COMPSTAT

Proceedings in
Computational Statistics

14th Symposium held
in Utrecht,
The Netherlands, 2000

Edited by
Jelke G. Bethlehem
and Peter G.M. van der Heijden

With 117 Figures
and 96 Tables

Physica-Verlag
A Springer-Verlag Company

Prof. Jelke G. Bethlehem
Statistics Netherlands
Department of Statistical Methods
P.O. Box 4000
2270 JM Voorburg, The Netherlands

Prof. Peter G.M. van der Heijden
Utrecht University
Faculty of Social Sciences
Department of Methodology and Statistics
P.O. Box 80140
3508 TC Utrecht, The Netherlands

ISBN 3-7908-1326-5 Physica-Verlag Heidelberg New York

Cataloging-in-Publication Data applied for
Die Deutsche Bibliothek – CIP-Einheitsaufnahme
Proceedings in computational statistics: 14th symposium held in Utrecht, The Netherlands, 2000; with 96 tables / COMPSTAT. Ed. by Jelke G. Bethlehem and Peter G.M. van der Heijden. – Heidelberg: Physica-Verl., 2000
ISBN 3-7908-1326-5

Physica-Verlag Heidelberg a member of BertelsmannSpringer Science+Business Media GmbH

Printed in Germany

Softcover Design: Erich Kirchner, Heidelberg

SPIN 10775102 88/2202-5 4 3 2 1 0 – Printed on acid-free paper

Preface

This Volume contains the Keynote, Invited and Full Contributed papers presented at COMPSTAT 2000. A companion volume (Jansen & Bethlehem, 2000) contains papers describing the Short Communications and Posters. COMPSTAT is a one-week conference held every two years under the auspices of the International Association of Statistical Computing, a section of the International Statistical Institute.

COMPSTAT 2000 is jointly organised by the Department of Methodology and Statistics of the Faculty of Social Sciences of Utrecht University, and Statistics Netherlands. It is taking place from 21-25 August 2000 at Utrecht University. Previous COMPSTATs (from 1974-1998) were in Vienna, Berlin, Leiden, Edinburgh, Toulouse, Prague, Rome, Copenhagen, Dubrovnik, Neuchatel, Vienna, Barcelona and Bristol. The conference is the main European forum for developments at the interface between statistics and computing. This was encapsulated as follows on the COMPSTAT 2000 homepage http://neon.vb.cbs.nl/rsm/compstat.

Statistical computing provides the link between statistical theory and applied statistics. As at previous COMPSTATs, the scientific programme will range over all aspects of this link, from the development and implementation of new statistical ideas through to user experiences and software evaluation. The programme should appeal to anyone working in statistics and using computers, whether in universities, industrial companies, research institutes or as software developers.

At COMPSTAT 2000 there is a special interest in the interplay with official statistics. This is evident from papers in the area of computerised data collection, survey methodology, treatment of missing data, and the like.

Selection of the Full Contributions from over 250 submitted abstracts presented a challenging but very interesting task for the Scientific Programme Committee (SPC): Prof. Jelke Bethlehem (The Netherlands), Prof. Andrej Blejec (Slovenia), Prof. Wolfgang Härdle (Germany), Prof. Peter van der Heijden (The Netherlands, Chair), Prof. Roger Payne (United Kingdom), Prof. Christian P. Robert (France), and Prof. Roberta Siciliano (Italy).

The papers were all refereed by the SPC, and we would like to thank our colleagues on the SPC for all their help. The resulting Proceedings provide a broad overview of the currently active research areas in statistical computing.

We are also especially grateful to the members of the Local Organising Committee: Wim Jansen (Chair) and Leoniek Wijngaards of Utrecht University, Anco Hundepool of Statistics Netherlands, Douwe van der Sluis and Gerrit Stemerdink of the Statistical Software Section of the Dutch Society for Statistics and Operations Research, and Marcelle Buma of the FBU Congress Bureau. We gratefully acknowledge the careful editing by Sandra Vogelesang, and the support of Maarten Bethlehem in maintaining the Conference website.

Jelke Bethlehem
Peter van der Heijden
May 2000

Contents

Keynote papers

The broad role of multiple imputation in statistical science 3
D.B. Rubin

Official Statistics: an estimation strategy for the IT-era 15
P. Kooiman, A.H. Kroese & R.H.Renssen

Invited papers

Bayesian model selection methods for nonnested models 29
F. Bertolino, E. Moreno & W. Racugno

Spatio-temporal hierarchical modeling of an infectious disease from 41
(simulated) count data
N. Cressie & A.S. Mugglin

GBMs: GLMs with bilinear terms 53
A. de Falguerolles

Generalized calibration and application to weighting for non-response 65
J.-C. Deville

Methodological issues in data mining 77
D.J. Hand

Practical data mining in a large utility company 87
G. Hebrail

HGLMs for analysis of correlated non-normal data 97
Y. Lee & J.A. Nelder

Bootstrapping impulse responses in VAR analyses 109
H. Lütkepohl

An application of TRAMO-SEATS; model selection and out-of-sample 121
performance. The Swiss CPI series
A. Maravall & F.J. Sánchez

Spreadsheets as tools for statistical computing and statistics education 131
E. Neuwirth

An algorithm for deepest multiple regression 139
P.J. Rousseeuw & S. van Aelst

Non-proportional hazards models in survival analysis 151
H.C. van Houwelingen & P.H.C. Eilers

A spatio-temporal analysis of a field trial 161
A. Verbyla, M. Lorimer & R. Stevens

Contributed papers

Principal component logistic regression 175
A.M. Aguilera & M. Escabias

Sieve bootstrap prediction intervals 181
A.M. Alonso, D. Peña & J. Romo

Clustering by maximizing a fuzzy classification maximum likelihood criterion 187
C. Ambroise & G. Govaert

Tree-based algorithms for missing data imputation 193
M.J. Bárcena & F. Tusell

MiPy: a system for generating multiple imputations 199
J. Barnard

A linear approximation to the wild bootstrap in specification testing 205
K. Bartels

The size of the largest nonidentifiable outlier as a performance criterion for multivariate outlier identification: the case of high-dimensional data 211
C. Becker

The influence of data generation and imputation methods on the bias of factor analysis of rating scale data 217
C.A. Bernaards

Model on a population and prediction on another one: a generalized discriminant rule 223
C. Biernacki, F. Beninel & V. Bretagnolle

Disclosure control on multi-way tables by means of the shuttle algorithm: extensions and experiments 229
L. Buzzigoli & A. Giusti

An MLE strategy for combining optimally pruned decision trees ... 235
C. Cappelli & W.D. Shannon

Semi-parametric models for data mining ... 241
C. Conversano & F. Mola

Preliminary estimation of ARFIMA models ... 247
M. Corduas

A collection of applets for visualizing statistical concepts ... 253
P. Darius, J.-P. Ottoy, A. Solomin, O. Thas, B. Raeymaekers & S. Michiels

Non-parametric regression and density estimation under control of modality ... 259
P.L. Davies & A. Kovac

Multivariate approaches for aggregate time series ... 265
C. Davino & V. Esposito

Some space-time models: an application to NO_2 pollution in an urban area ... 271
S. De Iaco & D. Posa

SLICE: generalised software for statistical data editing and imputation ... 277
T. de Waal

Improved PCB inspection: computational issues ... 283
D. Denteneer

Optimization of the antithetic Gibbs sampler for Gaussian Markov random fields ... 289
J.M. Dreesman

Computing zonoid trimmed regions of bivariate data sets ... 295
R. Dyckerhoff

Outlier resistant estimators for canonical correlation analysis ... 301
P. Filzmoser, C. Dehon & C. Croux

Graphical and phase space models for univariate time series ... 307
R. Fried

The use of the Tweedie distribution in statistical modelling ... 313
R. Gilchrist & D. Drinkwater

Predictive dimension: an alternative definition to embedding dimension ... 319
D. Guègan & F. Lisi

Post-stratification to correct for nonresponse: classification of ZIP code areas ... 325
M. Huisman

Approximate Bayesian inference for simple mixtures ... 331
K. Humphreys & D.M. Titterington

Correlated INAR(1) process ... 337
N. Kolev & D. Paiva

Confidence regions for stabilized multivariate tests ... 343
S. Kropf

Comparison of stationary time series using distribution-free methods ... 349
E.A. Maharaj

Generation of Boolean classification rules ... 355
R.J. Marshall

A statistical package based on Pnuts ... 361
J. Nakano, T. Fujiwara, Y. Yamamoto & I. Kobayashi

Generalized regression trees ... 367
M. Núñez

Generalized linear mixed models: an improved estimating procedure ... 373
J.-X. Pan & R. Thompson

The stochastic dimension in a dynamic GIS ... 379
E.J. Pebesma, D. Karssenberg & K. de Jong

A robust version of principal factor analysis ... 385
G. Pison, P.J. Rousseeuw, P. Filzmoser & C. Croux

TESS: system for automatic seasonal adjustment and forecasting of time series ... 391
A. Prat, V. Gomez, I. Solé & J.M. Catot

The BADAME project ... 397
H. Rezanková & L. Marek

Improving Statistics Canada's cell suppression software (CONFID) ... 403
D.A. Robertson

The multimedia project MM*STAT for teaching statistics ... 409
B. Rönz, M. Müller & U. Ziegenhagen

Projection pursuit approach to robust canonical correlation analysis 415
M. Rosário de Oliveira & J.A. Branco

A fast algorithm for highly robust regression in data mining 421
P.J. Rousseeuw & K. van Driessen

Optimal classification trees 427
P. Savicky, J. Klaschka & J. Antoch

GAM spline algorithms: a direct comparison 433
M.G. Schimek

Markov Chain Monte Carlo methods for handling missing covariates in 439
longitudinal mixed models
E. Schuster

Robust Bayesian classification 445
P. Sebastiani & M. Ramoni

Social science measurement by means of item response models........................ 451
K. Sijtsma

Multivariate DLMs for forecasting financial time series, with application 457
to the management of portfolios
A. Simpson & D.J. Wilkinson

An algorithm for the multivariate Tukey median 463
A. Struyf & P.J. Rousseeuw

Analyzing and synthesizing information from a multiple-study database 469
The Diverse Populations Collaborative Group

Bootstrapping neural discriminant model 475
M. Tsujitani & T. Koshimizu

An improved algorithm for robust PCA 481
S. Verboven, P.J. Rousseeuw & M.Hubert

Applying techniques of dynamic programming to sequential mastery 487
testing
H.J. Vos

The introduction of formal structure into the processing of statistical 493
summary data
A. Westlake

Validation of association rules by interactive mosaic plots 499
A. Wilhelm & H. Hofmann

Optimality models for PRAM 505
L. Willenborg

Dealing with real ordinal data: recent advances in analyzing tied, censored, and multivariate observations 511
K. M. Wittkowski

Testing for differences in location: a comparison of bootstrap methods in the small sample case 517
K. Wolf-Ostermann

Two principal points for location mixtures 523
W. Yamamoto & N. Shinozaki

Vector splines and other vector smoothers 529
T.W. Yee

Index of authors 535

Index of keywords 537

Keynote papers

The broad role of multiple imputation in statistical science

Donald B. Rubin

Department of Statistics, Harvard University, Cambridge, MA 02138, USA

Abstract. Nearly a quarter century ago, the basic idea of multiple imputation was proposed as a way to deal with missing values due to nonresponse in sample surveys. Since that time, the essential formulation has expanded to be proposed for use in a remarkably broad range of empirical problems, from many standard social science and biomedical applications involving missing data in surveys and experiments, to nonstandard survey and experimental applications, such as preserving confidentiality in public-use surveys and dealing with noncompliance and "censoring due to death" in clinical trails, to common "hard science" applications such as dealing with below-threshold chemometric measurements, to other scientific or medical applications such as imaging brains for tumors, and exploring the genetics of schizophrenia. The purpose of this presentation is to provide some links to this broad range of applications and to indicate the associated computing requirements, primarily using examples in which I am currently involved.

Keywords Bayesian methods, censored data, chemometrics, clinical trials, confidentiality, experiments, genetics, image reconstruction, latent variables, mixture models, noncompliance, surveys

1 Introduction

Missing data are a pervasive problem in almost all areas of empirical research. It is now nearly a quarter century since multiple imputation was first proposed as a general solution to the problem of missing data. The first formally published seeds were in Rubin, 1977, p. 539): "One can think of the method given here as simply summarizing the results of simulations, where one uses the respondents to generate 'reasonable' hypothetical responses for nonrespondents." A year later, the idea was explicitly proposed (Rubin, 1978). A decade later, the basic reference was published (Rubin, 1987).

Multiple imputation has now become a major applied approach to the general problem of obtaining valid statistical inferences when faced with missing data, and may become the dominant approach in practice in the near future. A recent article with many references is Rubin (1996). It is critically important to recognize that multiple imputation does not try to create information that is not in the data. The goal of multiple imputation is to represent the information you do have in such a way that you can analyze it using standard complete-data software and get valid statistical inferences.

One reason for the growth of multiple imputation in the last decade is its suitability for modern computing environments, with the division of the two tasks of creating and analyzing a multiply-imputed data set. The task of analyzing an already-created multiply-imputed data set is relatively easy, requiring only (a) standard complete-data statistical routines to be applied to each complete data set created by the imputations, and (b) entirely general purpose macros for combining the repeated complete-data analyses to reach one inference. The task of creating a multiply-imputed data set is far more demanding — here is where modern computing is really essential.

The complexity of the task of creating multiple imputations varies greatly with the problem. In "standard" problems in surveys and experiments, commercial or software already exists. In less standard problems in surveys and experiments (e.g., involving the preservation of confidentiality in public-use files or dealing with noncompliance or censoring due to death in clinical trials), the conceptual formulation using multiple imputation has lead to important insights, but widely available software is not yet available. In some basic missing data problems outside the simple context, such as below threshold measurements in chemometric data bases or medical imaging, some software does exist, but may not be available in commercial or well supported packages. In other situations, only exploratory or special-purpose "academic" software may exist.

This article briefly indicates the broad range of applications, with some links to literature, and provides notes on the underlying effort needed to create the multiple imputations. The selection of the topics and the literature is, admittedly, a bit idiosyncratic, in the sense of focusing primarily on applications in which I am currently involved.

2 Standard applications

The definition of what constitutes a "standard" missing data problem is, of course, somewhat ambiguous, but a useful one can be given by the following "algorithmic" description. The problem is standard if the missing values for a nonresponding unit in a rectangular units by variables data set can be realistically predicted, with appropriate recognition of uncertainty, by examining the corresponding observed outcomes for "similar" responding units, where similar is defined by variables that are observed for both the nonresponding and responding units. The formalization of this idea is done in Rubin (1976).

When the process that creates missing data is ignorable, as defined in this article, the effort of creating multiple imputations is basically one of prediction and so is relatively straightforward. Two main approaches can be distinguished, both of which have advantages and limitations.

The first approach is a theme for SOLAS 2 (Statistical Solutions, 2000) and is particularly suited, though not limited to, cases with monotone patterns of missingness. When such a pattern occurs, fully principled and flexible statistical modeling for imputation can be done without extraneous assumptions (Little and Rubin, 1987, Chapter 6; Rubin, 1987, Section 5.4). A monotone pattern of missingness arises when the data matrix can be divided into observed and missing parts with a "staircase" line dividing them — e.g., variable 1 is observed on all the units (1-100), variables 2 and 3 are observed on units 1-75, but missing on units 76-100; variable 4 is observed on units 1-50 but missing on units 51-100. The great statistical advantage of a monotone pattern is that the creation of multiple imputations in a multivariate data set is reduced to a series of single variable imputations, which allows tremendous modeling flexibility. In principle, with a monotone pattern, we have all the flexibility that we have with complete-data modeling (e.g., transformations of the outcome variable and predictor variables, interactions of predictor variables, nonlinear regressions [even for conditional variances], etc.). There is also a great computational advantage with a monotone pattern because the existence of missing data does not impose any need for iterative algorithms, either for maximum likelihood or Bayesian posterior estimation.

Of course, the problem in practice is that it is rare that any real data set has a perfectly monotone pattern of missingness. As a result, to retain all the great advantages of the monotone pattern, in practice something less than fully principled often needs to be done to fill in missing values to create a monotone pattern. This suggestion to create a monotone pattern by filling in values (or even deleting observed values, which SOLAS 2 does not do) is, in fact, even older than the idea of multiple imputation (or EM — Dempster, Laird, Rubin, 1977, for that matter); it appears in Rubin, 1974, Section 6.2. Moreover, the recognition of the importance of monotone patterns predates Rubin (1974), which focused on such special patterns; in fact, that article can be viewed as full-fledged extension of Anderson (1957), which itself was a neat derivation of maximum likelihood estimates for the simple case of bivariate normal monotone missing data done first by Wilks (1932). An early application of this idea of reducing a missing data problem to one with a monotone pattern is given in Marini, Olsen and Rubin (1980).

The way SOLAS 2 gets to a monotone pattern is by first sorting to get close to a monotone pattern and then multiply-imputing the values that destroy the monotone pattern using a series of carefully selected available-case regressions. Other options are available in SOLAS 2, and these could be especially useful when much is known about the reasons for missing values. There is some "art" needed to do this reduction to a monotone pattern well, but once accomplished, the principled flexibility for creating multiple imputations is very rewarding, especially if the extra percent of missing information due to the non-monotone missingness beyond the monotone missingness is not large.

The second currently available general method for creating multiple imputations in standard missing data problems is to specify one encompassing multivariate model for the entire data set (at least conditional on completely observed variables), and then to use fully principled likelihood/Bayesian techniques for analysis under that model. This generates a posterior distribution for the parameters of the model and a posterior predictive distribution for the missing values (given the model specifications and the observed data values).

The primary example of this approach is Joseph Schafer's freeware (Schafer:www) based on Schafer (1997), which involves iterative Markov chain Monte Carlo (MCMC) computations. A more limited version is Gary King's freeware (King:www), which involves iterative maximization using EM and draws under an asymptotic normal approximation refined using SIR (Rubin, 1987, p. 192). The theoretical advantage of this second general approach is that it is technically correct, no matter what the pattern of missing data, but only if the specified model is correct and the Bayesian/likelihood analysis method is correctly implemented.

The practical disadvantages of this second general approach, unfortunately, are just as clear as its theoretical advantage. First, despite the availability of some collections of multivariate models (e.g., the normal general location model in Schafer's software and extensions to the t-family in Liu and Rubin, 1998), these collections are extremely limited — at least limited relative to the enormous collection of models available with the monotone-missingness approach. This limitation can be a serious concern in practice unless the person doing the multiple imputation is a real expert, not only at statistical modeling with complete data and missing data, but also at making the particular software work in clever and often unanticipated ways. The second major disadvantage of using the "one encompassing model" approach is that, even if one of the limited models that is available in the software is appropriate for the data at hand, making an iterative program work can be a nightmare in practice, especially without the resources of Ph.D. level experts and the patience to deal with potentially misleading "nonconvergent" converging Markov chain Monte Carlo output, or to evaluate the propriety of large sample asymptotic approximations refined with SIR, which despite my earlier hopes, may not work very well in large dimensions, especially when applied to large sample approximations where the underlying likelihoods can be multi-modal.

My own assessment is that unless a user has the resources to use Schafer's software with data that conform pretty well to the underlying assumptions (or the statistical resources to enhance those specifications), the iterative versions of software for creating multiple imputations in standard missing data problems are not yet ready for reliable application by the typical user dealing with standard problems of missing data. At this point, I think that the SOLAS 2 attack of pushing the data to be monotone and then being fully principled and flexible is a safer path for the typical user dealing with standard problems of missing data. This is a subjective assessment, which could be studied in real applications. Also, of course, there is the important consideration of having a validated and supported computing environment that is nearly always more associated with commercial products than freeware.

The future, of course, holds the promise of methods for the multiple imputation of standard missing data that combine the best features of both general approaches. For example, use the SOLAS 2 type of routine to sort the data matrix as closely as possible to conform to a monotone pattern; use Schafer's type of software (or perhaps the "incompatible Gibbs" software of T.E. Raghunathan at the Institute for Social Research (Raghunathan:www) to multiply-impute the missing values that destroy the monotone pattern; and then use the SOLAS 2 type of routine to multiply-impute the monotonely missing values. Or perhaps even better, iterate the entire process, including the filling-in of the monotonely missing values. This latter approach is one I am currently directing with a major data set of substantial importance (the U.S. National Medical Expenditure Survey), using a very knowledgeable third party statistical consulting company, but this effort is extremely time-consuming to set up and very expensive to implement (well into six figures of U.S. dollars and counting), and the creating of just one imputed data set takes more than week of CPU time on a modern minicomputer and days of diagnostic checking — clearly this approach is not yet ready for the average user even with today's fastest PCs.

Lest all these comments make the task of multiple imputation in standard missing data problems appear too daunting to attempt, I must add that even doing multiple imputation relatively crudely, using simple methods, is very likely to be inferentially far superior to any other equally easy method to implement (e.g., complete- cases, available cases, single imputation, LVCF) because the multiple copies of the data set allow the uncertainty about the values of the missing data to be incorporated into the final inferences; Heitjan and Rubin (1990) provides some evidence for this statement, as does Raghunathan and Paulin (1998). And the use of SOLAS 2 (or Schafer's software or King's) to create a multiply-imputed data set is nearly certain to be far superior to any other generally feasible approach.

Major examples of the use of multiple imputation in standard examples are discussed in Rubin and Schenker (1991) and Ezzati-Rice, Johnson, Khare, Little, Rubin and Schafer (1995). Other examples are in the references in Rubin (1996).

3 Non-standard situations

Multiple imputation also shows substantial promise is situations where simple prediction will not work to create the posterior predictive distribution of the missing values. A few examples will be used to illustrate how multiple imputation for missing values, properly conceived, can successfully address difficult problems not normally viewed as missing data problems. Of course, it is not surprising that each nonstandard situation requires special software to create the multiple imputations. Still, the analysis of a multiply-imputed data set requires only standard complete-data analyses and generic combining rules to reach the final inference.

3.1 Missing data in clinical trials

There are distinctly different kinds of missing data that arise in clinical trials (i.e., randomized experiments involving people in medical/pharmaceutical contexts). These different types of missing data cannot be treated as if they were all standard and still get valid inferences. Of critical importance, all experiments have missing data by design: the potential outcomes for subjects under the treatments they were not assigned. These are the quantities that, with the observed data, define the causal effects, which are the objects of inference (e.g., Rubin, 1990). The fact that most of the potential outcomes are missing by design is not problematic; in fact, the analysis of a perfect randomized experiment with no complications is well understood. But other kinds of missing data in experiments can be difficult to handle well in practice because the unavoidable missing potential outcomes mean that there is more than one kind of missing data in the study. The complications are especially severe when various types of unintended missing data are simultaneously present in a clinical trial.

Missing covariate values are typically the simplest type of missing data to handle, because unless the covariates were part of the design, and thus blocking variables, their role is primarily to increase precision. In this role, the pattern of missing data itself is a valid covariate as well, and pattern mixture models can often be used (e.g., Rubin, 1987, Section 6.4; Little, 1993). Non-blocking covariates can be ignored if the extra precision, or subgroup analyses, are not needed, and this is often the best practical approach. When blocking covariate values are missing, they must be modeled in order to reach valid inferences, and then multiple imputation is an excellent approach. A correct analysis, however, ensures that the underlying distribution of covariates is the same across all treatment groups, as required by randomization. This requirement can create computational complications. I know of no generally available software that enforces this requirement on the covariate distribution in any generality.

Missing outcomes when all patients stay alive and on their assigned protocol can create problems but are conceptually straightforward to address. Considering each treatment group separately, this situation can typically be handled in the standard way, that is, assuming an ignorable missing data process and multiply-imputing the unintended missing data. This approach essentially treats each arm of the experiment as if it were a survey.

When a clinical trial has noncompliance with protocol, but outcomes are recorded for all subjects and all remain living, the crucial missing data are the compliance behaviors of the subjects under the treatments they were not assigned, that is, their "true" compliance statuses. For instance, with one treatment group, one control group, and all-or-none compliance, there are four possible compliance statuses: the true compliers are those who would adhere to their assigned treatment no matter how assigned; the always-takers are those who would receive the treatment no matter how assigned; the never-takers would receive the control treatment no matter how assigned; and the defiers would do the opposite of their assignment. This is the situation for which traditional econometric "instrumental variables" techniques can be applicable when "monotonicity" and "exclusion" restrictions are asymptotically appropriate (Angrist, Imbens and Rubin, 1996).

A more general and more satisfactory approach is implemented using multiple imputation for the missing true compliance statuses, as developed by Imbens and Rubin (1997) and illustrated by Hirano, Imbens, Rubin and Zhou (2000) in the context of a randomized "encouragement" design studying the efficacy of flu shots, where the multiple imputations are created by MCMC under specific models. In these situations, the primary estimand of interest is the effect of treatment for the subgroup of true compliers. These types of analyses are needed in order to go beyond standard intention-to-treat analyses, which not only may lack power under reasonable alternatives (Rubin, 1998) but also generally address a scientifically confounded question involving both the action of the drug and compliance behavior (Sheiner and Rubin, 1994). There is some development of software here, both free and commercial, but it is quite limited at present.

When some subjects die before the end of the trial, but otherwise there is full compliance and there are no missing covariates or outcomes, Y, the missing data are the survival statuses of the subjects under the treatments they were not assigned (not the outcomes that were not measured on the dead subjects). Thus, in analogy with noncompliance, the missing data in a two treatment trial are the "true" survival statuses of the subjects at the end of the experiment under both treatments: alive/alive, alive/dead, dead/alive, dead/dead. The estimands of primary interest concern the effect of treatment on survival, which is straightforward to estimate, and the effect of treatment on Y, which is only well-defined for the true survivors (i.e., the alive/alive group of subjects), which is more difficult to estimate (Rubin, 1998; 2000). The approach of multiply-imputing the true survival statuses has been successfully applied to this situation with real data, but to my knowledge, as yet no generally available software exits for such analyses.

Of course, all of these types of missing data can occur simultaneously, and then valid analyses become more complex and require more assumptions. For example, with both missing outcomes and noncompliance, an additional assumption, called "latent ignorability" by Frangakis and Rubin (1999), is needed to reach valid inferences even for the intention to treat estimand. Multiple imputation can validly address these complicated situations, and can do so, I believe, with more flexibility than any other competing technique, because of its reliance on modern computational environments rather than asymptotic or closed-form analysis. Software needs to be developed to address these kinds of situations.

3.2 Below threshold data

Many chemical and environmental data sets are complicated by the existence of fully missing values or censored values known to lie below detection thresholds. For example, week-long samples of airborne particulate matter were obtained at Alert, N.W.T. Canada between 1980 and 1991, where some of the concentrations of 24 particulate constituents were coarsened in the sense of being either fully missing or below detection limits. This situation is a special case of "coarsened data". To facilitate scientific analysis, it is very appealing to create complete data by filling in missing values so that standard complete-data methods can be applied. In Hopke, Liu and Rubin (2000), three statistical models are developed for multiply-imputing the missing values of airborne particulate matter making different assumptions about the incomplete multivariate time series data. All models assume the missing data are "coarsened at random" (Heitjan and Rubin, 1991), a generalization of "missing at random" (Rubin, 1976); an early application of this idea to age heaping is Heitjan and Rubin (1990).

The first of the three models for the Alert data is the standard multivariate normal but allowing both censored data with known censoring points and fully missing data. The second model is an extension to include a multivariate moving average time series structure. The third model is a further extension that allows seasonal components within the time series.

All models are fit using MCMC, and are used to create multiple imputations for the fully missing and below threshold values. Diagnostic analyses suggest that for the Alert data set, the seasonal time series model provides the most realistic imputations. Although not a standard missing data problem in the sense of Section 3.1, this problem is so common that these three models and their associated software could be considered prospects for general distribution.

3.3 Preserving confidentiality in public-use files

A rather unusual topic, more legal and ethical than scientific, is how to release valuable data paid for by public funding without fear that confidentiality assurances made to survey respondents will be maintained. Rubin (1993) proposed preserving confidentiality in public-use data bases by releasing only multiply-imputed synthetic microdata. This proposal has recently attracted more attention.

For simplicity, consider an actual microdata set of size n drawn using design D from a much larger population of N units, where X represents background variables, Z represents outcome variables with no confidentiality concerns, and Y represents outcome variables with some confidentiality concerns; X is, in principle, observed for all N units in the population, whereas Z and Y are only observed for the n sampled units and are missing for the $N-n$ unsampled units. A multiply-imputed population consists of the actual X data for all N units, the actual (Z,Y) data for the n units in the survey, and M (the number of multiple imputations, typically between 3 and 10) matrices of (Z,Y) data for the $N-n$ nonsampled individuals. The variability in the imputed values ensures, theoretically, that valid inferences for population estimands can be obtained, at least for population estimands up to some level of detail. A standard predictive model for predicting (Z,Y) from X is used to multiply-impute (Z,Y) in the population. The choice of model is influenced by the X data, and design D, scientific understanding of how to predict (Z,Y) from X, and judgement about the range of questions users may ask of the data. In the setting of preserving confidentiality, the model should exclude certain detailed or interactive terms that could lead to the identification of individuals from the subsequent multiply-imputed data set.

Given such a multiply-imputed population and a new survey design D^* for the released microdata (possibly, but not necessarily, the survey design actually used to collect the data), we can draw a sample of n^* units ($n^* \ll N$), which looks structurally just like an actual microdata base of size n^* drawn from the population using design D^*, and we can do this M times to create M replicates of (Z, Y) values. The result is a multiply- imputed synthetic data base. Of critical importance, this data set releases only synthetic confidential (i.e., Y) data because $N \gg n$ and $N \gg n^*$. To make sure that no actual microdata are released, the samples can be drawn from the population excluding the n actual units; also to guard against all M imputations of a Y variable being identical, M can be made large enough so that this event is very rare.

The resultant multiply-imputed data base creates M microdata bases, each of which is analyzed using complete-data methods of analysis appropriate to design D^*. These repeated complete-data analyses are then combined to create one valid inference for the population using slightly modified generic combining rules, which in initial simulation work appear to be quite satisfactory. (Raghunathan and Rubin, 2000).

The software needed to implement this scheme really does not have to multiply-impute the entire population, but rather only the subset of units chosen by design D^* for each of the M microdata sets. New software, which is a straightforward extension of that for standard missing data, needs to be written. Moreover, software for the modified generic combining rules needs to be written, but this too is straightforward.

3.4 Latent variables model

In a variety of contexts, as when studying schizophrenia, the use of mixture models arises. Such models posit hidden groups, for instance, types of subjects with and without an unobserved genetic deficit. A latent (i.e., missing) variable is used to indicate the hidden groups. Analyses proceed by multiply-imputing the missing indicators, and then doing complete-data analyses as if the memberships in the hidden group were known.

For example, Belin and Rubin (1995) and Rubin and Wu (1997) show how missing indicators can be used to model schizophrenic reaction times and eye-tracking movement, respectively. In these applications, some schizophrenics are "susceptible" and others are not, and person-specific indicators for susceptibility to reaction time and eye-tracking deficiencies are treated as missing variables. Moreover, to reflect the characteristic behaviors of susceptible schizophrenics, there are also missing values for each trial indicating whether each reaction-time or eye-tracking measurement arose from the "usual" or "dialyptic" component. Highly specific and complex MCMC software is needed to multiply-impute the missing indicators in these models; see especially Rubin and Wu (1997).

Extensions in the context of schizophrenia involve the inclusion of relatives of schizophrenics, who are viewed as a mixture of two genotypes: those susceptible to the behavior being studied (e.g., short-term spatial memory deficits), and those who are not susceptible (Lo, Matthysee and Rubin, 2000).

3.5 Image reconstruction

The problem of image reconstruction, for example, of a brain, using indirect information, such as provided by a PET scan, can also usefully be viewed as a problem of missing data where the image is the estimand. Here, each complete-data analysis is a picture or image, for example of the brain, and the variations in these images across the data sets completed by imputation represent the uncertainty in the image reconstruction (e.g., see Gelman and Price, 1999).

The following description follows that in Rubin's (1985) discussion of Vardi, Shepp and Kaufman (1985) on PET scans. There are several important conceptual steps that in combination lead to multiple imputations for image reconstruction, where the observed data are the counts in D detectors, and the estimand is the image of the brain represented by pixel intensities on a grid.

1. Formulate a pixel model of the brain with a distinct Poisson intensity parameter characterizing each pixel, say $\lambda(b), b = 1, 2, \ldots, B$, where B is the number of pixels. Typically, there are 10,000 pixels on a 100 by 100 grid.
2. Relate the B pixels to the D detectors via a $D \times B$ matrix of conditional probabilities, $\Pr(detector = d|pixel = b$, derived from physical considerations; $\Pr(d|b)$ gives the probability that an emission from pixel b is recorded in detector d, and it is regarded as a known a priori for all d and b.
3. View the estimation problem as one involving missing data, where the complete data (which we do not fully observe) consist of a $D \times B$ indicator matrix for each emission, $\boldsymbol{\delta} = \{\delta_i(d,b)\}, i = 1, \ldots, n$, where $\delta_i(d,b) = 1$ if the ith emission emanated from pixel b and was recorded in detector d, and $\delta_i(d,b) = 0$ otherwise. Notice that if $\boldsymbol{\delta}$ were known, the estimation of $\boldsymbol{\lambda}$ would be trivial after summing the column margins of $\boldsymbol{\delta}$ across all n emissions to obtain the emission counts for each pixel. The observed (or incomplete) data consist of the n row margins of $\boldsymbol{\delta}$, which are $D \times 1$ vectors indicating the detectors underlying the emissions.
4. Multiply impute the entries of $\boldsymbol{\delta}$, and use the completed emission counts in the pixels to create the image.

Another very recent example of the potential of applying multiple imputation to imaging concerns spectral imaging of galaxies using data from the recently launched Chandra X-ray Observatory. Here, a particularly difficult problem is that of photon "pile-up", which confounds one high energy photon with two or more closely arriving photons with the reported energy equal to their total energy. A recent proposal, arising in the context of a project at the Harvard-Smithsonian Center for Astrophysics, is to attack the pile-up problem using multiple imputation, rather than the enormously computationally expensive fully model-based analyses (van Dyk, Connors, Kashyap and Siemiginowska, 2000). The computation effort is substantial, but still much less than without the use of multiple imputation.

4 Conclusions

There are many other examples where the idea of multiply-imputing some essential features of a data set leads to flexible computation (e.g., Tanner and Wong, 1987; Kong, Liu and Wong, 1994; Liu and Wu, 1999). The key idea in many of these examples is to think of Bayesian inference for a missing data problem. In fact, in some ways, many of the currently existing algorithmic techniques for biologic and genetic problems (e.g., Liu, Neuwald and Lawrence, 1999), can be thought of as fundamentally applying the essential ideas of multiple imputation in creative ways. In these examples, the computational demands are specialized and extremely substantial, but the resulting algorithms address problems that appear to be essentially hopeless without this perspective.

There is indeed a broad role for multiple imputation!

References

Anderson, T.W. (1957). Maximum likelihood estimates for a multivariate normal distribution when some observations are missing. *Journal of the American Statistical Association*, **52**, 200–203.

Angrist, J.D., Imbens. G.W. and Rubin, D.B. (1996). Identification of causal effects using instrumental variables. *Journal of the American Statistical Association*, **91**, 434, as Applications Invited Discussion Article with discussion and rejoinder, 444–472.

Belin, T.R. and Rubin, D.B. (1995). The analysis of repeated-measures data on schizophrenic reaction times using mixture models. *Statistics in Medicine*, **14**, 8, 747–768.

Dempster, A.P., Laird, N. and Rubin, D.B. (1977). Maximum likelihood from incomplete data via the EM algorithm. *Journal of the Royal Statistical Society, Series B*, **39**, 1, 1–38, with discussion and reply.

Ezzati-Rice, T., Johnson, W., Khare, M., Little, R., Rubin, D. and Schafer, J. (1995). A simulation study to evaluate the performance of model-based multiple imputations in NCHS Health Examination Surveys. *Bureau of the Census Eleventh Annual Research Conference*, 257–266.

Frangakis, C. and Rubin, D.B. (1999). Addressing complications of intention-to-treat analysis in the combined presence of all-or-none treatment-noncompliance and subsequent missing outcomes. *Biometrika*, **86**, 2, 366–379.

Gelman, A. and Price, P.N. (1999). All maps of parameter estimates are misleading. *Statistics in Medicine*, **18**, 3221–3234.

Heitjan, D.F. and Rubin, D.B. (1990). Inference from coarse data via multiple imputation with application to age heaping. *Journal of the American Statistical Association*, **85**, 410, 304–314.

Heitjan, D.F. and Rubin, D.B. (1991). Ignorability and coarse data. *The Annals of Statistics*, **19**, 4, 2244–2253.

Hirano, K., Imbens, G., Rubin, D.B. and Zhou, X.H. (2000). Estimating the effect of an influenza vaccine in an encouragement design. *Biostatistics*, **1**, 69–88.

Hopke, P.K., Liu, C. and Rubin, D.B. (2000). Multiple imputation for multiple time series of the chemical concentrations of air pollution in the Arctic. Revision submitted to *Biometrics*.

Imbens, G. and Rubin, D.B. (1997). Bayesian inference for causal effects in randomized experiments with noncompliance. *The Annals of Statistics*, **25**, 1, 305–327.

King, G.: `gking.harvard.edu/amelia`

Kong, A., Liu, J.S. and Wong, W.H. (1994). Sequential imputations and Bayesian missing data problems. *Journal of the American Statistical Association*, **89**, 425, 278–288.

Little, R.J.A. (1993). Pattern-mixture models for multivariate incomplete data. *Journal of the American Statistical Association*, **88**, 125–134.

Little, R.J.A. and Rubin, D.B. (1987). *Statistical Analysis with Missing Data.* New York: John Wiley and Sons. Translated into Russian in 1991: Finansy and Statistika Publishers, Moscow, Andrei Nikiforev, translator.

Liu, C.H. and Rubin, D.B. (1998). Ellipsoidally symmetric extensions of the general location model for mixed categorical and continuous data.

Biometrika, **85**, 3, 673–688.

Liu, J.S. Neuwald, A.F. and Lawrence, C.E. (1999). Markovian structures in biological sequence alignments. *Journal of the American Statistical Association*, **94**, 445, 1–15.

Liu, J.S. and Wu, Y.N. (1999). Parameter expansion for data augmentation. *Journal of the American Statistical Association*, **94**, 448, 1264–1274.

Lo, Y., Matthysee, S. and Rubin, D.B. (2000). A permutation test for detecting the presence of mixtures of working memory performances in relatives of schizophrenia patients on spatial and object tasks. Unpublished.

Marini, M.M., Olsen, A.R. and Rubin, D.B. (1980). Maximum likelihood estimation in panel studies with missing data. *Sociological Methodology 1980*, Chapter 11, 314–357.

Raghunathan, T.E.: `www.isr.umich.edu/src/smp/ive`

Raghunathan, T.E. and Paulin, G.D. (1998). Multiple imputation in the Consumer Expenditure Survey: evaluation of statistical inference. *Proceedings of the Business and Economics Section of the American Statistical Association*, 1–10.

Raghunathan, T.E. and Rubin, D.B. (2000). Preserving confidentiality in public-use files via multiple imputation. Invited presentation, International Society for Bayesian Analysis. May 2000, Crete.

Rubin, D.B. (1974). Characterizing the estimation of parameters in incomplete data problems. *Journal of the American Statistical Association*, **69**, 346, 467–474, Section 6.2.

Rubin, D.B. (1976). Inference and missing data. *Biometrika*, **63**, 3, 581–592. With discussion and reply.

Rubin, D.B. (1977). Formalizing subjective notions about the effect of nonrespondents in sample surveys. *Journal of the American Statistical Association*, **72**, 359, 538–543.

Rubin, D.B. (1978). Multiple imputations in sample surveys — A phenomenological Bayesian approach to nonresponse. *The Proceedings of the Survey Research Methods Section of the American Statistical Association*, 20–34. With discussion and reply. Also in *Imputation and Editing of Faulty or Missing Survey Data*, U.S. Department of Commerce, 1–23.

Rubin, D.B. (1985). EM for PET. Discussion of "A Statistical Model for Positron Emission Tomography" by Vardi, Shepp and Kaufman, *Journal of the American Statistical Association*, **80**, 31–32.

Rubin, D.B. (1987). *Multiple Imputation for Nonresponse in Surveys.* New York: John Wiley and Sons.

Rubin, D.B. (1990). Neyman (1923) and causal inference in experiments and observational studies. *Statistical Science*, **5**, 4, 472–480.

Rubin, D.B. (1993). Satisfying confidentiality constraints through the use of synthetic multiply-imputed micro-data.— Discussion of statistical disclosure limitation. *Journal of Official Statistics*, **9**, 2, 461–468.

Rubin, D.B. (1996). Multiple imputation after 18+ years. *Journal of the American Statistical Association*, **91**, 434, 473–489, with discussion 507–515, rejoinder 515–517, and extensive references 486–489.

Rubin, D.B. (1998). More powerful randomization-based p-values in double-blind trials with noncompliance. *Statistics in Medicine*, **17**, 371–385, with discussion by D.R. Cox, 387–389.

Rubin, D.B. (2000). The utility of counterfactuals for causal inference. Comment on A.P. Dawid, "Causal inference without counterfactuals". *Journal of the American Statistical Association*, **95**, 450, 435–438.

Rubin, D.B. and Schenker, N. (1991). Multiple imputation in health-care data bases: An overview and some applications. *Statistics in Medicine*, **10**, 585–598.

Rubin, D.B. and Wu, Y. (1997). Modelling schizophrenic behavior using general mixture components. *Biometrics*, **53**, 1, 243–261.

Schafer, J.L. (1997). *Analysis of Incomplete Multivariate Data.* London: Chapman and Hall.

Schafer, J.L.: `www.stat.psu.edu/~jls`

Sheiner, L.B. and Rubin, D.B. (1994). Intention-to-treat analysis and the goals of clinical trials. *Clinical Pharmacology and Therapeutics*, **87**, 1, 6–15.

Statistical Solutions (2000): `www.statsol.ie`

Tanner, M.A. and Wong, W.H. (1987). The calculation of posterior distributions by data augmentation. *Journal of the American Statistical Association*, **82**, 528–540.

van Dyk, D.A., Connors, A., Kashyap, V.L. and Siemiginowska, A. (2000). Analysis of energy spectrum with low photon counts via Bayesian posterior simulation. *Technical Report, Harvard University, Department of Statistics.*

Vardi, Y., Shepp, L.A. and Kaufman, L. (1985). A statistical model for positron emission tomography. *Journal of the American Statistical Association*, **80**, 8–20.

Wilks, S.S. (1932). Moments and distributions of estimates of population parameters from fragmentary samples. *Annals of Mathematical Statistics*, **2**, 163–195.

Official Statistics: an estimation strategy for the IT-era

P. Kooiman, A.H. Kroese and R.H. Renssen*)

Statistics Netherlands, P.O. Box 4000, 2270 JM Voorburg, The Netherlands

Abstract. This paper gives an overview of the efforts at the Department of Statistical Methods of Statistics Netherlands to develop a design based strategy aimed towards full numerical consistency of its statistical estimates, given a matched data file containing all available register and survey data about the target population. The key element of the estimation procedure is repeated re-calibration of survey data sets, i.e. a new set of raising weights is derived each time additional estimates are being produced. This allows us to take account of any related statistics obtained in earlier rounds of estimation.

Keywords. Repeated weighting, calibration estimator, generalized regression estimator, numerical consistency, official statistics

1 Introduction

National Statistical Institutes (NSI's) prepare official statistics about a broad variety of aspects of socio-economic and socio-cultural behaviour, demographics, the natural environment and other features relevant for decision making, scientific research and policy evaluation. In doing so NSI's aim to present a coherent and complete quantitative description of the state of society and its environment. Technically speaking this represents a huge estimation problem. Indeed, in many cases statistical concepts to be measured are not directly observable; available data, be it from secondary sources or from own survey taking activities, are incomplete, inflicted with measurement errors, and, sometimes, even contradictory. Through balance and identity restrictions the various features to be estimated are often intimately related, both within and between individual statistical data collections. Although NSI's typically do not perceive themselves primarily as pursuing a huge simultaneous statistical estimation exercise, they in fact overwhelmingly are.

The estimation problem indicated is much too big to be tackled by a one shot full information estimation strategy, processing all available information and all a priori restrictions simultaneously in the calculation of some optimal joint estimator. In fact the estimation approach taken is necessarily much more fragmented, partial and sequential. Individual survey data collections are treated in isolation. By edit checking influential measurement errors are located and repaired, missing items are imputed, and the available records are raised, using a few auxiliary variables to

---

*) The views expressed in this paper are those of the authors and do not necessarily reflect the policies of Statistics Netherlands

control for nonresponse selectivity, reduce sampling variance, and enforce coherence with known totals for these variables. Remaining inconsistencies between different surveys are mostly taken for granted and incidentally reconciled *ex post* in the context of data integration frameworks such as the National Accounts system. This leads to revised outcomes as compared to the individual surveys themselves. This is an example of the fact that for many important figures NSI's disseminate multiple estimates, each stemming from a different data collection or survey programme. For the users of such output this is very confusing, even if such differences can technically be justified by slight differences in definitions of concepts, slight differences in reference populations, implicit differences in loss functions (optimally estimating levels or changes) or conditioning on different information sets.

Mainstream estimation strategy of NSI's is *design based*. Data are collected by means of a well designed sample of the population as represented in a sampling frame. The design based nature of the selection of the individuals to be observed allows for unbiased population estimates, free from selectivity bias due to untrue model assumptions. In the spirit of post stratification auxiliary information can be exploited to improve the precision of the estimates, using regression or calibration type models, without violating the fundamental (approximate) design unbiasedness of the resulting estimator. This provides a strong mathematical operationalisation of a fundamental property official statistics should satisfy in a democratic society: estimates should not reflect any particular interest or presumption. Official statisticians therefore have always been reluctant to employ *model based* statistical methods, and have tried to marginalize their use. Nevertheless such methods cannot be avoided when dealing with deviations from the survey design: frame errors, item and unit non-response, measurement errors, mismatches, as well as in reconciling conflicting outcomes from various surveys. In order to let the design based strategy dominate NSI's typically devote large amounts of professional resources to reduce non-sampling errors to such an extent, that after (model based!) treatment their impact of the final estimates can be judged negligible.

Due to the IT-revolution the amount of secondary data sources available for NSI's has increased enormously. Contrary to earlier times, when we knew almost nothing about households and enterprises so that we had to go out to find out ourselves, we now have vast amounts of data from administrative records and government registrations, and get these almost for free. Numbers of records of these data sources are typically several orders of magnitude larger than the sample surveys we are accustomed to. Some registers cover the whole population; by matching these registers we obtain vastly richer population frames. Many registers only cover specific subpopulations, other subpopulations being completely absent. Some registers have high quality data, others are inflicted with substantial errors through lack of maintenance. In many cases the variables in public or private administrations are not identical to the statistical concepts NSI's wish to report about, so that they have to be translated or transformed. Obviously, administrative data do not derive from a well-designed sampling experiment. The availability of vast amounts of non-experimental data not only offers an opportunity, but also severely complicates the estimation problem. The question then is: how can NSI's make use of these secondary data sources, without violating their commitment to high quality design unbiased estimation?

In the past few years methodologists of the department of Statistical Methods of Statistics Netherlands have tried to develop a comprehensive estimation strategy combining three features:

- (approximately) design unbiased estimators
- exploitation of register data
- complete numerical consistency of all outcomes produced

The present paper reports about our advances in this area. It summarizes a number of research reports and publications, notably Renssen & Nieuwenbroek (1997), Renssen (1998), Kroese & Renssen (1999), Renssen & Martinus (2000), Kroese et al. (2000), and Renssen et al. (2000).

Depending on the quality and relevance of the register data these can either be used as a direct source for tabulation, as auxiliary information to reweight sample surveys against, or both. It is demonstrated below that typically full coherence of the outcomes cannot be obtained by reweighting each sample survey only once, and fixing the weights for all subsequent tabulations, as is customary practice. Instead we advocate a calibration strategy, where the weighting of survey data is repeated for each subsequent tabulation, such that any related outcome obtained in earlier phases of the overall estimation process will be matched. In the remainder of this paper we present this strategy in some detail.

In section 2 we briefly introduce the general regression and calibration estimator on which the whole exercise rests. In section 3 we sketch our sequential reweighting procedure by means of an artificial example. In section 4 we discuss a number of complications, and indicate solutions.

2 Generalized regression and calibration estimators

The customary way of constructing estimates based on a sample survey is *weighting*. Each element of the sample is assigned a raising weight. Given these weights population totals of the survey variables can be calculated by weighted counting, i.e. the values in the sample are multiplied by their weight and subsequently added. Primarily the weight represents the inclusion probability of the population element in the survey. Using the inverse of this inclusion probability as the weight we obtain the so-called Horvitz-Thompson, or direct estimator.

In many cases the direct estimator can be improved upon when for some variables observed in the survey or in the frame underlying the survey population totals are known. If, for instance, we know the number of females in the population, then we can adjust the Horvitz-Thompson weights such that weighted counting of the number of females *from the survey* exactly reproduces the known population total. In an elementary form the technique of reweighting, or calibrating estimators to match +known population totals is known as *post-stratification*, *(generalized) regression estimation*, or *calibration estimation*, see e.g. Bethlehem and Keller (1987), Särndal et al. (1992), Deville and Särndal (1992) and Deville et al. (1993). In the remainder of this section we sketch the procedure.

Starting from the Horvitz-Thompson weights, or any other set of weights properly representing the sampling design, a new set of weights is calculated that is as close as possible to the set of starting weights, but satisfies the restriction that a set of population totals is exactly matched. The new weights are used to derive estimators of population totals by weighted counting. The requirement that the new weights are "close" to the starting weights makes this an optimization problem. To solve it the concept of "closeness" has to be made operational. A convenient possibility is to conceive of a set of weights as an n-dimensional vector, n being the sample size, and then minimize the (weighted) squared Euclidian distance between the vector of starting weights and the vector of calibration weights. If this choice is being made the calibration weights obtained are identical to the weights associated with the generalized regression estimator, which we describe in the next paragraph.

Suppose that for all elements i of a sample S from a population U we have data on a target variable y_i as well as on a k-vector $x_i = (x_{1i}, x_{2i},\dots,x_{ki})^t$ of auxiliary variables. The direct estimator $Y_{HT} = \sum_{i\in S} d_i y_i$ of the population total Y of y_i is obtained using the Horvitz-Thompson weights d_i. Although not our target variable, we can use the same weights to derive a direct estimate X_{HT} of the population total X of x_i. Now suppose that from another source we know the true value of X, which almost certainly will not be equal to the estimate X_{HT}. The basic idea of the generalized regression estimator is to use the observed estimation error $(X - X_{HT})$ to improve upon the estimation error $(Y - Y_{HT})$ of the direct estimator Y_{HT}. Therefore we concentrate on the least squares regression $y_i - Y_{HT} = b^t(x_i - X_{HT})$, $i\in U$, b being a k-vector of unknown population regression coefficients representing the (multiple) correlation between the estimation errors of Y_{HT} and X_{HT}. Once an (approximately) unbiased estimate B of b has been obtained from the available sample a hopefully better estimate for Y obtains as $Y_{HT} + B^t(X - X_{HT})$. Working out the regression formalism it appears that the generalized regression estimator has a representation in terms of raising weights as follows

$$w_i = d_i[1 + x_i^t G_s (X - X_{HT})], \qquad (1)$$

where G_s is the (generalized)[1] inverse of the weighted design matrix $\sum_{i\in S} d_i x_i x_i^t$. This shows that by using the generalized regression estimator we convert the starting weights d_i in a new set of weights w_i satisfying the calibration conditions $\sum_{i\in S} w_i x_i = X$.

Under mild conditions the resulting regression/calibration type estimators are (approximately) design unbiased. This also holds true when it is not a known population total X to which the estimator is calibrated, but another design unbiased estimator thereof, e.g. an estimate which has been obtained from another survey, or from the same survey at an earlier estimation stage; see Renssen and Nieuwenbroek (1997).

3 A fictitious example: Samplonia

The country of *Samplonia* has exactly 1000 inhabitants; for all of them the population frame lists *place of residence* (Wheaton / Greenham / Newbay / Oakdale / Smokely / Crowdon / Mudwater), *age* (young / middle / old) and *sex* (male / female). A simple random sample of size 100 provides information on

[1] Traditionally the generalized regression estimator is defined for non-singular design matrices. In the present context we prefer to extend it to singular design matrices, as long as the resulting estimator is invariant to the choice of the generalized inverse. Sufficient conditions are discussed in Renssen and Martinus (2000).

employment (yes/no). The communities of Wheaton, Greenham and Newbay jointly constitute the *province* of Agria, the other four communities jointly constitute the province of Indusia. From these data we want to estimate several tables such that all outcomes are numerically consistent. Tables 1 and 2 can straightforwardly be tabulated from the population frame, so these tables constitute a possible point of departure. Notice that the marginal *sex* x *province* table is numerically identical in both tables, which obviously, is due to the fact that we tabulate from one and the same population list.

Table 1: Place of residence x sex

	Agria				Indusia					total
	Wheat	Green	Newb	total	Oakd	Smoke	Crowd	Mudw	total	
male	70	44	31	145	36	128	80	122	366	511
female	74	50	24	148	25	116	67	133	341	489
total	144	94	55	293	61	244	147	255	707	1000

Table 2: Sex x province x age

	male			female			total		
	Agria	Indusia	total	Agria	Indusia	total	Agria	Indusia	total
young	80	146	226	61	148	209	141	294	435
middle	47	156	203	57	135	192	104	291	395
old	18	64	82	30	58	88	48	122	170
total	145	366	511	148	341	489	293	707	1000

We now consider the weighting model for the available sample survey. In its fullest detail the population frame defines 7*3*2 = 42 population cells, which is too many in view of the sample size of 100; typically the weighting model cannot possibly exploit all available population counts available from the register. Assume that for the purpose of non-response correction and/or variance reduction we opt for a parsimonious incomplete post stratification model (*province* + *sex*), i.e. we calibrate our sample survey estimates to match the marginal population counts of *province* and *sex* separately. This guarantees that in subsequent tabulations from the weighted survey these marginal population counts will always be reproduced, *but other population counts will not*. This is demonstrated in tables 3 through 5.

Table 3: Place of residence x employment

	Agria				Indusia					total
	Wheat	Green	Newb	total	Oakd	Smoke	Crowd	Mudw	total	
job	66	25	32	123	32	65	76	65	238	361
no job	64	73	33	170	22	120	120	207	469	639
total	130	98	65	293	54	185	196	272	707	1000

Table 4: Sex x province x employment

	male			female			total		
	Agria	Indusia	total	Agria	Indusia	total	Agria	Indusia	Total
job	49	205	254	74	33	107	123	238	361
no job	105	152	257	65	317	382	170	469	639
total	154	357	511	139	350	489	293	707	1000

Table 5: Employment x age

	young	middle	old	total
job	38	312	11	361
no job	342	126	171	639
total	380	438	182	1000

A few features of these tables are worth mentioning. First, the set of tables 3-5 taken by itself is numerically consistent, since they represent the same underlying weighted micro data set. Once we stick to the weights derived, any subsequent tabulation gives outcomes that are numerically consistent with all earlier tabulations. This is the main reason for the common, and convenient practice to use just one single set of weights for a given survey data set, no matter the tabulation involved[2]. We emphasize this point, since in seeking a more far reaching numerical consistency, it is precisely this common practice that we have to abandon. Second, due to the weighting model employed the *sex* and *province* margins in tables 3-5 are numerically identical to the same margins in tables 1-2. Third, it can easily be checked that other common margins between the two sets of tables are numerically different. Indeed, comparing the last rows of table 1 and 3 we see that the counts for *place of residence* are not identical (the *province* subtotals are). Similarly, by comparing the last rows of tables 2 and 4 we see that the body of the *sex* x *province* subtable differs (the margins are identical). Also, by comparing the last column of table 2 to the last row of table 5 we see that the counts for *age* are different.

Overall numerical consistency between tables 1 through 5 could be obtained by using the less parsimonious weighting model *place of residence* + (*province* x *sex*) + *age*. However, the resulting weights would give inconsistent results for other tabulations from the population register again. Due to a fundamental lack of degrees of freedom it is generally impossible to use a weighting model which is sufficiently rich to adjust estimates against all possible future tabulations. The increased availability of administrative data sources supports the construction of rich population registers, from which very large numbers of subpopulation counts, i.e. potential tabulations, can be obtained. The inevitable conclusion is that no single set of weights per survey data set will bring about the overall consistency that we are looking for[3]. The solution we advocate in this paper is a sequential

---

[2] From the point of view of estimation theory *per se* there is no need to use the same reweighting model for different target variables. Indeed, it is quite likely that the optimal model differs per variable, so that "better" individual estimates will be obtained using models tailored to the variable at hand.

[3] Overall consistency can easily be obtained by a mass imputation strategy, where we impute values for the complete population, given the information from registers and sample

reweighting procedure, which uses a calibration approach to assure overall consistency with all statistical figures that have been released before. In our simple example the procedure runs as follows.

First we re-estimate table 3, forcing numerical consistency with the *place of residence* margin that has already been obtained in table 1. This can be done[4] by weighting the sample from scratch using the extended weighting model (*province* + *sex*) + (*place of residence*). Since *province* is contained in *place of residence* this amounts to using the model (*place of residence* + *sex*) to weight the sample in view of the estimation of table 3. The resulting table 3 will be numerically consistent with table 1. Secondly, and similarly, we re-estimate table 4, taking account of the *sex* x *province* subtable contained in tables 1-2, and the *province* x *employment* subtable contained in table 3, using the extended weighting model (*sex* x *province*) + (*province* x *employment*). Note that the *province* and *sex* margins of the original weighting model are contained in the (*sex* x *province*) term of the extended model now used to re-estimate table 4. Finally we re-estimate table 5, taking account of the *age* margin of table 2 and the *employment* margin of table 3, using the extended weighting model (*province* + *sex* + *age* + *employment*).

Using the sequential reweighting procedure indicated we obtain a fully consistent set of tables 1-5. The estimates are (approximately) design unbiased, and design variances of the estimates can be obtained. They cannot be claimed to be optimal in the sense of minimizing the (expected) design variance, though, and the figures obtained are dependent upon the order in which the tables are estimated. If, for instance, we had re-estimated table 5 before tables 3-4, the weighting model would have been extended with the *age* margin of table 2 only, thus giving a different weighting model, and therefore different estimates. This could be considered as a drawback of our procedure. However, if we estimate the most reliable tables first, and the least reliable last, some sort of overall optimality might occur.

In our simple example it is quite easy to identify the weighting model minimally to be employed in order to obtain estimates consistent with all earlier tabulations. However, in a more realistic setting, this is not so trivial. In the next section we discuss some complications due to more complicated data structures as well as more complicated dependencies between variables.

4 Some complications

The procedure we advocate involves three steps. First we match all available register and survey data for a given population, consisting of a well-defined finite set of elements, say persons, households or businesses. Secondly, if we want to estimate some cross tabulation we select the largest subset of records containing

---

surveys. As long as we tabulate from the virtual census thus obtained we get consistent outcomes. However, these outcomes are model based, and are likely to be severely biased in many respects, due to the fact that the imputation model employed is necessarily very restrictive in view of the richness of the available register data; see Kroese and Renssen (1999).

[4] Alternatively we can start from the calibration weights obtained from the original (*province* + *sex*) reweighting, and perform a second round of reweighting simply using the model (*place of residence*).

information about this cross tabulation, and we perform a first round of weighting for this subset. At this stage consistency requirements are not yet predominant: we select the weighting model primarily to reflect the sampling design, a correction for possible nonresponse bias, and variance reduction. Some consistency may be obtained, though, since already at this stage weighting is likely to involve calibration towards either some known population totals, or towards estimates obtained in earlier rounds of estimation. Thirdly, and finally, we derive the complete set of calibration restrictions the tabulation has to satisfy, given all earlier estimations that have been disseminated, and we perform a second round of weighting to match this set. This last step involves a careful examination of all relationships the variables involved in the tabulation have with the complete body of statistical output already published by the statistical office.

Matching problems are not the topic of this paper: therefore we shall simply assume that all data sources can be correctly matched at the level of the individual population elements. So we start with the second step. Consider figure 1 that depicts a situation where we have matched a complete register R_1 (*age, sex, etc*), an incomplete register R_2 (*wages*), and two disjoint samples: a labour survey S_1 (*profession, education*) and a health survey S_2 (*education, health*), with overlapping *education* variables. As the figure indicates we derive eight record sets from this matched file, each representing a different combination of variables. For any cross tabulation we want to estimate we select the largest of these eight record sets containing all the information needed. Table 6 lists the record sets to be used for the various possible tabulations.

Table 6: Feasible tabulations per record set.

	Persons	Variables *)	Aggregates *)
1	R_1	V_{R1}	V_{R1}
2	R_2	$V_{R1} \cup V_{R2}$	$V_{R2}, V_{R1} x V_{R2}$
3	$S_1 \cup S_2$	$V_{R1} \cup (V_{S1} \cap V_{S2})$	$V_{S1} \cap V_{S2}, (V_{S1} \cap V_{S2}) x V_{R1}$
4	$(S_1 \cup S_2) \cap R_2$	$V_{R1} \cup (V_{S1} \cap V_{S2}) \cup V_{R2}$	$(V_{S1} \cap V_{S2}) x V_{R2}$, $(V_{S1} \cap V_{S2}) x V_{R1} x V_{R2}$
5	S_1	$V_{R1} \cup V_{S1}$	$V_{S1} \backslash V_{S2}, (V_{S1} \backslash V_{S2}) x V_{R1}$
6	$S_1 \cap R_2$	$V_{R1} \cup V_{S1} \cup V_{R2}$	$(V_{S1} \backslash V_{S2}) x V_{R2}, (V_{S1} \backslash V_{S2}) x V_{R1} x V_{R2}$
7	S_2	$V_{R1} \cup V_{S2}$	$V_{S2} \backslash V_{S1}, (V_{S2} \backslash V_{S1}) x V_{R1}$
8	$S_2 \cap R_2$	$V_{R1} \cup V_{S2} \cup V_{R2}$	$(V_{S2} \backslash V_{S1}) x V_{R2}, (V_{S2} \backslash V_{S1}) x V_{R1} x V_{R2}$

*) $V_1 \cup V_2$: variables in set 1or in set 2
$V_1 \cap V_2$: variables jointly in set 1 and set 2
$V_1 \backslash V_2$: variables in set 1 but not in set 2

When deriving starting weights two complications arise. First the incomplete register R_2 has no known inclusion probabilities: unavailable data are not *missing by design*. Some assumption about the missing data mechanism has to be made, just as we do in case of nonresponse. For each of the record sets involving R_2 variables this inevitably brings us firmly within the realm of *model based* methods. Since we have no new solution to offer we will not elaborate on this point, and refer to the relevant literature on missing data; see e.g. Gelman e.a. (1995). Second, when we consider record set 3, i.e. the union of the labour and health surveys we have to derive joint inclusion probabilities, i.e. the probability that a person is sampled for either one or the other survey. We offer the following practical solution; see Renssen et al. (2000).

Figure 1. Decomposition of a matched micro data set into eight record sets

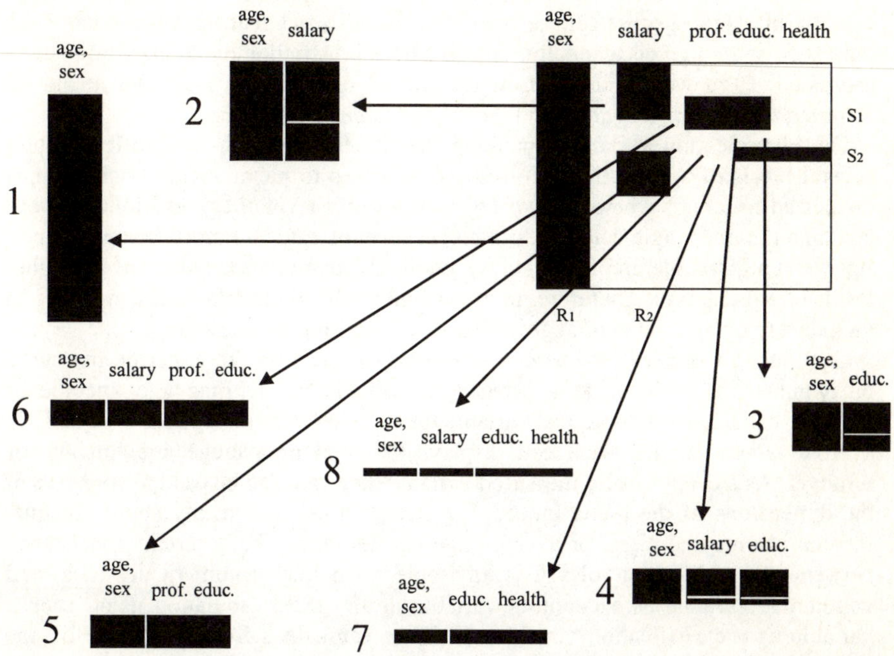

Let π_{1i} and π_{2i} denote the first order inclusion probabilities of the i-th element with respect to S_1 and S_2. Using equation (1) we calculate new calibration weights as follows. Define $d_i = \lambda\pi_{1i}^{-1}$ for $i \in S_1$ and $d_i = (1-\lambda)\pi_{2i}^{-1}$ for $i \in S_2$, as starting weights, with $\lambda \in [0,1]$. Define the associated pooled Horvitz-Thompson estimator of X as

$$\hat{X}_{HT} = \lambda \sum_{i \in S_1} x_i / \pi_{1j} + (1-\lambda) \sum_{i \in S_2} x_i / \pi_{2i}$$

and let G_s denote a generalized inverse matrix of the associated pooled design matrix

$$\lambda \sum_{i \in S_1} \frac{x_i x_i^t}{\pi_{1i}} + (1-\lambda) \sum_{i \in S_2} \frac{x_i x_i^t}{\pi_{2i}}.$$

Inserting these expressions in equation (1) we obtain a new set of weights satisfying the calibration equation $\sum_{i \in (S_1 \cup S_2)} w_i x_i = X$, where X is either a known population total or an estimate obtained at an earlier round of estimations.

Although the mixing parameter λ can be chosen freely, it ideally reflects the total survey error, i.e. the relative degree of confidence one has in each one of the two surveys.

We now turn to the third step in our sequential reweighting procedure, i.e. the determination of the calibration restrictions an estimated table has to satisfy in order to obtain full numerical consistency of the total output of a statistical office. Clearly, in case a specific cell, or range of cells such as a margin, has already been published at another occasion, this entails a trivial restriction on the new table to be estimated. Here we concentrate on less trivial ones, in particular those due to different *levels of aggregation* and those due to *consistency edits*.

The classification variables spanning the dimensions of a table often feature several levels of aggregation. We restrict ourselves to hierarchical classifications, i.e. nested structures where every category at a lower level of aggregation is always contained in one single category at a higher level of aggregation. When publishing figures at a given hierarchical level we implicitly also publish figures at all higher levels of aggregation. Therefore, in case of hierarchical classifications, we have to include any upward level of aggregation in our calibration restrictions.

Continuous variables such as *age* or *income* or *turnover* can either be measured on a numerical scale or as a categorical variable by defining age, income or turnover classes. As a numerical variable they can serve as the counting variable in a cross tabulation, i.e. each cell displays the total or average age, income or turnover. As a categorically measured variable they can also be used to span one of the dimensions of the table. Indeed, we can even report sensibly about the total income per income class, or average age per age class. To ascertain consistency between these different roles it is advisable to include a numerically measured continuous variable as a counting variable in any cross tabulation using such a variable as a classification variable. So if we estimate a frequency table of *age class* crossed by *income class*, we should not only count numbers of persons per table cell, but also the average age and total income and include these as estimates to be published. This guarantees that at a next occasion these figures will be taken into account as potential calibration restrictions.

A special relationship between variables exists in case of consistency edits, such as "IF (age < 18) THEN *driver's license* = no", or numerical edits of the type "*turnover* minus *costs* equals *profit*". Aggregates of these variables should properly reflect such edits. In searching for calibration restrictions we therefore have to inspect not only the set of published data, but also the set of consistency edits we want the published aggregates to satisfy. We elaborate a bit on IF-THEN edits for categorical variables. Suppose we want to estimate *driver's license* ownership by *municipality*, while the age distribution by municipality is known. Neglecting the age distribution in estimating license ownership we could easily obtain municipalities for which the proportion of license ownership is larger than the proportion of *age* ≥ 18, which is inconsistent under the stated edit rule. A solution is to inflate our target cross tabulation with the extra dimension *age* in two classes (<18, ≥18), so that we derive our calibration restrictions from the extended table *driver's license* x *municipality* x *age*. This leaves us with the two-way table *municipality* x *age* as our reweighting model. Using this model the resulting calibration weights will satisfy the known age distribution per municipality, which under the edit rule excludes estimates of license ownership larger than the proportion of *age* (≥ 18).

Inflating the target cross tabulation by an extra dimension seems to be a rather crude way to account for IF-THEN categorical edits, since this could easily result in unstable estimates due to insufficient degrees of freedom. However, such unstable estimates, if they arise, should be interpreted as a warning rather then as a drawback. Consider, again, the license ownership example. Suppose that in some municipality all *sampled* elements did own a license. Obviously, in this case the sample merely contains persons of age 18 or older. If we neglect the edit rule, the population fraction of license ownership is probably estimated as unity, which contradicts the register counts for age. If we use *age* x *municipality* as our reweighting model we end up with the post-stratification estimator, which is not well defined, though, since there are no persons younger than 18 in the sample. So the regression estimator collapses. Nevertheless it seems sensible to use the population fraction of persons of age 18 or older as an estimate of the population fraction of *license* ownership, given the outcome of the sample.

More formally, let $P(license)$ and $P(\geq 18)$ denote the population fraction of persons owning a license and persons of age 18 or over. Then by elementary probability theory we have

refer to the relevant literature on missing data; see e.g. Gelman e.a. (1995). Second, when we consider record set 3, i.e. the union of the labour and health surveys we have to derive joint inclusion probabilities, i.e. the probability that a person is sampled for either one or the other survey. We offer the following practical solution; see Renssen et al. (2000).

$$P(license) = P(license \mid \geq 18)P(\geq 18) + P(license \mid <18)[1 - P(\geq 18)]. \tag{2}$$

In case of post-stratification $P(license \mid \geq 18)$ and $P(license \mid <18)$ are both estimated from the sample, while $P(\geq 18)$ is obtained from a register. A precondition for post-stratification is that neither of the two sample strata should be empty. When this is the case, and the individual data satisfy the edit rule, the sample fraction of license ownership is zero in the below 18 stratum, which gives an estimate of zero for $P(license \mid <18)$, as it should be. However, given the edit rule, we do not need the sample to draw this conclusion: it is directly immanent that equation (2) reduces to

$$P(license) = P(license \mid \geq 18)P(\geq 18), \tag{3}$$

so that it is sufficient to estimate $P(license \mid \geq 18)$ from the sample. As long as the regression estimator is well defined this is automatically accomplished when we inflate the target table, including the variable *age* as an extra dimension. Otherwise we should explicitly concentrate on directly estimating the conditional proportion $P(license \mid \geq 18)$.

References

Bethlehem, J. and Keller, W. (1987). Linear weighting of sample survey data, *Journal of Official Statistics*, **3**, pp. 141-153.

Deville, J.C. and Särndal, C.E. (1992). Calibration estimators in survey sampling, *Journal of the American Statistical Association*, **87**, pp. 376-382.

Deville, J.C., Särndal, C.E. and Sautory, O. (1993). Generalized raking procedures in survey sampling, *Journal of the American Statistical Association*, **88**, pp. 1013-1020.

Gelman, A., Carlin, J.B., Stern, H.S. and Rubin, D.B. (1995). *Bayesian data analysis*, New York: Chapman and Hall.

Kroese, A.H. and Renssen, R.H. (1999). Weighting and imputation at Statistics Netherlands, paper presented at the International Associations of Survey Statisticians Satelite Conference on Small Area Estimation, Riga, Latvia.

Kroese, A.H., Renssen, R.H. and Trijssenaar, M. (2000). Weighting or imputation? Constructing a set of estimates based on data from different sources, to appear in *Netherlands Official Statistics*.

Renssen, R.H. (1998). Use of statistical matching techniques in calibration estimation, *Survey Methodology*, **24**, pp. 171-183.

Renssen, R.H. and Martinus, G.H. (2000). On the use of generalized inverse matrices in sampling theory; with references to Bascula 3.0, Research report, Department of Statistical Methods, Statistics Netherlands, Voorburg/Heerlen.

Renssen, R.H. and Nieuwenbroek, N.J. (1997). Aligning estimates for common variables in two or more sample surveys, *Journal of the American Statistical Association*, **92**, pp. 369-374.

Renssen, R.H., Kroese, A.H. and Willeboordse, A.J. (2000). Aligning estimates by repeated weighting, Research Report, Department of Statistical Methods, Statistics Netherlands, Voorburg/Heerlen.

Särndal, C.E., Swensson, B. and Wretman, J.H. (1992). *Model Assisted Survey Sampling*, New York: Springer Verlag.

Invited papers

Bayesian model selection methods for nonnested models

Francesco Bertolino[1], Elias Moreno[2] and Walter Racugno[1]

[1] Department of Mathematics, University of Cagliari, Italy
[2] Department of Statistics, University of Granada, Spain

Abstract. In the Bayesian approach to model selection and prediction, the posterior probability of each model under consideration must be computed. In the presence of weak prior information we need using default or automatic priors, that are typically improper, for the parameters of the models. However this leads to ill-defined posterior probabilities.

Several methods have recently been proposed to overcome this difficulty. Of particular interest are the intrinsic and fractional methodologies introduced by Berger and Pericchi (1996) and O'Hagan (1995), respectively. In the specific case of comparing nested models, a significant feature of these methods is that they allow to derive proper priors to be used in the analysis.

Unfortunately, the above methods do not apply to nonnested models: either the priors derived depend on the label we assign to models under comparison, and this may give two possible values to the posterior probability of each model, or they depends on the form of "encompassing".

We first consider a particular nonnested model selection problem: the one-sided testing problem. The encompassing approach for converting the nonnested problem into a nested one is discussed, and an alternative solution is proposed. Its behavior is illustrated on exponential distributions.

For more general nonnested model selection problems we argue that an accessible piece of prior information on the observable random variable helps to find the posterior probability of the models. The way to deal with such a prior information is developed and illustrated on the comparison of separate normal and double exponential families of distributions.

Keywords. Bayes factor, Intrinsic prior, Nonnested models

1 Introduction

Suppose that data $\mathbf{x} = (x_1, ..., x_n)$ are independently drawn from either model M_1 or from M_2. Under M_i the density is $f_i(x|\theta_i)$, where $\theta_i \in \Theta_i$, and prior $\pi_i(\theta_i)$, $i = 1, 2$. Given a prior P on the set $\{M_1, M_2\}$, the Bayes theorem provides the posterior probabilities

$$P(M_i|\mathbf{x}) = \frac{m_i(\mathbf{x})P(M_i)}{m_1(\mathbf{x})P(M_1) + m_2(\mathbf{x})P(M_2)}, \; i = 1, 2,$$

where $m_i(\mathbf{x}) = \int_{\Theta_i} f_i(\mathbf{x}|\theta_i)\pi_i(\theta_i)d\theta_i$ is the marginal density of $\mathbf{x}$ under model M_i.

When we want to choose one of the above models, as in testing problems, it is easily seen that the optimal decision under 0-1 loss function is to choose

M_1 if the inequality $\frac{P(M_1|\mathbf{x})}{P(M_2|\mathbf{x})} \geq 1$, holds. This ratio can also be written:

$$\frac{P(M_1|\mathbf{x})}{P(M_2|\mathbf{x})} = B_{12}(\mathbf{x})\frac{P(M_1)}{P(M_2)},$$

where $B_{12}(\mathbf{x}) = m_1(\mathbf{x})/m_2(\mathbf{x})$ is known as the Bayes factor of model M_1 *versus* M_2, and contains all the information the data provides on the posterior odds.

Other interesting loss functions for model choice can be found in San Martini and Spezzaferri (1984) and Bernardo and Smith (1994).

When we are concerned with the density of a future observation y, the posterior density $f(y|\mathbf{x})$ is given by

$$f(y|\mathbf{x}) = f(y|\mathbf{x}, M_1)P(M_1|\mathbf{x}) + f(y|\mathbf{x}, M_2)P(M_2|\mathbf{x}).$$

In other words both for the decision problem of choosing a model and for predictive purposes, the posterior probability of each model must be computed.

When subjective elicitation of priors is not available automatic or default priors are often assumed. That is, the prior P is set as $P(M_1) = P(M_2)$, and $\pi_i(\theta_i)$ as the reference prior (Berger and Bernardo, 1992). The reference prior is typically improper and hence it is written $\pi_i^N(\theta_i) = c_i h(\theta_i)$, where c_i is an indeterminate positive constant. In this situation,

$$P(M_1|\mathbf{x}) = (1 + B_{21}^N(\mathbf{x}))^{-1}, \tag{1}$$

is not well-defined since $B_{21}^N(\mathbf{x}) = m_2^N(\mathbf{x})/m_1^N(\mathbf{x})$ depends on the arbitrary constant c_2/c_1. Various solutions for overcoming this difficulty have been proposed in the literature, which are briefly review here. These solutions differ significantly depending on whether the models are nested or nonnested.

The sampling model $\{f_1(x|\theta_1), \theta_1 \in \Theta_1\}$ is said to be nested in $\{f_2(x|\theta_2), \theta_2 \in \Theta_2\}$, if

$$(i)\ \Theta_1 \subset \Theta_2, \qquad (ii)\ f_2(x|\theta_2) \equiv f_1(x|\theta_1),\ \text{for } \theta_2 = \theta_1.$$

Otherwise, the models are said to be nonnested. For instance, nonnested models arise in the important one-sided testing problem which we will consider.

1.1 The Bayesian information criterion

The so-called *Bayesian information criterion* uses an asymptotic approximation and replaces $B_{21}^N(\mathbf{x})$ in (1) with

$$B_{21}^S(\mathbf{x}) = \frac{f_2(\mathbf{x}|\hat{\theta}_2(\mathbf{x}))}{f_1(\mathbf{x}|\hat{\theta}_1(\mathbf{x}))} \exp\{\frac{\log n}{2}\ (d_1 - d_2)\},$$

where $\hat{\theta}_i(\mathbf{x})$ is the maximum likelihood estimator of θ_i, and d_i the dimension of Θ_i. The approximation is obtained by expanding $m_i^N(\mathbf{x})$ at point $\hat{\theta}_i(\mathbf{x})$ and deleting the dependency of $\pi_i^N(\hat{\theta}_i)$. It has been argued, however, that $\pi_2^N(\hat{\theta}_2)/\pi_1^N(\hat{\theta}_1)$ can be a dominant factor for small sample size n, so that it cannot be ignored in practice (Berger and Pericchi, 1996). Further, $B_{21}^S(\mathbf{x})$ does not correspond to a Bayesian analysis except for special nested models.

1.2 Intrinsic Bayes factor

The basic idea is to split the sample $\mathbf{x}$ into two parts $(x(\ell), x(n-\ell))$. Part $x(\ell)$, called *training sample*, is designed to convert the improper prior into a proper posterior. That is,

$$\pi_i^N(\theta_i|x(\ell)) = \frac{f_i(x(\ell)|\theta_i)\pi_i^N(\theta_i)}{m_i^N(x(\ell))}, \; i=1,2, \tag{2}$$

where $x(\ell)$ is such that $0 < m_i^N(x(\ell)) < \infty$. With the remaining part $x(n-\ell)$ the Bayes factor is computed by using (2) as the prior. This gives

$$B_{21}^P(\mathbf{x}) = \frac{\int f_2(x(n-\ell)|\theta_2)\pi_2^N(\theta_2|x(\ell))d\theta_2}{\int f_1(x(n-\ell)|\theta_1)\pi_1^N(\theta_1|x(\ell))d\theta_1}, \tag{3}$$

which is called partial Bayes factor (PBF) by O'Hagan (1995). Equivalently, (3) can be written as

$$B_{21}^P(\mathbf{x}) = \frac{m_2^N(\mathbf{x})}{m_1^N(\mathbf{x})}\,\frac{m_1^N(x(\ell))}{m_2^N(x(\ell))} = B_{21}^N(\mathbf{x})\; B_{12}^N(x(\ell)). \tag{4}$$

The PBF corrects $B_{21}^N(\mathbf{x})$ with the correction term $B_{12}^N(x(\ell))$, and by so doing the arbitrary constants c_1 and c_2 cancel out in (4).

To avoid the dependence of the training sample, Berger and Pericchi (1996) suggested first considering all possible subsamples $x(\ell)$ for which there is no proper subsample satisfying the inequalities $0 < m_i^N(x(\ell)) < \infty$ for any c_i. They termed this subsample a *minimal training sample*. Second, they take the arithmetic mean of the PBFs for all minimal training samples. This produces the so-called arithmetic intrinsic Bayes factor (AIBF), defined as

$$B_{21}^{AI}(\mathbf{x}) = B_{21}^N(\mathbf{x})\frac{1}{L}\sum_{\ell=1}^{L} B_{12}^N(x(\ell)),$$

where L is the number of minimal training samples contained in the sample. Other ways of "averaging" PBFs are in Berger and Pericchi (1996, 1998).

1.3 Fractional Bayes factor

An alternative way of avoiding the arbitrariness of choosing the training sample for which the PBF is computed, was developed by O'Hagan (1995). He replaces the correction term $B_{21}^N(x(\ell))$ in (4) with

$$F_{21}(\mathbf{x}) = \frac{\int\{f_2(\mathbf{x}|\theta_2)\}^{\ell/n}\pi_2^N(\theta_2)d\theta_2}{\int\{f_1(\mathbf{x}|\theta_1)\}^{\ell/n}\pi_1^N(\theta_1)d\theta_1}.$$

In this way, he defines the fractional Bayes factor (FBF) as

$$B_{12}^F(\mathbf{x}) = B_{12}^N(\mathbf{x})\; F_{21}(\mathbf{x}).$$

Other fractions apart from ℓ/n can be considered. In fact, O'Hagan(1995) argues that a larger fraction would reduce sensitivity to the prior and he also proposes using the fractions $\log n/n$ or $\sqrt{n}/n$. However, compelling arguments exist in fovour of the fraction ℓ/n, (Berger and Mortera 1995, Moreno 1997).

The above "Bayes factors" have been extensively studied (see, for instance, Berger and Pericchi (1996, 1998), Bertolino and Racugno (1996), De Santis and Spezzaferri (1999), O'Hagan (1995, 1997), Sansó, Pericchi and Moreno (1995)).

1.4 Intrinsic priors

Note that the intrinsic and fractional Bayes factors are not *actual* Bayes factors. Further, stability and coherence of the AIBF are a matter of concern (Bertolino and Racugno, 1996). Conceivably, for a given sample $\mathbf{x}$, the number of minimal training samples might be small and small changes in the data could make that number vary substantially. Moreover, the equality $B_{21}^{AI}(\mathbf{x}) = 1/B_{12}^{AI}(\mathbf{x})$ is not necessarily satisfied, so that the coherent equality $P(M_1|\mathbf{x}) = 1 - P(M_2|\mathbf{x})$ does not hold.

To be coherent, an important question is to know whether $B_{21}^{AI}(\mathbf{x})$ corresponds to an actual Bayes factor for sensible priors. If so, it provides a Bayesian justification for the method. With the so-called *intrinsic* priors, this question was asymptotically answered by Berger and Pericchi (1996). These are priors $\pi_1^I(\theta_1)$ and $\pi_2^I(\theta_2)$ for which the corresponding Bayes factor

$$B_{21}(\mathbf{x}) = \frac{\int_{\Theta_2} f_2(\mathbf{x}|\theta_2)\pi_2^I(\theta_2)d\theta_2}{\int_{\Theta_1} f_1(\mathbf{x}|\theta_1)\pi_1^I(\theta_1)d\theta_1},$$

and $B_{21}^{AI}(\mathbf{x})$ are asymptotically equivalent under the two models M_1 and M_2. Note that using the Bayes factor for the intrinsic priors in place of the improper priors we started with, coherency is assured.

By equating the limit of $B_{21}^{AI}(\mathbf{x})$ and $B_{21}(\mathbf{x})$ as $n \to \infty$ under the two models $B_{21}(\mathbf{x}) = B_{21}^{AI}(\mathbf{x})(1+o_P(1))$, Berger and Pericchi (1996) showed that the intrinsic priors satisfy the functional equations

$$E^{M_1}_{x(\ell)|\theta_1} B_{12}^N(x(\ell)) = \frac{\pi_2^I(\psi_2(\theta_1))}{\pi_2^N(\psi_2(\theta_1))}\frac{\pi_1^N(\theta_1)}{\pi_1^I(\theta_1)},$$

$$E^{M_2}_{x(\ell)|\theta_2} B_{12}^N(x(\ell)) = \frac{\pi_2^I(\theta_2)}{\pi_2^N(\theta_2)}\frac{\pi_1^N(\psi_1(\theta_2))}{\pi_1^I(\psi_1(\theta_2))}.$$

The expectations in these equations are taken with respect to $f_1(x(\ell)|\theta_1)$ and $f_2(x(\ell)|\theta_2)$, respectively; $\psi_2(\theta_1)$ denotes the limit of the maximum likelihood estimator $\hat{\theta}_2(\mathbf{x})$ under model M_1 at the point θ_1, and $\psi_1(\theta_2)$ the limit of $\hat{\theta}_1(\mathbf{x})$ under model M_2 at the point θ_2 (Cox, 1961 and Dmochowsky, 1996).

For nested models the above functional equations collapse into a single one. Although the solution (π_1^I, π_2^I) to this single equation is not unique, a sensible selection is the pair

$$\pi_1^I(\theta_1) = \pi_1^N(\theta_1),\ \pi_2^I(\theta_2) = \pi_2^N(\theta_2)E^{M_2}_{x(\ell)|\theta_2} B_{12}^N(x(\ell)). \tag{5}$$

These are a unique limit of proper intrinsic priors based on a fully automatic procedure (Moreno, Bertolino and Racugno, 1998b). The Bayes factor for the intrinsic priors in (5) becomes

$$B_{21}(\mathbf{x}) = \frac{\int_{\Theta_2} f_2(\mathbf{x}|\theta_2)\pi_2^I(\theta_2)d\theta_2}{\int_{\Theta 1} f_1(\mathbf{x}|\theta_1)\pi_1^N(\theta_1)d\theta_1},$$

which does not depend on the arbitrary constants c_1, c_2. This is a limit of actual Bayes factors and it has been proved to behave well when applied to some important nested testing problems; see Berger and Pericchi

(1996, 1998), Moreno, Bertolino and Racugno (1999a), Bertolino, Moreno and Racugno (1999), Moreno and Liseo (1998), among others.

A similar construction to that of intrinsic priors but using the fractional methodology is possible (Moreno, 1997).

1.5 Preview

In nonnested situations, however, the above two functional equations do not collapse into a single one. It is not yet known what the solutions are, or even if a solution exists.

For some models an *encompassing* approach allows us to convert the nonnested problem into a nested one (Cox, 1961, Berger and Pericchi, 1996). This is the case of the one-sided testing problem which is considered in Section 2, where the encompassing model approach is discussed and an alternative solution is proposed and illustrated for a simple example.

Section 3 deals with nonnested models for which the solution proposed in Section 2 does not apply. It is shown that by adding weak prior information on the observable random variable X, a procedure for model selection can be constructed. We illustrate the procedure with separate normal and double exponential families. Concluding remarks are made in Section 4.

2 One-sided testing

Consider the sampling model $f(x|\theta)$, where θ is an unknown real parameter. The one-sided testing problem consists in testing the null hypothesis $H_0 : \theta \leq \theta_0$ *versus* the alternative $H_1 : \theta \geq \theta_0$, where θ_0 is a specified value. Rewriting the problem in the form used in the preceding section we have the models

$$M_1 : f(x|\theta_1),\ \pi_1^N(\theta_1) = \pi^N(\theta_1) 1_{(-\infty,\theta_0)}(\theta_1),$$
$$M_2 : f(x|\theta_2),\ \pi_2^N(\theta_2) = \pi^N(\theta_2) 1_{(\theta_0,\infty)}(\theta_2),$$

where $\pi^N(\theta)$ is the reference prior associated to $f(x|\theta)$, and 1_A represents the indicator function of set A.

This is a nonnested problem for which the procedure for finding the intrinsic priors in subsection 1.4 does not apply. The encompassing approach proposes embedding M_1 and M_2 in the larger model $M_3 = M_1 \cup M_2 : f(x|\theta), \pi^N(\theta) 1_{(-\infty,\infty)}(\theta)$. Thus, M_1 and M_2 are nested in M_3 so that intrinsic priors can be constructed to compute $B_{31}(\mathbf{x})$ and $B_{23}(\mathbf{x})$. Then, $B_{21}(\mathbf{x})$ is defined as $B_{21}(\mathbf{x}) = B_{23}(\mathbf{x}) B_{31}(\mathbf{x})$. However, there are no grounds for accepting this equality. This is due to the fact that when computing $B_{23}(\mathbf{x})$ the intrinsic prior for the parameter θ in M_3 comes from the comparison of M_2 *versus* M_3. This prior is not necessarily the same as the one given by computing $B_{31}(\mathbf{x})$ which is obtained comparing M_1 *versus* M_3. Hence, the denominator of $B_{23}(\mathbf{x})$ does not cancel out with the numerator of $B_{31}(\mathbf{x})$ and consequently $B_{21}(\mathbf{x})$ does not correspond with an actual Bayes factor.

The solution we propose consists in considering the auxiliary model $M_0 : f(x|\theta),\ \pi(\theta) = 1_{\{\theta_0\}}(\theta), f(x|\theta)$, defined by the singular point θ_0, which is nested in M_3. Then, according to (5) the intrinsic priors for comparing M_0 *versus* M_3 are

$$\pi_0^I(\theta) = 1_{\{\theta_0\}}(\theta),\ \pi^I(\theta|\theta_0) = \pi^N(\theta) E_{x(\ell)|\theta}^{M_3} B_{03}^N(x(\ell)), \tag{6}$$

where $B_{03}^N(x(\ell)) = f(x(\ell)|\theta_0) / \int f(x(\ell)|\theta) \pi^N(\theta) d\theta$.

Note that $\pi^I(\theta|\theta_0)$ is a proper prior. Therefore, we can consider its restriction to the set $\{\theta \le \theta_0\}$ and $\{\theta \ge \theta_0\}$. So we have the well-defined models

$$M_1 : f(x|\theta_1),\ \pi_1^I(\theta_1|\theta_0) = \frac{1}{k_1}\pi^I(\theta_1|\theta_0)1_{(-\infty,\theta_0)}(\theta_1),$$

$$M_2 : f(x|\theta_2),\ \pi_2^I(\theta_2|\theta_0) = \frac{1}{k_2}\pi^I(\theta_2|\theta_0)1_{(\theta_0,\infty)}(\theta_2),$$

where $k_1 = \int_{-\infty}^{\theta_0} \pi^I(\theta|\theta_0)d\theta$, $k_2 = \int_{\theta_0}^{\infty} \pi^I(\theta|\theta_0)d\theta$.

Now the coherent equality $B_{12}(\mathbf{x}) = B_{10}(\mathbf{x})\ B_{02}(\mathbf{x})$ holds as long as we are dealing with actual Bayes factors.

Example 1. Let X be a random variable with exponential density $f(x|\theta) = \theta^{-1}\exp\{-x/\theta\}$. Suppose that we want to test $H_0 : \theta \le 1$ *versus* $H_1 : \theta \ge 1$. The two default models under comparison are

$$M_1 : f(x|\theta_1),\ \pi_1^N(\theta_1) = \frac{c_1}{\theta_1}1_{(0,1)}(\theta_1),\quad M_2 : f(x|\theta_2),\ \pi_2^N(\theta_2) = \frac{c_2}{\theta_2}1_{(1,\infty)}(\theta_2).$$

The encompassing model is

$$M_3 : f(x|\theta_3),\ \pi_3^N(\theta_3) = \frac{c_3}{\theta_3}1_{(0,\infty)}(\theta_3).$$

The minimal training sample is a single replication of X. From (5) it follows that the intrinsic priors for comparing M_1 *versus* M_3 are

$$\pi_1^I(\theta_1) = \frac{c_1}{\theta_1}1_{(0,1)}(\theta_1),\ \pi_{13}^I(\theta_3) = \frac{c_1}{\theta_3(1+\theta_3)}.$$

Similarly, the intrinsic priors for comparing M_2 *versus* M_3 are

$$\pi_2^I(\theta_2) = \frac{c_2}{\theta_2}1_{(1,\infty)}(\theta_2),\ \pi_{23}^I(\theta_3) = \frac{c_2}{(1+\theta_3)}.$$

Note that $\pi_{13}^I(\theta_3)$ is not the same as $\pi_{23}^I(\theta_3)$ so that $B_{31}(\mathbf{x})B_{32}(\mathbf{x})$ is not a Bayes factor but a product of Bayes factors.

The solution we propose considers the auxiliary model

$$M_0 : f(x|\theta) = \frac{1}{\theta}\exp\{-\frac{x}{\theta}\},\ \pi_0(\theta) = 1_{\{1\}}(\theta).$$

Then, from (6) the intrinsic priors for M_1 and M_2 turn out to be

$$\pi_1^I(\theta_1|\theta_0 = 1) = \frac{2}{(1+\theta_1)^2}1_{(0,1)}(\theta_1),\ \pi_2^I(\theta_2|\theta_0 = 1) = \frac{2}{(1+\theta_2)^2}1_{(1,\infty)}(\theta_2).$$

By construction, both intrinsic priors are probability densities.

For the data set $\mathbf{x}$, the Bayes factor $B_{21}(\mathbf{x})$ for these intrinsic priors is

$$B_{21}(\mathbf{x}) = \frac{\int_1^{\infty} \theta_2^{-n}(1+\theta_2)^{-2}\exp\{-n\,\bar{x}/\theta_2\}\,d\theta_2}{\int_0^1 \theta_1^{-n}(1+\theta_1)^{-2}\exp\{-n\,\bar{x}/\theta_1\}\,d\theta_1},$$

which only depends on $\mathbf{x}$ through the sufficient statistic $(n, \bar{x})$. For different values of the statistic $(n, \bar{x})$ and prior $P(M_1) = P(M_2) = 1/2$ the corresponding values of the posterior probability of M_1 are shown in Table 1.

Table 1. Posterior probabilities of model M_1

$(n,\bar{x})$	$P(M_1\|n,\bar{x})$	$(n,\bar{x})$	$P(M_1\|n,\bar{x})$
$(5,0.1)$	0.999	$(10,0.1)$	≈ 1
$(5,0.6)$	0.806	$(10,0.6)$	0.911
$(5,1.0)$	0.448	$(10,1.0)$	0.461
$(5,2.0)$	0.035	$(10,2.0)$	0.006

The numbers in Table 1 show that for small values of $\bar{x}$, that favour M_1, the posterior probability of M_1 is high, moreso as the sample size increases. As the sample mean increases the posterior probability decreases. For a large sample size and $\bar{x}=1$, the posterior probability of M_1 is 0.5 which is sensible.

2.1 Prediction

Suppose that we want to predict a future value y. The predictive density of y, conditional on $\mathbf{x}$, is

$$f(y|\mathbf{x}) = f(y|\mathbf{x},M_1)P(M_1|\mathbf{x}) + f(y|\mathbf{x},M_2)P(M_2|\mathbf{x}),$$

where $f(y|\mathbf{x},M_i) = \int f(y|\theta_i)\pi_i^I(\theta_i|\mathbf{x})d\theta_i$.

Example 1 (continued). The predictive density conditional on $(n,\bar{x})$ is

$$f(y|n,\bar{x}) = f(y|n,\bar{x},M_1)P(M_1|n,\bar{x}) + f(y|n,\bar{x},M_2)P(M_2|n,\bar{x}),$$

where

$$f(y|n,\bar{x},M_i) = \int \theta_i^{-1}\exp\{-y/\theta_i\}\pi_i^I(\theta_i|n,\bar{x},M_i)d\theta_i,\ i=1,2,$$

$$\pi_1^I(\theta_1|n,\bar{x},M_1) = \frac{\theta_1^{-n}(1+\theta_1)^{-2}\exp\{-n\,\bar{x}/\theta_1\}}{\int_0^1 \theta_1^{-n}(1+\theta_1)^{-2}\exp\{-n\,\bar{x}/\theta_1\}\,d\theta_1}1_{(0,1)}(\theta_1),$$

$$\pi_2^I(\theta_2|n,\bar{x},M_i) = \frac{\theta_2^{-n}(1+\theta_2)^{-2}\exp\{-n\,\bar{x}/\theta_2\}}{\int_1^\infty \theta_2^{-n}(1+\theta_2)^{-2}\exp\{-n\,\bar{x}/\theta_2\}\,d\theta_2}1_{(1,\infty)}(\theta_2).$$

Predictive densities for $(n,\bar{x})=(5,0.6)$ and $(n,\bar{x})=(5,1.2)$ are plotted in Figure 1 with continuous and dashed line respectively.

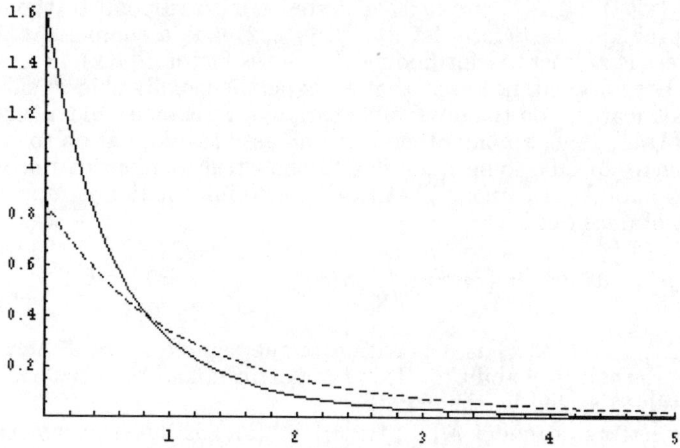

Figure 1: Predictive densities

This figure shows that for a sample mean value that favours M_1 the predictive density has a thinner tail than that exhibited for a sample mean value that favours M_2. This is sensible. Further it can be seen that their behavior is quite robust with respect to the sample size.

2.2 Extensions

The procedure here proposed, generalizes to the case of the one-sided testing problem in presence of a nuisance parameter, assuming that the parameter of interest and the nuisance parameter are *a priori* independent. That is, consider the model $f(x|\theta,\psi)$ and suppose we want to test $H_1 : \theta \leq \theta_0$ *versus* $H_2 : \theta \geq \theta_0$, where ψ is a nuisance parameter.

We consider the auxiliary model $M_0 : f(x|\theta_0,\psi_0)$, where ψ_0 is an arbitrary but fixed coordinate, and compute the intrinsic prior for the parameter (θ,ψ) that results from comparing M_0 *versus* $M_3 = M_1 \cup M_2$. This prior is given by

$$\pi^I(\theta,\psi|\theta_0,\psi_0) = \pi^N(\theta,\psi) E^{M_3}_{x(\ell)|\theta,\psi} \frac{f(x(\ell)|\theta_0,\psi_0)}{\int_{(\psi)}\int_{(\theta)} f(x(\ell)|\theta,\psi)\pi^N(\theta,\psi)d\theta d\psi},$$

where $\pi^N(\theta,\psi)$ is the reference prior for the sampling model $f(x|\theta,\psi)$. The restriction of $\pi^I(\theta,\psi|\theta_0,\psi_0)$ to $\{\theta \leq \theta_0\}$ and $\{\theta \geq \theta_0\}$ are denoted respectively as $\pi^I_i(\theta,\psi|\theta_0,\psi_0)$, $i = 1, 2$.

Hence, for the dataset $\mathbf{x}$ the Bayes factor is

$$B_{21}(\mathbf{x}) = \frac{\int_{(\psi_0)}\int_{(\psi)}\int_{\theta\geq\theta_0} f(\mathbf{x}|\theta,\psi)\pi^I_1(\theta,\psi|\theta_0,\psi_0)\pi^N(\psi_0)d\theta d\psi d\psi_0}{\int_{(\psi_0)}\int_{(\psi)}\int_{\theta\leq\theta_0} f(\mathbf{x}|\theta,\psi)\pi^I_2(\theta,\psi|\theta_0,\psi_0)\pi^N(\psi_0)d\theta d\psi d\psi_0},$$

where the arbitrary coordinate ψ_0 has been integrated out with the reference prior $\pi^N(\psi_0)$. Note that this Bayes factor is well-defined as long as the arbitrary constant involved in $\pi^N(\psi_0)$ cancels out in the ratio.

3 General nonnested models

Suppose that the models under consideration, $M_1 : \{f_1(x|\theta_1), \pi^N_1(\theta_1)\}$ and $M_2 : \{f_2(x|\theta_2), \pi^N_2(\theta_2)\}$ are neither nested nor correspond to the one-sided testing problem. As before, let $\mathbf{x} = (x_1, ..., x_n)$ be a sample. As far as we know there is no method for finding the Bayes factor $B_{12}(\mathbf{x})$.

It has been argued, however, that an expert is usually able to subjectively elicit some features on the observable variable X (Kadane and Wolfson, 1998 and O'Hagan, 1998, among others). An accessible information to the expert would be to specify some quantiles of the "true" unknown unconditional distribution of X, say $m(x)$. That is, the prior information is represented by the class of densities

$$\mathfrak{M}_k(\mathbf{p}) = \{m(x) : \int_{\mathfrak{X}_i} m(x)dx = p_i,\ i = 1, ..., k\},$$

where $\{\mathfrak{X}_i,\ i = 1, ..., k\}$ is a partition of the sample space $\mathfrak{X}$ and $\{p_i,\ i = 1, ..., k\}$ are their probabilities. Both the partition and their probabilities are subjectively specified by the expert.

In this setting, a model $M_i : \{f(x|\theta_i), \pi^N(\theta_i)\}$ is but a parametric representation of the "true" unknown distribution $m(x)$. A first question arises: is model M_i compatible with the expert class? In other words: is the marginal $m_i^N(x) = \int f_i(y|\theta_i)\pi_i^N(\theta_i)d\theta_i$ contained in the expert class $\mathfrak{M}_k(\mathbf{p})$? The answer is clearly no since $m_i^N(x)$ is not well-defined as it depends on the arbitrary constant c_i involved in $\pi_i^N(\theta_i)$.

One way of overcoming this difficulty is to convert $\pi_i^N(\theta_i)$ into a proper prior by using a minimal theoretical training sample. More precisely, suppose that a single replication of X, say y, is such that

$$\pi_i(\theta_i|y) = \frac{f_i(y|\theta_i)\pi_i^N(\theta_i)}{\int_{\Theta_i} f_i(y|\theta_i)\pi_i^N(\theta_i)d\theta_i},\ y \in \mathfrak{X},$$

is a proper density. Then integrating out y we obtain a proper prior for θ_i. The choice of the distribution for integrating out y has been discussed by Perez and Berger (1999). For instance, under no prior information they suggest of using the empirical distribution function of the sample.

In our situation where prior information is available, Moreno, Bertolino and Racugno (1999b) argue that y should be integrated out by using any $m(y) \in \mathfrak{M}_k(\mathbf{p})$, since this is the "true" marginal density of y. By so doing we obtain $\pi_i^I(\theta_i|m) = \int_{\mathbf{X}} \pi_i(\theta_i|y)m(y)dy$. Note that there are a multiplicity of priors $\pi_i^I(\theta_i|m)$, one for each $m \in \mathfrak{M}_k(\mathbf{p})$.

The compatibility of the model and the prior information can be asserted if $m_i^I(x|m) = \int_{\Theta_i} f_i(y|\theta_i)\pi_i^I(\theta_i|m)d\theta_i$ is a member of the class $\mathfrak{M}_k(\mathbf{p})$ for some $m \in \mathfrak{M}_k(\mathbf{p})$. We also want to characterize the subclass of marginal $m(y)$ for which compatibility is achieved. This characterization is given in the next theorem.

Theorem *The model $\{f(x|\theta), \pi^I(\theta|m)\}$, where $m \in \mathfrak{M}_k(\mathbf{p})$, matches the expert prior information $\mathfrak{M}_k(\mathbf{p})$, if and only if the equations*

$$p_i = \sum_{j=1}^{k} p_j P^f(\mathfrak{X}_i|y_j),\ i = 1, ..., k-1, \tag{7}$$

have at least one solution $y_j \in \mathfrak{X}_j,\ j = 1, ..., k$, where

$$P^f(\mathfrak{X}_i|y) = \int_\Theta P^f(\mathfrak{X}_i|\theta)\pi^f(\theta|y)d\theta, \text{ and } P^f(\mathfrak{X}_i|\theta) = \int_{\mathbf{X}_i} f(x|\theta)dx.$$

Proof. See Theorem 1 in Moreno, Bertolino and Racugno, 1999b.

Consider the subclass of densities of $\mathfrak{M}_k(\mathbf{p})$ defined as

$$\mathfrak{M}_k^*(\mathbf{p}) = \{m(y) : m(y) = \sum_{i=1}^{k} p_j 1_{\{y_j^*\}}(y)\},$$

where $\{y_j^*, j = 1, ..., k\}$ denotes a generic solution to (7). Under model $\{f(x|\theta), \pi^I(\theta|m)\}$, where m is such that compatibility is achieved, the data $\mathbf{x}$ have marginal values in the interval $(\underline{m}^I(\mathbf{x}), \overline{m}^I(\mathbf{x}))$ given by

$$\underline{m}^I(\mathbf{x}) = \inf_{m \in \mathrm{M}_k^*(\mathbf{p})} m^I(\mathbf{x}|m), \text{ and } \overline{m}^I(\mathbf{x}) = \sup_{m \in \mathrm{M}_k^*(\mathbf{p})} m^I(\mathbf{x}|m).$$

It is to be hoped that this interval is small enough; this would mean that $m^I(\mathbf{x}|m)$ is robust as m varies over the subclass of $\mathfrak{M}_k(\mathbf{p})$ for which compatibility is achieved.

Applying the above procedure to model M_1 and M_2 separately, the range of the Bayes factor $B_{21}(\mathbf{x})$ is $(\underline{B}_{21}(\mathbf{x}), \overline{B}_{21}(\mathbf{x}))$, where

$$\underline{B}_{21}(\mathbf{x}) = \frac{\underline{m}_2^I(\mathbf{x})}{\overline{m}_1^I(\mathbf{x})}, \ \overline{B}_{21}(\mathbf{x}) = \frac{\overline{m}_2^I(\mathbf{x})}{\underline{m}_1^I(\mathbf{x})}.$$

An important point is how large is the interval $(\underline{B}_{21}(\mathbf{x}), \overline{B}_{21}(\mathbf{x}))$. This is a question that in general cannot be answered. It depends not only on the models and prior information but also on the collected data. However, we feel that for models that contain a small number of parameters, a few quantiles are enough to obtain robustness of the Bayes factor. Let us illustrate the method with a simple example.

Example 2. Suppose we want to assess the precision of a measuring instrument. A sample $\mathbf{x}$ is collected and there are two possible sampling models: normal and double exponential with known location parameter. Without loss of generality the location is fixed at zero. Then, the models are

$$M_1 : f_1(x|\tau) = \frac{\tau^{1/2}}{(2\pi)^{1/2}} \exp\{-\frac{\tau\, x^2}{2}\}, \ \pi_1^N(\tau) = \frac{c_1}{\tau} 1_{R^+}(\tau),$$

$$M_2 : f_2(x|\lambda) = \frac{\lambda}{2} \exp\{-\lambda\, |x|\}, \ \pi_2^N(\lambda) = \frac{c_2}{\lambda} 1_{R^+}(\lambda),$$

where $\pi_1^N(\tau)$ and $\pi_2^N(\lambda)$ are the reference priors and c_1, c_2 arbitrary positive constants. We assume that $P(M_1) = P(M_2) = 1/2$. Suppose also that the prior expert information on X is

$$\begin{aligned} \mathfrak{M}(\mathbf{p}) = \{m(x) : m(x) \text{ is symmetric around zero}, \\ \int_0^1 m(x)dx = 0.3, \int_1^3 m(x)dx = 0.1, \int_3^\infty m(x)dx = 0.1\}. \end{aligned}$$

The aim is to find $P(M_1|\mathbf{x})$. The training sample for each model is a single replication of X. Then, simple calculations give

$$\pi_1^I(\tau|m) = 2\frac{\tau^{-1/2}}{\sqrt{2\pi}} \int_0^\infty y \exp\{-\tau y^2/2\} m(y)dy,$$

$$\pi_2^I(\lambda|m) = 2 \int_0^\infty y \exp\{-\lambda y\} m(y)dy.$$

The compatibility conditions (7) for the normal model turn out to be

$$\frac{.6}{\pi} \arctan(\frac{1}{y_1}) + \frac{.2}{\pi} \arctan(\frac{1}{y_2}) + \frac{.2}{\pi} \arctan(\frac{1}{y_3}) = .6,$$

$$\frac{.6}{\pi} \arctan(\frac{3}{y_1}) + \frac{.2}{\pi} \arctan(\frac{3}{y_2}) + \frac{.2}{\pi} \arctan(\frac{3}{y_3}) = .8,$$

where $0 \leq y_1 \leq 1, 1 \leq y_2 \leq 3, 3 \leq y_3$. Likewise, for the double exponential model the conditions are

$$\frac{.6}{1+y_1} + \frac{.2}{1+y_2} + \frac{.2}{1+y_3} = .6,$$
$$\frac{1.2y_1}{1+y_1} + \frac{.4y_2}{1+y_2} + \frac{.4y_3}{1+y_3} = .2,$$

where y_i varies over the region stated above for $i = 1, 2, 3$.

For the specific sample $\mathbf{x} = (0, 0.1, 0.3, 0.8, 2, 4)$, the marginal when using the normal model varies over the range

$$(\underline{m}_1^I(\mathbf{x}), \overline{m}_1^I(\mathbf{x})) = (0.770 \times 10^{-6}, 0.837 \times 10^{-6}),$$

and for the double exponential model,

$$(\underline{m}_2^I(\mathbf{x}), \overline{m}_2^I(\mathbf{x})) = (4.728 \times 10^{-6}, 5.305 \times 10^{-6}).$$

Therefore, the range of the Bayes factor $B_{12}(\mathbf{x})$ is

$$(\underline{B}_{12}(\mathbf{x}), \overline{B}_{12}(\mathbf{x})) = (0.145, 0.177).$$

Hence, the conclusion is that this dataset favours the double exponential *versus* the normal model. In fact, the posterior probability of the double model ranges over $(\underline{P}(M_1|\mathbf{x}), \overline{P}(M_1|\mathbf{x})) = (0.850, 0.873)$,while for the normal model ranges over $(\underline{P}(M_2|\mathbf{x}), \overline{P}(M_2|\mathbf{x})) = (0.127, 0.150)$.

Although the above example concerns models that have a real valued parameter, it is possible to apply the above method to models that have a vector-valued parameter (see, Moreno, Bertolino and Racugno, 1999b).

4 Concluding remarks

Objective Bayesian inference proposes that for a given parametric sampling model the prior distribution for the parameters involved is "automatically" determined. In this context an important problem is that of finding the posterior probability of two nonnested models. We have shown that when the models have the same structure but their parameter spaces are a partition of a given space, as occurs in the one-sided testing problem, the posterior probabilities of the models are coherently found.

We have also argued that in most real problems it is possible to consider partial prior information on the observable random variable. In this context, we have seen that by incorporating this information into the default parametric model we can find the posterior probability of the nonested model, even for separate families of sampling distributions.

Acknowledgments. This research has been funded by the grant PB97-1403-C03-01 from the Spanish Ministry of Education and by MURST, Italy.

References

Berger, J.O. and Bernardo, J.M. (1992). On the development of the reference prior method. In *Bayesian Statistics 4*, (J.M. Bernardo et al. eds.), London: Oxford University Press, pp. 35-60.

Berger, J. O. and Mortera, J. (1995). Discussion of O'Hagan. *Journal of the Royal Statistical Society*, B, **57**, pp. 99-138.

Berger, J.O. and Pericchi, L.R. (1996). The intrinsic Bayes factor for model selection and prediction. *Journal of the American Statistical Association*, **91**, pp. 109-122.

Berger, J.O. and Pericchi, L.R. (1998). On Criticism and Comparison of Default Bayes Factors for Model Selection and Hypothesis Testing. In *Proceedings of the Workshop on Model Selection*, (Ed. W. Racugno), Pitagora Ed., Bologna, pp. 1-50.

Bernardo, J.M. and Smith, A.F.M. (1994). *Bayesian Theory*. N.Y.: Wiley.

Bertolino, F. and Racugno, W. (1996), Is the intrinsic Bayes factor intrinsic?, *Metron*, **1**, pp. 5-15.

Bertolino, F., Moreno, E., and Racugno, W. (1999). Bayesian Model Selection Approach to Analysis of Variance under Heteroscedasticity. *Technical Report*. University of Granada.

Cox, D.R. (1961). Test of separate families of hypothesis. In *Proceedings of the Fouth Berkeley Symposium*, **1**, pp. 105-123.

De Santis, F. and Spezzaferri, F. (1999). Methods for Default and Robust Bayesian Model Comparison: the Fractional Bayes Factor Approach. *International Statistical Review*. **67**, pp. 26-286.

Dmochowski, J. (1996). Intrinsic Bayes factor via Kullback-Leibler geometry. In *Bayesian Statistics* 5, (Eds. J.M. Bernardo et al.), London: Oxford University Press.

Kadane, J.B. and Wolfson, L.J. (1998). Experiences in Elicitation. *The Statistician*, **47**, pp. 3-19.

Moreno, E. (1997). Bayes Factor for Intrinsic and Fractional Priors in Nested Models: Bayesian Robustness, in *L_1-Statistical Procedures and Related Topics*, (Ed. D. Yadolah), Hayward, CA: IMS, **29**, pp. 257-270.

Moreno, E., Bertolino, F. and Racugno, W. (1998a). Default approaches to compare means of normal populations (with discussion). In *Proceedings of the Workshop On Model Selection*, (Ed. W. Racugno), Pitagora Ed., Bologna, pp. 133-155.

Moreno, E., Bertolino, F. and Racugno, W. (1998b). An Intrinsic Limiting Procedure for Model Selection and Hypotheses Testing. *Journal of the American Statistical Association*, **93**, pp. 1451-1460.

Moreno, E., Bertolino, F. and Racugno, W. (1999a). Default Bayesian Analysis of the Behrens-Fisher Problem. *Journal of Statistical Planning and Inference*, **81**, pp. 323-333.

Moreno, E., Bertolino, F. and Racugno, W. (1999b). Inference under partial information on observable random variables. *Technical Report*. University of Granada.

Moreno, E. and Liseo, B. (1999). Default priors for testing the number of components of a mixture. *Technical Report*. University of Granada.

O'Hagan, A. (1995). Fractional Bayes Factor for Model Comparison (with discussion). *Journal of the Royal Statistical Society*, B, **57**, pp. 99-138.

O'Hagan, A. (1997). Properties of Intrinsic and Fractional Bayes Factors. *Test*, **6**, pp. 101-118.

O'Hagan, A. (1998). Eliciting Expert Beliefs in Substantial Practical Applications. *The Statistician*, **47**, pp. 21-35.

San Martini, A. and Spezzaferri, F. (1984). A predictive model selection criterion. *Journal of the Royal Statistical Society*, B, **46**, pp. 296-303.

Sansó, B,. Pericchi, L.R. and Moreno, E. (1995). On the robustness of the intrinsic Bayes factor for nested models (with discussion). In *Bayesian Robustness*, (Eds. J.O. Berger et al.), Hayward CA: IMS, **29**, pp. 157-176.

Spatio-temporal hierarchical modeling of an infectious disease from (simulated) count data

Noel Cressie and Andrew S. Mugglin

Department of Statistics, The Ohio State University, 1958 Neil Avenue, Columbus OH 43210, USA

Abstract. An infectious disease spreads through "contact" between an individual who has the disease and one who does not. However, modeling the individual-level mechanism directly requires data that would amount to observing (imperfectly) all individuals' disease statuses along their space-time lines in the region and time period of interest. More likely, data consist of spatio-temporal aggregations that give small-area counts of the number infected during successive, regular time intervals. In this paper, we give a spatially descriptive, temporally dynamic hierarchical model to be fitted to such data. The dynamics of infection are described by just a few parameters, which can be interpreted. We take a Bayesian approach to the analysis of these space-time count data, using Markov chain Monte Carlo to compute Bayes estimates of all parameters of interest. As a "proof of concept," we simulate data from the model and investigate how well our approach recovers important hidden features.

Keywords: small-area counts, multivariate autoregression, Markov chain Monte Carlo, Markov random field

1 Introduction

It is often the case that disease-incidence data come in the form of spatial aggregation into small-area counts, observed over a series of time periods. When the disease is non-infectious, like lung cancer, the time period might be a year or more, and the goal of the study is to find dependence on a covariable(s) that might inform on public policy to help prevent it. Allowing for spatio-temporal dependence in the model is important for efficient inference on regression coefficients of the covariables.

By contrast, when the disease is infectious, like influenza, the time period is usually short (e.g., several days or a week for influenza). Here the disease is much more dynamic, and it is of interest to find out how last week's counts will influence this week's counts. Covariables are still important in order to explain heterogeneities, but the emphasis has shifted toward those parameters governing the temporal dynamics. In the past, epidemiological analyses of infectious diseases have emphasized temporal and demographic heterogeneities (e.g., Rhodes and Anderson, 1996; Keeling, Rand, and Morris, 1997). The purpose of this article is to feature spatio-temporal heterogeneity in a way that incorporates much of the sophisticated Bayesian hierarchical technology used for modeling non-infectious diseases; Lawson and Leimich (2000) take the same basic approach, as do Mugglin, Cressie, and Gemmell (2000). As a "proof of concept" of the analysis given in Mugglin et al. (2000), we simulate (in space and time) disease-incidence counts, and then analyze the data using the same model.

The Bayesian hierarchical modeling strategy acts on essentially three collections of variables: (i) data, (ii) processes, and (iii) parameters. Then the three basic components of the model are the probability distributions of (I) data given processes and parameters, (II) processes given parameters, and (III) parameters. The joint distribution of all variables is just the product of these three probability distributions. In the non-infectious, space-time context, (I) is usually a product of independent Poisson distributions; (II) is usually a temporally autoregressive and spatially Markov log Gaussian random field; and (III) is usually a product of independent, noninformative parameter distributions. Such models can be found in Waller et al. (1997) and Knorr-Held and Besag (1998); for a comparison of these and other spatio-temporal models, see Cressie, Stern, and Wright (2000).

Once the modeling process is complete, Bayes' Theorem yields the posterior distribution of the processes and parameters given the data. Historically, computing the posterior has been too complicated without major simplification of (I), (II), and (III). However, more recently, these modeling constraints have been relaxed enormously due to the discovery of numerically intensive simulation approaches, such as Markov chain Monte Carlo (MCMC); see, for example, Gilks, Richardson, and Spiegelhalter (1996).

Our model is the most interesting for the distributions in (II), where the dynamic features are built in. We are guided by a qualitative expression of the way infections spread. We suppose that any small area's log disease intensity at time t depends (linearly) on the same for the small area, its nearest neighbors, and its second nearest neighbors, all at time $(t-1)$. The coefficients of this multivariate autoregression have special interest for us, even though the dynamics do not refer to individual contact between those who have the disease and those who do not. Further, the model has predictive power for determining health-care-delivery needs as the epidemic runs its course.

In Section 2, we build the statistical models described in (I), (II), and (III). Section 3 contains the MCMC algorithm needed to compute the posterior distribution. The methodology is applied to a simulated epidemic in Section 4. There we simulate from the model in Section 2 and use the methods in Section 3 to recover the "truth" about the parameters and to predict the "hidden" log Gaussian process. Section 5 discusses further research we plan to pursue, in particular for the application of our technology to influenza epidemics.

2 Dynamic model of small-area counts of infecteds

Consider for the moment a situation where the variable of interest is a Gaussian time series $\{s_t : t = 1, 2, \ldots\}$. We wish to construct a model for $\{s_t\}$ that has the flexibility to show growth, recession, and stationary behavior over time. Consider the first-order autoregression, abbreviated as $AR(1)$,

$$(s_t - \mu_t) = \eta(s_{t-1} - \mu_{t-1}) + \nu_t\,, \tag{1}$$

where $\{\mu_t\}$ is the sequence of means of $\{s_t\}$, the "shock" term ν_t is independent of the previous period's value s_{t-1}, and $\{\nu_t\}$ is a time series of independent Gaussian random variables with zero mean. Usually it is assumed that $\{\nu_t\}$ also has constant variance σ^2, in which case stationary variances and covariances of $\{s_t\}$ are achieved by choosing $|\eta| < 1$. Then we see that

$$\mathrm{var}(s_t) = \sigma^2/(1-\eta^2)$$
$$\mathrm{corr}(s_t, s_{t-h}) = \eta^h \,.$$

Unfortunately, the choice of $|\eta| > 1$ does not allow us to turn growth or recession "on" and "off" in a controlled way. However, *instability* is easy, simply by ensuring that $|\eta| > 1$. For example, suppose that $\eta > 1$. Then $\mathrm{var}(s_t) = \eta^2 \mathrm{var}(s_{t-1}) + \mathrm{var}(\nu_t)$, which implies the variance is growing. Further, $\mathrm{corr}(s_t, s_{t-1}) = \{1 - \mathrm{var}(\nu_t)/\mathrm{var}(s_t)\}^{1/2}$, which increases to 1 as time goes on. The situation is clearest in the case of constant mean $\mu_t \equiv \mu$. A moderate positive value of autoregressive component $\eta(s_{t-1} - \mu)$ could be dominated by a negative realization ν_t, in which case $(s_t - \mu)$ would start an excursion into negative "territory" that, once started, would be difficult to leave. The converse is also true. Furthermore, a little thought reveals that if $\eta < -1$, the instability becomes even worse.

Notice that the occurrence of growth or recession in these autoregressions is not something that can be controlled. Consider instead the model,

$$s_t = \eta s_{t-1} + \varepsilon_t \,, \tag{2}$$

where $\{\varepsilon_t\}$ is a time series of independent Gaussian random variables with distribution $\mathrm{Gau}(\beta, \sigma^2)$, and β may *not* be zero. If we impose stationarity on $\{s_t\}$, we see that for $|\eta| < 1$, the stationary mean must be $E(s_t) = \beta/(1-\eta)$, the stationary variance must be $\mathrm{var}(s_t) = \sigma^2/(1-\eta^2)$, and $\mathrm{corr}(s_t, s_{t-h}) = \eta^h$.

Now, we can create growth at t_0 by specifying $\beta = \beta_0$ for $t < t_0$, and $\beta = \beta_1 > \beta_0$ for $t \geq t_0$; eventually, a new stationary mean will be reached by growth to $E(s_t) = \beta_1/(1-\eta)$. Recession from this mean back to $\beta_0/(1-\eta)$ (or another value less than $\beta_1/(1-\eta)$) can be achieved by specifying $\beta = \beta_0$ (or $\beta < \beta_1$) again, for $t \geq t_1 > t_0$. Notice that (2) is actually a special case of (1) by putting $\mu_t - \eta\mu_{t-1} \equiv \beta$.

Now we shall generalize (2) to a multivariate (actually spatial) model, which is "hidden" behind Poisson counts. Consider disease counts aggregated in space and time such that y_{it}, the count in region i at time interval t, is Poisson-distributed with mean parameter $E_i e^{s_{it}}$, $i = 1, 2, \ldots, I$, and $t = 1, 2, \ldots, T$. Here E_i is the expected count for region i, obtained through direct or indirect adjustment (e.g., Lilienfeld and Stolley, 1994), and s_{it} is the log relative risk that measures deviation from the expected count and is where the spatial and temporal dependence is modeled. In particular, writing $(s_{1t}, s_{2t}, \ldots, s_{It})'$ as $\mathbf{s}_t$, let

$$\mathbf{s}_t = H\mathbf{s}_{t-1} + \boldsymbol{\varepsilon}_t \,, \tag{3}$$

a multivariate $AR(1)$ process with autoregressive coefficient matrix H and shock term $\boldsymbol{\varepsilon}_t$, which we take to be a realization from a Gaussian Markov random field. Specifically, $\boldsymbol{\varepsilon}_t \sim MVN(\beta_{\rho(t)}\mathbf{1}, \Sigma)$, where $MVN(\,\cdot\,,\,\cdot\,)$ denotes the multivariate normal distribution, $\mathbf{1}$ is a vector of ones, $\rho(t) = 0, 1$, or 2, indicating the stage of disease at time t (0 = stable, no epidemic; 1 = growth; 2 = recession), and Σ is the variance-covariance matrix for a Markov random field. Following Stern and Cressie (1999), we take

$$\Sigma = \sigma^2(I - \phi C)^{-1} M, \tag{4}$$

where $M = \text{diag}(E_i^{-1} : i = 1, \ldots, I)$, $c_{ij} = (E_j/E_i)^{1/2}$ for site j in the neighbor set of site i and 0 elsewhere, and $\phi \in (\phi_{\min}, \phi_{\max})$, where $\phi_{\min}$ and $\phi_{\max}$ are determined from the eigenvalues of C such that $M^{-1}(I - \phi C)$ is positive definite (Cressie, 1993, p. 559).

We take the autoregressive matrix H to be parameterized by η_0, η_1, and η_2, such that

$$h_{ij} = \begin{cases} \eta_0 & \text{if } j = i \\ \eta_1 & \text{if } j \in N_i\,, \text{ that is, site } j \text{ is a neighbor of site } i \\ \eta_2 & \text{if } j \in N_i^2\,, \text{ that is, site } j \text{ is a second-order neighbor of site } i \\ 0 & \text{otherwise}\,. \end{cases}$$

Site j is a neighbor of site i if some proximity criterion is satisfied (e.g., share a common boundary), and we define site j to be a second-order neighbor of site i if $j \notin N_i$ but $\exists k$ such that $k \in N_i$ and $j \in N_k$. We can assign prior distributions to the η_ℓ ($\ell = 0, 1, 2$) directly, but it may be more sensible to apply a transformation, $\theta_\ell = f(\eta_\ell)$, and assign a prior to θ_ℓ. In the one-dimensional case, the time series (1) is stationary if $|\eta| < 1$. Now $\theta = \log\left(\frac{1+\eta}{1-\eta}\right)$ is defined on $(-\infty, \infty)$, and hence a Gaussian prior on θ would be reasonable. For the matrix H, we make a similar transformation on its entries: we choose $\theta_\ell = \log\left(\frac{1+\eta_\ell}{1-\eta_\ell}\right)$, $\ell = 0, 1, 2$. In future work, we shall consider what transformation and what prior would guarantee that the variances and covariances of the $AR(1)$ process (3) are stationary.

Using $[Y]$ as generic notation for the density of Y, we can write the joint posterior of all parameters as being proportional to the product of the likelihood and the priors:

$$\begin{aligned} &[\beta_0, \beta_1, \beta_2, \sigma^2, \phi, \theta_0, \theta_1, \theta_2, \{s_{it}\} \mid \{y_{it}\}] \\ &\propto \left(\prod_{t=1}^{T} \prod_{i=1}^{I} [y_{it} \mid s_{it}]\right) [\mathbf{s}_1 \mid \beta_{\rho(1)}, \sigma^2, \phi] \\ &\qquad \times [\mathbf{s}_2 \mid H\mathbf{s}_1, \beta_{\rho(2)}, \sigma^2, \phi] \cdots [\mathbf{s}_T \mid H\mathbf{s}_{T-1}, \beta_{\rho(T)}, \sigma^2, \phi] \\ &\qquad \times [\beta_0][\beta_1][\beta_2][\sigma^2][\phi][\theta_0][\theta_1][\theta_2]\,. \end{aligned} \tag{5}$$

Finally, we specify priors, which completes the Bayesian hierarchical model:

$$\begin{aligned} \beta_\ell &\sim \text{Gau}(\mu_\ell, \tau_\ell^2)\,,\ \ell = 0, 1, 2 \\ \theta_\ell &\sim \text{Gau}(0, \delta_\ell^2)\,, \ell = 0, 1, 2 \\ \sigma^2 &\sim \text{Inverse Gamma } (a, b) \\ \phi &\sim \text{Uniform } (\phi_{\min}, \phi_{\max})\,. \end{aligned} \tag{6}$$

For a particular application, the parameter values of these prior distributions will be specified (e.g., see Section 4).

3 MCMC algorithm

Obtaining the posterior distribution in (5) as well as the marginal posterior distributions of individual parameters requires evaluation of integrals that are analytically intractible and numerically infeasible. However, a simulation-based method known as Markov chain Monte Carlo (MCMC) is possible. To accomplish this, we must write out first the full conditional distribution for each parameter. In what follows, "rest" means all other parameters in the model, as well as the data $\{y_{it}\}$.

- $\{\beta_\ell\}$:

$$[\beta_0, \beta_1, \beta_2 \mid \text{rest}] \propto [\mathbf{s}_1 \mid \beta_{\rho(1)}, \sigma^2 \phi] \prod_{t=2}^{T} [\mathbf{s}_t \mid H\mathbf{s}_{t-1}, \beta_{\rho(t)}, \sigma^2, \phi] \cdot [\beta_0][\beta_1][\beta_2]$$

- σ^2:

$$[\sigma^2 \mid \text{rest}] \propto [\mathbf{s}_1 \mid \beta_{\rho(1)}, \sigma^2 \phi] \prod_{t=2}^{T} [\mathbf{s}_t \mid H\mathbf{s}_{t-1}, \beta_{\rho(t)}, \sigma^2, \phi] \cdot [\sigma^2]$$

- ϕ:

$$[\phi \mid \text{rest}] \propto [\mathbf{s}_1 \mid \beta_{\rho(1)}, \sigma^2 \phi] \prod_{t=2}^{T} [\mathbf{s}_t \mid H\mathbf{s}_{t-1}, \beta_{\rho(t)}, \sigma^2, \phi] \cdot [\phi]$$

- $\{\theta_\ell\}$: For $\ell = 0, 1, 2$,

$$[\theta_\ell \mid \text{rest}] \propto \prod_{t=2}^{T} [\mathbf{s}_t \mid H\mathbf{s}_{t-1}, \beta_{\rho(t)}, \sigma^2, \phi] \cdot [\theta_\ell]$$

- $\{s_{it}\}$: For $i = 1, \ldots, I$,

$$[s_{it} \mid \text{rest}] \propto \begin{cases} [y_{it} \mid s_{it}][\mathbf{s}_t \mid H\mathbf{s}_{t-1}, \beta_{\rho(t)}, \sigma^2, \phi] \\ \quad \times [\mathbf{s}_{t+1} \mid H\mathbf{s}_t, \beta_{\rho(t+1)}, \sigma^2, \phi] & \text{if } t = 2, 3, \ldots, T-1 \\ [y_{i1} \mid s_{i1}][\mathbf{s}_1 \mid \beta_{\rho(1)}, \sigma^2, \phi] \\ \quad \times [\mathbf{s}_2 \mid H\mathbf{s}_1, \beta_{\rho(2)}, \sigma^2, \phi] & \text{if } t = 1 \\ [y_{iT} \mid s_{iT}][\mathbf{s}_T \mid H\mathbf{s}_{T-1}, \beta_{\rho(T)}, \sigma^2, \phi] & \text{if } t = T. \end{cases}$$

The simulation process entails specifying starting values for all parameters, drawing an observation from each full conditional distribution, and cycling through each parameter while conditioning on the most recent values of the others. This results in a Markov chain whose stationary distribution is the posterior distribution of all parameters. Once we are satisfied that the process has converged, we can easily obtain a large sample from the posterior and use it to answer virtually any inferential question of interest. The price we pay is a combination of statistical computing challenges and coding effort.

Several of the full conditionals are recognizable distributional forms. For example, $[\beta_0, \beta_1, \beta_2 \mid \text{rest}]$ is proportional to a product of Gaussian distributions; thus, it too must be Gaussian. After some algebra, we can write this distribution explicitly, and sampling from it becomes easy. The same can be said for $[\sigma^2 \mid \text{rest}]$, which follows the Inverse Gamma distribution. This type of simulation is known as Gibbs updating. However, $[\phi \mid \text{rest}]$ and $[s_{it} \mid \text{rest}]$

are not recognizable forms, so a rejection-type algorithm (known as Metropolis updating) must be employed. The essence here is to propose a candidate value randomly, and then decide whether to keep it as the next value in the chain. This decision is based upon the ratio of the full conditional evaluated at the candidate value (s_{it}^{cand}), to the full conditional evaluated at the previous value (s_{it}^{prev}) in the chain. This is generally not as efficient as Gibbs updating, and it is somewhat tricky to propose good candidate values, but the method works well once it is calibrated properly; see Carlin and Louis (1996) and Gilks et al. (1996). Simulating from $[\theta_\ell \mid \text{rest}]$ can be accomplished via Gibbs updating, but we chose not to for two reasons: the algebraic derivation of the Gaussian full conditional is tedious, and if we leave it as a Metropolis step, it is a straightforward matter to change the transformation function that relates θ_ℓ (the scale on which we apply priors) to η_ℓ (the parameters that comprise H).

Careful attention must be paid to computational speed. Although there is considerable computaton required in updating $\{\beta_\ell\}$, σ^2, ϕ, and $\{\theta_\ell\}$, these are only done once per iteration of the sampler. In contrast, updating $\{s_{it}\}$ involves repeating similar calculations $I \cdot T$ times for each iteration, so every effort must be made to increase computational efficiency. In the analysis of Section 4, $I = 56$ and $T = 26$, so the s_{it} updating routine must be executed 1456 times at each iteration, and any computational efficiencies in this routine pay big dividends in overall running time. In this case it is well worth the effort to simplify the acceptance ratio $[s_{it}^{\text{cand}} \mid \text{rest}]/[s_{it}^{\text{prev}} \mid \text{rest}]$ algebraically before coding it, rather than writing a function that evaluates $[s_{it} \mid \text{rest}]$ and evaluating it at both s_{it}^{cand} and s_{it}^{prev}. This simplification alone improves overall running time in Section 4 by almost a factor of 2.

4 Simulated epidemic and its statistical analysis

We simulate an epidemic in Scotland via the model of Section 3. In this analysis we do not stratify on age or sex, so $E_i \propto n_i$, the population of region i. Scotland has $I = 56$ local government districts (LGDs), and we use neighborhood definitions given in Stern and Cressie (1999) to form the matrix C in (4). The eigenvalues of C then induce $\phi_{\min} = -.325$ and $\phi_{\max} = .175$, and we select $\phi = 0.1$ and $\sigma^2 = 0.2$ as the true values. Selecting $\theta_0 = .2$, $\theta_1 = .10$, and $\theta_2 = .02$ yields a 56-dimensional $AR(1)$ process whose variances and covariancs are stationary; after transforming from the θ-scale to the η-scale, these autoregressive parameters are interpretable as the effect of autoregressing $s_{i,t}$ on $s_{i,t-1}$, on $s_{j,t-1}$, where LGD j is a first-order neighbor of LGD i, and on $s_{k,t-1}$, where LGD k is a second-order neighbor or LGD i.

To simulate the epidemic growth, we specify that weeks 1–10 are in disease stage 0 (stable), weeks 11–15 are in stage 1 (growth), weeks 16–20 are in stage 2 (recession), and weeks 21-26 are again in stage 0. To achieve disease growth and decline, we assign $\beta_0 = -.5$, $\beta_1 = .5$, and $\beta_2 = -.3$. We then simulate the $\{s_{it}\}$-process and finally generate disease counts $\{y_{it}\}$ using the Poisson distribution. These counts represent the *data* inputted into the MCMC algorithm described in Section 2.

Figure 1 shows $\sum_i s_{it}$ and $\sum_i y_{it}$ as functions of time. The epidemic in weeks 11 through 20 is clearly seen. We also note that in no week does the number of cases exceed 100, while in weeks of no epidemic it is typically less than 20. This is deliberate, in an attempt to mimic what might actually occur in Scotland with an infectious disease like influenza, and it induces many zero values for particular y_{it}.

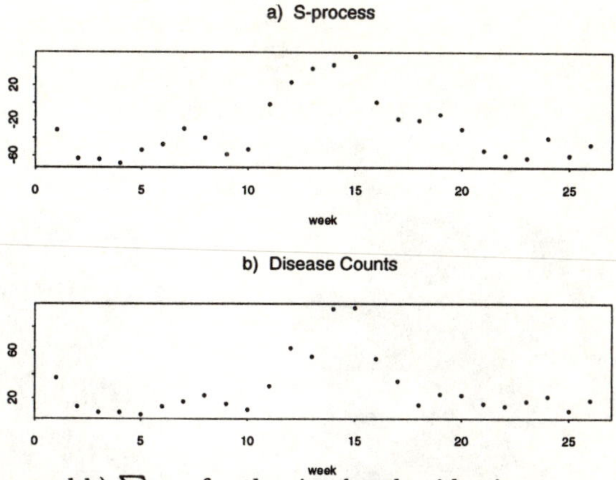

Fig. 1. a) $\sum_i s_{it}$ and b) $\sum_i y_{it}$ for the simulated epidemic

Regarding prior-parameter specification in (6), we specify fairly vague priors by choosing $\delta_\ell^2 = \tau_\ell^2 = 4$, $\mu_\ell = 0$, and $a = 6$, $b = 1$ (so that the prior mean and standard deviation of σ^2 are .2 and .1, respectively). We implement the MCMC algorithm of Section 3 using five parallel chains; using the convergence diagnostic of Gelman and Rubin (1992) and visual inspection of parallel chains, we find that the sampler has converged sufficiently by 200 iterations; however, sample lag-1 autocorrelations in some parameters (particularly σ^2, ϕ, and the $\{\theta_\ell\}$) are quite high, so we ran the sampler for 11,000 iterations. We used the last 10,800 iterations $\times$ 5 independent chains = 54,000 posterior observations from the Gibbs sampler to estimate the posterior distributions. The algorithm, implemented in C and run on a Sun Ultrasparc workstation, takes about 5 hours to run.

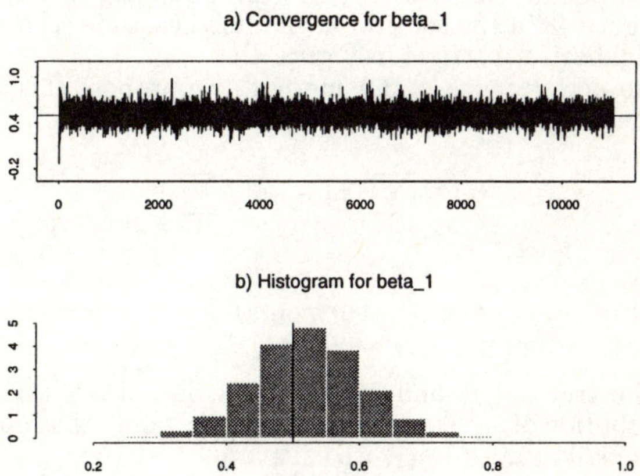

Fig. 2. Posterior distribution of β_1, shown with a) a convergence plot of 5 parallel chains, and b) a histogram. The dark lines in each represent the true value 0.5.

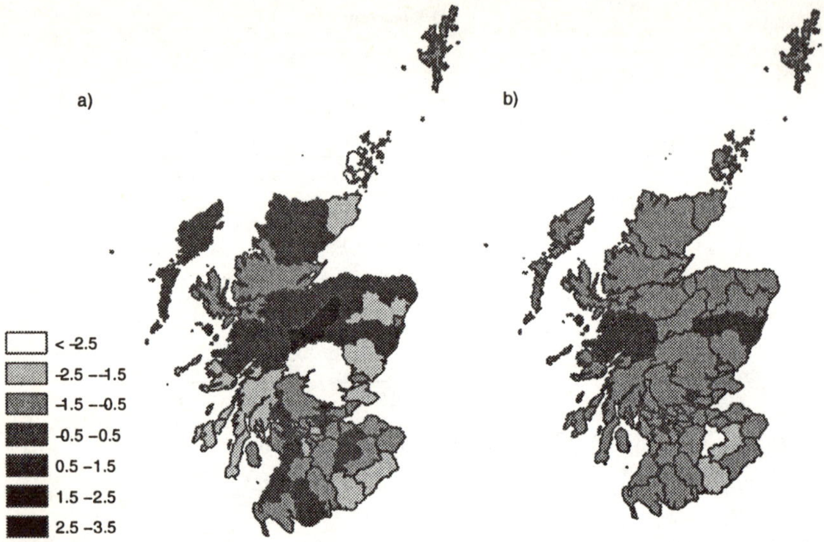

Fig. 3. A comparison of a) the true $s_{i,10}$ and b) $s_{i,10}$ estimated through posterior medians.

Figure 2 shows the convergence behavior of β_1 as a representative parameter. Because the data are simulated, we have the opportunity to compare the posterior realizations with the true value $\beta_1 = 0.5$; this is shown as the dark line, and we find the agreement to be excellent.

Figure 3 exhibits maps of the log relative risk $s_{i,10}$ at week 10, just before the epidemic begins; Figure 3a shows the true value, while Figure 3b shows the posterior medians from our analysis. As expected, the posterior medians are oversmoothed. Figure 4 displays posterior medians for $s_{i,11}$ and $s_{i,12}$, and here we see the beginning of the epidemic. Week 11 is the first week of the new stage, and already we note the overall darkening of the shading. The hotspots in Figure 4a in the southwest and northeast are seen to expand and migrate to neighboring districts in Figure 4b.

To assess the performance of our method, we propose the following measures of goodness-of-fit:

$$G_i \equiv \sum_t |s_{it} - \tilde{s}_{it}| \,/ \sum_t 1$$

$$F_t \equiv \sum_i |s_{it} - \tilde{s}_{it}| \,/ \sum_i 1,$$

where s_{it} is the true value, and $\tilde{s}_{it}$ is the median of our samples from the posterior distribution of s_{it}. As a measure of uncertainty that does not depend on knowing the true values $\{s_{it}\}$, we calculate

$$W_i \equiv \sum_t w_{it} / \sum_t 1,$$

Fig. 4. Posterior medians for a) $s_{i,11}$ and b) $s_{i,12}$, showing the beginning of the epidemic

where w_{it} is the width of the 95% Bayesian credible interval from the posterior distribution of s_{it}. Figure 5 shows plots of a) $\{(E_i, G_i) : i = 1, \ldots, 56\}$ and b) $\{(W_i, G_i) : i = 1, \ldots, 56\}$. We see from Figure 5a that there is a definite inverse relationship between $E_i (\propto n_i)$ and G_i: the model fits better where the population is larger, and worse where the population is smaller. We note that the two values with $E_i > 2$ are Edinburgh are Glasgow, while the highest G_i value represents a district with the second smallest E_i (and where there happened to be no disease cases in the entire 26-week period). We do not usually know the truth, but we can always find the credible interval widths w_{it}, and Figure 5b shows a strong linear relationship between the variability measure W_i and the goodness-of-fit measure G_i. Consequently, W_i and E_i will share the same inverse relationship as G_i shares with E_i. Put another way, it is very difficult to predict accurately the relative risk of disease in areas where population (and thus case counts) are small. A plot of F_t versus time indicates the same general concept, only we observe that the best fit occurs during the epidemic of weeks 11–20, when case counts are generally the highest.

The hierarchical Bayesian approach to modeling this process not only allows the construction of a rich, complicated model based on simple pieces, but it also enables inference on almost any question of interest. One subject of interest to epidemiologists may be the rate at which this disease spreads. Some information related to this question is available directly through the posteriors of η_ℓ. Although our sampler has produced observations from the posterior distributions of $\theta_\ell, \ell = 0, 1, 2$, it is a straightforward matter to transform these to the η-scale, which can in turn be interpreted as described in the first paragraph of this section. But other inferences are also possible. Consider a measure of disease spread, developed as follows: Recall that $\mathbf{s}_t = H\mathbf{s}_{t-1} + \boldsymbol{\varepsilon}_t$. Then from (3), a stationary mean $\boldsymbol{\mu}$ would satisfy $\boldsymbol{\mu} = H\boldsymbol{\mu} + \beta\mathbf{1}$,

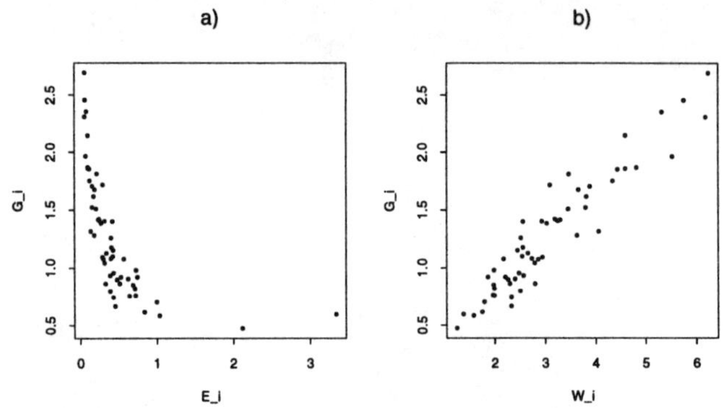

Fig. 5. Goodness-of-fit statistic G_i versus a) expected cases E_i and b) uncertainty measure W_i

where $E(\boldsymbol{\varepsilon}_t) = \beta \mathbf{1}$, and $\mathbf{1}$ is a vector of ones. Solving for $\boldsymbol{\mu}$, we obtain

$$\boldsymbol{\mu} = (I - H)^{-1}\mathbf{1} \cdot \beta, \tag{7}$$

assuming $(I - H)^{-1}$ exists. Now suppose we consider a time point such as week 11, where the stage of the disease has changed suddenly. The linear relation between $\boldsymbol{\mu}$ and β, given by (7), implies that for incremental change $\Delta\beta$,

$$\Delta\boldsymbol{\mu} = (I - H)^{-1}\mathbf{1} \cdot \Delta\beta. \tag{8}$$

The quantity $\Delta\boldsymbol{\mu}$ could be interpreted as a vector velocity of disease change, and can be easily calculated from the posterior samples generated by the MCMC sampler. We did this for $\Delta\beta = \beta_1 - \beta_0$, the change that actually occured at week 11. It should be noted that $(I - H)^{-1}$ is not guaranteed to exist for every realization from the posterior of (η_0, η_1, η_2); in our case it always did exist, though there were a few values for which one eigenvalue of $I - H$ was close to 0. However the number of these was small enough to hardly effect the posterior medians and 95% credible intervals.

Having produced a sample from the posterior of $\Delta\boldsymbol{\mu}$ via (8), we map the posterior medians as shown in Figure 6. There is a definite spatial structure here, with the regions of highest velocity in the geographical center, and generally decreasing toward the periphery. We can interpret the dark regions as the places where, given a β-change that uniformly affects all of Scotland (as perhaps induced by weather), the introduction of a pathogen will have the fastest and most dramatic effect on overall disease incidence. Perhaps special precautions might be taken to remediate disease early in these zones. Given the geography of Scotland, that these hotspots should be in the center makes intuitive sense, partly because the disease can spread outward to many more locations than it could from, say, the Shetland Islands in the northeast.

5 Discussion

The Bayesian hierarchical model defined around (3) expresses the dynamic structure of the (log) expected counts in terms of spatial proximities. Although this may not capture the individual-level dynamics very well, it does allow some epidemiological interpretation at a local, small-area level, and it

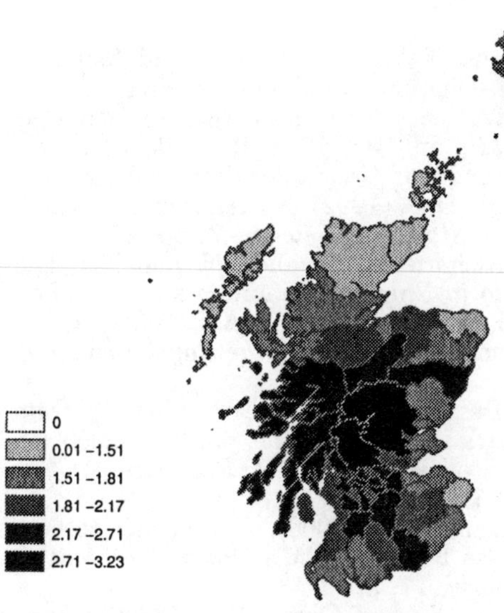

Fig. 6. Velocity of disease spread per unit increase in β

does allow predictions of how the epidemic will spread between the small areas. Nevertheless, in future research, we shall consider other parameterizations of H that might take into account such things as an area's population density.

Other enhancements to the hierarchical model presented could include the incorporation of explanatory variables exogenous to the probability model; the use of a stochastic process (in time) to determine the switching of $\rho(t)$ from 0 to 1, 1 to 2, and so forth; and the incorporation, through $\{E_i\}$, of age and sex strata in the analysis. These will be explored in Mugglin et al. (2000).

With the natural evolution of spatial models to spatio-temporal models comes the need to produce maps that evolve in time. Whether this is done to track the spread of infectious diseases in a population (epidemiological application) or to track the movement of enemy forces in a battlefield (military command-and-control application), the use of a Geographic Information System (GIS) is crucial to the quality and consistency of the maps from one time point to the next. Our Bayesian hierarchical statistical analysis has produced optimal estimates, and the GIS is a powerful tool for presenting the results. The maps shown in Figures 3, 4, and 6 were made using a GIS; in future research, we shall explore its use further for presenting dynamic spatial information.

Acknowledgements

The authors would like to thank Islay Gemmell of the Medical Research Council, Social & Public Health Sciences Unit, Glasgow, Scotland for her help with the geography of Scotland and her insights into the nature of influenza. This research was supported by the Office of Naval Research under grant number N00014-99-1-0001.

References

Carlin, B.P. and Louis, T.A. (1996). *Bayes and Empiricial Bayes Methods for Data Analysis.* Chapman and Hall, London.

Cressie, N. (1993). *Statistics for Spatial Data*, revised edition. Wiley, NY.

Cressie, N., Stern, H.S., and Wright, D.R. (2000). Mapping rates associated with polygons. *Journal of Geographical Systems*, **2**, forthcoming.

Gelman, A. and Rubin, D.B. (1992). Inference from iterative simulation using multiple sequences. *Statistical Science*, **7**, pp. 457-511.

(ed.) Gilks, W.R., Richardson, S., and Spiegelhalter, D.J. (1996). *Markov chain Monte Carlo in Practice.* Chapman and Hall, London.

Keeling, M.J., Rand, D.A., and Morris, A.J. (1997). Correlation models for childhood epidemics. *Proceedings of the Royal Society of London*, **264**, pp. 1149-1156.

Knorr-Held, L. and Besag, J.E. (1998). Modelling risk from a disease in time and space. *Statistics in Medicine*, **17**, pp. 2045-2060.

Lilienfeld, D.E. and Stolley, P.D. (1994). *Foundations of Epidemiology*, third edition. Oxford, NY.

Lawson, A.B. and Leimich, P. (2000). Approaches to space-time modelling of infectious disease behavior. *IMA Journal of Mathematics Applied to Medicine and Biology*, **17**, pp. 1-13.

Mugglin, A.S., Cressie, N., and Gemmell, I. (2000). Hierarchical statistical modeling of influenza-epidemic dynamics in space and time (Tentative title), in preparation.

Rhodes, C.J. and Anderson, R.M. (1996). Power laws governing epidemics in isolated populations. *Nature*, **381**, pp. 600-602.

Stern, H.S. and Cressie, N. (1999). Inference for extremes in disease mapping in A. Lawson, A. Biggeri, D. Bohning, E. Lesaffre, J-F Viel, and R. Bertollini, eds. *Disease Mapping and Risk Assessment for Public Health.* Wiley, Chichester, pp. 63-84.

Waller, L.A., Carlin, B.P., Xia, H., and Gelfand, A.E. (1997). Hierarchical spatio-temporal mapping of disease rates. *Journal of the American Statistical Association*, **92**, pp. 607-617.

GBMs: GLMs with bilinear terms

Antoine de Falguerolles

Laboratoire de Statistique et Probabilités, UMR C5583, Université Paul Sabatier, F-31062 Toulouse, France

Abstract. Generalized bilinear models can hide under a variety of denominations. These correspond broadly to two types of statistical activities which are often combined: exploratory analysis and explicit modeling. It turns out that the wide diversity of correlative methods can be considered as minor variations of a single model which is introduced in the standard framework set for generalized linear models. The paper presents this unifying approach and illustrates some its strengths.

Keywords. Biplot, Generalized Bilinear Models, Generalized Linear Models, Generalized Non-Linear Models, Matched Tables, Singular Value Decomposition, Square Tables.

1 Introduction

Generalized bilinear models (GBMs) can hide under a variety of names: biadditive models, reduced rank interaction models, linear-bilinear models, additive main effects and multiplicative interaction models, generalized singular value decomposition, correspondence analysis, $R \times C$ association models, correlation models, principal component analysis, log multiplicative models and so on. Some denominations evoke immaculate descriptive methods, others orthodox modeling. To some extent, the aficionados of the descriptive methods and the aficionados of the modeling approach seem reluctant to accept a unifying framework. Indeed, there are several reasons for maintaining the *status quo*. One reason is specialization or background. This can be illustrated by papers presented at former COMPSTAT meetings. In the COMPSTAT'92 proceedings, "Algorithmic Approaches for Fitting Bilinear Models" by Falguerolles and Francis and "Computational Procedures for Estimation and Hypothesis Testing in Analyzing Two-way Tables with Interaction and No Replication" by Seyedsadr and Cornelius do not have any common reference. The same situation prevails in the COMPSTAT'98 proceedings for "LLAMA: an Object-Oriented System for Log Multiplicative Models" by Firth and "Using Singular Value Decomposition in Non-Linear Regression" by Brain.

Another reason is that there are instances where modeling and exploratory methods are used jointly. For example, correspondence analysis (CA) efficiently extracts information from the residuals of a log-linear model in situations where the residuals are meaningfully structured as a two-way table (van der Heijden & de Leeuw, 1985, van der Heijden *et al.*, 1989). Here, the simple but powerful superimposition of these tools may reinforce the idea that the modeling and the descriptive methods are alien to each other.

However recent papers by Choulakian (1996), Falguerolles and Francis (1992, 1995), Gabriel (1998) and van Eeuwijk (1995) advocate the use of the generalized linear model framework to present seemingly unrelated methods. In this respect they follow the suggestion of Gower when discussing the paper by van der Heijden *et alii* (1989): " I have long felt that the generalized linear models (GLMs) could be beneficially further generalised to include bilinear terms". Here I am more than 10 years later, still in need of help! In particular, I will not discuss standard errors of parameters, degrees of freedom, and testing order m against order $m+1$.

2 From description to modeling (and vice versa)

2.1 Additive main effects and multiplicative interaction

Let us start with the simple situation where a quantitative response variable is cross-classified according to two factors I and J with respectively n_I and n_J categories. There is exactly one observation y_{ij}^{IJ} per cell (i,j) and it is further assumed that the two factors are on an equal footing.

A standard descriptive analysis of such a data matrix tries to gauge an average value for the response, and specific effects for factor I or factor J. This can be done by computing the grand mean ($\overline{y}$), the specific mean deviations for I (the $\overline{y}_i^I - \overline{y}$), and for J (the $\overline{y}_j^J - \overline{y}$). Other forms for these summaries can be envisaged on robustness grounds and this will be discussed later (see subsection 3.11).

The analysis then proceeds to compute the deviations from the identified effects:

$$r_{ij}^{IJ} = y_{ij}^{IJ} - (\overline{y}_i^I - \overline{y}) - (\overline{y}_j^J - \overline{y}) - \overline{y}.$$

These are viewed as resulting from the interaction of the two factors and significant patterns in the interaction are looked for.

It is clear that, having already considered an additive decomposition for the effects of the factors, some other form of decomposition is to be considered. Following Gollob (1968), a simple method consists in investigating a multiplicative decomposition. Taking advantage of the matrix form of the r_{ij}^{IJ} (cross-classified according to I and J), a singular value decomposition (SVD) gives:

$$r_{ij}^{IJ} = \sum_{k=1}^{M} d_k u_{k,i}^I u_{k,j}^J.$$

The $\mathbf{u}_k^I$ and the $\mathbf{u}_k^J$ are the right and left singular vectors, which are centered and orthonormed; the d_k are the singular values, positive and arranged in decreasing order. Moreover, $M \leq \min(n_I, n_J) - 1$.

This procedure provides reduced order approximations of the response variable:

$$\tilde{y}_{ij}^{IJ}(m) = \overline{y} + (\overline{y}_i^I - \overline{y}) + (\overline{y}_j^J - \overline{y}) + \sum_{k=1}^{m} d_k u_{k,i}^I u_{k,j}^J$$

which correspond to a fixed value of m ($m < M$).

This descriptive approach can be reformulated within a modeling framework. The y_{ij}^{IJ} are considered as realizations of independent random variables, with normal distribution, means equal to

$$\mu_{ij}^{IJ}(m) = \lambda^{\emptyset} + \lambda_i^I + \lambda_j^J + \sum_{k=1}^{m} \phi_k \xi_{k,i}^I \xi_{k,j}^J, \tag{1}$$

common unknown variance σ^2, and fixed value for m. Obviously the model needs some identification constraints for the linear parameters ($\lambda^{\emptyset}$, λ_i^I, and

λ_j^J), for the score vectors (ξ_k^I, and ξ_k^J, k=1,...,m), and for the association parameters (ϕ_k). Those currently set for linear models and for SVD are usually retained. Interestingly, a maximum likelihood approach for model 1 gives exactly the estimates already obtained by SVD.

The history of this model is discussed by Gower (1996) under the heading "biadditive models" in subsection 5.1 of his paper on "multivariate and multidimensional analysis". But straightforward extensions can be envisaged. One may wish to bring into the picture different weights for the observations. One may also wish to consider a more complex relationship between the righthand and the lefthand side in formula 1: a non-identity link function in the terminology of GLMs. It turns out that the introduction of weights is precisely considered by Gabriel & Zamir (1979) and that CA (Benzécri, 1973, and Greenacre, 1984) implicitly introduces a non-identity link function and specific weights in SVD.

2.2 Correspondence analysis

Let us consider a two-way table of counts. An established exploratory approach for such data is CA. CA (reinvented several times under different names, such as reciprocal averaging, dual scaling ...) leads to the so-called reconstitution formula for the data:

$$y_{ij}^{IJ}(m) = \frac{y_{i+}^I y_{j+}^J}{y_{++}}(1 + \sum_{k=1}^{M} d_k u_{k,i}^I u_{k,j}^J)$$

where the left and right singular vectors are orthonormed with respect to the I and the J marginal proportions ($\frac{y_{i+}^I}{y_{++}}$ and $\frac{y_{j+}^J}{y_{++}}$).

CA can be presented in various ways, each emphasizing different aspects of the method. The simplest, in the context of this paper, is to present it as a generalized SVD (Greenacre, 1984). This amounts to performing the (ordinary) SVD of the matrix of Pearson residuals from a trivial model:

$$r_{ij}^{IJ} = \frac{y_{ij}^{IJ} - \frac{y_{i+}^I y_{j+}^J}{y_{++}}}{\sqrt{\frac{y_{i+}^I y_{j+}^J}{y_{++}}}} = \sum_{k=1}^{M} \tilde{d}_k \tilde{u}_{k,i}^I \tilde{u}_{k,j}^J.$$

With $d_k = \frac{\tilde{d}_k}{\sqrt{y_{++}}}$, $u_{k,i}^I = \tilde{u}_{k,i}^I\sqrt{\frac{y_{++}}{y_{i+}^I}}$, $v_{k,i}^I = \tilde{v}_{k,j}^J\sqrt{\frac{y_{++}}{y_{j+}^J}}$, the desired decomposition is obtained.

In a modeling approach, the counts y_{ij}^{IJ} can be viewed as realizations of independent random variables normally distributed with means μ_{ij}^{IJ}, which, for a given value of m ($m < M$), satisfy the following relationship:

$$\frac{y_{++}}{y_{i+}^I y_{j+}^J}\mu_{ij}^{IJ} - 1 = \sum_{k=1}^{m} \phi_k \xi_{k,i}^I \xi_{k,j}^J. \tag{2}$$

Again, a maximum likelihood approach to estimation can be considered. But now prior weights for the observations need to be introduced. Not surprisingly, weights $\frac{y_{i+}^I y_{j+}^J}{y_{++}}$ do the trick. It may be of interest to note that, in this model, the main effects are absorbed by the non-identity and data-driven link function which is sometimes called the inverse independence link function.

Since CA is most frequently used for count data, obvious extensions are to consider a Poisson distribution with either the log link or the *ad hoc* link function above. These extensions are respectively known in the literature as the

Goodman 's $R \times C$ association model and the correlation model (Goodman, 1991).

2.3 Data transformation

We still consider a two-way table of counts. In an attempt to go "beyond correspondence analysis", Escoufier (1988) advocates pre-processing the data and using the log or the square root transformation. We consider here the square root transformation. Singular value decomposition then gives:

$$\sqrt{y_{ij}^{IJ}} = \sum_{k=1}^{K} \tilde{d}_k \tilde{u}_{k,i}^I \tilde{u}_{k,j}^J.$$

Since the entries of the transformed data matrix are non-negative, the Perron-Frobenius theorem shows that the coordinates of the first right and left singular vectors are positive (Baccini *et al.*, 1991). It then follows that:

$$\sqrt{y_{ij}^{IJ}} = \tilde{d}_1 \tilde{u}_{i,1}^I \tilde{u}_{j,1}^J (1 + \frac{1}{2} \sum_{k=2}^{K} 2 \frac{\tilde{d}_k}{\tilde{d}_1} \frac{\tilde{u}_{k,i}^I}{\tilde{u}_{i,1}^I} \frac{\tilde{u}_{k,j}^J}{\tilde{u}_{j,1}^J})$$

or that:

$$\frac{(\frac{y_{ij}^{IJ}}{w^{\emptyset} w_i^I w_j^J})^{\frac{1}{2}} - 1}{\frac{1}{2}} = \sum_{k=2}^{K-1} d_k u_{k,i}^I u_{k,j}^J.$$

where the "singular vectors" $\mathbf{u_k^I}$ and $\mathbf{u_k^J}$ are orthonormed with respect to data-driven weights. This formula is nothing else than the Box-Cox transformation for the somehow standardized counts.

The above formula suggests straightforward modeling extensions for the means:

$$\frac{(\frac{\mu_{ij}^{IJ}}{\lambda^{\emptyset} \lambda_i^I \lambda_j^J})^{\lambda} - 1}{\lambda} = \sum_{k=1}^{m} \phi_k \xi_{k,i}^I \xi_{k,j}^J \qquad (3)$$

where the link function (including the parameter λ, $\lambda \neq 0$) is to be estimated. Note that the main effects are again absorbed. For $\lambda = 0$, the model is:

$$\log(\mu_{ij}^{IJ}) = \lambda^{\emptyset} + \lambda_i^I + \lambda_j^J + \sum_{k=1}^{m} \phi_k \xi_{k,i}^I \xi_{k,j}^J. \qquad (4)$$

The class of "power models" is discussed by Rom & Sarkar (1992), Baccini *et al.* (1993), Falguerolles & Francis (1994), Goodman (1996).

But a natural question in this modeling context is to select a distribution for the y_{ij}^{IJ}: a normal or a Poisson?

2.4 A confusing example: principal component analysis

Principal component analysis (PCA) is often used to explore structures in multivariate continuous data. Let $\mathbf{x_i'} = (x_i^1, \ldots, x_i^j, \ldots, x_i^{n_J})$, $i = 1, \ldots, n_I$, be n_I independent realizations of a random vector with sample mean $\overline{\mathbf{x}}' = (\overline{x}^1, \ldots, \overline{x}^{n_J})$ and sample covariance matrix $\mathbf{S}$.

On the one hand, an eigenanalysis of $\mathbf{S}$, with eigenvectors $\mathbf{v}_k$ and eigenvalues d_k^2, gives a multiplicative decomposition $\mathbf{S} = \mathbf{VD^2V'} = \mathbf{VD(VD)'}$ which permits a biplot vizualization of the n_J variables. On the other hand, the standardized principal component vectors $\mathbf{u_k}$, the $\mathbf{v}_k$, and the d_k provide a bilinear reconstruction formula:

$$x_i^j = \overline{x}^j + \sum_{k=1}^{n_J} d_k \mathbf{u}_{k,i} \mathbf{v}_{k,j}.$$

Then, the units are often clustered in view of their scores on the first m principal components ($d_k \mathbf{u_k}$, $k = 1, \ldots, m$). But the clustering somehow contradicts the hypothesis of equal means and can be viewed as another example of SVD analysis of the two-way table of residuals $r_i^j = x_i^j - \overline{x}^j$ (see subsection 2.1).

A possible reconciliation is to consider that the x_i^j are independent Gaussian observations with unknown variance σ^2 and means $\mu_i^j = \lambda^j + \sum_{k=1}^m \phi_k \xi_{k,i}^I \xi_{k,j}^J$. This is a particular case of the fixed effect model for PCA considered by Caussinus (1986). Again, a link function other than identity, a distribution other than the normal, and weights for the variances can be envisaged.

3 Generalized bilinear models

3.1 Generalized linear models

Previous situations have shown the need to accommodate non-normal response distributions and transformations of the means ($g(\mu_s) = \eta_s$). This is exactly what GLMs do. Comprehensive references are the books of Aitkin *et al.* (1989) or McCullagh and Nelder (1989).

GLMs can be conveniently extended by incorporating prior weights for the cases and by adding fixed quantities, called offsets, in each predictor. Note that, for each response distribution in the exponential family, there is a canonical link function (g) which guarantees some properties. However non-canonical link functions can be used as previously seen.

3.2 Generalized bilinear models

A generalized bilinear model (GBM) is a statistical model that includes bilinear parameters for two-way interactions, but is otherwise in the form of a GLM. Assuming that the covariates include two factors (I and J), each having a large number of categories (at least two), the following partial hierarchy of models for the predictor illustrates how GLMs and GBMs are interleaved:

$$\eta_s = \lambda^\emptyset + \lambda_{I(s)}^I + \lambda_{J(s)}^J + \cdots \quad (5)$$

$$\eta_s = \lambda^\emptyset + \lambda_{I(s)}^I + \lambda_{J(s)}^J + \sum_{k=1}^m \phi_m \xi_{k,I(s)}^I \xi_{k,J(s)}^J + \cdots \quad (6)$$

$$\eta_s = \lambda^\emptyset + \lambda_{I(s)}^I + \lambda_{J(s)}^J + \lambda_{I(s)J(s)}^{IJ} + \cdots \quad (7)$$

The cumbersome notation used here dramatizes the vectorization of the data matrix considered in the motivating examples from section 1. Moreover, the "dots" represent other covariates than the two factors of interest which can be incorporated in the predictor. But, in such a situation, a strict hierarchical point of view forbids interactions involving factor I and J and a third variable.

3.3 Identification constraints

Clearly, all the above models need identification constraints for their parameters. These are standard in the first and last model. For the intermediate model, the most popular constraints are borrowed from SVD: the m score vectors associated with each factor (the m vectors ξ_k^I and the m vectors ξ_k^J)

are centered (in most cases) and orthonormed with respect to weights w^I and w^J. In some situations the weights are suggested by practical or theoretical reasons (see subsection 4.1).

3.4 Flexibility

GBMs are quite flexible: a variety of distributions for the response variable, a variety of link functions (canonical and non-canonical). Like for GLMs, unbalanced data and structural values are easily handled. Quasi-likelihood methods can also be easily entertained (van Eeuwijk, 1995).

3.5 Estimation

In particular settings, direct maximum likelihood estimation has been considered (Becker, 1990). Another strategy could be to carry out nested non-linear optimisation as suggested by Lane (1996) for generalized non-linear models (GNLMs), which include GBMs. The facilities offered by some statistical software for fitting non-linear models have been also used (Green, 1986, Francis & Saint-Pierre, 1986 with GLIM 4).

However, there is quite a tradition (Wold, 1966, Gabriel & Zamir, 1979) for using a sequence of alternated generalized linear regressions. In particular, Gabriel & Zamir (1979) coined the name "criss-cross regressions" for this sequence. This is greatly facilitated if the software environment allows products and indexing in regression formulas. The procedure is motivated by the following straightforward remarks:

- **Criss**: given the m J-score vectors ξ_k^J $(k = 1, \ldots, m)$, the generalized regression formula $I + J + \sum_{k=1}^m \xi_k^J(J) * I$ estimates the $\phi_k \xi_{k,i}^I$.
- **Cross**: given the m I-score vectors ξ_k^I $(k = 1, \ldots, m)$, the generalized regression formula $I + J + \sum_{k=1}^m \xi_k^I(I) * J$ estimates the $\phi_k \xi_{k,j}^J$.
- **End**: given the m J-score vectors (the ξ_k^J) and the m I-score vectors (the ξ_k^I), the generalized regression formula $I+J+\sum_{k=1}^m \xi_k^I(I)\xi_k^J(J)$ estimates the ϕ_k.

More details on alternating "criss" and "cross" generalized regressions can be found in Falguerolles & Francis (1992, 1995).

Note that the problem of local optima may be handled by using several random starting values for the J-score vectors (ξ_k^J).

3.6 Choice of dimension m

The choice of dimensionality for the bilinear term is widely considered in particular situations: PCA and CA. In the more general context of GBMs, this is nothing else than a problem of finding parsimoniously parameterized models. A simple solution can be based on using some form of penalized likelihood (AIC, BIC ...) An early reference on stepwise tests is Yochmowitz & Cornell (1978). But more work should be done in this area.

Note that, except for special models, the score vectors of same order obtained for two different values of m are not identical.

3.7 Multiple bilinear terms

Multiple multiplicative interaction terms in the predictor can be considered. As an example, consider the case of three factors (H, I and J). The predictor for the all-two-way interaction model is given by:

$$\eta_s = \lambda^\emptyset + \lambda_{H(s)}^H + \lambda_{I(s)}^I + \lambda_{J(s)}^J + \lambda_{H(s)I(s)}^{HI} + \lambda_{H(s)J(s)}^{HJ} + \lambda_{I(s)J(s)}^{IJ}.$$

Indeed, some of the HI, HJ and IJ interaction terms can be constrained to multiplicative forms of given order. With some computational efforts, identical score vectors for H in the HI and HJ interaction terms can also be envisaged. The number of possibilities becomes rather large and this matter will not be further pursued. Note that multiple correspondence analysis (MCA) and joint correspondence analysis (JCA) provide some sort of approximation to the all-two-way interaction model.

3.8 Visualization

Given its multiplicative structure, the bilinear term in the predictor lend itself to biplot visualisations (Gabriel, 1971) where the categories of factor I and of factor J can be represented as points in a low dimensional space with coordinates computed from the coordinates of the score vectors. A state-of-the-art review of the topic is given in Gower and Hand (1996) and this matter will not be discussed here.

Note that the first score vectors can be used to order the row and column categories as suggested by Bertin (Falguerolles *et al.*, 1997). In this spirit, some mosaic displays are of interest (Friendly, 1994), where the pattern of association is shown by the shading of tiles according to residuals from a model, and the H, I, J ... categories are permuted according to the first score vectors, in order to make the pattern of association most coherent.

3.9 Implementation

These ideas have been implemented in several environments. Some of them are specific to particular instances of the model, others are more general. To name but a few, there are a FORTRAN program (Becker, 1990), LLAMA (Firth, 1998), Genstat macros (van Eeuwijk, 1995), a GLIM macro (Falguerolles and Francis, 1992, 1995). The LEM package allows a wide variety of models (Vermunt, 1997) to be entertained (see subsection 3.7).

3.10 Incorporating linear external information

This problem has been primarily investigated as a problem of SVD under linear constraints on the rows and the columns. In the more general context of GBMs, the I-score vectors or the J-score vectors can be further constrained to belong to linear spaces generated by external covariates (R^r, $r = 1, \ldots, n_R$ for factor I and C^c, $c = 1, \ldots, n_C$ for factor J).

It turns out that the "criss-cross" approach facilitates the implementation task since linearity is preserved:

$$\textbf{Criss: } \lambda^{\emptyset} + \lambda^I_{I(s)} + \lambda^J_{J(s)} + \sum_{k=1}^m \phi_m \xi^I_{k,I(s)} \xi^J_{k,J(s)} = \lambda^{\emptyset} + \lambda^I_{I(s)} + \lambda^J_{J(s)} + \sum_{k=1}^m \phi_m \left(\sum_{r=1}^{n_R} \hat{\gamma}^I_{k,r} R^r_{I(s)}\right) \xi^J_{k,J(s)}$$

$$\textbf{Cross: } \lambda^{\emptyset} + \lambda^I_{I(s)} + \lambda^J_{J(s)} + \sum_{k=1}^m \phi_m \xi^I_{k,I(s)} \xi^J_{k,J(s)} = \lambda^{\emptyset} + \lambda^I_{I(s)} + \lambda^J_{J(s)} + \sum_{k=1}^m \phi_m \left(\sum_{c=1}^{n_C} \hat{\gamma}^J_{k,c} C^c_{J(s)}\right) \xi^I_{k,I(s)}$$

The case where different sets of constraints are necessary to characterize each dimension is discussed in Takane *et al.* (1995).

3.11 Influence and robustness

Most standard statistical methods are prone to influential observations. This may happen with GBMs as well: Critchley (1985) and Tanaka (1989) have studied influence for PCA and Pack & Jollife (1992) for CA.

The flexibility of GBMs allows case deletion (as opposed to row or column deletion in previous works). For any statistic T of interest such as the deviance, the difference $\hat{T} - \hat{T_{(s)}}$ (where (s) indicates the omission of the sth observation), measures the influence of the sth observation.

Another strategy is to take advantage of the estimation process by alternated regressions. This can done in two different ways. Firstly, at convergence of each of the alternated generalized regressions, some by-products such as leverages and Cook's distances provide measures of influence; but these may be crude since obtained conditionally upon other estimated parameters. Secondly, robust regressions can be used in the alternating procedure. The latter have been considered by Ukkelberg & Borgen (1993) for PCA and Croux & Filzmoser (1998) in the context of a two-way analysis of variance.

4 In support of bilinear models for count data

4.1 Link function and identification weights

A multinomial (or Poisson) distribution is considered for a two-way table cross-classified according to two factors I and J with ordered categories. To explain the ordering, it is often assumed that the I and J categories arise from the discretization of a continuous bi-variate distribution.

The bi-variate normal and the Eyraud-Farlie-Gumbel-Morgenstern copula are of special interest (Baccini *et al.*, 1993, 1994). In particular, these bi-variate distributions provide some theoretical basis for the choice of weights in the identification constraints and for the choice of link function. It turns out that both distributions advocate marginal proportions for identification weights and give rise to a bilinear model of order $m = 1$. However, the former induces the log link while the latter the inverse independence link. Moreover, in both instances, link mis-specification leads to spurious dimensions and to horseshoe or Guttman effects.

Note that other choices for identification weights can be made on other grounds. For example, non-symmetrical correspondence analysis (NSCA) uses marginal proportions for one factor and equal weights for the other (Lauro & D'Ambra, 1984).

4.2 Dimension m

Each marginal variable in the bi-variate situation considered in the previous subsection can be thought of as being itself multivariate with dimension m ($m > 1$). The bi-variate distributions in subsection 4.1 extend nicely (Baccini *et al.*, 1993, 1994, Eshima & Tabata, 1997). Both give rise to bilinear models of order m with link functions as in subsection 4.1. Again, link mis-specification generates spurious dimensions, but now with more intricate patterns.

4.3 Latent variables and bilinear models

Other situations than the discretization of continuous bivariate distributions give rise to bilinear models for a multinomial distribution.

An interesting situation is provided by the bivariate latent class model which assumes the existence of a categorical latent variable in order to explain the association between the manifest factors I and J. Relations between CA and latent class analysis (LCA) have been widely investigated. In particular, if an LCA model of order m holds, then a CA model of order m also holds (de Leeuw & van der Heijden, 1991).

There is also a connection between Goodman's $R \times C$ association model and some mixed graphical models involving latent conditional normal variables along with the manifest factors. For a particular structure of the conditional independence, the (marginal) joint distribution of I and J has a bilinear structure with log link function (Whittaker, 1989) and order m depending on the number of latent continuous variables.

5 Reduced three-way interaction

5.1 Trilinear interaction

Up to now, only two-way interactions λ_{ij}^{IJ} have been considered and modeled. A straightforward extension of GBMs for modeling a three-way interaction λ_{hij}^{HIJ} is the generalized trilinear model where

$$\lambda_{hij}^{HIJ} \simeq \sum_{k=1}^{m} \phi_k^{HIJ} \, \xi_{k,h}^{H} \, \xi_{k,i}^{I} \, \xi_{k,j}^{J}.$$

The identification constraints introduced for the bilinear score vectors apply again. Moreover, the "criss-cross" estimation procedure can be easily adapted by including an extra step. The discussion of these models goes far beyond the scope of this paper: an early reference is Pannekoek (1985), a recent one is Carlier & Kroonenberg (1998). Here, we restrict our attention to the simple case where H has two categories (Falguerolles & Greenacre, 2000).

5.2 Matched tables

A set of two two-way contingency tables cross-classified by the same two factors (I and J) and stratified according to gender (H) typifies the situation that we have in mind. If there is a three-way interaction, then the two-way interaction parameter $\lambda_{i,j}^{IJ}$ represents the common part of the matched tables and the three-way interaction parameter $\lambda_{1,i,j}^{HIJ} = -\lambda_{2,i,j}^{HIJ}$ their specific parts. Accordingly and omitting other possible covariates, the hierarchy of models for the matched two-way tables looks as follows:

$$\begin{aligned}
\eta_{hij}^{HIJ} &= \lambda^{\emptyset} + \lambda_h^H + \lambda_i^I + \lambda_j^J \\
\eta_{hij}^{HIJ} &= \lambda^{\emptyset} + \lambda_h^H + \lambda_i^I + \lambda_j^J + [\lambda_{hi}^{HI} + \lambda_{hj}^{HJ}] + \sum_{k=1}^{m_1} \phi_k^1 \xi_{k,i}^{1I} \xi_{k,j}^{1J} \\
\eta_{hij}^{HIJ} &= \lambda^{\emptyset} + \lambda_h^H + \lambda_i^I + \lambda_j^J + \lambda_{hi}^{HI} + \lambda_{hj}^{HJ} + \lambda_{ij}^{IJ} \\
\eta_{hij}^{HIJ} &= \lambda^{\emptyset} + \lambda_h^H + \lambda_i^I + \lambda_j^J + \lambda_{hi}^{HI} + \lambda_{hj}^{HJ} + \lambda_{ij}^{IJ} + (-1)^h \sum_{k=1}^{m_2} \phi_k^2 \xi_{k,i}^{2I} \xi_{k,j}^{2J} \\
\eta_{hij}^{HIJ} &= \lambda^{\emptyset} + \lambda_h^H + \lambda_i^I + \lambda_j^J + \lambda_{hi}^{HI} + \lambda_{hj}^{HJ} + \lambda_{ij}^{IJ} + \lambda_{hij}^{HIJ}
\end{aligned}$$

with m_1 and m_2 fixed. The second model gives a parsimonious bilinear description of the common part λ_{ij}^{IJ} of the matched tables. In this model, $[\lambda_{hi}^{HI} + \lambda_{hj}^{HJ}]$ denotes that the corresponding interaction terms may be excluded. The fourth model gives a parsimonious bilinear description of the specific part of each table $\lambda_{1,i,j}^{HIJ} = -\lambda_{2,i,j}^{HIJ}$.

If I and J are two homologous factors ($I \equiv J$) and if $y_{1ij}^{HIJ} = y_{2ji}^{HIJ}$ these models can be further simplified as will be seen below.

5.3 Square tables

The modeling of square tables addresses several substantive questions such as symmetry versus departure from symmetry, inclusion of the diagonal counts versus exclusion, marginal homogeneity versus marginal heterogeneity and so on. It turns out that several models can be fitted as particular cases of matched tables by pairing the square table and its transpose. This is known in the literature as the three-dimensional representation of square tables. In this context, the common part then describes the symmetry, while the specific part describes skew-symmetry. The models above become simplified since the third dimension (H) is only technical and the hierarchy simplifies to:

$$\eta_{ij}^{IJ} = \lambda^{\emptyset} + \lambda_i + \lambda_j$$

$$\eta_{ij}^{IJ} = \lambda^{\emptyset} + \lambda_i^{[I]} + \lambda_j^{[J]} + \sum_{k=1}^{m_1} \phi_k^1 \xi_{k,i}^1 (-1)^{\epsilon_k} \xi_{k,j}^1 \quad \text{(with } \epsilon_k = 0 \text{ or } 1\text{)}$$

$$\eta_{hij}^{IJ} = \lambda^{\emptyset} + \lambda_i^I + \lambda_j^J + \lambda_{ij}^{IJ} \quad \text{(with } \lambda_{ij}^{IJ} = \lambda_{ji}^{IJ}\text{)}$$

$$\eta_{ij}^{IJ} = \lambda^{\emptyset} + \lambda_i^I + \lambda_j^J + \lambda_{ij}^{IJ} + \sum_{k=1}^{m_2} \phi_k^2 (\xi_{2k-1,i}^2 \xi_{2k,j}^2 - \xi_{2k,i}^2 \xi_{2k-1,j}^2)$$

$$\eta_{ij}^{IJ} = \lambda^{\emptyset} + \lambda_i^I + \lambda_j^J + \lambda_{ij}^{IJ} \quad \text{(with unrestricted } \lambda_{ij}^{IJ}\text{)}$$

Here, the second model and the fourth model respectively give a parsimonious description of symmetry and skew-symmetry. With an identity link and a normal distribution, they recover the SVD of the symmetric and of the skew symmetric part obtained in the Gower (Hermitian) decomposition of square matrices. With a log link and a Poisson distribution, they recover bilinear models considered by van der Heijden & Mooijaart (1995).

References

Aitkin, M., Anderson, D., Francis, B. and Hinde, J. (1989). *Statistical Modelling in Glim.* Oxford: Clarendon Press.

Baccini, A., Caussinus, H., and Falguerolles, A. de. (1991). Discussion of Goodman. *Journal of the American Statistical Association*, **86**, pp. 1085-1138.

Baccini, A., Caussinus, H., and Falguerolles, A. de. (1993). Analysing Dependence in Large Contingency Tables: Dimensionality and Patterns in Scatter Plots. In: *Multivariate Analysis: Future Directions 2.* C.M. Cuadras and C.R. Rao (eds.), pp. 245-263, Elsevier B.V. (North-Holland).

Baccini, A., Caussinus, H., and Falguerolles, A. de. (1994). Diabolic Horseshoes. In: *Proceedings of the 9th International Workshop on Statistical Modelling.* Exeter, 11-15 July.

Becker, M.P. (1990). Algorithm AS 253. Maximum Likelihood Estimation of the RC(M) Association Model. *Applied Statistics*, **39**, pp. 152-167.

Benzécri, J.P. (1973). *L'analyse des données*, vol. **2**, Paris: Dunod.

Brain, Ph. (1998). Using Singular Value Decomposition in Non-Linear Regressions. In: *COMPSTAT98 Proceedings in Computational Statistics*, pp. 197-202. Heidelberg: Physica-Verlag.

Carlier, A. and Kroonenberg, P.M. (1998). The Case of the French Cantons: An Application of Three-Way Correspondence Analysis. In: *Vizualization of Categorical Data*, J. Blasius & M. Greenacre (eds.), pp. 253-276. Academic Press.

Caussinus, H. (1986). Models and Uses of Principal Components Analysis. In: *Multidimensional Data Analysis*, J. de Leeuw *et al.* (eds.), pp. 149-170. Leiden: DSWO Press.

Choulakian, V. (1996). Generalized Bilinear Models. *Psychometrika*, **61**, pp. 271-283.

Critchley, F. (1985). Influence in Principal Components Analysis, *Biometrika,* **72**, pp. 627-636.

Croux, C. and Filzmoser, P. (1998). Robust factorization of a Data Matrix. In: *COMPSTAT98 Proceedings in Computational Statistics*, pp. 245-250.

de Leeuw and van der Heijden. (1991). Reduced Ranks Models for Contingency Tables, *Biometrika*, **78**, pp. 229-232.

Escoufier, Y. (1988). Beyond Correspondence Analysis. In: *Classification & Related Methods of Data Analysis*, H.H. Bock (eds.), pp. 505-514. Elsevier B. V. (North-Holland).

Eshima, N. and Tabata, M. (1997). The $R \times C(M)$ Association Model and Canonical Correlation Analysis. *Journal of the Japan Statistical Society*, **27**, pp. 109-120.

Falguerolles, A. de and Francis, B. (1992). Algorithmic Approaches for Fitting Bilinear Models. In: *COMPSTAT92, Proceedings in Computational Statistics*, **Vol. 1**, pp. 77-82. Heidelberg: Physica-Verlag.

Falguerolles, A. de and Francis, B. (1994). Fitting Power Models to Two-Way Tables. In: *COMPSTAT94 Proceedings in Computational Statistics*, pp. 470-475. Heidelberg: Physica-Verlag.

Falguerolles, A. de and Francis, B. (1995). Fitting Bilinear Models in GLIM, *GLIM Newsletter*, **25**, pp. 9-20.

Falguerolles, A. de and Greenacre, M. (2000). Statistical Modelling for Matched Tables. (to appear in: *Proceedings of the 15th International Workshop on Statistical Modelling*. Bilbao, 17-21 July.

Falguerolles, A. de, Friedrich, F. and Sawitzki, G. (1997). A Tribute to Bertin's Graphical Data Analysis In: *SOFTSTAT'97*, pp. 11-20. Stuttgart: Lucius & Lucius.

Firth, D. (1998). LLAMA: an Object Oriented System for Log Multiplicative Models. In: *COMPSTAT98 Proceedings in Computational Statistics*, pp. 305-310. Heidelberg: Physica-Verlag.

Francis, B., and Saint-Pierre, J. (1986). Correspondence Analysis Models Using GLIM. In:*Compstat86, Short communications and posters*, pp. 255-256. Roma: Università "La Sapienza".

Friendly, M. (1994). Mosaic Displays for Multi-way Contingency Tables. *Journal of the American Statistical Association*, **89**, pp. 190-200.

Gabriel, K.R. (1971). The Biplot-Graphic Display of Matrices with Application to Principal Component Analysis, *Biometrika*, **54**, pp. 453-467.

Gabriel, K.R. (1998). Generalised Bilinear Regression. *Biometrika*, **85**, pp. 689-700.

Gabriel, K.R. and Zamir, S.(1979). Lower Rank Approximation of Matrices by Least Squares with any Choice of Weights, *Technometrics*, **21**, pp. 489-498.

Gollob, H.C. (1968). A Statistical Model which Combines Features of Factor Analysis and Analysis of variance Techniques. *Psychometrika*, **33**, pp. 73-115.

Goodman, L. (1991), Measures, Models, and Graphical Displays in the Analysis of Cross-Classified Data (with discussion). *Journal of the American Statistical Association*, **86**, pp. 1085-1138.

Goodman, L. (1996). A Single General Method for the Analysis of Cross-Classified Data: Reconciliation and Synthesis of some Methods of Correspondence Analysis. *Journal of the American Statistical Association*, *91*, pp. 408-428.

Gower, J.C. (1989). Discussion of van der Heijden, P.G.M., Falguerolles, A. de, & de Leeuw, J. *Applied Statistics*, **38**, pp. 249-292.

Gower, J.C. (1996). Multivariate and Multidimensional Analysis. In: *Advances in Biometry*, P. Armitage & H.A. David (eds.), pp. 149-175. Wiley.

Gower, J.C. and Hand, D. (1996). *Biplots*. London: Chapman and Hall.

Green, M. (1986). Generalized Multiplicative Models. In: *Compstat86 Proceedings in Computational Statistics*, pp. 102-107. Heildeberg: Physica-Verlag.

Greenacre, M. (1984). *Theory and Applications of Correspondence Analysis.* Academic Press.

Lane, P. (1996). Generalized Nonlinear Models. In: *COMPSTAT96 Proceedings in Computational Statistics*, pp. 331-336. Heidelberg: Physica-Verlag.

Lauro, N.C., and D'Ambra, L. (1984). L'Analyse non symétrique des correspondances. In: *Data Analysis and Informatics, III*, E. Diday *et al.* (eds.), pp. 433-446. Amsterdam: North-Holland.

McCullagh, P. and Nelder J.A. (1989). *Generalized Linear Models,* (2nd Edition). London: Chapman and Hall.

Pack, P. and Jolliffe I.T. (1992) Influence in Correspondence Analysis, *Applied Statistics,* **41**, pp. 365-380.

Pannekoek, J. (1985). Log-Multiplicative Models for Multiway Tables. *Sociological Methods and Research*, **14**, pp. 137-153.

Rom, D., and Sarkar, S.K. (1992). A Generalized Model for the Analysis of Association in Ordinal Contingency Tables, *Journal of Statistical Planning and Inference*, **33**, pp. 205-212.

Seyesadr, M. and Cornelius, P.L.(1992). Computational Procedures for Estimation and Hypothesis Testing in Analyzing Two-Way Tables with Interaction and No Replication. In: *COMPSTAT92 Proceedings in Computational Statistics*, vol.**1**, pp. 89-94. Heidelberg: Physica-Verlag.

Takane, Y., Kiers, H.A.L., and de Leeuw, J. (1995). Component Analysis with Different Sets of Constraints on Different Dimensions. *Psychometrika*, **60**, pp. 259-280.

Tanaka Y. (1989). Influence Functions Related to Eigenvalue Problems which Appear in Multivariate Analysis. *Communication in Statistics*, **18**, pp. 3991-4010.

Ukkelberg, Å. and Borgen, O. (1993). Outlier Detection by Robust Alternating Regression, *Analytica Chimica Acta,* **277**, pp. 489-497.

van der Heijden, P.G.M. and de Leeuw, J. (1985). Correspondence Analysis Used Complementary to Loglinear Analysis, *Psychometrika*, **50**, pp. 429-447.

van der Heijden, P.G.M., Falguerolles, A. de, and de Leeuw, J. (1989). A Combined Approach to Contingency Table Analysis Using Correspondence Analysis and Log-Linear Analysis (with discussion). *Applied Statistics*, **38**, pp. 249-292.

van der Heijden, P.G.M., and Mooijaart, A. (1995). Some New-Bilinear models for the Analysis of Symmetry in a Square Contingency Table. *Sociological methods and research*, **24**, pp. 7-29.

van Eeuwijk, F.A. (1995). Multiplicative Interaction in Generalized Linear Models, *Biometrics,* **51**, pp. 1017-1032.

Vermunt, J.K. (1997). *LEM: A general program for the analysis of categorical data.* Department of Methodology/WORC, Tilburg University.

Whittaker, J. (1989). Discussion of van der Heijden, P.G.M., Falguerolles, A. de, & de Leeuw, J. (1989). *Applied Statistics*, **38**, pp. 249-292.

Wold, H. (1966). Nonlinear Estimation by Iterative Least Squares Procedures. In: *Research papers in statistics, Festschrift for J. Neyman*, H. David (ed), pp. 411-444. Wiley.

Yochmowitz, M.G. and Cornell, R.G. (1978). Stepwise Tests for Multiplicative Components of Interaction, *Technometrics*, **20**, pp. 79-84.

Generalized calibration and application to weighting for non-response

Jean-Claude Deville

Laboratoire de Statistique d'Enquête, ENSAI/crest,Campus de Ker-Lann, 35170-Bruz, France

Abstract. A generalised theory for calibration is developed distinguishing two set of variables and leading to instrumental regression estimation in the linear case. The dissymmetry of the variables receives a very interesting application when we apply generalised calibration to the problem of weighting for non-response: one set of variables is connected to factors inducing nonresponse, the second one to variables correlated to the variable of interest. A calibration principle is proposed as an estimation method for the parameters of the response model. Its advantage is to produce a reduction of the variance bound to the calibration. A complete treatment is given in the case of an exhaustive survey, and some indication for the general case. We show also that imputation « weighting-like » can be performed by using of balanced sampling techniques.

Keywords : calibration estimators, non-response, weighting, imputation, exhaustive surveys

1 Some general ideas on nonresponse correction

Correction for nonresponse in surveys relies generally on two different techniques : weighting which is the principal tool to deal with complete nonresponse, and imputing whose purpose is to correct partial nonresponse.

Weighting (or reweighting) procedures are based on some model describing the response mechanism. This mechanism consist in selecting a sample r of respondents according to an unknown "pseudo" sampling design $q(r|s)$. This pseudo sampling design is described by an hypothetical model containing a vector of parameters β with p dimensions. Then we seek for an estimation of this β parameter. Then we admit that the true mechanism of selection of the respondent sample is generated by the model with the estimated parameter. That is the reason why this approach of the nonresponse problem is called « pseudorandomization ». At the same time, estimation theory for sampling surveys received in the last years interesting developments. Calibration theory allows to synthesise and to generalise almost all kinds of linear estimators used before and to define new ones (Deville, Särndal (1992), Deville, Särndal, Sautory (1993)). In this theory, weights are corrected by a factor having the form $F(x_k^{'}\lambda)$ where F is some regular increasing function, x_k a p-vector of auxiliary variable and λ a p-vector of parameters allowing to have an exact estimate of the total of the auxiliary variables. This form of the weights is very similar to the form given to a probability (or to the inverse of a probability) in the generalised linear model set-up. Actually it could be shown that this similarity is more than a coincidence and that the analogy is in fact a rather deep property leading to an interesting way to describe formally the response mechanism. This lead to a synthesis between the two approaches already outlined in (Deville, Dupont (1993)). A new extension is presented here, based on a generalised calibration theory. This leads to new weighting procedures which seem to be easy to use and efficient. In particular, it will be seen that some reductions of variance are obtained by using a convenient estimation procedure.

Standard ideas about imputation are based on models allowing to predict the variable of interest y_k using cofactors. Generally a model is estimated using data available only for respondents. The estimated model provides an estimated probability for the variable y_k conditionally to the cofactors. There are now two commonly used possibilities :

- Impute the best prediction for y_k
- Impute a value at random using the estimated probability law for y_k.

Both methods present severe drawbacks. The first one distort completely the distribution of the *y* variable and cannot be used if some quantiles of *y* need to be estimated (income distribution for instance). The second one produce an artificially increased variance because of the randomness of the imputation procedure. This extra variance could be reduced by multiple imputation, but, unhappily, this procedure cannot be used in the context of large scale government surveys. However we shall see that we can mix the two possibilities using balanced sampling methods. This provides an interesting trade-off between the two variations.

In this paper, we deal essentially with nonresponse correction in the case of exhaustive surveys. For a correct treatment of the matter, we need an extension of calibration techniques surveyed in part 2. In part 3 we shall see how it works in the case of an exhaustive survey with nonresponse. Some indications on the general case are given in part 4. Imputation weighting-like is outlined in part 5.

2 Generalised calibration theory: some basic indications

2.1 What do we know?

Let us recall some major idea of the standard theory of calibration for sample surveys. We start with an unbiased (or at least convergent) estimator of the total *Y* of the y_k values. Generally one uses the Horvitz-Thompson estimator, $\hat{Y} = \sum_s d_k y_k$. We seek for new weights w_k , near of the d_k for some distance function and verifying the calibration equations: $X = \sum_s w_k x_k$ where *X* is the p-vector of totals for *p* variables which are the co-ordinates of the vector x_k. It is found that the new weights are given by $w_k = F(q_k x_k' \lambda)$ where the positive numbers q_k are introduced to deal with some heteroskedasticity in the generalised regression model bounding y_k and x_k, λ is a p-vector and F a regular increasing function from $\boldsymbol{R}$ to $\boldsymbol{R}$ verifying $F(0)=1$ and $F'(0)=1$.

The advantage of this class of estimators is that it eliminates the variance induced by the x_k variables. More precisely we have $Var\left(\sum_s w_k y_k\right) = Var\left(\sum_s w_k e_k\right)$, where the e_k are the residuals in the regression $y_k = x_k' \beta + e_k$ obtained by the least square method *using the weights* q_k.

2.2 Outline of the generalised theory

To every k in the population is associated a calibration function $F_k : \boldsymbol{R}^p \to \boldsymbol{R}$ verifying $F_k(0)=1$, regular in a domain sufficiently large containing *0*. Of course those functions are constructed by using auxiliary variables associated to each individual *k*, as it will be clear from the following examples, but for the moment it does not matter .Then we solve the calibration equations :

$$X = \Phi(\lambda) = \sum_s d_k x_k F_k(\lambda) \quad (1).$$

2.2.1 The linear case

Let us begin with the most simple case and assume that we take simply $F_k(\lambda) = 1 + z_k' \lambda$. Vector z_k is a vector of « instrumental variables ». The choice of this denomination will become clear from the following. If we try to solve the calibration equations (1) we get :

$$\lambda = \left(T_{szx}'\right)^{-1}\left(X - \hat{X}\right) \quad (2)$$

where the design matrix $T_{szx} = \sum_s d_k z_k x_k'$ is supposed to have full rank.

Geometrically, this condition can be interpreted as follows. Let L_X be the linear subspace of $\mathbb{R}^N$ generated by the p variables which are the co-ordinates of x_k . In other words, if $X_\bullet$ is the *pxN* matrix of the x_k , L_X is the linear subspace generated by the columns of $X_\bullet'$. L_z is defined in the same way from the z_k. The condition of inversibility can be written $L_X \cap L_Z^\perp = 0$. and is of course necessary to solve (1). Actually we need the same condition for the sample. However, if the condition is fulfilled, T_{szx} has full rank with a very high probability.

A straightforward calculation show that we have for the calibrated estimator $\hat{Y}_C = \sum_s d_k F_k(\lambda) y_k$ the two following equivalent expressions :

$$\hat{Y}_C = \hat{Y} + \left(X - \hat{X}\right)' T_{szx}'^{-1} \sum_s d_k z_k \, y_k \; = \hat{Y} + \left(X - \hat{X}\right)' \tilde{\beta},$$

where $\tilde{\beta}$ is solution of the set of normal equations: $\sum_s d_k z_k \left(y_k - x_k' \hat{\beta}\right) = 0.$

We get the following conclusions :

- $\tilde{\beta}$ is the p-vector of the instrumental regression coefficients (Fuller (1987) for example)) of y on x using the z_k as instruments.

- Geometrically, it is interpreted as the projection, in $I\!R^N$, of the vector having the y_k as co-ordinates on L_X along the direction $L_Z^{\perp}$.
- It can be shown that the new weights can be obtained by the minimisation of a quadratic distance between the old and the new weights.
- Moreover, the approximate variance of $\hat{Y}_C$ *can be computed with the residual technique*. This means that $Var(\hat{Y}_C) \cong Var(\sum_s d_k \tilde{e}_k)$ where $\tilde{e}_k = y_k - x_k'\beta$ and β is the vector of coefficients of the instrumental regression. The proof is easy and standard.
- Variance estimation follows the same principle.
- The "instruments" z_k needs to be known only for the sample : they are *not* an external auxiliary information.

2.2.2 General case

In the general case, equations (1) can be solved numerically without difficulty using for example Newton's method. Exactly like in the habitual case of calibration estimators, we can obtain general results in having a closer look to this equation.

Let $z_k = grad\ F_k(0) \in I\!R^p$, be the vector of the partial derivative of F_k at *0*. Exactly like in the habitual case we get, with a design matrix $T_{szx} = \sum_s d_k z_k x_k'$:

$$\lambda = \left(T_{szx}'\right)^{-1}\left(X - \hat{X}\right) + O\left(\left\|X - \hat{X}\right\|^2\right).$$

It follows that, up to terms in n^{-2}, the calibrated estimator using a non-linear method has the same variance as the estimator using the linear method with $z_k = grad\ F_k(0)$. Therefore the approximate variance is also obtained with the residual technique and the variance can be estimated on the same basis.

2.3 Examples

2.3.1 Ratio estimator

X et x_k are one dimensional. The instrumental variable is the "free of charge" variable $z_k = 1$. The calibration equation has the expression : $X = \sum_s d_k x_k (1 + z_k'\lambda)$

and therefore $\tilde{\beta} = \dfrac{\hat{Y}}{\hat{X}} = \hat{R}$ with residuals $\tilde{e}_k = y_k - \hat{R}x_k$.

2.3.2 Weighted regression estimator

The instrument are the variables : $z_k = q_k x_k$.

2.3.3 Optimal regression estimator (Montanari, 1987)

This estimator is obtained by solving the optimality problem : find the estimator having the form $\hat{Y} - B'\hat{X}$ and a minimum variance (the estimators in the formula are Horvitz-Thompson). It is easily shown that, for any variable, the optimal estimator is obtained by taking $B = Var(\hat{X})^{-1}\ Cov(\hat{X},\hat{Y})$. Then one can replace the unknown vector B by some estimator. As $Var(\hat{X}) = \sum_U \sum_U (\pi_{k\ell} - \pi_k \pi_\ell) \frac{x_k}{\pi_k} \frac{x'_\ell}{\pi_\ell}$ we can take as instruments the variables $z_k = \sum_{\ell \in s} \Delta_{k\ell} x_\ell$ with $\Delta_{k\ell} = \frac{\pi_{k\ell}}{\pi_k \pi_\ell} - 1$.

In the case of a stratified design with simple random sampling for each strata we get simply $z_k = \left(x_k - \overline{X}_h\right) \frac{N_h^2}{n_h} \left(1 - \frac{n_h - 1}{N_h - 1}\right)$ with the usual notations.

2.3.4 A non-linear example

Let u_k be a strictly positive variable known for every unit k in the sample s, $\lambda = (a,b,c)$ and. $F_k(\lambda) = a + \exp(bu_k) u_k^c$. It is clear that $F_k(O) = 1$ and we have:

$$\frac{\partial F_k}{\partial a} = 1$$

$$\frac{\partial F_k}{\partial b} = u_k \exp(bu_k) u_k^c \qquad => \frac{\partial F_k}{\partial b}(0) = u_k$$

$$\frac{\partial F_k}{\partial c} = Log\, u_k \exp(bu_k) u_k^c \qquad => \frac{\partial F_k}{\partial c}(0) = Log\, u_k$$

Actually this example shows that when we can use a numerical variable, it is always possible to construct other variables by taking some functions of the original one. Therefore the fact that vectors z_k and x_k must have the same dimension is not a practical limitation.

3 Weighting for non-response in the case of an exhaustive survey

3.1 Response model

The response mechanism is modelled as a sampling design $q(r\,;\beta)$ giving the probability to get the sample r of respondents given s. In this expression β is an unknown parameter in R^P. The model provide us extrapolation weights "à la Horvitz-Thompson", $\pi_k^{-1} = F_k(\beta)$. Eventually we get also from the model the second order inclusion probabilities if we need them for computing or estimating the variance.

The most simple model, and, in some respects the most natural one, is Poisson sampling. In this scheme, there are independent draws for each individual with given probabilities $\pi_k = 1 / F_k(\beta)$ where the F_k are functions determined by the model and taking their value from 1 to infinity. Therefore the probability to get the sample r is given by :

$$q(r;\beta) = \prod_{k \in r} F_k^{-1}(\beta) \prod_{k \in U-r} (1 - F_k^{-1}(\beta))$$

It is possible, of course, to introduce more sophisticated models. We could desire, for instance, to take into account some clustering effects if we suspect some influence of the interviewers on the non-response pattern.

3.2 Estimation problems: the calibration principle

An important question seems to be now how to estimate β.

The answer is a bit surprising : it doe's not matter ! Maximum likelihood, moments method, minimum -Chi 2, it's nearly as you want ! The only important fact is the following : anyway, we will have to solve a set of p estimating equations. And they will be solved *for any feasible sample of respondent r.* This means that we get a set of p expressions equated to *0 for every sample r and therefore a set of p statistics for which variance is equal to 0*

Example : Suppose we want to estimate the β parameter of the Poisson model just above. If we apply the maximum likelihood method, we obtain the estimating score-equations $\sum_r F_k(\beta) z_k^* = \sum_U z_k^*$ with $z_k^* = \dfrac{grad\, F_k(\beta)}{F_k(\beta)(F_k(\beta)-1)}$.

If we specify a generalised linear model $F_k(\beta) = F(z_k' \beta)$ with a link function F and its derivative $\dot{F}$, then $z_k^* = z_k \; \dot{F} / F(F-1)$. If $F = 1 - exp$ (log-linear model), we have simply $z_k^* = z_k$.

Actually, we shall see that it is much better to rely upon a *calibration principle.* It consists in using the following unbiased estimating equations :

$$\sum_r F_k(\beta) x_k = \sum_U x_k$$

or, when using a Generalised Linear Model : $\sum_r F(z_k' \beta) x_k = \sum_U x_k$

It is clear that the solution of this equation has a variance of the order of 1/n, n being the size of U. However we can also interpret the equations in the following way :

$$X = \sum_r x_k \, F_k(\beta_0) \, G_k(\lambda).$$

where β_0 is the true value of the parameter and $G_k(\lambda) = \dfrac{F_k(\beta_0 + \lambda)}{F_k(\beta_0)}$.

Observe that $G_k(0) = 1$. Actually, it is easy to see that those equations are nothing else than calibrations equations since $F_k(\beta_0)$ is, under the model, the exact Horvitz-Thompson weight. In particular the G_k are calibration functions as defined in part 2 of this paper. With this interpretation we can make the following observations :

- $\beta = \beta_0 + \lambda$ is an estimator of β_0.
- If, by chance, we have at hand another estimator $\hat{\beta}_0$ of β_0, we can write $\beta = \beta_0 + \left(\hat{\beta}_0 - \beta_0\right) + \lambda$, and the interpretation remain the same as before because λ and the estimation error $\hat{\beta}_0 - \beta_0$ have the same order of magnitude, *1/n*.

- We need to know the functions F_k (or variables contained in it) only for the respondents.

We get the same effect on the variance than with the usual calibration techniques. Therefore we can estimate the variance with the residual technique.

3.3 A closer look

From the theory in section 2, this estimator has the same variance as the "linear estimator" of the same family, that is :

$$\hat{Y} = \sum_r y_k\, F_k(\beta_0)(1 + z_k'\lambda)$$

with λ such that

$$X = \sum_r x_k\, F_k(\beta_0)(1 + z_k'\lambda) \text{ and } z_k = grad\ G_k(0) = grad\ Log\, F_k(\beta_0).$$

If the model is a Generalised Linear Model, we have $F_k(\beta) = F\left(z_k^{*'}\beta\right)$ and

$$z_k = z_k^{*}\frac{\dot{F}(z_k^{*'}\beta_0)}{F(z_k^{*'}\beta_0)} = z_k^{*}(Log\, F(z_k^{*'}\beta_0)\dot{)} = z_k^{*}q_k .$$

Therefore we have : $Var(\hat{Y}) = Var(\sum_r \tilde{e}_k F_k(\beta_0))$ with $\tilde{e}_k = y_k - \tilde{B}'\, x_k$,where $\tilde{B}$ is the solution of the normal equations : $\sum_r z_k\, F_k(\beta_0)(y_k - \tilde{B}'\, x_k) = 0$.

In the case of a GLM, the equations take the form : $\sum_r q_k z_k^{*} F_k(\beta_0)\,(y_k - \tilde{B}'\, x_k) = 0$.

If $z_k^{*} = x_k$ we get $\sum_r q_k x_k\, F_k(\beta_0)\,(y_k - \tilde{B}'\, x_k) = 0$, and the instrumental regression resumes to the weighted least squares regression.

3.4 Some examples

3.4.1 Ratio weighting for non-response

Here x_k is a positive variable and $z_k = 1$, the « free of charge » variable. In other words, responses are missing at random with the same probability, but calibration is performed on the sum X of the x_k . It is easily seen that we get a commonly used estimator : $\hat{Y} = X\frac{Y_r}{X_r}$.

If our response model is the Poisson model we get the following expression for the variance : $Var(\hat{Y}) = \frac{X}{X_r}(\frac{X}{X_r} - 1)\sum_r (y_k - Rx_k)^2$. With $x_k = 1$ we get the classical and natural result :

$$Var(\hat{Y}) = \frac{N}{n}(\frac{N}{n} - 1)\sum_r (y_k - \bar{y})^2 .$$

3.4.2 Poststratification and formal raking-ratio

For the poststratification case, $x_k = z_k$ is the vector of the indicator variables : if k belongs to strata h, the hth co-ordinate of x_k is equal to 1, and the other to 0. It is a special degenerate case of GLM.

For the formal raking case, x_k=z_k is the vector of the indicator variables of the different margins on which calibration is performed (with exclusion of one modality for the margin of rank two or more). The response model takes the form $F_k(\beta)=\exp(x_k^{'}\beta_0)=\prod_i P_{h_i}$ where i is an index for the ith margin.

3.4.3 An original and useful example

Vector x_k the indicator vector of a qualitative variable having I modalities $i = 1$ *à* I. The total number N_i of the ith class is assumed to be known. Vector z_k is the indicator vector of another qualitative variable having the same number of modalities indexed by a. In a concrete application, x_k is a situation known in the sampling frame (occupation of a household head, for example) and z_k the effective situation observed in the survey (occupation of the household head in the sampled dwelling). Let $R_{i\,a}$ be the total number of respondents classified in the modality i of x_k and a of z_k,. The vector $\beta=(\ldots,\beta_a,\ldots)$ of parameters of the response model verifies the equations : $N_i=\sum_a R_{ia}(1+\beta_a)$. Those equations are estimating equations for β. The variance of the calibrated estimator has the expression :

$$\sum_a \beta_a(1+\beta_a)\sum_i \sum_{k\in r_{ia}}\left(y_k-\hat{\bar{Y}}_i\right)^2$$

where $\hat{\bar{Y}}_i$ is the estimation of the mean of the y_k for k in the ith modality of x.

3.4.4 Remark on the nature of the z and x variables

Variables z_k are introduced because we believe that they explain the pattern of the non-response. Actually if the response model is correct, their influence is to reduce the bias. Variables x_k are introduced because we have already some information about them. The influence of this use is to reduce the variance of the estimates in proportion of the amount of information they carry.

Observe that the z variables have to be known only for the respondents. Nevertheless, they can be themselves variable of interest. This remark shows that generalised calibration applied to non-response can supply a solution for the case of non-response caused by a variable of interest (non ignorable mechanism).

4 Non-response after sampling

In this paragraph, to make it short, we only state problems and results.

4.1 Use of an information available in the sample

A sampling design p on the population U provides a sample s and its extrapolation weights d_k. Conditional to s, non-response is modelled by the design $q(r|s\,;\beta)$ providing the respondent sample r. The extrapolation weights from r to s are $F_k(\beta_0)$ as in paragraph 3. If the x_k are known on s, we can estimate β in accordance with the calibration principle. As $\hat{X}_s=\sum_s d_k x_k$ is assumed to be known, we are led to the estimating equations : $\hat{X}_s=\sum_s d_k x_k=\sum_r d_k x_k F_k(\beta_0)$.

As it is well known (Särndal, Wretman (1987)), we have to deal with a two phases sampling. Variance has two terms which can be estimated separately . In the following we suppose that a software is available for computing those two quadratic forms (like POULPE, Caron, (1998), Petit (1998)) :

$$V\hat{a}r\left(\hat{Y}\right) = Q_1\left(y_r\right) + Q_2\left(y_r\right), \text{ where } y_r = \left\{y_k \,; k \in r\right\}.$$

If β_0 is estimated as above, we estimate the variance of $\hat{Y}_1$ by: $V\hat{a}r\left(\hat{Y}_1\right) = Q_1\left(y_r\right) + Q_2\left(e_r^{zx}\right)$ where e_r^{zx} is the set of residuals in the instrumental regression in s with the normal equations :

$$\sum_s z_k (d_k y_k - d_k x_k' B) = 0 \quad (z_k = \frac{grad\ F_k(\beta_0)}{F_k(\beta_0)}).$$

These equations estimate the instrumental regression in the population characterised by $\sum_U z_k(y_k - x_k' B) = 0.$

The estimation in r is obtained by solving: $\sum_r d_k F_k(\beta)\, z_k(\underbrace{y_k - x_k' B}_{ré siduals}) = 0$.

4.2 Use of an external information

In many cases, the total X is known from external sources and $\hat{X}_s$ is not available. The typical case is formal poststratification.In the general case, it is sufficient to solve : $X = \sum_r d_k x_k F_k(\beta)$. and the application of the results in paragraph 3 leads to an (estimated) variance having the expression : $\hat{V}ar(\hat{Y}_2) = Q_1(e_r^{zx}) + Q_2(e_r^{zx})$.

If $F_k(\beta) = F(z_k^{*'} \beta)$ then $z_k = z_k^* \dfrac{\dot{F}(z_k^{*'})}{F(z_k^{*'})} = z_k^* q_k$. If $F_k(\beta) = F(x_k' \beta)$ one get the standard calibrated estimator with regression weights q_k. Moreover if $F=exp(.)$, then simply $q_k = 1$.

4.3 Use of information coming from the sample and from an external source

Suppose now that we have auxiliary information at both level.

The total of a first set of auxiliary variable is known for the whole population : $X_o = \sum_U x_{ok}$, $\dim X_o = p_o$. A second set of auxiliary variables is available on s and allows to compute the Horvitz-Thompson estimator $\hat{X}_1 = \sum_s d_k x_{1k}$, $\dim \mathrm{X}_1 = p_1$.

For $k \in r$, the calibration/non-response functions $F_k\left(\beta\right)$ are known, where β is a parameter in $R^p, p = p_0 + p_1$. It can be shown that the best use of this information is to solve the following set of calibration equations :

$$\begin{pmatrix} X_0 \\ \hat{X}_1 \end{pmatrix} = \sum_r d_k \begin{pmatrix} x_{0k} \\ x_{1k} \end{pmatrix} F_k(\beta) = \sum_r d_k x_k F_k(\beta_0)\, G_k(\lambda)$$

$$\text{with} \quad G_k(\lambda) \cong 1 + z_k' \lambda \quad \text{and} \quad z_k = \frac{grad\, F_k(\beta_0)}{F_k(\beta_0)}$$

The weighted estimator $\hat{Y}_W$ obtained with this equations has for (estimated) variance : $\hat{V}ar\left(\hat{Y}_w\right) = \sum_r d_k F_k(\beta) y_k = Q_1\left(e_r^{\$}\right) + Q_2(\tilde{e}_r)$,where :

$$---- \tilde{e}_k = y_k - x_{ok}\hat{\beta}_o - x_{1k}^{'}\tilde{\beta}_1$$

verify the normal (instrumental) equations :

$$\sum_r d_k F_k(\beta)\, z_k(y_k - \begin{pmatrix} x_{0k}^{'} & x_{1k}^{'} \end{pmatrix}\begin{pmatrix} \tilde{B}_0 \\ \tilde{B}_1 \end{pmatrix} = 0$$

and : $e_r^{\$} = y_k - x_{0k}^{'} B_0^{\$}$ verify the equations $\sum_r d_k F_k(\beta)\, x_{0k}^{*}(y_k - x_{0k}^{'} B_0^{\$}) = 0$ with the instruments : $x_{0k}^{*} = T_{rx_0z}\, T_{rzz}^{-1}\, z_k$ (and $T_{ruv} = \sum_r d_k F_k(\beta) u_k v_k^{'}$).

When $x_{o\,k}$ is a linear transformation of z_k. $B_o^{\$}$ is nothing else as the mean square regression vector and, moreover $x_{ok}^{*} = x_{ok}$.

If we have $z_k = \begin{pmatrix} x_{ok} \\ x_{1k} \end{pmatrix}$ then $\begin{pmatrix} \tilde{B}_o \\ \tilde{B}_1 \end{pmatrix} = \begin{pmatrix} B_o \\ B_1 \end{pmatrix}$ because we are in a least square setting. In the same way $B_o^{\$} = B_{oo}$, least square estimate of the regression of y on x_o. This last result is a slight improvement for F.Dupont (1995).

5 Imputation « weighting-like » for missing values

5.1 General considerations on the imputation methodology

The usual way of imputing for a missing value y_k consists in estimating a probability law for y_k conditionally to some cofactor x_k. The model is estimated using data where x_k and y_k are together available. Two opposite methods are commonly used. One of them consist in the imputation of the expectation of y_k .It provides the best predictor for $\sum_U y_k$ or $\sum_s w_k y_k$. It has the drawback to give a bad restitution of dispersion of the y_k and therefore to provide inconsistent estimates for the distribution function of y. The other way is to impute a value at random in the estimated law of y_k. . The drawback is to add an arbitrary extra-variance which is sometimes large (see for instance Oh and Scheuren (1983)). When one uses non-parametric models, they resume often to a donor imputation . Formally, it is always the case when y_k is a (0-1)-variable or a qualitative variable.

None of these drawbacks arises when non-response is compensated by weighting methods. That's the reason of the theoretical superiority of weighting upon imputation methods. However, imputation is often necessary because of practical constraints. We indicate below a method aiming to restrict the drawbacks of random imputation. It makes uses of balanced sampling with the cube method (Deville and Tillé (2000)).

5.2 An elementary example

The most simple example is the following :y_k is a (0-1) binary variable. Some models have given to us some estimated probabilities P_k for $y_k = 1$. If the sample is drawn with equal probabilities (or if the survey is exhaustive like in paragraph 3) the best prediction is based on the fact that $\sum_o P_k = m$, assumed to be an integer (o is the sample of non-respondents). Thus the question is to sample the value 1 m times and the value 0 ($|o| - m$) times. This is exactly a problem of fixed size unequal probability sampling, which is the basic case of balancing a sample. This example can be generalised to many situations and it introduces a very rich connection between imputation and advanced sampling methods.

5.3 A more complex example

Let us finish with another example coming from the practice. The situation is the one described in paragraph 4-2 : the total of the vector of auxiliary variables is given by an external source, but the values of x_k, z_k and y_k are not available for the (eventually partial) non-respondents. We are able to find new weights by calibration. They have the form $w_k = g_k d_k$ and the g_k can be interpreted as inverses of response probabilities. With some precautions they can be taken greater than 1 and rarely greater than 2. For simplicity, we treat only the case of sampling with equal probabilities (covering the exhaustive case as well as the stratified self-weighted sampling which is very frequent in population surveys). Thus $d_k = N/n$ where n is the sample size. Therefore the calibration estimating equations (4-2) can be stated as the following :

$$X = \frac{N}{n} \sum_r g_k x_k \quad (5)$$

However, it happens that we cannot use these weights because of practical constraints. If we deal with partial non-response, it is not possible to use a set of weights specific for each variable, or for each pair of variables for cross tabulations. Some software do not work with unequal weights. The following trick allows to perform an imputation system and to keep important properties of weighting.

The g_k are estimated by the set (5) of estimating equations. In the non-response set-up, the « free of charge » variable, equal to 1 for every individual, must always belong to (5). It is written $\sum_r g_k = n$. The procedure is the following :

- Duplicate $[g_k]$ - 1 times unit k ($[g_k]$ is the integer part of de g_k.). It is clear that $\sum_r g_k - [g_k] = m$, size of o.
- Very often $[g_k]=1$. However, in the general case, $\sum_o g_k - [g_k]$ is an integer q.

  One samples q units k in r with unequal probabilities equal to g_k - $[g_k]$ and subject to the constraints defined by (5). The cube method (Deville, Tillé (2000)) can be used for this purpose.

  Including the duplicate units of the first step, we have get exactly m units in our « donor basin » and we can impute these values at random to the non-respondent for variables x_k, z_k and y_k . It is easily seen that the completed data set verify (5) (if balancing is perfect) and share for variables x_k , z_k and y_k the same statistical properties (variance reduction included) with the data set that could have been obtained by weighting.

References

Deville, J.C, and Sarndal, C.E. (1992). Calibration estimators in survey sampling. *Journal of the American Statistical Association*, Vol **87**, pp. 376-382 .

Deville, J.C, Sarndal, C.E and Sautory, O.(1993). Generalised raking procedures in survey sampling. *Journal of the American Statistical Association,* Vol **88**, pp. 1013-1020 .

Deville, J-C. and Tillé, Y.(2000). *Balanced Sampling by Means of the Cube Method.* Manuscript submitted for publication.

Fuller, W.A. (1987). *Measurement Error Models*, Wiley .

Caron, N. (1998). Le logiciel POULPE : aspects méthodologiques. In: *INSEE:Actes des Journées de Méthodologie (1998).*

Petit, J.N. (1998). Le logiciel POULPE: modélisation informatique. In: *INSEE:Actes des Journées de Méthodologie (1998).*

Dupont, F. (1995). Redressements alternatifs en présence de plusieurs niveaux d'information auxiliaire. *Techniques d'enquête*, Vol. **21**, n° 2, pp.141-150 .

Montanari, G.E.(1987). Post sampling efficient prediction in large scale surveys. *International statistical review*, Vol. **55**, pp. 191-202 .

Sarndal, C.E. and Swensson, B. (1987): A general view of estimation for two phases of selection with application to two-phase sampling and non-response. *International statistical review*, Vol **55**, pp. 279-294.

Deville, J-C. and Dupont, F. (1996). Non-réponse : Principes et Méthodes. In: *INSEE : Actes des Journées de Méthodologie Statistique de 1993.*

Oh, H.L and Scheuren, F.J. (1983). Weighting Adjustments for Unit Non-response. In: *Incomplete Data in Sample Surveys*, Vol **2**, pp. 143-184 .

Methodological issues in data mining

David J. Hand

Department of Mathematics, Imperial College, London, SW7 2BZ, UK

Abstract. The nature of the new science of data mining is examined, drawing attention to the concepts and ideas it has inherited from areas such as statistics, machine learning, and database technology. Particular issues looked at include the emphasis on algorithms as well as models, and the fundamental importance of data quality to data mining exercises.

Keywords. Large data sets, models, algorithms, data quality, statistics.

1 Introduction

I shall define data mining as the discovery of interesting, unexpected, or valuable structures in large data sets. Others may have slightly different definitions, but I believe this captures the essence of the discipline. There are also further features which are typically characteristic of data mining exercises, perhaps most notably the fact that the analysis is often *secondary*. This means that the data set has already been collected, and is ready to be examined. Data mining is thus generally not concerned with methods for the efficient collection of data such as experimental and survey design.

The analysis is secondary because the data will have been initially collected for some other purpose, and then retained in the database. For example, clinical trials data will have been collected for the specific purpose of answering the question of the trial, but will be kept in the company's database in case queries arise. Once there, the data can be used for further investigation – and this can be in the context of similar data sets from other trials. A rather different example arises in retail sales: supermarkets automatically collect data from electronic point of sale transaction machines, initially so that the customer's bill can be quickly worked out, but once collected and stored the data may be subjected to detailed analysis and investigation. Sometimes the data are collected without any specific question in mind, though with the background idea that there are interesting structures concealed within its mass. This would be the case for much scientific data, for example. In yet other cases, the data are collected for a specific 'local' purpose, but then form part of a much larger database which may reveal interesting things about 'global' aspects. For example, the obvious role of automatic teller machines is to allow one to interrogate one's bank account and withdraw money from it – a specific local use relating to the individual customer. However, once stored in a database, this individual information becomes part of the overall collection of data describing customer behaviour – which can then be used to look for particular patterns of behaviour, with the ultimate aim being to improve customer service (or perhaps increase the bank's profits).

These examples demonstrate why data mining, as a discipline in its own right, has appeared now. Much of the data capture is automatic: it relies on sophisticated electronic measuring instruments and computer technology. Complementary to this is the development of large storage devices, which can store the huge data sets being collected. Statisticians have always sifted through data sets seeking 'interesting, unexpected, or valuable' structures, but when manual search was required the scope was limited to at most a few thousand records. The adjective 'large' in the opening sentence can nowadays mean many billions of data points. I have been involved in data mining a 150 gigabyte banking data set, and this is as nothing compared to some of the data sets which are being compiled (see Hand *et al*, 2000, for some examples).

Much of the incentive for the development of data mining comes from commercial interests. It is possible, even likely, that the large databases rapidly accumulating on a company's computers, describing transaction patterns, customer behaviour, and so on, contain information which can be used to increase profits and reduce losses. For example, in an analysis of credit card repayment behaviour, we have been able to distinguish between different classes of behaviour which allow the bank to predict who is likely to be about to have repayment problems. Likewise, one can characterise certain classes of transaction behaviour as being linked to possible fraud and examine these customers in more detail. An important use of data mining in marketing applications is to identify those customers most likely to respond to a mailshot: if this can avoid the need to mail the entire customer database substantial savings may be made.

Of course, commercial concerns have always analysed their databases. What is different now are the sizes of the databases, and the fact that a wider range of methods is being brought to bear to examine them, but it still remains to be seen whether the intense commercial interest in data mining will prove economically justifiable. There are relatively few cases where one can stand up and say, unequivocally, that a discovery was due to the application of data mining technology. Perhaps this is understandable, firstly since commercial concerns will naturally want to preserve any market edge they may have, and hence keep things secret, and secondly because data mining is a process, not simply a one-off analysis. That is, data mining is a cyclic and iterative process in which the data are examined, any possibly interesting discoveries are reported back to the database owner, who considers whether they are worth pursuing, and then the data are re-examined in the light of these conclusions. In data mining, like data analysis in general, it is seldom possible to put one's finger on a single step of the process and say 'this is the point at which the discovery occurred.' There is also a feeling that if other companies are doing it perhaps we should too. This, of course, is not unique to data mining, and I have seen the same thing with other data analytic technologies in commercial environments.

If the commercial concerns have stimulated the growth of interest in data mining, scientific bodies have not been far behind. In some areas (the human genome work is an obvious example) commercial and scientific interests seem to be closely entwined, but in others (e.g., the analysis of electronic catalogues of distant astronomical objects) there seem to be few commercial applications.

Although new computer-dependent technology for data mining is being developed, the notion of data mining is not a new one. People have always trawled through data sets hoping to make interesting discoveries. Indeed, terms like 'data dredging' and 'fishing expeditions' have long been in use amongst statisticians - generally in a derogatory sense, appealing to the truth that there are certain to be apparent structures in large enough data sets, arising from chance conjunctions of data points. What is new is an appreciation that, although this pessimistic perspective is true, it is also very likely that there is valuable information in the databases. The trick is to locate the valuable information.

Although I referred to 'unexpected' structures in my opening sentence, this should not be interpreted too literally. The better one's understanding of the data, the more chance one has of finding something of interest: luck, as they say, favours the prepared mind.

There are other views of data mining, rather different from that which I am presenting, but which I believe are rather specialised and over-restrictive. For example, econometricians sometimes view data mining as simply a very extensive model search (see, for example, Hoover and Perez, 2000), often based on fairly basic models such as linear regression. Others stress the role of cluster analysis and segmentation in data mining, believing it to be mainly concerned with partitioning bodies of data. Yet others focus on the predictive classification models, especially using recursive partitioning tree classifiers. All of these are data mining tools, but they are not all of the data mining tools. Data mining is properly regarded as a term for the heterogeneous collection of tools for extracting the potentially valuable information in data mountains. It is a new discipline, emerging from the interface of others, addressing new problems arising from the new challenges of modern data sets. It overlaps with statistics, database theory, machine learning, pattern recognition, artificial intelligence, and other areas, but it already possesses its own unique flavour. An illustration of the overlap - and of the differences - between data mining and statistics is given in the next section. This section makes the distinction between two kinds of structure which are sought in data mining activities: I call them *models* and *patterns*.

2 Models versus patterns

I shall define a model as a large scale data summary. That is, it tries to convey the main features of a data set. All modern statisticians are familiar with the notions of models. Building models is the main focus of much, arguably the major part, of modern statistical work. Examples include a regression function, a cluster analytic partition of a data set, a Bayesian belief network, and a neural network.

In contrast, patterns are small scale features of a data set. That is, a pattern is a characteristic structure exhibited by just a few data points, a few variables, or a small part of the data in some other sense. Examples of patterns are characteristic local waveforms in time series (often called *episodes*), anomalous customer behaviour in a retail database, or outliers. Outliers are unusual, firstly because they are an example of a pattern which statisticians have examined and secondly because they are defined relative to a model. While patterns may be defined as departures from a model, it is not clear that this is always the best way to think of things. For example, in technical analysis a gradual decline followed by a sudden sharp rise in a stock's price may be regarded as a good omen, but it seems rather contrived to think of this as a departure from a model. Pattern detection is particularly important in areas such as banking, credit card fraud detection, and intrusion detection in computer systems.

A pattern might be used to describe a model. For example, one might characterise a cluster model in terms of a single stereotype for each cluster, that is, a pattern which lies at the centre of each cluster. Despite this, the result of a cluster analysis is a model – it is a large scale summary of a body of data. This illustrates that patterns need not be 'unusual' configurations of data points. Adams *et al* (2000) contrast cluster analytic models with patterns characterising unusual densities of sample points in a retail banking data set.

Of course, it is inevitable that we will find patterns in data. Firstly, it is the very nature of human perception that we try to interpret visual or other stimuli in terms of known objects. Thus, for example, we see faces in clouds, we see archers and crabs in star patterns, and we see all sorts of things in Rorschach ink blots. The point is that we are matching an observed structure in data with a vast collection of *known* patterns (objects) until we find a good match. Without some prior restriction on what we mean by 'pattern', we can hardly fail to identify a structure as a pattern.

Secondly, it is *inevitable* that certain patterns will occur in data sets. For example, given 100 possible values that a particular variable can take, and 101 objects measured, it is *certain* that at least two of them will have the same value. Or, to take another example, if we order 10,001 objects according to the values they take on a variable, then it is *certain* that we can find 101 objects which have either increasing order or decreasing order on any other variable. These things are inevitable consequences of the properties of numbers.

Thirdly, if we have a large enough data set, then the probability of any given small pattern may be very large. Here, in contrast to the first situation above, we are matching known patterns (objects) with a vast collection of potential occurrences in the data. The probability that any particular record shows some pattern may be only 1 in a million, but if there are 100 million records we should not be too surprised to see the pattern.

Given the above, the challenges of deciding if the patterns are real (in the sense that they represent something beyond mere chance fluctuation, and also are not merely a consequence of human perceptual idiosyncrasies) and are also of value can be seen to be very demanding. One important property is that one should be able to provide an explanation for the apparent patterns: if one cannot see how or why a particular pattern should have arisen, then one might doubt its 'reality'. (The word 'reality' is in inverted commas there to indicate that it depends on whether we are referring to the actual observed data (when an observed pattern is necessarily real) or to a population from which a sample – the data – have been drawn). However, although post hoc rationalisation is a necessary condition for a pattern to be taken seriously, it is not a sufficient one: as noted above in the contexts of clouds and ink blots, humans are very good at this, even to the extent of seeing patterns where none exist. Further, the fact that a pattern seems 'obvious' *after* the fact should not detract from its value. What matters is whether it was thought of before – and here the proof will generally be in terms of whether the discovered pattern (retrospectively obvious or not) is valuable (in a commercial or scientific sense). If it does turn out to be valuable then one has a fairly convincing proof that, obvious or not, no-one thought of it before.

An important issue in the context of pattern detection is how to decide whether a pattern is sufficiently interesting or unusual to be worth reporting. Statisticians have tackled this in the modeling context via significance and hypothesis tests based on probability. Given the large pattern space and the huge number of potential data configurations, this may not be feasible in data mining. Because of this, researchers have tended to ignore the probability interpretation, defining a measure of 'unusualness', or similarity, or whatever, and simply reporting those configurations which have high values (letting a domain expert decide which, if any, are of value). In the data mining context these are called *score functions*. It is doubtful that they represent anything new – one might argue that score functions are simply measures used without the additional merits of a probabilistic interpretation.

That the distinction between models and patterns is a real one is illustrated by the role of sampling in data mining. First let us distinguish between three situations: sometimes the data set we are presented with is a sample, sometimes it is a notional sample from a superpopulation (for example, today's set of transactions – there is only one such set, but we might wish to make inferences to tomorrow's, so we imagine that today's could have been different), and sometimes it is the entire population (for example, descriptions of all of the possible molecules in a given class of chemicals or all the employees in a corporation). If the data are a sample or notional sample, then we might want to make inferences – and statistics tells us how. If the data are the complete population (or even a sample, if it is large), then it may be useful, on computational grounds, to work with only a sample (subsample), again using statistical inferential ideas. (Data miners have not always appreciated the relevance of samples and notional samples and the role of inference – I hypothesise because of the close relationship to database technology.) However, statistical ideas are only relevant if one is building a model – a large scale summary of the data. If, in contrast, one is concerned with pattern detection then there may be no alternative to examining every individual data point. If, for example, we wish to know *which* taxpayers are evading tax, we must examine each person's records. We can estimate *how many* are guilty (that is, we can build a simple model) using a sample, but to draw conclusions about *individuals* we have to examine individuals.

It is also worth mentioning in this context that drawing a sample from a data set is not always straightforward. Often the data are dynamic, evolving over time (e.g. telephone call logs, weather data, disease data). Hand *et al* (2000) give some examples. In other contexts it might be very difficult to define the sampling frame. Mining text data from the World Wide Web is an important subdiscipline of data mining, and this is very much a dynamic population. Furthermore, even if one is presented with a sample to analyse, it may not be a nice probability sample of the kind required for valid statistical inference. This will be discussed further in Section 5.

3 Structures or algorithms?

I have already remarked on the central role of models in statistics. Data mining, however, descending from more than one discipline, places much more emphasis on algorithms. Salzberg (1997) illustrates this, when he says (p317): 'Researchers have a range of different data mining algorithms at their disposal, including nearest neighbour methods, decision tree induction, error back propagation, reinforcement learning, and rule learning.' Statisticians would focus on the structures derived by these procedures. The contrast is further illustrated by the titles of the two major books on recursive paritioning: Breiman *et al* (1984) (statistics) called their book *Classification and Regression **Trees***. That is, it is clearly about structures, about *models*. In contrast, Quinlan (1993) (machine learning) called his book *C4.5: **Programs** for Machine Learning*. That is, about programs or *algorithms*.

Perhaps the focus on algorithms is because of the central role they must necessarily play: manual analysis of a billion data points is not feasible, and the use of a computer inevitable. Indeed, given such large data sets, an ideal analysis, in which a complex model is fitted using an iterative estimation procedure may not be practicable. It might be necessary to resort to a cruder model, acknowledging its imperfections, but with the attractive feature that one can fit it in reasonable time. In general, scalability of algorithms is important. An algorithm which takes time proportional to the square of the number of data points might be feasible for a hundred or a thousand points, but may well not be practicable with a million. The econometric perspective on data mining referred to above, involving fitting a very large number of linear regression models may not be viable using logistic regression.

These sorts of points mean that many data mining exercises favour adaptive or sequential algorithms, developed by machine learning researchers, in which a single pass is made through the data and in which the data do not need to be stored in memory. These sorts of algorithms may also be applied with data which arrive over time, in on-line processing, an important subdomain of data mining. An example is the automatic updating of probability estimates so that customers can best be advised on the existence of products in which they might be interested.

Perhaps it is worth adding a cautionary remark about validation here. It is easy to invent a new algorithm, but it is less easy to establish in what ways (if any) they are superior to existing ones - at least, not without a great deal of rather tedious work. Data mining is rather lacking in this regard. This may be a consequence of its youth and novelty. Perhaps as more familiarity is gained with very large data sets so more careful testing and assessment will be undertaken.

4 Data mining tools

A huge number of tools are used in data mining. In principle, at least, any tool from statistics, machine learning, database management, or other areas may be used to find structures in data and hence might be used in data mining. This section illustrates with just a few examples, plucked at random from the literature.

4.1 Examples of tools for models

Many tools have been developed for mining time series, a very important class of data. For example, tools for clustering time series into traces possessing similar dynamics and tools for segmenting time series into a small number of segment types. Applications of such methods include stock analysis, speech processing, EEG and ECG analysis, and other areas. Clustering and partitioning methods are also important in commercial applications – for example in partitioning customers into different behaviour types. An example of a partitioning method involving non-numeric data is the generation of classification schemes for document collections, such as legal abstracts, case reports, and medical research reports.

Predictive models are widely used. These may involve a response variable which is numerical (as in regression, a basic workhorse tool which is widely used) or nominal (in supervised classification). Tree based classifiers seem to be particularly widely used in data mining.

In collaborative filtering, each of a number of users rate (probably different but overlapping) subsets of items and model is constructed which permits the prediction of which items any user will rate highly. A similar idea is used by some retailers: on the basis of a customer's transaction pattern, their attention can be drawn to other potential purchases in which they might be interested, where the predictive model is based on the transactions of previous customers.

4.2 Examples of tools for pattern detection

I commented on the use of models to segment time series such as EEG traces and computer log-on patterns, but one can also seek patterns or episodes in such traces. That is, no effort is made to build a global model describing the overall evolution of the series, but instead attention is focused on episodes which repeat themselves or which might be predictive of other behaviour patterns. In many situations it is change which is of interest, as a transition from one type of behaviour to another occurs (when a customer begins to find repayments difficult, for example). These sorts of ideas have also been extended to two (and higher) dimensional patterns such as occur in meteorology (high pressure centres, pressure troughs, thermal packings, jet streams, etc.) and in fingerprints (whorls, for example).

More generally, of course, one can seek patterns of behaviour which are indicative of fraud, not related to time series. Obvious examples are application characteristics in bank loan applications. This illustrates one of the difficulties: unlike nature, in many of these problems one is working against an intelligent opponent whose behaviour will alter as you learn to detect fraud. The situation is very much a dynamic one, and one which will never be adequately 'solved'.

Market basket analysis, in which one seeks to identify high conditional probabilities (of item A occurring given that item B does) is a classic pattern detection problem.

5 Data quality

In my view the issue of data quality is of fundamental importance to data mining. I find it convenient to distinguish between high and low level problems with data. High level problems refer to the entire data set: the data may be an opportunity or convenience sample (not a random and representative sample from a population to which an inference is to be made) or some other selection process may have been operating so that the data are distorted in the sense that some entire records are missing. This prompts one to ask questions such as which entire records are missing from the database, are they missing differentially across the population, whether the chosen records have been included because they were easy to obtain, and so on. Road accident statistics provide a nice example of the dangers. The more serious accidents, those resulting in fatalities, are recorded with great accuracy, but the less serious ones, those resulting in minor or no injury, are not recorded so rigorously. Indeed a high proportion are not recorded at all. This gives a distorted impression - and could lead to mistaken conclusions.

More generally, individual variables in each record may be missing, and this may not be in a random way. Likewise, individual variables may be distorted. When records within cases are missing or distorted, I refer to the problem as one of low level distortion. One of my PhD students, studying the relationship between meteorological conditions and telecommunications faults, discovered that curious spikes in some of the variables were caused by an electronic measuring instrument automatically resetting itself. This is just one example. In all of the large data sets we have examined we have found similar faults. Their main common characteristic is that they are just the sort of thing one might hope a data mining tool would detect: they are curious, anomalous, and interesting aspects of the data. However, and this is the important point, they are not of substantive importance. They relate to the way the data are measured, or the data collection process, or some other inadequacy of the data. This may in turn lead to improved data collection procedures but it does not shed light on any substantive question: in particular, it does not lead to valuable information for the database owners. The promise of data mining is not being fulfilled.

One might briefly contemplate automatic data cleaning (as in survey work) as a way to tackle this problem, but this risks the automatic erasure of the very patterns of interest. The difficulty seems to be a very real one, which will require some novel approaches.

It has been suggested (Jane Hutton, private communication) that one of the beneficial effects of data mining might be a raising of an awareness of the poor quality of much data, with the consequence that steps are made to improve it. Certainly, it is well attested in the literature of machine learning and statistics applied in medicine that superior performance of objective statistical procedures compared to (subjective) human approaches can often be attributed simply to the greater efforts which are taken over collecting the data for the former. This is, of course, similar to the Hawthorne effect observed in the social sciences: performance improves when people are being studied and an interest is being taken in them.

6 Conclusions

Data mining is a new discipline, which has arisen in response to the challenges and opportunities provided by the very large data sets which are increasingly common. It represents a merger of several other disciplines, especially statistics, machine learning, and database management technology, but it is also leading to novel problems for which the solutions are far from obvious. The cautious statistical view of data mining is partly justified – but just because a problem is difficult does not mean a solution should not be attempted. This is a trap into which statisticians have fallen before, with the consequence that others (typically, various flavours of computer scientist) have gained intellectual ownership of the discipline. Instead, the statisticians should join with other disciplines in bringing their joint understanding to bear on finding the best solutions possible – even if 'best' may not be very good.

Further discussions of data mining, as well as illustrative examples, can be found in Fayyad *et al* (1996), Glymour *et al* (1996), Elder and Pregibon (1996), Heckerman *et al* (1997), Agrawal *et al* (1998), Hand (1998), and Hand *et al* (2000).

References

Adams, N.M., Hand, D.J. and Till, R.J. (2000). Mining for classes and patterns in behavioural data. Working paper, Department of Mathematics, Imperial College, London.

Agrawal, R., Stolorz, P., and Piatetsky-Shapiro, G. (eds.) (1998). *Proceedings of the Fourth International Conference on Knowledge Discovery and Data Mining*. Menlo Park: AAAI Press.

Breiman, L., Freidman, J.H., Olshen, R.A., and Stone, C.J. (1984). *Classification and Regression Trees*. Belmont, California: Wadsworth.

Elder, J., IV and Pregibon, D. (1996). A statistical perspective on knowledge discovery in databases. In Fayyad, U.M., Piatetsky-Shapiro, G., Smyth, P. and Uthurusamy, R. (eds.) *Advances in Knowledge Discovery and Data Mining*. Menlo Park, California: AAAI Press., pp. 83-113

Fayyad, U..M., Piatetsky-Shapiro, G., Smyth, P. and Uthurusamy, R. (eds.) (1996). *Advances in Knowledge Discovery and Data Mining*. Menlo Park, California: AAAI Press.

Glymour, C., Madigan, D., Pregibon, D. and Smyth, P. (1996). Statistical inference and data mining. *Communications of the ACM*, **39**, pp. 35-41.

Hand, D.J., (1998). Data mining: statistics and more? *The American Statistician*, **52**, pp. 112-118.

Hand, D.J., Mannila, H. and Smyth, P. (2000). *Principles of Data Mining*, MIT Press. In press.

Hand, D.J., Blunt, G., Kelly, M.G. and Adams, N.M. (2000). Data mining for fun and profit. To appear in *Statistical Science*, May.

Heckerman, D., Mannila, H., Pregibon, D. and Uthurusamy, R. (eds.) (1997). *Proceedings of the Third International Conference on Knowledge Discovery and Data Mining*. Menlo Park: AAAI Press.

Hoover, K.D. and Perez, S.J. (2000). Data mining reconsidered: encompassing and the general-to-specific approach to specification search'. To appear in *Econometrics Journal*.

Salzberg, S.L. (1997). On comparing classifiers: pitfalls to avoid and a recommended approach. *Data Mining and Knowledge Discovery*, **1**, pp. 317-328.

Quinlan, J.R. (1993). *C4.5: Programs for Machine Learning*. San Mateo, CA: Morgan Kaufmann.

Practical data mining in a large utility company

Georges Hébrail

Electricité de France, R&D Division, 1 Av. du Général de Gaulle, 92141 Clamart, France

Abstract. We present in this paper the main applications of data mining techniques at Electricité de France, the French national electric power company. This includes electric load curve analysis and prediction of customer characteristics. Closely related with data mining techniques are data warehouse management problems: we show that statistical methods can be used to help to manage data consistency and to provide accurate reports even when missing data are present.

Keywords. Data mining, statistical methods, data warehouses, applications, load curve analysis, utility company.

1 Introduction and short presentation of EDF

Electricité de France (EDF) is the French national electric power company which is in charge of the generation, transmission and distribution of electric power in France. In 1999, the electric power market was deregulated in France so that now EDF does not have anymore the monopoly on the market of large customers. In parallel with the deregulation, EDF develops its activity in other countries in Europe and in the world. EDF is a large company, gathering nearly 115 000 employees with around 30 million customers in France.

Acting in a deregulated environment is quite different from the previous situation where EDF had a monopolistic position on the French electric power market. The company currently develop new tools based on data mining techniques to be able to face this competition. Actually, EDF developed in the past many tools in order to optimize its activity and reduce the cost of generation, transmission and distribution of electric power: many of these tools designed to solve technical problems are adapted to solve commercial problems.

In this paper, we present the main developments of data mining techniques at EDF in the area of customer relationship management. Data mining techniques are also used in other domains, like power plant maintenance, electrical network operation, human resource management, but are not described here. In the paper, we will not make a special distinction between the terms *'data mining'* and *'statistical'* methods since in our opinion most of efficient data mining techniques are based on well-known statistical methods, the main difference being related to their scalability to large volumes of data.

In Section 2, we present the development of data mining techniques for dealing with electric power load curves of individual customers, i.e. curves representing their electric power consumption. In particular, we present a software developed in our Division which performs clustering of such curves interactively: this software is based on the Self Organizing Map (SOM) approach (see Kohonen (1990)). In this section, we also present a new approach we are studying to analyze long time series associated with the consumption of one customer: pattern extraction techniques are applied to build a symbolic description of the time series.

Section 3 is devoted to the presentation of data mining techniques applied to predict missing data in customer databases. Some fields in customer databases are partially filled, like for instance the possible use of electric power for water heating. Logistic regression methods are applied to predict missing information, from customers with non missing data.

In Section 4, applications related to interactions between statistical methods and data warehouses are described. We show that standard statistical methods can be used for two very useful goals: detection and characterization of records which fail consistency checking of the data warehouse, and production of accurate reports even when data is missing, by statistical adjustment.

2 Electric load curve analysis

2.1 Typical available data

Electric power sales for a customer are represented by a time series (also called a *load curve*) figuring the electric power consumption for every period of time. Availability of such data deeply depends on the type of customer. General public small customers (like residential ones) are poorly described since a communicating meter is too expensive regarding to their consumption: for these customers there are only a few points of the curve every year.

For larger customers, a communicating meter is often available for many reasons: the billing is done every month, the consumption is high and justifies the communicating meter investment, a detailed record of consumption is necessary because prices depend on the period. For these customers, a load curve is available figuring points every 10 minutes. Figure 1 shows an example of such a curve for a period of one day. The curve is generally available all over the year. In the rest of this section, we describe applications when such data is available.

Figure 1 : Example of a daily load curve

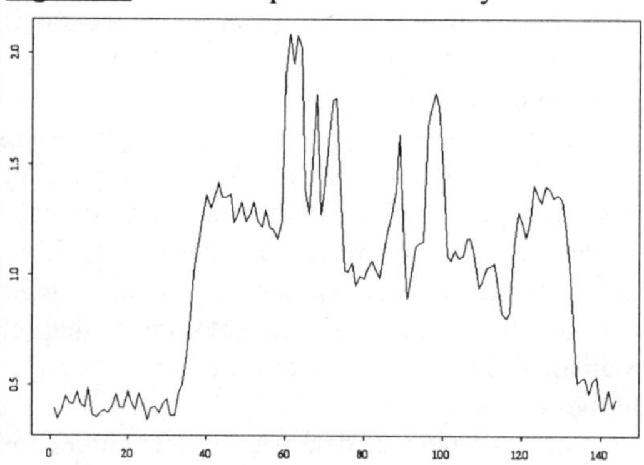

2.2 Goals of load curve analysis

Load curves have been analyzed at EDF for more than 20 years, with different goals of study. Analyzed curves may correspond to individual customer curves or to aggregates over a geographical sector. Applications may be either understanding the behavior of customers in relation to some external characteristics, or predicting consumption in a short or long term perspective.

In this paper, we focus on the exploratory data analysis of a set of load curves, each curve being associated with a single customer. The goal is to understand customer consumption behavior all along the year, in relation to his usage of electric power, to the pricing policy, and to the characteristics of the customer (mainly its activity).

2.3 The "Courboscope" software

The "Courboscope" (*courbe* is the French word for curve) is a software which has been developed a few years ago in our R&D Division. It is described in detail in Debrégeas and Hébrail (1998). This software is a tool for clustering sets of curves. Curves are here assumed to be described by the same number of points (for instance 24 points for a daily hourly measured curve). Each curve is thus considered as a point in a p-dimensional euclidian space (3^{24} for daily curves). The euclidian distance is used as a measure of dissimilarity between curves. If this distance is not accurate, a preliminary process normalizes the curves so that the euclidian distance becomes appropriate.

Curves are also described by additional attributes, such the name of the customer, his activity, the day in the week, the month, the season, the pricing option, ...

Figure 2: Main screen of the 'Courboscope' software

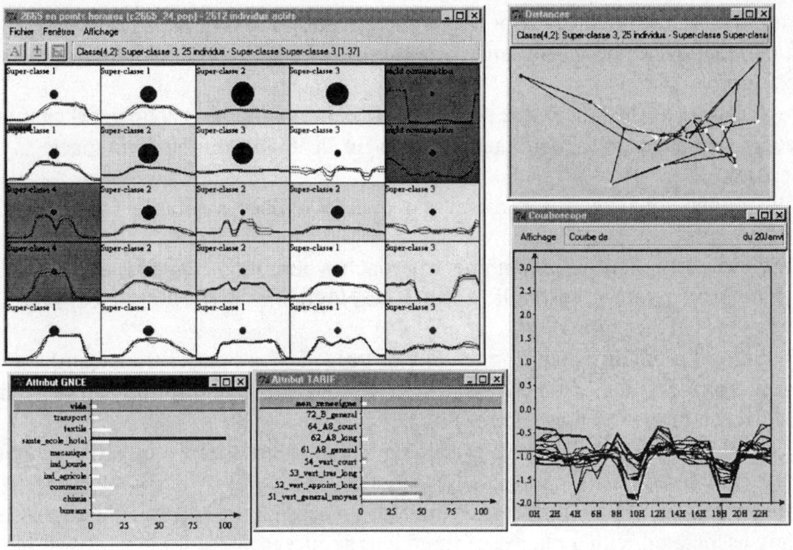

The user's task associated with this software is to build clusters of curves, where curves belonging to the same clusters have similar shapes. The clustering method is the SOM (Self-Organized Map) method proposed by Kohonen, see Kohonen (1990). This method, based on a neural network approach, produces clusters organized in a spatial way so that clusters which are close in the map have similar global shapes (see the upper left window in Figure 2). This special arrangement of clusters is quite interesting because it allows to perform clustering with a large number of clusters (for instance 10 x 10 = 100 clusters): the result is still readable.
Beyond construction of clusters, the software helps the user to give an interpretation to each cluster. A sophisticated and easy to use interface enables the user both to examine details of each cluster (lower right window in Figure 2), and to characterize each cluster with external attributes (lower left windows).

This software is used by load curve analysts who bring out typical shapes of curves (usually on a basis of daily or weekly curves), associated with some external information, typically a tariff, an activity, or an electric power usage. The result of this process is what we call an *interpreted map*, where a label is associated with each cluster cell.

Interpreted maps can be used by other (operational) people through another module of the software which is designed to classify new curves into the interpreted map. Main applications here are default detection and pricing advice. Two kinds of expertise are thus mixed together: the analyst expertise by the means of the interpreted map, and the expertise of people on the field through this classification module, which enables them to check if the use of the euclidian distance does not get completely crazy.
The Courboscope software is not dedicated to electric power load curves but is generic. It can be used with any type of curves described by any additional attributes.

2.4 Towards a symbolic analysis of curves

The approach described in the previous section shows several restrictions:
- all curves have to be of the same length and described by points at the same time positions,
- long curves (for instance time series describing one month or one year) cannot be treated because the euclidian distance in a 8000 dimension space is not meaningful,
- multi-dimensional curves (i.e. several distinct values available for every time tick) cannot be processed easily.

We are currently working on another approach where long curves (i.e. time series) are transformed into a symbolic form (see Hugueney and Hébrail (1999)). The basic process is the following:
- selection of a window width (manually or guided by a Fourier transform),
- generation of a set of fixed length curves obtained by sliding a fixed length window all along the long curve,
- clustering of the fixed length curves with the Courboscope, in order to obtain clusters of fixed length curves,
- transformation of the long curve into a succession of symbols, each symbol being associated with a cluster of fixed length curves.

Figure 3: Example of a symbolic representation of curves

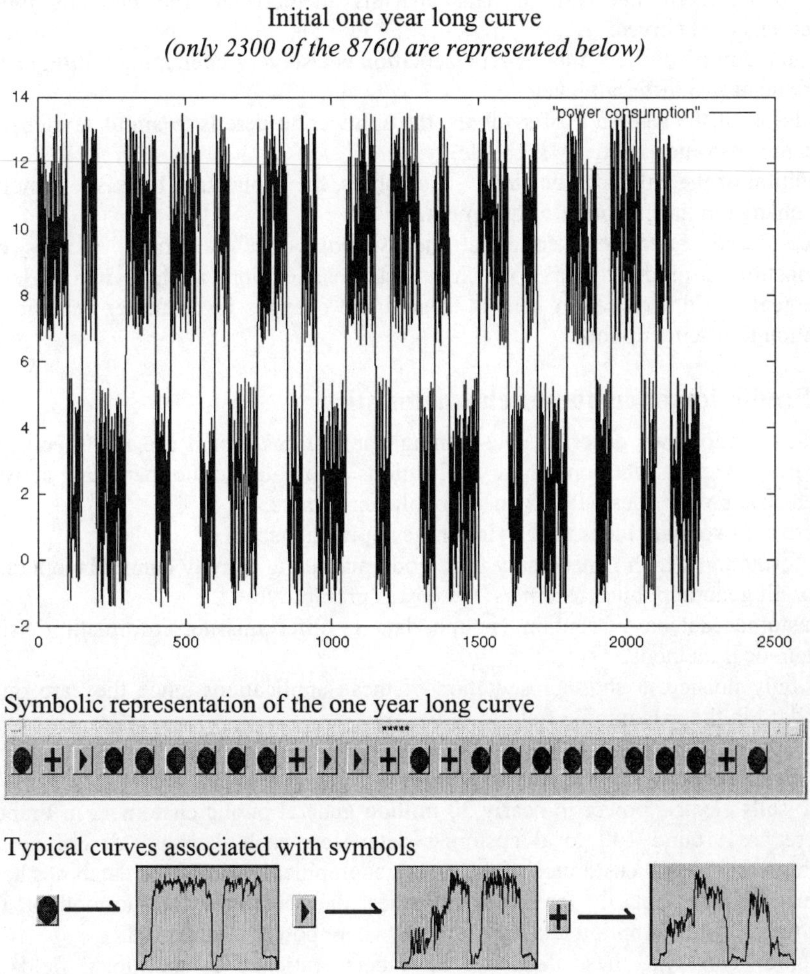

Figure 3 shows an example where a curve of one year long (8760 hourly points) is replaced by a succession of 26 from a set of 3 different symbols. Here, each symbol represents a two week typical curve.

Once a long curve has been transformed into such a symbolic form, several methods can be used, in an exploratory perspective:

- visualization of the symbolic representation: this provides a very synthetic view of the long curve,

- application of data mining methods such as as sequence analysis, in order to discover frequent subsequences (see Agrawal and Srikant (1995)),
- application of multivariate data analysis methods in the case of multi-dimensional curves.

Manual editing of the symbolic representation is also very interesting. Editing can be done at two different levels:

- edition of the sequence of symbols, the application here is to simulate a change in the customer's activity schedule,
- edition of the curves associated with symbols, the application here is to simulate a change in the customer's equipment.

Once curves have been edited at the symbolic level, a simple process can reconstitute a predicted curve which reflects changes done at the symbolic level. The real-world application here is to simulate changes in customer behavior in relation to pricing options.

3 Prediction of customer characteristics

In this section, we describe data mining applications which are not specific to electric power distribution. These applications are related to the marketing activity of electric power sales within general public customers.

There are two main ideas underlying these applications:

- information which is necessary for a good marketing activity cannot be obtained for all general public customers: the cost is prohibitive,
- customer databases contain enough data to infer missing information using statistical methods.

We only do here a short presentation of these applications since they are quite standard in the data mining field.

3.1 Brief overview of customer databases

EDF sells electric power to nearly 30 million general public customers in France. There are around 100 local customer databases, each database describing an average of 300 000 customers of the same geographical area. These databases have been designed primarily for billing purposes: they contain information about the contract, the consumption and the premises where power is delivered.

Some years ago, these databases have been enriched by additional fields to improve customer relationship and to allow marketing actions to be more efficient. This fields are filled out with the stream of contacts between the company and its customers. Consequently, they are currently partially filled.

In parallel, national surveys are carried out in order to acquire more information on a small number of customers. These surveys also contain a copy of information available in our databases for the corresponding customers (identity of each customer is lost for ethical reasons).

3.2 Prediction of customer characteristics

Data mining techniques are used to predict some customer characteristics when they are missing. The basic method we use is logistic regression.

Two different problems are addressed:
- application of logistic regression within a local database: partially filled fields define the prediction variables and the learning sample gathers customers with non missing values.
- application of logistic regression to predict values of fields available in surveys but not in customer databases. Values are predicted for all customers in each local database.

3.2.1 Prediction within local databases

As written before, local databases describe an average of 300 000 general public customers. Some fields describing them are partially filled because these fields have been introduced in the database only some years ago. Depending on the database, these fields are filled out for 10% to 50% of the customers. This is usually enough to build a prediction model from a sample gathering the customers with non missing values.

Several fields – like for instance the presence/absence of electric heating of water – have been predicted successfully with error rates from 10 to 30%, depending on the field and on the local database. This precision is enough to feed marketing campaigns.

It is important to note that one model is built for each local database, in order to take into account specific local characteristics of the populations. A special attention has to be given to the fact that in this case the sample is biased. A reweighting is necessary and is based on a stratification of the local database.

3.2.2 Projection of national survey data to local customer databases

The other application is prediction (within each local customer database) of fields which are available only in national surveys. Survey data contain both answers to questionnaires and customer data extracted from our databases: this enables to build prediction models applicable to local customer databases.

In this case, the sample is not made of customers from the local database on which the prediction is performed. Actually, customers of national surveys are picked up randomly from all local databases. So, there is only a very small number of customers from each local database in the survey. This number is not sufficient to build a model for each database. A model is built from the whole survey and applied to the different local databases.

We have done preliminary experiments which show that prediction is possible but with an error rate which is still too high for our needs. We are currently running more experiments on this problem to improve the method.

4 Statistics and data warehouses

In this section, we examine the joint use of statistical techniques and data warehouse facilities. We are currently working in two directions:
- quality of data in *relational* data warehouses,
- estimation of missing data in *multi-dimensional* data warehouses.

All applications described in this section correspond to very basic statistical methods but are really useful. The only difficulty is that the volume of data may become very large: algorithms must be scalable.

4.1 Quality of data warehouses

In this section, we focus on data warehouses where the database model is the *relational* model, i.e. the database is a collection of standard data tables.

First, simple statistics are systematically computed on every table of the data warehouse in order to give a diagnosis of the quality of data. The following basic statistics appear to be very useful to detect errors: number of values, number of different values, and number of missing values for every column. Barcharts, boxplots and histograms on some specified variables (like the contract type or the power consumption) are also of good help, especially to detect outliers which often correspond to errors in the database.

Secondly, we carry out a project addressing the problem of detecting errors or inconsistencies in relational data warehouses. This project is based on a database point of view: a set of constraints which should be satisfied by the database is defined. A program generates for each table of the database the set of records which violate the constraints. Here, statistical methods (see Morineau (1984) for instance) are used to characterize error records against others, using all columns of the table. This helps the user to understand the reason of errors: for instance, it may reveal that records in error correspond to bills of October or to customers of some geographical area.

4.2 Estimation of missing data in data warehouses

In Section 3.2.1, we have seen that logistic regression is used to predict values for customers showing missing values on some partially filled fields. This is prediction of missing data at the detailed level. In this section, we address the problem of estimation of aggregated data from a database containing missing values at the detailed level.

Data warehouses now offer a *multi-dimensional* model of data in addition to the relational model. This new model allows to define data structures which are matrices with several dimensions. In real-world applications, usual dimensions are *Product*, *Store*, and *Time*. For each product, store, and date, data usually consist in sales defined by several numerical values such as the number of sold items, the amount of the transaction, ... For each dimension, hierarchies are defined in order to build syntheses of data, for instance: products are grouped into types of products, brands, or store departments; stores are grouped into geographical areas and regions; time is aggregated at the day, month, quarter, and year levels.

Once sales data have been collected from stores, the user can use the dimension hierarchies to build interactively any aggregated array of data: for instance the total sales of food by region and month for year 1999. The database system ensures that the response time is good even if the volume of data is huge (millions of transactions for a large corporate department store). This facility assumes that all data are available, i.e. there are no missing values in the detailed data.

Within our local databases, we are currently building such a multi-dimensional database centered on electric power sales. This database is intended to be used by marketing people in order to have an idea of the sales by geographical area, time period, and type of customer. As mentioned in Section 3, there are many missing values in the customer characteristics which are not necessary for billing activity. This is the case for information about usages of electric power (heating, water-heating, type of appliances, ...). At the detailed level of data, prediction can be done with a logistic regression. At aggregated levels, prediction of aggregate values can also be done with statistical methods but the approach is different: it is related to sample adjustment to take into account the fact that non missing values constitute a biased sample.

We are currently working on the problem of adding statistical inference capabilities to multi-dimensional databases, in order to be able to query the database as if there were no missing values. An error has to be defined and evaluated for each query, depending on the level of aggregation of the query.

5 Conclusion

In this paper, we have presented several applications of data mining and statistical methods in the context of a large utility company.

Some of them are rather specific to electric power distribution, or by extension to utilities. The methods in this case are related to the analysis of load curves of large customers. The Courboscope software allows to analyze short period curves (day or week) to understand customer behavior. Symbolic curve analysis is intended to analyze longer curves (one year for instance): it is an active research area for us. There are also other applications of statistical methods related to load curves, in order to predict the consumption demand in the future. These applications were not described in this paper and are based on ARIMA and neural network models (see for instance Mangeas and Muller (1997)).

Other applications presented in the paper are related to general public customers. Better knowledge of these customers is achieved by analysis of customer databases and national surveys. Here statistical methods allow prediction of customer characteristics either at a detailed or aggregated level. Work on projecting survey data on customer databases is an active research area for us.

References

Agrawal, R. and Srikant, R. (1995). Mining Sequential Patterns. In: *Proceedings of 11th Int'l Conf. on Data Engineering (DE'95)*, Taipei, Taiwan.

Debrégeas, A. and Hébrail, G. (1998). Interactive Interpretation of Kohonen Maps Applied To Curves. In: *KDD'98, Proceedings of the 4th International Conference on Knowledge Discovery and Data Mining*, New-York, pp.179-183, AAAI Press.

Hugueney, B. and Hébrail, G. (1999). Construction de descriptions symboliques de courbes numériques. EDF R&D Division internal report n° HI-23/99-025.

Kohonen, T. (1995). *Self-organizing maps*, Berlin: Springer.

Mangeas, M. and Muller, C. (1997). An automatic search of feed forward neural network architecture based on genetic algorithms: application to the short term load forecasting. In: *ISAP'97, Proceedings of the International Conference on Intelligent System Applications for Power Systems*, Korea.

Morineau, A. (1984). Note sur la Caractérisation Statistique d'une Classe et les Valeurs-test. *Bulletin du CESIA*, Vol. **2**, N°1-2, Paris.

HGLMs for analysis of correlated non-normal data

Youngjo Lee[1] and John A. Nelder[2]

[1] Seoul National University, Seoul, Korea
[2] Imperial College of Science, Technology and Medicine, London, UK

Abstract. Hierarchical generalized linear models (HGLMs) are developed as a synthesis of (i) generalized linear models (GLMs) (ii) mixed linear models, (iii) joint modelling of mean and dispersion and (iv) modelling of spatial and temporal correlations. Statistical inferences for complicated phenomena can be made from such a HGLM, which is capable of being decomposed into diverse component GLMs, allowing the application of standard GLM procedures to those components, in particular those for model checking.

Keywords. Generalized linear model, GLM, hierarchical generalized linear model, HGLM, joint modelling of mean and dispersion, spatial correlation, temporal correlation

1 Introduction

There have been many models and many methods proposed for the description and analysis of correlated non-normal data. One general approach is via the use of random effect models. In 1996 we introduced a class of models called HGLMs, and recently Lee & Nelder (2000a) presented them as a synthesis of three widely-used existing model classes, (i) GLMs (McCullagh & Nelder, 1989), (ii) mixed linear models having both fixed and random effects (Longford, 1993) and (iii) models with structured dispersions as used in the analysis of data from quality improvement experiments (Nelder & Lee, 1991). In Lee & Nelder (2000b) we introduced an extended HGLM for modelling and analysing spatial and temporal correlations for correlated non-normal data. This covers a broad class of models, and we show that many previously developed models appear as instances of our model.

Rich classes of correlated patterns in non-Gaussian models can be produced via HGLMs without requiring explicit multivariate generalizations of non-Gaussian distributions. We (2000a) have shown that HGLMs can be decomposed into component GLMs, which means that standard procedures such as model-checking in GLMs (McCullagh & Nelder, 1989), can be applied directly to HGLMs. An alternative approach for the analysis of correlated non-normal data is by means of generalized estimating equations (GEEs, Zeger et al. 1988). Because the corresponding quasi-likelihood (QL) is usually not uniquely determined for correlated errors the GEE approach in general lacks a probabilistic interpretation. Without such a description, model checking may not be straightforward. In this paper, we describe basic attributes for GLMs, and show how joint modelling of mean and dispersions and HGLMs can be described as an integrated model, decomposable into a set of component GLMs. Each component GLM can be described by basic attributes of GLMs. We give an example of analysis, using gas consumption data in UK from Durbin & Koopman (2000).

2 GLMs and quasi-likelihood

GLMs were introduced by Nelder & Wedderburn (1972). They extend classical linear models in two ways: (i) the assumption of the normal distribution to describe the random part of the model is extended to one-parameter exponential families and (ii) additivity of the effects of explanatory variables is assumed to hold on some transformed scale defined by a monotone function called the link function.

The first extension allows the distribution to come from, among other possibilities, the Poisson, binomial, gamma, and inverse gamma distributions. For them the variance of a response variable y with mean μ can be written in the form

$$var\,(y) = \phi V\,(\mu)$$

where ϕ is called the dispersion parameter, and $V()$ is the variance function. Within one-parameter exponential families, these distributions are characterized by their variance functions. Thus, the assumption of constant variance is equivalent to the normal distribution in such families.

The second extension implies that the mean μ can be written in the form

$$\eta = g\,(\mu) = X\beta,$$

where η is called the linear predictor, and $g()$ is the link function. Within the GLM framework classical linear models have normal errors with an identity link, log-linear models Poisson errors with a log link, and logistic regressions binomial errors with logit link. In addition, a GLM allows an offset and a prior weight for more flexible data analysis

Quasi-likelihood (QL) was introduced by Wedderburn (1974) as a way of extending GLMs to the analysis of data which have errors beyond one-parameter exponential families. His starting point was the score equations for GLMs, which have the kernel

$$\partial l\,/\partial\mu = \frac{y-\mu}{\phi V\,(\mu)}.$$

If $V\,(\mu) = \mu$ the corresponding one-parameter exponential family is Poisson, for which ϕ must be unity. Often we need an over-dispersed Poisson distribution with $\mathrm{var}(y) = \phi\mu$ where $\phi > 1$. Wedderburn defined a QL (strictly a quasi-log-likelihood) q by

$$\partial q\,/\partial\mu = \frac{y-\mu}{\phi V\,(\mu)},$$

and showed that maximum QL estimators have many properties similar to those of ML estimators. We can thus extend the range of exponential families by allowing quasi-GLMs with the same pattern of score equations.

As with a GLM, a QL is characterized by its variance function. Basic attributes for GLMs are summarized as in Table 1. The QL is defined by the first two moments only. It is sometimes said that the QL uses the mean-variance relationship only, but this is not so. There are further implications. If a QL is converted into a distribution by appropriate normalization, it will be found that the third and fourth cumulants approximate those predicted from the formula for cumulants of a one-parameter exponential family if one existed. Lee & Nelder (1998) illustrated that the QL estimator is robust against misspecification of the third and fourth cumulants. By robustness we mean that a possible information gain using other estimators could be offset by a larger information loss from misspecification. For model checking for overdispersion see Lee & Nelder (in press).

Table 1. Attributes for GLM

Response	y
Mean	μ
Variance	$\phi V(\mu)$
Link	$g(\mu)$
Linear Predictor	$X\beta$
Offset	arbitrary
Prior Weight	arbitrary

3 Joint modelling of mean and dispersion

Quality improvement experiments, introduced by Taguchi, often aim to minimize variation among products while holding the mean at target. The design is usually a fractional factorial, which may or may not have any replication. The analysis of such experiments requires the modelling of both mean and dispersion. For response variable y having $\mathrm{var}(y) = \phi V(\mu)$ Nelder & Lee (1991) introduced the joint model
mean GLM

$$\eta = g(\mu) = X\beta$$

dispersion GLM

$$\zeta = h(\phi) = G\gamma.$$

A QL, like a GLM likelihood, produces a deviance with component

$$d_i = 2\int_{\mu_i}^{y_i} (y_i - s)/V(s)\, ds.$$

This deviance can be used to compare different link functions and linear predictors on the same data. It cannot, however, be used to compare different variance functions. For this we require the extended quasi-likelihood (EQL) of Nelder & Pregibon (1987). Independently, Efron (1986) introduced the double exponential family for structured dispersions, and this leads to identical inferences to those from the EQL (Lee & Nelder, 2000c). Nelder & Lee (1992) in a numerical study showed that the EQL provides good dispersion estimators. Lee & Nelder (1998) and Nelder & Lee (1998) showed that the EQL leads to the following fitting algorithm: Given dispersion parameters ϕ, the mean GLM

$$\eta = g(\mu) = X\beta \text{ with } E(y) = \mu \text{ and } var(y) = \phi V(\mu)$$

can be fitted using responses y. Let $d_i^* = d_i/(1-p_i)$, where p_i is the GLM leverage. Given mean parameters μ, the dispersion GLM defined by

$$\zeta = h(\phi) = G\gamma \text{ with } E(d^*) = \phi \text{ and } var(d^*) = 2\phi^2$$

can be fitted using response d^*. In consequence the joint models can be fitted by using two interconnected GLMs, one for the mean and one for the dispersion. The links between two models are (i) that the mean GLM provides the response d^* for the dispersion GLM, and (ii) that reciprocal of the fitted value from the dispersion GLM provides the prior weight for the mean GLM. In classical linear models, having $E(y_i) = x_i\beta$, and $var(y_i) = \phi_i$,we have

$$d_i^* = \left(y_i - x_i\widehat{\beta}\right)^2 / (1 - p_i),$$

where $p_i = x_i(X^tX)x_i^t$. Note that the d_i^* are squares of standardized residuals. This method is unbiased because $E(d^*) = \phi$, and forms an extension of the restricted maximum likelihood (REML) estimators of Patterson & Thompson (1971) to the joint model. In general, the deviance components d_i can be thought as an extension of squares of residuals, so that REML techniques can be extended to a wider classes of models (Lee & Nelder, 2000a).

Basic GLM attributes for joint modelling of mean and dispersions are summarized as in Table 2.

Table 2. GLM attributes for joint modelling of mean and dispersion

	Mean GLM	Dispersion GLM
Response	y	d^*
Mean	μ	ϕ
Variance	$\phi V(\mu)$	$2\phi^2$
Link	$g(\mu)$	$h(\phi)$
Linear Predictor	$X\beta$	$G\gamma$
Offset	arbitrary	arbitrary
Prior Weight	$1/\phi$	$1-p$

4 HGLMs

Normal linear models with additional random effects have long been used; they occur both in designed experiments such as split-plot experiments, and in models for survey data such as cattle-breeding surveys, for example, where sire effects are treated as looking like a sample from a normal distribution. GLMs assume a single random component, and can be extended by allowing extra normal random components in the linear predictor of a GLM, called generalized linear mixed models (GLMMs). Inferential procedures for GLMMs have been studied by many authors including Schall (1991) and Breslow & Clayton (1993). Lee & Nelder (1996) further extended GLMMs to HGLMs, where the random components may have non-normal distributions, in particular the conjugate distributions of the GLM distributions, giving rise to Poisson-gamma, binomial-beta, etc. models.

We (1996) introduced the HGLM in the following form:

(i) $y|u$ has a GLM form with

$$E(y|u) = \mu \text{ and } var(y|u) = \phi V(\mu).$$

The linear predictor takes the form

$$\eta = g(\mu) = X\beta + Zv,$$

where $v = g_1(u)$ for some monotone function $g_1()$.

(ii) The random components u are assumed to have arbitrary distribution. We (2000a) showed that in the conjugate HGLM the distribution for v has the GLM form with

$$E(u) = \psi \text{ and } var(u) = \lambda V(\psi)$$

and $v = \theta(u)$, where $\theta()$ is the GLM canonical link.

We further showed that various combinations of GLM distributions and links for $y|v$ and any conjugate GLM distribution and link for v can be used to construct HGLMs. Once the data y are observed, the random effects v become fixed but unknown. Assuming the inverse relationship, $var(\psi) = \lambda V_1(u)$ with $u = E(\psi)$, for pseudo data ψ, we extended our 1996 augmented-variable techniques for fitting HGLMs as follows: Augmented data $y_a = (y^t, \psi^t)^t$ have a GLM form with

$$E(y_a) = \mu_a = (\mu^t, u^t)^t, var(y) = \phi V(\mu), var(\psi) = \lambda V_1(u),$$

and linear predictor

$$\eta = \left(g(\mu)^t, g_1(u)^t\right)^t \text{ with } g(\mu) = X\beta + Zv, \text{ and } g_1(u) = v.$$

We use a subscript 1, if necessary, to emphasize that the distribution for v can have a different link $g_1()$ and a different variance function $V_1()$. For example, in Poisson GLMMs, $g(\mu) = \log\mu$ and $V(\mu) = \mu$, but $v = g_1(u) = u$ and $V_1(u) = 1$. This framework allows structured dispersion as follows: for dispersion parameters ϕ_i and λ_i, satisfying

$$var(y_i|v) = \phi_i V(\mu_i) \text{ and } var(u_i) = \lambda_i V_1(\psi_i),$$

we assume GLM linear predictors

$$h(\phi) = G\gamma \text{ and } h_1(\lambda) = G_1\gamma_1.$$

To extend the REML we (2000a) extended deviance components as follows: For the $y|u$ GLM,

$$d_i = 2\int_{\widehat{\mu}_i}^{y_i} (y_i - s)/V(s)\, ds,$$

and for the v GLM,

Table 3. GLM attributes for HGLMs

	Mean model	Dispersion model
	y|v component	
Response	y	d^*
Mean	μ	ϕ
Variance	$\phi V(\mu)$	$2\phi^2$
Link	$g(\mu)$	$h(\phi)$
Linear Predictor	$X\beta + Zv$	G_γ
Offset	arbitrary	arbitrary
Prior Weight	$1/\phi$	$1-q$
	v component	
Response	ψ	d_1^*
Mean	u	λ
Variance	$\lambda V_1(u)$	$2\lambda^2$
Link	$g_1(u)$	$h_1(\lambda)$
Linear Predictor	v	$G_1\gamma_1$
Offset	arbitrary	arbitrary
Prior Weight	$1/\lambda$	$1-q$

$$d_{1i} = 2\int_{\widehat{u}_i}^{\psi} (\psi - s)/V_1(s)\, ds.$$

This allows the class of HGLMs to be characterized by separate variance functions for the $y|v$ and the v components. For an extension of the leverage q to HGLMs see Lee & Nelder (2000a). Basic GLM attributes for HGLMs are summarized as in Table 3.

5 Extended HGLMs

When the data consist of a series of repeated measurements made on each of a set of subjects, it is usual to find that the observations on any one subject or those in neighbouring areas are correlated, and these temporal and spatial correlations must be reflected in the model used in the analysis. For this purpose we (2000b) introduced an extended HGLM as follows:

(i) $y\,|u$ has a GLM form with

$$E(y\,|u) = \mu \text{ and } var(y\,|u) = \phi V(\mu)$$

(ii) $v = L(\rho)\,r$, where $r \sim$MVN$(0, \Lambda)$ with $\Lambda = \text{diag}(\lambda_i)$; MVN stands for the multivariate normal distribution. Here ρ denotes parameters for temporal and spatial correlations. Because the normal distribution is unique in the sense it remains normal for any linear transformation, we assume normality for v.

We (2000b) showed that our representation covers a broad class of models: When $\Lambda = \lambda I$ and $L(\rho) = I$, we have Lee & Nelder's (1996) HGLM, and with arbitrary Λ Lee & Nelder's (2000a) HGLM with structured dispersion components. The class $\Lambda = \lambda I$ and $L(\rho) = L$, a matrix with fixed elements not depending upon parameters ρ, includes many state-space models of Harvey

(1989) and those of Besag et al. (1995) and Besag & Higdon (1999). The models of Laird & Ware (1982) and Diggle et al. (1994) have $var(v) = \lambda C(\rho)$, where $C(\rho) = L(\rho) L(p)^t$. Finally, we may choose to model the precision matrix $[var(v)]^{-1}$instead of $var(v)$; the setting $[var(v)]^{-1} = P(\rho)/\lambda$ gives us the models of Cressie (1991) in which we choose $L(\rho)$ to satisfy $P(\rho) = \left(L(p)^t\right)^{-1} L(\rho)^{-1}$. Some of the above models, e.g. that of Laird & Ware and Cressie, are multivariate normal. Our generalization, however, is to the wider class of generalized linear mixed models, themselves a subset of HGLMs.

6 Example

We analyse the quarterly demand for gas in the UK from 1960 to 1986, given in Harvey (1989). The purpose of the investigation was to study the seasonal pattern in the data with a view to seasonally adjusting the series. It is known that for most of the series the seasonal component changes smoothly over time, except during a disruption in the gas supply in the third and fourth quarters of 1970. Durbin & Koopman (2000) studied a heavy-tailed distribution for the $y|v$ component in their state-space models. Originally, they used the normal distribution for log transformed data. In discussion Lee (2000) showed that the outliers disappear if structured dispersions are used, with the variance changing with quarter. With HGLMs we can now fit the log-linked gamma model for the original data. Lee's (2000) final model can be written as a gamma HGLM with $var(y_t|v) = \phi_t \mu_t^2$, where

$$\log \mu_t = \alpha + t\beta + \alpha_i + t\beta_i + \delta_1 I(t = 43) + \delta_2 I(t = 44)$$
$$+\gamma_1 \sin(2\pi t/104) + \gamma_2 \cos(2\pi t/104) + v_t,$$

and,

$$\zeta_t = \log \phi_t = k + \omega_i,$$

here $i = 1,...,4$ represents quarters, δ_1 and δ_2 are for the third and fourth quarters of 1970, γ_1 and γ_2 are for the seasonal effect, $r_t = v_t - v_{t-1} \sim N(0, \lambda)$ and ω_i are the quarter main effects for the dispersion. Note that $\log \phi_t$ increases roughly linearly with quarter.

The results are in Table 4 and the model checking plots of Lee & Nelder (2000a) in Fig. 1. Fig. 1 does not indicate much need for a heavy-tailed distribution for the $y|v$ random component. Lee & Nelder (2000b) illustrated that a good model with a standard distribution often gets rid of most of the outliers. Care is required in using a heavy-tailed distribution for the errors if its use is not supported by other considerations, because it may be used to justify the fit of a model that is bad in other respects.

Table 4. Parameter estimates for gas consumption data

mean model		dispersion model	
coefficient	est	coefficient	est
α_2	−0.094	ω_2	0.718
α_3	−0.493	ω_3	1.229
α_4	−0.358	ω_4	1.683
β	0.020	$\log \lambda$	−9.382
β_2	−0.006		
β_3	−0.009		
β_4	0.001		
δ_1	0.474		
δ_2	−0.391		
γ_1	−0.140		
γ_2	0.034		

Software

We have developed a set of Genstat 5.4.1. procedures to fit HGLMs in this paper. The procedures are built on those of the K-system (Nelder, 1993). Software is available by electronic mail from j.nelder@ma.ic.ac.uk.

Acknowledgement

This work was supported by the Brain Korea 21 Project.

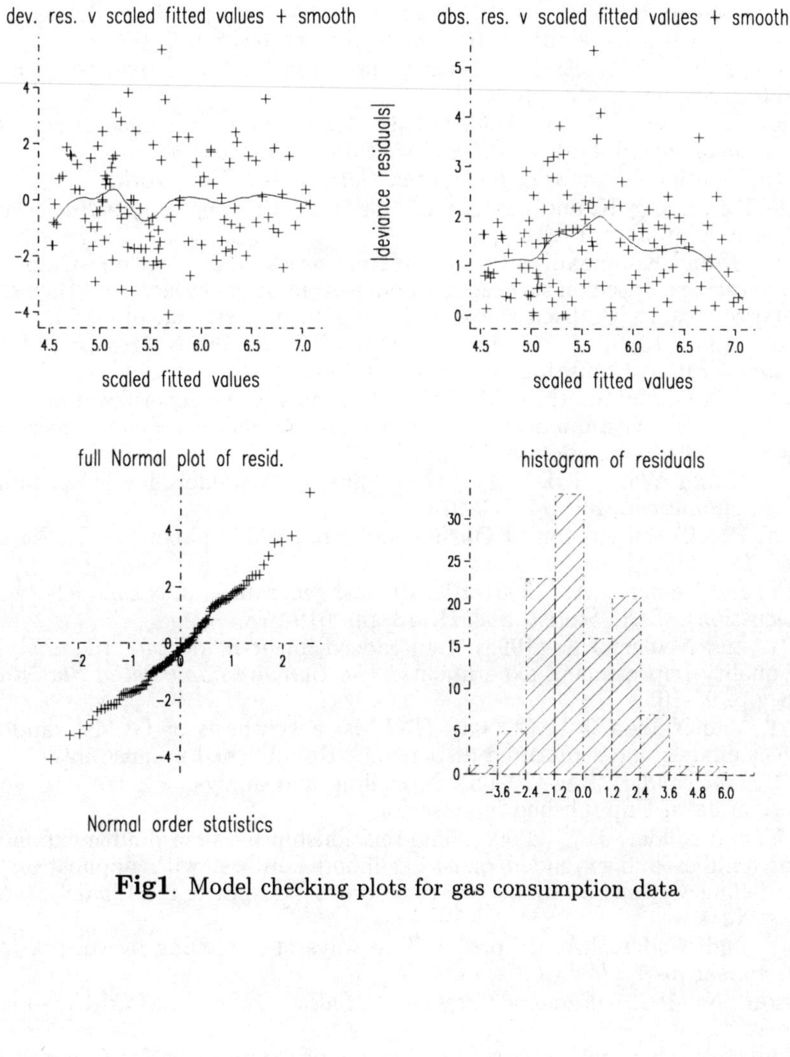

Fig1. Model checking plots for gas consumption data

References

Baltagi, B.H. (1995). *Economic Analysis of Panel Data.* New York: Wiley.

Besag, J., Green, P. Higdon, D. and Mengersen, K. (1995). Bayesian computation and stochastic systems (with discussion). *Statistical Science*, **10**, pp. 3-66.

Besag, J. and Higdon, P. (1999). Bayesian analysis of agriculture field experiments (with discussion). *J. R. Statist. Soc.* **B**, **61**, pp. 691-746.

Breslow, N.E. (1984). Extra-Poisson variation in log-linear models. *J. Roy. Statist. Soc.Ser.C*, **33**, pp. 38-44.

Breslow, N.E. and Clayton, D.G. (1993). Approximate inference in generalized linear mixed models. *J.Am.Statist.Ass.*, **88**, pp. 9-25.

Cressie, N. (1991). *Statistics for Spatial Data.* Wiley: New York.

Diggle, P.J., Liang, K. and Zeger, S.L. (1994). *Analysis of Longitudinal Data.* Oxford: Clarendon Press.

Durbin, J. and Koopman, S.J. (2000). Time series analysis of non-Gaussian observations, based on state space models from both classical and Bayesian perspectives, (with discussion). *J. R. Statist. Soc.* **B.**, **62**, pp. 3-56.

Harvey, A.C. (1989). *Forecasting, Structural Time Series Models and the Kalman Filter.* Cambridge University Press.

Kenward, M.G. and Smith, D.M. (1995). Computing the generalized estimating equations with quadratic covariance estimation for repeated measurements. *Genstat Newsletter*, **32**, pp. 50-62.

Laird, N. and Ware, J.H. (1982). Random-effects models for longitudinal data. *Biometrics*, **38**, pp. 963-974.

Lee, Y. (2000). Discussion of Durbin and Koopman's paper, *J. R. Statist. Soc*, **B**.

Lee, Y. and Nelder, J.A. (1996). Hierarchical generalized linear models (with discussion). *J. R. Statist. Soc.* **B**, **58**, pp. 619-678.

Lee, Y. and Nelder, J.A. (1998). Generalized linear models for the analysis of quality-improvement experiments. *The Canadian Journal of Statistics*, **26**, pp. 95-105.

Lee, Y. and Nelder, J.A. (2000a). HGLMs: a synthesis of GLMs, random effect models and structured dispersions. Unpublished manuscript.

Lee, Y. and Nelder, J.A. (2000b). Modelling and analysing correlated non-normal data. Unpublished manuscript.

Lee, Y. and Nelder, J.A. (2000c). The relationship between double exponential families and extended quasi-likelihood families, with application to modelling Geissler's human sex ratio data. To appear in *Applied Statistics*, No.**4**.

Lee, Y. and Nelder, J.A. (in press). Two ways of modelling overdispersion. To appear in *Applied Statistics.*

Longford, N. (1993). *Random Coefficient Models.* Oxford: Oxford University Press.

McColloch, C. E. (1997). Maximum Likelihood Algorithms for Generalized Linear Mixed Models, *J. Am. Statist. Ass*, **92**, pp. 162-170.

McCullagh P. and Nelder, J.A. (1989). *Generalized Linear Models, 2nd edn.* London: Chapman and Hall.

Nelder, J.A. (1993). *The K system for GLMs in Genstat.* Technical report TRI/93. Oxford: NAG.

Nelder, J.A. and Lee, Y. (1991). Generalized linear models for the analysis of Taguchi-type experiments. *Applied Stochastic Models and Data Analysis*, **7**, pp. 107-120.

Nelder, J.A. and Lee, Y. (1992). Likelihood, quasi-likelihood and pseudo-likelihood: Some comparisons. *J. R. Statist. Soc.* **B**, **54**, pp. 273-284.

Nelder, J.A. and Lee, Y. (1998). Letters to the editor. *Technometrics*, **40**, pp. 168-175.

Nelder, J.A. and Pregibon, D. (1987). An extended quasi-likelihood function. *Biometrika*, **74**, pp. 221-231.

Nelder, J.A. and Wedderburn, R.W.M. (1972). Generalized linear models. *J.R. Statist. Soc.* **A**, **135**, pp. 370-384.

Patterson, H.D. and Thompson, R. (1971). Recovery of interblock information when block sizes are unequal. *Biometrika*, **58**, pp. 545-554.

Schall, R. (1991). Estimation in generalized linear models with random effects. *Biometrika*, **78**, pp. 719-727.

Pierce, D.A. and Sands, B.R. (1975). Extra-Bernoulli Variation in binary data. Department of Statistics, Oregon State University, bfTR46.

Pierce, D.A. and Schafer, D.W. (1986). Residuals in generalized linear models. *J. Am. Statist. Ass.*, **81**, pp. 977-986.

Reid, N. (1991). itApproximations and asymptotics. Statistical Theory and Modelling, edn. by D. V. Hinkely, N. Reid and E. J. Snell. London: Chapman and Hall.

Robinson, G.K. (1991). That BLUP is a good thing: the estimation of random effects. *Statist. Sci*, **6**, pp. 15-51.

Wedderburn, R.W.M. (1974). Quasi-likelihood functions, generalized linear models and the Gauss-Newton method. *Biometrika*, **61**, pp. 439-447.

Zeger, S.L., Liang, K.Y. and Albert, P.S. (1988). Models for longitudinal data: a generalized estimating equation approach. *Biometrics*, **44**, pp. 1049-1060.

Bootstrapping impulse responses in VAR analyses

Helmut Lütkepohl[1]

Humboldt-Universität, Institut für Statistik und Ökonometrie, Spandauer Straße 1, D-10178 Berlin, Germany,

Abstract. Because the parameters of vector autoregressive processes are often difficult to interpret directly, econometricians use quantities derived from the parameters to disentangle the relationships between the variables. Bootstrap methods are often used for inference on the derived quantities. Alternative bootstrap methods for this purpose are discussed, some related problems are pointed out and proposals are presented to overcome the difficulties at least partly. Some remaining problems are presented.

Keywords. Impulse response, bootstrap, vector autoregression, confidence interval

1 Introduction

In econometric analyses vector autoregressive (VAR) models are established tools for describing the generation process of a set of time series variables. The advantage of these models is that their statistical analysis can be done with standard methods. On the other hand, the direct interpretation of the VAR parameters is often difficult. Therefore, to reveal the interaction between the variables involved, impulse responses and related quantities are used. They are usually determined from the estimated process parameters and are therefore also estimates. Their estimation uncertainty is often visualized by plotting confidence intervals (CIs) together with the impulse response coefficients. These CIs are frequently determined by bootstrap methods. The objective of this lecture is to discuss this practice, present some alternative bootstrap methods and also draw attention to the problems related to the application of these methods in the present context.

Before a more formal treatment of the objects of interest is presented in the next section it may be useful to discuss some characteristics of the variables for which the models are constructed. Economic time series typically have nonzero means, stochastic and/or deterministic trends and seasonal components. Such components have to be allowed for in the models used to describe the data generation process (DGP). For simplicity I will occasionally call variables $I(0)$ if they are generated by a process with stationary stochastic part. If they are not $I(0)$ but become $I(0)$ upon taking first differences, they will be called $I(1)$ variables. In principle $I(0)$ and $I(1)$ variables can have polynomial trends and other deterministic components. Because these components

[1] I am grateful to Alexander Benkwitz for helping with the computations and for comments on an earlier version of this paper. I also thank Michael Neumann for comments. The Deutsche Forschungsgemeinschaft, SFB 373, and the European Commission under the Training and Mobility of Researchers Programme (contract No. ERBFMRXCT980213) provided financial support.

do not affect the quantities of interest in the present context, I will ignore them for simplicity in most of the discussion. Also, some economic variables cannot be reduced to $I(0)$ variables upon differencing once. Such variables are not considered here again for simplicity, although some of the problems are equally valid or even more so for variables with higher order of integration.

The lecture is organized as follows. I will first present some quantities of interest. Then I will consider alternative bootstrap methods that have been proposed in the recent literature and I will also point out some related problems which are relevant in the present context. Moreover, attempts to overcome the problems will be considered. Most of this discussion is based on systems of $I(0)$ variables generated by purely stochastic stationary processes. Admittedly, this set-up is quite unrealistic. The advantage is, however, that the main problems are most easily presented in this framework. Moreover, it will become apparent that the problems are not mainly due to unusual properties of the variables but arise for the simplest cases as well. In the last section extensions will be discussed.

A word on the notation: The symbol $\mathcal{L}(X)$ will be used to denote the distribution of the random variable or random vector X. The symbol $\stackrel{d}{\to}$ signifies convergence in distribution. The $(K \times K)$ identity matrix is denoted by I_K and vec is the column stacking operator which stacks the columns of a matrix in a column vector.

2 Quantities of interest

For a given set of K time series variables $y_t = [y_{1t}, \dots, y_{Kt}]'$, I consider a VAR process of the form

$$A_0 y_t = A_1 y_{t-1} + \cdots + A_p y_{t-p} + u_t \tag{2.1}$$

as the basic model in the following. Here the A_i $(i = 0, 1, \dots, p)$ are $(K \times K)$ coefficient matrices which are restricted in a suitable way so that they can be estimated. The disturbances $u_t = [u_{1t}, \dots, u_{Kt}]'$ are assumed to be white noise, that is, the u_t are serially independent or at least uncorrelated with zero mean and positive definite covariance matrix Σ_u.

As mentioned in the introduction, in practice there will usually also be other components necessary to describe the DGP of a given set of time series in a suitable manner. For example, there may be deterministic terms such as intercepts, seasonal dummy variables or polynomial trend terms. In addition there may be unmodelled stochastic explanatory variables. For the moment I will not consider those terms because the issues of central importance in the following can be discussed without taking such additional terms into account.

The model (2.1) summarizes the instantaneous and intertemporal relations between the variables. The interactions between the variables are usually difficult to infer directly from the coefficient matrices. Therefore impulse responses and other quantities are often computed from the A_i and Σ_u which help in the interpretation of the model. The following examples are of interest in this context:

- *Forecast error impulse responses*
  Defining

$$\Phi_h = A_0^{-1} \sum_{k=1}^{h} A_k \Phi_{h-k}, \quad h = 1, 2, \dots, \tag{2.2}$$

with $\Phi_0 = A_0^{-1}$. The ijth elements of the matrices Φ_s, regarded as a function of s, trace out the expected response of $y_{i,t+s}$ to a unit change in y_{jt} holding constant all past values of y_t. If $A_0 = I_K$, (2.1) is a reduced form model. In that case the change in y_{it} given $\{y_{t-1}, y_{t-2}, \ldots\}$ is measured by the innovation u_{it} and the elements of Φ_s represent the impulse responses of the components of y_t with respect to the u_t innovations. These impulse responses are sometimes called *forecast error impulse responses* because, for a reduced form model, the u_t are the 1-step ahead forecast errors. Note that generally the $\phi_{ij,h}$ are sums of products of the elements of the A_i. For instance, if $p = 1$ and $A_0 = I_K$ it is easily seen that $\Phi_h = A_1^h$.

- *Orthogonalized impulse responses*
  Denoting by P a lower triangular matrix such that $\Sigma_u = PP'$, the orthogonalized shocks are given by $\varepsilon_t = P^{-1}u_t$ which have unit covariance matrix, that is, $E(\varepsilon_t\varepsilon_t') = I_K$. The orthogonalized impulse responses are the elements of the $\Psi_i = \Phi_i P$ $(i = 0, 1, 2, \ldots)$.
- *Structural impulse responses*
  Since many matrices P exist which satisfy $PP' = \Sigma_u$, considering the orthogonalized impulse responses is to some extent arbitrary. If a priori hypotheses are used to choose a matrix P such that $u_t = P\varepsilon_t$ results in so-called structural impulse responses $\Psi_i = \Phi_i P$ $(i = 0, 1, 2, \ldots)$.
- *Forecast error variance decomposition*
  Let $y_{T+h|T}$ be the optimal h-step forecast in period T. Then the corresponding forecast error can be written as

$$y_{T+h} - y_{T+h|T} = \Psi_0\varepsilon_{T+h} + \Psi_1\varepsilon_{T+h-1} + \cdots + \Psi_{h-1}\varepsilon_{T+1}.$$

Denoting the ijth element of Ψ_s by $\psi_{ij,s}$, the kth element of the forecast error vector becomes

$$y_{k,T+h} - y_{k,T+h|T} = \sum_{s=0}^{h-1}(\psi_{k1,s}\varepsilon_{1,T+h-s} + \cdots + \psi_{kK,s}\varepsilon_{K,T+h-s}).$$

Using that the ε_{kt} are contemporaneously and serially uncorrelated and have unit variances by construction, it follows that the corresponding forecast error variance is

$$\sigma_k^2(h) = \sum_{s=0}^{h-1}(\psi_{k1,s}^2 + \cdots + \psi_{kK,s}^2) = \sum_{j=1}^{K}(\psi_{kj,0}^2 + \cdots + \psi_{kj,h-1}^2).$$

The term $(\psi_{kj,0}^2 + \cdots + \psi_{kj,h-1}^2)$ is interpreted as the contribution of variable j to the h-step forecast error variance of variable k. This interpretation makes sense if the ε_{it} can be interpreted as shocks in variable i. Dividing the above terms by $\sigma_k^2(h)$ gives the percentage contribution of variable j to the h-step forecast error variance of variable k.

For detailed discussions of these and further quantities of interest in the present context see Sims (1980, 1981, 1992), Lütkepohl (1990, 1991), Watson (1994), Lütkepohl & Breitung (1997). A crucial property of these quantities from the point of view of our analysis is that they are particular nonlinear functions of the parameters of the model (2.1), say,

$$\phi_{ij,h} = \phi_{ij,h}(A_0, A_1, \ldots, A_p) \quad \text{or} \quad \psi_{ij,h} = \psi_{ij,h}(A_0, A_1, \ldots, A_p, \Sigma_u), \quad (2.3)$$

where $\phi_{ij,h}$ or $\psi_{ij,h}$ represents the response of variable i to an impulse in variable j, h periods ago.

3 Bootstrap inference

Usually the coefficients of model (2.1) are estimated by some standard procedure such as (pseudo) maximum likelihood (ML), feasible generalized least squares (GLS) or generalized method of moments (GMM). Estimators of the impulse responses are then, for example, obtained as

$$\hat{\phi}_{ij,h} = \phi_{ij,h}(\hat{A}_0, \ldots, \hat{A}_p), \tag{3.1}$$

where the $\hat{A}_0, \ldots, \hat{A}_p$ are the estimated VAR parameter matrices. The properties of the estimator $\hat{\phi}_{ij,h}$ then follow from those of $\hat{A}_0, \ldots, \hat{A}_p$. To assess the sampling uncertainty of the quantities of interest, confidence intervals (CIs) are sometimes determined on the basis of the asymptotic distributions. Alternatively, bootstrap methods are often used for this purpose, because these methods occasionally lead to more reliable small sample inference than CIs based on standard asymptotic theory. It may be worth noting, however, that the bootstrap is also justified by asymptotic arguments only. In the following, alternative bootstrap approaches will be considered for setting up CIs for impulse responses. They are all residual based bootstraps using the following general algorithm which assumes a sample of size T plus some presample values.

(1) Estimate the parameters of the model in (2.1) and denote the estimation residuals by $\hat{u}_t$.
(2) Compute centered residuals $\hat{u}_1 - \bar{u}_., \ldots, \hat{u}_T - \bar{u}_.$, where $\bar{u}_. = T^{-1}\sum \hat{u}_t$, and generate bootstrap residuals $u_1^*, \ldots, u_T^*$ by randomly drawing with replacement from the centered residuals.
(3) Compute bootstrap time series recursively as

$$y_t^* = \hat{A}_0^{-1}\left(\hat{A}_1 y_{t-1}^* + \cdots + \hat{A}_p y_{t-p}^* + u_t^*\right), \quad t = 1, \ldots, T,$$

where $(y_{-p+1}^*, \ldots, y_0^*) = (y_{-p+1}, \ldots, y_0)$.
(4) Reestimate the parameters $A_0, \ldots, A_p$ based on the bootstrap time series.
(5) Based on the parameter estimates obtained in Stage (4), calculate a bootstrap version of the statistic of interest, say $\hat{\phi}^*$.
(6) Repeat Steps (2) - (5) a large number of times (say N times).

This algorithm is based on an assumed VAR(p) model. If there is uncertainty with respect to specific aspects of the model like, for instance, the VAR order, a block bootstrap may be used which is based on the original observations rather than the model residuals (see, e.g., Li & Maddala (1996) for details). Because construction of a suitable model for the DGP is often the main purpose of an analysis, using the block bootstrap is not an obvious procedure in the present context, however, and will not be discussed further.

In the following I use the symbols ϕ, $\hat{\phi}$ and $\hat{\phi}^*$ to denote some general impulse response coefficient, its estimator implied by the estimators of the model coefficients and the corresponding bootstrap estimator, respectively. The following bootstrap CIs have, for instance, been considered in the literature:

– *Standard percentile interval*
The most common method in setting up CIs for impulse responses in practice is to use the interval

$$CI_S = \left[s^*_{\gamma/2}, s^*_{(1-\gamma/2)} \right],$$

where $s^*_{\gamma/2}$ and $s^*_{(1-\gamma/2)}$ are the $\gamma/2$- and $(1-\gamma/2)$-quantiles, respectively, of the bootstrap distribution $\mathcal{L}(\hat{\phi}^* \mid y_{-p+1}, \ldots, y_0, \ldots, y_T)$. The interval CI_S is the percentile confidence interval described, e.g., by Efron & Tibshirani (1993).
– *Hall's percentile interval*
Hall (1992) uses the usual bootstrap analogy,

$$\mathcal{L}(\hat{\phi} - \phi) \approx \mathcal{L}(\hat{\phi}^* - \hat{\phi} \mid y_{-p+1}, \ldots, y_0, \ldots, y_T),$$

to derive the interval

$$CI_H = \left[\hat{\phi} - t^*_{(1-\gamma/2)}, \hat{\phi} - t^*_{\gamma/2} \right],$$

where $t^*_{\gamma/2}$ and $t^*_{(1-\gamma/2)}$ are the $\gamma/2$- and $(1-\gamma/2)$-quantiles of $\mathcal{L}(\hat{\phi}^* - \hat{\phi} \mid y_{-p+1}, \ldots, y_0, \ldots, y_T)$, respectively.
– *Hall's studentized interval*
In some situations, using a studentized statistic $(\hat{\phi} - \phi)/(\widehat{\text{var}}(\hat{\phi}))^{1/2}$ for constructing confidence intervals may be advantageous. In that case bootstrap quantiles from the distribution of $(\hat{\phi}^* - \hat{\phi})/(\widehat{\text{var}}(\hat{\phi}^*))^{1/2}$ are used to construct an interval

$$CI_{SH} = \left[\hat{\phi} - t^{**}_{(1-\gamma/2)}(\widehat{\text{var}}(\hat{\phi}))^{1/2},\ \hat{\phi} - t^{**}_{\gamma/2}(\widehat{\text{var}}(\hat{\phi}))^{1/2} \right].$$

In this approach the variances are estimated by a bootstrap within each bootstrap replication.

A number of refinements and modifications of these intervals exist. Some of them will be mentioned in the next section where important properties of the present intervals are discussed.

4 Properties and problems

A desirable property of a CI is that it has the nominal confidence content at least asymptotically. The bootstrap CIs mentioned in the previous section have this property under certain conditions. Roughly speaking, if $\mathcal{L}(\sqrt{T}(\hat{\phi} - \phi))$ converges as $T \to \infty$, $\mathcal{L}(\sqrt{T}(\hat{\phi}^* - \hat{\phi}) \mid y_{-p+1}, \ldots, y_0, \ldots, y_T)$ converges to the same limit distribution under suitable conditions (e.g., Hall (1992)). It follows immediately that in this case CI_H has the correct size asymptotically, that is, $Pr\,(\phi \in CI_H) \to 1 - \gamma$ as $T \to \infty$ and, hence, Hall's percentile method is asymptotically correct. Similar arguments can be used to establish the asymptotic validity of the CI_{SH} interval. The same can also be shown for CI_S if the limiting distribution corresponding to $\mathcal{L}(\sqrt{T}(\hat{\phi} - \phi))$ is symmetric about zero, for instance, if it is zero mean normal. Roughly speaking, CI_S works with an implicit unbiasedness assumption for $\hat{\phi}$. However, if the distribution of $\hat{\phi}$ is not centered at ϕ, CI_S will generally not have the desired confidence content even asymptotically (see also Efron & Tibshirani (1993) and Benkwitz, Lütkepohl & Neumann (2000) for a more detailed discussion of this point). Hence, we have to consider the distribution of the quantities of interest.

Assuming that the $\hat{A}_i$ have an asymptotic normal distribution,

$$\sqrt{T}\mathrm{vec}([\hat{A}_0,\ldots,\hat{A}_p]-[A_0,\ldots,A_p]) \xrightarrow{d} N(0,\Sigma_{\hat{A}}), \tag{4.1}$$

using the delta method (e.g., Serfling (1980)), $\hat{\phi}$ can be shown to have an asymptotic normal distribution as well,

$$\sqrt{T}(\hat{\phi}-\phi) \xrightarrow{d} N(0,\sigma^2_{\hat{\phi}}), \tag{4.2}$$

where

$$\sigma^2_{\hat{\phi}} = \frac{\partial\phi}{\partial\alpha'}\Sigma_{\hat{A}}\frac{\partial\phi}{\partial\alpha} \tag{4.3}$$

with $\alpha = \mathrm{vec}[A_0,\ldots,A_p]$, and $\partial\phi/\partial\alpha$ denotes a vector of first order partial derivatives. The result (4.2) holds if $\sigma^2_{\hat{\phi}}$ is nonzero which is a crucial condition for asymptotic inference to work in the usual way. Note that $\Sigma_{\hat{A}}$ may be singular if there are constraints on the coefficients or if the variables are integrated and/or cointegrated (see Lütkepohl (1991, Chapter 11)). However, even if $\Sigma_{\hat{A}}$ is nonsingular, $\sigma^2_{\hat{\phi}}$ may be zero because the partial derivatives in (4.3) may be zero. In fact, they will usually be zero in parts of the parameter space because an impulse response coefficient ϕ generally consists of sums of products of the VAR parameters and, hence, the partial derivatives will also be sums of products of such coefficients which may be zero, of course.

To see this consider a simple, stable, univariate AR(1) process $y_t = \alpha y_{t-1} + u_t$ with $|\alpha| < 1$. In that case, forecast error impulse responses are of the form $\phi_h = \alpha^h$. If $\hat{\alpha}$ is the usual LS estimator, it is well-known that $\sqrt{T}(\hat{\alpha}-\alpha) \xrightarrow{d} N(0,1-\alpha^2)$ under general conditions (Anderson (1959)) and, hence, $\sqrt{T}(\hat{\phi}_h-\phi_h) \xrightarrow{d} N(0,h^2\alpha^{2(h-1)}(1-\alpha^2))$ for $\alpha \neq 0$. If $\alpha = 0$, however, $\sqrt{T}(\hat{\phi}_h-\phi_h) = \sqrt{T}\hat{\alpha}^h \xrightarrow{d} 0$ for $h > 1$. In other words, $(\hat{\phi}_h - \phi_h)$ converges to zero more rapidly than $T^{-1/2}$ so that the convergence rate is not constant for all points in the parameter space. This problem is aggravated for integrated processes with $\alpha = 1$ and for higher order or higher dimensional processes.

In fact, in the simple AR(1) example it is easy to see that CI_S has zero coverage probability for $\alpha = 0$ and even h, because in that case, denoting by $\hat{\alpha}^*$ the estimator of α based on the bootstrap time series, we have that $(\hat{\alpha}^*)^h$ will be strictly positive with probability 1. Consequently, using simply the relevant lower and upper percentage points of the bootstrap distribution results in an interval which cannot include zero. Hence, $\alpha^h = \phi_h = 0$ is not included. Although the other bootstrap CIs do not have this problem, they may not have the correct coverage probability asymptotically (see Benkwitz, Lütkepohl & Neumann (2000) for details).

The latter authors have also performed a Monte Carlo experiment to investigate the performance of bootstrap CIs in finite samples. Some of their

Table 4.1. Relative coverage frequencies of nominal 95% CIs for impulse responses ϕ_h

	$\alpha = 0$	$\alpha = .20$	$\alpha = .50$	$\alpha = .90$	$\alpha = .99$
			$h = 1$		
CI_S	.954	.953	.953	.891	.776
CI_H	.933	.929	.937	.890	.929
CI_{SH}	.942	.942	.945	.915	.928
			$h = 2$		
CI_S	.000	.982	.953	.891	.776
CI_H	.982	.703	.876	.882	.925
CI_{SH}	.979	.788	.956	.920	.933
			$h = 3$		
CI_S	.954	.953	.953	.891	.776
CI_H	1.000	.676	.821	.866	.919
CI_{SH}	1.000	.764	.958	.924	.935
			$h = 4$		
CI_S	.000	.982	.953	.891	.776
CI_H	.976	.620	.786	.855	.905
CI_{SH}	.979	.764	.962	.929	.937

(CI_{SH} figures extracted from Tables III - VI of Benkwitz, Lütkepohl & Neumann (2000). Other figures simulated based on the same generated time series.)

simulation results are replicated in Table 4.1 together with own results. The figures are based on the AR(1) example, a sample size $T = 100$ and $N = 2000$ bootstrap replications in each simulation run. The relative coverage frequencies shown in Table 4.1 are obtained for 1000 replications of the experiment.

Obviously, CI_S fails completely for $\alpha = 0$ and $h = 2$ and 4, that is, the true ϕ_h never falls in CI_S in these cases. For odd h or $\alpha > 0$, the coverage of CI_S improves considerably. It deteriorates, however, if α approaches 1 and, hence, the instability region. In contrast, CI_H and CI_{SH} perform well in terms of coverage for $\alpha = 0$ and $h = 1, 2, 3, 4$. They sometimes even have a larger coverage probability than desired. They do not perform very well in some of the other cases, however. For example for $h = 4$ and $\alpha = .20$, the actual coverage frequencies of around 60% for CI_H and of 76.4% of CI_{SH} are not satisfactory, although it is apparent that CI_{SH} performs better than CI_H in general. For α close to 1, both CI_H and CI_{SH} have a better coverage than CI_S.

In summary, the three types of CIs considered here suffer from both asymptotic as well as small sample deficiencies. In particular, there are points in the parameter space where the asymptotic theory fails and, thus, it cannot be used to justify the bootstrap. Moreover, some intervals perform poorly in small samples even for asymptotically unproblematic true parameter values. For cases where the coverage is correct asymptotically, small sample adjustments have been proposed in the literature. For instance, Hall (1992) argues that an iterative bootstrap procedure may result in more accurate CIs. In that method a further layer of bootstrap samples is drawn from each original bootstrap sample. Then CIs are computed from each of the second stage bootstrap samples and these CIs are used to estimate the actual coverage by checking how often the original estimate falls within these intervals. Then the original CI is furnished with a correction factor to adjust the coverage. This procedure can be used iteratively by repeating it more than once.

Small sample distortions of CI_S may also reflect biased estimation procedures. In those cases a bias correction may be used. A possible procedure for the presently considered cases was proposed and investigated by Kilian (1998a). Although these modifications help to improve the coverage probabilities in some small sample situations which are such that the standard asymptotic theory works, they do not overcome the problems for those cases where the usual $\sqrt{T}$-asymptotics break down.

5 Possible solutions

There are at least three possible strategies to overcome the problems resulting from the different rates of convergence in the parameter space. First, one may consider bootstrap procedures that adjust to the kind of singularity in the asymptotic distribution which we have to deal with here. A couple of different proposals of this sort are discussed in Benkwitz, Lütkepohl & Neumann (2000). For instance, subsampling may be used to estimate the convergence rate in addition to the model parameters. These and other methods were shown to have drawbacks, however, in empirical applications. Either they are not very practical for processes of realistic dimension and autoregressive order or they do not perform well in samples of typical size.

A second possibility to tackle the singularity problem is to single out and eliminate the points where problems occur before an impulse response analysis is carried out. In the present case this proposal amounts to determining all zero coefficients in a first stage of the analysis and enforcing the resulting zero restrictions in the next stage where the model underlying the impulse response analysis is estimated and used for computing impulse responses. This solution is, for instance, considered by Benkwitz, Lütkepohl & Wolters (2000). Problematic in this approach may be the uncertainty with respect to the actual zero restrictions.

A third way out of the singularity problem is to consider a different type of modelling approach based on the assumption of a potentially infinite VAR order. So far it has been assumed that a given finite order VAR model is considered. In practice, it is usually unknown a priori which model provides a suitable representation of the DGP. Therefore the model order and other restrictions are chosen by some kind of statistical procedure. Kilian (1998b) proposes to use a bootstrap procedure at this stage as well. In any case, such a preliminary analysis introduces additional uncertainty into the analysis. An alternative approach is therefore to explicitly allow the original model to be an infinite order VAR process and to integrate the model choice directly into the estimation procedure by assuming that the model order is increased when more sample information becomes available. In other word, the model order is assumed to approach infinity with the sample size. A suitable asymptotic theory is developed by Lütkepohl (1988, 1996), Lütkepohl & Poskitt (1991, 1996), Lütkepohl & Saikkonen (1997) and Saikkonen & Lütkepohl (1995, 1996) based on work by Lewis & Reinsel (1985) and Saikkonen (1992). It turns out that this asymptotic theory avoids the kind of singularity in the asymptotic distribution which causes the failure of the bootstrap procedures. On the other hand, the greater generality of the model results in an inefficiency relative to the model with finite fixed order.

6 Extensions and open problems

As mentioned earlier, in practice deterministic terms are often used in VAR processes. Such terms can be added without affecting the previous results because they do not enter the computation of the impulse response coefficients, that is, impulse responses are computed for the stochastic part only anyway. For $I(0)$ processes, the estimates of the deterministic part will be asymptotically independent of the stochastic part if the model is specified in a suitable way. Hence, from the point of view of asymptotic theory no new problems arise. Of course, in small samples the inclusion of deterministic terms can matter to some extent. Moreover, additional stochastic exogenous variables and their impact on the endogenous variables may be of interest. In that case, dynamic multipliers may be considered which are closely related to impulse responses. In principle, the analysis of the former quantities can proceed in a similar way as for impulse responses.

A different situation arises if $I(1)$ variables are also included in the system under consideration. In that case there may be singularities in the limiting distribution of the VAR parameters due to unit roots. Therefore, the singularity problem is aggravated. Unfortunately, in that case it cannot be overcome simply by imposing zero restrictions on the parameters because it arises from the superefficient estimation of cointegration parameters or, alternatively, of certain linear combinations of VAR parameters. One possible way to get around the resulting problems for bootstrapping impulse responses may be a detailed modelling of the cointegration relations and taking into account the implied singularities in the asymptotic distribution. In fact, in that case it may be possible to estimate the cointegration relations in a first round and fix the associated parameters in the bootstrap procedure. Such an approach is discussed by Benkwitz, Lütkepohl & Wolters (2000) who also present detailed examples. Unfortunately, a complete asymptotic justification of such a procedure seems to be missing to date.

Sims & Zha (1999) point out that the individual impulse responses are not independent. Consequently, reporting CIs for individual impulse response coefficients may give a misleading impression of the actual uncertainty in the estimated impulse response function. In other words, the band obtained by drawing individual $(1-\gamma)100\%$ CIs around each coefficient separately is not a confidence band with confidence content $(1-\gamma)100\%$. Sims & Zha propose a method that can help in getting a better picture of the overall uncertainty in the estimated impulse response functions.

So far I have discussed linear models only. In these models impulse responses are especially easy to analyze because the response to an innovation does not depend on the state of the system. Moreover, the direction or the size of a shock do not have an impact on the shape of the response. Therefore it is easy to represent the essential features of the dynamic interactions between the variables in a relatively small set of impulse response functions. On the other hand, it is clear that these invariance properties are not realistic in all situations of practical interest. As a consequence, nonlinear dynamic models and corresponding extensions of impulse response functions have been proposed in the literature (e.g., Gallant, Rossi & Tauchen (1993), Pesaran & Shin (1998)). In general the inference problems observed in the simpler linear models will be aggravated if more complicated models are considered.

References

Anderson, T. W. (1959). On asymptotic distribution of estimates of parameters of stochastic difference equations. *The Annals of Mathematical Statistics*, **30**, pp. 676–687.

Benkwitz, A., Lütkepohl, H. and Neumann, M.H. (2000). Problems related to confidence intervals for impulse responses of autoregressive processes. *Econometric Reviews*, **19**, pp. 69–103.

Benkwitz, A., Lütkepohl, H. and Wolters, J. (2000). Comparison of bootstrap confidence intervals for impulse responses of German monetary systems. *Macroeconomic Dynamics*, forthcoming.

Efron, B. and Tibshirani, R.J. (1993). *An Introduction to the Bootstrap*. New York: Chapman & Hall.

Gallant, A.R., Rossi, P.E. and Tauchen, G. (1993). Nonlinear dynamic structures. *Econometrica*, **61**, pp. 871–907.

Hall, P. (1992). *The Bootstrap and Edgeworth Expansion*. New York: Springer.

Kilian, L. (1998a). Small-sample confidence intervals for impulse response functions. *Review of Economics and Statistics*, **80**, pp. 218–230.

Kilian, L. (1998b). Accounting for lag order uncertainty in autoregressions: The endogenous lag order bootstrap. *Journal of Time Series Analysis*, **19**, pp. 531–548.

Lewis, R. and Reinsel, G.C. (1985). Prediction of multivariate time series by autoregressive model fitting. *Journal of Multivariate Analysis*, **16**, pp. 393–411.

Li, H. and Maddala, G.S. (1996). Bootstrapping time series models. *Econometric Reviews*, **15**, pp. 115–158.

Lütkepohl, H. (1988). Asymptotic distribution of the moving average coefficients of an estimated vector autoregressive process. *Econometric Theory*, **4**, pp. 77–85.

Lütkepohl, H. (1990). Asymptotic distributions of impulse response functions and forecast error variance decompositions of vector autoregressive models. *Review of Economics and Statistics*, **72**, pp. 116–125.

Lütkepohl, H. (1991). *Introduction to Multiple Times Series Analysis*. Berlin: Springer–Verlag.

Lütkepohl, H. (1996). Testing for nonzero impulse responses in vector autoregressive processes. *Journal of Statistical Planning and Inference*, **50**, pp. 1–20.

Lütkepohl, H. and Breitung, J. (1997). Impulse response analysis of vector autoregressive processes, in: C. Heij, J.M. Schumacher, B. Hanzon and C. Praagman (eds.), *System Dynamics in Economic and Financial Models*, Chichester: John Wiley, pp. 299–326.

Lütkepohl, H. and Poskitt, D.S. (1991). Estimating orthogonal impulse responses via vector autoregressive models. *Econometric Theory*, **7**, pp. 487–496.

Lütkepohl, H. and Poskitt, D.S. (1996). Testing for causation using infinite order vector autoregressive processes. *Econometric Theory*, **12**, pp. 61–87.

Lütkepohl, H. and Saikkonen, P. (1997). Impulse response analysis in infinite order cointegrated vector autoregressive processes. *Journal of Econometrics*, **81**, pp. 127–157.

Pesaran, H.H. and Shin, Y. (1998). Generalized impulse response analysis in linear multivariate models. *Economics Letters*, **58**, pp. 17–29.

Saikkonen, P. (1992). Estimating and testing of cointegrated systems by an autoregressive approximation. *Econometric Theory*, **8**, pp. 1–27.

Saikkonen, P. and Lütkepohl, H. (1995). Asymptotic inference on nonlinear functions of the coefficients of infinite order cointegrated VAR processes. Discussion Paper 66, SFB 373, Humboldt University Berlin.

Saikkonen, P. and Lütkepohl, H. (1996). Infinite order cointegrated vector autoregressive processes: Estimation and inference. *Econometric Theory*, **12**, pp. 814–844.

Serfling, R.J. (1980). *Approximation Theorems of Mathematical Statistics.* New York: John Wiley.

Sims, C. A. (1980). Macroeconomics and reality. *Econometrica*, **48**, pp. 1–48.

Sims, C. A. (1981). An autoregressive index model for the U.S. 1948-1975, in: J. Kmenta & J.B. Ramsey (eds.), *Large-Scale Macro-Econometric Models*, Amsterdam: North-Holland, pp. 283-327.

Sims, C.A. (1992). Interpreting the macroeconomic time series facts: The effects of monetary policy. *European Economic Review*, **36**, pp. 975–1000.

Sims, C.A. and Zha, T. (1999). Error bands for impulse responses. *Econometrica*, **67**, pp. 1113–1155.

Watson, M. W. (1994). Vector autoregressions and cointegration, in: Engle, R.F. and D.L. McFadden (eds.), *Handbook of Econometrics*, Vol. IV, New York: Elsevier.

An application of TRAMO-SEATS; model selection and out-of-sample performance. The Swiss CPI series

Agustín Maravall and Fernando J. Sánchez

Servicio de Estudios, Banco de España, Alcalá 50, 28014 Madrid, Spain

Keywords. Time series, ARIMA models, Seasonal adjustment, Trend estimation, Outliers

1 Introduction

The programs TRAMO, "Time Series Regression with ARIMA Noise, Missing Observations and Outliers", and SEATS, "Signal Extraction in ARIMA Time Series" (Gomez and Maravall, 1996) have experienced an explosion in their use by data producing agencies and short-term economic analysts. TRAMO is a program for estimation and forecasting of regression models with possibly nonstationary ARIMA errors and any sequence of missing values. The program interpolates these values, identifies and corrects for several types of outliers, and estimates special effects such as Trading Day and Easter and, in general, intervention-variable type effects. Fully automatic procedures are available. SEATS is a program for estimation of unobserved components in time series following the so-called ARIMA-model-based (AMB) method. The trend-cycle, seasonal, irregular and perhaps transitory components are estimated and forecasted with signal extraction techniques applied to ARIMA models. The two programs are structured so as to be used together both for in-depth analysis of a few series or for automatic routine applications to a large number of series. When used for seasonal adjustment, TRAMO preadjusts the series to be adjusted by SEATS. The two programs are officially used (and recommended) by Eurostat and, together with X12 ARIMA, by the European Central Bank.

The AMB methodology for seasonal adjustement was originally proposed by Burman (1980) and Hillmer and Tiao (1982). A more complete description of the methodology behind TRAMO and SEATS can be found in Gómez and Maravall (1994,1996,2000a,2000b) and Maravall (1995). In essence, given the vector of observations $y = (y_{t1}, \ldots ,y_{tm})$ where $0< t1< \ldots < tm$, TRAMO fits the regression model $y_t = z'_t\beta + x_t$, where β is a vector of regression coefficients, z'_t denotes a matrix of regression variables, and x_t follows the stochastic general ARIMA process $\phi(B)\delta(B)\, x_t = \theta(B)a_t$, where B is the backshift operator, a_t is assumed a n.i.i.d. (0,Va) white-noise innovation, and $\phi(B)$, $\delta(B)$, $\theta(B)$ are finite polynomials in B that have the multiplicative form: $\delta(B)=(1-B)^d(1 - B^s)^D$; $\phi(B) = (1+ \phi_1B+ \ldots +\phi_PB^P)(1+\Phi_1B^s)$; $\theta(B) = (1+\theta_1B+ \ldots +\theta_qB^q)(1+\Theta_1B^s)$, where s denotes the number of observations per year. SEATS decomposes x_t as in $x_t = p_t + s_t + c_t + u_t$, where p_t, s_t, c_t, u_t, are the trend-cycle, seasonal, transitory, and irregular components, which also follow ARIMA-type models, possibly with deterministic effects added.

This paper illustrates application of the programs to the monthly CPI Swiss series for the period December 1982 - May 1999 (198 observations), displayed in Figure 1. It is shown how the results of the purely automatic procedure can be improved with two simple modifications, one that emerges from the TRAMO-SEATS diagnostics, and another that uses *a priori* information. We shall describe some relevant considerations in the comparison of different models. The final section contains an analysis of the out-of-sample performance of the selected procedure.

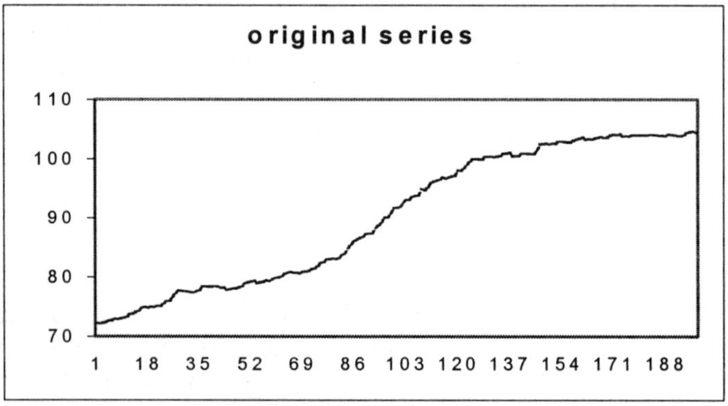

Figure 1

2 Model choice and series decomposition

Model 1: With the purely automatic procedure, the model $(1,1,0)(0,1,1)_{12}$ for the levels detailed in Table 1 is obtained. It is very parsimonious and provides a good fit, and hence it should perform well in forecasting. From Table 2, the diagnostics are acceptable, except for the Normality test, and looking at the skewness and kurtosis, the nonnormality is seen to be associated mostly with the latter (therefore the effect on point forecasts will be small). But when decomposing the model in SEATS, the AR root (1-.438B) is assigned by default to an additional "transitory component" and the model does not accept an admissible decomposition into 4 components. SEATS automatically transforms the above AR root into an MA root, changes the model thus to a $(0,1,1)\ (0,1,1)_{12}$ model, reestimates the parameters and decomposes the series.

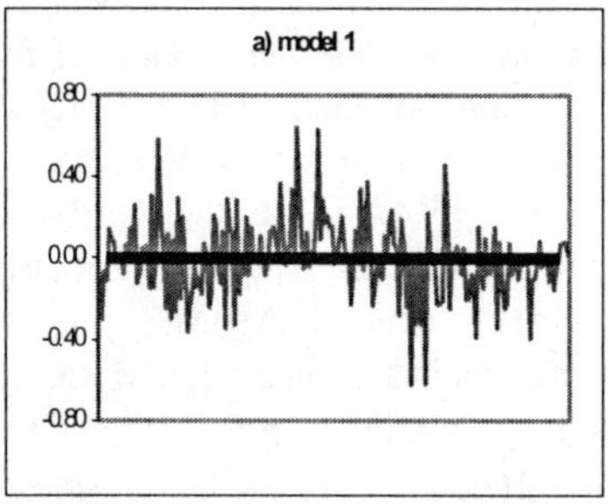

Figure 2a

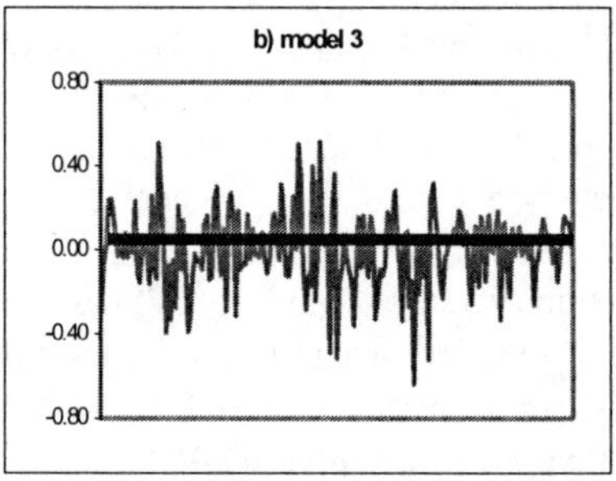

Figure 2b

Figure 2a presents the residuals and they seem to display a moving mean level, which is an indication of possible nonstationarity. The model contains 6 outliers, a relatively high number, all concentrated in the period 1990-1992, at the end of the first half of the series. Looking at TRAMO for this last model, it is seen that the ACF of the residuals has significant autocorrelation (Q(24) = 55) and, as seen in Figure 3a, the first 23 autocorrelations are > 0. This may well be interpreted as an additional indication of nonstationarity. As before, the model residuals show high kurtosis.

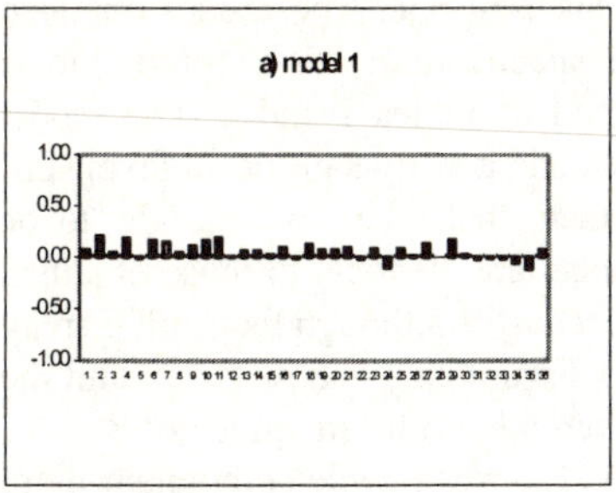

Figure 3a

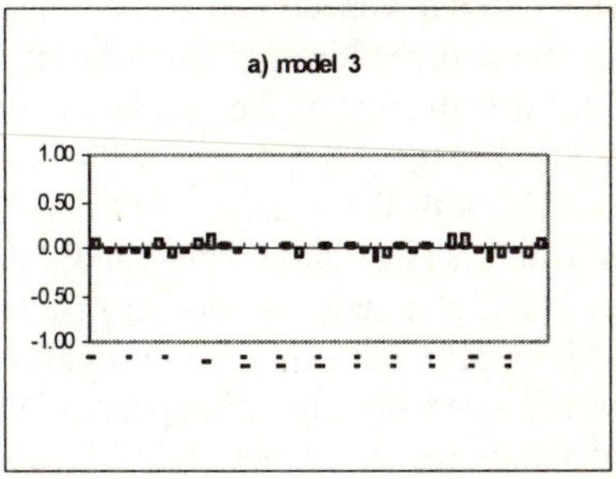

Figure 3b

SEATS decomposes the series using the approximated model. A summary of the results is contained in Table 3. The component innovation variances show a highly stochastic trend, a considerably moving seasonal, and a heteroscedastic (small) irregular component (figures 4a, 5a and 6a). The stochasticity of the seasonal component produces a not very precise estimation of seasonality. As seen in Table 2, for historical estimation, seasonality is clearly significant; for the last year, it is borderline significant; and for the one-year ahead forecasts seasonality is not worth forecasting. Altogether, one would like to obtain more stable, better behaved components, that can be estimated and forecast with more precission.

Model 2: Without considering any additional information, from the TRAMO results we have seen that the residuals give some indications of possible nonstationarity. Besides, Figure 4a showed a nonconstant mean for the series rate of growth. This would point towards an increase in the regular difference to d = 2 and, in order to protect against possible overdifferencing, to increase q to 2. Proceeding in this way the results are improved. In TRAMO, nonnormality disappears, only 3 outliers, are automatically detected, and the diagnostics are passed with no problem (see Table 2).

The model is decomposable, and the SEATS results are summarized in Table 3 . The decrease in the variance of the trend and seasonal innovations imply more stable components. Both the trend and SA series are estimated with smaller error, and subject to smaller revisions. Seasonality now is highly significant also for preliminary estimators and for the next year of forecasts. For both models, convergence of the SA series to its final estimator is relatively slow (it requires about 6 years of revisions), and there is little gain from using concurrent instead of once-a-year adjustment. The slow convergence, however, is a minor problem, since the standard desviation (SD) of the full revision in the concurrent estimator of the SA series is equal to .06 percent points, and hence of small importance.

Model 3: Some a-priori information, however, is available. Namely, that: a) in May 1993 the actual weights were introduced; b) in January 1995 there was an increase in the VAT rate; c) in January 1999 there was another increase in the tax rate. We tested for these possible effects by introducing them as regression variables. Several possible models were used, starting with the fully automatic one, with different specifications for the regression-variable effect. It became obvious that the VAT increase in January 1999 and the 1993 weights setting had insignificant effects. As for the January 1995 effect, it was significant as a level shift (t always larger than 2). Given that, on occasion, the effect of legislative issues takes more than one month to be fully captured, we tried a specification that allowed the new level to be reached in 2 or 3 months. The 2-months specification was clearly the best one. The model obtained with the automatic procedure and the 2-month VAT specification will be referred to as Model 3, and a summary of the results is contained in Tables 1 and 2. The model obtained is a $(0,2,1)\ (0,1,1)_{12}$ process, with 3 outliers and a fairly significant level shift effect associated with the January 1995 VAT change, that takes two months to be completed. The orders of differencing and the outliers are identical to those obtained with Model 2. The residuals (Figure 2b) are homocedastic, although they still display a moderate amount of kurtosis. The residual ACF (Figure 3b) is now clean, and the sequence of positive values has disappeared. Altogher, the results are quite good.

SEATS decomposes the model (Table 3), and yields a trend-cycle component more stable that models 1 or 2, a more stable seasonal component, and a larger irregular. The estimation errors of the trend-cycle and seasonally adjusted (SA) series are smaller. Seasonality is detected well for the last two years of the sample and for the one-year ahead forecasts, and the SEATS diagnostics are all passed. As seen in Figures 4, 5 and 6, the trend-cycle of the model is smoother, the seasonal component is markedly more stable, and the irregular component larger and homoscedastic. If models 1, 2 and 3 are compared, it is clear than Model 1 should be discarded. As for models 2 and 3, the latter outperforms the former on practically all accounts.

Model 4: We considered a further modification. Because the residuals of Model 3 displayed some non-normality, the critical value for outlier detection was lowered, and (as mentioned before) to protect against overdifferencing, we set $q = 2$. A summary of the results is presented in Tables 2 and 3. The model contains 6 outliers, the 3 main ones identical to those of models 2 and 3, and the 5 main ones identical to those of Model 1. The diagnostics in TRAMO are excellent, but the decomposition achieved by SEATS is less apealing than the one obtained with Model 3: the trend-cycle and seasonal components are less stable, the precission of the component estimators is smaller, and the revisions are larger. Besides, Model 4 is cleary less parsimonious than Model 3 (it has 3 more outliers and one more MA parameter), and parsimony is an important feature for model stability and out-of-sample forecasting. Thus the best choise seems to be Model 3. The choice implies that we are willing to pay the price of some excess kurtosis in order to substantially increase parsimony and the stability of the trend-cycle and seasonal components.

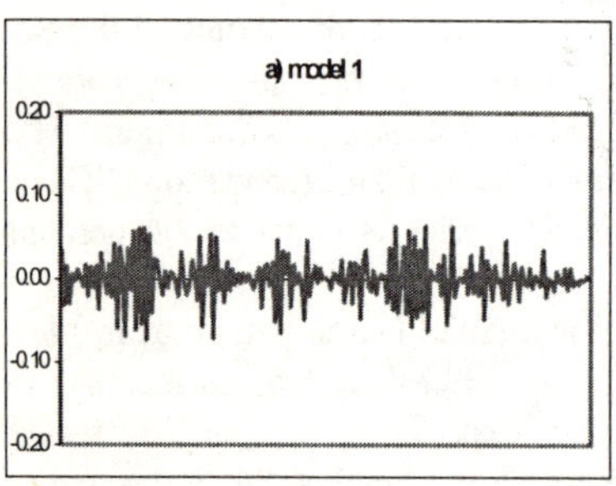

Figure 4a

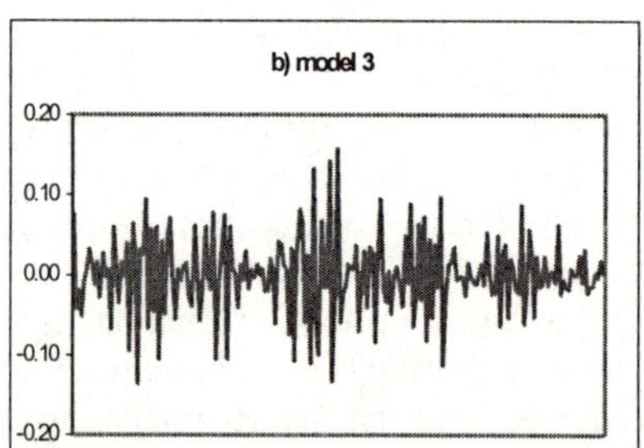

Figure 4b

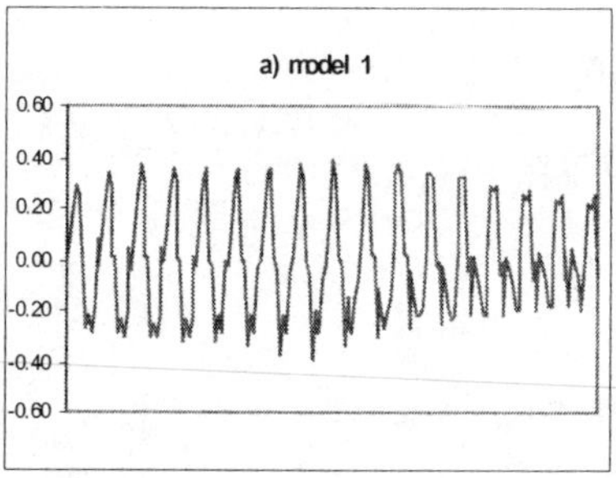

Figure 5a

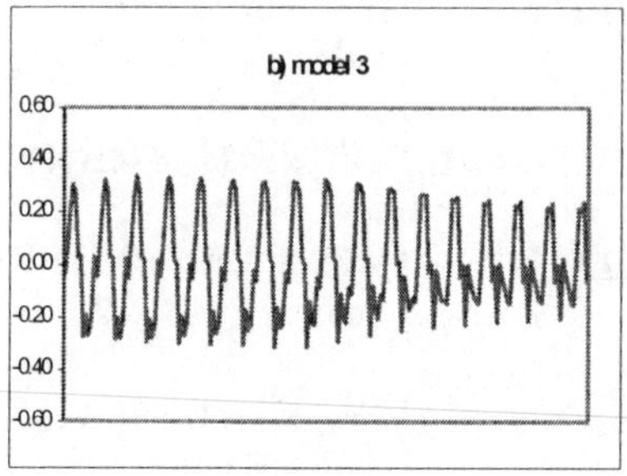

Figure 5b

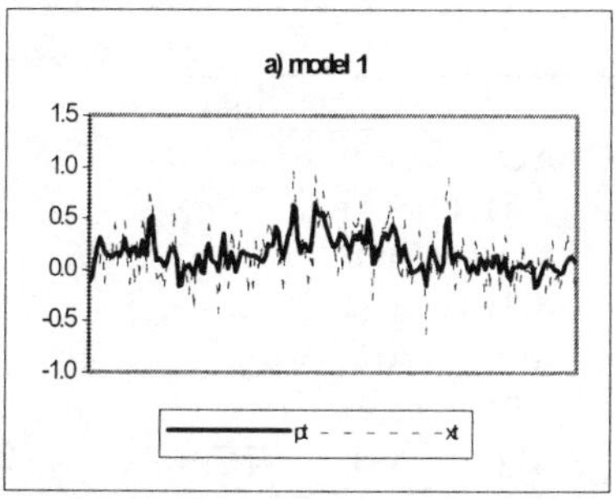

Figure 6a

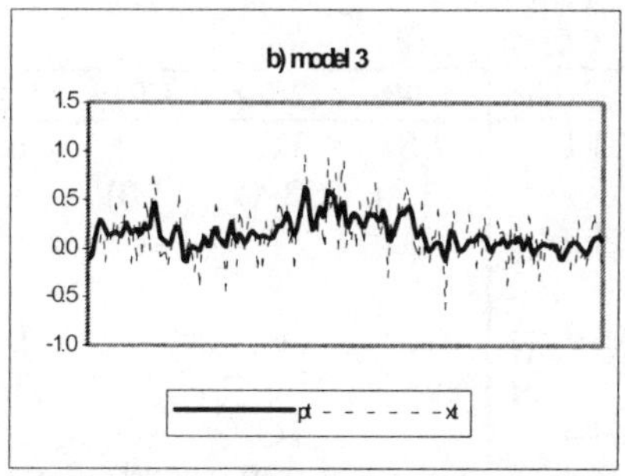

Figure 6b

Model 3 implies models for the SA series (n_t) and the seasonal component of the type $\nabla^3 n_t = (1\text{-}1.815B\text{+}.821B^2)a_{nt}$; $Ss_t = \theta_s(B)a_{st}$, where $S = 1+B+ \dots +B^{11}$ and θ_s (B) is of order 11. The variables a_{nt} and a_{st} are orthogonal component innovations. The I(3) nature of the trend model may seem high but, if the MA polynomial is factorized, one of the roots is equal to (1- .99B), and hence indistinguishable from ∇. Therefore, in practice, the model simplifies into $\nabla^2 n_t = (1\text{-}.834B)a_{nt} + \eta$, a sensible specification.

3 The decomposition in tables

Table 1. Specification of the models

MODEL 1 y_t = 6 outliers + x_t

$$(1\text{-}.438B)\,\nabla\nabla_{12}\,x_t = (1\text{-}.719B^{12})\,a_t.$$

The model does not accept an admissible decomposition and is replaced by

$$\nabla\nabla_{12}\,x_t = (1\text{+}.343B)\,(1\text{-}.688B^{12})\,a_t.$$

MODEL 2 y_t = 3 outliers + x_t

$$\nabla^2\nabla_{12}\,x_t = (1\text{-}.706B\text{ -}.155B^2)\,(1\text{-}.777B^{12})\,a_t.$$

MODEL 3 $y_t = 3$ outliers $+ 0.541$ REG $+ x_t$
t-value: 4.2

$$\nabla^2\nabla_{12}\, x_t = (1-.834B)\,(1-.771B^{12})\, a_t.$$

MODEL 4 $y_t = 6$ outliers $+ 0.571$ REG $+ x_t$
t-value: 4.2

$$\nabla^2\nabla_{12}\, x_t = (1-.677B-.150B^2)\,(1-.736B^{12})\, a_t.$$

OUTLIERS

S.E. of date	5/90	12/90	2/91	5/91	11/91	11/92	5/94
Model 1	LS	TC	AO	LS	AO	AO	-
	(-3.3)	(-3.4)	(3.0)	(4.7)	(6.4)	(5.1)	-
Model 2	-	-	-	LS	AO	AO	-
				(3.8)	(5.2)	(4.1)	
Model 3	-	-	-	LS	AO	AO	-
				(3.9)	(4.9)	(3.9)	
Model 4	-	TC	AO	LS	AO	AO	TC
		(-3.2)	(3.2)	(.39)	(6.2)	(5.0)	(-3.1)

t-values are in parenthesis

Table 2. Summary statistics; TRAMO

MODEL	1	2	3	4
1. S.E. of residuals	.193	.202	.195	.182
2. BIC	-3.11	-3.06	-3.14	-3.18
3. Normality	12.71	5.07	9.03	4.73
4. Skewness	-.30	-.12	-.32	-.06
5. Kurtosis	4.17	3.78	3.88	3.79
6. Ljung-Box Q-Statistics (24 autoc.)	26.06	19.11	25.81	19.80
7. Pierce Q_s-stat (2 autoc.)	1.67	1.16	2.10	3.31
8. Q-stat for squared res.	18.3	30.3	25.2	30.7
9. Run of residuals	.75	.75	.60	.60

Criteria: 1. As small as possible; 2. As small as possible;
3. < 6 (95 % Chi-squared); 4. | . | < 2 x .184;
5. | . | < 3+2 x .365; 6. | . | < 34 (95 % Chi-squared);
7. | . | < 6 (95 % Chi-squared); 8. | . | < 34 (95 % Chi-squared);
9. | . | < 2.

(Note: | . | denotes absolute value)

Table 3. Summary statistics; SEATS

S.E. of MODEL	1(*)	2	3	4
10. Component Innovation Variance (in units of Va)				
Trend	.324	.276	.200	.264
Seasonal	.038	.017	.016	.023
Irregular	.077	.119	.165	.110
SA series	.724	.802	.797	.767
11. E Estimation Error Variance (in units of Va)				
Trend	.451	.320	.305	.351
Seasonal	.394	.228	.189	.266
12. SD of revision in concurrent estimator (in % points)				
Trend	.089	.082	.078	.080
SA series	.084	.071	.062	.070
13. Significance of seasonality (95%): # significant months per year				
Historical Estimation	4	6	6	6
Preliminary Estimation	2	5	5	6
Next year forecast	0	5	4	4
14. Convergence (% decrease in variance of revision error after five years)				
Trend	86	78	82	83
SA series	85	72	73	78
15. Gain from concurrent adjustment (% decrease in RMSE)	12	7	7	7
16. Diagnostics: Comparation of ACF and CCF	OK	OK	OK	OK

Criteria: 10. As small as possible; 11. As small as possible;
12. As small as possible; 13. As many as possible;
14. As large as possible.

(*) The SEATS results are for the approximation to Model 1

Comments to tables 2 and 3

1)ROW 2-9: BIC denotes the Bayesian Information Criterion. The normality, skewness, kurtosis and the Ljung-Box Q-statistics (which tests the hypothesis of no residual autocorrelation) are described, for example, in Harvey (1993). The Pierce-Q_s-statistics tests for the lack of seasonal autocorrelation in the residuals, and is described in Burman (1980). The Q-statistics for the squared residuals is a test for linearity of the process; see, for example, Mills (1990). Finally, the residual-run-statistics tests for randomness in the signs of the residuals.

2)ROWS 10-11: In order to standarize measurements, variances are expressed as a fraction of the variance of the residuals a_t.

3)ROW 10: The innovations in the components are the cause of their stochastic behavior (i.e, their "moving" features). The larger is the variance, the more volatile the component will be. The trend and seasonal component should be as smooth and stable as possible, and hence the irregular should absorb as much noise as possible.

4)ROW 11: The error includes the one in the final estimator and the revision error (see, Maravall, 1995)

5)ROW 12: The revision error variances, from the SEATS output, are multiplied by Va in order to express the SD in the series units (% points). The revisions are of moderate size, in all cases smaller than one-tenth of a percent point. Confidence intervals, based on revisions, around the concurrent estimators can be immediately obtained.

6)ROW 13: The significance of seasonality is determined for the historical and preliminary (last year) estimators and for the one-year-ahead forecasts of the seasonal component. A seasonal component estimate is significant if (in absolute terms) its value is > 2 x SE of the estimate. It is seen that, for Model 1, although historical estimation detects significant seasonality in the series, preliminary and concurrent estimation will be fairly unreliable, and forecasts useless. On the contrary, seasonality is significant for historical and preliminary estimators, and also for forecasts, in the other three models.

7)ROWS 14-15: Row 14 shows the percentage of the revision error variance that has disappeared after 5 more years of data have become available. Row 15 shows how much is gained if concurrent (instead of once-a-year) adjustment is performed. Convergence is seen to be in all cases slow. Accordingly, the improvement in MSE that can be achieved by moving from a once-a-year adjustment to a concurrent one is also small.

8)ROW 16: The SEATS diagnostics are those associated with the comparison of the ACF and CCF of the stationary transformation of the components, their theoretical estimators, and their empirical estimates. The variance of the theoretical component should be larger than those of the theorical estimator and estimate actually obtained; theoretical estimators and empirical ones should be close. These requirements, as indicated in the output of SEATS, are satisfied in all four cases.

4 Out-of-sample month-to month performance

The previous exercise was made following a consultation from Nicolas Fasel (Swiss Statistical Office) and, at that time, the series ended in May 1999. Since then, 10 more observations have become available, and we shall use them to test out-of-sample performance of the procedure.

As explained in Gómez and Maravall (1998), month-to-month routine application of TRAMO-SEATS should consist of, once an initial identification of the model has been selected, to mantain it for a year unless some drastic error becomes apparent, after which a new re-identification should be made. The procedure implies fixing for the next 12 months: the orders $(p,d,q)(P,D,Q)_{12}$, as well as the log/level and mean/no-mean choices, the type and location of outliers already identified, and regression variables included in the original identification, while maintaining during the year the automatic outlier detection. Then, every month one reestimates simply the values of the parameters. This means that, for the 10 extra months, the model " $(0,2,1)(0,1,1)_{12}$ in the levels and no mean ", the LS outlier for 5/91, and the AOs for 11/91 and 11/92, are maintained, as well as the VAT-effect regression variable (with 2 months to reach the level). Figure 7 compares the monthly forecast with the observations, and displays the forecast errors. The mean of these error is 0.023, and the SD is equal to 0.157. The zero mean hypothesis can be accepted, and the variance is not significantly different from the in-sample value, it is, in fact, a bit smaller. The forecast errors show no evidence of misspecification. In fact, when the model identified for the short sample is applied to the extended one, the results barely change, as shown in Table 4. No new outliers are detected, and the only difference worth pointing out is a slight decrease in the significance of seasonality for historical estimation, associated with the slight increase in seasonal volatility.

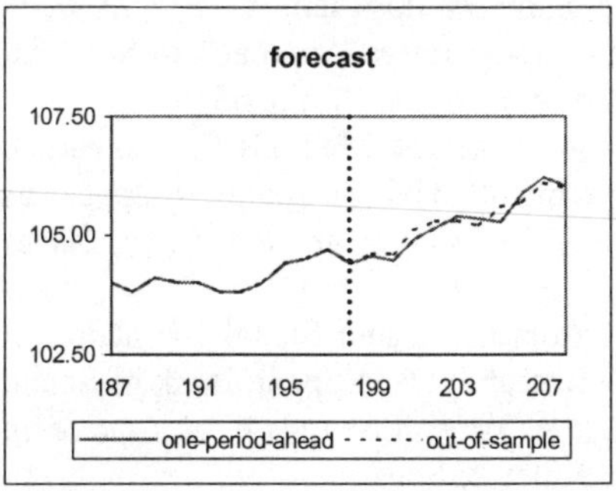

Figure 7a

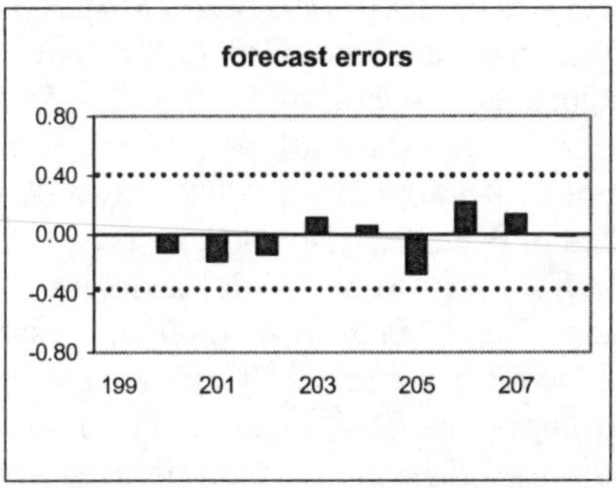

Figure 7b

Further, if the fully automatic procedure (with the VAT effect included) is applied, the model obtained is a $(1,1,1)(0,1,1)_{12}$ model, with the regular AR polynomial equal to (1-.96B). Given that it is reasonable to replace this root by (1-B), the model reidentified after the new 10 months of observations is the same as Model 3 with very similar parameters estimates. The behaviour of TRAMO-SEATS has been, as a consequence, very stable over the out-of-sample period.

Table 4. Reidentification of model

TRAMO		SEATS	
Model identified	: Model 3	var (innov. trend) :	.195
θ_1	: -.833	var (innov. seasonal) :	.019
θ_{12}	: -.749	var (innov. irregular) :	.161
Outliers	: As model 3	SD (rev. in conc. trend) :	.077
SE (residuals)	: .193	SD (rev. in conc. trend) :	.077
BIC	: -3.14	Significance of seasonality: Historical	5
BL-Q-stat	: 25.0	Significance of seasonality: Preliminary	4
Kurtosis	: 3.81	Significance of seasonality: Forecast	4

6 Conclusion

The paper applies TRAMO-SEATS to the CPI monthly swiss series, and illustrates how the programs output can be used to discriminate among alternative models and decompositions. In particular, it is seen how the SEATS output can be used to choose a model for seasonal adjustment among the ones that are equally compatible with the sample. The identification and estimation of the model is done for data finishing in May 1999, and 10 months of additional observations are used to illustrate the out-of-sample performance. The model selected (the result of the automatic procedure, with a regression variable capturing an increase in the VAT) is seen to perform very well, and remains unchanged when reidentification is made after incorporation of the out-of-sample values.

References

Burman, J.P. (1980). Seasonal Adjustment by Signal Extraction. *Journal of Royal Statistical Society A,* **143**, pp. 321-337.

Harvey, A.C. (1993). *Time Series Models.* Deddington: Philip Allan.

Hillmer, S.C. and Tiao, G.C. (1982). An Arima-Model Based Approach to Seasonal Adjustment. *Journal of the American Statistical Association* **77**, pp. 63-70.

Gómez, V. and Maravall, A. (2000a). Automatic Modelling Methods for Univariate Series . *Working Paper* ***9808***, Research Department, Bank of Spain. To appear as Ch.8 in Peña, D., Tiao, G.C. & Tsay, R.S (eds.) *A Course in Advanced Time Series Analysis,* New York: J.Wiley and Sons.

Gómez,V. and Maravall,A. (2000b). Seasonal Adjustment and Signal Extraction in Economic Time Series. *Working Paper* ***9809*** Research Department, Bank of Spain. To appear as Ch.7 in Peña, D., Tiao, G.C. & Tsay, R.S. (eds.) *A Course in Advanced Time Series Analysis,* New York: J.Wiley & Sons.

Gómez, V. and Maravall, A. (1998). Guide for Using the Programs TRAMO and SEATS. *Working Paper* ***9805***, Research Department, Bank of Spain.

Gómez, V. and Maravall, A. (1996). Programs TRAMO (Time Series Regression with Arima noise, Missing observations, and Outliers) and SEATS (Signal Extraction in Arima Time Series). Instruction for the User. *Working Paper* ***9628***, Research Department, Bank of Spain.

Gómez, V. and Maravall, A. (1994). Estimation, Prediction and Interpolation for Nonstationary Series with the Kalman Filter, *Journal of the American Statistical Association* **89**, pp. 611-624.

Maravall, A. (1995). Unobserved Components in Economic Time Series. In: Pesaran, H. and Wickens, M. (eds), *The Handbook of Applied Econometrics*, vol. **1**, Oxford: Basil Blackwell.

Mills, T.C. (1990). *Time Series Techniques for Economist,* Cambridge: Cambridge University Press.

Spreadsheets as tools for statistical computing and statistics education

Erich Neuwirth

Department of Statistics and Decision Support Systems, University of Vienna, Austria, A-1010

Abstract. Spreadsheets are an ubiquitous program category, and we will discuss their use in statistics and statistics education on various levels, ranging from very basic examples to extremely powerful methods. Since the spreadsheet paradigm is very familiar to many potential users, using it as the interface to statistical methods can make statistics more easily accessible.

Keywords. Spreadsheets, Interfacing software, Statistical Education

1 Spreadsheets as tools

Except for word processing software, spreadsheet programs probably are the most popular software category. In a certain sense spreadsheet programs are the paradigm for numerical software for most users of desktop PCs. As a consequence, many people becoming "clients" for statistical education already know how to handle this kind of software, and therefore statistical education can build upon the pre-existing knowledge of learners.

In this paper, we will deal with different ways of using spreadsheet programs as the central tool for very different kinds of statistical work. Some examples are

- Didactical applications (including animation)
- Probability and combinatorics
- Simple database applications
- Elementary statistical methods (including cross table analysis)
- Extending spreadsheets programs through built in programming languages
- Interfacing spreadsheet programs with heavy duty statistical software (locally and over the Internet)

2 Didactical applications

Spreadsheets are quite powerful tools for numerical work and additionally have easy to use graphical tools (e.g. for plotting functions, histograms). Using these facilities, one can easily design interactive didactical applications to illustrate statistical concepts or mathematical tools for statistics. These range of complexity for these applications extends from simple to possibly rather complex. A more complex example can be seen at http://sunsite.univie.ac.at/Projects/demography.

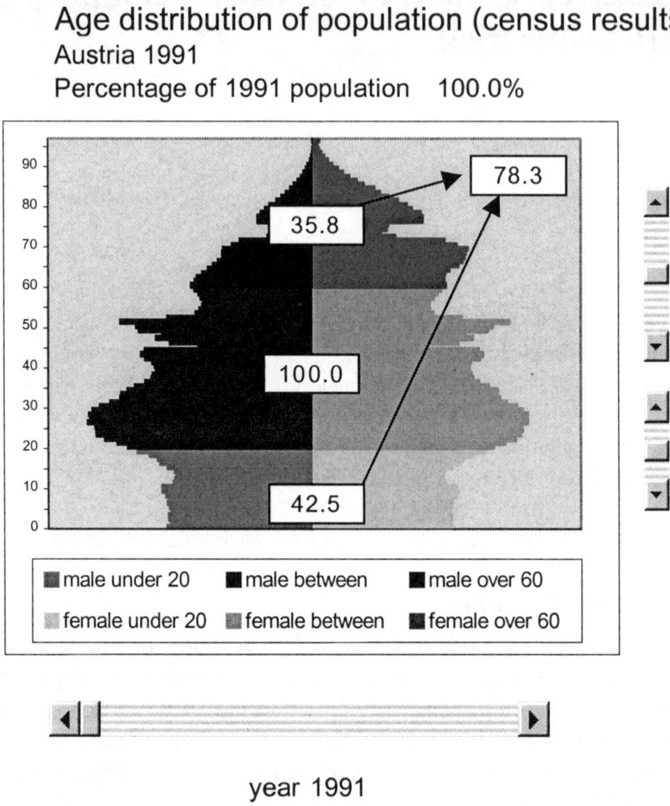

This example calculates a prediction of the age structure of populations over time, and it presents results as a user controlled "movie", i.e. as a graphic where the user can use the "slider" at the bottom of the graphic to move through time. As an aside, it is worth mentioning that this model has been implemented in Microsoft Excel by using "in sheet" spreadsheet techniques only, no VBA programming code was needed.

This application tries to demonstrate that spreadsheets can be used as powerful didactical devices allowing users to interact with models. It is downloadable from the WWW, and so we see that spreadsheet programs can be used as numerical engines for distributing didactical applications or "special purpose statistical labs" on the Internet. The advantage of this approach is that teachware authors do not have to learn new tools for Web programming. It is quite easy to configure a Web server in way that when the user clicks a link, the spreadsheet program on the local PC immediately starts, and so the concept of an interactive textbook can easily be implemented using software tools which are well established and for which a large knowledge base exists.

3 Probability and combinatorics

When teaching statistics, giving the learners an introduction to the basic concepts of probability and combinatorics is one of the harder tasks for teachers of statistics. One of the problems is that quite often the learners do not feel comfortable with mathematical techniques and notations, and therefore they concentrate on technical details instead of coping with the difficulties of the statistical concepts by themselves. Here, spreadsheets can play an important role in helping with easier access to concepts.

Let us illustrate this by an example:

One of the basic combinatorics problems is counting the number of combinations without repetitions. As a concrete example, we want to calculate the number of possible subsets of 4 elements out of a set of 6 elements. With some additional didactical explanations we can show that these subsets can be represented as increasing sequences of integers of length 4 with the last element of the sequence being not larger than 6. Therefore, the number of these sequences is the number of combinations we want to calculate. Calling these sequences 4-6-sequences we can show by a simple argument that their number is equal to the sum of the number of 4-5-sequences and the number of 3-5-sequences. This principle holds for any n-k-sequence (with n>1 and k>1), and therefore, when we write the numbers down in a spreadsheet table, we see that we can describe the structure of the table in the following graphical way:

n \ k	1	2	3	4	5	6
1	1	0	0	0	0	0
2	2					
3	3					
4	4					
5	5					
6	6			+		

This graphical representation expresses the recursive nature of the binomial coefficients in a way which is much easier to understand than the usual notational conventions, and recursion (which is a key concept in combinatorics) becomes very natural when expressed in sentences like "each cell except the ones in the first row and the first column contains the sum of the cell above and the cell above and to the left".

This approach to the basic mathematical tools for statistics is quite helpful because most students already are used to work with spreadsheets, and so the mathematical ideas are less separated from the tools they normally use for any type of calculation.

4 Simple database applications

If statistics is used as a tool, almost any statistical project has to deal with data in tabular form. Since spreadsheets are "tables with smarts", they are natural for dealing with data in the pre-analysis stage. Modern spreadsheets support database work by integrating special tools for operating on data sets.

Programs like Microsoft Excel know the concept of a data list in the sense that a table with a first row consisting of text labels only and columns (all of the same length) of values of equal type below the header row usually represent names of variables and values of these variables for different cases. If one has such a structure in a spreadsheet, Excel will offer Sorting and Filtering in the Data menu item. Filtering will allow to easily construct boolean conditions for subsets directly on the sheet, and the data list interface will also allow to work with the data in a "questionnaire" view on a case by case basis in addition to the table view for the database.

When dealing with statistical data, ordering and selecting subsets are easily done with a "point and select" user interface, so basic operations on data are done without using a programming language (which is what is necessary when using any of the major statistical packages like SPSS or S-Plus).

Transforming and recoding data also can easily be accomplished by using standard spreadsheet techniques.

5 Elementary statistical methods (including cross table analysis)

Modern spreadsheet programs usually have quite a lot of statistical methods already built into them. Frequency counts, histograms, univariate and multiple regression, analysis of variance for one and two factor are part of the standard set of tools offered by these programs. So again, for not too complicated problems, spreadsheets offer quick ways of performing statistical analyses. It is well known, however, that some of the methods used in spreadsheet programs are ill behaved numerically. The matrix inversion algorithm in Excel for example does a very bad job when the matrix to be inverted is near singular, and therefore one should not rely on results from Excel when doing very sensitive analyses. In a later section we will discuss how to connect a spreadsheet program with special purpose statistics software to use numerically more reliable statistical algorithms.

The set of statistical tools offered by spreadsheet programs and especially by Excel is not as complete as a professional statistician would wish. Excel, for example, cannot do histograms with unequal interval widths.

Excel also has all the common statistical distribution functions and their inverse functions implemented as spreadsheet functions. Therefore significance calculations for statistical tests can be done directly in the spreadsheet. These functions also do have shortcomings, and they give imprecise results at the tails of the distributions. But by connecting spreadsheet programs with statistical software we can avoid the shortcomings of the all purpose spreadsheet program by integrating precise methods from specialized statistics software packages, and still use the well known interface of the spreadsheets.

The WWW offers information about using spreadsheets for statistics, e.g. see the following web sites:

- http://www.mailbase.ac.uk/lists/assume/files/ and
- and http://sunsite.univie.ac.at/Spreadsite

One of the highlights for using Excel for statistical work is Pivot Tables. This tool allows multi-dimensional cross table analyses with an extremely comfortable user interface. Since Excel knows about statistical data lists, a contingency table can be created by just dragging labels with the names of the appropriate variables in the right places on a template for the table, and one will immediately get the corresponding tables. The "point and click" interface even goes one step further.

Clicking on any cell in a contingency table will automatically create a data table with the subset of cases corresponding to the stratification variables defining the clicked cell. This tool makes it very easy to identify outliers in data sets.

6 Extending spreadsheets programs through built in programming languages

Since spreadsheet programs and especially Excel also contain a programming language which allows one to extend the functionality, it is possible to implement statistical methods not available in the standard version of the program. Modules with such code, often called addins, can be loaded into Excel and this way the spreadsheet program can become the host application for a much enhanced statistical set of tools. The power of the embedded programming language (VBA = Visual Basic for Applications in the case of Excel) is comparable to C or Pascal or FORTRAN, therefore the language itself does not restrict the choice of algorithms to be implemented. The implementation of the language, however, is through an interpreter, not a compiler, and therefore execution of the added code is not very fast. The web sites already mentioned offer links to statistical addins for Excel. The option of extending Excel's functionality has been used by a few software companies also, there are quite a few commercial addin packages offering advanced statistical methods in Excel.

As stated previously, these packages try to build on the user's knowledge of spreadsheet programs, thereby making statistical methods convenient to use by allowing the user to access them through the familiar interface of spreadsheet programs.

7 Interfacing spreadsheet programs with heavy duty statistical software (locally and over the Internet)

As noted already, spreadsheet programs do have their shortcomings when it comes to heavy duty statistical calculations. Methods like cluster analyses or linear models with many parameters are beyond the scope and the numerical methods offered by spreadsheets. On the other hand, it would still be convenient to keep data in a spreadsheet and just transfer the data to a statistical package and then get back the results into the spreadsheet program. Spreadsheets offer very convenient ways of formatting data, therefore bringing results of analyses back into the spreadsheet can help to reduce the work needed to produce reports quite dramatically.

The integration of spreadsheet programs and statistical packages can be done on various levels.

The most simple way is having the two programs exchange data in a format understood by both programs. Column oriented ASCII data files are understood by most programs, so this method can be used. It is, however, rather clumsy and therefore should only be used as a last resort.

Modern operating systems allow interprocess communication, an therefore the spreadsheet program and the statistics package can communicate more directly. In such a situation, one of the two programs has to play the host for the other one. Since we already discussed the convenience of spreadsheets as the statistical user interface, it seems very reasonable to have the user interact with the spreadsheet, and have the spreadsheet program use the statistics package as its "numerical library".

This kind of connection between spreadsheets and specialized statistics software has been implemented for Microsoft Excel on one side and R and XploRe on the other side, and the conference presentation will show some demonstrations of these "software bundles in action".

R is an open source statistical package (modeled somewhat after the statistical package S) developed as a joint project of a few universities and research facilities in Europe, New Zealand, and North America, available from
http://www.ci.tuwien.ac.at/R/.

XploRe is a commercial statistical package developed under the auspices of Humboldt University of Berlin, and it is available from http://www.xplore-stat.de/.

When setting up software connections between these programs, the degree of visibility for the statistics package is of key importance. Depending on the level of expertise of the intended user, one might want to hide the details of the statistics package completely from the user. In this model, the added statistical methods could appear in additional menus in the spreadsheet program, and the user would not even notice that by using a method offered by one of the menus an additional software package (different from the software package he interacts with) is called behind the scenes.

On the other end of the spectrum, the statistical method developer needs full access to the statistics package. Therefore, the interface package connecting the spreadsheet and the statistics packages has to be scalable as far as exposing the statistics package explicitly to the user is concerned.

The basic elements for such a connection package are transfers of data between the spreadsheet and the statistics package and calling statistical routines within the statistics package from Excel. So we need data transfer in both directions, and facilities for transferring code and starting execution of code for the statistics packages from within Excel. Results of calculation in the statistics program then can be transferred back to the spreadsheet via the mechanism for transferring data.

All these mechanisms have been implemented for Microsoft Excel and R, and the packages needed are freely available from http://www.ci.tuwien.ac.at/R/ as contributed extensions. The 3 example spreadsheets in
http://www.ci.tuwien.ac.at/R/contrib/extra/excel/
demonstrate 3 different user interaction models.

In developer mode, the user is able to select rectangular areas in the spreadsheet and transfer these data to R and assign the values to a matrix object in R. The developer can use spreadsheet regions as scratchpad for R code, write the code, and then transfer it to R and execute it in R. Getting data from R is done by creating an object with all the data needed within R (with assignment statements) and then transferring all the data in contained in the R object into a rectangular region in Excel.

In end user mode, the transfer mechanism can be hidden completely. In this mode, Excel offers some additional toolbars or menu items, the user selects data in a spreadsheet region, clicks the menu item, and then the analyses are performed and the results put in appropriate places in the spreadsheet.

There is a third mode of connecting Excel and R. Excel allows to define functions which call external functions and routines, but are executed as part of the automatic recalculation procedure of Excel. This way, one can define functions which look indistinguishable from functions built in Excel, but use the much more reliable algorithms in R. This way Excel can access distribution functions of noncentral distributions in R which are not available in Excel directly, but for the user it still will look like these functions are functions supplied by Excel itself. This mechanism has very important didactical applications, it allows us to combine "the best of both worlds". We can use the well known spreadsheet interface, and enhance the method base by the full range of algorithms supplied by the statistics package. Especially, we can shield the learners from the intricacies of the programming language of the statistics package. Of course, it is not mandatory to hide the statistics package completely. It is the strength of this approach that the visibility of the statistics package is scalable, and we can set up the system in a way close to developer mode, so the learner can see and even change the code for the statistics package.

The package we just described uses two programs, Excel and R, which run on the same computer. Additional possibilities arise when we separate these two programs. As a interface package between Excel and XploRe we have implemented a setup where Excel and XploRe reside on different machines, but XploRe still seems to be part of Excel. The implementation uses socket technology available in Microsoft Windows and in UNIX. On a UNIX server, XploRe runs as a server process listening to a port and reacting to calls from users from other machines. Technically, this can be implemented by using inetd on the UNIX server to pipe incoming XploRe calls to an XploRe process and send the results back to the calling program (in our case Excel). From the user's point of view, this is not very much different from what we have seen for the connection between R and Excel. Like in the one-computer model there is a developer mode, and there is an end user mode. The difference to the one-computer model is that data are transferred to a program running on a different computer, and that the program code transferred to a different program is also executed on a computer different from the one running Excel. Developer mode and end user mode behave exactly like in the case with two programs running on the same machine. The only difference is timing. Since the data transfer is done over the internet, transfer time may vary, and therefore one has to expect much longer answering times. Therefore, executing remote calls of statistical procedures in XploRe as part of Excel's automatic recalculation is theoretically possible, but extremely infeasible and has not been implemented.

Separation of the Excel side and the XploRe side makes this setup a client server configuration. Therefore, XploRe can be considered as a statistical method server with Excel as the user front end. Since XploRe resides on a different machine, all the details of the code can be completely shielded from the user. The user himself does not even have to have access to the UNIX machine. This way, a complete method base can be set up on the server machine (which does not have to be UNIX, this is just a detail of the current implementation), and access to this method base is controlled by the facilities the Excel interface package (implemented as a VBA-based addin) offers the user.

Similar to the very first example in our paper such an Excel sheet can be embedded in a web page. Therefore the user can go to a web page, get the details and explanations of a statistical method in written form (like in a textbook), and then by just clicking a link on the page open an Excel sheet which connects to the remote statistics server on the internet. This spreadsheet then executes statistical programs on the internet, and transfers the results back to Excel on the user's machine.

It is important to note that the computational architecture we have been describing is quite adaptable to different needs. The one fixed point in our configurations is that the spreadsheet program is the hub for the user to interact with the statistical data and with the specialized statistics program. There is a choice of how much of the work is to be done by the spreadsheet program, and how much by the statistics program. There also is a choice for having the statistics program either reside on the same machine as the spreadsheet program, or on a different machine. Finally, there is a choice on how much of the statistics program to expose to the user, and how much convert into an integral part of the spreadsheet program (at least for the spreadsheet user).

References

Evans, I G (1997). A Note on p-values. *Teaching Statistics* **19**, 22-23.

Donald Piele (1990). *Introductory Statistics With Spreadsheets*. Reading: Addison-Wesley.

Erich Neuwirth (1990). Visualizing Correlation with Spreadsheets. *Teaching Statistics* **12**

An algorithm for deepest multiple regression

Peter J. Rousseeuw [1] and Stefan Van Aelst[2]

[1] Department of Mathematics and Computer Science, Universitaire Instelling Antwerpen (UIA), Universiteitsplein 1, B-2610 Wilrijk, Belgium

[2] Research Assistant with the FWO-Belgium, Department of Mathematics and Computer Science, Universitaire Instelling Antwerpen (UIA), Universiteitsplein 1, B-2610 Wilrijk, Belgium

Abstract. Deepest regression (*DR*) is a method for linear regression introduced by Rousseeuw and Hubert (1999). The *DR* is defined as the fit with largest regression depth relative to the data. *DR* is a robust regression method. We construct an approximate algorithm for fast computation of *DR* in more than two dimensions. We also construct simultaneous confidence regions for the true unknown parameters, based on bootstrapped estimates.

Keywords. Regression depth, robustness

1 Introduction

Consider a dataset $Z_n = \{\mathbf{z}_i = (x_{i1}, \cdots, x_{i,p-1}, y_i); i = 1, \ldots, n\} \subset I\!R^p$. In linear regression we want to fit a hyperplane of the form $y = \theta_1 x_1 + \ldots + \theta_{p-1} x_{p-1} + \theta_p$ with $\boldsymbol{\theta} = (\theta_1, \ldots, \theta_p)^t \in I\!R^p$. We denote the x-part of each data point $\mathbf{z}_i$ by $\mathbf{x}_i = (x_{i1}, \ldots, x_{i,p-1})^t \in I\!R^{p-1}$. The residuals of Z_n relative to the fit $\boldsymbol{\theta}$ are denoted as $r_i = r_i(\boldsymbol{\theta}) = y_i - \theta_1 x_{i1} - \cdots - \theta_{p-1} x_{i,p-1} - \theta_p$. To measure the quality of a fit, Rousseeuw and Hubert (1999) introduced the notion of *regression depth.*

Definition 1 *The regression depth of a candidate fit $\boldsymbol{\theta} \in I\!R^p$ relative to a dataset $Z_n \subset I\!R^p$ is given by*

$$rdepth\,(\boldsymbol{\theta}, Z_n) = \min_{\mathbf{u},\mathbf{v}} \{\#(r_i(\boldsymbol{\theta}) \geq 0 \text{ and } \mathbf{x}_i^t\mathbf{u} < \mathrm{v}) + \#(\mathrm{r_i}(\boldsymbol{\theta}) \leq 0 \text{ and } \mathbf{x}_i^t\mathbf{u} > \mathrm{v})\} \quad (1)$$

where the minimum is over all unit vectors $\mathbf{u} = (u_1, \ldots, u_{p-1})^t \in I\!R^{p-1}$ and all $v \in I\!R$ with $\mathbf{x}_i^t\mathbf{u} \neq v$ for all $(\mathbf{x}_i^t, y_i) \in Z_n$.

The regression depth of a fit $\boldsymbol{\theta} \subset I\!R^p$ relative to the dataset $Z_n \subset I\!R^p$ is thus the smallest number of observations that need to be passed when tilting $\boldsymbol{\theta}$ until it becomes vertical. Therefore, we always have $0 \leq rdepth(\theta, Z_n) \leq n$.

In the special case of $p = 1$ there are no $\mathbf{x}$-values, and Z_n is a univariate dataset. For any $\theta \in I\!R$ we then have

$$\mathrm{rdepth}(\theta, \mathrm{Z_n}) = \min\,(\#\{\mathrm{y_i}(\theta) \geq 0\}, \#\{\mathrm{y_i}(\theta) \leq 0\})$$

which is the 'rank' of θ when we rank from the outside inwards. For any $p \geq 1$, the regression depth of $\boldsymbol{\theta}$ measures how balanced the dataset Z_n is about the linear fit determined by $\boldsymbol{\theta}$. It can easily be verified that regression depth is scale invariant, regression invariant and affine invariant according to the definitions in Rousseeuw and Leroy (1987, page 116).

Based on the notion of regression depth, Rousseeuw and Hubert (1999) introduced the deepest regression estimator (DR) for robust linear regression. In Section 2 we give the definition of DR and its basic properties. Section 3 proposes the fast approximate algorithm MEDSWEEP to compute DR in higher dimensions ($p \geq 3$). In Section 4, confidence regions for the unknown parameters $\theta_1, \ldots, \theta_p$ are constructed.

2 Definition and properties of deepest regression

Definition 2 *In p dimensions the deepest regression estimator $DR(Z_n)$ is defined as the fit $\boldsymbol{\theta}$ with maximal rdepth $(\boldsymbol{\theta}, Z_n)$, that is*

$$DR(Z_n) = \underset{\boldsymbol{\theta}}{\operatorname{argmax}}\ rdepth\,(\boldsymbol{\theta}, Z_n). \tag{2}$$

Since the regression depth of a fit $\boldsymbol{\theta}$ can only increase if we slightly tilt the fit until it passes through p observations (while not passing any other observations), it suffices to consider all fits through p data points in Definition 2. If several of these fits have the same (maximal) regression depth, we take their average. Note that no distributional assumptions are made to define the deepest regression estimator of a dataset. The DR is a regression, scale, and affine equivariant estimator. For a univariate dataset, the deepest regression is its median. The DR thus generalizes the univariate median to linear regression.

In the population case, let $(\mathbf{x}^t, y)$ be a random p-dimensional variable with distribution H on $I\!R^p$. Then $rdepth(\boldsymbol{\theta}, H)$ is defined as the smallest amount of probability mass that needs to be passed when tilting $\boldsymbol{\theta}$ in any way until it is vertical. The deepest regression $DR(H)$ is the fit $\boldsymbol{\theta}$ with maximal depth. The natural setting of deepest regression is a large semiparametric model $\mathcal{H}$ in which the functional form is parametric and the error distribution is nonparametric. Formally, $\mathcal{H}$ consists of all distributions H on $I\!R^p$ that satisfy the following conditions:

$$\text{H has a strictly positive density and there exists a } \tilde{\boldsymbol{\theta}} \in I\!R^p \text{ with}$$
$$med_H(y \mid \mathbf{x}) = (\mathbf{x}^t, 1)\tilde{\boldsymbol{\theta}}. \tag{H}$$

Note that this model allows for skewed error distributions and heteroscedasticity. Van Aelst and Rousseeuw (2000) have shown that the DR is a Fisher-consistent estimator of $med(y|\mathbf{x})$ when the data come from the natural semiparametric model $\mathcal{H}$. The asymptotic distribution of the deepest regression was obtained by He and Portnoy (1998) in simple regression, and by Bai and He (1999) in multiple regression.

Figure 1 shows the Educational Spending data, obtained from the DASL library at `http://lib.stat.cmu.edu/DASL`. This dataset lists the expenditures per pupil versus the average salary paid to teachers, for $n = 51$ regions in the US. The fits $\boldsymbol{\theta_1} = (0.17, 0.6)^t$ and $\boldsymbol{\theta_2} = (-0.3, 12)^t$ both have regression depth 2, and the deepest regression $DR(Z_n) = (0.17, -0.51)^t$ is the average of fits with depth 23. Figure 1 illustrates that lines with high regression depth fit the data better than lines with low depth. The regression depth thus measures the quality of a fit, which motivates our interest in the deepest regression $DR(Z_n)$.

The breakdown value ε_n^* of an estimator T is defined as the smallest fraction of contamination that can cause the estimator T to become useless (see

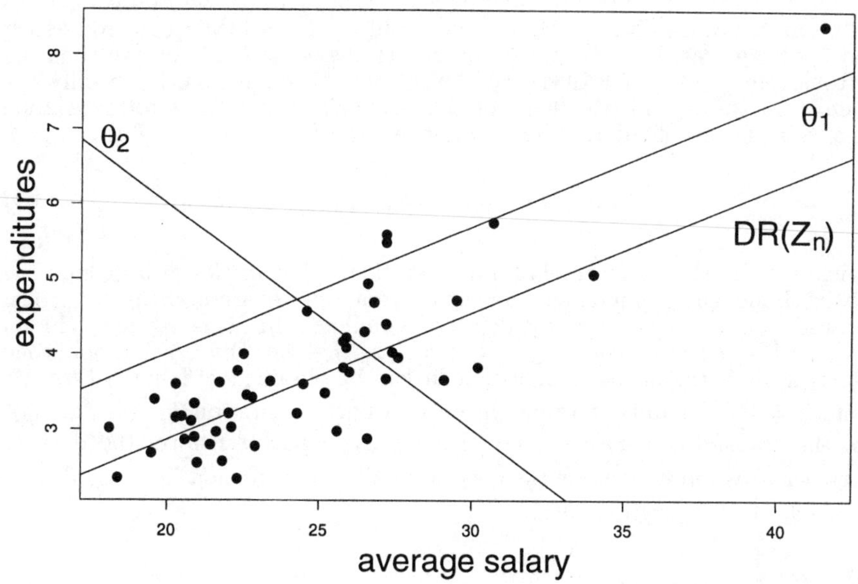

Fig. 1. Educational spending data, with $n = 51$ observations in $p = 2$ dimensions. The lines `$_1$ and `$_2$ both have depth 2, and the deepest regression $DR(Z_n)$ is the average of fits with depth 23.

Hampel et al. 1986). The breakdown value of the deepest regression is always positive, but it can be as low as $1/(p+1)$ when the original data are themselves peculiar (Rouseeuw and Hubert 1999). Fortunately, Van Aelst and Rousseeuw (2000) have shown that the deepest regression functional has breakdown value 1/3 in any dimension p, for all distributions from the model $\mathcal{H}$. This means that the deepest regression does not break down when at least 67% of the data are generated from the semiparametric model $\mathcal{H}$ while the remaining data (i.e., up to 33% of the points) may be anything. This result holds in any dimension. The DR is thus robust to leverage points as well as to vertical outliers. Moreover, this illustrates that the deepest regression is different from L^1 regression, which is defined as

$$L^1(Z_n) = \underset{\boldsymbol{\theta}}{\text{argmin}} \sum_{i=1}^{n} |r_i(\boldsymbol{\theta})|.$$

Note that L^1 is another generalization of the univariate median to regression, but with zero breakdown value due to its vulnerability to leverage points.

In simple regression, Van Aelst and Rousseeuw (2000) derived the influence function of the DR for elliptical distributions and computed the corresponding sensitivity functions. The influence functions of the DR slope and intercept are piecewise smooth and bounded, meaning that an outlier cannot affect DR too much, and the corresponding sensitivity functions show that this already holds for small sample sizes.

The deepest regression also inherits a monotone equivariance property from the univariate median, which does not hold for L^1 or other estimators such as least squares, least trimmed squares (Rousseeuw 1984) or S-estimators (Rousseeuw and Yohai 1984). By definition, the regression depth only depends on the $\mathbf{x}_i$ and the signs of the residuals. This allows for monotone transformations of the response y_i: Assume the functional model is

$$y = g(\theta_1 x_1 + \cdots + \theta_{p-1}x_{p-1} + \theta_p) \tag{3}$$

with g a strictly monotone link function. Typical examples of g include the logarithmic, the exponential, the square root, the square and the reciprocal transformation. The regression depth of a nonlinear fit (3) is defined as in (1) but with $r_i(\boldsymbol{\theta}) = y_i - g(\theta_1 x_1 + \cdots + \theta_{p-1}x_{p-1} + \theta_p)$. Due to the monotone equivariance, the deepest regression fit can be obtained as follows. First we put $\tilde{y}_i = g^{-1}(y_i)$ and determine the deepest linear regression $\hat{\boldsymbol{\theta}} = (\hat{\theta}_1, \ldots, \hat{\theta}_p)^t$ of the transformed data $(\mathbf{x}_i^t, \tilde{y}_i)$. Then we can backtransform the deepest linear regression $\hat{\boldsymbol{\theta}}$, yielding the deepest nonlinear regression fit $y = g(\hat{\theta}_1 x_1 + \cdots + \hat{\theta}_p)$ to the original data.

3 Computation

In $p = 2$ dimensions the regression depth can be computed in $O(n \log n)$ time with the algorithm described in (Rousseeuw and Hubert 1999). To compute the regression depth of a fit in $p = 3$ or $p = 4$ dimensions, Rousseeuw and Struyf (1998) constructed exact algorithms with time complexity $O(n^{p-1} \log n)$. For datasets with large n and/or p they also give an approximate algorithm that computes the regression depth of a fit in $O(mp^3 + mpn + mn \log n)$ time. Here m is the number of $(p-1)$-subsets in x-space used in the algorithm. The algorithm is exact when all $m = \binom{n}{p-1}$ such subsets are considered.

A naive exact algorithm for the deepest regression computes the regression depth of all $O(n^p)$ fits through p observations and keeps the one(s) with maximal depth. This yields a total time complexity of $O(n^{2p-1} \log n)$ which is very slow for large n and/or high p. Even if we use the approximate algorithm of Rousseeuw and Struyf (1998) to compute the depth of each fit, the time complexity remains very high. In simple regression, collaborative work with several specialists of computational geometry yielded an exact algorithm of complexity $O(n \log^2 n)$, i.e. little more than linear time (van Kreveld et al. 1999). To speed up the computation in higher dimensions, we will now construct the fast algorithm MEDSWEEP to approximate the deepest regression.

The MEDSWEEP algorithm is based on regression through the origin. For regression through the origin, Rousseeuw and Hubert (1999) defined the regression depth (denoted as rdepth_0) by requiring that $v = 0$ in Definition 1. Therefore, the $\text{rdepth}_0(\boldsymbol{\theta})$ of a fit $\boldsymbol{\theta} \subset I\!R^p$ relative to a dataset $Z_n \subset I\!R^{p+1}$ is again the smallest number of observations that needs to be passed when tilting $\boldsymbol{\theta}$ in any way until it becomes vertical. Rousseeuw and Hubert (1999) have shown that in the special case of a regression line through the origin ($p = 1$), the deepest regression (DR_o) of the dataset $Z_n = \{(x_1, y_1), \ldots, (x_n, y_n)\}$

is given by the slope

$$DR_o(Z_n) = \operatorname*{med}_{i=1}^{n} \frac{y_i}{x_i} \tag{4}$$

where observations with $x_i = 0$ are not used. This estimator has minimax bias (Martin, Yohai and Zamar 1989) and can be computed in $O(n)$ time.

We propose a sweeping method based on the estimator (4) to approximate the deepest regression in higher dimensions. Suppose we have a dataset $Z_n = \{(x_{i1}, \ldots, x_{i,p-1}, y_i); i = 1, \ldots, n\}$. We arrange the n observations as rows in a $n \times p$ matrix $\mathbf{X} = [X_1, \ldots, X_{p-1}, Y]$ where the X_i and Y are n-dimensional column vectors.

Step 1: In the first step we construct the sweeping variables $X_1^S, \ldots, X_{p-1}^S$. We start with $X_1^S = X_1$. To obtain X_j^S $(j > 1)$ we successively sweep $X_1^S, \ldots, X_{j-1}^S$ out of the original variable X_j. In general, to sweep X_k^S out of X_l $(k < l)$ we compute

$$\hat{\beta}_{lk} = \operatorname*{med}_{j \in J} \frac{x_{jl} - \operatorname*{med}_{i=1}^{n} x_{il}}{x_{jk}^S - \operatorname*{med}_{i=1}^{n} x_{ik}^S} \tag{5}$$

where J is the collection of indices for which the denominator is different from zero, and then we replace X_l by $X_l - \hat{\beta}_{lk} X_k$. If $k < l - 1$ we can now sweep the next variable X_{k+1}^S out of this new X_l. If $k = l - 1$ then $X_l^S = X_l$. Thus we obtain the sweeping variables

$$\begin{aligned}
X_1^S &= X_1 \\
X_2^S &= X_2 - \hat{\beta}_{21} X_1^S \\
&\vdots \\
X_{p-1}^S &= X_{p-1} - \hat{\beta}_{p-1,1} X_1^S - \cdots - \hat{\beta}_{p-1,p-2} X_{p-2}^S.
\end{aligned}$$

Step 2: In the second step we successively sweep $X_1^S, \ldots, X_{p-1}^S$ out of Y. Put $Y^0 = Y$. For $k = 1, \ldots, p-1$ we now compute

$$\hat{\beta}_k = \operatorname*{med}_{j \in J} \frac{y_j^{k-1} - \operatorname*{med}_{i=1}^{n} y_i^{k-1}}{x_{jk}^S - \operatorname*{med}_{i=1}^{n} x_{ik}^S} \tag{6}$$

with J as before, and we replace the original Y^{k-1} by $Y^k = Y^{k-1} - \hat{\beta}_k X_k^S$. Thus we obtain

$$Y^S = Y - \hat{\beta}_1 X_1^S - \cdots - \hat{\beta}_{p-1} X_{p-1}^S. \tag{7}$$

The process (6)-(7) is iterated until convergence is reached. In each iteration step all the coefficients $\hat{\beta}_1, \ldots, \hat{\beta}_{p-1}$ are updated. Usually only a few iterations are needed, and in any case the maximal number of iterations has been set to 100. After the iteration process, we take the median of Y^S to be the intercept $\hat{\beta}_p$.

Step 3: By backtransforming $\hat{\beta}_1, \ldots, \hat{\beta}_p$ we obtain the regression coefficients $(\hat{\theta}_1^S, \ldots, \hat{\theta}_p^S)^t$ corresponding to the original variables $X_1, \ldots, X_{p-1}, Y$. The obtained fit $\hat{\boldsymbol{\theta}}^S$ is then slightly adjusted until it passes through p observations, because we know that this can only improve the depth of the fit. We start by making the smallest absolute residual zero. Then for each of the directions $X_1, \ldots, X_{p-1}$ we tilt the fit in that direction until it passes an observation while not changing the sign of any other residual. This yields the fit $\hat{\boldsymbol{\theta}}$.

Step 4: In the last step we approximate the depth of the final fit $\hat{\boldsymbol{\theta}}$. Let $\mathbf{u}_1^S, \ldots, \mathbf{u}_{p-1}^S$ be the directions corresponding to the variables $X_1^S, \ldots, X_{p-1}^S$, then we compute the minimum over

$$\mathbf{u} \in \{\mathbf{e}_1, -\mathbf{e}_1, \ldots, \mathbf{e}_{p-1}, -\mathbf{e}_{p-1}, \mathbf{u}_1^S, -\mathbf{u}_1^S, \ldots, \mathbf{u}_{p-1}^S, -\mathbf{u}_{p-1}^S\}$$

instead of over all unit vectors $\mathbf{u} \in I\!R^{p-1}$ in the right hand side of expression (1).

Since computing the median takes $O(n)$ time, the first step of the algorithm needs $O(p^2 n)$ time and the second step takes $O(hpn)$ time where h is the number of iterations. The adjustments in step 3 also take $O(p^2 n)$ time, and computing the approximate depth in the last step can be done in $O(pn \log n)$ time. The time complexity of the MEDSWEEP algorithm thus becomes $O(p^2 n + hpn + pn \log n)$ which is very low.

To measure the performance of our algorithm we carried out the following simulation. For different values of p and n we generated $m = 10{,}000$ samples $Z^{(j)} = \{(x_{i1}, \ldots, x_{i,p-1}, y_i); i = 1, \ldots, n\}$ from the standard Gaussian distribution. For each of these samples we computed the deepest regression $(\hat{\theta}_1^{(j)}, \ldots, \hat{\theta}_p^{(j)})^t$ with the MEDSWEEP algorithm and measured the total time needed for these 10,000 estimates. For each n and p we also computed the bias of the intercept, which is the average of the 10,000 intercepts, and the bias of the vector of the slopes, which we measure by

$$b(\hat{\theta}_1, \ldots, \hat{\theta}_{p-1}) = \sqrt{\frac{1}{p-1}((\operatorname*{ave}_j \hat{\theta}_1^{(j)})^2 + \cdots + (\operatorname*{ave}_j \hat{\theta}_{p-1}^{(j)})^2)}. \tag{8}$$

We also give the mean squared error of the vector of the slopes, given by

$$\mathrm{MSE}(\hat{\theta}_1, \ldots, \hat{\theta}_{p-1}) = \frac{1}{m} \sum_{j=1}^{m} \frac{1}{p-1} \sum_{i=1}^{p-1} (\hat{\theta}_i^{(j)} - \theta_i)^2 \tag{9}$$

where the true values $\theta_i; i = 1, \ldots, p-1$ equal zero, and the mean squared error of the intercept, given by $\frac{1}{m}\sum_{j=1}^{m}(\hat{\theta}_p^{(j)})^2$. Table 1 lists the bias and mean squared error of the vector of the slopes, while the bias and mean squared error of the intercept are given in Table 2. Note that the bias and

Table 1. Bias (8) and mean squared error (9) of the *DR* slope vector, obtained by generating $m = 10,000$ standard Gaussian samples for each n and p. The *DR* fits were obtained with the MEDSWEEP algorithm. The results for the bias have to be multiplied by 10^{-4} and the results for the MSE by 10^{-3}.

		p				
n		2	3	4	5	10
50	bias	18.01	18.27	28.53	22.47	21.42
	MSE	54.55	52.32	52.23	52.54	58.65
100	bias	7.56	8.41	11.16	8.98	12.68
	MSE	24.59	25.32	25.99	26.15	27.17
300	bias	8.11	5.98	6.29	7.39	10.89
	MSE	8.11	8.27	8.26	8.41	8.42
500	bias	0.44	1.68	2.91	9.10	5.49
	MSE	4.93	4.89	5.02	4.93	4.97
1000	bias	6.34	3.58	6.77	3.54	3.98
	MSE	2.47	2.51	2.46	2.46	2.50

mean squared error of the slope vector and the intercept are low, and that they decrease with increasing n. From Tables 1 and 2 we also see that the mean squared error does not seem to increase with p. Table 3 lists the average time the MEDSWEEP algorithm needed for the computation of the *DR*, on a Sun SparcStation 20/514. We see that the algorithm is very fast.

To illustrate the MEDSWEEP algorithm we generated 50 points in 5 dimensions, according to the model $y = x_1 - x_2 + x_3 - x_4 + 1 + e$ with x_1, x_2, x_3, x_4 and e coming from the standard Gaussian distribution. The *DR* fit obtained with MEDSWEEP is $y = 0.98x_1 - 1.01x_2 + 0.97x_3 - 1.00x_4 + 1.14$ with approximate depth 21. The algorithm needed 14 iterations till convergence. In a second example, we generated 50 points according to the model $y = 100x_1 + x_2 - 2x_3 + 3x_4 - 4 + e$ with standard Gaussian x_1, x_2, x_3, x_4 and e. After 25 iterations the MEDSWEEP algorithm yielded the fit $y = 99.99x_1 + 0.99x_2 - 2.03x_3 + 3.00x_4 - 3.86$ with approximate depth 21. Note that in both cases the coefficients obtained by the algorithm approximate the true parameters in the model very well. The MEDSWEEP algorithm is available from our website `http://win-www.uia.ac.be/u/statis/` where its use is explained.

4 Confidence regions

In order to construct a confidence region for the unknown true parameter vector $\tilde{\boldsymbol{\theta}} = (\tilde{\theta}_1, \ldots, \tilde{\theta}_p)^t$ we use a bootstrap method. Starting from the dataset $Z_n = \{(\mathbf{x}_i^t, y_i); i = 1, \ldots, n\} \in I\!R^p$, we construct a bootstrap sample by randomly drawing n observations, with replacement. For each bootstrap sample $Z_n^{(j)}$, $j = 1, \ldots, m$ we compute its deepest regression fit

Table 2. Bias and mean squared error of the *DR* intercepts, obtained by generating $m = 10,000$ standard Gaussian samples for each n and p. The *DR* fits were obtained with the MEDSWEEP algorithm. The results for the bias have to be multiplied by 10^{-4} and the results for the MSE by 10^{-3}.

n		p				
		2	3	4	5	10
50	bias	-3.04	-48.70	14.38	13.23	-19.64
	MSE	32.89	32.62	35.23	36.72	44.35
100	bias	5.75	-3.32	12.21	9.92	-1.70
	MSE	15.78	16.01	16.92	16.77	18.14
300	bias	-2.71	-7.99	-2.37	3.49	4.54
	MSE	5.25	5.31	5.22	5.23	5.47
500	bias	1.35	-5.53	-4.33	-4.26	2.39
	MSE	3.09	3.15	3.21	3.21	3.18
1000	bias	-3.76	-3.40	-5.10	0.75	-0.01
	MSE	1.56	1.56	1.55	1.58	1.62

Table 3. Computation time (in seconds) of the MEDSWEEP algorithm for a sample of size n with p dimensions. Each time is an average over 10,000 samples.

	p				
n	2	3	4	5	10
50	0.017	0.023	0.040	0.057	0.20
100	0.027	0.071	0.15	0.25	0.51
300	0.056	0.21	0.42	0.67	1.38
500	0.093	0.39	0.73	1.12	2.24
1000	0.20	0.66	1.43	2.18	4.42

$\hat{\boldsymbol{\theta}}^{(j)} = (\hat{\theta}_1^{(j)}, \ldots, \hat{\theta}_p^{(j)})^t$. Note that there will usually be a few outlying estimates $\hat{\boldsymbol{\theta}}^{(j)}$ in the set of bootstrap fits $\{\hat{\boldsymbol{\theta}}^{(j)}; j = 1, \ldots, m\}$, which is natural since some bootstrap samples contain disproportionally many outliers. Therefore we don't construct a confidence ellipsoid based on the classical mean and covariance matrix of the $\{\hat{\boldsymbol{\theta}}^{(j)}; j = 1, \ldots, m\}$, but we use the robust minimum covariance determinant estimator (MCD) proposed by (Rousseeuw 1984,1985).

The MCD looks for the $h \geq n/2$ observations of which the empirical covariance matrix has the smallest possible determinant. Then the center

(a)

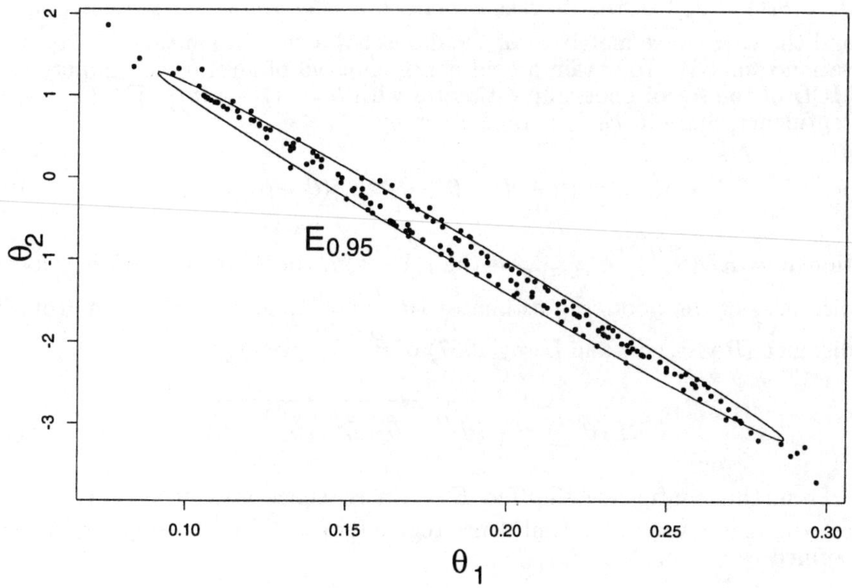

(b)

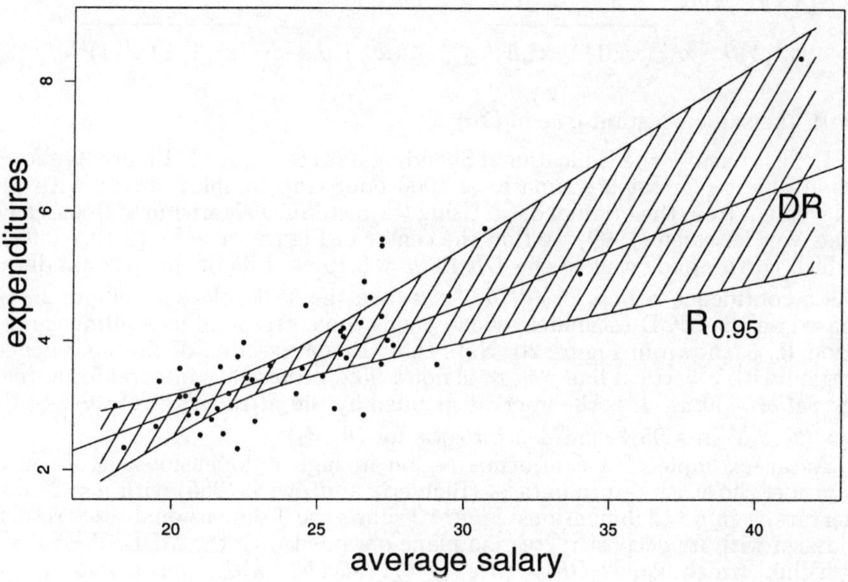

Fig. 2. (a) *DR* estimates of 1,000 bootstrap samples from the Educational Spending data with the 95% confidence ellipse $E_{0.95}$; (b) Plot of the data with the *DR* line and the 95% confidence region $R_{0.95}$ for the fitted value.

$\hat{\boldsymbol{\theta}} = (\hat{\theta}_1, \ldots, \hat{\theta}_p)^t$ of the dataset is defined as the average of these h points, and the covariance matrix $\hat{\Sigma}$ of the dataset is a certain multiple of their covariance matrix. To obtain a confidence ellipsoid of level α we compute the MCD of the set of bootstrap estimates with $h = \lceil (1-\alpha)m \rceil$. The $(1-\alpha)\%$ confidence ellipsoid $E_{1-\alpha}$ is then given by

$$E_{1-\alpha} = \{\boldsymbol{\theta} \in I\!R^p; (\boldsymbol{\theta} - \hat{\boldsymbol{\theta}})^t \hat{\Sigma}^{-1} (\boldsymbol{\theta} - \hat{\boldsymbol{\theta}}) \leq c^2\}. \tag{10}$$

Here $c := RD(\hat{\boldsymbol{\theta}}^{(j)})_{\lceil (1-\alpha)m \rceil : m}$ is the $\lceil (1-\alpha)m \rceil$ order statistic of the robust distances of the bootstrap estimates $\{\hat{\boldsymbol{\theta}}^{(j)}; j = 1, \ldots, m\}$, where the robust distance (Rousseeuw and Leroy 1987) of $\hat{\boldsymbol{\theta}}^{(j)}$ is given by

$$RD(\hat{\boldsymbol{\theta}}^{(j)}) = \sqrt{(\hat{\boldsymbol{\theta}}^{(j)} - \hat{\boldsymbol{\theta}})^t \hat{\Sigma}^{-1} (\hat{\boldsymbol{\theta}}^{(j)} - \hat{\boldsymbol{\theta}})}. \tag{11}$$

From this confidence ellipsoid $E_{1-\alpha}$ in fit space we can also derive the corresponding regression confidence region for $\hat{y} = \tilde{\theta}_1 x_1 + \cdots + \tilde{\theta}_{p-1} x_{p-1} + \tilde{\theta}_p$ defined as

$$R_{1-\alpha} = \{(\mathbf{x}^t, y)^t \in I\!R^p; \mathit{min}((\mathbf{x}^t, 1)\boldsymbol{\theta}) \leq y \leq \mathit{max}((\mathbf{x}^t, 1)\boldsymbol{\theta}) \ \mathit{where} \ \boldsymbol{\theta} \in E_{1-\alpha}\}.$$

This region $R_{1-\alpha}$ is given by

$$\begin{aligned} &\{(\mathbf{x}^t, y)^t \in I\!R^p; \\ &\quad (\mathbf{x}^t, 1)\hat{\boldsymbol{\theta}} - c\sqrt{(\mathbf{x}^t, 1)\hat{\Sigma}(\mathbf{x}^t, 1)^t} \leq y \leq (\mathbf{x}^t, 1)\hat{\boldsymbol{\theta}} + c\sqrt{(\mathbf{x}^t, 1)\hat{\Sigma}(\mathbf{x}^t, 1)^t}\} \end{aligned} \tag{12}$$

with the same constant c as in (10).

Let us consider the Educational Spending data of Figure 1. Figure 2a shows the deepest regression estimates of 1000 bootstrap samples, drawn with replacement from the original data. Using the fast MCD algorithm of Rousseeuw and Van Driessen (1999) we find the center in Figure 2a to be $(0.19, -0.95)^t$ which corresponds well to the DR fit $y = 0.19x - 1.05$ of the original data. As a confidence region for $(\tilde{\theta}_1, \tilde{\theta}_2)^t$ we take the 95% tolerance ellipse $E_{0.95}$ based on the MCD estimates, which yields the corresponding confidence region $R_{0.95}$ shown in Figure 2b. Note that the intersection of this confidence region with a vertical line $x = x_0$ is not a 95% probability interval for an *observation* y at x_0. It is the interval spanned by the *fitted values* $\hat{y} = \theta_1 x_0 + \theta_2$ for $(\theta_1, \theta_2)^t$ in a 95% confidence region for $(\tilde{\theta}_1, \tilde{\theta}_2)^t$.

As an example of a confidence region in higher dimensions, let us now consider the stock return dataset (Benderly and Zwick 1985) with $n = 28$ observations in $p = 3$ dimensions. Figure 3 shows the 3-dimensional stock return dataset with its deepest regression plane obtained with the MEDSWEEP algorithm, which equals $DR = (3.41, -2.22, 6.66)^t$ with approximate depth 11. The 95% confidence region shown in Figure 3 was based on $m = 1,000$ bootstrap samples.

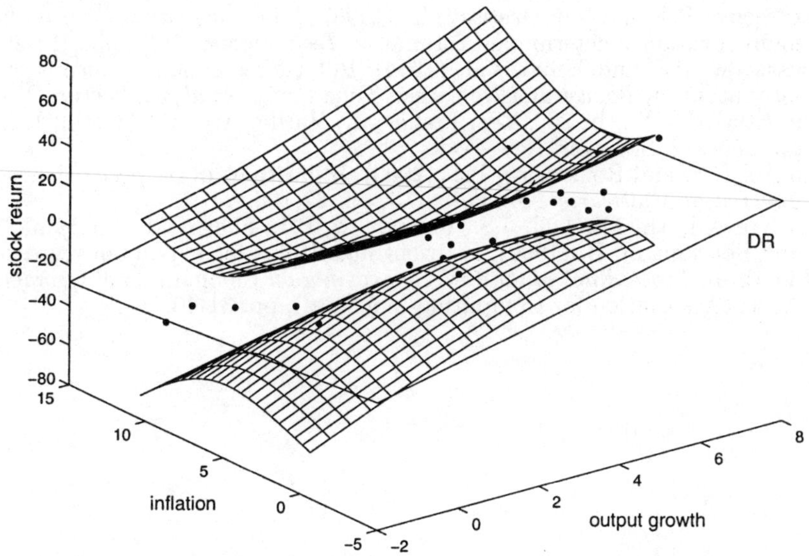

Fig. 3. Stock return dataset with its deepest regression plane, and the upper and lower surface of the 95% confidence region $R_{0.95}$ based on $m = 1,000$ bootstrap samples.

References

Bai, Z. and He, X. (1999). Asymptotic distributions of the maximal depth estimators for regression and multivariate location, *The Annals of Statistics*, pp. 1616-1637.

Benderly, J. and Zwick, B. (1985). Inflation, real balances, output, and real stock returns. *American Economic Review*, pp. 1117.

Hampel, F.R., Ronchetti, E.M., Rousseeuw, P.J. and Stahel, W.A. (1986). *Robust Statistics: the Approach based on Influence Functions.* New York: John Wiley.

He, X. and Portnoy, S. (1998). Asymptotics of the deepest line. In *Applied Statistical Science III: Nonparametric Statistics and Related Topics* (Eds. Ahmed, S.E., Ahsanullah, M. and Sinha, B.K.), pp. 71-81, New York: Nova Science Publishers Inc.

Martin, R.D., Yohai, V.J. and Zamar, R.H. (1989). Min-max bias robust regression, *The Annals of Statistics,* **17**, pp. 1608-1630.

Rousseeuw, P.J. (1984). Least median of squares regression. *Journal of the American Statistical Association*, **79**, pp. 871-880.

Rousseeuw, P.J. (1985). Multivariate estimation with high breakdown point. In *Mathematical Statistics and Applications*, Vol. **B** (Eds. Grossmann, W., Pflug, G., Vincze, I. and Wertz, W.), pp. 283–297, Dordrecht: Reidel.

Rousseeuw, P.J. and Hubert, M. (1999). Regression depth. *Journal of the American Statistical Association*, **94**, pp. 388-402.

Rousseeuw, P.J. and Leroy, A.M. (1987). *Robust Regression and Outlier Detection.* New York: Wiley-Interscience.

Rousseeuw, P.J. and Struyf, A. (1998). Computing location depth and regression depth in higher dimensions. *Statistics and Computing*, **8**, pp. 193-203.

Rousseeuw, P.J. and Van Driessen, K. (1999). A fast algorithm for the minimum covariance determinant estimator. *Technometrics*, **41**, pp. 212-223.

Rousseeuw, P.J. and Yohai, V.J. (1984). Robust regression by means of S-estimators. In *Robust and Nonlinear Time Series Analysis,* Lecture Notes in Statistics No. bf 26, (Eds. Franke, J., Härdle, W. and Martin, R.D.), pp. 256-272, Springer, New York.

Van Aelst, S. and Rousseeuw, P.J. (2000). Robustness of deepest regression. *Journal of Multivariate Analysis*, **73**, pp. 82-106.

Van Kreveld, M., Mitchell, J.S., Rousseeuw, P.J., Sharir, M., Snoeyink, J., and Speckmann, B. (1999). Efficient algorithms for maximum regression depth. In *Proceedings of the 15th Symposium on Computational Geometry,* ACM (Association for Computing Machines), pp. 31-40.

Non-proportional hazards models in survival analysis

Hans C. van Houwelingen and Paul H.C. Eilers

Department of Medical Statistics, Leiden University Medical Center, P.O. Box 9604, 2300 RC Leiden, The Netherlands

Abstract. Cox' proportional hazard model is usually the model of choice in survival analysis. It is shown that this model can be embedded in a GLM-model by proper discretization of the time axis. That approach easily allows non-proportional hazard models, that are special cases of time-varying coefficients models. It is shown how the effective dimension of the general non-proportional hazards model can be controlled by either reduced rank regression methods of P-splines methodology.

Keywords. Poisson regression, B-splines, penalized estimation

1 Survival analysis

1.1 Some general notions

In medical statistics the outcome of interest is often the "survival time" T after diagnosis or after start of treatment. "Survival time " should not be taken too literally: it can also mean time until relapse, progression or even time till recovery after surgery. The general interest is to describe the relation between the distribution of T and a (vector of k) covariate(s) X. The relation between T and X can either be described through the survivor function

$$S(t|X) = P(T > t|X)$$

with corresponding density

$$f(t|X) = \frac{d(1 - S(t|X))}{dt},$$

or the hazard function, defined by

$$h(t|X)dt = P(T < t + dt|T \geq t, X),$$

or

$$h(t|X) = -\frac{d\ln(S(t|X))}{dt}.$$

The link between the two concepts is through the cumulative hazard function

$$H(t|X) = \int_0^t h(s|X)ds = -\ln(S(t|X))$$

or via

$$f(t|X) = S(t|X)h(t|X).$$

A typical phenomenon in survival analysis is right censoring. Since follow-up is limited, it is impossible to observe the "death" of all patients. For some patients we only know that T is larger than some value C (censoring time). The survival information is usually summarized in the pair (T, δ) where T is survival time and $\delta = 1$ in case of a true "death" and T is censoring time and $\delta = 0$ in case of a censored observation. We assume that censoring in non-informative, that is stochastically independent of T and, hence, that the likelihood of the observed pair (T, δ) is given by

$$L = f(T|X)^{\delta} S(T|X)^{1-\delta} = S(T|X) h(T|X)^{\delta}.$$

1.2 Accelerated failure time (AFT) models

The modelling of the relation between T and X through the survivor function S stays very close to classical statistical modelling. Since T is positive, it is kind of natural to take the logarithm of T and model it as

$$\ln(T) = X\beta + \sigma e$$

(here, the constant is included in X). The parameter σ describes the shape of the distribution and the linear predictor $X\beta$ the length of survival. In terms of the survivor function we have

$$S(t|X, \beta, \sigma) = S_{\sigma}(t/\exp(X\beta)).$$

Well-known examples are the log-normal distribution, where e has a normal distribution, and the Weibull distribution, where e has cumulative distribution function $1 - \exp(-\exp(x))$, and, hence

$$S(t|X, \beta, \sigma) = \exp(-\exp(\frac{\ln(t) - X\beta}{\sigma}))$$

or

$$H(t|X, \beta, \sigma) = t^{1/\sigma} \exp(-X\beta/\sigma).$$

In many cases the log-normal is a reasonable choice and classical regression models can be used. The only problem is the handling of the censored data. Iterated multiple imputation could be a practical solution.

1.3 Cox' proportional hazards regression

The analysis of choice in medical applications is through the so-called Cox regression (Cox, 1972). The model used is the proportional hazards (PH) model

$$h(t|X) = h_0(t) \exp(X\beta) \quad (1)$$

(no constant contained in X). Let the cumulative hazard corresponding with the baseline $h_0(t)$ be given by

$$H_0(t) = \int_0^t h_0(s) ds$$

and the corresponding baseline survival function by

$$S_0(t) = \exp(-H_0(t)).$$

The general survivor function is given by

$$S(t|X) = S_0(t)^{RR(X)}$$

with

$$RR(X) = \exp(X\beta).$$

The approach is to estimate β first by Maximal Partial Likelihood and use that estimate of β in the estimation of the baseline survival function. To explain the notion of partial likelihood, assume that the survival times are ordered $t_1 < t_2 < \ldots < t_n$ (no ties) and that (δ_1, X_1), (δ_2, X_2), ... are the corresponding values of the censoring variable and the covariate. The partial likelihood is given by

$$PL(\beta) = \prod_{\delta_i=1} \frac{\exp(X_i\beta)}{\sum_{j\geq i}\exp(X_j\beta)}.$$

It is the product of the conditional probabilities that at time t_i, with $\delta_i = 1$, the person dies that actual has died. The advantage of the partial likelihood approach is that the baseline hazard does not appear in the partial likelihood. The baseline cumulative hazard is estimated in the second stage by the Nelson-Aalen estimator

$$\hat{H}_0(t) = \sum_{t_i\leq t} \frac{\delta_i}{\sum_{j\geq i}\exp(X_j\beta)}.$$

The hazard itself can be obtained by smoothing and differentiation of $\hat{H}_0(t)$. The main practical disadvantage of the Cox regression as used in standard software packages is its emphasis on the relative risk and its neglect of estimating the baseline (cumulative) hazard and survivor function. The PH model needs not to be linear in X as in (1). The only requirement is that $h(t|X_1)/h(t|X_2)$ is independent of t. The more general PH model could be written as

$$\ln(h(t|X)) = \ln(h_0(t) + f(X|\beta).$$

1.4 Non-proportional hazards; time-dependent effects

One way of stating that the Proportional Hazards Assumption does not hold true is by saying that the regression coefficients depend on the time t, that is by replacing (1) by

$$h(t|X) = h_0(t)\exp(X\beta(t)). \tag{2}$$

The standard choice is a simple model for each component of $\beta(t)$, such as

$$\beta_i(t) = \beta_{i0} + \beta_{i1}\ln(t) + \beta_{i2}t. \tag{3}$$

This can be squeezed into standard software if the data set is not too big, but usually the baseline hazard gets lost in the process. Moreover, the number of unknown parameters increases very rapidly.

This non-proportional hazard model can be seen as very (too) flexible model to describe survival data. In the rest of the paper we will discuss how parsimonious (non-)proportional hazards models can be fitted by turning survival analysis into a simple Generalized Linear Model (GLM) and estimating time effects, covariate effects and their interaction simultaneously.

2 Survival as a GLM

2.1 Back to basics

Long before the Cox regression model was introduced, epidemiologists estimated hazards by counting number of deaths and person-years in the age-period-cohort model. Let us divide the time axis in small intervals of equal length Δ and number them from 1 to m. The probability that individual i dies in interval j, given that he is alive at the beginning of interval j, is given by

$$\mu_{ij} = h(t_j|X_i)\Delta$$

with $t_j = (j-1)\Delta$, the start point of interval j. Let

$$Y_{ij} = 1 \text{ if person } i \text{ is at risk at time } t_j,\ Y_{ij} = 0 \text{ if not,}$$

and

$$\delta_{ij} = 1 \text{ if person } i \text{ dies in interval } j,\ \delta_{ij} = 0 \text{ if not.}$$

If Δ is very small

$$T_i = \Delta \sum_{j=1}^{m} Y_{ij}, \text{ and } \delta_i = \sum_{j=1}^{m} Y_{ij}\delta_{ij}$$

and the contribution of the i-th person to the log-likelihood is given by

$$\sum_{j=1}^{m} Y_{ij}(\delta_{ij}\theta_{ij} - \exp(\theta_{ij})), \text{ with } \theta_{ij} = \ln(\mu_{ij}).$$

This Poisson approximation was already used when no software for Cox regression was available yet. The total log-likelihood is given by

$$l = \sum_{i=1}^{n}\sum_{j=1}^{m} Yij(\delta_{ij}\ln(\mu_{ij}) - \mu_{ij}). \tag{4}$$

This log-likelihood is the starting point for further analysis. If the sample size n is not too big and Δ not too small, it is helpful to reorganize the data in records for each combination of person and interval for which $Y_{ij} = 1$. That asks for at most nm records. In this set-up, the data analysis can be done by standard GLM-software. However, the data set can become prohibitively big if organized this way and, conceptually, it is also more convenient to think of Y, δ, θ and μ as matrices.
The approach to model the hazard function directly is very much in line with the work of Kooperberg et al. (1995) on hazard regression.

2.2 Building blocks for modelling

In modelling the data we are dealing with two design matrices: X, the n by k matrix with covariate vectors as rows, the covariates as columns, and Z, the m by p matrix with p time functions to model time-effects. The constant is neither included in X nor in Z. technically we assume the rank$(X,1) = k+1$ and rank$(Z,1) = p+1$. The design matrix X is dictated by the available covariate information. It can be extended by categorizing, introducing polynomials in covariates and interactions. The choice of the design matrix Z is up to the analyst. Possible choices for the base functions to be included in Z are (fractional) polynomial or B-splines. For the time being we take X and Z as given and fixed. In section 5 B-splines and penalties are used.

3 Fitting the proportional hazards model

3.1 The technicalities

In the notation of section 2.1 the proportional hazards model is simply an additive model in X and Z. To be more precise, the cell-wise model is

$$\theta_{ij} = \alpha + \sum_{u=1}^{k} X_{iu}\beta_u + \sum_{v=1}^{p} Z_{jv}\gamma_v \tag{5}$$

In vector notation that can be written as

$$\theta_{ij} = \alpha + X_i\beta + Z_j\gamma. \tag{6}$$

Here, X_i is the i-th row of X, the covariate information of person i, Z_j is the j-th row of Z, the values of the base functions at t_j. The parameters are α (scalar), β (k-vector) and γ (p-vector). Matrix-wise the model could be written as

$$\theta = \alpha 1_{nm} + (X\beta)1'_m + 1_n(Z\gamma)'.$$

The maximum likelihood estimator (MLE) is obtained by maximizing (4), leading to the score equations

$$\begin{aligned} &\sum_{i=1}^{n}\sum_{j=1}^{m} Y_{ij}(\delta_{ij} - \mu_{ij}) = 0, \\ &\sum_{i=1}^{n}\sum_{j=1}^{m} Y_{ij}(\delta_{ij} - \mu_{ij})X_i = 0, \\ &\sum_{i=1}^{n}\sum_{j=1}^{m} Y_{ij}(\delta_{ij} - \mu_{ij})Z_j = 0, \end{aligned} \tag{7}$$

with $\mu_{ij} = \exp(\theta_{ij})$. The information matrix is given by

$$H = \sum_{i=1}^{n}\sum_{j=1}^{m} Y_{ij}\mu_{ij}(1, X_i, Z_j)'(1, X_i, Z_j). \tag{8}$$

The estimates are easily derived by Newton-Raphson. The covariance matrix of the estimated parameters is a direct by-product.

3.2 Example

We analyzed a data set of 358 patients with advanced ovarian cancer. The covariates are:

- DIAM: diameter of residual tumor after surgery, coded as 0, 1, 2, 3, 4 (small to large);
- FIGO: stage of the cancer, coded as 0 (FIGO = III) and 1 (FIGO = IV);
- KARN: Karnovsky status at start of therapy, coded as an integer from 0 (100% functional) to 4 ($\leq$ 60% functional).

Cox regression by SPSS gave $\hat\beta_{\text{DIAM}} = 0.200$ ($se = 0.050$), $\hat\beta_{\text{FIGO}} = 0.542$ ($se = 0.135$) and $\hat\beta_{\text{KARN}} = 0.172$ ($se = 0.053$). To apply the GLM approach we took intervals of one week and "time" functions $\ln(j/52)$, $j/52$ and $(j/52)^2$. The estimated β-vector and the standard errors were very similar to the SPPS result. The log-likelihood was $l = -1616.43$. The baseline hazard and survivor function are shown in Figure 1. Confidence bands around these functions are easily obtained.

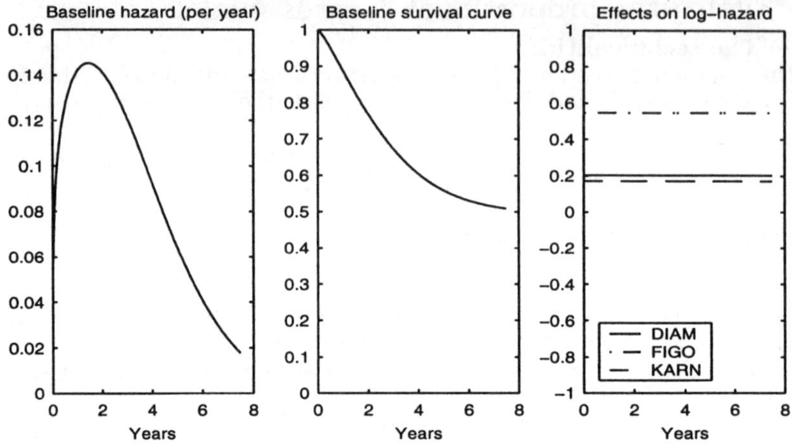

Fig. 1. Parametric estimates of baseline hazard (left), survival function (middle), and (constant) covariate effects for the parametric additive model.

4 Non-proportional hazards models

4.1 Fitting the full model

The time-dependent coefficient model of (2) and (3) can easily be imbedded in the GLM set up. Allowing interactions between all columns of X and Z we obtain the full non-proportional hazards model as an extension of (5)

$$\theta_{ij} = \alpha + \sum_{u=1}^{k} X_{iu}\beta_u + \sum_{v=1}^{p} Z_{jv}\gamma_v + \sum_{u=1}^{k}\sum_{v=1}^{p} \lambda_{uv} X_{iu} Z_{jv} \tag{9}$$

This equation can be written again in vector or matrix notation with Λ denoting the $k \times p$ matrix of all interaction parameters. The score equations (7) are expanded with a term

$$\sum_{i=1}^{n}\sum_{j=1}^{m} Y_{ij}(\delta_{ij} - \mu_{ij})M_{ij}, \tag{10}$$

where $M_{ij} = Z_j' X_i$ is the Kroneckerproduct of X_i and Z_j. The information matrix is easily obtained by adding an appropriate vectorisation of M_{ij} in (8). Fitting this model to the ovarian cancer data set resulted in a log-likelihood $l = -1604.24$, a significant improvement of the PH model. ($\chi^2 = 24.38$, $df = 9$ and $p = 0.004$.) The baseline survivor function derived from the baseline hazard $\exp(\alpha + \sum_{v=1}^{p} Z_{jv}\gamma_v)$ and the time-dependent coefficients $\beta_u + \sum_{v=1}^{p} \lambda_{uv} Z_{jv}$ are shown in Figure 2. The curves in Figure 2b are a bit wild, but the findings are more or less in line with Verweij and Van Houwelingen (1992), in the sense that the effect of KARN vanishes after about one year. Within the GLM set-up the non-proportional model is just a special case of the varying coefficient model (VCM) of Hastie and Tishirani (1993)

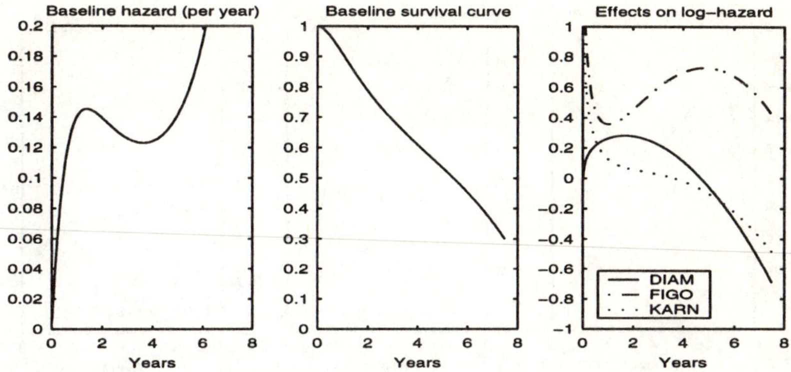

Fig. 2. Baseline hazard (left), baseline survival function (middle) and the effects on the log-hazard of three covariates for the full-rank model (right).

4.2 Parsimonious models

Parsimony can be achieved by putting restrictions on the interaction matrix Λ. A general approach is reduced rank regression as practiced in Van Eeuwijk (1995). In the spirit of (6) a general rank= r model can be written as

$$\theta_{ij} = \alpha + X_i\beta_0 + Z_j\gamma_0 + \sum_{w=1}^{r}(X_i\beta_w)(Z_j\gamma_w). \tag{11}$$

The simplest case is r=1 and that could even be simplified by taking $\beta_1 = \beta_0$. That is the kind of model that arises in a natural way in frailty models. See Aalen(1994). The general model (11) is not fully identified. If a model of rank= r is fitted, there are r^2 redundant parameters. Since the models are fitted by alternating between β's and γ's, the lack of identifiability does not constitute a serious problem. We fitted a sequence of models bridging the gap between the PH model of (5) and the full non-PH model of (9). The results are given in Table 1.

To compare models, we correct the likelihood by the dimension of the model, giving the information criterion IC. From this point of view, the model of choice is the rank= 1 model. The graphs for this model are presented in Figure 3.

Model	Parameters	Dimension	Log-likelihood	IC
Additive	μ, β, γ	7	-1616.43	-1623.4
"Frailty"	$\mu, \beta_0 = \beta_1, \gamma_0, \gamma_1$	7 + 3 = 10	-1609.11	-1619.1
Rank = 1	$\mu, \beta_0, \beta_1, \gamma_0, \gamma_1$	7 + 5 = 12	-1606.96	**-1619.0**
Rank = 2	$\mu, \beta_0, \beta_1, \beta_2, \gamma_0, \gamma_1, \gamma_2$	7 + 8 = 15	-1604.5	-1619.50
Full rank	$\mu, \beta, \gamma, \Lambda$	7 + 9 = 16	-1604.24	-1620.2

Table 1. An overview of five models, their parameters, dimension and log-likelihood. The column marked IC presents the log-likelihood minus the dimension.

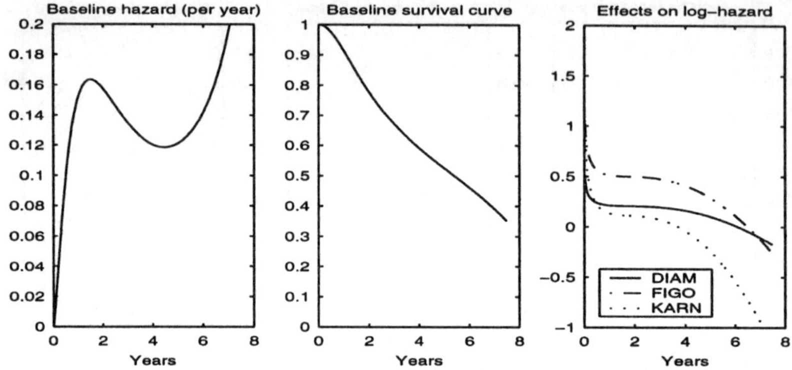

Fig. 3. Baseline hazard (left), baseline survival function (middle) and effects of the covariates for the parametric rank= 1 model.

The curves are better behaved in than the ones in Figure 2. The curves for DIAM and FIGO show a very similar rather constant behaviour, while the KARN effect again vanishes after a while.

5 Smooth non-parametric components

Up till here we modeled time varying coefficients with a small set of parametric functions $(t, t^2, \log t)$. As a more flexible set we choose B-splines, with a roughness penalty on their coefficients, for continuous control over smoothness (Eilers and Marx, 1996). For example, when only a baseline hazard is to be estimated, we put

$$\ln(h_0(t_j)) = \sum_{v=1}^{p} b_{jv}\gamma_v = \sum_{v=1}^{p} B_v(t_j)\gamma_v, \tag{12}$$

and we maximise the penalized log-likelihood

$$LL^* = LL - \frac{\omega}{2} \sum_{v=d+1}^{p} (\Delta^d \gamma_v)^2, \tag{13}$$

where Δ^d is the differencing operator of order d. The higher ω, the smoother the curve. With more curves in the model, additional sets of coefficients and the corresponding penalties are introduced.

As the effective dimension of a non-parametric model we use the trace of $(H + \omega D'D)^{-1}H$, where H is the information matrix as defined in (8). More details of this approach to hazard modeling were presented by Eilers (1998).

We compare parametric and non-parametric approaches for the additive and the full-rank model. We used eight cubic B-splines and a second order penalty. The optimum of ω was searched on a geometric grid (steps 1, 2, 5). Figure 4 shows results for the additive model (ω = 0.02; IC = −1621.2). When these are compared to Figure 1, we see that the coefficient are practically the same and that the baseline survival curves don't differ much. But the baseline hazard now shows a conspicuous peak between 5 and 6 years.

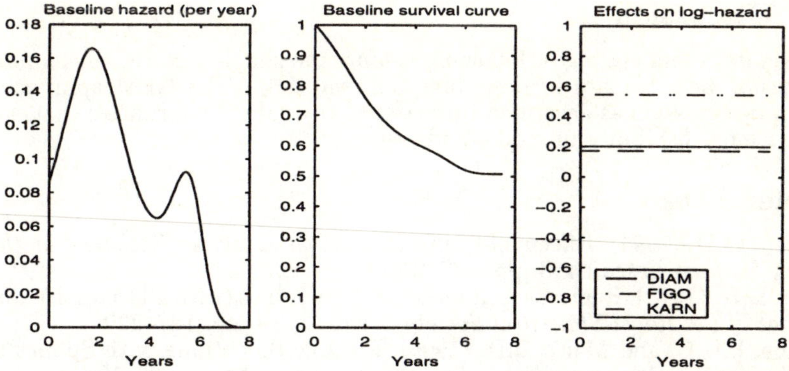

Fig. 4. Parametric estimates of baseline hazard (left), survival function (middle), and (constant) covariate effects for the non-parametric additive model.

Figure 5 shows the results for the non-parametric full-rank model. There is a complete change of the baseline hazard, as it now increase monotonically. On the other hand, the curves of the time dependencies of the covariate effects are rather wiggly. As there are four time effects (for the baseline and for each coefficient), there are four penalties and thus four smoothing parameters: $\omega_1 \ldots \omega_4$. To avoid a four-dimensional grid search, we took $\omega_2 = \omega_3 = \omega_4$. The optimal values appeared to be $\omega_1 = 100$ and $\omega_2 = 0.01$, with $IC = -1617.3$. The minimum of IC is rather flat.

Note that the two non-parametric models have IC about 2 lower than the corresponding parametric models. Apparently the extra flexibility allows a better fitting model. However, the optimization with respect to ω has not been taken into account in the value of IC.

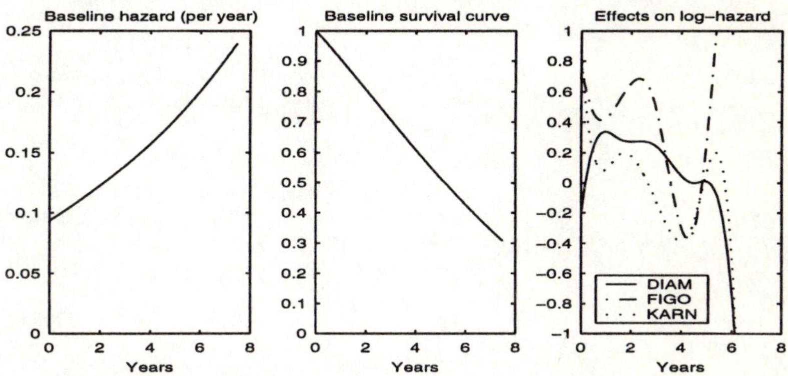

Fig. 5. Parametric estimates of baseline hazard (left), survival function (middle), and covariate effects for the non-parametric full rank model.

6 Discussion

When data sets are big or follow-up is long, the simple Cox regression model may not be adequate in describing survival data. The GLM approach to hazard regression as described turns out to be a useful alternative that offers much more flexibility in model building.

References

Aalen, O.O. (1994). Effects of frailty in survival analysis, *Statistical Methods in Medical Research*, **3**, pp. 227–243.

Cox, D.R. (1972). Regression Models and Life Tables (with Discussion). *Journal of the Royal Statistical Society Series B*, **39**, pp. 187–220.

Eilers, P.H.C. and Marx, B.D. (1996). Flexible Smoothing with Splines and Penalties (with Discussion). *Statistical Science*, **11**, pp. 89–121.

Eilers, P.H.C. (1998). Hazard Smoothing with B-splines. In: *Statistical Modeling. Proceedings of the 13th International Workshop on Statistical Modeling, New Orleans.*, Marx, B. D. and Friedl, H., (eds.), pp. 200–207.

Hastie, T. and Tibshirani, R. (1990). *Generalized Additive Models.* London: Chapman and Hall.

Kooperberg, C., Stone, C.J. and Truong Y.K. (1995). Hazard Regression *J. American Statistical Association*, **90**, pp. 78-94.

McCullagh, P. and Nelder, J.A. (1989). *Generalized Linear Models* (2nd ed.). London: Chapman and Hall.

van Eeuwijk, F.A. (1995). Multiplicative Interaction in Generalized Linear-Models *Biometrics,* **51**, pp. 1017–1032.

Verweij P.J.M. and van Houwelingen, H.C. (1995). Time-Dependent Effects of Fixed Covariates in Cox Regression *Biometrics* **51**, pp. 1550–1556.

A spatio-temporal analysis of a field trial

Arūnas Verbyla[1], Michelle Lorimer[1] and Robert Stevens[2]

[1] BiometricsSA, The University of Adelaide and South Australian Research and Development Institute, Urrbrae, SA, 5064, Australia

[2] Loxton Research Centre, South Australian Research and Development Institute, Loxton, SA, 5333, Australia

Abstract. A study involving the growth of Australian Eucalypts under irrigation regimes with varying levels of salinity and nutrients was conducted in Loxton South Australia. The field experiment was conducted over a period of six years. The aim of the study was to determine the impact of salinity on the growth of eucalypts and to provide recommendations on the commercial suitability of growing eucalypts using saline drainage water. The data has both spatial and temporal aspects which are examined in this paper. Spatial modelling follows current methods for field trials while the temporal modelling involves smoothing splines. A joint mixed model is developed which uses a mixed model representation of the smoothing spline.

Keywords. Best linear unbiased prediction; Mixed models; Residual maximum likelihood; Separable process; Smoothing spline; Spatio-temporal modelling; Variogram

1 Introduction

1.1 Woodlot experiment

In the Murray River region of Australia, irrigation is a common and necessary practice for the effective growth of horticultural crops. Excess application of irrigation either to leach salt from the root zone, or because of poor irrigation management, produces saline drainage water throughout the year. This water is unsuitable for reuse on crops because of high salinity and, among other things, the presence of boron at high concentrations. Australian Eucalypts can be tolerant of high salinity and high boron.

A study was conducted in Loxton, South Australia, to investigate the effect of irrigation water salinity and nutrition on the growth and water use of a *Eucalyptus camaldulensis* woodlot. Five month old samplings were planted in January 1990 on a regular grid. A split-plot design was used with 8 replicates of five main plots, each main plot having two sub-plots of 35 trees (at a spacing of 2m within rows and 3m between rows). The layout is given in Table 1. Five irrigation salinity treatments were used for the main plots, and are given in Table 2.

The sub-plot treatment (Table 2), the addition or otherwise of nutrients as superphosphate, with N and P respectively 200 and 100 kg/ha/yr, was included to examine the benefits of addition of nutrients in the presence of salinity.

Tree growth is the focus of the analysis presented here. The butt circumference at a height of 0.3m was measured on all trees at 9 (unequally spaced) measurement times, over a 5 year period. These were converted to an area (assuming a circular trunk) so that tree growth could be represented on a

Table 1. Field layout for Woodlot experiment

3+	2+	5−	4+	1+
3−	2−	5+	4−	1−
4+	5−	3+	2+	1−
4−	5+	3−	2−	1+
5−	1−	4−	3−	2+
5+	1+	4+	3+	2−
1−	4−	3−	5+	2−
1+	4+	3+	5−	2+
2−	4+	5+	3+	1−
2+	4−	5−	3−	1+
3+	1−	2−	4+	5−
3−	1+	2+	4−	5+
5+	1−	2+	4−	3−
5−	1+	2−	4+	3+
1−	3+	4−	5−	2−
1+	3−	4+	5+	2+

Table 2. Treatments in Woodlot trial

Salinity treatment			Nutrition treatment	
Code	Substance	Salinity (dS/m)	Code	Nutrition level
1	river water	0.5	+	+N+P
2	drainage water	2.5	−	Nil
3	ground drainage water	5.0 (to 10.0)		
4	ground drainage water	7.5 (to 14.0)		
5	ground drainage water	10.0 (to 18.0)		

basal area basis. The mean basal area in each plot is the response analysed. Late in 1994, the salinity levels for treatments 3 to 5 were raised to the levels indicated in Table 2. It was expected this change would have a marked effect on the growth of trees; measurements for treatments 3 to 5 in months 46 and 58 (the last two times) may therefore exhibit a change in trend and possibly variation.

Figure 1 is a trellis plot (Becker & Chambers, 1997) of the basal area over time for the 5x2 factorial experiment. Each line in Figure 1 represents a plot in the field. The data is typical of growth studies with the increase in the mean accompanied by an increase in the variability over time. This type of behaviour is often modelled using random intercepts and slopes (random coefficients) in models advocated by Laird & Ware (1982). There are plots which are consistently higher or lower than other plots in their trace over time and random coefficients models may capture this aspect. Spatial effects, such as fertility gradients, may be present at the plot level and should be accounted for in any analysis. However, the spatial features of the data, trends in rows and columns of the field, are hidden in the temporal representation of Figure 1, and their possible determination requires some initial model fitting.

Nonetheless, Figure 1 suggests that increasing the level of salinity reduces growth, but the impact of actual level is unclear. The reduction in growth appears to be dampened by the addition of nutrients.

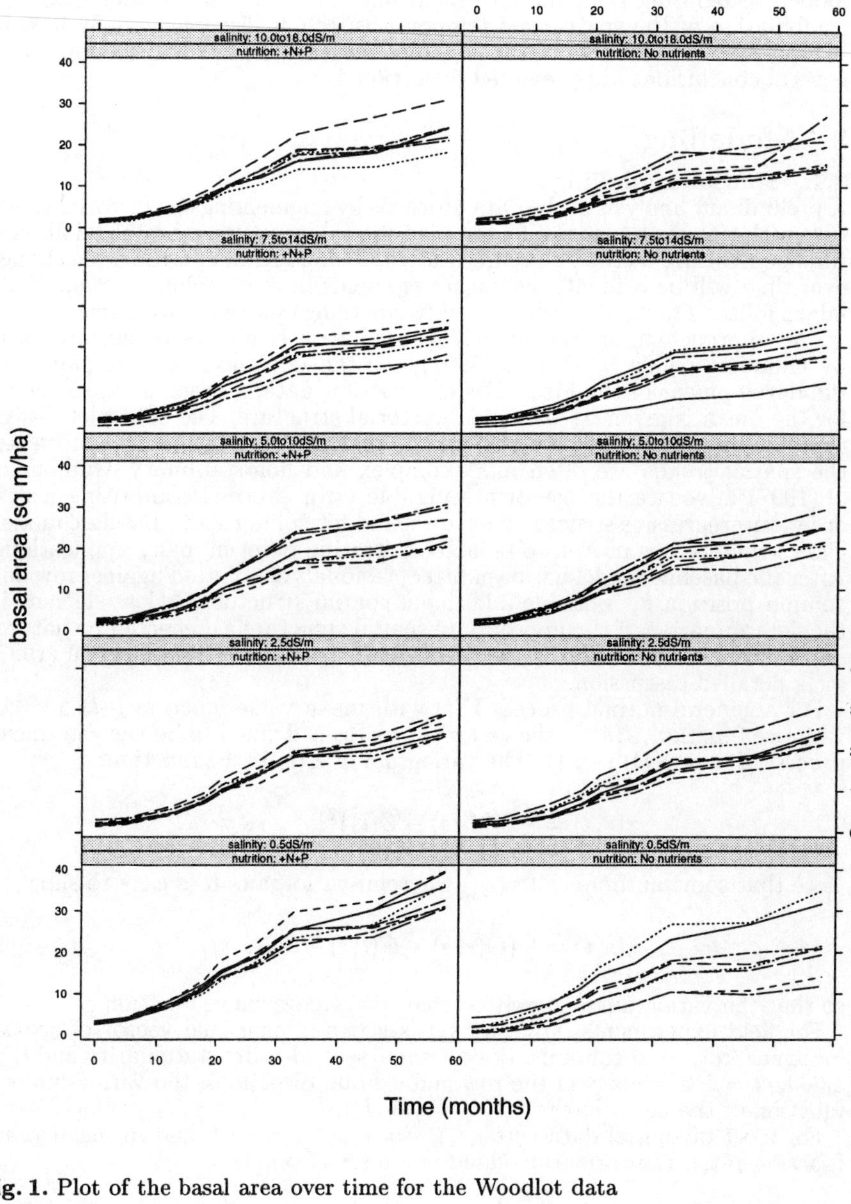

Fig. 1. Plot of the basal area over time for the Woodlot data

1.2 Outline of the paper

In this paper we consider an approach to modelling the Woodlot experiment. Section 2 presents an outline of the modelling approach. We begin by considering spatial models that are used for field trials in sub-section 2.1. These models recognise the regular row and column placement of the basic units and allow for various sources of variation. The temporal modelling approach is discussed in sub-section 2.2. Spline smoothers are used in an hierarchical model, recognising both the treatment and unit structure in the data. The combination of the spatial and temporal aspects is discussed briefly in sub-section 2.3. The analysis of the Woodlot data is presented in section 3 and general conclusions are presented in section 4.

2 Modelling

2.1 Spatial modelling

A preliminary analysis of the data proceeds by augmenting the factorial structure with time as taken as a factor. The spatial structure is examined at each time point using methods discussed briefly below. Although the scale change over time will be evident, the nature of trends in rows, columns or spatially might follow similar (but also possibly emerging) patterns over time.

The approach to spatial modelling used here is discussed most recently by Gilmour, Cullis & Verbyla (1997); see the references in that paper for the development of the ideas. For the woodlot data, a baseline linear model for the mean is given by the 5x2x9 factorial structure. The split-plot design could be used to provide a model for the spatial trends in the field. However, the spatial trends are often more complex and non-stationary. Gilmour et al (1997) advocate the use of a separable error process, comprising a first order autoregressive structures for rows and for columns as a baseline model. This structure has proven to be a good starting point in many applications. After the baseline model has been fitted, residuals are plotted against row and column position for possible additional spatial structure. A key element in the determination of the appropriate spatial structure as reflected by natural and extraneous variation is the (semi-)variogram; see Gilmour et al (1997) for a detailed discussion.

For a general normal process $Y(t)$ with mean value function $\mu(t)$ and covariance function $G(s,t)$, the residual process is defined to be the zero-mean process $Z(t) = Y(t) - \mu(t)$. The variogram of $Z(t)$ is the function

$$\gamma(s,t) = \frac{1}{2}\mathrm{E}[\{Z(s) - Z(t)\}^2], \qquad s \neq t.$$

Note that some authors call $\gamma(\cdot,\cdot)$ the semi-variogram. It is easy to show,

$$\gamma(s,t) = \frac{1}{2}\{G(s,s) + G(t,t) - 2G(s,t)\}.$$

so that the variogram is closely related to the covariance function.

For field experiments, $t = (t_1,\ t_2)$ is a two-dimensional vector of spatial locations, rows and columns. If we assume second-order stationarity and $l_i = s_i - t_i$, $i = 1, 2$ are lags in the row and column directions, the variogram is a function of the lags, that is $\gamma(s,t) = \gamma(l_1, l_2)$.

For a set of spatial data $(y(t_i), t_i)$, $i = 1, 2, \ldots, n$ with known mean value function $\mu(t_i)$, the variogram cloud is the set of points,

$$(t_i, t_j, v(t_i, t_j)), \quad i, j = 1, \ldots, n;\ i \neq j$$

where

$$v(t_i, t_j) = \frac{1}{2}[\{y(t_i) - \mu(t_i)\} - \{y(t_j) - \mu(t_j)\}]^2.$$

Under the normality assumption, these quantities are scaled chi-square variates on 1 degree of freedom.

Under second-order stationarity, $v(t_i, t_j) = v(l_{ij1}, l_{ij2})$ where $l_{ijk} = t_{ik} - t_{jk}$ for $k = 1, 2$, and we can average the elements of the cloud at common pairs of lags to provide a smoother representation of $\gamma(\cdot, \cdot)$. A plot of $\gamma(l_{ij1}, l_{ij2})$ against l_{ij1} and l_{ij2} is a standard diagnostic tool in field trial analysis. Estimates of $\mu(t_i)$ are required in practice and bias correction to allow for estimation of the mean process is discussed by Gilmour et al (1997) for the mixed model case.

For a separable AR(1) x AR(1) spatial process, the variogram increases smoothly from the origin, rising to a plateau. As Gilmour et al (1997) show, the AR(1) x AR(1) structure is inherited in the residuals and hence the sample variogram. Thus the sample variogram plotted against row and column displacement should exhibit a similar pattern to the theoretical variogram. Of course, the sample variogram summarises second-order properties of the process and will exhibit considerable sampling variation, so that some care is required in interpretation and fitting of model terms.

2.2 Temporal modelling

Verbyla, Cullis, Kenward & Welham (1999) present an approach to the analysis of designed experiments and longitudinal data. The approach involves using the cubic smoothing spline both as a vehicle for modelling nonlinear trends and as a generator of covariance structures, thereby (often) solving several difficult issues in a modelling context. Speed (1991) pointed out that the cubic smoothing spline is a Best Linear Unbiased Predictor (BLUP) and makes links with residual maximum likelihood (REML). Hence the cubic smoothing spline has a mixed model representation (see Verbyla, 1995), and this is the basic result used in Verbyla et al (1999) (see also Wang (1998a, 1998b), Brumback & Rice (1998) and Zhang, Lin, Raz and Sowers (1998)). This algebraic result allows nonlinear models to be included in standard mixed linear models, allowing ease of fitting and flexibility in modelling, and places the cubic smoothing spline in a parametric setting. Verbyla et al (1999) consider an analysis of variance representation involving spline terms, random coefficients and lack of fit random effects terms and discuss applications and extensions.

The fundamental result is not new, but is expressed in a manner that may be more familiar to practitioners. Indeed, the basic idea goes back to Kimeldorf & Wahba (1970) who develop a stochastic model with a Bayesian interpretation of the cubic smoothing spline. This is equivalent to the finite dimensional representation discussed by Silverman (1985) and so elegantly presented in Green & Silverman (1994).

2.3 Spatio-temporal models

Cullis, Gogel, Verbyla & Thompson (1998) extend spatial analysis to the multi-site situation. For the woodlot data, the measurement times could be taken as sites in what then would be a multi-site analysis. This combination of the spatial and temporal aspects then occurs through an additive structure with separable covariance structures. An analysis of variance type decomposition is utilised, with a decomposition into (losely speaking) strata of variation. The number of parameters, fixed effects plus variance/covariance components, is quite small. The analysis is however powerful and allows the assignment of sources of variation to inherent aspects of the experiment. The treatment effects can then be assessed more appropriately.

3 Analysis

3.1 Preliminary analysis

A preliminary factorial model for the mean and separable AR(1) x AR(1) spatial processes at each time point led to the sample variograms given in Figure 2. These variograms consist of means of points in the variogram cloud at the same lag (row and column displacement) that have been standardized by the estimated error variance at each time point. Table 3 gives the spatial parameter estimates as well as the estimated error variances.

Both the sample variograms and the estimates of the spatial parameters are reasonably consistent over time and the use of a single AR(1) x AR(1) spatial process seems sensible. The sample variograms do suggest additional structure may be present, and developing over time, and this may be part of row, column and plot effects that have not been included at this point. The analysis proceeds using a common AR(1) x AR(1) structure.

Table 3. Spatial parameter estimates

Variance	row AR(1)	column AR(1)
0.47	0.45	0.27
0.50	0.31	0.26
1.35	0.27	0.23
2.39	0.20	0.19
4.26	0.22	0.21
7.24	0.23	0.30
11.18	0.20	0.24
14.13	0.23	0.26
19.90	0.24	0.32

Residual plots at each time point were also examined but are not presented here. The patterns observed were reasonably consistent over time, again suggesting a single spatial process may be appropriate.

3.2 AOV decomposition

Table 4 gives an analysis of variance table for the maximal model fitted, although row and column random coefficients were also fitted but are not explicitly listed in the table. Interaction effects are given using the notation of Wilkinson & Rogers (1993). The term lin(time) is a slope parameter in time, spl(time) is the random component of the smoothing spline on time, ran(time) is a random effect taking time as a factor, and RC stands for random coefficients (random intercepts and slopes which may be correlated). AR stands for autoregressive process.

ASREML (Gilmour et al, 1996) was used to fit this and reduced models. ASREML implements residual or restricted maximum likelihood (REML, Patterson & Thompson, 1971; Harville, 1977) using an average information algorithm (Gilmour, Thompson & Cullis, 1995). A sequence of submodels was fitted and non-significant random terms removed. The usual asymptotic theory fails for likelihood ratio tests involving parameters on the boundary of their parameter space; see Stram & Lee (1994). In addition, if fixed effects are added or removed, the space of the residual likelihood changes. Wald tests are used to test the significance of fixed effects. See Verbyla et al (1999) for a discussion of these issues.

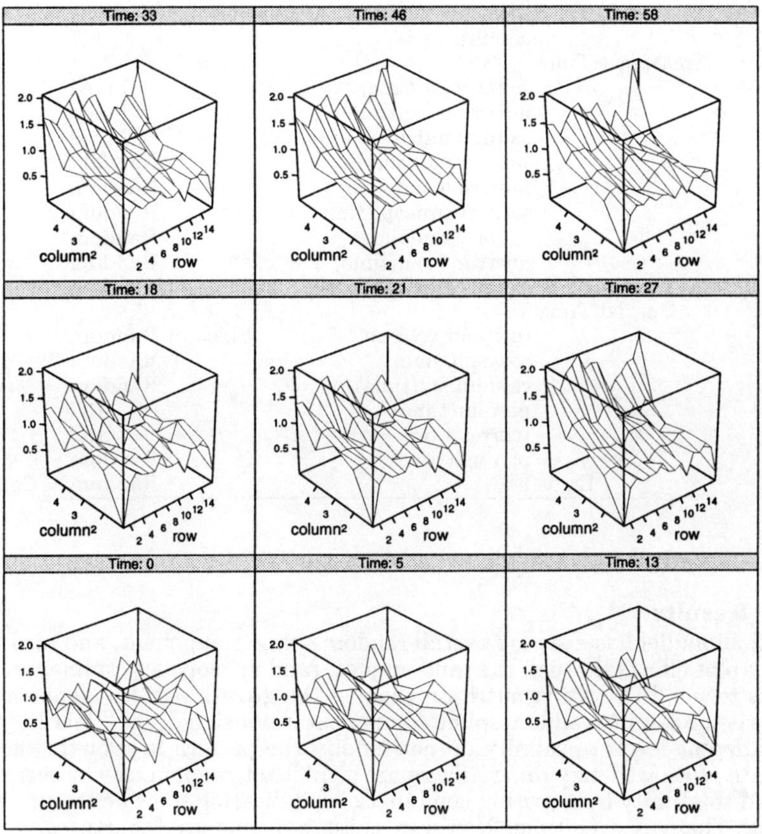

Fig. 2. Scaled sample variograms for the spatial pattern at each time point

Table 4. Woodlot data: AOV-type Decomposition

Term	Decomposition	Type
Constant	1	Fixed
Time		
	lin(time)	Fixed
	spl(time)	Random
	ran(time)	Random
Treatments		
	Salinity	Fixed
	Nutrition	Fixed
	Salinity.Nutrition	Fixed
Spatial		
	row and column	Fixed or Random
	plot	RC
	row	AR
	column	AR
Treatment.Time		
	salinity.lin(time)	Fixed
	nutrition.lin(time)	Fixed
	salinity.nutrition.lin(time)	Fixed
	salinity.spl(time)	Random
	nutrition.spl(time)	Random
	salinity.iron.spl(time)	Random
	salinity.ran(time)	Random
	nutrition.ran(time)	Random
	salinity.iron.ran(time)	Random
Spatial.Time		
	row and column	Fixed or Random
	row.spl(time)	Random
	column.spl(time)	Random
	plot.lin(time)	RC
	(correlation)	RC
	plot.spl(time)	Random
Error		Random

3.3 Results

The final model has a strong overall random spline component, and also very strong plot effects through the random plot, random slope and spline random effects terms. In addition nutrition changes the growth pattern significantly, whereas the salinity random spline component is not significant. This suggests the different levels of salinity do not modify the pattern in growth, though the rate of growth may differ. There are significant random effects over time related to salinity by nutrition combinations, indicating some departure from smooth changes over time. However, these non-smooth departures do not appear related to the change in salinity treatment levels. Table 5 gives the variance component estimates for the final model.

Table 6 gives the fixed effect slope estimates for combinations of salinity and nutrition. Wald tests of equality of slopes were carried out on the non-river water salinity treatments within the nutrition levels. There are no significant differences for the treatments with additional nutrients (approximate chi-square of 3.8 on 3 degrees of freedom). There is a significant difference for

Table 5. Variance parameter estimates for Woodlot data

spl(time)	4.63
nutrition.spl(time)	0.11
salinity.nutrition.ran(time)	0.025
row.spl(time)	0.057
row.ran(time)	0.010
column	0.18
unit.spl(time)	0.29
Variance	0.052
Row AR(1)	-0.024
Column AR(1)	0.38
random unit intercept	0.36
(correlation)	0.36
random unit.lin(time)	0.038

treatments without additional nutrients (chi-square of 11.14 on 3 degrees of freedom). This latter effect is mainly due to the two highest salinity levels which is interesting remembering the change in the level of salinity of these treatments during the experiment.

A Wald test of the hypothesis of no difference between the river water with no nutrients and the non-river water with nutrient treatments was not significant (chi-square of 5.49 on 4 degrees of freedom). This has important implications discussed below.

Figure 3 gives the smooth growth patterns over time for the 10 treatment combinations. The similarity of growth for all but two treatment combinations, river water with nutrients and the highest salinity with no additional nutrients is apparent.

Table 6. Slope estimates for salinity by nutrition combinations (SE = 0.024)

Salinity	Nutrition	Estimate
0.5	+N+P	0.58
0.5	Nil	0.33
2.5	+N+P	0.44
2.5	Nil	0.40
5.0 to 10	+N+P	0.43
5.0 to 10	Nil	0.40
7.5 to 14	+N+P	0.39
7.5 to 14	Nil	0.33
10.0 to 18	+N+P	0.42
10.0 to 18	Nil	0.30

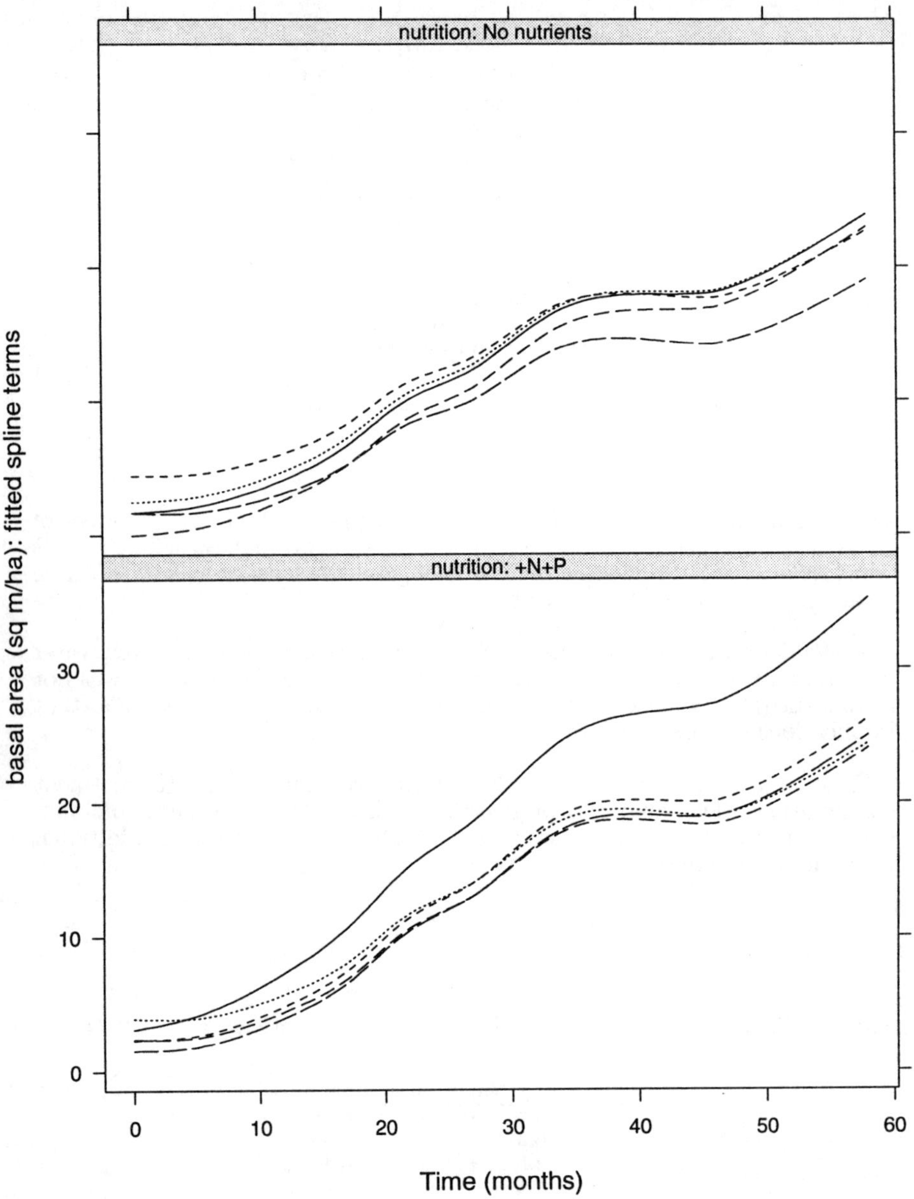

Fig. 3. Fitted trends for Woodlot omitting unit and other non-smooth effects

4 Conclusions

The results of this analysis provide important information for the scientist. Firstly, nutrition alters the growth rate and curvature. Secondly, over the range of salinity levels examined, nutrition helps to keep the growth much the same as river water without nutrients, although rather less than in river water with nutrients. Without nutrients, there is a high dose at which further reduction in growth is apparent. These results suggest that drainage water of high salinity can be used with Eucalypts, and although a subsequent decline in growth will result, the addition of nutrients can alleviate the decline to some extent.

References

Becker, R. and Chambers, J. (1996). *S-PLUS Trellis Graphics User's Manual.* MathSoft Inc. and Bell Labs.

Brumback, and Rice, J.A. (1998). Smoothing spline models for the analysis of nested and crossed samples of curves. *Journal of the American Statistical Association* **93**, pp. 961-994.

Cullis, B.R, Gogel, B.J., Verbyla, A.P. and Thompson, R. (1998). Spatial analysis of multi-environment early generation trials. *Biometrics* **54**, pp. 1-18.

Green, P.J. and Silverman, B. (1994). *Nonparametric regression and generalized linear models.* Chapman and Hall, London.

Gilmour, A.R., Cullis, B.R. and Verbyla, A.P. (1997). Accounting for natural and extraneous variation in the analysis of field experiments. *Journal of Agricultural, Biological and Environmental Statistics* **2**, pp. 269-293.

Gilmour, A.R., Cullis, B.R., Welham, S.J. and Thompson R. (1996). *AS-REML, an average information REML program.* Biometric Bulletin, NSW Agriculture.

Gilmour, A.R., Thompson, R. and Cullis, B.R. (1995). Average Information, an efficient algorithm for REML estimation in linear mixed models. *Biometrics* **51**, pp. 1440-1450.

Harville, D.A. (1977). Maximum likelihood estimation of variance components and related problems. *Journal of the American Statistical Association* **72**, pp. 320-340.

Kimeldorf, G. and Wahba, G. (1970). A correspondence between Bayesian estimation of stochastic processes and smoothing by splines. *Annals of Mathematical Statistics* **41**, pp. 195-502.

Laird, N.M. and Ware, J.H. (1982). Random-effects models for longitudinal data. *Biometrics* **38**, pp. 963-974.

Patterson, H.D. and Thompson, R. (1971). Recovery of inter-block information when block sizes are unequal. *Biometrika* **58**, pp. 545-554.

Silverman, B. (1985). Some aspects of the spline smoothing approach to non-parametric regression curve fitting (with discussion). *Journal of the Royal Statistical Society, Series* B **47**, pp. 1-52.

Speed, T.P. (1991). Comment on: That BLUP is a good thing: The estimation of random effects, by G. K. Robinson, *Statistical Science* **6**, pp. 44.

Stram, D.O. and Lee, J.W. (1994). Variance component testing in longitudinal mixed effects models. *Biometrics* **50**, 1171-1177.

Verbyla, A.P. (1995). A mixed model formulation of smoothing splines and testing linearity in generalized linear models. *Research Report* **95/5**, Department of Statistics, The University of Adelaide.

Verbyla, A.P., Cullis, B.R., Kenward, M.G. and Welham, S.J. (1999). The analysis of designed experiments and longitudinal data by using smoothing splines (with discussion). *Applied Statistics* **48**, pp. 269-311.

Wang, Y. (1998a). Mixed effects smoothing spline analysis of variance. *Journal of the Royal Statistical Society, Series* B **60**, pp. 159-174.

Wang, Y. (1998b). Smoothing spline models with correlated errors. *Journal of the American Statistical Association* **93**, pp. 341-348.

Wilkinson, G.N. and Rogers, C.E. (1973). Symbolic description of factorial models for analysis of variance. *Applied Statistics* **22**, pp. 392-399.

Zhang, D., Lin, X., Raz, J. and Sowers, M. (1998). Semi-parametric stochastic mixed models for longitudinal data. *Journal of the American Statistical Association* **93**, pp. 710-719.

Contributed papers

Principal component logistic regression

A.M. Aguilera[1] and M. Escabias[2]

[1] Department of Statistics and O.R., University of Granada, 18071-Granada, Spain
[2] Department of Statistics and O.R., University of Jaén, 23071-Jaén, Spain

Abstract. The objective of this paper is to develop an extension of principal component regression for multiple logistic regression with continuous covariates. A practical application with simulated data will be included where the accuracy of the proposed principal component logistic regression model will be evaluated starting from the estimated parameters and probabilities.

Keywords. Linear regression, logistic regression, principal components

1 Introduction

The purpose of this paper is to develop a generalization of the principal component regression (PCR) model for the case of a binary response variable.

This type of response variables frequently appears in fields like medicine, epidemiology, and many others. The logistic regression model, that has been applied and studied by many authors in the last years, solves the problem of explaining a binary random variable (with only two possible outcomes) in terms of a set of non random variables. For example, the problem of predicting if a person is likely to develop heart disease from a set of risk factors.

PCR performs least squares linear estimation of a quantitative response variable on the principal components (p.c.'s) associated with a set of predictor variables. There are two principal reasons for using principal component analysis (PCA) of the predictor variables. First, we may have a large number of highly correlated explanatory variables (multicollinearity). In this case, the p.c.'s, that are uncorrelated, may be used in place of the original variables in order to obtain an accurate estimation of the regression coefficients. We should remember that multicollinearity is a problem if we are using a regression equation for description, estimation, or control, but it may or may not be a problem if we are using it for prediction. Second, the p.c.'s may be used to reduce the dimension of the problem and to regress the response variable in terms of a small set of p.c.'s of interest.

2 Basic theory on principal component regression

In order to formulate the multiple regression problem let us consider a vector $Y = (Y_1, ..., Y_n)^T$ of responses on a dependent random variable Y. Let X be a $n \times k$ matrix of predictor variables. That is, x_{ij} is the *ith* observation on the *jth* independent variable X_j and x_i denotes the *ith* row of such a matrix. We will also consider without loss of generality that the independent variables are centered so that $\sum_i x_{ij} = 0$.

The multiple linear regression model is then given by

$$Y = \beta_0 \mathbf{1} + X\beta + \varepsilon, \tag{1}$$

where $\mathbf{1}$ is an $n \times 1$ vector of ones, β_0 is an unknown constant, β is a $k \times 1$ vector of unknown constants and ε is an $n \times 1$ vector of uncorrelated errors with expectations zero and common variance, σ^2.

The least squares linear estimation of this model gives the following results:

$$\widehat{\beta}_0 = \overline{Y}; \quad \widehat{\beta} = \left(X^T X\right)^{-1} X^T Y; \quad SSE = \left(Y - \widehat{Y}\right)^T \left(Y - \widehat{Y}\right),$$

where $\widehat{\beta}_0$ and $\widehat{\beta}$ are the estimated parameters, and SSE is the residual sum of squares.

In order to obtain PCR, let us define $W = XV$ as an $n \times k$ matrix whose columns are the p.c.'s of X, with V being a $k \times k$ matrix whose columns are the unit length eigenvectors of the covariance matrix of X given by $\Sigma_X = (X^T X)/(n-1)$. Let us denote by $\lambda_1 \geq \lambda_2 \geq \cdots \geq \lambda_k$ the decreasing sequence of eigenvalues of the positive definite Σ_X matrix and by $v_1, v_2, \ldots, v_k$ the corresponding eigenvectors. It can be easily proved that $V^T V = V V^T = I$ (a $k \times k$ identity matrix) and $W^T W/(n-1) = \Lambda$ (the $k \times k$ diagonal matrix of eigenvalues). Let us remember that the p.c.'s so defined are uncorrelated variables with means zero and variances equal to the associated eigenvalues.

The model (1) can be written in terms of the p.c.'s as

$$Y = \beta_0 \mathbf{1} + W\gamma + \varepsilon,$$

where $\gamma = V^T \beta$. The least squares estimator of γ is $\widehat{\gamma} = ((n-1)\Lambda)^{-1} W^T Y = V^T \widehat{\beta}$, and the prediction equation is $\widehat{Y}(x) = \overline{Y} + \sum_{j=1}^{k} w_j(x)\widehat{\gamma}_j$, where $w_j(x) = x^T v_j$ is the jth p.c. value for a point x at which prediction is desired.

The PCR model deletes the r p.c.'s associated with the r smallest eigenvalues (Mansfield et al., 1977) giving the following predictor:

$$\widehat{Y}^s(x) = \overline{Y} + \sum_{j=1}^{s} w_j(x)\widehat{\gamma}_j \quad (s + r = k), \tag{2}$$

that is equivalent to the least squares linear estimator of Y in terms of the first s p.c.'s. Let us observe that the multiple linear regression model for Y can be reduced by taking as covariates the first s p.c.'s as follows

$$Y = \beta_0 \mathbf{1} + W_s \gamma^s + \varepsilon^s,$$

where $W_s = XV_s$, V_s is a $k \times s$ matrix whose columns are the eigenvectors associated with the s largest eigenvalues, and γ^s is the vector of parameters associated with the first s p.c.'s contained in the W_s matrix . Then the parameters $\widehat{\gamma}_j$ used in PCR model (2) can be obtained as the components of the least squares estimator of γ^s given by $\widehat{\gamma}^s = ((n-1)\Lambda_s)^{-1} W_s^T Y$, with Λ_s being the diagonal matrix of the s largest eigenvalues.

On the other hand, the residual sum of squares for $\widehat{Y}^s$ can be expressed as

$$SSE^s = (Y - \bar{Y})^T (Y - \bar{Y}) - \widehat{\gamma}^{s^T} ((n-1)\Lambda_s) \widehat{\gamma}^s = SSE + Y^T W_r ((n-1)\Lambda_r)^{-1} W_r^T Y,$$

where Λ_r is the diagonal matrix of the r smallest eigenvalues and W_r has as columns the corresponding r p.c.'s.

Finally, it is known that p.c.'s with the largest variances are not necessarily the best predictors because p.c.'s with the smallest variances could be highly correlated with the response variable. Therefore, the p.c.'s are entered in the PCR model in order of magnitude of the squares of their correlations with the response variable (Jackson, 1991).

3 Logistic regression model

Logistic regression is used when the response variable Y is binary. The multiple logistic regression model is given by

$$Y = \pi(X) + \varepsilon, \tag{3}$$

where $\pi(X)$ is an $n \times 1$ vector whose components are defined as

$$\pi_i = E[Y/X = x_i] = P[Y_i = 1] = \exp\left(\beta_0 + x_i\beta\right) / \left(1 + \exp\left(\beta_0 + x_i\beta\right)\right), \tag{4}$$

and X and β defined as in previous section. Now the errors are assumed to be independent with Bernouilli probability distribution so that the variances are not equal. The most common method used for estimating the parameters in model (3) is maximum likelihood (Hosmer and Lemeshow, 1989).

The logistic model (3) can be equivalently written in linear form as

$$L(X) = \beta_0\mathbf{1} + X\beta, \tag{5}$$

where $L(X)$ is an $n\times 1$ vector whose components are the logit transformations $L_i = \ln\left[\pi_i/(1-\pi_i)\right]$. That is, the logistic model is a generalized linear model with the logit transformation as link function (McCullagh and Nelder, 1983).

The maximum likelihood estimates verify the following matrix equation that is generally solved by the Newton-Raphson method (Ryan, 1997):

$$\mathbf{1}^T\left(Y - \widehat{\pi}\right) = 0, \quad X^T\left(Y - \widehat{\pi}\right) = 0,$$

where $\widehat{\pi}$ is the vector of predicted probabilities with components $\widehat{\pi}_i$ obtained by replacing in expression (4) the parameters by their likelihood estimators.

4 Principal component logistic regression

As in linear regression, the logit transformation (5) can be written in terms of the p.c.'s of the predictor variables in the form

$$L(X) = \beta_0\mathbf{1} + W\gamma, \tag{6}$$

with $\gamma = V^T\beta$. As a consequence of the invariance property of maximum likelihood estimates we obtain $\widehat{\gamma} = V^T\widehat{\beta}$, and the prediction equation

$$\widehat{Y}(x) = \widehat{\pi}(x), \tag{7}$$

with $\widehat{\pi}(x) = \exp\left(\widehat{\beta}_0 + \sum_{j=1}^k w_j(x)\widehat{\gamma}_j\right)/(1 + \exp\left(\widehat{\beta}_0 + \sum_{j=1}^k w_j(x)\widehat{\gamma}_j\right))$.

By analogy with the PCR model, we are now going to eliminate those p.c.'s with the smallest variances. Taking into account that the logit transformation (6) can be decomposed as $L(X) = \beta_0\mathbf{1} + W_s\gamma^s + W_r\gamma^r$, the principal component logistic regression (PCLR) model in terms of the first s p.c.'s is

$$Y = \pi^s(X) + \varepsilon^s,$$

where the corresponding logit transformation, $L^s(X)$, with components $L_i^s = \ln[\pi_i^s/(1-\pi_i^s)]$, is defined as $L^s(X) = \beta_0\mathbf{1} + W_s\gamma^s$.

The maximum likelihood parameter estimates of this PCLR model obey the following matrix equation:

$$\mathbf{1}^T(Y - \widehat{\pi}^s) = 0, \quad W_s^T\left(Y - \widehat{\pi}^s\right) = 0,$$

where the associated logit transformation is estimated as $\widehat{L}^s = \widehat{\beta}_0^s \mathbf{1} + W_s \widehat{\gamma}^s$.

Then the vector, β, of parameters in terms of the original variables can be reconstructed in the form $\widehat{\beta}^s = V_s \widehat{\gamma}^s$.

The main difference between PCR and PCLR is that in the latter, the estimator $\widehat{\gamma}^s$ with the first s p.c.'s is not the vector of the first s components of the estimator $\widehat{\gamma}$ with all the p.c.'s. Therefore the predicted probabilities $\widehat{\pi}^s(x)$ are different to the ones obtained by removing in the maximum likelihood predictor given by expression (7) the last r p.c.'s. As a consequence the PCLR model has to be readjusted every time you remove or enter a p.c.

Let us finally observed that, as in PCR, the p.c.'s will be entered in the PCLR model in order of their ability to model the response variable rather than in order of their explained variances.

5 Simulated example

In order to illustrate the proposed PCLR model we have simulated five hundred probabilities of a logistic model with ten covariates whose parameters, which have been randomly simulated from an uniform distribution on the interval $[-3, 10]$, are given by $\beta_0 = -1.18$; $\beta = (2.61, 4.69, 0.78, -0.98, 6.09, 1.5, 2.79, -2.32, -1.66, -1.16)^T$. The design matrix X corresponds to five hundred values randomly simulated from a multivariate normal distribution with highly correlated components. In fact X has been obtained as the linear transformation $X = X^\star A$ with A being a 10×10 matrix with one hundred values of an uniform distribution on $[0, 1]$, and $X^\star$ being a 500×10 matrix of independent values of a standard normal distribution. After that, the simulated probabilities, π_i, have been used for generating the values Y_i of a Bernouilli distribution of parameter π_i, $i = 1, \ldots, 500$.

The first step is to obtain the PCA of the X matrix. The percentages of variance explained by the p.c.'s are shown in Table 1. As we can see the first three p.c.'s explain more than a 92% of the total variability, and with seven of them more than 99% is achieved.

Table 1. Variance explained by the p.c.'s

p.c.	Variance	% of variance	Accumulated %
1	33.773	83.949	83.949
2	2.255	5.606	89.555
3	1.384	3.352	92.907
4	1.046	2.599	95.506
5	0.990	2.460	97.967
6	0.395	0.982	98.949
7	0.279	0.695	99.643
8	0.099	0.247	99.890
9	0.040	0.100	99.990
10	0.004	0.010	100.00

The second step is to adjust the PCLR models by entering the p.c.'s in order of their explained variances (method I), and to study the evolution of the predicted probabilities and the reconstructed parameters for the original variables. All computations have been performed by using the statistical package SPSS 9.0.

The reconstructed parameters appear in Table 2. Let us observe that the parameters estimated from a PCLR model with ten p.c.'s (the same as that with all the original covariates) are very different to the real ones, with some of them being outside the definition interval $[-3, 10]$. This fact can be caused by the multicollinearity of the data (very high pairwise correlations between the predictor variables). It can be seen that the parameters estimated with PCLR are more accurate than the ones estimated with the original x-variables.

Table 2. Parameters reconstructed by the PCLR models (method I)

s	1	2	3	4	5	6	7	8	9	10
$\widehat{\beta}_0^s$	-0.153	-0.196	-0.170	-0.163	-0.161	-0.168	-0.427	-0.544	-0.600	-0.953
$\widehat{\beta}_1^s$	0.054	0.039	0.019	-0.037	-0.019	-0.062	2.460	5.440	7.204	-1.794
$\widehat{\beta}_2^s$	0.052	0.110	0.174	0.108	0.104	0.363	4.046	3.719	1.869	-2.377
$\widehat{\beta}_3^s$	0.040	0.014	0.209	-0.041	-0.058	0.081	0.624	0.309	1.755	-7.302
$\widehat{\beta}_4^s$	0.053	0.003	0.089	0.109	0.076	0.003	-0.838	0.051	-3.045	1.659
$\widehat{\beta}_5^s$	0.043	-0.053	-0.029	-0.037	-0.015	0.147	1.216	0.614	1.513	17.339
$\widehat{\beta}_6^s$	0.050	0.111	0.078	0.011	0.003	0.221	-1.833	-1.255	1.181	15.903
$\widehat{\beta}_7^s$	0.038	0.076	0.022	0.140	0.170	0.142	-2.795	-3.207	-7.287	-9.970
$\widehat{\beta}_8^s$	0.042	0.089	0.159	0.290	0.290	-0.013	2.912	3.151	5.705	20.425
$\widehat{\beta}_9^s$	0.050	0.040	0.257	0.282	0.283	0.254	-2.279	-3.755	-2.402	-19.758
$\widehat{\beta}_{10}^s$	0.047	0.039	-0.145	-0.320	-0.323	-0.633	-2.622	-4.545	-5.659	-9.897

Moreover the parameters estimated with the first seven p.c.'s are the best and some of them are really similar to the original ones.

In order to compare the probabilities predicted by the PCLR models we have computed the residual sum of squares

$$SSE^s = \sum_{i=1}^{500} (\pi_i - \widehat{\pi}_i^s)^2, \quad s = 1, \ldots, 10.$$

The results are shown in Table 3 next to the correct classification rate (CCR^s) that will be used as a global measure of the goodness of fit of the model. CCR^s is defined as the percentage of subjects in the data set that are classified correctly by taking as cut-point $p = 0.5$. It can be observed that the SSE^s is drastically reduced when the seventh p.c. is introduced in the PCLR model.

Table 3. SSE and CCR for PCLR models

Method I			Method II		
s	SSE^s	CCR^s	s	SSE^s	CCR^s
1	87.23728	65.8	1	59.45250	76.2
2	86.31259	66.4	2	38.99664	83.4
3	81.37741	67.4	3	30.67354	86.2
4	79.24568	67.6	4	24.20960	88.2
5	79.01974	68.0	5	20.82932	88.8
6	75.17057	69.8	6	15.92241	91.0
7	19.02657	88.8	7	11.22945	92.8
8	13.05216	91.8	8	4.42280	94.0
9	7.17477	93.2	9	1.80470	94.6
10	1.71766	94.6			

The third step has been to choose the p.c.'s to be included in the PCLR model in the order given by forward stepwise selection based on conditional likelihood ratio chi-square tests (method II). At each step, the hypothesis is tested that the term entered or removed at that step significantly changes the prediction. The p.c.'s of the models selected at each step have been: step 1 (w_7), step 2 (w_7, w_1), step 3 (w_7, w_1, w_3), step 4 (w_7, w_1, w_3, w_6), step 5 (w_7, w_1, w_3, w_6, w_8), step 6 ($w_7, w_1, w_3, w_6, w_8, w_9$), step 7 ($w_7, w_1, w_3, w_6, w_8, w_9, w_{10}$), step 8 ($w_7, w_1, w_3, w_6, w_8, w_9, w_{10}, w_4$), step 9 ($w_7, w_1, w_3, w_6, w_8, w_9, w_{10}, w_4, w_2$), with w_j being the *jth* p.c. The parameters reconstructed by the PCLR model selected at each step are shown in Table 4. The SSE^s and CCR^s associated to these PCLR models appear in Table 3 (method II).

Let us observe that the SSE for the model PCLR with only the seventh p.c. (method II step 1) is significantly smaller than the SSE for the PCLR model with the first six p.c.'s (method I). This fact may explain the big reduction obtained in SSE when w_7 is entered in the corresponding model obtained with method I. In any case the SSE^s given by the PCLR models selected with method II is always smaller than with method I. Moreover, the parameters and probabilities given by the PCLR model with four p.c.'s selected by method II are very accurate and similar to the ones given by the PCLR model with the first seven p.c.'s given by method I.

From this example we can conclude that there are two important reasons for using PCLR. On one hand, a more accurate estimation of the original parameters in the case of multicollinearity has been obtained by removing p.c.'s. This is very useful because parameters have an important interpretation in terms of odds ratios. On the other hand an important dimension reduction is obtained by selecting the p.c.'s in terms of the improvement in the predicted probabilities (bigger with method II).

Table 4. Parameters reconstructed by the PCLR models (method II)

step	1	2	3	4	5	6	7	8	9	Real
$\widehat{\beta}_0^s$	-0.187	-0.329	-0.337	-0.444	-0.513	-0.542	-0.623	-0.800	-0.953	-1.18
$\widehat{\beta}_1^s$	-0.116	1.936	2.180	2.400	4.069	5.671	-1.051	-1.031	-1.824	2.61
$\widehat{\beta}_2^s$	-0.111	2.249	2.701	3.723	3.263	1.715	-1.597	-2.259	-2.390	4.69
$\widehat{\beta}_3^s$	-0.084	0.393	-0.001	0.330	0.028	0.933	-5.223	-5.616	-7.308	0.78
$\widehat{\beta}_4^s$	-0.114	-0.500	-0.399	-0.632	0.140	-2.118	1.261	1.530	1.685	-0.98
$\widehat{\beta}_5^s$	-0.092	0.674	0.831	1.351	0.838	1.470	11.648	14.441	17.356	6.09
$\widehat{\beta}_6^s$	-0.107	-1.602	-1.918	-1.687	-1.135	0.753	10.296	12.495	15.952	1.50
$\widehat{\beta}_7^s$	-0.081	-2.087	-2.514	-2.973	-3.233	-6.049	-6.921	-8.476	-10.015	2.79
$\widehat{\beta}_8^s$	-0.090	2.172	2.627	2.276	2.251	3.822	12.640	16.595	20.463	-2.32
$\widehat{\beta}_9^s$	-0.107	-1.946	-1.801	-2.145	-3.196	-2.029	-13.010	-16.130	-19.782	-1.66
$\widehat{\beta}_{10}^s$	-0.100	-0.626	-1.094	-2.005	-3.282	-3.748	-5.553	-8.080	-9.906	-1.16

References

Jackson, J.E. (1991). *A User's Guide to Principal Components*, New York: Wiley.

Ryan, T. P. (1997). *Modern Regression Methods.* New York: Wiley.

Mansfield, E. R., Webster, J. T. and Gunst, R. F. (1977). An analytic variable selection technique for principal component regression. *Applied Statistics*, **26**,34-40.

McCullagh, P. and Nelder, J.A. (1983). *Generalized Linear Models.* London: Chapman Hall.

Hosmer, D. W. and Lemeshow, S. (1989). *Applied Logistic Regression.* New York: Wiley.

Sieve bootstrap prediction intervals

Andrés M. Alonso, Daniel Peña and Juan Romo

Departmento de Estadística y Econometría, Universidad Carlos III de Madrid, 28903 Getafe, Madrid, Spain

Keywords. Sieve bootstrap, prediction intervals, time series, linear processes

1 Introduction

When studying a time series, one of the main goals is the estimation of forecast confidence intervals based on an observed trajectory of the process. The traditional approach of finding prediction intervals for a linear time series assumes that the distribution of the error process is known. Thus, these prediction intervals could be adversely affected by departures from the true underlying distribution.

Some bootstrap approaches have been proposed as a distribution free alternative to compute prediction intervals. Stine (1987) proposes a bootstrap method to estimate the prediction mean squared error of the estimated linear predictor of an AR(p) where p is known. Also, for an AR(p) process with known p, and relaxing the assumptions of Stine (1987), Thombs and Schucany (1990) propose a first backward and then forward bootstrap method to estimate prediction intervals. Cao *et al.* (1997) study a conditional bootstrap method alternative to Thombs and Schucany's proposal, which is computationally much faster. Pascual *et al.* (1998) generalize this conditional bootstrap to ARMA(p,q) processes with known p and q and they also include the parameter estimation variability.

This paper considers bootstrap methods to construct nonparametric prediction intervals for a class of linear processes that can be written as a one-sided infinite-order moving average process with at most a polynomial decay of the coefficients $\{\psi_j\}_{j=0}^{+\infty}$. This class includes the stationary and invertible ARMA(p,q) processes. Our approach uses the sieve bootstrap of Bühlmann (1997) based on residual resampling from a sequence of approximating autoregressions for $\{X_t\}_{t\in Z}$ with order $p = p(n)$ that increases as a function of the sample size n. This sieve bootstrap has a nice nonparametric property, being model-free within the considered class of linear processes. Thus, the proposed bootstrap prediction intervals could be applied to this more general class of linear models without specifying a finite dimensional model as in previous bootstrap proposals.

The paper is organized as follows. Section 2 introduces the sieve bootstrap for estimating forecast intervals. Section 3 establishes that the sieve bootstrap provides consistent estimators of the conditional distribution of future values X_{T+h}. Section 4 presents a Monte Carlo study comparing the finite sample properties of the sieve bootstrap with those of alternative methods. It shows that the average coverage is better when intervals are constructed incorporating the parameter uncertainty. Some real data examples could be found in Alonso *et al.* (2000).

2 Sieve bootstrap forecast intervals

Let $\{X_t\}_{t\in Z}$ be a real valued stationary process with expectation $E[X_t] = \mu_X$, that admits a MA(∞) representation. Under the additional assumption of invertibility we can represent $\{X_t\}_{t\in Z}$ as a one-sided infinite-order autoregressive process. This AR(∞) representation motivates Bühlmann's sieve bootstrap. The method proceeds as follows:

1. Given a sample $\{X_1, ..., X_n\}$, select the order $p = p(n)$ of the autoregressive approximation by AICC criterion (cf. Section 9.3 of Brockwell and Davis (1991)).
2. Construct some estimators of the autoregressive coefficients: $(\widehat{\phi}_1, \widehat{\phi}_2, \ldots, \widehat{\phi}_p)^t$. Following Bühlmann (1997) we take the Yule-Walker estimates.
3. Compute the residuals: $\widehat{\varepsilon}_t = \sum_{j=0}^{p} \widehat{\phi}_j (X_{t-j} - \bar{X})$, where $\widehat{\phi}_0 = 1$, and $t \in (p+1, \ldots, n)$.
4. Compute the empirical distribution function of the centred residuals: $\widehat{F}_{\tilde{\varepsilon}}(x) = (n-p)^{-1} \sum_{t=p+1}^{n} 1_{\{\tilde{\varepsilon}_t \le x\}}$, where $\tilde{\varepsilon}_t = \widehat{\varepsilon}_t - \widehat{\varepsilon}^{(\cdot)}$ and $\widehat{\varepsilon}^{(\cdot)} = (n-p)^{-1} \sum_{t=p+1}^{n} \widehat{\varepsilon}_t$.
5. Draw a resample ε_t^* of i.i.d. observations from $\widehat{F}_{\tilde{\varepsilon}}$.
6. Define X_t^* by the recursion:

$$\sum_{j=0}^{p} \widehat{\phi}_j (X_{t-j}^* - \bar{X}) = \varepsilon_t^*, \tag{1}$$

   where the starting p observations are equals to $\bar{X}$.

Up to this step, the resampling plan coincides with the sieve bootstrap, and is valid for bootstrapping some statistics defined as a functional of a m-dimensional distribution function. However, it is not effective for bootstrap prediction, because it does not replicate the conditional distribution of X_{T+h} given the observed data. But, if we proceed as do Cao *et al.* (1997) fixing the last p observations we can obtain resamples of the future values X_{T+h}^* given $X_{T-p+1}^* = X_{T-p+1}, \ldots, X_T^* = X_T$.

7. Compute the estimation of $(\widehat{\phi}_1^*, \ldots, \widehat{\phi}_p^*)^t$, as in step 2.
8. Compute future bootstrap observations by the recursion:

$$X_{T+h}^* - \bar{X} = -\sum_{j=1}^{p} \widehat{\phi}_j^* (X_{T+h-j}^* - \bar{X}) + \varepsilon_t^*, \tag{2}$$

   where $h > 0$, and $X_t^* = X_t$, for $t \le T$.

Finally, $F_{X_{T+h}^*}^*$ the bootstrap distribution of X_{T+h}^* is used to approximate the unknown distribution of X_{T+h} given the observed sample. As usual, a Monte Carlo estimate $\widehat{F}_{X_{T+h}^*}^*$ is obtained by repeating the steps 5 to 8 B times. The $(1-\alpha)$% prediction interval for X_{T+h} is given by

$$[Q^*(\alpha/2), Q^*(1-\alpha/2)], \tag{3}$$

where $Q^*(\cdot)$ are the quantiles of the estimated bootstrap distribution. Notice that, if we omit step 7 and use the $\widehat{\phi}_j$ in recursion (2), our resampling plan is similar though more general than the conditional bootstrap of Cao *et al.* (1997).

3 Asymptotic results

The asymptotic validity of the proposed intervals (3) depends on the limiting behavior of the distribution $F^*_{X^*_{T+h}}$, and it is sufficient to establish convergence in the conditional distribution of the bootstrap version X^*_{T+h} to X_{T+h}. We now consider the precise assumptions about the stationary process $\{X_t\}_{t\in Z}$ required to prove our results:

Assumption A1: $X_t - \mu_X = \sum_{j=0}^{+\infty} \psi_j \varepsilon_{t-j}$, $\psi_0 = 1$ $(t \in Z)$ with $\{\varepsilon_t\}_{t\in Z}$ stationary, ergodic and $E[\varepsilon_t|F_{t-1}] \equiv 0$, $E[\varepsilon_t^2|F_{t-1}] \equiv \sigma^2 < \infty$, $E[|\varepsilon_t|^s] < \infty$ for some $s \geq 4$, and F_{t-1} is the σ-field generated by $\{\varepsilon_s\}_{s=-\infty}^{t-1}$.

Assumption A2: $\Psi(z)$ is bounded away from zero for $|z| \leq 1$, and $\sum_{j=0}^{+\infty} j^r |\psi_j| < \infty$ for some $r \in N$.

Additionally, we impose the following assumption about the autoregressive approximation:

Assumption B: $p = p(n) \to \infty$, $p(n) = o(n)$ $(n \to \infty)$, and the $\widehat{\phi}_p = (\widehat{\phi}_{1,n}, \ldots \widehat{\phi}_{p,n})^t$ satisfy the Yule Walker equations $\widehat{\Gamma}_p \widehat{\phi}_p = -\widehat{\gamma}_p$, where $\widehat{\Gamma}_p = [\widehat{R}_{i-j}]_{i,j=1}^p$, $\widehat{\gamma}_p = (\widehat{R}_1, \ldots, \widehat{R}_p)^t$, and $\widehat{R}_j = n^{-1} \sum_{t=1}^{n-|j|} (X_t - \bar{X})(X_{t+|j|} - \bar{X})$.

The following theorem, gives the consistency of the conditional sieve bootstrap, i.e., without step 7 in our proposed algorithm, as in Cao *et al.* (1997).

Theorem 1 *Suppose that assumptions A1 with $s = 4$, A2 with $r > 1$ and B with $p = o((n/\log(n))^{1/4})$ and $\max_t |X_t| = O_P(\log(n)^{1/2})$ hold. Then*

$$X^*_{T+h} \xrightarrow{d} X_{T+h}, \text{ in probability.} \tag{4}$$

Proof: Without loss of generality, we could assume that $\mu_X = 0$, then we can write X_{T+h} and X^*_{T+h} as:

$$\begin{aligned} X_{T+h} &= -\sum_{j=1}^{+\infty} \phi_j X_{T+h-j} + \varepsilon_{T+h} \\ X^*_{T+h} &= -\sum_{j=1}^{+\infty} \widehat{\phi}_{j,n} X^*_{T+h-j} + \varepsilon^*_{T+h} \end{aligned} \tag{5}$$

where $\widehat{\phi}_{j,n}$ denote the estimates of ϕ_j with the sample $(X_{T-n+1}, \ldots, X_T)$, $\widehat{\phi}_{j,n} = 0$ for $j > p(n)$, and $X^*_t = X_t$ for $t \leq T$. For simplicity of notation we prove the theorem for $h = 1$. From Lemma 5.4 of Bühlmann (1997), we have $\varepsilon^*_{T+1} \xrightarrow{d} \varepsilon_{T+1}$, in probability. Then, only rest to prove that the difference of first terms in X^*_{T+1} and X_{T+1} goes to 0 in probability:

$$\begin{aligned} -\sum_{j=1}^{+\infty} (\widehat{\phi}_{j,n} - \phi_j) X_{T+1-j} &= -\sum_{j=1}^{p(n)} (\widehat{\phi}_{j,n} - \phi_j) X_{T+1-j} \\ &\quad + \sum_{j=p(n)+1}^{+\infty} \phi_j X_{T+1-j} = S_{1,1} + S_{2,1} \end{aligned} \tag{6}$$

First, we have $S_{2,1} = o_{a.s.}(1)$, since

$$|S_{2,1}| \leq \max_t |X_t| \sum_{j=p(n)+1}^{+\infty} |\phi_j| = O_{a.s.}(\log(n)^{1/2}) o_{a.s.}(p^{-r}) \tag{7}$$

and second, we establish that also $S_{1,1} = o_{a.s}(1)$. From Theorem 3.2 of Bühlmann (1995) we have that

$$\max_{1 \le j \le p} |\widehat{\phi}_{j,n} - \phi_j| = O_{a.s.}((\log(n)/n)^{1/2}) + o_{a.s}(p^{-r}). \quad (8)$$

Therefore,

$$\begin{aligned} |S_{1,1}| &\le \sum_{j=1}^{p(n)} |(\widehat{\phi}_{j,n} - \phi_j) X_{T+1-j}| \\ &\le p \cdot \max_{1 \le j \le p} |\widehat{\phi}_{j,n} - \phi_j \cdot | \max_t |X_t| = O_{a.s.}(n^{\frac{1-r}{2r+2}} \log(n)^{\frac{2r}{2r+2}}) \end{aligned} \quad (9)$$

and then,

$$-\sum_{j=1}^{+\infty} \widehat{\phi}_{j,n} X_{T+h-j} = -\sum_{j=1}^{+\infty} \phi_j X_{T+1-j} + O_{a.s.}(n^{\frac{1-r}{2r+2}} \log(n)^{\frac{2r}{2r+2}}) \quad (10)$$

Finally, $X^*_{T+1} \xrightarrow{d} X^*_{T+1}$ in probability. For general h, it is clear that we could write the difference of first terms in X_{T+h} and X^*_{T+h} as a sum of some function $f(\phi_1, \ldots, \phi_{h-1}, \widehat{\phi}_{1,n}, \ldots, \widehat{\phi}_{h-1,n})(S_{1,1} + S_{2,1})$, and a term similar to $S_{1,1} + S_{2,1}$. The second terms in X_{T+h} and X^*_{T+h} is a "linear" combination of the corresponding (and independent) errors $(\varepsilon_{T+1}, \ldots, \varepsilon_{T+h}, \varepsilon^*_{T+1}, \ldots, \varepsilon^*_{T+h})$.

4 Simulation results

First, we compare the sieve bootstrap approaches with the proposal (PRR) of Pascual *et al.* (1998). We use two sieve bootstrap prediction methods: (CS) a conditional sieve, and (VS) the complete algorithm. We report the results for the following model considered in Pascual *et al.* (1998):

Model 1: $X_t = 0.7X_{t-1} + \varepsilon_t - 0.3\varepsilon_{t-1}$.

Since method PRR use the correct model, we could interpret their results as benchmarks. In practice, having observed a sample of size n, the model, and particularly p and q, is invariably unknown.

The error distributions F_ε considered are the uniform (-1,1), a standard normal, and a t_6. We take sample sizes $n = 50$, and 100, leads $h = 1$, 2 and 3, and nominal coverages $1 - \alpha = 0.8$ and 0.95.

To compare the different prediction intervals, we use their mean coverage and length, and the following combined measure of coverage and length:

$$CQ_M = |1 - \bar{C}_M/\bar{C}_T| + |1 - \bar{L}_M/\bar{L}_T|, \quad (11)$$

where $\bar{L}_T$ is the "true" mean interval length, $\bar{C}_T = (1 - \alpha)\%$ is the nominal coverage, $\bar{L}_M$ and $\bar{C}_M$ are the estimated mean length and coverage for method M ($M \in \{$ CS, VS, PRR, $\}$), and $|\cdot|$ denotes the absolute value.

In Table 1, we present the results for Model 1 and t_6 distribution (which is the less favorable distribution, since it does not satisfy the condition on $\max_t |X_t|$), using the two sample sizes, nominal coverage 95%, and lead times $h = 1$ and 3. The other possible combinations of parameters are available on request to the authors.

Method PRR have a better performance than CS and VS in terms of mean coverage, as expected. Notice that in this case, the sieve approach never uses the correct model. In terms of mean length PRR tends to over-estimate the nominal lengths, but for sieve bootstrap the opposite is observed. For $h = 1$, similar results are obtained for both methods in terms of CQ_M, but for $h = 3$ the PRR have better results. Bhansali (1996) shows that the AIC type criterion is not efficient for multistep prediction. The better behavior of VS compared to CS is evident, revealing the importance of parameter estimation variability. Also, as expected, PRR, CS and VS improve coverage with the sample size.

Table 1. Simulation results for Model 1, with t_6 errors.

Lag	Sample size	Method	$\bar{C}_M$ (se)	Cov. (below/above)	$\bar{L}_M$ (se)	CQ_M
k	n	Theoretical	95%	2.50% / 2.50%	4.87	0.00
1	50	CS	90.63(0.46)	4.58/ 4.78	4.70(0.08)	0.09
		VS	92.05(0.31)	3.82/ 4.13	4.88(0.08)	0.03
		PRR	92.94(0.28)	3.49/ 3.58	4.94(0.07)	0.03
	100	CS	93.20(0.24)	3.50/ 3.30	4.84(0.05)	0.03
		VS	93.48(0.21)	3.28/ 3.24	4.82(0.04)	0.03
		PRR	94.05(0.18)	3.01/ 2.94	4.95(0.05)	0.02
k	n	Theoretical	95%	2.50% / 2.50%	5.39	0.00
3	50	CS	91.53(0.35)	4.01/ 4.47	5.23(0.09)	0.07
		VS	91.64(0.34)	4.18/ 4.18	5.26(0.10)	0.06
		PRR	92.66(0.31)	3.71/ 3.64	5.44(0.08)	0.03
	100	CS	93.06(0.23)	3.57/ 3.36	5.28(0.05)	0.04
		VS	93.23(0.22)	3.49/ 3.28	5.31(0.05)	0.04
		PRR	93.68(0.20)	3.19/ 3.13	5.36(0.05)	0.02

As mentioned in the Introduction, the proposed method is applicable to linear models other than ARMA's. In Tables 2 and 3, we present the results for the following models:

Model 2: X_t is a Gaussian process with autocovariance generating function equal to $G(z) = \sum_{k=-\infty}^{+\infty} \gamma_k z^k$, where $\gamma_k = 1/(|k|+1)^2$.

Model 3: X_t is a Gaussian ARFIMA(0, d, 0) process, with $d = 0.25$.

Both models are simulated using the Cholesky decomposition of the autocovariance matrix (cf. Beran (1990)). Note that Model 2 satisfies Assumption *A2* with $r > 1$, but the fractional process does not satisfy *A2* with any $r \geq 0$. However, sieve bootstrap methods perform reasonably well in Model 3 and similarly for Model 2.

Table 2. Simulation results for Model 2.

Lag	Sample size	Method	$\bar{C}_M$ (se)	Cov. (below/above)	$\bar{L}_M$ (se)	CQ_M
k	n	Theoretical	95%	2.50% / 2.50%	3.79	0.00
1	50	CS	91.31(0.36)	4.42/ 4.27	3.61(0.04)	0.08
		VS	92.06(0.31)	3.96/ 3.98	3.68(0.04)	0.05
	100	CS	93.16(0.22)	3.44/ 3.40	3.72(0.03)	0.04
		VS	93.53(0.20)	3.22/ 3.26	3.75(0.03)	0.03
k	n	Theoretical	95%	2.50% / 2.50%	3.90	0.00
3	50	CS	91.54(0.33)	4.44/ 4.02	3.72(0.04)	0.09
		VS	91.78(0.31)	4.12/ 4.09	3.74(0.04)	0.08
	100	CS	93.55(0.22)	3.23/ 3.22	3.86(0.03)	0.03
		VS	93.57(0.21)	3.16/ 3.27	3.85(0.03)	0.03

Table 3. Simulation results for Model 3.

Lag	Sample size	Method	$\bar{C}_M$ (se)	Cov. (below/above)	$\bar{L}_M$ (se)	CQ_M
k	n	Theoretical	95%	2.50% / 2.50%	3.51	0.00
1	50	CS	90.38(0.36)	4.20/ 5.42	3.26(0.03)	0.12
		VS	91.41(0.31)	3.70/ 4.89	3.34(0.03)	0.09
	100	CS	92.93(0.22)	3.72/ 3.35	3.40(0.02)	0.05
		VS	93.27(0.21)	3.48/ 3.25	3.42(0.02)	0.04
k	n	Theoretical	95%	2.50% / 2.50%	3.75	0.00
3	50	CS	91.29(0.30)	3.86/ 4.85	3.49(0.03)	0.11
		VS	91.32(0.30)	3.83/ 4.85	3.49(0.03)	0.11
	100	CS	93.00(0.22)	3.72/ 3.28	3.62(0.02)	0.05
		VS	93.06(0.20)	3.62/ 3.31	3.62(0.02)	0.05

Acknowledgements
We would like to thank Mike Wiper for carefully reading that greatly improved the paper. This research was partially supported by the Dirección General de Educación Superior project DGES PB96-0111.

References

Alonso, A.M., Peña, D. and Romo, J. (2000) Forecasting time series with sieve bootstrap, Working Paper 00-07, Universidad Carlos III de Madrid.

Beran, J. (1994) *Statistics for Long-Memory Processes*, New York: Chapman & Hall.

Bhansali, R.J. (1996) Asymptotically efficient autoregressive models selection for multistep prediction, *Ann. Inst. Statist. Math.*, **48**, 577-602.

Brockwell, P.J. and Davis, R.A. (1991) *Time Series: Theory and Methods*, New York: Springer-Verlag.

Bühlmann, P. (1995) Moving-average representation of autoregressive approximations, *Stochastic Process. Appl.*, **60**, 331-342.

Bühlmann, P. (1997) Sieve bootstrap for time series, *Bernoulli*, **3**, 123-148.

Cao, R., Febrero-Bande, M., González-Manteiga, W., Prada-Sánchez, J.M. and García-Jurado, I. (1997) Saving computer time in constructing consistent bootstrap prediction intervals for autoregressive processes, *Comm. Statist. A*, **26**, 961-978.

Pascual, L., Romo, J. and Ruiz, E. (1998) Bootstrap predictive inference for ARIMA processes, Working Paper 98-86, Universidad Carlos III de Madrid.

Stine, R.A. (1987) Estimating properties of autoregressive forecasts, *J. Amer. Statist. Assoc.*, **82**, 1072-1078.

Thombs, L.A. and Schucany, W.R. (1990) Bootstrap prediction intervals for autoregression, *J. Amer. Statist. Assoc.*, **85**, 486-492.

Clustering by maximizing a fuzzy classification maximum likelihood criterion

Christophe Ambroise and Gérard Govaert

UMR CNRS 6599 Centre de Recherches de Royallieu, BP 20259, 60205 Compiègne, cedex France

Abstract. Basing cluster analysis on mixture models has become a classical and powerful approach. In this paper we propose an extension of this approach to fuzzy clustering. We define a fuzzy clustering criterion which generalizes both the maximum likelihood and the classification maximum likelihood. Finally using a generalization of the well-known EM and CEM algorithms we design an algorithm to optimize this criterion.

Keywords. Fuzzy Clustering, EM algorithm, mixture model

1 Introduction

Cluster analysis aims to partition a set of objects into subgroups where the objects inside a cluster show a certain degree of "similarity". Classical clustering, also called hard clustering in a fuzzy context, considers that each object belongs to only one cluster and assumes well separated clusters. This model is sometimes unrealistic.

Fuzzy clustering tackles this problem by assigning each object to all clusters with different degrees of membership. This model allows a more precise description of the underlying structure of the datasets. One of the better known fuzzy clustering algorithm is the fuzzy c-means algorithm (Bezdeck, 1974). This algorithm is based on a fuzzy extension of the least square criterion.

We advocated here a probabilistic approach of fuzzy clustering using mixture models.

2 Mixture models

A frequently appearing model in probabilistic based clustering considers that the observed data set $\mathbf{x} = (\mathbf{x}_1, ..., \mathbf{x}_N)$ is a sample of a mixture of K subpopulations with class conditional densities $f_k(\mathbf{x}_i|\theta_k)$, where θ_k is a parameter vector. In this framework, the density of the population is a mixture of the class densities

$$f(\mathbf{x}_i|\Phi) = \sum_{k=1}^{K} p_k f_k(\mathbf{x}_i|\theta_k), \tag{1}$$

where $p_1, ..., p_K$ denote the mixing proportions.

In this context, two commonly used maximum likelihood (m.l.) approaches have been proposed: the mixture approach and the classification approach. Loosely speaking, the mixture approach is aimed to maximize the likelihood over the mixture parameters, whereas the classification approach is aimed to maximize the likelihood over the mixture parameters and over the identifying labels of the mixture component origin for each point.

In the mixture approach, the parameter $\Phi = (p_1, \ldots, p_{K-1}, \theta_1, \ldots, \theta_K)$ is chosen to maximize the log-likelihood

$$L(\Phi|\mathbf{x}_1, \ldots, \mathbf{x}_N) = \sum_{i=1}^{N} \ln \left[\sum_{k=1}^{K} p_k f_k(\mathbf{x}_i|\theta_k) \right]. \tag{2}$$

The EM algorithm (Dempster, Laird and Rubin, 1977) is generally used to maximized this criterion (2).

In the classification approach, the parameter Φ and the partition $\mathbf{c}$ are chosen to maximize the Classification Maximum Likelihood

$$\mathrm{CL}(\Phi, \mathbf{c}_1, \ldots, \mathbf{c}_N|\mathbf{x}_1, \ldots, \mathbf{x}_N) = \sum_{k=1}^{K} \sum_{\mathbf{x}_i \in P_k} \ln \left[p_k f_k(\mathbf{x}_i|\theta_k) \right] \tag{3}$$

where $P = (P_1, \ldots, P_K)$ is the partition of $\mathbf{x}_1, \ldots, \mathbf{x}_N$ associated to the indicator vectors $c_1, \ldots, c_N : P_k = \{\mathbf{x}_i / c_{ik} = 1\}$. This criterion can be optimized by making use of a classification version of the EM algorithm, the so-called CEM algorithm (Celeux and Govaert, 1992).

3 Relation between the EM algorithm and fuzzy clustering

Estimating the parameters of a mixture allows to compute the posterior probabilities

$$t_k(\mathbf{x}_i) = \frac{p_k f_k(\mathbf{x}_i|\theta_k)}{f(\mathbf{x}_i)}, \tag{4}$$

which can be considered as fuzzy membership coefficients describing a fuzzy partition.

Moreover Hathaway (1986) has shown that the EM algorithm can be interpreted as a fuzzy clustering algorithm. Considering the posterior probabilities $t_k(\mathbf{x}_i)$ as variables denoted c_{ik}, the log likelihood of the parameters $\Phi = (p_1, \ldots, p_{K-1}, \theta_1, \ldots, \theta_K)$ can be expressed as a function of all the c_{ik}'s and Φ

$$L(\mathbf{c}, \Phi) = \sum_{i=1}^{N} \sum_{k=1}^{K} c_{ik} \log p_k f_k(\mathbf{x}_i|\theta_k) - \sum_{i=1}^{N} \sum_{k=1}^{K} c_{ik} \log c_{ik}, \tag{5}$$

where $\mathbf{c} = \{c_{ik} : 0 \leq c_{ik} \leq 1, \sum_{k=1}^{K} c_{ik} = 1, \sum_{i=1}^{N} c_{ik} > 0 (1 \leq i \leq N, 1 \leq k \leq K)\}$.

An alternate optimization between $\mathbf{c}$ and Φ for maximizing the criterion gives exactly the two steps of the EM algorithm for mixture distributions.

Notice that the maximized criterion is the sum of two terms:

- the first term is the classification likelihood with fuzzy c_{ik},
- the second term can be considered as an entropy which measures the fuzziness of the partition. This term is maximum when all the objects belong to all clusters with the same membership coefficient and minimum when the partition is hard.

From this point of view the EM algorithm can be considered as a fuzzy clustering algorithm which optimized a fuzzy classification criterion penalized by an entropy.

4 A fuzzy CEM algorithm

The relation existing between the classification likelihood and the Hathaway's criterion (5) leads us to the idea of an intermediate criterion

$$L(\mathbf{c}, \Phi) = \sum_{i=1}^{N} \sum_{k=1}^{K} c_{ik} \log p_k f_k(\mathbf{x}_i|\theta_k) - \alpha \cdot \sum_{i=1}^{N} \sum_{k=1}^{K} c_{ik} \log c_{ik}, \tag{6}$$

whose entropy term is weighted by α varying between 0 and ∞. Notice that for $\alpha = 0$ this criterion is equivalent to the classification likelihood and for $\alpha = 1$ it is equivalent to the likelihood.

This criterion has the disadvantage to have no statistical justifications, but allows to generalize the EM and CEM algorithms. Consequently it is able to deal with intermediary situations where the CEM classification is too hard and the EM classification too fuzzy. In this sense losing the statistical background may have some advantages.

In the alternate optimization of this criterion the first step is obtained by solving the following system:

$$\begin{cases} \log\left(p_k f_k(\mathbf{x}_i|\theta_k)\right) - \alpha - \alpha \log c_{ik} + \lambda_i = 0; \\ \sum_{k=1}^{K} c_{ik} = 1; \end{cases}$$

where the λ_i stand for the Lagrange coefficients. This system can be expressed as,

$$\begin{cases} c_{ik} = \left((p_k f_k(\mathbf{x}_i|\theta_k) \cdot \exp(-\alpha + \lambda_i)\right)^{\frac{1}{\alpha}}; \\ \sum_{k=1}^{K} c_{ik} = 1; \end{cases} \tag{7}$$

Thus the new values of c_{ik} are:

$$c_{ik} = \frac{(p_k f_k(\mathbf{x}_i|\theta_k))^{\frac{1}{\alpha}}}{\sum_{\ell=1}^{K} (p_\ell f_\ell(\mathbf{x}_i)|\theta_\ell)^{\frac{1}{\alpha}}}. \tag{8}$$

Using these new values of the c_{ik}'s, the second step corresponds exactly to the M step of the classical EM algorithm.

5 Numerical experiments

In this section, we report a summary of extensive Monte Carlo simulations. Data sets have been generated in $\mathbf{R}^2$ from two different models: a two-component Gaussian mixture with identical spherical variance matrices and a two-component mixture of uniform distribution drawn within two disks with identical radius. For each data type we considered different overlapping degrees which are measured by a criterion varying between 0 (well-separated) and 1 (very-mixed):

$$\delta_z = \frac{1}{2N} \sum_{i=1}^{N} \sum_{k=1}^{K} (z_{ik} - t_{ik})^2,$$

where

- z_{ik} identifies the mixture component origin: $z_{ik} = 1$ or 0 according as $\mathbf{x}_i$ $(1 \leq i \leq n)$ has been drawn from the kth component or from another one,
- $t_{ik} = \frac{p_k f_k(\mathbf{X}_i|\theta_k)}{f(\mathbf{X}_i)}$ are the true posterior probabilities.

We have compared four different algorithms: CEM, EM, FuzzyCEM (with $\alpha = 0.5$) and Fuzzy c-means with $m = 2$ (Bezdeck, 1974). For each experiment 30 simulations were used. We ran 10 times the four algorithms from random initial position and we selected the best solution out of the 10 runs.

We used a generalization of the error rate to fuzzy partition in order to evaluate the quality of the results:

$$\delta_c = \frac{1}{2N} \sum_{i=1}^{N} \sum_{k=1}^{K} (c_{ik} - r_{ik})^2,$$

where

- c_{ik} are the estimated posterior probabilities,
- r_{ik} is a reference partition.

This criterion can be considered as a distance between two partitions: a reference partition and the partition estimated by a given algorithm. Notice that we also tested the Kullback-Leibler divergence as comparison criterion, and we obtained similar results.

The simulation results are summarized in Tables 1-3. In these tables are displayed the sample mean and, between parentheses, the standard deviation of the fuzzy error rate over the 30 trials for each data structure and for each model.

δ_z	CEM	EM	$FuzzyCem$	Fcm
0.255(0.057)	0.256(0.044)	0.094(0.052)	0.192(0.052)	0.145(0.053)
0.133(0.051)	0.152(0.032)	0.061(0.037)	0.108(0.023)	0.096(0.024)
0.083(0.037)	0.085(0.022)	0.044(0.026)	0.061(0.020)	0.095(0.024)
0.035(0.027)	0.043(0.013)	0.018(0.010)	0.030(0.009)	0.083(0.010)
0.023(0.022)	0.019(0.012)	0.010(0.009)	0.014(0.010)	0.078(0.011)
0.006(0.011)	0.007(0.005)	0.003(0.004)	0.005(0.004)	0.066(0.009)

Table 1. Error rates (δ_c) for two-component Gaussian mixture models with identical spherical variance matrices mixed in the same proportion, using the true posterior probabilities t_{ik} as reference partition.

The main points arising from these tables are the followings.

- In Table 1, the EM algorithm always produces the best result. This may be explained by the fact that the EM algorithm hypothesis did exactly fit the simulation model. Comparison between the fuzzy algorithms shows

δ_z	CEM	EM	$FuzzyCem$	Fcm
0.255(0.057)	0.267(0.075)	0.328(0.059)	0.272(0.064)	0.341(0.052)
0.133(0.051)	0.141(0.053)	0.204(0.064)	0.151(0.050)	0.236(0.042)
0.083(0.037)	0.091(0.042)	0.127(0.045)	0.097(0.037)	0.188(0.034)
0.035(0.027)	0.037(0.027)	0.054(0.025)	0.039(0.024)	0.130(0.022)
0.023(0.022)	0.023(0.022)	0.031(0.022)	0.025(0.022)	0.102(0.020)
0.006(0.011)	0.006(0.011)	0.010(0.010)	0.007(0.010)	0.074(0.012)

Table 2. Error rates (δ_c) for two-component Gaussian mixture models with identical spherical variance matrices mixed in the same proportion, using the true initial hard partition as reference partition.

	δ_z	CEM	EM	$FuzzyCem$	Fcm
	0.158(0.046)	0.339(0.099)	0.300(0.039)	0.302(0.079)	0.297(0.053)
	0.097(0.043)	0.237(0.112)	0.258(0.069)	0.217(0.092)	0.254(0.082)
a	0.066(0.036)	0.137(0.090)	0.197(0.088)	0.132(0.078)	0.208(0.069)
	0.026(0.020)	0.071(0.065)	0.135(0.102)	0.071(0.064)	0.151(0.065)
	0.019(0.018)	0.039(0.077)	0.058(0.098)	0.037(0.075)	0.119(0.079)
	0.002(0.006)	0.007(0.018)	0.021(0.068)	0.006(0.017)	0.076(0.023)
	0.164(0.047)	0.189(0.056)	0.114(0.029)	0.144(0.047)	0.188(0.041)
b	0.073(0.039)	0.099(0.055)	0.074(0.030)	0.075(0.046)	0.145(0.038)
	0.021(0.019)	0.039(0.048)	0.051(0.030)	0.033(0.031)	0.130(0.029)
	0.000(0.000)	0.001(0.005)	0.018(0.010)	0.005(0.006)	0.104(0.011)

Table 3. Error rates for two-component Gaussian mixture models with different proportions (a) and for two-component uniform mixture model (b), using the true posterior probabilities t_{ik} as reference partition.

that Fcm performs better than FuzzyCEM when the clusters are very mixed but FuzzyCEM proves to be very robust and exhibits better performance when the two clusters become separated.

- Table 2 displays the fuzzy error rate considering the true underlying hard partition as the reference partition. In this configuration, the CEM algorithm exhibits the best behaviour and FuzzyCEM obtains the second best results. This shows that the FuzzyCEM is really a compromise between EM and CEM. It performs better than EM when searching a hard partition and better than CEM when seeking for a fuzzy partition.
- Table 3 displays results obtained when the underlying model used by the algorithms is different from the one used to simulate data sets. Two different divergences to the previous model have been considered: Gaussian mixture model with different mixing proportions (whereas the algorithms

implicitly assume equal proportions) and mixture of uniform distributions. In this context, when the degree of overlapping is not too high, the EM algorithm shows poor performance, while FuzzyCEM remains stable and always gives a better criterion than Fcm and CEM.

To summarize the preceding observations, we can state that the FuzzyCEM algorithm offers a good compromise between EM and CEM. Moreover it is a robust algorithm which performs really well when clusters show an intermediate degree of overlapping. A possible explanation consists in considering that the FuzzyCEM algorithm assumes that the partition has a certain degree of fuzziness: actually the second term of the minimized criterion (6) can be considered as a "prior knowledge".

6 Concluding remarks

The presented algorithm inherits all the advantages of the mixture model approach for clustering which is a fruitful line for dealing with various type of data, from binary data using Bernoulli distribution to continuous data using Gaussian or Laplace distributions. Moreover Gaussian mixture models have been proved to be efficient for handling many clustering situations in a simple and elegant manner since they allow easy specification of the volume, the orientation, the shape and the size of the clusters (Banfield and Raftery, 1993, Celeux and Govaert, 1995 and Rousseeuw, Kaufman and Trauwaert, 1996).

The performance of the new algorithm has been compared to the EM, CEM and fuzzy c-means algorithms on synthetic datasets using numerical experiments and it shows very encouraging results.

References

Banfield, J.D. and Raftery, A.E. (1993). *Model-based gaussian and non gaussian clustering.* Biometrics, **49**, 803–821.

Bezdek, J.C. (1974). *Numerical taxonomy with fuzzy sets.* Journal of mathematical biology, **1**, 57–71.

Celeux, G. and Govaert, G. (1992). *A classification EM algorithm for clustering and two stochastic versions.* Computational Statistics and Data Analysis, **14**, 315–332.

Celeux, G. and Govaert, G. (1995). *Gaussian parsimonious clustering models.* Pattern Recognition, **28**, 781–793.

Dempster, A.P., Laird, N.M. and Rubin, D.B. (1977). *Maximum likelihood from incomplete data via the EM algorithm.* Journal of the Royal Statistical Society, **39**, 1–38.

Hathaway, R.J. (1986). *Another interpretation of the EM algorithm for mixture distributions.* Journal of Statistics & Probability Letters, **4**, 53–56.

Rousseeuw, P.J., Kaufman, L. and Trauwaert, E. (1996). *Fuzzy clustering using scatter matrices.* Computational Statistics & Data Analysis, **23(1)**, 135–151.

Tree-based algorithms for missing data imputation

M.J. Bárcena and F. Tusell

Facultad de CC.EE. y Empresariales, Universidad del País Vasco / Euskal Herriko Unibertsitatea, Avda. Lehendakari Aguirre, 83, 48015 BILBAO, Spain. [1]

Abstract. Let **X** be a $N \times (p+q)$ data matrix, with entries partly missing in the last q columns. A problem of practical relevance is that of drawing inferences from such an incomplete data set. We propose to use a sequence of trees to impute missing values. Essentially, the two algorithms we introduce can be viewed as predictive matching methods. Among their advantages, is their flexibility, which makes no assumptions about the type or distribution of the variables.

Keywords. missing data; multiple imputation; binary trees; file matching; predictive matching.

1 Introduction

Let **X** be a $N \times (p+q)$ data matrix, with entries partly missing in the last q columns. This includes: i) Scattered missing entries which were never recorded, or were subsequently lost; ii) A full block missing, say the $n_{\text{mis}} \times q$ block consisting of the last n_{mis} rows and q columns. The second situation may arise, for instance, if **X** contains data collected in two surveys given to n_{obs} and n_{mis} subjects respectively ($N = n_{\text{obs}} + n_{\text{mis}}$), and complete data is available for the n_{obs} subjects interviewed in the first survey while the last q questions were not asked to the n_{mis} subjects interviewed in the second survey. Only case ii) is dealt with in this paper, although generalization is possible.

A problem of practical relevance is that of drawing inferences from such an incomplete data set, and a considerable body of literature exist on this issue. A landmark is the monograph Little and Rubin(1987), setting up a methodology and advocating the use of multiple imputation. A recent monograph is Schafer(1997) which develops algorithms for imputation based on the EM algorithm and data augmentation.

2 Goals

The methods in Schafer(1997) require the specification of a parametric model and a (possibly non-informative) prior on the parameters. Our intent has been to produce a good all-purpose nonparametric method, capable of coping with situations where little is known about the underlying data generation mechanism.

Our research was initially motivated by the problem of completing a partially observed sample with regular structure (problem ii in the Introduction).

---

[1] Research supported through grant PB98-0149 from CICYT.

For instance, we might have a file with $N = n_{\text{obs}} + n_{\text{mis}}$ subjects. The first n_{obs} subjects have been totally observed on $(p+q)$ variables. For the remaining n_{mis} subjects, only the first p variables have been observed. We would like to impute the missing q variables on these n_{mis} subjects with a method:

1. making as little assumptions as feasible on the joint distribution of the $(p+q)$ variables;
2. allowing for multiple imputation, and
3. taking into account the structure of the q variables that are imputed.

This last point was of particular interest to us. We had an application where the q variables to impute where the times devoted by each subject to different tasks, and were required to add up to twenty four hours (see Bárcena and Tusell(1998,1999)). It was clear that those q variables had to be jointly imputed to ensure mutual compatibility.

Binary trees are a flexible tool to capture the relationship between a response and a set of predictors. However, neither in the seminal work Breiman et al.(1984) nor in the large body of literature that followed could we find examples in which the response was multivariate[2] and jointly imputed.

Next section describes two algorithms built around univariate response binary trees and designed to meet the three goals above.

3 Algorithms

We use a collection of ordinary (scalar response) binary trees. They are built with the methodology described by Breiman et al.(1984) as implemented by Therneau and Atkinson(1997); but a different strategy can be used (e.g., Murthy et al.(1994)). We denote by $\mathcal{Y}_{x|\mathbf{z}}$ a tree "regressing" x on the predictors in $\mathbf{z}$ —the response x might just as well be qualitative, and the tree a classification rather than a regression tree. Assume we have a training sample of n_{obs} subjects, fully observed in $(p+q)$ variables, while for the remaining n_{mis} subjects we only observe the first p variables. Call $\mathbf{X}_{\text{obs}}$ the vector of the p variables fully observed (for all $n_{\text{obs}} + n_{\text{mis}}$ subjects) and $\mathbf{X}_{\text{mis}}$ the vector of the q variables incompletely observed (missing for the last n_{mis} subjects). The case where observations are missing irregularly can also be handled (using surrogate splits), but we will only deal with case ii) of the Introduction in the following. We propose the following imputation strategies.

3.1 The forest climbing algorithm

It can be summarized as follows:

1. Build trees $\mathcal{Y}_{X_{p+1}|\mathbf{X}_{\text{obs}}}, \ldots, \mathcal{Y}_{X_{p+q}|\mathbf{X}_{\text{obs}}}$ using the CART methodology and the n_{obs} complete observations.
2. Drop each of the n_{mis} incomplete cases down the q trees constructed. Let case i fall in the terminal nodes labelled $(\ell_{i,1}, \ldots, \ell_{i,q})$ of (respectively) trees $\mathcal{Y}_{X_{p+1}|\mathbf{X}_{\text{obs}}}, \ldots, \mathcal{Y}_{X_{p+q}|\mathbf{X}_{\text{obs}}}$. Call $(\ell_{i,1} \cap \ldots \cap \ell_{i,q})$ the subset of the n_{obs} complete cases which also end in said leaves. If $(\ell_{i,1} \cap \ldots \cap \ell_{i,q}) \neq \emptyset$, impute the missing values of case i by those of one complete case which also ends in $(\ell_{i,1} \cap \ldots \cap \ell_{i,q})$. If multiple imputation is desired, sample k cases out of that intersection.

[2] Note the work Ciampi(1991): it does require the specification of a likelihood, though. Recently, Siciliano and Mola(2000) address the problem of constructing trees with multivariate response in a non-parametric way.

3. If $(\ell_{i,1} \cap \ldots \cap \ell_{i,q}) = \emptyset$, iteratively replace leaves by their ancestors ("climb the trees") until a non empty intersection is found from which one or more complete cases can be drawn.

The idea is disarmingly simple. Take any tree $\mathcal{Y}_{X_k|\mathbf{X}_{\text{obs}}}$, $p < k \leq p+q$. The leaves of that tree are classes of a partition of the predictor space such that, within each class, knowledge of $\mathbf{X}_{\text{obs}}$ cannot help us in further refining our prediction of X_k (otherwise, the leave would have been splited). It then makes sense that if subject i with unknown X_k ends in leave $\ell_{i,k}$ when dropped down the tree $\mathcal{Y}_{X_k|\mathbf{X}_{\text{obs}}}$, its X_k value be predicted by a function of the X_k values of subjects in the training sample which ended in the same leave. This function can be the mean, median or other summary statistic; or else we can sample from that leave if multiple imputation is desired.

Since we want to jointly impute all values in $\mathbf{X}_{\text{mis}}$ for the subject at hand, we would like to use complete cases in $(\ell_{i,1} \cap \ldots \cap \ell_{i,q})$, and this is exactly what the algorithm above does. The only additional caveat is that the relevant intersection might be empty —not one of the subjects in the training sample ended in exactly the same leaves as the subject to impute. If that is the case, the algorithm replaces nodes by their ancestors ("climbs the trees"), until a non empty intersection is found. The order of climbing is governed by the deviance —we climb first the tree where the replacement of a node by its ancestor leads to the least possible increase in deviance. Inasmuch as the deviance is scale-dependent, this is only an *ad-hoc* device.

We can think of the forest climbing algorithm as a nearest neighbour method in which "nearness" is defined as "falling in the same leaves than". Similar ideas exist in the literature, under the name of predictive mean matching.

3.2 The cascade algorithm

The cascade algorithm is directed at finding subjects in the training sample that are simultaneously "close" to the subject to impute in the metrics defined by all trees, obviating the need to climb.

Again the idea is quite simple. Jointly imputing $\mathbf{X}_{\text{mis}}$ given the values in $\mathbf{X}_{\text{obs}}$ is easy as soon as we have the conditional distribution $f(\mathbf{X}_{\text{mis}}|\mathbf{X}_{\text{obs}})$: we only have to draw from that distribution to perform multiple imputation. By successively conditioning, we can write

$$\begin{aligned} f(\mathbf{X}_{\text{mis}}|\mathbf{X}_{\text{obs}}) = \; & f(X_{p+q}|\mathbf{X}_{\text{obs}}, X_{p+1}, \ldots, X_{p+q-1}) \\ & \times f(X_{p+q-1}|\mathbf{X}_{\text{obs}}, X_{p+1}, \ldots, X_{p+q-2}) \\ & \times \cdots \\ & \times f(X_{p+1}|\mathbf{X}_{\text{obs}}) \end{aligned}$$

We can regard a tree as a mechanism generating observations with a given conditional distribution. For instance, if we construct the tree $\mathcal{Y}_{x|\mathbf{X}_{\text{obs}}}$ we can generate approximate random drawings from $f(X|\mathbf{X}_{\text{obs}})$ by dropping $\mathbf{X}_{\text{obs}}$ down $\mathcal{Y}_{x|\mathbf{X}_{\text{obs}}}$ and sampling from the leave where it ends.

To generate observations with approximate distribution $f(\mathbf{X}_{\text{mis}}|\mathbf{X}_{\text{obs}})$ we can do the following:

1. Construct trees $\mathcal{Y}_{X_{p+1}|\mathbf{X}_{\text{obs}}}, \mathcal{Y}_{X_{p+2}|\mathbf{X}_{\text{obs}}X_{p+1}}, \ldots, \mathcal{Y}_{X_{p+q}|\mathbf{X}_{\text{obs}}X_{p+1},\ldots,X_{p+q-1}}$.
2. For each incomplete observation with observed $\mathbf{X}_{\text{obs}}$,

(a) Drop $\mathbf{X}_{\text{obs}}$ down the first tree. Sample the leave where it ends to obtain a value X_{p+1}.

(b) For $j = 2, \ldots, q$ do likewise: drop $\mathbf{X}_{\text{obs}}, \ldots, X_{p+j-1}$ down the j-th tree and sample the leave where it ends to obtain a vector of imputed values $X_{p+1}, \ldots, X_{p+j}$.

Note that while a joint distribution can be factored in any order, in the tree cascade algorithm the just sketched order does matter. The ideal would be to reorder variables $X_{p+1}, \ldots, X_{p+q}$ in such a way that we have first those which can be best predicted from $\mathbf{X}_{\text{obs}}$ and last those which cannot be predicted well from $\mathbf{X}_{\text{obs}}$ yet are closely related to previously predicted variables. These are potentially conflicting criteria, and there is no clear choice. We have investigated two different alternatives: *best first* and *best last*. In the first case, the trees are used in order of decreasing goodness of fit; the rationale being that, since each imputed variable can be input in subsequent trees, we want the values imputed earlier to be of the best possible quality.

On the other hand, in order to ensure consistency of the imputed variables, the whole vector $X_{p+1}, \ldots, X_{p+q}$ is imputed at the last step, which makes desirable a high quality tree at the end of the cascade.

4 Implementation and simulated results

We have written functions to implement our methods in the statistical language R (see Venables et al.(1997) for a description). We have used the functions in the `rpart` package (see Therneau and Atkinson(1997)) as building blocks. For the purpose of comparison, we used the routines in the package `norm`, a port[3] to R of the programs of the same name described in Schafer(1997).

We have generated data from a multivariate normal distribution $N_{15}(\mathbf{0}, \Sigma)$ with Σ exhibiting moderate correlation among variables. The variables were standardized to have variance equal to one. Each of the two hundred replications generated contains $N = 500$ observations. The last $n_{\text{mis}} = 50$ observations of the last $q = 5$ variables were deleted and then their values imputed using the remaining $n_{\text{obs}} = 450$ complete observations as the training sample.

We have simulated the behaviour of the forest climbing algorithm (FC) and the cascade algorithm, both with best first (BF) and best last (BL) orderings and joint imputation (that is, all of the missing values are imputed at once, thus ensuring compatibility of the imputed values). We have also simulated the behaviour of the cascade algorithm with BF order and individual imputation of each variable (IND). Finally, we have simulated Schafer's method (NOR), using the EM algorithm to find the maximum likelihood estimates of the parameters conditional on $\mathbf{X}_{\text{obs}}$ and subsequently drawing random observations from that conditional distribution $f(\mathbf{X}_{\text{mis}}|\mathbf{X}_{\text{obs}})$.

Figure 1 shows the mean square error (MSE) of imputation for four of the variables, averaged over the 200 replications of the experiment (the fifth variable, not shown for lack of space, behaves similarly). Notice that a naïve strategy of imputing with a random complete subject from the sample *(cold deck)* would give a MSE of 2. Naïve imputation using the mean would give a MSE of 1. Since data is generated following a multivariate normal model, we can expect the parametric method (NOR) to perform best, and this is indeed the case. What is interesting is that the forest climbing algorithm is always a

[3] By Alvaro Novo, and available at CRAN, `http://cran.ar.r-project.org`.

Fig. 1. Imputation results for four variables and two hundred replications with $n_{\text{obs}} = 450$, $n_{\text{mis}} = 50$, $p = 10$, $q = 5$ and multivariate normally distributed data. See text for description of the methods.

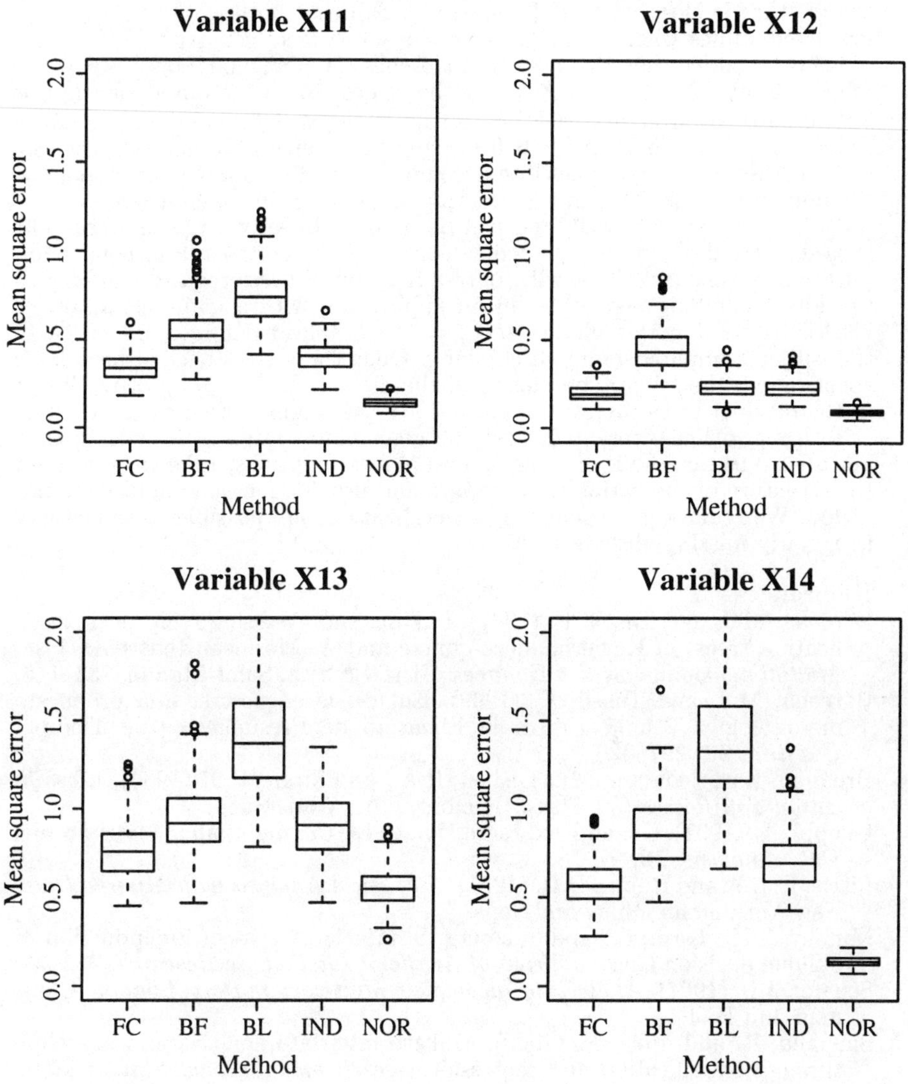

second best. When imputing using the cascade algorithm, the minimum MSE is of course obtained imputing each variable separately. Of the remaining two cascade algorithms, neither order BF or BL seems uniformly better (see for example the results for variable X11 and X12 in Figure 1). Additional more extensive results are available from the authors.

5 Some remarks

As mentioned previously, both the forest climbing and cascade algorithms can be seen as *ad hoc* methods of predictive matching: they replace in block the missing values of a subject with those of another subject in the training sample that is close. "Close" is taken to mean that both would have similar predicted values when dropped down the set of trees constructed. It is important to notice that this notion of closeness is ambiguous, because we are jointly imputing $X_{p+1}, \ldots, X_{p+q}$. If the scales vary widely and/or there is strong correlation, it makes sense to rescale the variables and/or transform them to principal components before using the intersection method. The cascade method explicitly takes into account the relation among the responses: the ambiguity resurfaces in the ordering of the trees in the cascade.

Both methods scale well, and can be used with fairly large samples. The largest portion of time is devoted to constructing the trees. Subsequent imputation is very fast. Typically, only a fraction of cases require climbing in the forest climbing algorithm: in an application with a training sample of 2521 subjects $p = 5$ predictors and $q = 24$ variables to impute, under 2% of the subjects imputed required climbing. Once the q trees needed have been constructed, the (time) complexity of the algorithms is $O(qn_{\text{mis}})$, i.e. linear in the product of variables to impute times the number of cases to impute.

Both algorithms presented meet the goals enumerated in Section 2: they are all-around methods making almost no assumptions, take into account the structure of the variables to impute and provide for easy multiple imputation. We remark in closing that generalizations are possible to the case of irregularly missing observations.

References

Bárcena, M.J. and Tusell, F. (1997). Linking surveys using reciprocal classification trees. In K. Fernández-Aguirre and A. Morineau (eds.) *Analyses Multidimensionnelles des Données*, Cisia-Ceresta, Saint-Mandé, 133–148.

Bárcena, M.J. and Tusell, F. (1999). Enlace de encuestas: una propuesta metodológica y aplicación a la Encuesta de Presupuestos de Tiempo. *Qüestiió*, **23**, 297–320.

Breiman, L., Friedman, J.H., Olshen, R.A., and Stone, C.J. (1984). *Classification and Regression Trees.* Belmont, CA.: Wadsworth.

Ciampi, A. (1991). Generalized regression trees. *Computational Statistics and Data Analysis*, **12**, 57–78.

Little, R.J.A. and Rubin, D.B. (1987). *Statistical Analysis with Missing Data.* New York: John Wiley and Sons.

Murthy, S.K., Kasif, S., and Salzberg, S. (1994). A system for induction of oblique decision trees. *Journal of Artificial Intelligence Research*, **2**, 1–32.

Schafer, J.L. (1997). *Analysis of Incomplete Multivariate Data.* London: Chapman and Hall.

Siciliano, R. and Mola, F. (2000). Multivariate data analysis and modeling through classification and regression trees. *Computational Statistics and Data Analysis*, **32**, 285–301.

Therneau, T.M. and Atkinson, E.J. (1997). An introduction to recursive partitioning using the RPART routines. Technical Report, Mayo Foundation.

Venables, B., Smith, D., Gentleman, R., and Ihaka, R. (1997). *Notes on R: A Programming Environment for Data Analysis and Graphics.* Auckland: Dept. of Statistics, University of Adelaide and University of Auckland. Available at `http://cran.at.r-project.org/doc/R-intro.pdf`.

MiPy: a system for generating multiple imputations

John Barnard

Department of Statistics, Harvard University, Cambridge, MA 02138, USA

Abstract. Multiple imputation has proven to be a useful mode of inference in the presence of missing data. It is a Monte-Carlo based methodology in which missing values are imputed multiple times by draws from a (typically explicit) imputation model. Generating "proper" imputations under reasonable models for the missing-data and complete-data mechanisms is often a computationally challenging task. The lack of software for generating multiple imputations is a serious impediment to the routine adoption of multiple imputation for handling missing data. Several groups have developed software for generating imputations, most of which is model specific, but none of this software is open, flexible, or extensible. In this paper I will introduce a computer software system, called MiPy, for generating multiple imputations under a wide variety of models using several computational approaches. The system is constructed from a combination of Python, an object-oriented and portable high-level interpreted language, and compiled modules in C, C++, and Fortran. MiPy features a clean syntax, simple GUI, open source, and portability to all of the major operating system platforms. In MiPy, Python can be viewed as the glue language that ties together computationally intensive modules written in lower-level languages.

Keywords. Missing Data, Object Oriented, Python, Simulation

1 Introduction

Multiple imputation (Rubin, 1987) has proven to be a useful approach for obtaining valid inferences in the presence of missing data for a variety of fields, especially in the context of public-use data files (Rubin, 1996). It is a Monte-Carlo based methodology in which missing values are imputed multiple times by draws from a (typically explicit and Bayesian) imputation model. The major aim of multiple imputation is to allow users, who generally have little information about the nonresponse mechanism, to obtain valid statistical inferences in the presence of nonresponse using only valid complete-data analysis techniques and simple combining rules. Generating "proper" imputations under reasonable models for the missing-data and complete-data mechanisms is often a computationally challenging task because the implied predictive distribution from which the imputations are to be drawn is often nonstandard. To meet these computational challenges, imputers have used a combination of modern sampling methods including Monte Carlo Markov Chain algorithms such as the Gibbs sampler or the Metropolis-Hastings algorithm, Bayesian bootstrap, and sampling importance resampling (see Schafer, 1997 for more details). In addition, computationally intensive methods such as EM and modifications to EM, e.g., ECM and PXEM, provide very useful and important information to imputers. Unfortunately, there has been and still is a lack of generally available and flexible software for implementing these methods to meet the computational challenges faced by creators of multiply imputed data.

The lack of software for generating multiple imputations is a serious impediment to the routine adoption of multiple imputation for handling missing data. Several groups have developed software for imputation, most of which is model specific (e.g., see Schafer, 1997), but none of this software is open, flexible, and extensible. We review the current state of software for handling missing data problems in Section 2. In Section 3, I introduce a computer software system, called MiPy, for generating multiple imputations under a wide variety of models using several computational approaches. The system is constructed from a combination of Python (van Rossum, 2000), a portable high-level interpreted language, and compiled modules in C, C++, and Fortran. The Python language and some of its useful extensions are described in Section 4, and the reasons for building an imputation system with Python are given in Section 4.4.

2 State of software for missing data problems

2.1 General statistical software

Several general purpose statistical packages have extensions for handling missing data and for generating imputations, although none of these extensions is flexible or extensible. Joseph Schafer (Schafer, 1997) has developed routines for S-PLUS (primarily in Fortran) that generate multiple imputations under several common models (using MCMC), and these have been ported to R. SPSS has a missing data module (SPSS, 1998) that allows for the generation of imputations and for obtaining maximum likelihood estimates under normality and ignorable missing data mechanisms.

Many programs have facilities for obtaining maximum likelihood (ML) estimates for general likelihoods, e.g., SAS, GAUSS, and Stata, but few of these are tailored to obtaining ML estimates in the presence of missing data (Mplus (Muthen and Muthen, 1998) is an exception), a task often best tackled with EM-based algorithms.

2.2 Specialized software

Several programs are specialized for generating multiple imputations. Joseph Schafer (Schafer, 1998) has developed several stand-alone Windows programs that produce imputations for various common models. These programs provide a GUI and limited diagnostic procedures for detecting convergence of MCMC methods. Amelia (Honaker et al., 1998), a freely available program, generates multiple imputations for a small class of models using a combination of sample importance resampling and asymptotic approximations. Solas (Statistical Solutions Ltd., 1997) is the first and currently only commercial program for generating multiple imputations. It uses a propensity-based approach for generating imputations (thus avoiding the need for MCMC) that is best suited for longitudinal settings. It provides a nice GUI, basic statistical analysis procedures, and is available only for Microsoft Windows.

BUGS (Spiegelhalter et al., 1996) and its Microsoft Windows implementation, WinBUGS (Spiegelhalter et al., 1999), provide a powerful and easy to use environment for generating draws from a Bayesian model and can be used to generate multiple imputations. However, BUGS suffers from several restrictions, most notably the inability to link code written in other programming languages, that make it unsuitable for handling large and complex missing data problems.

3 What is MiPy?

3.1 Design

Although the design of MiPy is still under development, several design aspects are fixed. MiPy will have a GUI, which will be optional in that the user will be able to use MiPy from a command line. It will provide high quality graphics based on the recently proposed SVG (Scalable Vector Graphics) (W3C, 1999) specification. A rich data framework, similar to the data frame idea of R and S-PLUS, will be provided. MiPy will contain interfaces to a variety of databases and will be able to seamlessly pass data back and forth from MiPy to the data sources. These design descriptions give some idea about the general look of MiPy. However, they are not directly related to the core goal of MiPy – to provide a flexible and efficient structure for conducting iterative procedures.

The design of the heart of MiPy aims to provide a collection of reusable model components (e.g., a linear regression component or a general location model component) that can be combined in almost arbitrary ways. New model components can be easily added, being implemented in pure Python or in a lower-level language (such as C, C++, or Fortran) or in a combination of languages. Graphical models, which are at the heart of BUGS/WinBUGS, will be used to help organize and connect the components; graphical models have proven to be very useful for structuring complex models, particularly from a Bayesian modeling perspective. Unlike BUGS/WinBUGS, the goal of MiPy may be to find a maximum of a model function instead of or in conjunction with trying to simulate from a model. Hence, the requirements of the modeling components and their structure will be minimal. In summary, the core design is loosely fashioned after that of BUGS (without the compilation step) but allows for greater flexibility and for differing goals (maximization vs. intergration/simulation).

3.2 Implementation

MiPy is in the prototype stage of development. However, the core libraries on which MiPy will be developed have been chosen. The GUI will be provided by wxPython (Dunn, 2000), the Python bindings to the cross-platform C++ GUI library wxWindows. wxPython/wxWindows provides native look and feel on Microsoft Windows, Linux/Unix (using GTK), Macintosh OS, and other operating platforms. Plotting and graphics will be provided by the Python packages PIL (Lundh, 2000) (for images), Pingo (for SVG graphics format) and Reportlab (Reportlab, 2000) (for PDF output and tables). The primary data structure will be provided by Metakit (Equi4 Software, 2000), a cross-platform C++ embedded database with tight integration to Python via the Mk4Py package. Interfaces to databases will be provided by the Python package mxODBC, which uses the standard ODBC to communicate with a variety of databases.

The core numerical services will be provided by NumPy (Ascher et al., 1999), a Python package that adds extensive and powerful numerical capabilities to Python, in conjunction with several C, C++, and Fortran libraries. For example, Linpack will be used to provide linear algebra routines, and drandlib will provide random number generators. NumPy has been used in various large scale numerical projects, e.g., see Beazley and Lomdahl (1997).

The plan is to have an alpha version of MiPy for release by the end of June 2000, with a beta release by the end of the summer. MiPy will be completely open source, freely distributed, and, hopefully, a community endeavor.

4 The what and why of Python

Python is a freely available, multiplatform, interpreted, interactive, object-oriented programming language. It has modules, classes, exceptions, threading, very high level data types, dynamic typing, and automatic memory management.

4.1 Core features

Python combines remarkable power with a clean syntax and design. It was invented by Guido van Rossum in the early 1990s and has close to half a million users and developers. Python is freely available for all of the major operating systems, including all of the Microsoft Windows variants, Macintosh OS, Linux, Be-OS, OS/2, VMS, and all flavors of Unix. Its source code is also freely available, and it can be used and modified for commercial and noncommercial applications. Support includes extensive on-line documentation with a tutorial, a newsgroup, comp.lang.python, numerous mailing lists and special interest groups (e.g., numerical computing, plotting and graphing, imaging, XML, and distributed computing), conferences, and several introductory and advanced books.

Python has a minimalist syntax; it uses white space to indicate blocking structure and uses very few special symbols (e.g., $ or @). A snippet of Python code is given below:

```
def lgeometricmean (inlist):
    mult = 1.0
    one_over_n = 1.0/len(inlist)
    for item in inlist:
        mult = mult * pow(item,one_over_n)
    return mult
```

Everything in Python is an object, including functions, and has a type. It is easy to add a new type of object (e.g., a tree type). Several highly useful types that are built in to Python are strings, lists, and dictionaries (sometimes called hash tables). Like many high-level languages (e.g., S), Python is dynamically typed, so it is not necessary (nor allowed) to declare the type of an object (unlike C, C++, or Java). Dynamic typing simplifies programming and allows for quick prototyping.

Python is an object-oriented programming (OOP) language. Its class model supports standard object-oriented ideas such as encapsulation, polymorphism, and multiple inheritance, and it allows for operator overloading. OOP has proven highly useful in implementing statistical models, as demonstrated in S-PLUS, R, and XLisp-Stat. Python's object-oriented nature makes it natural and easy for it to serve as a scripting language for lower-level OOP languages such as C++ and Java.

Exceptions are helpful for handling error conditions in programs and make for more robust and stable code. Python has exceptions built in to the language (like Java but unlike almost all statistical languages). In addition, Python has threading, which allows for multiple Markov chains to run (almost) simultaneously along with a GUI and an interactive session.

4.2 Extensions and modules

Python comes with an extensive library of modules, which are collections of functions, classes, and commands, ranging from object persistence to regular-expression matching to networking. Although most of these library modules are not directly useful for performing numerical calculations, they make it easy to interact with and manipulate data and other programs, which is often an important component of statistical computing in the broad sense.

In addition to the standard library modules, there are many third-party extensions and modules for Python that are useful for statistical computing and simulation. Interfaces exist (either direct or via ODBC) to many databases, including Oracle, Sybase, MySQL, Microsoft Access, and many others. Additional packages and modules allow Python to perform distributed computing (via CORBA, ILU, and XML-RPC). Python bindings exist to many GUI libraries, making it very easy to add a GUI to Python programs.

4.3 Language interfaces

Several tools are available that make integrating C, C++, and Fortran code into Python extremely simple and easy. SWIG (Beazley, 1995), the simplified wrapper and interface generator, takes C and C++ code and automatically builds the required C/Python code to seamlessly integrate the C/C++ code into Python. It has been used extensively to integrate a variety of numerical libraries and code into Python. Other tools such as SIP, CXX, and SCXX support integrating C++ code that utilizes some of the more advanced features of that complex language. Two powerful but easy to use tools exist for integrating Fortran code into Python, PyFort (Dubois, 1999) and f2py (Peterson, 2000). I have used these tools to quickly make available in Python (it took about an hour) all of the routines in the extensive linear algebra library Linpack.

4.4 Why Python?

Why build an imputation system with Python instead of with more statistically oriented languages/frameworks such as R, SAS, or XLisp-Stat? There are several reasons why I chose Python, a language virtually unknown to statisticians, for implementing MiPy:

1. Availability and Openness – Does it run on many platforms and is the source code available? Can it be distributed for free?
2. Extensibility and Integration – Can the program be extended? Can it interface with other languages such as C, C++, and Fortran?
3. Interactivity – Does the program allow for easy prototyping?
4. Simplicity and Performance – Is the programming language easy to use and learn? Does the program have enough muscle to perform large simulations?
5. Connectivity and Communication – Is it easy to get data from other sources, such as databases? Can it perform distributed/threaded computing?

Reason (1) removes commercial packages (e.g., S-PLUS, SAS, Stata, Solas); I want MiPy to be available to as many imputers as possible regardless of their preferred platform or statistical package. Reason (2) removes BUGS, WinBUGS, and Amelia – none of them can be extended or interfaced to foreign languages. Reason (3) removes the standard programming languages C, C++, Fortran, and Java (they may be utilized, but as extensions, not as the core). At this point the major choices left are the open source programs R and XLisp-Stat. Reason (4) partially removes R, as its performance for doing iterations is often poor and it lacks a GUI, and XLisp-Stat, whose Lisp syntax is very foreign to most statistical programmers. Reason (5) further weakens the arguments for these two competitors – neither of them has strong interfaces to external data sources (a group has started adding an ODBC interface for R).

Python's major weakness is that it has very little statistically oriented code. I feel that this is easily overcome, especially since the aim of MiPy is not to be a general statistics package, which would require a large amount of coding, but more focused on simulation and iterative methods. Overall, Python's strengths, which are in some sense the opposite of those of R and XLisp-Stat, far outweigh its lack of currently available statistical code. I suspect that once Python is "discovered" in the statistical world, which I hope will be aided by MiPy, it will become a popular alternative to the more mainstream languages and programs for developing statistical software.

References

Ascher, D., Hinsen, K., Oliphant, T. and Dubois, P. (1999). Numerical Python. UCRL-12859, Part 4, Lawrence Livermore National Laboratory, Livermore, CA. Available online only at http://numpy.sourceforge.net

Beazley, D.M. (1995). Simplified Wrapper and Interface Generator. See http://swig.sourceforge.net

Beazley, D.M. and Lomdahl, P.S. (1997). Feeding a large-scale physics application to Python. In: *Proceedings of the 6th International Python Conference*, pp. 21-29. San Jose, CA.

Dubois, P. (1999). Python-Fortran Connection Tool. See http://pyfortran.sourceforge.net

Dunn, R. (2000). wxPython. See http://alldunn.com/wxPython

Equi4 Software (2000). Metakit. See http://www.equi4.com

Honaker, J., Joseph, A., King, G., Scheve, K. and Singh, N. (1998). Amelia: A Program for Missing Data. See http://gking.harvard.edu/amelia

Lundh, F. (2000). Python Imaging Library. See http://www.python.org/sigs/image-sig/Imaging.html

Muthen and Muthen (1998). Mplus. See http://www.statmodel.com

Peterson, P. (2000). f2py: Fortran to Python Interface Generator. See http://cens.ioc.ee/projects/f2py2e

van Rossum, G. (2000). *Python Reference Manual.* Reston: Corporation for National Research Initiatives. Available online at http://www.python.org

Reportlab Inc. (2000). Reportlab. See http://www.reportlab.com

Rubin, D.B. (1987). *Multiple Imputation for Nonresponse in Surveys.* New York: Wiley.

Rubin, D.B. (1996). Multiple imputation after 18+ years (with discussion). *Journal of the American Statistical Association*, **91**, pp. 473-489.

Schafer, J.L. (1997). *Analysis of Incomplete Multivariate Data.* New York: Chapman and Hall.

Schafer, J.L. (1998). Norm. See http://www.stat.psu.edu/~jls

Spiegelhalter, D.J., Thomas, A. and Best, N.G. (1999). WinBUGS Version 1.2 User Manual. MRC Biostatistics Unit. See http://www.mrc-bsu.cam.ac.uk

Spiegelhalter, D.J., Thomas, A., Best, N.G. and Gilks, W.R. (1996). BUGS: Bayesian inference Using Gibbs Sampling, Version 0.5, (version ii). MRC Biostatistics Unit. See http://www.mrc-bsu.cam.ac.uk

SPSS (1998). Missing Value Analysis. See http://www.spss.com

Statistical Solutions Ltd. (1997). SOLAS for Missing Data Analysis. See http://www.statsol.ie

W3C (1999). Scalable Vector Graphics (SVG) 1.0 Specification. See http://www.w3.org/TR/SVG.

A linear approximation to the wild bootstrap in specification testing

Knut Bartels

University of Potsdam, Faculty of Economics and Social Sciences, Chair for Statistics and Econometrics, August-Bebel-Str. 89, D-14482 Potsdam, Germany

Abstract. The specification of a nonlinear regression model $\mathsf{E}[Y|X=x] = f(x,\theta)$ for $(Y,X) \sim \mathsf{D}$ for a known function $f: \mathbb{R}^d \times \Theta \to \mathbb{R}$ is to be tested. A possible test statistic is $\hat{T}_n = \frac{1}{n}\sum_{1\leq i<j\leq n}^{n} \hat{U}_i K_{ij} \hat{U}_j$, where $\hat{U}_i$ denote parametrically estimated residuals, and K_{ij} are kernel weights. Usually the wild bootstrap algorithm (Wu, 1986) is used for for deriving the critical values. Using the structure of an U-statistic inherent to $\hat{T}_n$, it is possible to approach its limiting distribution directly. The resulting Monte-Carlo-approximation can be viewed as a linear approximation to the wild bootstrap that consumes substantially less computer time for nonlinear models. In simulations this Monte-Carlo-approximation was demonstrated to be applicable. The theoretical foundation lies in an asymptotic consideration that differs from the usual assumptions: The kernel weights K_{ij} depend on a bandwidth h that is held fixed here, in contrary to the usual setting $h = h_n \to 0$. Thus the effects for $n \to \infty$ and $h_n \to 0$ are separated.

Keywords. specification tests, nonlinear regression, resampling methods, simulation studies

1 Test problem

Given is an i.i.d. random sample $Z_1 = (Y_1, X_1), \ldots, Z_n = (Y_n, X_n)$ from a distribution D on $\mathbb{R} \times \mathbb{R}^d$. We want to test if the continuous regression function $\phi(x) := \mathsf{E}[Y|X=x]$ belongs to a parametric family of known real functions $f(x,\theta)$ on $\mathbb{R}^d \times \Theta$ where $\Theta \subseteq \mathbb{R}^p$ denotes a compact parameter space and $(Y,X) \sim \mathsf{D}$. For a measurable function $g: \mathbb{R}^d \to \mathbb{R}$ let $\mathsf{D}\{g\}$ denote the set of all distributions on $\mathbb{R} \times \mathbb{R}^d$ with $\mathsf{E}[Y^2] < \infty$ and $\mathsf{P}\{\mathsf{E}[Y|X] = g(X)\} = 1$. Thus we consider the null hypothesis

$$\mathbf{H_0}: \quad \mathsf{D} \in \mathcal{D}_0 := \bigcup_{\theta\in\Theta} \mathsf{D}\{f(\cdot,\theta)\} \quad . \tag{1}$$

versus

$$\mathbf{H_1}: \quad \mathsf{D} \in \mathcal{D}_1 := \bigcup_{g\in\mathcal{B}(\mathbb{R}^d,\mathbb{R})} \mathsf{D}\{g\} \setminus \mathcal{D}_0 \quad , \tag{2}$$

where the union is taken over all measurable functions $g: \mathbb{R}^d \to \mathbb{R}$.

As a test statistic we use

$$\hat{T}_n = \frac{1}{n} \sum_{1 \le i < j \le n} \hat{U}_i K_{ij} \hat{U}_j, \tag{3}$$

where $\hat{U}_i := u(Z_i, \hat{\theta}_n) = Y_i - f(X_i, \hat{\theta}_n)$ are parametrically estimated residuals and $K_{ij} := k(X_i, X_j)$ are weights derived from a positive definite symmetric kernel k that may depend on a bandwidth h. Any strongly consistent and asymptotically normal parameter estimator $\hat{\theta}_n$ may be applied (see assumption **A3**).

It can be shown that (3), or its version including the diagonal

$$\hat{T}_n^{(v)} = \frac{1}{n} \sum_{1 \le i,j \le n} \hat{U}_i K_{ij} \hat{U}_j = 2\hat{T}_n + \frac{1}{n} \sum_{i=1}^{n} \hat{U}_i^2 K_{ii} \quad , \tag{4}$$

are general forms of the L_2-test statistics that cover, for example the tests of Bierens (1990), Härdle and Mammen (1993), Zheng (1996) or even Stute (1997). In the following the special but main case $K_{ij} = h^{-d} k(\frac{X_i - X_j}{h})$ with a symmetric and bounded kernel function $k = k_h : \mathbb{R}^d \to \mathbb{R}$ is considered. Then the test statistics (3) and (4) are U-statistics, respectively V-statistics, with U-kernel $s_n(z_1, z_2) := u(z_1, \hat{\theta}_n) k(x_1, x_2) u(z_2, \hat{\theta}_n)$ that are asymptotically degenerate under $\mathbf{H_0}$ and nondegenerate under $\mathbf{H_1}$. Further (3) and (4) are asymptotically normal distributed for $n \to \infty$ and $h = h_n \to 0$ with $h_n^{-d} n \to \infty$ (Härdle and Mammen, 1993; Zheng, 1996).

In contrary to this point of view, that is guided by the nonparametric heuristics leading to the test statistics, here the bandwidth h is assumed to be fixed. Then under $\mathbf{H_0}$ and assumptions **A1**, **A2**, **A3** for fixed bandwidths (3) has a development

$$\hat{T}_n = \frac{1}{n} \sum_{1 \le i < j \le n} q(Z_i, Z_j, \theta_0) + \frac{1}{n} \sum_{i=1}^{n} c(Z_i, \theta_0) + o_p(1) \tag{5}$$

with $\mathsf{E}[q(Z_1, Z_2, \theta_0) \,|\, Z_2] = 0$. By a limit law for degenerate U-statistics by Gregory (1977) it has an asymptotic distribution as

$$\mathcal{L}\Big(\sum_{j=1}^{\infty} \lambda_j (\chi_{1j}^2 - 1) + \gamma\Big) \tag{6}$$

where $\gamma \in \mathbb{R}$ is a constant, $\chi_{11}^2, \chi_{12}^2, \ldots$ independent χ_1^2-distributed random variables and λ_j the eigenvalues of the linear integral operator with kernel $q(\cdot, \cdot, \theta_0)$. Under $\mathbf{H_1}$ we have $\hat{T}_n \to +\infty$ in probability for all deviations in the positive eigenspaces of the integral operator with kernel k. Thus for a positive definite kernel function k, e.g. the Gauss-kernel, the test is consistent. Analogous results hold for $\hat{T}_n^{(v)}$. For a detailed description see Bartels (1999).

2 Derivation of critical values

Since the limiting distribution (6) depends on the unknown distribution D, it is case dependent and cannot be tabulated. Thus the critical values for the tests must be adaptively determined by resampling methods, e.g. bootstrap methods which try to estimate unknown quantities by evaluating samples from the given empirical distribution. In the context of nonlinear regression or U-statistics naive methods do not work, since the sum of the empirical residuals in general is not zero. Thus more sophisticated methods like the wild bootstrap or the following Monte-Carlo-approximation must be employed.

Using a theorem of Dehling and Mikosch (1994, Theorem 3.1) on the bootstrap of degenerate U-statistics, resampling methods for approximating the critical values of the tests can be derived. To implement this theorem, first a version on the U-kernel q_n that includes all effects of parameter estimation, denoted by $\hat{q}_n$, must be determined. This is varied by independent standardized random variables $\eta_1, \ldots, \eta_n$ through

$$Q_n^* := \sum_{1 \leq i < j \leq n} \eta_i \eta_j \hat{q}_n(z_i, z_j, \hat{\theta}_n) \,. \tag{7}$$

Then, for almost every sequence $\{z_n\}$ of realizations of $\{Z_n\}$, Q_n^* is (in η-probability) conditional on the given sample $z_1, \ldots, z_n$ asymptotically distributed as $\hat{T}_n - \gamma$. This leads directly to a Monte-Carlo-approximation of the limiting distribution. Only the estimator $\hat{q}_n$ must be derived and then the distribution of (7) can be approximated by the empirical distribution of Q_n^* calculated for sufficiently many sets of random numbers $\eta_1, \ldots, \eta_n$. These methods include the wild bootstrap of Wu (1986), if the additional assumption **A4** is satisfied. Versions with variable bootstrap sample size and resampling Z_{ni}^* from $z_1, \ldots, z_n$ are possible. Also additional assumptions like homoscedasticity or parametric error distribution can be integrated to the Monte-Carlo-approximation. Detailed results are proven in Bartels (1999).

The U-kernel q in (5) depends on the first order effects w of the parameter estimation. Thus only a linear approximation to the effect of parameter estimation is implemented in the Monte-Carlo-approximation, while for the wild bootstrap an estimator $\hat{\theta}_n^*$ must be calculated for every resample. Since the calculation of these estimators takes most of the time in computing the critical values for nonlinear models, with the new Monte-Carlo-approximation this time is reduced substantially.

For applying the Monte-Carlo-approximation the function w from (5) must be known. E.g. for least squares estimation this is $w(z, \theta) = \Sigma^{-1} u(z, \theta) f'(x, \theta)$. In linear models the equation (5) is exact without the $o_p(1)$-terms and with $f'(\cdot, \theta) = const.$ Therefore for linear models and least squares estimation the wild bootstrap and the Monte-Carlo-approximation give identical results, if applied to the test statistic $\hat{T}_n^{(v)}$.

3 Simulations

The nonlinear model $f(x, \theta) = (1 + \theta_2 \exp(-\theta_1 x))^{-1}$ was considered. For the simulations 5000 sets of data with sample size $n = 25$

$$x_{i,1}, \ldots, x_{i,25} \,, \quad \varepsilon_{i,1}, \ldots, \varepsilon_{i,25} \tag{8}$$

with $x_{i,j} \sim U[1-\pi, 1+\pi]$ und $\varepsilon_{i,j} \sim N(0,1)$ for $i \in \{1,\ldots,5000\}$ and $j \in \{1,\ldots,25\}$ were generated. With a sine-disturbance

$$f_a(x_{i,j}, (\mathbf{1},\mathbf{4})) = (1 + \mathbf{4} \cdot e^{-\mathbf{1} \cdot x_{i,j}})^{-1} + a_1 \cdot \sin(a_2 \cdot 2\pi(x_{i,j} - 1)) + \sigma \cdot \varepsilon_{i,j} \quad (9)$$

the final data sets for the simulations for different constellations of disturbance parameters a_1, a_2 but otherwise identical $x_{i,j}$ and $\varepsilon_{i,j}$ were produced; see figure 1.

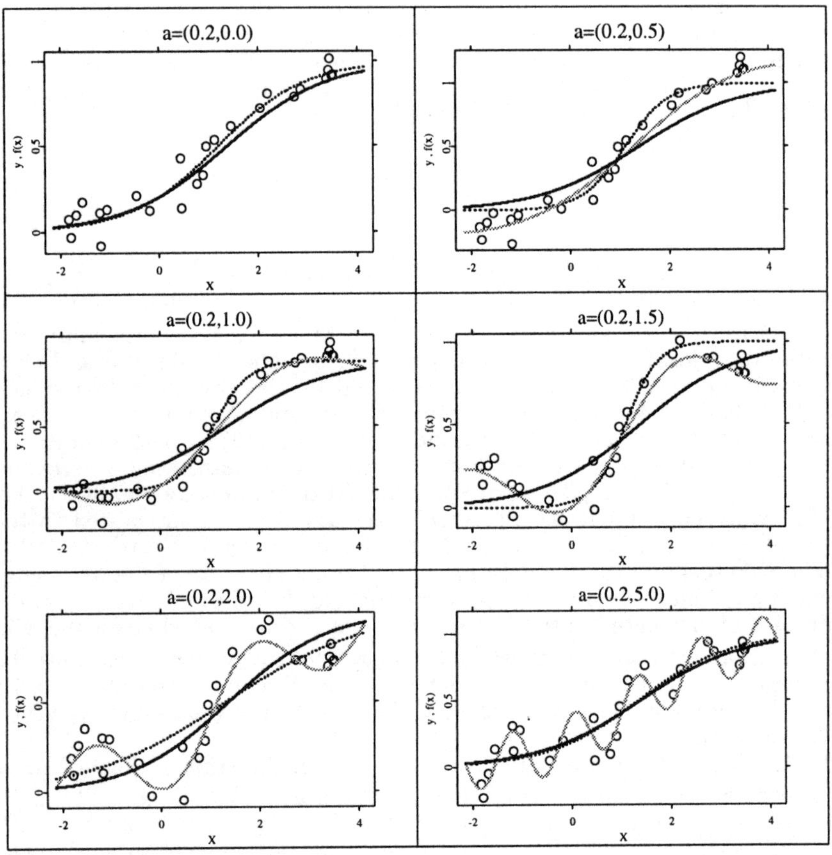

Fig. 1. Nonlinear model (9)(black), disturbed model (grey) and estimation (dashed) for a sample (circles) with $\sigma^2 = 0.01$ and different disturbance parameters

Table 1 lists the percentages of rejection calculated with the wild bootstrap (WBS) and three versions of the Monte-Carlo-approximation: MCA uses η_j's with a distribution satisfying $\mathsf{E}[\eta^3] = 1$, MCN uses standard normal η_j's and MCH uses η_j's drafted from residuals from an constrained least squares

Empirical power at nominal level 0.05 Gaussian kernel, $h = 1.0$, $n = 25$, $a_1 = 0.2$, $\sigma^2 = 0.1$								
method	a_2							
	0.0 ($\mathbf{H_0}$)	0.5	1.0	1.5	2.0	3.0	5.0	9.0
$\hat{T}_n$, WBS	**0.030**	0.076	0.202	0.319	0.088	0.033	0.040	0.025
$\hat{T}_n^{(v)}$, WBS	**0.031**	0.063	0.188	0.310	0.093	0.039	0.044	0.030
$\hat{T}_n$, MCA	**0.063**	0.130	0.285	0.447	0.183	0.077	0.063	0.056
$\hat{T}_n$, MCA, $\hat{c}_\eta$	**0.044**	0.116	0.255	0.392	0.139	0.054	0.046	0.041
$\hat{T}_n^{(v)}$, MCA	**0.047**	0.092	0.226	0.388	0.144	0.060	0.046	0.041
$\hat{T}_n$, MCN	**0.063**	0.132	0.287	0.457	0.188	0.079	0.066	0.058
$\hat{T}_n$, MCN, $\hat{c}_\eta$	**0.047**	0.120	0.264	0.417	0.151	0.061	0.051	0.047
$\hat{T}_n^{(v)}$, MCN	**0.033**	0.074	0.189	0.346	0.118	0.047	0.037	0.031
$\hat{T}_n$, MCH	**0.047**	0.106	0.239	0.405	0.152	0.057	0.053	0.044
$\hat{T}_n$, MCH, $\hat{c}_\eta$	**0.056**	0.134	0.273	0.427	0.160	0.064	0.058	0.050
$\hat{T}_n^{(v)}$, MCH	**0.037**	0.089	0.200	0.350	0.121	0.044	0.042	0.033
upper bounds	**0.005**	0.009	0.043	0.126	0.024	0.007	0.005	0.003
normal ($h \to 0$)	**0.000**	0.000	0.000	0.000	0.000	0.000	0.000	0.000

Table 1. Test for nonlinear model, data according to (9), variation of disturbance parameter a_2 and bandwidth h

Empirical power at nominal level 0.05 Gaussian kernel, $n = 25$, $a_1 = 0.2$, $\sigma^2 = 0.01$, method $\hat{T}_n^{(v)}$-MCA											
	a_2										
h	0.0($\mathbf{H_0}$)	0.5	1.0	1.5	2.0	2.5	3.0	3.5	4.0	5.0	9.0
0.2	0.023	0.236	0.521	0.831	0.819	0.781	0.745	0.694	0.634	0.478	0.089
0.5	0.043	0.507	0.871	0.985	0.935	0.842	0.701	0.352	0.223	0.083	0.045
1.0	0.053	0.521	0.946	0.995	0.659	0.228	0.145	0.075	0.073	0.057	0.053
1.5	0.054	0.483	0.959	0.992	0.245	0.111	0.094	0.062	0.066	0.056	0.054
2.0	0.051	0.451	0.962	0.987	0.166	0.105	0.084	0.064	0.065	0.056	0.056
2.5	0.052	0.384	0.967	0.984	0.149	0.106	0.080	0.065	0.066	0.056	0.056
3.0	0.052	0.291	0.973	0.984	0.131	0.105	0.079	0.065	0.066	0.057	0.056
3.5	0.053	0.206	0.976	0.984	0.117	0.105	0.076	0.067	0.064	0.057	0.055
4.0	0.055	0.147	0.979	0.985	0.104	0.106	0.073	0.068	0.063	0.057	0.056

Table 2. Test for nonlinear model, data according to (9), variation of disturbance parameter a_2 and bandwidth h

estimator assuring $\sum_{i=1}^{n} \hat{U}_i = 0$ (Bartels, 1999). *Upper bounds* refer to the bounds for the critical values according to Bierens and Ploberger (1997) and *normal* to the normal limit distibution for $h = h_n \to 0$ without bootstrapping.

Despite the very small sample size of $n = 25$ and the large variance $\sigma^2 = 0.1$ (chosen to abbreviate calculations) the results are close to the asymptotically expected ones. The results from Monte-Carlo-approximation are not worse than those obtained by wild bootstrap. On a Sun Ultra 300Mhz the calculation of the percentage of rejection for one constellation of parameters took about 5 days for the wild bootstrap and only 4 hours for the Monte-Carlo-approximation.

In table 2 results for different bandwidths are compared to different frequencies (a_2) of the disturbation. The variance $\sigma^2 = 0.01$ is smaller here than for the results in table 1. This, of course, leads to higher rates of rejection. The effect of bandwidth choice arises clearly: The smaller the bandwidth, the better alternatives with high frequency deviation can be detected. But on the other hand the empirical power decreases against low frequency alternatives, and bandwidths cannot be arbitrarily small due to the finite sample size.

4 Appendix

Assumptions

A1 The function $f : \mathbb{R}^d \times \Theta \to \mathbb{R}$ is measurable for every fixed $\theta \in \Theta$ and twice continuously differentiable w.r.t. θ. For $X \sim \mathsf{D}_X$ and every fixed $\theta \in \Theta$ it holds $\mathsf{E}\left[f(X,\theta)^2\right] < \infty$ und $\mathsf{E}\left[[f'(X,\theta)]_\alpha^2\right] < \infty$ with $\alpha \in \{1,\ldots,p\}$. Further for every $\theta \in \Theta$ exists a neighborhood $\Psi = \Psi(\theta) \subseteq \Theta$ with $\mathsf{E}\left[\sup_{\tau\in\Psi}[f''(X,\tau)]_{\alpha\beta}^2\right] < \infty$ for all $\alpha, \beta \in \{1,\ldots,p\}$.

A2 The kernel function $k : \mathbb{R}^d \times \mathbb{R}^d \to \mathbb{R}$ is measurable, symmetric and bounded.

A3 Under $\mathbf{H_0}$ with $\mathsf{D} \in \mathsf{D}\{f(\cdot,\theta)\}$ the estimator $\hat{\theta}_n = \hat{\theta}(Z_1,\ldots,Z_n)$ has a development $n^{\frac{1}{2}}(\hat{\theta}_n - \theta) = n^{-\frac{1}{2}} \sum_{i=1}^n w(Z_i,\theta) + o_p(1)$ for some measurable and w.r.t. θ continuous function $w : \mathbb{R} \times \mathbb{R}^d \times \Theta \to \mathbb{R}^p$ with $\mathsf{E}[w(Z_1,\theta)|X] = 0$. Further for every $\theta \in \Theta$ exists a neighborhood $\Psi = \Psi(\theta) \subseteq \Theta$ with $\mathsf{E}\left[\sup_{\tau\in\Psi}[w(Z,\tau)]_\alpha^2\right] < \infty$ for all $\alpha \in \{1,\ldots,p\}$.

A4 The estimator $\hat{\theta}_n$ is strongly consistent and the function w in **A3** satisfies $w(z,\theta) = w(y,x,\theta) = \tilde{w}(x,\theta) \cdot u(z,\theta)$ for some measurable function $\tilde{w}(\cdot,\theta) : \mathbb{R}^d \times \Theta \to \mathbb{R}^p$.

References

Bartels, K. (1999). *Tests zur Modellspezifikation in der nichtlinearen Regression.* Dissertation Universität Potsdam.

Bierens, H. J. (1990). A consistent conditional moment test of functional form. *Econometrica*, **58** (6), pp. 1443–1458.

Bierens, H. J. and Ploberger, W. (1997). Asymptotic theory of integrated conditional moment tests. *Econometrica*, **65** (5), pp. 1129–1152.

Dehling, H. and Mikosch, T. (1994). Random quadratic forms and bootstrap for U-statistics. *Journal of Multivariate Analysis*, **51**, pp. 392–413.

Gregory, G. G. (1977). Large sample theory for U-statistics and tests of fit. *The Annals of Statistics*, **5** (1), pp. 110–123.

Härdle, W. and Mammen, E. (1993). Comparing nonparametric versus parametric regression fits. *The Annals of Statistics*, **21** (4), pp. 1926–1947.

Stute, W. (1997). Nonparametric model checks for regression. *The Annals of Statistics*, **25** (2), pp. 613–641.

Wu, C.-F. (1986). Jackknife, bootstrap and other resampling methods in regression analysis. *The Annals of Statistics*, **14**, pp. 1261–1350.

Zheng, J. X. (1996). A consistent test of functional form via nonparametric estimation techniques. *Journal of Econometrics*, **75**, pp. 263–289.

The size of the largest nonidentifiable outlier as a performance criterion for multivariate outlier identification: the case of high-dimensional data

Claudia Becker

Department of Statistics, University of Dortmund, D-44221 Dortmund, Germany

Abstract. Various procedures exist for identifying outliers in multivariate data. To decide which identification rule should be chosen, several performance criteria can be used. We investigate here the problem of multivariate simultaneous outlier identification and concentrate on the criterion of the size of the largest nonidentifiable outlier. Four outlier identification rules are compared with respect to this criterion. Our main focus is on a comparison of the rules in high-dimensional data situations, and we present the results of a simulation study in an accordingly chosen 10-dimensional setting.

Keywords. MVE, MCD, outlier detection, robustness, S-estimators

1 Multivariate outlier identification

In the literature on statistical data analysis, there exists a variety of procedures for the detection of outliers with respect to several null models. Such procedures include for example graphical methods, block procedures, or consecutive testing (see Gather and Becker, 1997). In this paper, we deal with simultaneous outlier identification procedures which are designed to detect an unknown number of outliers in a data set in one step with respect to a normal reference model. Procedures of this kind have been suggested by various authors for the situation of univariate data as well as for the multivariate situation (e.g. Barnett and Lewis, 1994, Davies and Gather, 1993, Hampel, 1985, Rousseeuw and van Zomeren, 1990, Wilks, 1963). We investigate methods for the multivariate setting as considered by Gather and Becker (1997) who adapt the concept of univariate α_N outlier identifiers of Davies and Gather (1993).

Taking the p-variate normal $N(\boldsymbol{\mu}, \boldsymbol{\Sigma})$, $\boldsymbol{\mu} \in I\!R^p$, $\boldsymbol{S} \in I\!R^{p\times p}$ positive definite, as model distribution, the so-called α_N outlier region $\boldsymbol{out}(\alpha_N, \boldsymbol{\mu}, \boldsymbol{\Sigma})$ of $N(\boldsymbol{\mu}, \boldsymbol{\Sigma})$ is defined as

$$\boldsymbol{out}(\alpha_N, \boldsymbol{\mu}, \boldsymbol{\Sigma}) = \{\boldsymbol{x} \in I\!R^p : (\boldsymbol{x} - \boldsymbol{\mu})^T \boldsymbol{\Sigma}^{-1} (\boldsymbol{x} - \boldsymbol{\mu}) > \chi^2_{p;1-\alpha_N}\}.$$

The choice of $\alpha_N = 1 - (1 - \alpha)^{1/N}$ ensures that

$$P_{N(\boldsymbol{\mu},\boldsymbol{\Sigma})}(\text{no observation in a sample of size } N \text{ lies in } \boldsymbol{out}(\alpha_N, \boldsymbol{\mu}, \boldsymbol{\Sigma})) = 1-\alpha$$

for some given value of $\alpha \in (0, 1)$. Any observation lying in $\boldsymbol{out}$ is then called an α_N outlier.

To get an α_N outlier identifier, we estimate the α_N outlier region by

$$\boldsymbol{OR}(\underline{\boldsymbol{x}}_N, \alpha_N) = \{\boldsymbol{x} \in \mathbb{R}^p : (\boldsymbol{x} - \boldsymbol{m})^T \boldsymbol{S}^{-1}(\boldsymbol{x} - \boldsymbol{m}) \geq c(p, N, \alpha_N)\},$$

where $(\boldsymbol{m}, \boldsymbol{S})$ denotes a pair of affine equivariant estimators for $(\boldsymbol{\mu}, \boldsymbol{\Sigma})$, $\underline{\boldsymbol{x}}_N = (\boldsymbol{x}_1, \dots, \boldsymbol{x}_N)$, $\boldsymbol{x}_i \in \mathbb{R}^p$, a given sample of size N in which we want to detect the outliers. The normalizing constant $c(p, N, \alpha_N)$ is calculated according to

$$P_{N(\boldsymbol{\mu},\boldsymbol{\Sigma})}(\text{no } \alpha_N \text{ outliers identified in a sample of size } N) = 1 - \alpha$$

with $\alpha_N = 1 - (1 - \alpha)^{1/N}$ as before. An observation $\boldsymbol{x}_i \in \boldsymbol{OR}(\underline{\boldsymbol{x}}_N, \alpha_N)$ is then detected as an outlier (see Davies and Gather, 1993, Gather and Becker, 1997, or Becker and Gather, 1999, for details).

We take outlier identifiers which are all derived according to the concept given above, but differ in the way they estimate $\boldsymbol{\mu}$ and $\boldsymbol{\Sigma}$. The identifiers we investigate are based on the following robust estimators of location and scatter:

- MVE estimators,
- MCD estimators,
- S-estimators using Tukey's biweight (BW),
- S-estimators using the translated biweight (TW)

(Rousseeuw, 1985, Rocke, 1996), all of which are solutions to certain minimization problems in $\text{PDS}(p) \times \mathbb{R}^p$. Here, $\text{PDS}(p)$ denotes the set of all positive definite symmetric $p \times p$ matrices. For the MVE estimators, construct the ellipsoid of minimum volume which contains a certain number h of the observations; the centre and shape matrix of this ellipsoid then yield estimators of location and covariance. The MCD estimators are derived similarly as sample mean and sample covariance of that subsample of a certain size h which leads to the covariance matrix with minimum determinant. Both covariance estimators have to be normalized to guarantee consistency at the normal distribution (Rousseeuw, 1985, Rousseeuw and van Driessen, 1999). S-estimators are solutions of the minimization problem $\min_{\text{PDS}(p)} \det(\boldsymbol{S})$ subject to some constraint $1/N \sum_{i=1}^N \rho\left(((\boldsymbol{x}_i - \boldsymbol{m})^T \boldsymbol{S}^{-1}(\boldsymbol{x}_i - \boldsymbol{m}))^{1/2}\right) = b_0$ for a suitable function ρ and a constant b_0. For the biweight S-estimators (Rocke, 1996), take

$$\rho(d) = \rho_{\text{BW}}(d) = \begin{cases} d^2/2 - d^4/(2c_0^2) + d^6/(6c_0^4) & , 0 \leq d \leq c_0, \\ c_0^2/6 & , d > c_0. \end{cases}$$

The translated biweight S-estimators are considered by Rocke (1996) in the context of analyzing high-dimensional data. They are derived from

$$\rho(d) = \rho_{\text{TW}}(d) = \begin{cases} d^2/2, & 0 \leq d \leq M, \\ M^2/2 - M^2(M^4 - 5M^2c_0^2 + 15c_0^4)/(30c_0^4) & \\ \quad + d^2\left(1/2 + M^4/(2c_0^4) - M^2/(c_0^2)\right) & \\ \quad + d^3\left(4M/(3c_0^2) - 4M^3/(3c_0^4)\right) & \\ \quad + d^4\left(3M^2/(2c_0^4) - 1/(2c_0^2)\right) & \\ \quad - d^5\left(4M/(5c_0^4)\right) + d^6/(6c_0^4), & M \leq d \leq M + c_0, \\ M^2/2 + c_0(5c_0 + 16M)/30, & d > M + c_0. \end{cases}$$

The parameters h, b_0, c_0, M can be chosen such that all estimators achieve the maximum possible finite-sample breakdown points (see Rousseeuw, 1985, Davies, 1987, Rocke, 1996, for details).

2 Performance criteria

For the comparison of outlier identification rules, various performance criteria may be used. Among them are worst-case criteria like masking- and swamping-breakdown points and average-case considerations like determining the proportion of correctly classified observations or the proportion of correctly identified outliers (cf. Davies and Gather, 1993, Gather and Becker, 1997, Gather and Hilker, 1997, Gather, Becker and Kuhnt, 2000). According to theoretical considerations, in multivariate simultaneous outlier identification rules, the use of robust estimators with high breakdown point in the sense of Donoho and Huber (1983) is recommended to achieve protection against masking and swamping effects (Becker and Gather, 1999). Several such robust estimators with maximum possible finite-sample breakdown point have been proposed, such as the estimators we use here, namely MVE and MCD estimators or S-estimators (Rousseeuw, 1985, Rocke, 1996). The resulting outlier identification procedures behave equally well with respect to masking breakdown, having the highest possible masking breakdown point (Becker and Gather, 1999).

Supplementary insight into the performance of the rules can be given by the results of simulation studies. For the univariate normal as reference distribution, Davies and Gather (1993) introduce the maximum asymptotic bias of an estimator in a given outlier situation and develop the performance criterion of the size of the largest nonidentifiable outlier. Becker and Gather (2000) transfer this concept to the multivariate situation. They define the size of the largest nonidentifiable outlier with respect to the p-variate standard normal distribution as

$$\sup_{\underline{x}_k^0}\{\|\boldsymbol{y}\| : \boldsymbol{y} \in \underline{x}_k^0 \text{ and } (\boldsymbol{y} - \boldsymbol{m}(\underline{x}_{N,k}))^T \boldsymbol{S}(\underline{x}_{N,k})^{-1}(\boldsymbol{y} - \boldsymbol{m}(\underline{x}_{N,k})) < c(p, N, \alpha_N)\}.$$

Here, $\underline{x}_{N,k}$ denotes a sample of size N from $N(\mathbf{0}, \mathbf{I})$, in which a number k of observations is replaced by δ_N outliers for some suitably chosen δ_N. That means, $\underline{x}_{N,k} = (\underline{x}_n^r, \underline{x}_k^0)$ with $\underline{x}_n^r = (\boldsymbol{x}_{i_1}, \ldots, \boldsymbol{x}_{i_n})$ "regular" observations from the original sample, coming i.i.d. from $N(\mathbf{0}, \mathbf{I})$, and $\underline{x}_k^0 = (\boldsymbol{y}_1, \ldots, \boldsymbol{y}_k)$ with $\boldsymbol{y}_i \in \boldsymbol{out}(\delta_N, \mathbf{0}, \mathbf{I})$, $N = n + k$. It suffices to restrict on $N(\mathbf{0}, \mathbf{I})$ because we only deal with identification rules based on affine equivariant estimators.

The size of the largest nonidentifiable outlier tells us how "far away" from the centre of the "good" data outliers can be located before they are detected. In this sense, it can also be interpreted as a measure of bias of an outlier identification rule.

3 Simulations

In Becker and Gather (2000), three of the simultaneous outlier identification rules mentioned above are compared by simulation with respect to the largest nonidentifiable outlier for the case of low to medium dimension p. In this paper, we investigate the case of higher dimension which is known to be essentially different due to the so-called "curse of dimensionality". Becker and Gather (2000) found that for lower dimensions ($p = 2, 3, 4$) the outlier identifier based on the BW S-estimators yields good results in most of the situations they investigated, and thus should be preferred. The TW identifier cannot compete with the other rules in situations with low to medium p and is thus excluded from their study. In this paper, we investigate the case of $p = 10$ as an example of a high-dimensional situation, where the TW estimators should do better. All calculations are made using the statistical package S-plus (version 4.5). The MVE and MCD estimators are derived by using the respective functions cov.mve and cov.mcd. The S-estimators are calculated according to the iterative algorithm given in Rocke (1996) with MVE or MCD as initial estimators (called BW/MVE, BW/MCD, TW/MVE, TW/MCD, respectively).

Our simulation design is as follows. We take the case of $p = 10$ and $N = 500$; for each of the four identifiers we first calculate the normalizing constant $c(p, N, \alpha_N)$ for $\alpha = 0.1$ (i.e. $\alpha_N = 0.0002107$) by simulation. We then generate samples of size N from $N(\mathbf{0}, \mathbf{I})$. Following the arguments of Becker and Gather (2000), to create a worst-case situation for the outlier identification rules investigated here, we replace a number k of the observations in each sample by putting them all on some point $\boldsymbol{y}$. We then investigate how large we can make $||\boldsymbol{y}||$ while these observations are not identified as outliers by the respective rules. The direction of $\boldsymbol{y}$ does not matter, because we are working with respect to $N(\mathbf{0}, \mathbf{I})$, thus we chose $\boldsymbol{y}$ to be of the form $(y_1, 0, \ldots, 0)$. The maximum value of $||\boldsymbol{y}||$ gives the size of the largest nonidentifiable outlier. We report the mean of these sizes out of 500 simulation runs for $k = 50(10)150$. The results are shown in Figures 1-3.

There are several conclusions we can draw from these results. First, we see from Figure 1 that for moderate numbers of outliers, the MCD identifier clearly outperforms the MVE identifier. The picture changes for large numbers of outliers, where now the MVE identifier leads to drastically smaller sizes of the largest nonidentifiable outlier.

The second result of the simulations concerns the use of the different S-estimators. In Figure 2, the results are shown for the case with MVE as initial estimators. The case of MCD is similar. Although the TW estimators were considered in the context of high-dimensional data, the resulting outlier identification procedure breaks down with respect to the size of the largest nonidentifiable outlier in the situation investigated for a sample of size $N = 500$ in $p = 10$ dimensions when there are $k \geq 80$ maliciously placed outliers present in the sample. The BW identifier does not break down in the same situation. Since 500 observations are not very many when the data are 10-dimensional, it seems to be necessary to supplement our findings by further simulations. First results of taking a larger sample size ($N = 2000$ for $p = 10$) show a similar behaviour of the BW and TW identifiers, but detailed investigations should follow.

The third result refers to the choice of the initial estimate for the BW iteration. From Figure 3 it can be seen that there is almost no difference between the mean results of the MCD and the BW/MCD identifier as well as between the MVE and the BW/MVE identifier. But as there is a remarkable difference between the behaviour of the MCD and the MVE identifier, we see that the BW identifier behaves different depending on the choice of the initial estimator in the iteration. In fact, we get two different approximations of the "true" S-estimator here. There are two possible reasons for this. It may be the case that in the high-dimensional situation, the iteration step of the S-estimator algorithm does not change the initial estimate very much. In this case, contrary to the situation of low to medium dimension

Figure 1: Largest nonidentifiable outliers for MVE and MCD identifier

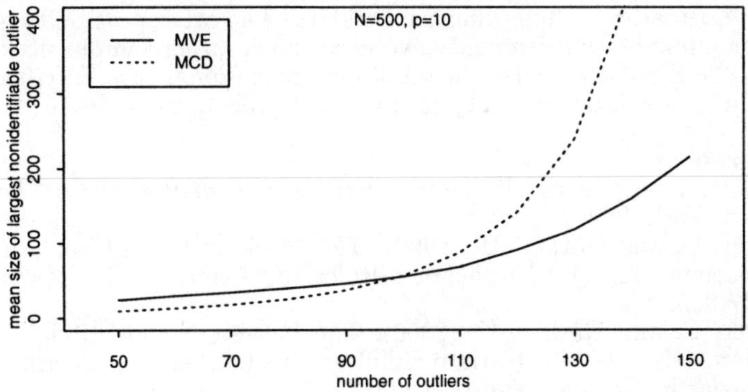

Figure 2: Largest nonidentifiable outliers for BW and TW identifier

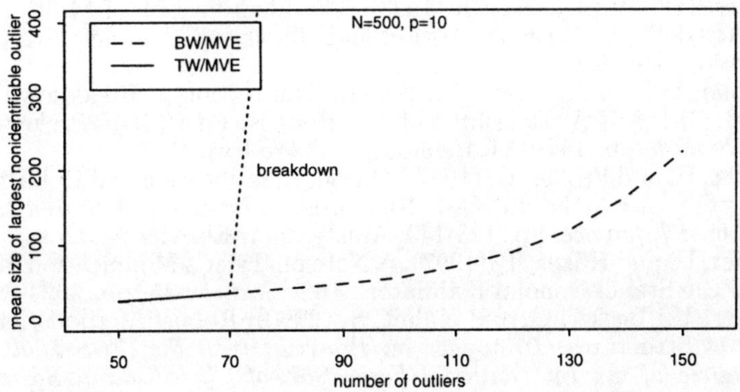

Figure 3: Largest nonidentifiable outliers for MVE, BW/MVE, MCD and BW/MCD identifier

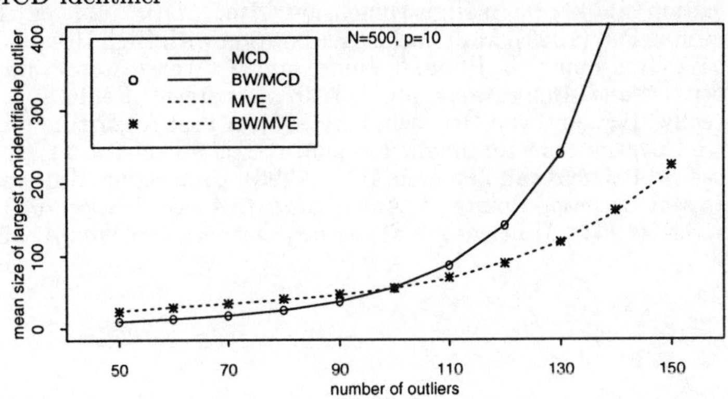

p (see Becker and Gather, 2000), the use of S-estimators according to the present algorithm will not lead to an improvement in simultaneous outlier identification. This would mean that we need special algorithms for calculating S-estimators in high-dimensional data. The other possibility is that, due to the curse of dimensionality, we need much larger sample sizes to benefit from the S-estimators like in the lower dimensional case. Further extensive simulations will be necessary to investigate this in more detail.

References

Barnett, V. and Lewis, T. (1994). *Outliers in Statistical Data: Third Edition.* New York: Wiley.

Becker, C. and Gather, U. (1999). The Masking Breakdown Point of Multivariate Outlier Identification Rules. *J. Amer. Statist. Assoc.*, **94**, pp. 947-955.

Becker, C. and Gather, U. (2000). The Largest Nonidentifiable Outlier: A Comparison of Multivariate Simultaneous Outlier Identification Rules. To appear in *Comput. Statist. Data Anal.*

Davies, P.L. (1987). Asymptotic Behaviour of S-Estimates of Multivariate Location Parameters and Dispersion Matrices. *Ann. Statist.*, **15**, pp. 1269–1292.

Davies, P.L. and Gather, U. (1993). The Identification of Multiple Outliers. Invited Paper With Discussion and Rejoinder. *J. Amer. Statist. Assoc.*, **88**, pp. 782-801.

Donoho, D.L. and Huber, P.J. (1983). The Notion of Breakdown Point. In: P.J. Bickel, K.A. Doksum and J.L. Hodges (Eds.), *A Festschrift for Erich Lehmann*, pp. 157–184. Belmont, CA: Wadsworth.

Gather, U. and Becker, C. (1997). Outlier Identification and Robust Methods. In: G.S. Maddala and C.R. Rao (Eds.), *Handbook of Statistics, Vol. 15: Robust Inference*, pp. 123-143. Amsterdam: Elsevier.

Gather, U. and Hilker, T. (1997). A Note on Tyler's Modification of the MAD for the Stahel-Donoho Estimator, *Ann. Statist.*, **25**, pp. 2024-2026.

Gather, U., Becker, C. and Kuhnt, S. (2000). Robust Methods for Complex Data Structures. To appear in: *Proceedings of the IFCS 2000, 7th Conference of the International Federation of Classification Societies, Data Analysis, Classification and Related Methods, Namur, July 11-14, 2000.*

Hampel, F.R. (1985). The Breakdown Points of the Mean Combined With Some Rejection Rules. *Technometrics*, **27**, pp. 95–107.

Rocke, D.M. (1996). Robustness Properties of S-Estimators of Multivariate Location and Shape in High Dimension. *Ann. Statist.*, **24**, pp. 1327-1345.

Rousseeuw, P.J. (1985). Multivariate Estimation with High Breakdown Point. In: W. Grossmann, G. Pflug, I. Vincze and W. Wertz (Eds.), *Mathematical Statistics and Applications*, pp. 283-297. Dordrecht: Reidel.

Rousseeuw, P.J., and van Driessen, K. (1999). A Fast Algorithm for the Minimum Covariance Determinant Estimator, *Technometrics*, **41**, pp. 212–223.

Rousseeuw, P.J. and van Zomeren, B.C. (1990). Unmasking Multivariate Outliers and Leverage Points, *J. Amer. Statist. Assoc.*, **85**, pp. 633-639.

Wilks, S.S. (1963), Multivariate Statistical Outliers, *Sankhya A*, **25**, pp. 407-426.

The influence of data generation and imputation methods on the bias of factor analysis of rating scale data

Coen A. Bernaards

Department of Methodology and Statistics, Utrecht University, P.O. Box 80140, 3508 TC Utrecht, The Netherlands.

Abstract. This paper focuses on the bias as a result of imputation methods applied to psychological questionnaire data. Multidimensional rating scale data were generated using three different models. A simulation was carried out with, among others, factors Method of Data Generation, and Imputation Method. It was found that imputation of the mean for each person separately had little bias whereas item mean imputation could result in severe underestimation of factor loadings.

Keywords. Factor analysis, imputation, missing data, multidimensional item response theory, psychometrics

1 Introduction

Psychological questionnaires are often used to measure traits or attitutes in a group of respondents. Typical examples are intelligence tests and opinion polls. Questions, usually called "items" in psychology, are mainly of the rating scale type, i.e. with five ordered answer categories such as "no preference at all", "no explicit dislike", "weak preference", "positive affection", and "strong preference". In this paper, ordered items with outcomes 1 (lowest category) through 5 (highest category) are used throughout. After data have been collected, the structure of the item list is often studied using factor analysis. Factor analysis uncovers the mutual relations between items. This way it is determined which item measures which trait. However, often item responses are the result of multiple traits. For example, an item on introversion could partly be determined by language skills. The phenomenom that more than one trait is measured by an item is known in psychology as "multidimensionality" (not to be confused with the mathematical definition of multidimensionality). Throughout this paper it was assumed that two traits underly the item responses; an item measured the first latent trait with a certain preset weight, and the second latent trait with some other preset weight. These weights are technically called Mixing configuration.

A problem in any statistical analysis is item nonresponse. Item nonresponse occurs when a respondent replied to at least one item but refused to reply to another. Item nonresponse, if not treated correctly, may seriously hamper consecutive analyses. In the present paper it is investigated how simple imputation methods influence the bias of factor loadings in psychological questionnaires. Bias is studied for three methods of data generation popular in simulation of psychological questionnaire data.

The two-dimensional factor analysis model with orthogonal factors is used to study the structure of rating scale data. This model contains parameters for factor loadings denoted by Λ, item means μ, unique variances $U \sim N(0, \Xi)$ where Ξ is diagonal, and person specific factor scores $f \sim N_2(0, I_2)$. The vector of k item responses X is decomposed in

$$X = \Lambda f + \mu + U. \tag{1}$$

The factor analysis model is studied extensively in the literature, see e.g. Harman (1976).

The regular starting point for simulations are population values of a quantity of interest, here factor loadings. In simulation of rating scale data factor loadings are commonly not the starting point because two methods regularly used in the generation of rating scale data are from the class of models called Item Response Theory (IRT) models. The mathematical relation between these IRT models and factor analysis is known only in a limited number of cases, see Takane and de Leeuw (1987). Therefore, in this study, it was not possible to start from the population values of factor loadings directly, but instead we started from the population values obtained by generating 100000 simulees for each cell using IRT models. Next, each set of 100000 draws was factor analyzed and the obtained loadings served as population factor loadings. For comparison, a third method of data generation was implemented. It used the factor analysis model as a starting point and discretized data generated from this model.

In simulation studies, an undesirable situation occurs if the method of data generation is an additional factor in the final results and the analyst is not aware of that. If similar data had been generated using an alternative method, results should not change dramatically, and conclusions should be the same. In this paper we aimed at studying the influence of three rating scale data simulation methods, and four imputation methods on the bias of factor loadings under nonresponse.

A simulation study was carried out with design factors Method of Data Generation, Mixing Configuration, Correlation between Traits, Nonresponse Mechanism, Percentage Nonresponse, and Imputation Method. Fixed levels were Maximum Likelihood Factor Extraction, Varimax Rotation, Sample Size 100, Two-dimensional Latent Traits, and 20 Items.

2 Method

2.1 Design

Each design factor is discussed separately.

2.1.1 Method of data generation

Three methods which are all applied in practice, were used to generate two-dimensional item response data. The first IRT method was of the logistic type; see e.g. Bernaards and Sijtsma (1999, in press). Weighting coefficients B with indices for the item and the latent traits, and category separation parameters Ψ with indices for items and categories were the model parameters. B can be explained as the discriminating power of an item, and B determined the mixing configuration of the latent traits (see subsubsection 2.1.2). Ψ can be explained as thresholds for the categories. Latent traits are denoted by θ_1 and θ_2. For respondent i having values for the latent traits θ_{i1} and θ_{i2}, the probability of responding in category x to item j, was defined as

$$P(X_{ij} = x | \theta_{i1}, \theta_{i2}) = \frac{\exp(\theta_{i1} B_{j1} x + \theta_{i2} B_{j2} x - \Psi_{jx})}{\sum_{y=1}^{5} \exp(\theta_{i1} B_{j1} y + \theta_{i2} B_{j2} y - \Psi_{jy})} \tag{2}$$

The second IRT method used for generation of two-dimensional item response data was based on the cumulative normal distribution (usually called "normal ogive function" in psychology). It used the same weighting coefficients B as the logistic based model. The category treshold parameters Ψ^* were different from the parameters Ψ in model (2). The probability of responding in category x or lower was defined as

$$P(X_{ij} \leq x|\theta_{i1}, \theta_{i2}) = \Phi\left(\theta_{i1}B_{j1} + \theta_{i2}B_{j2} - \Psi^*_{jx}\right) \tag{3}$$

where $\Phi(z)$ is the standard normal distribution function evaluated at z. By definition, $P(X_{ij} < 1|\theta_{i1}, \theta_{i2}) = 0$, and $P(X_{ij} \leq 5|\theta_{i1}, \theta_{i2}) = 1$.

The third method used population factor loadings for data generation, and was based on discretization of a multivariate normal distribution; a method used in, for example, Muthén and Kaplan (1985). Data were generated using fixed Λ, μ, and Ξ. First, for a particular design cell, Λ, μ, and Ξ were chosen to be the population values based on 100000 draws of the normal ogive IRT model (3). Next, for each sample, latent traits values θ were drawn from a bivariate normal distribution with variance 1 and preset correlation. The same sample θ was used in all three models for generating data. The drawn θs were substituted for f, and, together with Λ, μ, and a realization of $U \sim N(0, \Xi)$, were put in the model (1). The realizations $x_1, \ldots, x_{100}$ of the model (1) with 2 factors, were discretized as follows. For simulee i, having the same θ across all three data generation methods, the scores 1, 2, 3, 4, and 5 in the normal ogive IRT sample were tabulated. Next, in the realization from the factor analysis model for simulee i, x_i, there are as many 1's assigned to the lowest values in x_i as there were 1's in the normal ogive sample. An analogous procedure was followed to obtain scores 2, 3, 4, and 5.

2.1.2 Mixing configuration

Using each of these methods, data were generated with the following structure. Twenty two-dimensional items were simulated which were divided into two groups of ten items each. Each group of ten items consisted of three easy items (most respondents reply in category 4), four items of medium difficulty (most respondents reply in category 3), and three difficult items (most respondents reply in category 2). The first ten items mainly measured latent trait θ_1 and the last ten items mainly measured latent trait θ_2. The latent traits had mixing configuration as follows: if the first ten items measured the latent traits in the mixing configuration 1:3, then the last ten items measured the latent traits in the mixing configuration 3:1. This configuration was called Mix 1:3 throughout. Similarly for the mixing configuration 1:0. The mixing configuration 1:0 can be explained as two groups of ten items where each group measured only one trait. This mixing configuration was called Mix 1:0 throughout.

2.1.3 Correlation between traits

This simulation considered correlated multidimensional items. Correlations were fixed at -0.5, 0, and 0.5.

2.1.4 Nonresponse mechanism

Three nonresponse mechanisms were considered. In the first mechanism (denoted MCAR) the missing item scores were a random sample of all scores. For the second mechanism (denoted HIGH) the higher answer categories had a higher probability of being missing than the lower categories. More specifically, category 5 had probability twice as high as category 4 to be missing;

category 4 had probability twice as high as category 3 to be missing; category 3 had probability twice as high as category 2 to be missing. Category 1 was always observed. In the third mechanism (denoted BOTH) the extreme categories had a higher probability of being missing than the middle categories. In case BOTH, categories 1 and 5 had equal probability of being missing. Categories 2 and 4 had probability of being missing only half of categories 1 and 5. Category 3 was always observed.

2.1.5 Percentage nonresponse

The percentages of nonresponse used here, were 10 percent and 30 percent.

2.1.6 Imputation methods

Imputation methods are widely used as a method for dealing with missing data. Bernaards and Sijtsma (1999, in press) investigated a number of imputation methods for nonresponse in factor analysis of rating scale data under several nonresponse mechanisms. They reported that imputation of the mean for each person (PM) separately worked well with respect to the measure of comparison used by them. Imputation of the mean for each item separately (IM) did not work as well as PM, but is a standard option in commercial software packages such as SPSS. Also, two versions were considered which added a draw from the residual standard error. The residual standard error was calculated by subtracting the observed data matrix from the matrix based only on imputed scores (the matrix of scores under the imputation model). The residual variance was calculated across all differences (missing values were omitted). Thus, in this paper, four imputation methods were considered: PM, PM+error, IM, IM+error. Other imputation methods are widely available, however, these are either too naive (e.g. overall mean imputation and random imputation), or require a number of completely observed variables or subjects (e.g. regression imputation and hot-deck imputation).

2.2 Comparing simulation results

The method used here to compare the simulation results was the mean of the biases of the factor loadings. The estimated factor loadings were compared with the population values of the parameters, i.e. the factor analysis results based on the 100000 draws for the IRT models and the population factor loadings for the discretized normal distribution. Comparison was done for each of group of ten items separately and for each latent trait separately. Similarly, the standard deviations of the bias of the factor loadings were obtained across 50 replications for each group of ten items and each factor separately. Since no distributional assumptions underly factor loadings, tests on factor loadings were carried out assuming that asymptotic normality holded elementswise.

3 Results and discussion

3.1 Results

In general the bias seldomly exceeded twice the accompanying standard error with sample size fixed at 100. For each imputation method and each design factor, first the average bias is discussed, followed by the estimated standard error of the bias. Interactions between design factors are discussed if necessary.

3.1.1 Method of data generation

Keeping other design factors fixed, IM and IM+error, and PM and PM+error all had approximately equal bias across all data generation methods. However, the accompanying standard error of the bias was 1 to 3 times larger for the discretized normal data generation method than for the other two methods.

3.1.2 Mixing configuration

For IM and IM+error, generally factor loadings were less biased for Mix 1:0 than for Mix 1:3. However, the accompanying standard error was lower for Mix 1:3 than for Mix 1:0. For Mix 1:3, and 30 percent nonresponse with correlation 0.5 between traits, and normal ogive and logistic based data generation under all nonresponse mechnisms, the bias exceeded twice the standard error of the bias, already for sample size 100. Note that IM and IM+error do not take any relation between items into account. This is a drawback especially for Mix 1:3.

For PM and PM+error, and normal ogive and logistic based data generation the bias and its accompanying standard error were approximately equal for Mix 1:0 and Mix 1:3. For the discretized normal data generation method, factor loadings tended to be more biased and less dispersed for Mix 1:3 than for Mix 1:0. However, for PM and PM+error, for the discretized normal data generation method the bias small compared to IM and IM+error.

3.1.3 Correlation between traits

For IM and IM+error, under all data generation methods, for correlation 0.5 the bias was 1 to 2 times the bias for correlation -0.5. The reverse holded for the standard error of the bias; for correlation -0.5 the standard error was 1 to 2 times the standard error for correlation 0.5.

For PM and PM+error, for all data generation methods, the bias was approximately equal for all correlations between traits. The standard error for correlation 0.5 was 1 to 1.5 times the standard error for correlation -0.5.

Notably, for all correlations between traits, and keeping other design factors fixed, the bias and its standard error were generally smaller for PM and PM+error than for IM and IM+error.

3.1.4 Nonresponse mechanism

For IM and IM+error, for all data generation methods, the bias was smallest for nonresponse mechanism MCAR, and highest for nonresponse mechanism BOTH. The bias for BOTH was 1 to 2 times higher than for MCAR, and the bias for HIGH was 1 to 1.5 times higher than for MCAR. The standard error of the bias was approximately equal for all nonresponse mechanisms. Since MCAR does not rely on any missing data assumption, it was to be expected that MCAR had the smallest bias. BOTH resulted in peaked items, and HIGH resulted in skewed items. Peaked items result in smaller correlations than skewed items, and, therefore, factor analysis may be hampered more by BOTH nonresponse than by HIGH nonresponse.

For PM and PM+error, the bias was approximately equal across nonresponse mechanisms for all data generation methods. The standard error of the bias was approximately equal for the normal ogive data generation method across all nonresponse mechanisms. The bias of the logistic based data generation method was 1.5 to 2 times the standard error of the normal ogive method across all nonresponse mechanisms. The standard error of the discretized normal data generation method was 2 to 3 times the standard error of the normal ogive data generation method across all nonresponse mechanisms. Thus, based on bias, PM and PM+error is useful under all nonresponse mechanisms. This finding is in line with Bernaards and Sijtsma (1999, in press).

3.1.5 Percentage nonresponse

Keeping all other design factors fixed, bias and standard error were much lower for 10 percent nonresponse than for 30 percent nonresponse. Under 30 percent nonresponse the bias was 2 to 4 times the bias under 10 percent nonresponse. Under 30 percent nonreponse, the standard error was 2 to 3 times the standard error under 10 percent nonresponse.

3.1.6 Imputation methods

In general, IM and IM+error had a negative bias in almost all cells. IM and IM+error underestimated correlations, and, consequently, factor loadings were underestimated ranging from 0 to 0.23. PM and PM+error did not result in severe bias of estimated factor loadings.

Keeping other design factors fixed, IM+error was more biased than IM. Also, the accompanying standard error was bigger for IM+error than for IM across all data generation methods.

Keeping other design factors fixed, the bias was smaller for PM+error than for simply imputing PM. The standard errors of the bias were approximately equal for PM and PM+error.

3.2 Discussion

In the present paper it was investigated how bias of factor loadings was influenced by imputation methods applied to several methods used to simulate rating scale data.

The way in which psychological questionnaire data is simulated did not seriously affect the bias of factor loadings. In practical questionnaire simulation, this implies that a method of data generation can be chosen which suits the purpose of the study and the researcher most.

Although the average bias did seldomly exceed two times the population standard error of the factor loadings, IM and IM+error were often close to that. Therefore, it can be expected that, when the sample size increases, IM and IM+error result in severely underestimated factor loadings. Hence, the use of IM and IM+error is not advised in practice, whereas PM and PM+error did not bias factor loadings much. These findings are in line with Bernaards and Sijtsma (1999, in press). Thus, practitioners are advised to apply an imputation method that uses PM when dealing with multidimensional rating scale data suffering from item nonresponse, and factor analysis of the data is envisaged.

References

Bernaards, C.A., and Sijtsma, K. (1999). Factor analysis of multidimensional polytomous item response data suffering from ignorable item nonresponse. *Multivariate Behavioral Research*, **34**, 277-313.

Bernaards, C.A., and Sijtsma, K. (in press). Influence of imputation and EM methods on factor analysis when item monresponse in questionnaire data is nonignorable. *Multivariate Behavioral Research.*

Harman, H.H. (1976). *Modern Factor Analysis.* Chicago: The University Chicago Press.

Muthén, B. and Kaplan, D. (1985). A comparison of some methodologies for the factor analysis of non-normal Likert variables. *British Journal of Mathematical and Statistical Psychology*, **38**, 171-189.

Takane, Y, and de Leeuw, J. (1987). On the relationship between item response theory and factor analysis of discretized variables. *Psychometrika*, **52**, 393-408.

Model on a population and prediction on another one: a generalized discriminant rule

Christophe Biernacki[1], Farid Beninel[2] and Vincent Bretagnolle[3]

[1] Université de Besançon, UMR CNRS 6623, Besançon, 25030, FRANCE
[2] IUT département STID, Niort, 79000, FRANCE
[3] CEBC-CNRS, Beauvoir sur Niort, 79360, FRANCE

Keywords. model-based discriminant analysis, relationship between populations, biometrical variables, sex determination, model selection

1 Introduction

1.1 The problem

Traditionally, discriminant analysis in a decision purpose proceeds in the following manner (McLachlan 1992): A sample is drawn from a population and a partition of this sample in two classes, males and females, is known. Using some variables, an allocation rule is established in order to classify other elements of the previous population. An underlying assumption of this procedure is that the learning sample is representative of the population, i.e. its parameters about the predictive features are statistically not different from the ones of the population.

In this paper, we are interested in the statistical discriminant analysis problem where the learning sample comes from a different population of the entities to classify. The aim is to obtain a generalized discriminant rule, it means a solution halfway between standard discriminant analysis and cluster analysis. Concrete illustration is provided by mature birds which are measured through several biometrical features. Sex of these birds is known and the aim is to guess the sex of non-pubescent animals because their sex is impossible to obtain without killing them. However, in some cases, mature birds have not the same dimension as the young ones: for some species of birds, growth finished several years after the taking fly (Bretagnolle *et al.* 1998). So, in this case, the problem is to obtain the sex of young birds by using some features of the allocation rule learning from the mature birds.

1.2 The data

In the previous example, the difference between populations was age. Nevertheless, the problem is quite general and the difference may concern other features. For instance, geographical difference will be concerned in the experimentation below. So, data are composed by two samples: a sexed one, S, coming from a population P and a non-sexed one, S^*, coming from a population P^*.

The sexed sample S is composed by n couples $(x_1, z_1), \ldots, (x_n, z_n)$ where x_i is a biometrical vector of $\Re^d$ for the ith individual, and where z_i is its sex with $z_i = 0$ for female and $z_i = 1$ for male. Main biometrical features are generally distributed from a Gaussian distribution conditionally to the sex (Van Franecker and Ter Brack 1993). Consequently, each couple (x_i, z_i) is a i.i.d. realization of the random couple (X, Z) of distribution ($k \in \{0, 1\}$)

$$X_{|Z=k} \sim N(\mu_k, \Sigma_k) \qquad \text{and} \qquad Z \sim B(p)$$

with $N(\mu, \Sigma)$ the Gaussian distribution of mean $\mu \in \Re^d$ and of variance matrix $\Sigma \in \Re^{d\times d}$, and with $B(p)$ the Bernoulli distribution of parameter p. This parameter corresponds to the male proportion in the population P.

The non-sexed sample S^* groups n^* individuals whose only biometrical variables $x_1^*, \dots, x_{n^*}^*$ are known (variables are the same than in the sexed sample), the corresponding sex labels $z_1^*, \dots, z_{n^*}^*$ being unknown. Each couple (x_i^*, z_i^*) is a i.i.d. realization of (X^*, Z^*) with the following distribution:

$$X^*_{|Z^*=k} \sim N(\mu_k^*, \Sigma_k^*) \qquad \text{and} \qquad Z^* \sim B(p^*).$$

The aim is to estimate the n^* unknown sex labels $z_1^*, \dots, z_{n^*}^*$ by using information from both samples S and S^*. The challenge is to exhibit now some links between P and P^*.

2 Relationship between the two populations

2.1 Linear relationship between biometrical variables

We would like to exhibit a relationship ϕ_k ($\Re^d \to \Re^d$) between biometrical random vectors of same sex k but different populations: $X^*_{|Z^*=k} \sim \phi_k(X_{|Z=k})$. In this aim, we make two assumptions on ϕ_k. First, biological reasons avoid combining completely different biometrical variables as length and weight. So, the jth component ϕ_k^j of ϕ_k only depends on the jth component $X^j_{|Z=k}$ of $X_{|Z=k}$ and we have $\phi_k(X_{|Z=k}) = [\phi_k^1(X^1_{|Z=k}) \dots \phi_k^d(X^d_{|Z=k})]'$. The second assumption is that each ϕ_k^j ($\Re \to \Re$) is C^1. Consequently, functions ϕ_k^j are necessarily linear (De Meyer *et al.* 2000). So, we have relations

$$\text{female: } X^*_{|Z^*=0} \sim D_0 X_{|Z=0} + b_0 \quad , \quad \text{male: } X^*_{|Z^*=1} \sim D_1 X_{|Z=1} + b_1$$

with D_0 and D_1 diagonal matrices of $\Re^{d\times d}$ and b_0 and b_1 vectors of $\Re^d$.

Moreover, in order to keep a biological significance of these links, we choose $b_0 = b_1 = 0$ because these vectors have only an influence on the mean of the distribution but not on its dispersion. It is often not realistic in biology.

2.2 Consequence on the discriminant rule of P^*

If the two diagonal matrices D_0 and D_1 are known, it is easy to obtain means and variance matrices of the population P^* from the ones of the population P by the following classical formula: $\mu_k^* = D_k \mu_k$ and $\Sigma_k^* = D_k \Sigma_k D_k$ with $k = 0, 1$. Then, the discriminant rule of P^* is directly given by these parameters (see McLachlan 1992 for instance).

3 Models of constraints

3.1 Five models

The main idea of this work is to define some constraints on the transformation matrices D_0 (females) and D_1 (males) to explicitly express some particular links between both populations P and P^*. We suggest five, so-called, models of constraints M_1 to M_5 on D_0 and D_1. Definition of these models is:

M_1 : $D_0 = D_1 = I$, i.e. both populations are the same.
M_2 : $D_0 = D_1 = \alpha I$, i.e. transformation is variable and sex independent.
M_3 : $D_0 = D_1 = D$, i.e. transformation is only sex independant.

M_4 : $D_0 = \alpha_0 I$ and $D_1 = \alpha_1 I$, i.e. transformation is only variable independant.
M_5 : D_0, D_1, i.e. it is the most general situation.

Some models call for some interesting comments. Model M_1 corresponds to the classical discriminant analysis case. Models M_2 and M_3 preserve an eventual linearity of the rule: if $\Sigma_0 = \Sigma_1$ for P, then $\Sigma_0^* = \Sigma_1^*$ for P^*. Moreover, models M_4 and M_5 may transform a linear allocation rule on P to a quadratic rule for P^* with few parameters to estimate (see below).

3.2 Link with Van Franeker's work

We can establish a link between model M_2 and the work of Van Franeker and Ter Brack (1993). They assume homoscedasticity of the population P ($\Sigma_0 = \Sigma_1 = \Sigma$). In such a situation, the discriminant rule is linear: x is classed as male if $x'\Sigma^{-1}(\mu_0 - \mu_1) > c$ with $c = 1/2(\mu_0+\mu_1)\Sigma^{-1}(\mu_0-\mu_1)+\ln((1-p)/p)$. They suggest to keep the same discriminant rule for the population P^* by only changing the threshold c into a threshold c^*. Model M_2 has the same property as we show now but leads to a more justified new threshold c^* than this one proposed by Van Franeker and Ter Brack.

Model M_2 preserves the rule of linearity, so

$$x^* \text{ male if } x^{*'}\Sigma^{*-1}(\mu_0^* - \mu_1^*) > 1/2(\mu_0^* + \mu_1^*)\Sigma^{*-1}(\mu_0^* - \mu_1^*) + \ln((1-p^*)/p^*).$$

By using the fact that $\mu_k^* = \alpha\mu_k$ ($k = 0, 1$) and $\Sigma^* = \alpha^2\Sigma$, we obtain

$$x^* \text{ male if } x^{*'}\Sigma^{-1}(\mu_0 - \mu_1) > \underbrace{\alpha/2(\mu_0 + \mu_1)\Sigma^{-1}(\mu_0 - \mu_1) + \ln((1-p^*)/p^*)}_{c^*}.$$

4 Parameter estimation

Estimation of matrices D_0 and D_1 depends on the used model. For models M_2 and M_3, we use a mean square estimation method since transformation is sex independent. In such cases, only straightforward computations are needed. For models M_4 and M_5, the maximum likelihood estimator is used. We adapt the maximization step of the EM algorithm (Dempster *et al.* 1977) in this aim. Moreover and for any models, if no knowledge on proportion p^* is available, it is also defined by maximum likelihood and the EM algorithm.

4.1 Model M_2

We have the following global relationship between expectation of the two populations: $E[X^*] = DE[X]$. Estimating respectively $E[X^*]$ and $E[X]$ by their empirical mean $\bar{x}^*$ and $\bar{x}$, the mean square estimator of α is given by $\hat{\alpha} = \arg\min \|\bar{x}^* - \alpha\bar{x}\|_E^2$, with $\|\cdot\|_E$ the Euclidian norm. We obtain after some calculus $\hat{\alpha} = \bar{x}^{*\prime}\bar{x}/(\bar{x}'\bar{x})$.

4.2 Model M_3

Using the same global relationship about expectations, we solve the system $\bar{x}^* = D\bar{x}$. It leads to $\{D\}_{jj} = \{\bar{x}^*\}_j/\{\bar{x}\}_j$ with $j = 1, \ldots, d$.

4.3 Model M_4

Since relationship between biometrical variables is now sex dependent, global previous methods cannot be longer used. We suggest to maximize the likelihood on p^*, D_0 and D_1 by using the EM algorithm. At the M step, Celeux and Govaert (1995) show that maximizing the likelihood on D_0, D_1 is the same as minimizing the following function

$$f(D_0, D_1) = \sum_{k=0}^{1} \sum_{i=1}^{n^*} t_{ik} \{\ln |D_k \Sigma_k D_k| + (x_i^* - D_k \mu_k)' D_k^{-1} \Sigma_k^{-1} D_k^{-1} (x_i^* - D_k \mu_k)\},$$

with t_{ik} ($i = 1, \ldots, n^*$, $k = 0, 1$) the conditional probability that x_i^* belongs to the class k.

Minimizing the function f with $D_k = \alpha_k I$ leads to solve a two order equation with the only non-negative solution

$$\hat{\alpha}_k = \frac{1}{2} \left[-\frac{\bar{x}_k^{*\prime} \Sigma_k^{-1} \mu_k}{d} + \sqrt{\left[\frac{\bar{x}_k^{*\prime} \Sigma_k^{-1} \mu_k}{d}\right]^2 + 4 \frac{\sum_{i=1}^{n^*} t_{ik} x_i^{*\prime} \Sigma_k^{-1} x_i^*}{n_k d}} \right]$$

where we noted $n_k = \sum_{i=1}^{n^*} t_{ik}$ and $\bar{x}_k^* = \sum_{i=1}^{n^*} t_{ik} x_i^* / n_k$.

4.4 Model M_5

We also minimize the function f in order to maximize the likelihood. Noting u_k the vector of $\Re^d$ composed by the diagonal of the matrix D_k^{-1}, X_i^* the diagonal matrix $\Re^{d \times d}$ with the vector x_i^* on the diagonal, $\bar{X}^*$ the same thing with $\bar{x}^*$, we can write $f(D_0, D_1) = \sum_{k=0}^{1} \Psi_k(u_k)$ with

$$\Psi_k(u_k) = -2n_k \sum_{j=1}^{d} \ln \{u_k\}_j + u_k' (\sum_{i=1}^{n^*} t_{ik} X_i^* \Sigma_k^{-1} X_i^*) u_k - 2n_k \mu_k' \Sigma_k^{-1} \bar{X}^* u_k.$$

After some calculus, the Hessian matrix H_k of Ψ_k is expressed by $H_k = 2 \sum_{i=1}^{n^*} t_{ik} X_i^{*\prime} \Sigma_k^{-1} X_i^* + 2n_k D_k^2$. All matrices $X_i^* \Sigma_k^{-1} X_i^*$ are non-negative definite since $\forall a \in \Re^d, a' X_i^* \Sigma_k^{-1} X_i^* a = c' \Sigma_k^{-1} c > 0$ with $c = X_i^* a \in \Re^d$ (recall Σ_k is non-negative definite). Moreover, sum of non-negative definite matrices being non-negative definite, H_k is non-negative and the function Ψ_k is strictly convex on each subspaces delimited by $u_k = 0$. So, Ψ_k is a strictly convex function for the subpace of interest where each u_k component is non-negative. Moreover, it is easy to show that $\lim \Psi_k(u_k) \to \infty$ when $\|u_k\|_E \to \infty$.

In conclusion, there exists a unique minimum of Ψ_k in the subspace of interest. It can be computed by any numeric method as the Nelder-Mead simplex method by starting from the parameter $u_k = (1 \ldots 1)'$ for instance.

4.5 Estimation of the proportion p^*

If proportions are not preserved, there is a need to estimate p^*. This can be done by maximizing the likelihood using the EM algorithm again. The M step of EM gives the standard result $p^* = n_1/n^*$.

4.6 Choosing a model with the BIC criterion

There exists five models $M_1, \ldots, M_5$ on D_k matrices and two models on proportions ($p^* = p$ or p^* is unknown), so, by combination we obtain 10 models. We will note M_j for model M_j with $p^* = p$ and pM_j for model M_j with unknown p^*. One of these 10 models may be choosen by the biologist himself since models are easily understood. The Bayesian criterion BIC (Schwarz 1978) may also be used to automatically select one model. This criterion, to maximize, is given by BIC(model) $= \ln \ell - \nu/2 \ln n^*$, where ℓ is the maximum likelihood value and ν is the number of estimated parameters (see Table 1). Note that we will use the likelihood of the mean square estimators for models M_2 and M_3 instead.

Table 1. Number of estimated parameters for each model

model	M_1	M_2	M_3	M_1	M_5	pM_1	pM_2	pM_3	pM_4	pM_5
ν	0	1	d	2	2d	1	2	d+1	3	2d+1

5 Experiments

5.1 Data

Petrels are birds implanted in different geographical zones. They lead to different subspecies: borealis living in the Atlantic islands (the Azores, etc.), and diomedea living in the Mediterranean islands (Illes Balears, etc.) (Thibault *et al.* 1997). We have a borealis sample of size 206, each bird being measured on five biometrical features: culmen, tarsus, wing and tail length, and also culmen depth. We have also a diomedea sample of size 38 with same biometrical features.

We consider borealis birds as being the sexed sample S and diomedea birds as being the non-sexed sample S^*. In fact, in our data both samples are sexed but sex of diomedea will be only used to measure quality of results provided by the proposed method.

5.2 Results

First, the cross-validation criterion value is computed to choose between homoscedastic and heteroscedastic models for both samples by using sex information of each of them. Values are in Table 2 and homoscedastic model is selected twice. Consequently, parameters of borealis are estimated by the homoscedastic model.

Table 2. Cross-validation criterion value.

model	homoscedastic	heteroscedastic
borealis	**9.71**	13.59
diomedea	**15.79**	26.32

Second, from the borealis estimated parameters, the 10 models are successively applied to the non-sexed diomedea sample. The empirical error rate (deduced from the true partition of diomedea) and the BIC value, are displayed for each model in Table 3. Moreover, empirical error rate of the cluster analysis situation is reported. In this case, mixture parameters of the non-sexed sample diomedea are estimated with EM using the optimal model of diomedea (the homoscedastic model). This situation is of interest because no information comes from the borealis sample. Moreover, it gives an optimistic error estimation of the clustering procedure since the optimal model is used.

Table 3. Empirical error rate (error) and BIC value (BIC)

model	M_1	M_2	M_3	M_4	M_5	pM_1	pM_2	pM_3	pM_4	pM_5	clustering
error	42.11	28.95	21.05	28.95	21.05	42.11	42.11	**18.42**	42.11	21.05	34.21
BIC	-753	-503	**-452**	-504	-458	-725	-490	-453	-491	-460	–

Interest of our approach is exhibited by the poor error rate obtained with standard discriminant analysis (models M_1 and pM_1) and standard cluster analysis. The model selected by the empirical error rate is pM_3. This model preserves homoscedasticity, relevant property since both rules selected by cross-validation criterion were homoscedastic. The model selected by the BIC criterion is M_3 and the error rate is the second best value with M_5 and pM_5. Note that the BIC value for pM_3 is very close to this one of M_3.

6 Concluding remarks

Other experiments are required to confirm these encouraging first results. In another context, it is straightforward to apply the method with more than two groups. Moreover, an interesting extension of this work would be to consider a partially labelled second sample.

References

Bretagnolle, V., Genevois, F. and Mougeot F. (1998). Intra and Intersexual Function in the Call of a non Passerine Bird. *Behaviour*, **135**, 1161-502.

Celeux, G. and Govaert, G. (1995). Gaussian Parsimonious Models. *Pattern Recognition*, **28**, 5, 781-193.

De Meyer, B., Roynette, B., Vallois, P. and Yor, M. (2000). On Independent Times and Positions for Brownian Motion. *Les prépublications de l'Institut Élie Cartan*, **1**, 46-48.

Dempster, A.P., Laird, N.M. and Rubin, D.B. (1977). Maximum Likelihood from Incomplete Data (with discussion). *Journal of the Royal Statistical Society, Series B*, **39**, 1-38.

McLachlan, G.J. (1992). *Discriminant Analysis and Statistical Pattern Recognition.* New York: Wiley.

Schwarz, G. (1978). Estimating the Dimension of a Model. *Annals of Statistic*, 6: 461-464, 1978.

Thibault, J.C., Bretagnolle, V. and Rabouam, C. (1997). *Cory's Shearwater Calonectris diomedea.* In BWP update Vol 1 (2): 75-98. Oxford: Oxford University Press.

Van Franecker, J.A. and Ter Brack, C.J.F. (1993). A Generalized Discriminant for Sexing Fulmarine Petrels from External Measurements. *Auk*, **110**, 492-502.

Disclosure control on multi-way tables by means of the shuttle algorithm: extensions and experiments

Lucia Buzzigoli and Antonio Giusti

Dipartimento di Statistica "Giuseppe Parenti", Università di Firenze, Viale G. B. Morgagni, I 50134 Firenze, Italy

Abstract. In this paper we re-examine our algorithm for calculating the lower and upper bounds of an array given the complete set of its marginals, proposing some extensions. The algorithm has some interesting properties that can be useful in various fields of application, such as statistical disclosure control of count tables. These properties involve both theoretical and computational issues: in particular, the algorithm has relevant links with probabilistic and statistical aspects (e.g. Fréchet and Bonferroni bounds) and is particularly easy to implement, has a low storage requirement and is very fast.

Keywords: Statistical disclosure control, linked tables, shuttle algorithm, linear programming, zero restrictions.

1 Introduction

In a recent paper, we introduced the 'shuttle' algorithm, a methodology alternative to linear programming to calculate lower and upper bounds of the elements of an n-way array, starting from the complete set of its (n-1)-way marginals (Buzzigoli & Giusti, 1999). This algorithm, which is computationally simple and fast, could be useful for the evaluation of the disclosure risk connected with the release of 'linked' tables (de Vries, 1993), that is, tables derived from the same set of original data: the known marginals represent the released linked tables, while the unknown n-way array represents the original data (to be protected) collected by means of censuses or surveys.

The advantage over a linear programming strategy is clear when n is large, although some characteristics of our algorithm are still in need of exploration. For example, as we will discuss later, there are some simple cases of zero restrictions which call for some refinement.

Let us consider an n-way array X with generic element:

$$x_{i_1,i_2,\ldots,i_n}$$

and suppose to have the complete set of the n marginal values that belong to the (n-1)-way marginal distributions:

$$x_{+,i_2,\ldots,i_n}, x_{i_1,+,\ldots,i_n}, \ldots, x_{i_1,i_2,\ldots,+} \quad .$$

The algorithm is iterative and is based on two steps, which calculate, respectively, the upper and the lower bound of each element of the array. The lower bounds can be used to revise upper bounds (repetition of first step) and the revised upper bounds to recalculate new lower bounds (repetition of the second step), and so on until convergence.

In order to formalize this logic, let us consider two n-way arrays X^{U_i} and X^{L_i}, which are the arrays containing, respectively, the upper and the lower bounds of the elements of X computed at the i-th iteration. Their generic elements are $x^{U_i}_{i_1,i_2,\ldots,i_n}$ and $x^{L_i}_{i_1,i_2,\ldots,i_n}$.

In summary, the algorithm can be described as follows:

1st step:

$$x^{U_s}_{i_1,i_2,\ldots,i_n} = \min\Big(x^{U_{s-1}}_{i_1,i_2,\ldots,i_n}, x_{+,i_2,\ldots,i_n} - \sum_{i\neq i_1} x^{L_{s-1}}_{i,i_2,\ldots,i_n}, x_{i_1,+,\ldots,i_n} - \sum_{i\neq i_2} x^{L_{s-1}}_{i_1,i,\ldots,i_n}, \ldots, x_{i_1,i_2,\ldots,+} - \sum_{i\neq i_n} x^{L_{s-1}}_{i_1,i_2,\ldots,i}\Big)$$

2nd step:

$$x^{L_s}_{i_1,i_2,\ldots,i_n} = \max\Big(x^{L_{s-1}}_{i_1,i_2,\ldots,i_n}, x_{+,i_2,\ldots,i_n} - \sum_{i\neq i_1} x^{U_s}_{i,i_2,\ldots,i_n}, x_{i_1,+,\ldots,i_n} - \sum_{i\neq i_2} x^{U_s}_{i_1,i,\ldots,i_n}, \ldots, x_{i_1,i_2,\ldots,+} - \sum_{i\neq i_n} x^{U_s}_{i_1,i_2,\ldots,i}\Big)$$

where $x^{L_0}_{i_1,i_2,\ldots,i_n} = 0$ and $x^{U_0}_{i_1,i_2,\ldots,i_n} = x_{++\ldots+}$.

It obviously stops when $X^{U_s} = X^{U_{s-1}}$ or $X^{L_s} = X^{L_{s-1}}$.

As shown in Buzzigoli & Giusti (1999), the algorithm always converges in a finite number of steps and, for specific cases of relevance, the dimensions of the original array bound the number of steps needed to reach a stop.

We managed to prove that shuttle solutions coincide with linear programming solutions only for particular cases, but Roehrig (1999) showed that the shuttle algorithm is related to the dual of the linear programming approach. Therefore, it seems that the methods may have some general connections although they solve the problem of finding bounds with very different foundations.

It should be stressed that the shuttle algorithm seems to have relevant links also with some probabilistic and statistical issues: in fact, the problem of finding lower and upper bounds for the cells of a contingency table given a set of subtables can be converted into the problem of finding the probability distribution corresponding to some given marginal distributions. A vast literature is available on this subject (see, for instance, Dall'Aglio et al. 1990), often dealing with the issue of the existence of the probability distribution: in our case this is not a problem, because the various marginals are surely compatible, as they derive from the same original array.

In this context, the algorithm can be seen as a useful tool to implement the automated calculus of the combined Fréchet-Bonferroni bounds proposed by Fienberg (1999) in a rapid and efficient way (as done in Buzzigoli & Giusti, 1998), filling a gap in this field of application.

The purpose of this paper is to suggest a generalization of the algorithm. In particular, we are going to use the algorithm in situations where some of the internal values of the original n-way array are known (section 2).

From a computational point of view, we will present the results of an intensive simulation to test the performances of the shuttle algorithm (section 3). The simulation results will be useful in evaluating the potential of the shuttle algorithm in various applications related to statistical disclosure control, not only for the problem of linked tables but also for the monitoring of cell suppression methods.

2 Zero restrictions

An interesting field of application of the shuttle algorithm would be the monitoring of cell suppressions for disclosure control problems (Willenborg & de Waal, 1996). In this case we have a table where some of the entries have been deleted and the algorithm can be used to check whether it is possible to identify the suppressed cells using the known cells and the marginals of the table.

When some entries in the table are given, we can subtract their values from the corresponding marginals, fixing the original values to zero. For instance, if we suppose that the element x_{i_1,i_2,i_3} of a three-dimensional table is known, the shuttle will be applied to three sets of marginals (layer totals, row totals and column totals) where x_{+,i_2,i_3} is replaced by $x_{+,i_2,i_3} - x_{i_1,i_2,i_3}$, $x_{i_1,+,i_3}$ by $x_{i_1,+,i_3} - x_{i_1,i_2,i_3}$ and $x_{i_1,i_2,+}$ by $x_{i_1,i_2,+} - x_{i_1,i_2,i_3}$; therefore x_{i_1,i_2,i_3} is set to zero (i.e. $x^U_{i_1,i_2,i_3} = x^L_{i_1,i_2,i_3}$ =0).

This situation is known in disclosure control literature as zero-restriction (Cox, 1999).

The same condition arises when one of the marginals is zero, because in count tables this means that all the cells contributing to that marginal are identically zero.

In particular cases of zero restrictions, the algorithm is not able to find linear programming solutions. For instance, the application of the shuttle algorithm on the marginals in Fig. 2.1 (note that the layer totals equal to zero automatically fix some internal values to zero) produces the lower and upper bounds in Fig. 2.2, while linear programming is able to correctly identify the values $x_{123}=x_{223}=1$.

layer totals

2	2	0	0
2	2	2	0
0	0	2	2
0	0	2	2

row totals

2	2
3	3
2	2
2	2

column totals

2	2	3	2
2	2	3	2

Fig 2.1. Marginals of a 2×4×4 table.

The identifiability of x_{123} and x_{223} is easily proven: it is sufficient to note that the collapsed 2×2×2 array obtained from the original table summing the first row with the second one, the third row with the fourth one, the first column with the second one and the third column with the fourth one (Fig. 2.3) is a table where two cells are zero and two other cells contain only, respectively, x_{123} and x_{223}: this table is identifiable and, therefore, these two values are identified also with the shuttle algorithm.

X^U

2	2	0	0
2	2	2	0
0	0	2	2
0	0	2	2

2	2	0	0
2	2	2	0
0	0	2	2
0	0	2	2

X^L

0	0	0	0
0	0	0	0
0	0	0	0
0	0	0	0

0	0	0	0
0	0	0	0
0	0	0	0
0	0	0	0

Fig 2.2. Upper and lower bounds calculated with the shuttle algorithm.

original array

x	x	0	0
x	x	x	0
0	0	x	x
0	0	x	x

x	x	0	0
x	x	x	0
0	0	x	x
0	0	x	x

collapsed array

Σx	x
0	Σx

Σx	x
0	Σx

Fig 2.3. Original 2×4×4 array (x means unknown values) and corresponding collapsed 2×2×2 array.

3 Intensive simulation

Another issue worth exploring relates to the evaluation of the shuttle algorithm performance. To this end we ran a set of simulations on randomly generated three dimensional arrays in order to assess the computing time and the stability of the algorithm across the experiments.

The simulation was conducted on arrays with two layers: the various experiments differ by the number of rows and columns (3 sets: 2×4×4, 2×5×5, 2×6×6) composed by elements with different value limits (4 sets: 0-1, 0-10, 0-100, 0-1000). We broke down the evaluation of the performance in 10 groups of a varying number of arrays (2 sets, 1000 and 2000 each) in order to be able to compute an average running time across groups and a measure of variability to check the algorithm stability. Overall we performed 240 experiments (3×4×2×10).

For the simulation we used an APL2 interpreter running on a personal computer powered by a 600Mhz Pentium III. We used the APL2 language, despite the lower performances of an interpreter with respect to a compiler, because the translation of the algorithm in this kind of "matrix" language is easier and the resulting functions are more general and flexible. As is well known, APL2 permits the utilisation of nested generalised arrays; this fact contributes to reduce the use of iterative structures inside the code and allows for the reduction of the execution time (the total computing time to calculate lower and upper bounds of 360000 tree-way arrays is of 538.242 seconds).

We report the results of our experiments in Tab. 3.1 where the average computing time and the coefficient of variation (cv) are calculated over each set of 10 experiments: the mean is inversely related to the width of the value limits and is roughly double when more arrays are considered; the cv stresses how little the variation in mean computing time is, signalling the stability of the algorithm. Comparing the results by set of rows and columns for the same value limits the mean computing time decreases as the size of the array increases.

	2 layers 4 rows 4 columns			
value limits	10 experiments with 1000 arrays		10 experiments with 2000 arrays	
	mean	cv	mean	cv
0 - 1	1835.9	0.01170	3672.4	0.00791
0 - 10	1537.2	0.00945	3052.3	0.01610
0 - 100	1494.1	0.00575	2963.4	0.00529
0 - 1000	1486.2	0.00532	2944.1	0.00546

	2 layers 5 rows 5 columns			
value limits	10 experiments with 1000 arrays		10 experiments with 2000 arrays	
	mean	cv	mean	cv
0 - 1	1777.4	0.00877	3551.4	0.00360
0 - 10	1416.0	0.00647	2844.9	0.00573
0 - 100	1366.9	0.00497	2738.8	0.00505
0 - 1000	1355.0	0.00741	2722.0	0.00540

	2 layers 6 rows 6 columns			
value limits	10 experiments with 1000 arrays		10 experiments with 2000 arrays	
	mean	cv	mean	cv
0 - 1	1717.6	0.00784	3432.1	0.00508
0 - 10	1355.9	0.00888	2720.8	0.00659
0 - 100	1307.6	0.00688	2617.8	0.00486
0 - 1000	1300.9	0.00427	2613.5	0.01051

Tab 3.1. Computing time (in milliseconds): means and coefficients of variation for each group of ten experiments.

This last result shows that the shuttle algorithm converges faster and faster as the structure of the arrays becomes more complex. This property could be explained considering the two steps presented in the first section: as the number of layers/rows/columns increases, the probability that the first term in the min/max expressions will be enhanced decreases. This is clearly a remarkable characteristic of the algorithm, especially when we compare it with the simplex method in linear programming: when we apply the simplex method the computational effort increases with the number of layers/rows/columns of the tables because the constraints that must be satisfied in each minimisation/maximisation step increase as well.

Another advantage of the shuttle algorithm when used on large arrays is that it has limited storage and memory requirements, as the storage occupation for data, results and working data is mostly represented by the marginal distributions (given data) and by n+3 arrays of the same dimensions of the unknown array (two arrays represent the results, the other n+1 are working data).

4 Conclusions

The shuttle algorithm presents several appealing characteristics: in this paper we have explored some extensions and we have provided further evidence about the behavior of the algorithm in simulated conditions.

The theoretical issues that still need to be explored are related to the links between the shuttle algorithm and the linear programming method, in particular when some a priori knowledge on the elements of the array can be exploited.

The main advantage of the algorithm is in practical situations, especially in what the disclosure control of linked tables is concerned. Such a problem is felt at National Statistical Agencies in a particular way since the institutional duties related to disclosure may find it computationally burdensome to use a linear programming approach when released tables are large.

In this respect the shuttle algorithm with its limited computing time and convergence properties appears to be a promising solution.

References

Buzzigoli, L. & Giusti, A. (1998). Statistical disclosure control problems for linked tables. *Italian Journal of Applied Statistics*, **10**, 443-458.

Buzzigoli, L. & Giusti, A. (1999). An algorithm to calculate the lower and upper bounds of the elements of an array given its marginals. In: *Statistical data protection - Proceedings of the conference,* 131-147. Luxembourg: Eurostat.

Cox, L.H. (1999). Some remarks on research directions in statistical data protection. In: *Statistical data protection - Proceedings of the conference,* 163-176. Luxembourg: Eurostat.

Dall'Aglio, G., Kotz, S. & Salinetti, G. (1990). *Advances in Probability Distributions with Given Marginals*. Dordrecht: Kluwer Academic Publishers.

De Vries, R.E. (1993). Disclosure Control of Tabular Data Using Subtables. Voorburg: Netherlands Central Bureau of Statistics.

Fienberg, S.E. (1999). Fréchet and Bonferroni Bounds for Multi-way Tables of Counts With Applications to Disclosure Limitation. In: *Statistical data protection - Proceedings of the conference,* 115-129. Luxembourg: Eurostat.

Roehrig, S. F. (1999). Auditing Disclosure in Multi-Way Tables With Cell Suppression: Simplex and Shuttle Solutions. Paper presented at: *Joint Statistical Meeting 1999*, August 5-12, Baltimore.

Willenborg, L.C.R.J. & de Waal, A.G. (1996). *Statistical Disclosure Control in Practice*. Lecture Notes in Statistics. New York: Springer-Verlag.

An MLE strategy for combining optimally pruned decision trees

Carmela Cappelli [1] and William D. Shannon [2]

[1] Department of Mathematics and Statistics, Naples University Federico II, Via Cinthia, Monte S. Angelo, 80126 Napoli, Italy

[2] Division of General Medical Science, Washington University School of Medicine, Campus Box 8005, 660 S. Euclid Ave., S. Louis, MO 63110, USA

Abstract. This paper provides a maximum likelihood estimation strategy to identify a tree-based model which, being a function of a set of observed optimally pruned trees, represents the final classification model. The strategy is based on a probability distribution and it uses a metric based on structural differences among trees. An example on a real dataset is also presented to show how the procedure works.

Keywords: trees-based models, simplification, pruning, probability distribution, minimum mean distance

1 Introduction

Recursive partitioning methods, or tree-based models, can be used for discovering general classification rules from a set of examples. Briefly, the procedure begins with a *training set*, *i.e.*, a set of examples described in terms of a vector of features or attributes with each example belonging to a known class. The algorithm examines each attribute and a cut point along that attribute, choosing the "best" one on the basis of a *goodness of split* measure whose value reflects how well that attribute and cut point can discriminate classes into two mutually exclusive subsets. This procedure is recursively applied to each subset of the previous split in an effort to classify correctly as many of the training cases as possible.

When this *divide et conquer* approach is used to classify noisy data (*i.e.*, data that cannot be perfectly discriminated based on attribute measurements) the resulting tree tends to be very large and complex and, above all, many of the branches arise by chance and reflect sampling error present in the training set rather than representing underlying real classification rules. The consequences are difficulties of interpretation and *overfitting*. The standard approach to overfitting avoidance is simplification, i.e., to grow a very large (full) tree model and then *prune* it back by replacing sets of sibling terminal nodes by a common ancestor node. This results in removing entire branches of the full tree, according to some pruning criterion, which simplifies interpretation of the classification rules and produces fewer misclassifications of new data.

Several pruning methods have been proposed in literature however, empirical studies have shown pruning methods perform quite differently with respect to both the size and accuracy of the final optimal tree, (see for example Cappelli and Siciliano, 1998; Malerba *et* al., 1996; Mingers, 1989), so that a data analyst is faced with the problem of selecting a final pruned tree to be used as classification model.

In this paper we propose an MLE strategy for combining a set of optimally pruned trees produced by different pruning algorithms with the goal to address the variability across pruning methods. Some familiarity with pruning methods is assumed. In particular, we will refer to the pioneer pruning method proposed in the CART methodology (Breiman *et* al, 1984), the so called *cost-complexity pruning* (both the trees resulting from the 0-se rule and the 1-se rule will be considered), the *reduced error pruning* (rep) and the *pessimistic error pruning* (pep) of Quinlan (1985, 1993), the *minimum error pruning* (mep) provided by Niblett and Bratko (1986) and finally a variant of the cost-complexity pruning proposed by Cappelli *et* al (1998) that, opposed to the other methods uses the Gini index of heterogeneity instead of the error rate and it is based on statistical testing procedure (stp).

We will show how the proposed strategy works using a real dataset, in particular, the performances of the MLE tree compared to the optimally pruned trees will be investigated in terms of misclassification rate on independent validation data and distance to all the other trees.

2 The proposed methodology

The proposed MLE strategy is based on a probability distribution derived for graph-valued random object (Banks and Constantine, 1998) and applied to the study of instability of tree based classifiers (Shannon and Banks, 1999). The distribution is defined as follows:

$$P_{(g^*,\tau)}[g] = c(g^*,\tau) \times exp[-\tau d(g,g^*)], \tag{1}$$

where g^* is the central tree, $\tau \geq 0$ is a concentration parameter (which summarizes the concentration around the central tree), $c(g^*,\tau)$ is the normalizing constant, and d is a metric defined on the set G of all possible pruned subtrees of a given full tree. The central tree g^* can be viewed as the 'mean' while τ as a precision measure (inverse of dispersion). Actually, the highest the value of the concentration parameter, the highest the central tendency, suggesting that there exists a representative central tree. On the contrary, a low value of the concentration parameter meaning absence of central tendency, will be interpreted as equiprobability of the pruned trees.

Both the central tree and the concentration parameter are unknown parameters to be estimated from the data. The metric employed, based on the so called *discrepant paths*, is defined as follows. Let g_i and g_j be two trees in the set G, the distance between g_i and g_j is given by :

$d(g_i, g_j)$ = minimum # of nodes in g_i added or deleted to match the structure of g_j,

so that, adding or deleting a split to a tree node *i.e.*, for $d = 1$, generates the one discrepant path tree of the starting tree.

It can be easily verified that d satisfies the properties of a metric, being:

- $d(g_i, g_j) \geq 0$;
- $d(g_i, g_j) = 0$ if and only if g_i is identical to g_j;
- $d(g_i, g_j) \leq d(g_i, g_k) + d(g_k, g_j)$, $\forall g_i, g_j, g_k \in G$.

We propose to consider the set of optimally pruned trees obtained by different pruning methods as a random sample taken from the above probability distribution. We can then use this observed set to search for the maximum likelihood estimate of the central tree to be used as the classification model. Let $\{g_1, \ldots, g_i, \ldots, g_n\} \subset G$ be a random sample of observed trees, the log-likelihood of (1) is given by:

$$\begin{aligned} \ln L(g^*, \tau) &= n\ln c(g^*, \tau) - \tau \sum_{i=1}^{n} d(g_i, g^*) \\ &= -n\ln[\sum_{g \in G} exp(-\tau d(g, g^*))] - \tau \sum_{i=1}^{n} d(g_i, g^*), \end{aligned} \tag{2}$$

where $c(g^*, \tau) = [\sum_{g \in G} exp(-\tau d(g, g^*))]^{-1}$.

In order to maximize (2) a steepest ascent search algorithm is employed, which can be summarized as follows:

1. a sample observation g_i is choosen as $\hat{g}_0^*$;
2. estimate $\hat{\tau}$ given $\hat{g}_0^*$ and calculate the log-likelihood $\ln L(\hat{g}_0^*, \hat{\tau})$;
3. generate all the one discrepant path trees of $\hat{g}_0^*$, by adding or deleting one split; for the one discrepant path tree estimate $\hat{\tau}$ and calculate their associated log-likelihood, if the maximum of these is larger than the one calculated at step 2, choose the corresponding tree as the current candidate central tree $\hat{g}_j^*$ and return to step 3. Otherwise retain the current candidate MLE tree and select another observed tree as central tree and rerun the algorithm.

After each of the optimally pruned trees have been used as the starting point for the steepest ascent search, the candidate MLE trees are compared, and the one with the maximum likelihood is selected as the final tree. In our experience so far, each starting point has converged on the same candidate MLE tree.

The steepest ascent search is computationally expensive so simplifying approximations are used.

First, the solution of 2 involves the computation of the distance of the current candidate MLE tree $\hat{g}_j^*$ to all trees in G. This cannot be done due to the cardinality of G, so we limit the search to all trees within a small neighborhood around $\hat{g}_j^*$, *i.e.*, all $g \in G$ such that $d(g, \hat{g}_j^*) \leq k$ where k is a constant [1]. This is justified since the contribution to the log-likelihood for trees far from $\hat{g}_j^*$ is negligible .

Second, we consider another tree averaging approach similar in nature to the MLE tree: the minimum mean distance (MMD) tree that is simple to compute, and can provide a good starting point for the steepest ascent search for the MLE tree. The MMD tree is the tree whose mean distance to all the observed trees is minimum. A steepest ascent search for the MMD tree is similar to the MLE search algorithm presented above:

1. a sample observation g_i is choosen as $\hat{g}_0^*$;
2. calculate the mean distance of $\hat{g}_0^*$ to the n optimally pruned trees by $\frac{\sum_i d(g_i, \hat{g}_0^*)}{n}$;

[1] In the application we have considered $k = 3$, *i.e.*, all trees that are one, two or three discrepant paths away from the current candidate MLE tree $\hat{g}_j^*$.

3. generate all the one discrepant path trees of $\hat{g}_0^*$, by adding or deleting one split;
4. for each of the one discrepant path trees calculate the mean distance to all the observed pruned trees. If the minimum of these is smaller than the one calculated at step 2, choose the corresponding tree as the current candidate central tree and return to step 3. Otherwise retain the candidate MMD tree and select another observed tree as central tree and rerun the algorithm.

Each MMD tree obtained can be added to the set of optimally pruned trees to be used as the starting point for the MLE tree search. Our experience has been that the algorithm converges on one or a few MMD that are located very near or are equivalent to the MLE tree.

3 An application on a real data set

The proposed MLE strategy has been applied to a well known real data set: the Wisconsin diagnostic breast cancer dataset, which was contributed to the UCI repository of machine learning databases (http://www.uci.edu/~mlearn/MLRepository.htlm) by the University of Wisconsin Hospitals.

The dataset consists of 569 cellular samples, it has 30 real-valued features and a binary response variable; the two classes are malignant (212) and benign (357). Features were computed from a digitized image and they describe characteristics of the cell nuclei present in the image. These characteristics are: *radius*, *texture*, *smoothness*, *perimeter*, *area*, *compactness*, *concavity*, *concave points*, *symmetry*, *fractal dimension*. The mean, the standard error and the "worst" or largest (mean of the three largest values) of these features were computed for each image, resulting in 30 features. For instance, considering the radius, in the database there are: mean radius, radius SE, and worst radius (for more details see Mangasarian *et* al., 1995).

The dataset has been split in a training set, covering the 70% of cases, employed to fit the full tree, and a test set used to estimate the error rate.

By applying the MMD algorithm to the set of optimally pruned trees provided by the different pruning methods we have found two MMD trees, g^1_{mmd} and g^2_{mmd}. Then, each optimally pruned tree and the two MMD trees has been used as starting point for the MLE search algorithm.

In Figure 1 are shown the multidimensional scaling projections of the 6 optimally pruned trees, the two MMD trees and the MLE tree g_{mle}. All trees converged on the same final MLE tree. Note also that tree g^2_{mmd} is identical to tree g_{mle}, and it took only three iterations of the algorithm for the other MMD tree g^1_{mmd} to converge onto the MLE tree. The convergence of all the considered trees on the same final MLE tree suggests that this provides a stable combination of them.

The performances of the pruned trees, the MMD tree and the MLE tree have been investigated in terms of misclassifications on the test set and mean distance to all the other trees; the results are described in Table 1 where for each tree are reported: the terminal nodes (labeled according to the following notation: node t is split into node $2t$ and node $2t+1$, the root node is labeled $t = 1$), the test set error rate, the log-likelihood and the mean distance to all the other trees.

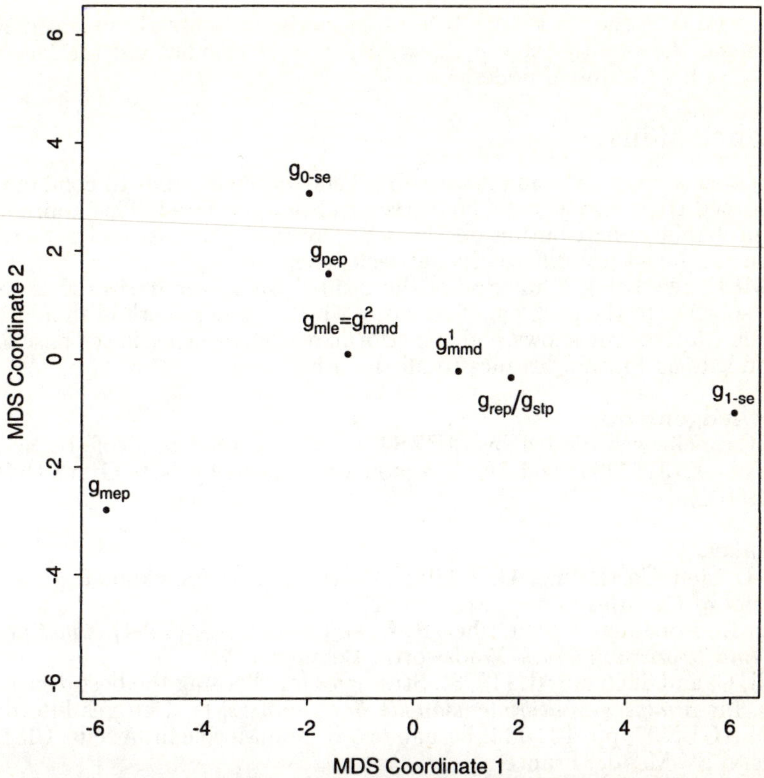

Figure 1. The mds projections of the pruned subtrees, the MMD trees and the MLE tree

Trees	Terminal nodes	% error	Log Likelihood	Mean distance
g_{1-se}	2, 3	13.78	-286.86	2.07
g_{0-se}	16, 17, 18, 38, 39, 10, 11, 12, 13, 14, 15	13.73	40.36	1.39
g_{pep}	16, 17, 18, 19, 10, 11, 12, 13, 14, 15	14.73	47.41	1.23
g_{rep}	4, 10, 11, 12, 13, 7	11.94	44.00	1.30
g_{mep}	32, 66, 67, 134, 135, 17, 18, 38 39, 10, 11, 12, 13, 14, 15	13.64	-6.34	1.92
g_{stp}	8, 9, 10, 11, 6, 7	12.67	44.00	1.30
g^1_{mmd}	8,9,10,11,12,13,7	12.66	50.20	1.18
$g_{mle} = g^2_{mmd}$	16,17,9,10,11,12,13,14,15	11.71	51.46	1.16

Table 1. Test set percent errors, log-likelihoods and mean distances to the other trees for the set of pruned trees, the MMD trees and the MLE tree

The g_{mle} tree outperforms the 6 pruned trees and g^1_{mmd} in terms of the percent error and it has also the smallest mean distance to the sample.

Finally, note that the MLE with 9 terminal nodes presents an intermediate size between the smallest tree g_{1-se} with 2 terminal nodes and the largest one g_{mep} with 14 terminal nodes.

4 Conclusions

In this paper a strategy using maximum likelihood estimation to combine a set of pruned trees into a single final tree has been presented. This approach uses a probability distribution on the set of pruned subtrees and a metric based on the structural differences between trees.

The MLE tree, being a function of the pruned trees, summarizes them offering a solution to the problem of selecting which pruning method should be employed. Moreover it shows higher performances presenting lower misclassification rate and mean distance to all the other trees.

Acknowledgements
Dr. C. Cappelli was funded by MURST Grant awarded to Prof. R. Siciliano (prot. 9913182289) and Dr. Shannon was funded by NIH Grant (RO1 GM61218-01).

References

Banks, D. and Costantine, G. (1998). Metric models for random graphs, *Journal of Classification*, **15**, pp. 199-223.

Breiman L., Friedman J. H., Olshen R. A. and Stone C. J. (1984). *Classification and Regression Trees*, Wadsworth, Belmont CA.

Cappelli, C. and Siciliano, R. (1998). Strategies for choosing the best decision trees. In: *Analyses Multidimensionelle des Donnees*, IV Congres International NGUS'97, pp. 101-111, Fernandez, K. and Morineau, A. eds, CISIA Ceresta, St. Mande, France.

Cappelli, C., Mola, F. and Siciliano, R. (1998). An alternative pruning method based on the impurity-complexity measure. In: *Proceedings in Computational Statistics*, pp. 221-226, Payne, R. and Green, P. eds, Heidelberg: Physica-Verlag.

Malerba, D., Esposito, F. and Semeraro, G. (1995). A further comparison of simplification methods for decision tree pruning. In: *Learning from Data: Artificial Intelligence and Statistics, Lecture Notes in Statistics*, Fisher, D. and Lanz, H. eds, Berlin: Springer-Verlag.

Mangasarian , O.L. et al. (1995). Breast cancer diagnosis and prognosis via linear programming. *Operations Research* , **43** (4), pp. 570-577.

Mingers, J. (1989), An empirical comparison of pruning methods for decision tree induction. *Machine Learning*, **4**, pp. 227-243.

Niblett, T., Bratko, I. (1986), Learning decision rules in noisy domains. In *Research and Development in Expert Systems, Proceedings of Expert Systems-86*,**3**, Bramer, M.A. Ed., Cambridge: Cambridge University Press.

Quinlan, J.R. (1987), Simplifying decision trees. *International Journal of Man-Machine Studies*, **27**, pp. 221-234.

Shannon, W.D. and Banks, D. (1999), Combining classification trees using MLE. *Statistics in Medicine*, **18**, pp. 727-740.

Semi-parametric models for data mining

Claudio Conversano[1] and Francesco Mola[2]

[1] Dipartimento di Matematica e Statistica, Universitá di Napoli Federico II, Via Cinthia Monte S. Angelo, I-80126, Napoli, Italia

[2] Dipartimento di Economia, Universitá di Cagliari, Viale Fra' Ignazio, I-90123, Cagliari, Italia

Abstract. In order to combine the exactness of a very large data set with the major predictability of statistical modeling, we introduce a two-step methodology that makes use of partitioning algorithms and semi-parametric models. The result is an alternative strategy for supervised classification and prediction problems when dealing with huge and complex data sets, in order to improve the predictability of the dependent variable on the basis of the previous detection of homogeneous sub-populations[1].

Keywords. Supervised classification and prediction, Generalized Additive Models, Recursive partitioning.

1 Introduction

Increasing amounts of data are collected in various fields of application, such as finance, market data analysis, astronomy, manufacturing, and education. Large data sets can lead to essential knowledge in decision-making, but knowledge discovery requires intelligent and largely automatic methods and systems supporting the analysts. The rapidly emerging fields of Knowledge Discovery in Databases (KDD) and Data Mining provides these tools. They also assimilate methods of machine learning and discovery, statistics, databases, and visualization. According to Hand (1998), Data Mining is the secondary analysis of large databases aimed at finding unsuspected relationships which are of interest or value to the database owners.

In the framework of supervised classification and prediction, i.e. in situations such that we dispose of a set of objects belonging to a known member of a set of possible classes, the aim is to formulate a rule which would permit the class of a new object to be predicted, based solely on its vector of measurements.

Many different kind of methods have been developed for such problems, ranging from simple linear methods through more sophisticated models (e.g. neural networks and projection pursuit regression), to non-parametric methods (e.g. recursive partitioning and nearest neighbor methods). They are reviewed in Ripley (1996) and Hand (1997).

The aim of this paper is to improve the predictability of the dependent variable in data mining. For this purpose, we introduce a methodology that makes use of partitioning algorithms and semi-parametric models.

[1] Research was supported by MURST funds 1999 (prot. 9913182289).

2 Generalized Additive Models

Among the class of semi-parametric models, reviewed in Härdle *et* al. (1997), our attention will be focused on Generalized Additive Models (GAM) of Hastie and Tibshirani (1990).

GAM are based on the sum of d non-parametric functions of the d predictors of $\mathbf{X}$ (plus an intercept term). In addition, they allow for a known parametric link function, $G(\cdot)$, that relates the sum of functions $f(\cdot)$ to the dependent variable Y. The link function might be the regression function in a multiple regression or the logistic transformation of the posterior probability in a logistic regression. The model is defined as follows:

$$E(Y|\mathbf{X}) = G\{\alpha + \sum_{j=1}^{d} f_j(X_j)\} \tag{1}$$

where $E(Y|\mathbf{X})$ denotes the usual expectation of the dependent variable given the set of predictors. These models aim to examine the effects of covariates one at a time, conditioned on the presence of the other covariates. GAM are an example of semi-parametric models, because they consist of a parametric and a non-parametric component (besides the error term). The response variable might depend on some covariates in a parametric (e.g. linear) fashion and on an additional (or several) covariate(s), not full-filling a parametric assumption(s). The functions f_j in (1) are smoothing functions (in the sense of having small derivatives), so fitting the model provides a *smoothing* of the data. An estimate of the function f_j is named *smoother*.

Although this approach provides a balance between the flexibility of the non-parametric techniques and the facility of interpretation of the linear regression, it cannot be used in a straightforward way when dealing with huge data sets.

3 The proposed methodology

The proposed methodology is in two steps, dealing successively with the dimensionality problem and the modeling strategy. The first step finds a partition of the N cases into 2^K disjoint and exhaustive groups, which are internally homogeneous with respect to the dependent variable. To this point, we consider a K-level recursive partitioning in order to obtain G sub-populations. Without loss of generality, we assume $G = 2^K$, namely at each level we split each sub-population in two groups. The result is a balanced binary tree-structure, in which terminal nodes are the sub-populations. The second step deals with each sub-population in a separate way, applying a recursive partitioning procedure based on semi-parametric modeling. As a result, we obtain G tree-structures to which it is possible to assign a hierarchy of nested models. In the following, we describe in detail the algorithms of the two steps procedure that are respectively summarized in Table 1 and 2.

The first algorithm, which identifies G homogeneous sub-populations, is made up of two loops. The inner loop allows us to find within each level k of the tree the best split $s_h{}^*$ of N_h cases (we label each node with h). The splitting rule is based on the minimization of the sum of the within-group variations $[W(2h) + W(2h+1)]$ of the dependent variable in the sub-nodes $2h$ and $(2h+1)$ of the node h. The outer loop updates the total within-group variation W_k of the k-th partition of the N cases as a sum of the within-group variations of each group at the k-th level. The usefulness of each partition is checked on the basis of the relative decrease of the total within-group variation, when passing from the initial sample to the k-th partition. We define an index $GAIN_0(k)$, which measures the gain in impurity reduction passing from the initial level ($k = 0$) to the k-th one. By definition $GAIN_0(k)$ increases with the level k so that as stopping rule we fix a threshold value α such that we stop the procedure as soon as the $GAIN_0(k)$ is greater than (1-α). It is worth noting that in any k-th partition there might be empty groups: this occurs when small groups at the $(k-1)$-th level cannot be split. The first-step yields to identify for some level K a restricted number $G = 2^K$ of homogeneous sub-populations which may be better described by statistical models.

Table 1. First Step: Identifying homogeneous sub-populations

1. *Initialize:* Set $k = 0$, $G = 1$, $h = 1$, $\bar{y} = \frac{1}{N}\sum_{i=1}^{N} y_i$, $W_0 = \sum_{i=1}^{N}(y_i - \bar{y})^2$, $GAIN_0(k) = 0$ and fix the threshold value α.

2. *Outer loop:* For $k = 0, 1, \ldots, K$:
   Inner loop: For $h = G, G+1, \ldots, 2G-1$:
   Define the set S_h of all possible splits of N_h cases;
   Find the best split $s_h{}^*$ of N_h cases such that:

$$\min_{s \in S_h}[W(2h) + W(2h+1)],$$

   where $W(2h) = \sum_{i \in 2h}(y_i - \bar{y}_{2h})^2$,
   and $W(2h+1) = \sum_{i \in 2h+1}(y_i - \bar{y}_{2h+1})^2$.
   Split the N_h cases according to $s_h{}^*$.

   Compute $W_k = \sum_{h=G}^{2G-1} W(h)$.

   Compute the gain measure at the k-th partition:

$$GAIN_0(k) = \frac{(W_0 - W_k)}{W_0}$$

   Set $G = 2^k$.

   Break the outer loop when $GAIN_0(k) \geq (1-\alpha)$ being $K = k$ the final level.

In the second step we consider for each sub-population g ($g = 1, \ldots, G$) a binary segmentation procedure based on a two-stage splitting criterion (Siciliano and Mola, 1994; Mola et al., 1996). The main idea is to assign the most suitable semi-parametric model to each node of the tree and, on the basis of the most significant predictors, to find the best split of the cases which maximizes the heterogeneity in regression between the two sub-groups.
For any sub-population g of N_g cases consider the N_g-vector $\mathbf{y}$ of the observations of the dependent variable Y and the matrix $\mathbf{X}$ of N_g row vectors

Table 2. Second Step: Semi-parametric modeling of homogeneous sub-populations

1. For each subpopulation g $(g = 1, \dots, G)$ of N_g cases:

2. Set $\mathcal{F}_0$ as the trivial model.

3. Fit the best semi-parametric model $\mathcal{F}_t^*$ at the current node t selecting the number d^* of best predictors through the statistic:

$$F_t = \frac{(RSS_{\mathcal{F}_t} - RSS_{\mathcal{F}_t^*})/(\nu_{\mathcal{F}_t} - \nu_{\mathcal{F}_t^*})}{RSS_{\mathcal{F}_t}/(N_t - \nu_{\mathcal{F}_t^*})}$$

4. Among all the d^* predictors included in $\mathcal{F}_t^*$, find the best split s^* of the N_t cases into two sub-groups t_l and t_r such that:

$$\max_{s^* \in S} F(s,t) = \frac{(RSS(t) - [RSS(t_l) + RSS(t_r)])/(d_t^* - 1)}{[RSS(t_l) + RSS(t_r)]/(N_t - 2d_t^* - 2)}$$

5. Repeat the partitioning procedure for each sub-node of t until the model $\mathcal{F}^*$ results the best in the child nodes t_l and t_r.

$\mathbf{x}_n = (x_{1n}, \dots, x_{pn})$ of the measurements of d predictors $(X_1, \dots, X_d)$ of a numerical or categorical type. The recursive partitioning algorithm finds at each node of the tree the most suitable semi-parametric model $\mathcal{F}_t{}^*$ for a given set of smoothing functions $F = \{f_{1,\nu}, \dots, f_{J,\nu}\}$ (where ν are their respective degrees of freedom) depending on some smoothing parameters to be estimated. Without loss of generality, we fix the function $f_{1,\nu}$ as the classical linear estimator, whereas the remaining $f_{j,\nu}$ $(j = 2, \dots, J)$ are scatterplot smoothers of the spline or kernel type. We perform *for each predictor* an F-test based on the residual sum of squares (RSS) for the hypothesis that its effect on the dependent variable is well explained by the parametric linear function $f_{1,\nu}$ (called the trivial model) against a non parametric alternative, in order to evaluate the significance of the nonlinearity exhibited by each of the smoothing functions $f_{j,\nu}$ $(j = 2, \dots, J)$. The result is a semi-parametric model $\mathcal{F}_t{}^*$, which includes the d^* significant predictors to be assigned to the node t. Notice that the non-significant predictors will not be considered in the successive splits. On the basis of the best model $\mathcal{F}_t{}^*$ we split the cases at node t according to the splitting criterion based on the F statistic for total homogeneity in regression: the best split is found maximizing the heterogeneity in the two sub-groups with respect to $\mathcal{F}_t^*$. The procedure is repeated for each sub-node until the model $\mathcal{F}_t^*$ results the best in the child nodes t_l and t_r.

4 An example on real world data

We show the usefulness of our methodology analyzing a data set concerning some economic indicators collected from a sample of 8089 Italian households by the Bank of Italy in a 1995 survey. Our aim was to determine how household savings (S) is influenced by some predictors, such as subordinate employment income (YL), retirement pension income (YT), self-employment income (YM), real estate income (YCR), financial income (YCF), durable consumption (CD), non-durable consumption (CN) and real estate (AR)[2]. In order to determine homogeneous sub-populations of the initial sample we performed the first step of our methodology with $\alpha = 0.05$. Main results are described in table 3, providing 8 sub-populations which identify different intervals of the values of the dependent variable.

Table 3. Identifying homogeneous sub-populations for household savings in the Bank of Italy data set

Partition level k	Groups g	Sum of the within group variation W_k	$GAIN_0(k)$
0	1	920.5	0
1	2	422.6	0.541
2	4	143.85	0.843
3	**8**	**44.135**	**0.952**
4	16	13.050	0.986
5	32	3.809	0.996
6	64	2.930	0.999

Table 4. Explaining different levels of savings through semi-parametric models in the Bank of Italy data set

Group	n	Average Savings	Significant Predictor
1	826	-15878	YL,s(YCR),lo(YCF),s(AR)
2	738	-11837	CD,s(CN), lo(YCR),lo(YCF)
3.1	1357	-1453	s (YT),s(YCF),s(AR),lo(CN)
3.2	1747	2774	YT,s(YCF),lo (CN)
4.1	1174	7572	s(YCF),lo (YCR),lo (CD)
4.2	781	12655	YL, s(CN)
5.1	415	16628	s(YL), s(YCF),lo(CD)
5.2	551	20844	s(CD), s(CN)
6	641	28139	YM, YT, s(YL),s(AR)
7	386	39281	lo(CD),s(YM)
8	211	54019	s(YCR),s(YCF)

Furthermore, we identified the most influencing predictors in the 8 sub-

[2] All measurements are expressed in Italian Lira

populations by means of semi-parametric models. On this purpose, we used for each predictor, as possible alternative estimators, the functions lowess (`lo`), smoothing spline (`s`) and the linear smoother, trying to find, at the same time, the most suitable degree of smoothness.

Starting from the third up to the fifth sub-population, a further binary partition was induced by the splitting criterion introduced in Table 2. At the end of the procedure we obtained in each group significant relationships between the dependent variable and the most important predictors.

Table 4 shows the final number of sub-populations and the number of their respective observations, the average level of savings in each group, the most powerful predictors and their respective smoothing functions.

5 Concluding remarks

The proposed methodology provides the identification of homogeneous sub-populations and the parsimonious description of their behaviours by means of semi-parametric models. As a matter of fact, the second step of the procedure may be also understood as a special case of a general methodology recently introduced by Conversano (2000).

We think main advantages of our approach are that our procedure for data mining allow us to:

- quickly identify sub-populations that are homogeneous with respect to a dependent variable;
- identify subsets of cases whose behaviour is well described by means of a parsimonious model detecting the most explicative predictors;
- not impose any constraint with respect to the nature of data (numerical or categorical variables can equally be treated).

References

Conversano, C. (2000). *Semi-parametric Models for Supervised Classification and Prediction. Some Proposals for Model Integration and Estimation Procedures* (in Italian), Ph.D Thesis in Computational Statistics and Data Analysis, Universitá di Napoli Federico II.

Fayyad, M., Piatesky-Shapiro, G. and Smith, N. (1996). *Advances in Knowledge Discovery and Data Mining*, AAAI Press/The MIT Press.

Härdle, W., Müller, M., Sperlich, S. and Werwatz, A. (1999) *A course on Non- and Semi-parametric Modeling*, Technical Report Hümbolt University of Berlin.

Hand, D.J. (1998) Data mining: statistics and more?, *The American Statistician*, **52**-2, pp. 112118.

Hand, D. J. (1997) *Construction and Assessment of Classification Rules*, Chichester: John Wiley and Sons.

Hastie, T. and Tibshirani, R. (1990). *Generalized Additive Models*, London: Chapman & Hall.

Mola, F., Klaschka, J. and Siciliano, R. (1996). Multinomial Logistic Regression for Classification Trees, in *COMPSTAT 96, Proceedings in Computational Statistics* (ed. A. Prat), pp. 373-378, Heidelberg: Physica-Verlag.

Ripley, B.D. (1996). *Pattern Recognition and Neural Networks.* Cambridge: Cambridge University Press.

Siciliano, R. and Mola, F. (1994). Modelling for Recursive Partitioning and Variable Selection, in *COMPSTAT 94, Proceedings in Computational Statistics* (ed. R. Dutter & R. Grossman), pp. 172-177, Heidelberg: Physica-Verlag.

Preliminary estimation of ARFIMA models

Marcella Corduas

Dipartimento di Scienze Statistiche, University of Naples Federico II, Via G. Sanfelice 47, 80134 Napoli, Italy

Abstract. In this article we propose a preliminary estimator for the parameters of an *ARFIMA(p,d,q)* model. The estimation procedure is based on the search of the element in the class of *ARFIMA* models *closest* to the estimated *ARMA* model which best fits the observed time series.

Keywords. Long memory, ARFIMA models, Preliminary estimate

1 Introduction

The class of *ARFIMA* (Autoregressive Fractional Integrated Moving Average) models offers a flexible tool to describe long memory series. Several estimation methods have been developed both in time and frequency domain (see Beran, 1995b, for a review).

A non parametric approach was first proposed by Janaceck (1982), Geweke and Porter-Hudack (1983). The procedure uses the log-periodogram to estimate the fractional differencing parameter. The approach is useful in the preliminary analysis of the time series in order to detect the presence of the fractional operator. Then the time series has to be pre-filtered to identify the order of *AR* and *MA* components from the filtered series.

Also, great attention has been paid to the use of the maximum likelihood method. In this respect, Sowell (1992) derived a procedure to compute the unconditional exact likelihood function which uses the hypergeometric function to evaluate the covariances of the process efficiently. Beran (1995a) developed an approximation based on conditional least squares whereas Hanslett and Raftery (1989) discussed the problem with reference to spatial models. Fox and Taqqu (1986) and Dahlhaus (1989) considered the estimation in frequency domain and discussed the asymptotic properties of the Whittle estimators and corresponding approximations.

The computation of the exact likelihood function may become time consuming because a large sample size is required to observe the long memory structures and since the autocorrelation function can not be neglected even for large lags. Moreover, the likelihood function may not be globally concave.

In such a context, it is generally recognized that a good selection of starting values for the model's parameters is a critical step in order to reduce the computational time and to avoid the final solution, produced by the maximum likelihood search, depending on that initial choice. The aim of this paper is to propose a new procedure to provide initial values of the *ARFIMA* parameters.

2 The preliminary estimator

According to the standard notation, let X_t be an invertible and stationary Gaussian process following an *ARFIMA(p,d,q)* model:

$$\phi(B)\nabla^d X_t = \theta(B)a_t$$

where a_t is a White Noise process with zero mean and constant variance σ_a^2, $\phi(B)$ and $\theta(B)$ are polynomials with no common factors and with roots outside the unit circle and, finally, $0 < d < 0.5$.

Then, X_t follows the *AR(∞)* representation: $\pi_F(B)X_t = a_t$ with $\pi_F(B) = \pi(B)(1-B)^d = 1 - \sum_{j=1}^{\infty} \pi_{Fj} B^j$. The weights $\{\pi_{Fj}\}$ are determined by the convolution of the polynomials: $\pi(B) = \theta^{-1}(B)\phi(B)$ and $(1-B)^d = \sum_{k=0}^{\infty} \alpha_k B^k$, where $\alpha_k = (-1)^k \Gamma(d+1)/[\Gamma(k+1)\Gamma(d-k+1)]$ and $\Gamma(y)$ denotes the Gamma function. The *AR(∞)* model truncated at a certain lag has been used for several purposes, for example, to produce estimates and forecasts as in Granger and Joyeaux (1980), Sutcliffe (1994).

Although the *ARMA* models can not fully capture the dynamics of a long memory series, they can still provide good approximations as discussed by Lawrance and Kottegoda (1977) and Hosking (1981, 1984).

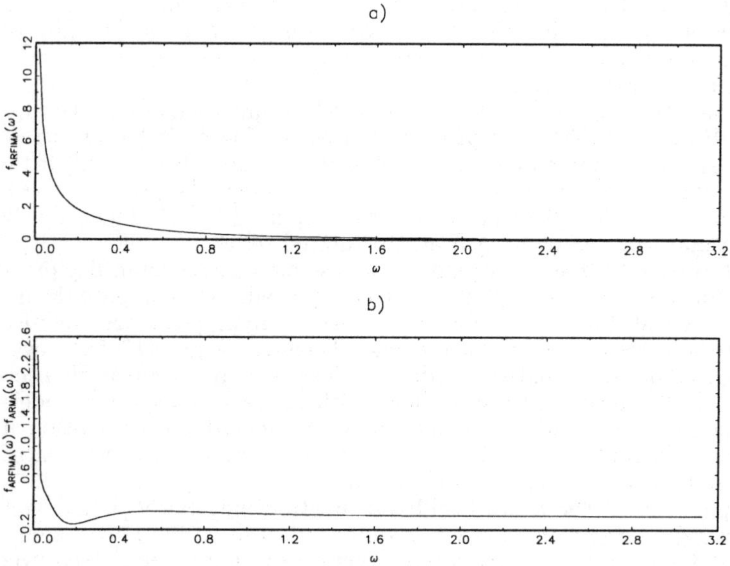

Figure 1. (a) Spectral density of the ARFIMA(1,d,0) model; (b) differences between the spectral density of the ARFIMA model and that of the proxy model

As an example, we consider the *ARFIMA(1,d,0)* process defined as:

$$(1 - 0.5B)\nabla^{0.35} X_t = a_t$$

and the *ARMA(2,1)*process:

$$(1 - 1.7409B + 0.7466B^2)X_t = (1 - 0.9445B)b_t.$$

The spectral densities of the processes (Figure 1) can not be easily distinguished. Of course, the difference between them is larger at the very low frequencies. But the corresponding time components are more difficult to detect in finite samples. They may not be observable unless the number of observations is very large. Moreover, the short memory components tends to interfere with the long memory effects and make them difficult to recognize in practice. For these reasons fitting an *ARMA* model to a long memory time series can provide some useful information on the underlying dynamic structure.

Recently, Tiao and Tsay (1994) suggested that *ARMA* models produce good forecasts for long memory series when an adaptive estimation procedure is applied. Moreover they showed that if the *true* model is *ARFIMA(0,d,0)* and instead we fit an *ARMA(1,1)* model: $(1-\lambda B)X_t = (1-\eta B)b_t$, the *MLE* estimators $\left\{\widehat{\lambda}, \widehat{\eta}\right\}$ are consistent for the parameters of the *wrong* model $\{\lambda, \eta\}$.

Extending this argument, we approximate an *ARFIMA(p,d,q)* model by an *ARMA(p+1,q+1) proxy* model: $\lambda(B)X_t = \eta(B)b_t$. Of course, if $X_t \sim ARFIMA(p,d,q)$, b_t is not a White Noise process but it is an *ARFIMA* process itself. Specifically, $b_t = \lambda(B)\eta^{-1}(B)\theta(B)\phi^{-1}(B)(1-B)^{-d}a_t$.

Then, from the maximum likelihood estimates $\left\{\widehat{\lambda}_1, ..., \widehat{\lambda}_{p+1}, \widehat{\eta}_1, ..., \widehat{\eta}_{q+1}, \widehat{\sigma}_b^2\right\}$ we compute the estimated *AR* weights $\{\widehat{\pi}_{Pj}\}$.

We propose to search the preliminary estimates $\widehat{\beta}$ of the parameter vector $\beta_0 = (\phi_1, ..., \phi_p, \theta_1, ..., \theta_q, d)^T$ of the *ARFIMA* models by minimizing:

$$\delta_m(\beta) = \sum_{j=1}^{m} \left(\pi_{Fj}(\beta) - \widehat{\pi}_{Pj}\right)^2$$

with respect to β. The square root of δ_m is the Euclidean distance between the $AR(\infty)$ weights of the fitted ARMA model and those of the *ARFIMA* model. This criterion was introduced by Piccolo (1990) in order to discriminate and classify time series. Here, we use that criterion with a different aim. Given the best *proxy* model for X_t within the class of *ARMA* models we search for the *closest* element in the class of *ARFIMA* models with given order p and q. We use the coefficients of that element as preliminary estimates of the parameters of interest.

3 Monte Carlo study

In order to assess the properties of the proposed estimator in finite samples, we present the result of a Monte Carlo study performed on a set of 1000 independent Gaussian time series with 900 observations. The number of π-weights to be included in the estimation was fixed to 100.

Table 1. ARFIMA(0,d,0) model

d	$ave(\widehat{d})$	$bias$	$r.m.s.e.$	$ave(\widehat{\lambda})$	$ave(\widehat{\eta})$	$ave(\widehat{\lambda - \eta})$
0.25	0.237	−0.013	0.049	0.810	0.587	0.223
0.35	0.331	−0.019	0.052	0.875	0.559	0.316
0.45	0.424	−0.026	0.047	0.931	0.516	0.415

In Table 1 we report the results for a simple *ARFIMA(0,d,0)* model. The preliminary estimator shows a small negative bias and a root mean squared error (rmse) of the expected order. Also, we report the average of the estimated *ARMA* parameters. Notice that a crude estimate of the fractional differencing parameter is given by the difference between the *AR* and *MA* term. We have compared these results with the approximated maximum likelihood estimates, derived in the frequency domain.

Specifically, let $I_n(\omega_j) = (2\pi n)^{-1} \left|\sum_t \widetilde{x}_t \exp(-i\omega_j t)\right|^2$ denote the periodogram of the centered series $\widetilde{x}_t = x_t - \overline{x}$, t=1,2,...,n, evaluated at the frequencies, $\omega_j = 2\pi j/n, \ j = 1, ..., [(n-1)/2]$. The maximum likelihood estimates are obtained by maximizing the approximated profile likelihood:

$$\widetilde{L}(d, \phi, \theta, \widehat{\sigma}_a^2) \propto -(n/2) \ln \left(\sum_j \frac{I_n(\omega_j)}{f(\omega_j)}\right) -(1/2) \sum_j \ln f(\omega_j)$$

where $f(\omega_j)$ is the parametric spectral density of the process and

$\widehat{\sigma}_a^2 = n^{-1} \sum_j I_n(\omega_j)/f(\omega_j)$.

The use of initial estimates in such a case reduces the computation time. The study has been implemented in GAUSS for Windows95 (version 3.2.37) on a Pentium II 300Mhz. We have used the function *hsec* included in that language to time the segment of code where the estimation is performed. Having initialized the iterative optimization with the preliminary estimate reduced on average the computation time from 33% (for $d = 0.25$) up to 43% (for $d = 0.45$) with respect to the use of the default value 0.1.

Moreover, we considered some *ARFIMA(1,d,0)* structures. These are more difficult to detect since the long memory effect can interfere with the short term *AR* component and make the estimation more difficult. The preliminary estimator shows a good performance with respect to the Whittle estimator (see Table 2.a and 2.b).

Table 2.a ARFIMA (1,d,0) model: preliminary estimates, d=0.35

$ave(\widehat{d})$	$bias(\widehat{d})$	$r.m.s.e(\widehat{d})$	ϕ	$ave(\widehat{\phi})$	$bias(\widehat{\phi})$	$r.m.s.e(\widehat{\phi})$
0.360	0.010	0.041	−0.8	−0.839	−0.039	0.047
0.340	−0.010	0.049	−0.5	−0.503	−0.003	0.044
0.326	−0.024	0.057	−0.3	−0.278	0.022	0.061
0.317	−0.033	0.130	0.3	0.336	0.036	0.150
0.283	−0.067	0.163	0.5	0.529	0.029	0.162
0.372	0.022	0.109	0.8	0.781	−0.019	0.092

Table 2.b ARFIMA (1,d,0) model: Whittle estimates, d=0.35

$ave(\widehat{d})$	$bias(\widehat{d})$	$r.m.s.e(\widehat{d})$	ϕ	$ave(\widehat{\phi})$	$bias(\widehat{\phi})$	$r.m.s.e(\widehat{\phi})$
0.350	0.000	0.0304	−0.8	−0.797	0.003	0.022
0.350	0.000	0.0346	−0.5	−0.499	0.001	0.035
0.348	−0.002	0.0372	−0.3	−0.298	0.002	0.044
0.346	−0.004	0.0732	0.3	0.302	0.002	0.084
0.330	−0.020	0.0983	0.5	0.515	0.015	0.100
0.348	−0.002	0.0847	0.8	0.791	−0.009	0.067

Only one situation seems to be extreme. When $d = 0.35$ and $\phi = 0.5$, the preliminary estimator is badly affected by the fact that the dynamic structure of the fractional model, in finite samples, can not be easily discriminated by that of a pure *AR* model with a larger coefficient. This implies a number of cases in which the fractional parameter d is estimated as zero and the whole dynamic structure is captured by the *AR* term.

In Figure 2, we present the box-plots of the simulated distribution of the two estimators for the differencing parameter. The difficulty of the preliminary estimator to identify the differencing parameter correctly in the case mentioned above is shown very clearly. In the other cases, the form of the distribution of the preliminary estimator seems to follow that of the Whittle estimator quite closely.

The time series were generated using the algorithm by Hosking (1984) and the Normal random number generator available in GAUSS. This is based on the fast rejection-acceptance algorithm proposed by Kindeman and Ramage (1976). The optimization was performed using the Newton-Raphson algorithm available in the CML module of GAUSS, written by Schoenberg (1995).

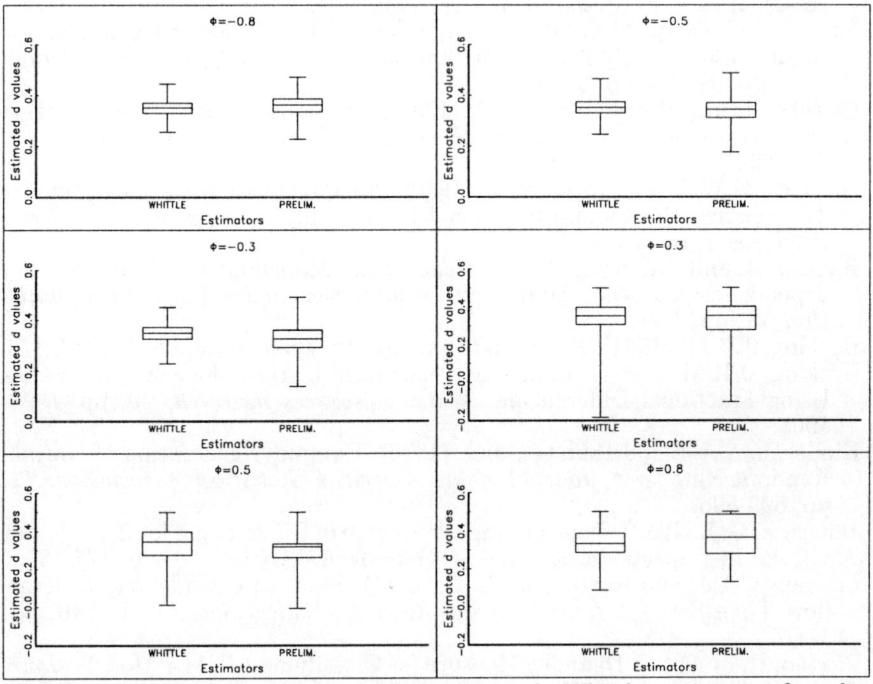

Figure 2. Box Plots of the preliminary and Whittle estimates for the ARFIMA(1,d,0) model with d=0.35

4 Discussion

In this article we have proposed a preliminary estimator for the parameters of an *ARFIMA(p,d,q)* model. The estimation procedure is based on the search of the element in the class of *ARFIMA* models *closest* to the estimated *ARMA* model which best fits the observed time series. The proposed procedure provides the initial values for the fractional differencing parameter and the *ARMA* coefficients at the same time. This aspect makes the proposal more appealing since many time series arising in practice present both the long-memory and short term components. Other estimators, such as those derived from the periodogram, can, in fact, only provide the initial estimate of d.

Acknowledgements: This research was supported by MURST funds.

References

Beran, J. (1995a). Maximum Likelihood Estimation of the Differencing Parameter for Invertible Short and Long Memory Autoregressive Integrated Moving Average Models. *Journal of the Royal Statistical Society*, B, **57**, pp. 659-672.

Beran, J. (1995b). *Statistics for Long Memory Processes.* New York: Chapman and Hall.

Dahlhaus, R. (1989). Efficient Parameter Estimation for Self-Similar Processes. *Annals of Statistics*, **17**, pp. 1749-1766.

Fox, R. and Taqqu M.S. (1986). Large Sample Properties of Parameter Estimates for Strongly Dependent Stationary Gaussian Time Series. *Annals of Statistics*, **14**, pp. 517-532.

Geweke, J. and Porter-Hudack, S. (1983). The Estimation and Application of Long-Memory Time Series. *Journal of Time Series Analysis*, **4**, pp. 231-238.

Granger, C.W.J. and Joyeaux, J. (1980). An Introduction to Long Memory Time Series Models and Fractional Differencing. *Journal of Time Series Analysis*, **1**, pp. 15-30.

Hanslett J. and Raftery A. (1989). Space-Time Modelling with Long-memory dependence: assessing the Ireland's wind power. *Journal of Applied Statistics*, **38**, pp. 1-21.

Hosking, J.R.M. (1981). Fractional Differencing. *Biometrika*, **68**, pp. 165-176.

Hosking, J.R.M. (1984). Modeling Persistence in Hydrological Time Series Using Fractional Differencing. *Water Resources Research*, **20**, pp. 1898-1908.

Kinderman, A.J. and Ramage, J.G. (1976). Computer generation of Normal Random Numbers. *Journal of the American Statistical Association*, **71**, pp. 893-896.

Janaceck, G.J. (1982). Determining the Degree of Differencing for Time Series via the Log Spectrum. *Journal of Time Series Analysis*, **3**, pp. 177-183.

Lawrance, A.J. and Kottegoda, N.T. (1977). Stochastic Modelling of Riverflow Time Series . *Journal of the Royal Statistical Society, A,* **140**, pp. 1-47.

Piccolo, D. (1990). A Distance Measure for Classifying ARIMA Models. *Journal of Time Series Analysis*, **11**, pp. 153-164.

Schoenberg, R. (1995). *CML -Maximum Likelihood Estimation with General Nonlinear Constraints on Parameters*, Maple Valley: Aptech System inc.

Sowell, F. (1992). Maximum Likelihood Estimation of Stationary Univariate Fractionally Integrated Time Series Models. *Journal of Econometrics*, **53**, pp. 165-188.

Sutcliffe, A. (1994). Time Series Forecasting using Fractional Differencing. *Journal of Forecasting*, **13**, pp. 338-344.

Tiao, G.C. and Tsay, R. (1994). Non Linear and Adaptive Modelling in Time Series. *Journal of Forecasting*, **13**, pp. 109-131.

A collection of applets for visualizing statistical concepts

P. Darius[1], J-P. Ottoy[2], A. Solomin[3], O. Thas[2], B. Raeymaekers[3] and S. Michiels[3]

[1] Faculty of Agricultural and Applied Biological Sciences, K.U. Leuven, Kard. Mercierlaan 92, B-3001 Leuven, Belgium
[2] Faculty of Agricultural and Applied Biological Sciences, Ghent University, Coupure links 653, B-9000 Gent, Belgium
[3] University Center of Statistics, K.U. Leuven, De Croylaan 52, B-3001 Leuven, Belgium

Abstract. This paper describes a set of didactic tools for statistical teaching, implemented as JAVA applets. The tools allow to visualize a number of statistical concepts, and to experiment with them interactively.

Keywords. Teaching, Statistics Education, JAVA, Applets

1 Background

Teachers of statistics encounter a number of didactic problems, especially when teaching the introductory statistics course. Many of the key concepts are defined in a quite abstract way: e.g. through complex formula's, based on the frequentist approach, based on asymptotic results,.. Some students experience parts of the material as too abstract, and have difficulty mastering the key concepts. It has been argued that supplementing the traditional material with tools based on a visual approach and a more active form of learning could improve the effectiveness of the teaching (Moore (1994), Marasinghe et al. (1996)).

This paper describes such a tool. It consists of a set of JAVA applets called VESTAC (Visualization of and Experimentation with STAtistical Concepts), covering selected topics useful for an introductory course and a second regression and/or anova course. Each applet gives a visual representation of the topic, with ample possibilities to see the asymptotic results build up, or to experiment with the data and the parameters and see the effects immediately. Those topics were selected where it was thought that an interactive visual applet could have didactic properties exceeding those attainable with more traditional didactic tools.

An implementation in the form of JAVA applets was chosen because of its apparent advantages in a teaching context: the software is kept and maintained at one place (the Internet server), and can be executed on any PC with an Internet connection and an Internet browser. No special software is needed on the client PC, yet all the computational work is done there, allowing the server to serve large quantities of clients at the same time. JAVA applets also have disadvantages, which are discussed at the end of section 3.

The VESTAC applets can be accessed over the web through one of the following URL's:

```
http://www.kuleuven.ac.be/ucs/java
http://fltbw2.rug.ac.be/iloapp/Applets_Home.html
```

2 Scope

The VESTAC collection currently consists of 32 applets. Additional ones are under construction. They cover selected topics from the following four areas: distributions and plots, tests and confidence intervals, regression and analysis of variance.

The distribution applets visualize the form and relevant tail areas of the usual univariate distributions and the bivariate normal. They also show how these change when the parameters are changed. The central limit theorem is illustrated by showing how the mean of samples from different distributions gradually approaches a normal distribution. Other applets let the user experiment with histograms, boxplots and QQ-plots, and visualize the concept of correlation.

The second group of applets deals with the concept of a confidence interval, illustrates one sample, two sample and some non-parametric tests and visualizes power and type I and type II errors.

The regression applets deal with the relation between population and sample, and show how the variation of the sample regression line is influenced by the choice of the x-values. It is also shown how fitting a curve according to the least squares criterion relates to finding the minimum of an SSerror surface, in the linear as well as in the non-linear case. The distributions of MSregression, MSerror and their quotient are built up through repeated sampling. The correlation between the estimators for slope and intercept is shown. Confidence bands and prediction intervals are illustrated, residual and normal probability plots, as well as the effect of influential points and outliers. The difference between regressing y on x and x on y is shown (in all these cases the data points can be dragged with the mouse and the effect immediately seen). Other applets illustrate the Box-Cox transformation and the effect of near-multicollinearity on the confidence intervals of the parameters. Finally, a couple of applets visualize permutation tests for small datasets.

The anova applets illustrate the relation between population and sample, and allow to see the effect of changes in the data on residual and normal probability plots. Another applet shows how the distributions of MStreatment, MSerror and their quotient are built up, when the usual null hypothesis is true, as well as in the case of an a user specified alternative hypothesis being true. There is also an applet that shows the mechanism of the permutation test and visualizes the exact randomization distribution. Finally, an applet illustrates the danger of using t-tests for testing hypotheses generated by data snooping. It shows both the distribution of pairwise differences between sample group means under repeated sampling, and the distribution of the difference between the largest and the smallest sample group mean.

3 Implementation

Although several interactive tools for statistical teaching have been described in the literature (e.g. Mathieson et al. (1995), Marasinghe et al. (1996), Dunn (1999)), or are available at several places on the Internet, the VESTAC collection is rather different in a number of important aspects.

First of all, every effort has been made to give the applets, as much as possible, a common "look and feel". The collection is based on an extensive class library specially written in JAVA to implement certain GUI aspects, repeated random simulation, construction of statistical graphs and statistical computation. All the applets use classes from this library.

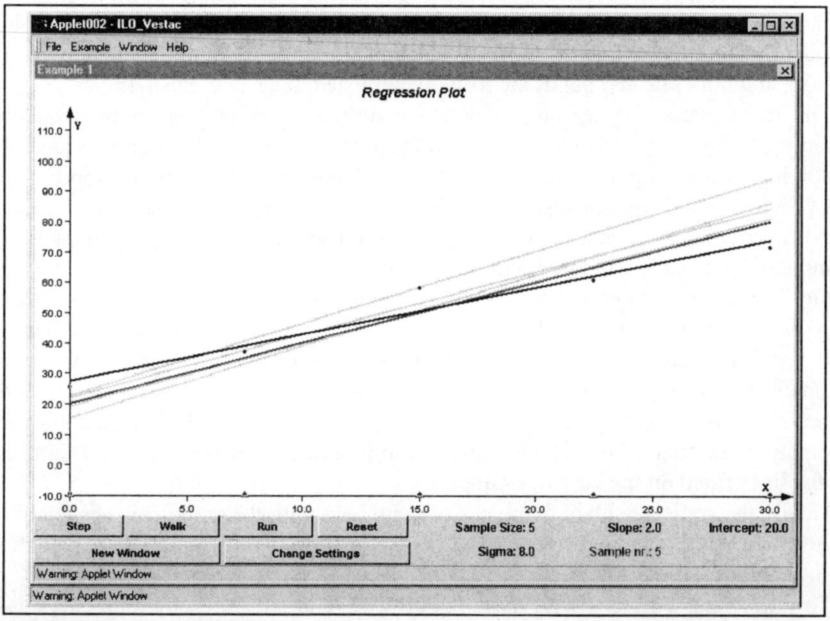

Fig 1. Variability of the sample regression line, as illustrated by repeated sampling from a population (see explanation in section 3).

Repeated sampling is implemented through "Step", "Walk" and "Run" buttons. The "Step" button generates one new sample. The "Walk" button generates one sample after the other until a specified limit is reached, and does this at a slow rate, allowing the user to see how the final result is built up. The "Run" button generates new samples at the fastest possible rate. Fig. 1 illustrates this for the applet visualizing the variability of the sample regression line. It shows the situation after hitting "Step" five times. Shown are: the population regression line (the dark line with intercept 20.0), and (in a different colour, but also dark in Fig 1) the current sample points and the fitted line. The gray lines are the lines fitted on the previous samples.

An applet originally appears in a separate window, with default parameter settings applied. This makes it possible to start the applet demonstration quickly (e.g. while teaching in a classroom). The visual elements (e.g. the points on a scatterplot) can, wherever applicable, be manipulated directly by dragging with the mouse. The most important parameters (e.g. the degrees of freedom of a distribution) can be adjusted directly on the screen. All the other parameters can be changed through a pop-up window. The applets not only execute simulations based on settings of population parameters, those for which it is relevant also allow import of an actual dataset. This makes it possible for the applets to use the same datasets as e.g. those in the textbook. However, due to security measures in Internet browsers, this is only possible with datasets stored on the web server.

Fig 2 hints at the available possibilities for interactive experimentation. This applet shows a set of points (either obtained by random sampling from a specified population, or from an available dataset), with the fitted regression line and/or prediction/confidence bands. Each of the points can be dragged around with the mouse, and both line and bands are instantly adjusted to the new position.

The most interesting learning experience occurs when two or more parameter settings can be compared side by side. To make this possible, all the applets have the following features: upon hitting a "new window" button, a new window appears and the previous windows are resized, so that they are all visible on the screen. Each window now behaves as a separate applet: parameters can be changed on each and computations in each window occur concurrently.

Fig 3 shows two windows with the same applet (the one from Fig 1). It illustrates how the x-values of the sample points influence the variability of the fitted line. In both windows, sampling is from the same population. But in each window points are sampled from different x-locations. The user can specify these by dragging the little triangles on the x-axis to the appropriate positions. As in Fig 1, the screen shows the population regression line, the last sample and its fitted regression line, and (in gray) all the lines fitted on the previous samples.

Finally the applets contain two types of help: help with the graphical user interface (when the mouse is moved over a part, the explanation is shown) and an explanation of the statistical background.

To implement a library with these features, several software environments could have been chosen. In VESTAC, JAVA applets were used because of their apparent advantages from an educational point of view, summarized at the end of section 1. A client server approach could have been used, with a statistical package on the server side. This might slow down the execution speed, especially when a substantial number of users requests access at the same time (a typical situation with guided exercises in a PC-class), and raises license fee issues. The download speed of the JAVA code will be a problem when the internet connection is slow. When its speed is reasonable, we found the download speed of the applets acceptable (only the needed parts of the class library are downloaded). The original JAVA version 1.1 implementation proved to be rather stable, and only minor adjustments were required to keep the applets fully operational with the newest versions of the major internet browsers.

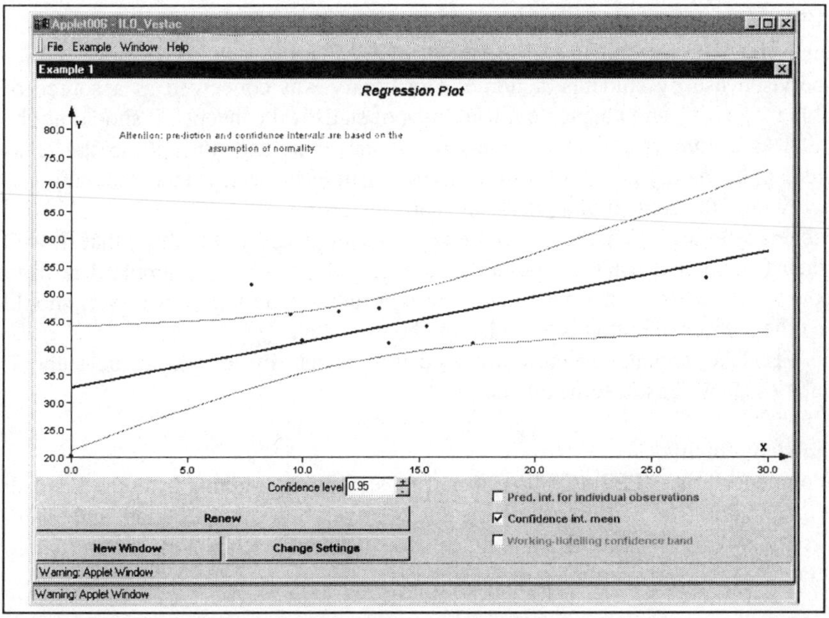

Fig 2. Sensitivity of the fitted line and the associated prediction/confidence bands for the location of particular points (see explanation in section 3).

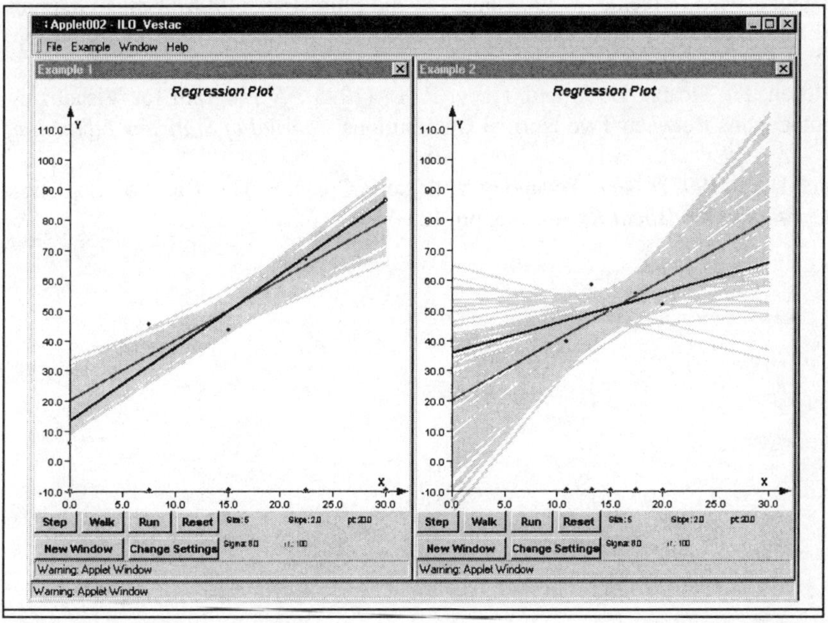

Fig 3. Comparing the variability of the fitted line, for samples from the same population but with different x-values (see explanation in section 3).

4 Role in teaching

The VESTAC applets are intended for use by the teacher in the classroom, as well as for supervised use by students during practical exercises in a PC classroom, and for unsupervised use by students at home. The library was conceived as a source of possible interactive and adaptable illustrations of statistical concepts. It should not be regarded as a comprehensive overview of all the important concepts in the areas covered, and certainly not as a learning environment of its own. Hence it should only be used within the context of a (classical) statistics course.

Teachers who want to make use of the applets should select only those that fit well into their course, and wherever possible use the applets with the example data from the course. Teachers who want to store certain datasets on the web server, should contact the authors (a button is provided on the web site).

The VESTAC applets are currently used in a variety of courses at the authors' universities, as well as in some others.

Acknowledgements

Development of the VESTAC library was financed through grants from the Flemish government (Ministry of Education), the K.U. Leuven Education Council, and Ghent University. Wim Moreau, Brian van de Noortgate and Ruqi Wang contributed substantially to the programming of the library.

References

Dunn, P.K. (1999). Three Tools for Interactively Visualizing Some Distribution Theory Concepts. *The American Statistician*, **53**, pp. 137-139.

Marasinghe, M.G., Meeker, W.Q., Cook, D., and Shin, T.-S. (1996). Using Graphics and Simulation to Teach Statistical Concepts. *The American Statistician*, **50**, pp. 342-351.

Mathieson, K., Doane, D. P. and Tracy, R. L. (1995). A Program for Visualizing Comparisons Between Two Normal Distributions. *Journal of Statistics Education*, **3**.

Moore, D.S. (1997). New Pedagogy and New Content: The Case of Statistics. *International Statistical Review,* **65**, pp. 123-165.

Non-parametric regression and density estimation under control of modality

P. L. Davies and A. Kovac
Department of Mathematics, University of Essen, 45117 Essen, Germany

Abstract. Most available methods in non-parametric regression and density estimation are not directly concerned with modality. New methods are presented that avoid artifacts and yield estimates that have asymptotically the correct modality.

Keywords. Non-parametric regression, density estimation, modality, strings, multiresolution analysis

1 Non-parametric regression and modality

Non-parametric regression is a topic of much current interest in statistics. The problem is to estimate a function f on the basis of observations $y_1, \ldots, y_n$ at time points $t_1, \ldots, t_n$ where

$$y_i = f(t_i) + \varepsilon_i$$

and the $\varepsilon_1, \ldots, \varepsilon_n$ are white noise.

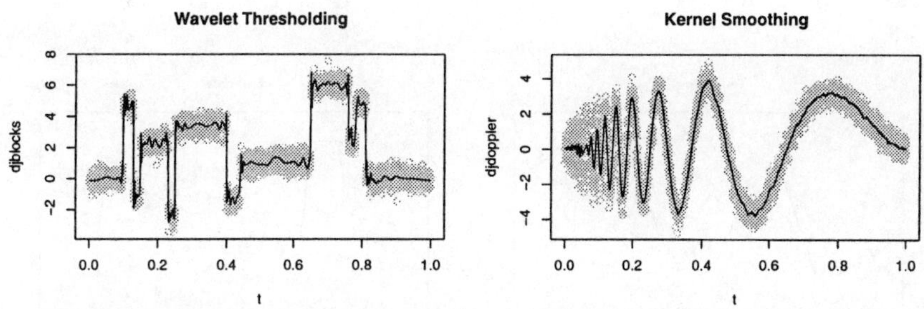

Fig. 1. Application of two classical smoothing methods to simulated data. The left panel shows Gibbs effects that occur often when wavelet thresholding is applied to functions with discontinuities. In the right panel a kernel estimator does not adapt locally to different degrees of smoothness.

Numerous methods have been suggested including kernel estimators (Nadaraya, 1964, Watson, 1964), spline smoothing (Silverman, 1985), local polynomial fitting (Fan and Gijbels, 1996) and wavelet thresholding (Donoho et al., 1995; Härdle et al., 1998). However, none of these methods controls the number of local extremes, and as a result they all tend to produce superfluous extremes or exhibit artifacts like Gibbs effects near discontinuities (see Figure 1). Several procedures have been proposed to decide for each local extreme whether it comes from the data or is an artifact (Marron et al., 1995). In the latter case the extreme value is eliminated.

2 A taut string procedure

We follow a different idea and attempt to minimize the modality among all functions which give an adequate approximation to the data. The notion of adequacy was introduced by Davies (1995). In the context of function estimation adequate functions are those such that the residuals look like white noise. Adequacy regions based on the length of the longest run of the signs of the residuals and related measures have been analysed by Davies (1995) and Dümbgen (1998). We will introduce a new definition of adequacy: a function f is adequate if the multiresolution coefficients $w_{j,k}$

$$w_{j,k} = 2^{-j/2} \sum_{i=k2^j+1}^{(k+1)2^j} (y_i - f(t_i)).$$

of the residuals are all smaller than $\sqrt{2\log(n)}\sigma$.

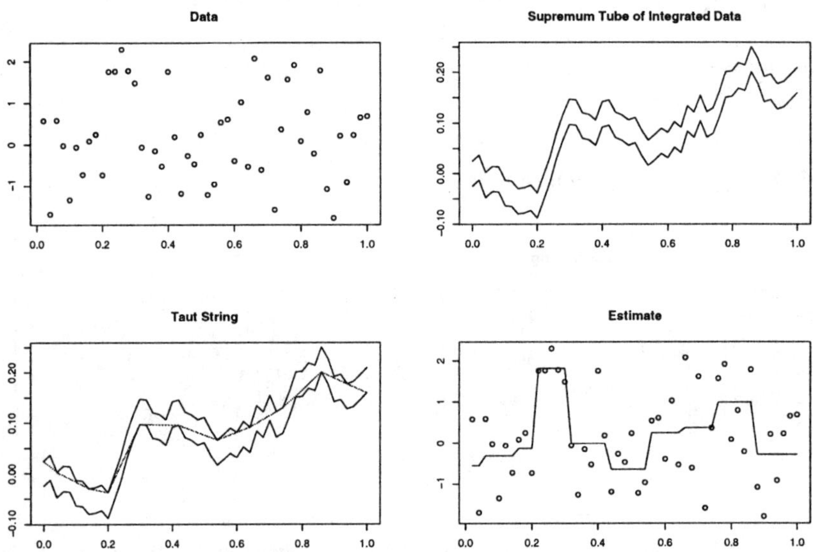

Fig. 2. The taut string procedure. The given data are integrated and a supremum tube of the integrated process is considered. A string in the tube is tightened and the derivative of the taut string is finally an estimate of the data.

For the sake of simplicity we assume that we have $n = 2^J$ time points, equally spread over the interval $[0,1]$. These assumptions can be relaxed using the methodology of Kovac and Silverman (2000). In order to obtain functions with minimal modality we consider the integrated process Y defined by

$$Y(j/n) = \frac{1}{n} \sum_{i=1}^{j} y_i, \quad j = 1, \ldots, n$$

with linear interpolation between the design points. Then we employ the method of a taut string $\tilde{F}$ constrained to lie in a supremum neighbourhood of Y, ie between $Y-\varepsilon$ and $Y+\varepsilon$. Differentiating $\tilde{F}$ yields a piecewise constant model for f. This procedure is illustrated in Figure 2. This method goes back to Hartigan and Hartigan (1985) and in the context of density estimation to Davies (1995). It has recently been adopted to non-parametric regression by Mammen and van de Geer (1997).

The choice of the width ε is important. Asymptotic results show that a width of order $C/\sqrt{n}$ attains the correct modality (Davies and Kovac, 1999). Moreover, the taut string procedure attains the optimal rate of convergence away from the extremes.

Another idea is to choose ε automatically from the data as the largest width such that the multiresolution coefficients of the residuals are all smaller than $\sqrt{2\log(n)}\sigma$. In an actual implementation this width can be approximated by an iterative procedure where one starts with a sufficiently large bandwidth which is subsequently being reduced until all multiresolution coefficients are small enough.

3 Local squeezing

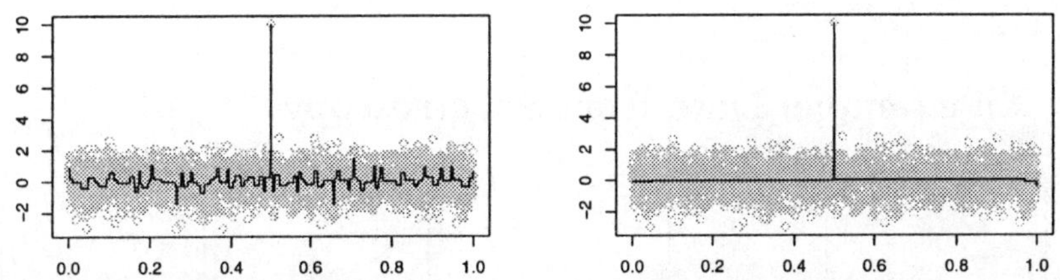

Fig. 3. Global and local squeezing. The taut string procedure was applied to white noise with an exceptional large data point in the center of the data using both squeezing methods. In order to approximate the small peak global squeezing has to reduce the bandwidth on the whole interval, introducing many additional extremes. Local squeezing avoids this effect and attains the correct modality.

Usually the taut string procedure leads to reasonable estimates for f. However, the theoretical results mentioned above show that the procedure is not optimal at local extremes. Figure 3 shows an example where the global squeezing procedure picks up many artifacts while reducing the bandwidth until the peak in the center of the data is approximated sufficiently well.

Such artefacts can be avoided by further reducing the radius ε only in the neighbourhood of local extreme values. The localization of a multiresolution decomposition in time can be used to determine the regions where the bandwidth must be reduced. In practice we start with a sufficiently large global bandwidth which is gradually decreased only in those intervals where the corresponding multiresolution coefficients are larger than $\sqrt{2\log(n)}$. Applied to the data set from Figure 3 local squeezing retains the peak in the center of the data without introducing further extremes.

Figure 4 shows the application of the local squeezing method to two simulated data sets using Donoho and Johnstone's (1995) often used test functions. Number and positions of the local extremes are estimated correctly for the Blocks data and for the right half of the Doppler data. The high-frequency phenomena at the left-hand side are also approximated very well. This empirical result is supported by theoretical results (Davies and Kovac, 1999) that show that the local squeezing method attains asymptotic consistency of the modality and optimal rate of convergence.

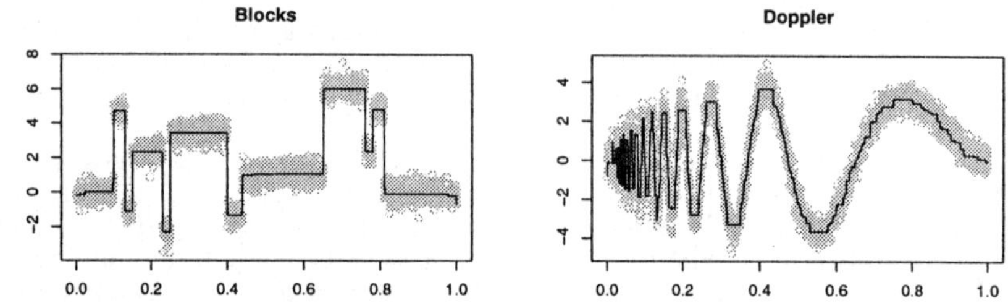

Fig. 4. Local squeezing. The taut string procedure with local squeezing is applied to simulated data sets using Donoho and Johnstone's test functions.

4 An example: Data from Spectroscopy

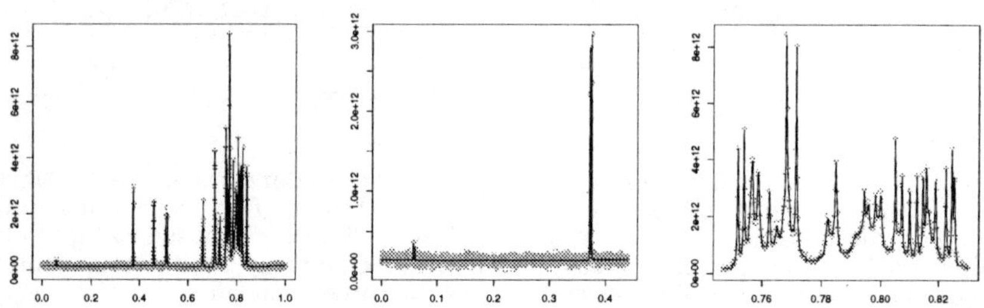

Fig. 5. Spectroscopical data from a gall stone. The left panel shows the entire data set. The other panels show two excerpts.

An important application arises in Spectroscopy where local extremes play a crucial part in the analysis of data. The positions of the extremes determine bonds that are present in the substance which is analyzed. The power of a peak represents the quantity of a specific bond. Therefore an appropiate smoothing method must take into account the number and positions of local extreme values. Figure 5 shows that the taut string method works very well on this kind of data and retains the peaks while it is constant in between.

5 Density estimation

A closely related topic is density estimation where modality often plays a crucial part in the analysis of the data. Local maxima in the density estimate reflect certain effects in the mechanism that generated the data. Again, a

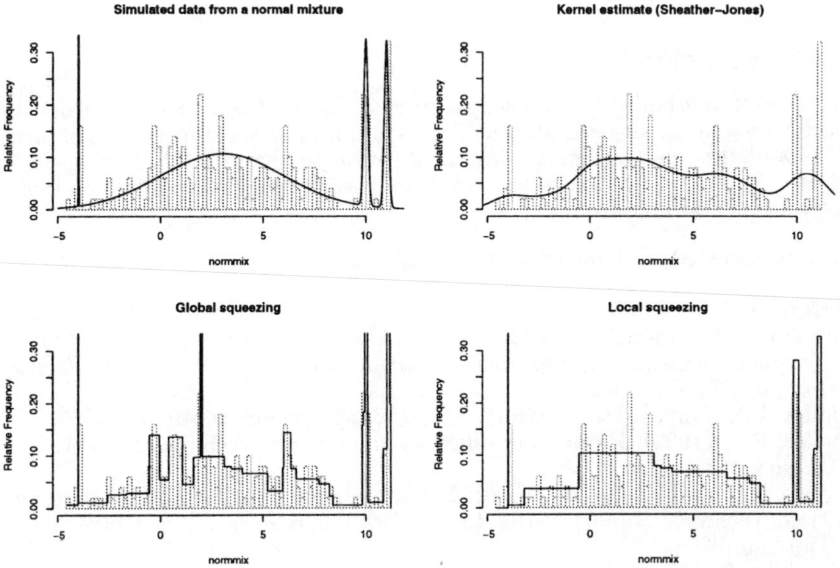

Fig. 6. Density estimation. A sample of a mixture of four normal distribution was analysed with a kernel estimator using a Sheather-Jones bandwidth and with the taut string procedure using global and local squeezing.

taut string estimator based on the empirical distribution function instead of the partial sum process attains asymptotically the correct modality if the bandwidth is chosen in a suitable way. The taut string estimator can be enhanced by an adaptation of local squeezing using a multiresolution-like technique. Assume that data $x_1, \ldots x_n$ are to be analysed and that $\hat{F}$ is the taut string through the Kolmogorov tube with some bandwidth. Then we consider coefficients

$$w(I_{j,k}) = \#\{i : \hat{F}(x_i) \in I_{j,k}\}$$

and decrease the bandwidth in those intervals where the corresponding coefficients exceed some theoretical bound.

Figure 6 shows the application of this methodology to 250 observations from a mixture Q of four normal distributions

$$Q = 0.04 \cdot \mathcal{N}(-4, 0.01) + 0.8 \cdot \mathcal{N}(3, 3) + 0.08 \cdot \mathcal{N}(10, 0.1) + 0.08 \cdot \mathcal{N}(11, 0.1).$$

Kernel estimators do not produce estimates with the correct modality because all bandwidths that separate the peaks at the right-hand end of the data lead to additional modes in the middle of the data. This problem was still present with another sample of the same density containing 2500 observations. The figure shows the estimate for 250 observations using the plug-in method by Sheather and Jones (1991) for choosing the bandwidth. Global squeezing also leads to spurious local extreme values in the center of the data due to the sharp peak at the left-hand side of the data and the problems described in the preceding paragraph. However, applied to the larger sample the estimate showed the correct modality. The local squeezing method found the modes of the distribution even in the small sample

6 Conclusion

Taut string methods can be used to control the modality in non-parametric regression and density estimation. Their computational complexity is of order $O(n)$ and they are automatic. An application to the estimation of spectral density functions in time series analysis was developed by Davies (1999). Software is available from our web page at

`http://wwwstat.mathematik.uni-essen.de`

References

Donoho, D.L., Johnstone, I.M., Kerkyacharian, G. and Picard, D. (1995). Wavelet shrinkage: asymptopia?. *Journal of the Royal Statistical Society, Ser. B*, **57**, 371–394.

Davies, P.L. (1995). Data features. *Statistica Neerlandica*, **49**, 185–245.

Davies, P.L. (1999). Hidden periodicities and strings. Technical Report, University of Essen.

Davies, P.L. and Kovac, A. (1999). Modality, Runs, Strings and Multiresolution. Technical Report 16/99, Sonderforschungsbereich 475, University of Dortmund.

Dümbgen, L. (1998). New goodness-of-fit tests and their application to nonparametric confidence sets. *The Annals of Statistics*, **26**, 288–314.

Fan, J. and Gijbels, I. (1996). *Local Polynomial Modelling and its Applications.* London: Chapman and Hall.

Härdle, W., Kerkyacharian, G., Picard, D. and Tsybakov, A. (1998). *Wavelets, Approximation, and Statistical Applications.* New York: Springer-Verlag.

Hartigan, J.A. and Hartigan, P.M. (1985). The dip test of unimodality. *The Annals of Statistics*, **13**, 70–84.

Kovac, A. and Silverman, B. W. (2000). Extending the Scope of Wavelet Regression Methods by Coefficient-dependent Thresholding *Journal of the American Statistical Association*, to appear.

Mammen, E. and van de Geer, S. (1997). Locally adaptive regression splines. *The Annals of Statistics*, **25**, 387–413.

Marron, J.S. and Tsybakov, A.B. (1995). Visual error criteria for qualitative smoothing. *Journal of the American Statistical Association*, **90**, 499–507.

Nadaraya, E. A. (1964). On estimating regression. *Theory of Probability and its Applications*, **10**, 186–190.

Sheather, S. J. and Jones, M. C. (1991). A reliable data-based bandwidth selection method for kernel density estimation. *Journal of the Royal Statistical Society, Ser. B*, **53**, 683–690.

Silverman, B. W. (1985). Some aspects of the spline smoothing approach to non-parametric regre ssion curve fitting. *Journal of the Royal Statistical Society, Ser. B*, **47**, 1–52.

Watson, G. S. (1964). Smooth regression analysis. *Sankhyā, Ser. A*, **26**, 101–116.

Multivariate approaches for aggregate time series[1]

Cristina Davino and Vincenzo Esposito

Dipartimento di Matematica e Statistica - Università degli Studi di Napoli Federico II - Via Cintia, 80126 Napoli, Italy

Abstract. The problem of choosing between the direct and the indirect method for the seasonal adjustment of aggregate time series is addressed. Different multivariate approaches are proposed in order to define properly optimised aggregation weights for the component time series. The new weighing systems help in discriminating between the classical methods and may perform better than the weights assigned *a-priori*. Applications on real economic series are shown.

Keywords. Seasonal adjustment, aggregation weights, coherence, compromise components.

1 Introduction

The need of methods for forecasting and seasonally adjust economic and social time series is clear for a broad spectrum of real-world problems and, in particular, for the production of official statistics by national and international statistical institutes.

Very often, the methodological approach to these needs has to cope with the problem of defining a suitable aggregate series from component ones. This is the case, for example, of most macroeconomic series such as those related to the consumer prices index, import/export at constant prices indexes, employment rate and so on. Generally, the aggregate is obtained by combining the components on the basis of a set of weights preassigned by official institutions that, unfortunately, may not be optimal for either seasonal adjustment or forecasting purposes.

The seasonal adjustment of aggregate time series is usually carried out by means of an indirect or a direct method. In the *indirect method*, the component series are firstly seasonally adjusted (Modelling phase) and then combined in order to form the aggregate series (Aggregation phase). On the contrary, in the *direct method*, the raw elementary data are firstly combined to form the aggregate series (Aggregation phase) which is then seasonally adjusted (Modelling phase). Direct and indirect methods do not necessarily produce the same results. It is therefore advisable to evaluate the difference between the two aggregate series with respect to their capability to represent the components.

Actually, the direct method may naturally provide the users with a more meaningful aggregate from the point of view of its economic interpretability

[1] **Acknowledgments**: This research was supported by an ESPRIT (29.741 TESS) grant. We are very much indebted to Prof. Carlo Lauro for providing us with the main ideas behind this paper as well as for strongly supporting and encouraging the further developments.

as it is a weighted sum of the raw component series. On the other side, the indirect method seems to be much more consistent from the statistical point of view. In fact, statistically speaking, each component series has its own characteristics (models and parameters) which must be taken into account when modeling for forecasting.

In order to help in choosing between these methods, we propose three approaches (hereinafter called *unsupervised*, *supervised* and *hybrid*) by suitably adapting multivariate techniques. The first two approaches intervene in determining *alternative* systems of weights for the different components to form the aggregate with better performances in terms of both *coherence* with the raw series and revisions in the seasonal adjustment. On the other side, the hybrid approach aims at defining *compromise* component series.

The performance of the newly defined aggregates from the three proposed approaches are evaluated in terms of revisions and forecasting errors in the application section.

2 The unsupervised approach

Let us define $\mathbf{C}$=[C_1, C_2, ..., C_p] the original component series matrix and $\tilde{\mathbf{C}}$=[$\tilde{C}_1$, $\tilde{C}_2$, ..., $\tilde{C}_p$] the seasonally adjusted component series matrix. Each component series is usually assigned a weight (w_1, w_2, ..., w_p) in order to form the aggregate series either by the direct A^{dir}=w_1C_1+w_2C_2+ ... +w_pC_p or the indirect method A^{ind}=$w_1\tilde{C}_1$+$w_2\tilde{C}_2$+ ... +$w_p\tilde{C}_p$.

Given the seasonally adjusted component series from the indirect method, the proposed unsupervised approach (Fig. 1) aims at finding an alternative aggregate series which is the most coherent with the models for the single components. We refer to Principal Component Analysis (PCA, Joliffe, 1986) and Clustering of Variables (CLU, Hartigan, 1975) in order to find the *compromise* aggregate series defined as a linear combination of (all or some) the component series.

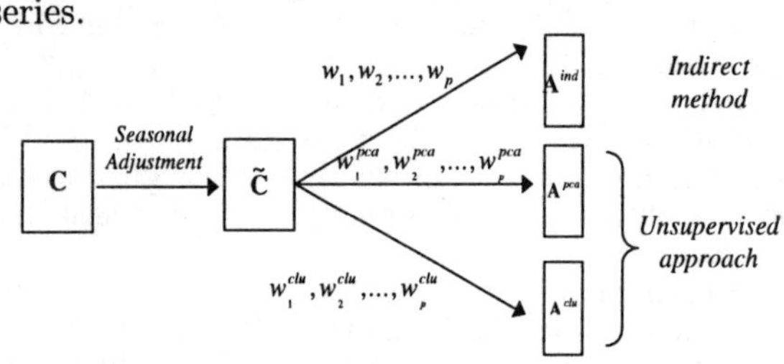

Figure 1: The unsupervised approach

Namely, a PCA on the matrix $\tilde{\mathbf{C}}$ furnishes the eigenvectors, i.e. the weights of the original values obtained in the linear combination, whose elements represent the weights (w_1^{pca}, w_2^{pca}, ..., w_p^{pca}) to be assigned to the component series in order to get the newly defined aggregate A^{pca}=$w_1^{pca}\tilde{C}_1$+$w_2^{pca}\tilde{C}_2$+ ... +$w_p^{pca}\tilde{C}_p$. Actually, the retained eigenvector is the first one which, in the usual case of very highly correlated component series, accounts for more than 90% of the total variability.

With respect to the Clustering of Variables, the similarity measure that allows to merge the component series at each step of the clustering procedure furnishes the weights (w_1^{clu}, w_2^{clu}, ..., w_p^{clu}) assigned to them in order to form an alternative aggregate A^{clu}=$w_1^{clu}\tilde{C}_1$+$w_2^{clu}\tilde{C}_2$+ ... +$w_p^{clu}\tilde{C}_p$. After defining

A^{pca} and A^{clu}, their coherence with the components in $\tilde{\mathbf{C}}$ need to be evaluated so as to assess their performance with respect to the original aggregate A^{ind}.

We propose to do that in terms of *reliability* of the aggregate series with respect to the seasonally adjusted components by referring to the Cronbach's α index. This coefficient was introduced (Cronbach, 1984) in the framework of the item analysis to test the reliability of a measure scale obtained from the answers to a set of items in a questionnaire. In order to make this coefficient feasible to our context, we need to consider that the aggregate series is a weighted sum of the single components so that we use the following adjusted version (in the case of PCA):

$$\alpha^{pca} = \frac{p}{p-1}\left[1 - \frac{\sum_{j=1}^{p} \sigma^2_{\tilde{C}_j^{pca}}}{\sigma^2_{A^{pca}}}\right]$$

where $\sigma^2_{\tilde{C}_j^{pca}}$ is the variance of the j-th seasonally adjusted component weighted by w_j^{pca} and $\sigma^2_{A^{pca}}$ is the variance of A^{pca}. The index α^{clu} may be analogously defined. These indexes can be interpreted as mean correlation coefficients so that if the components represent accidental elements and are not correlated, they will be equal to 0. The higher is the value of α-indexes, the better is the reliability of the aggregate series with respect to its components.

There exist other versions of the index that allow to evaluate the role played by each component in the definition of the aggregate (e.g. mean and variance if deleted, component-aggregate correlation). Considerations on these versions might even lead to drop some component series. In real analyses, however, it is never advisable to completely drop a series (i.e. assign a weight equal to zero) so that these peculiar versions are hereinafter neglected.

3 The supervised approach

Given the seasonally adjusted aggregate series from the direct method A^{dir}, considered as a *target*, this approach aims at finding its maximal agreement with the seasonally adjusted components in $\tilde{\mathbf{C}}$. We propose to refer to special regressions for multicollinearity situations. In fact, the component series are naturally highly correlated and, in this framework, play the role of explanatory variables with respect to the response variable defined by the aggregate series A^{dir}.

In particular, we propose (Fig. 2) to define a system of weights for the seasonally adjusted components in $\tilde{\mathbf{C}}$ which is more coherent with, and therefore *supervised* by, A^{dir}. This task is accomplished by means of Regressions on *factorial syntheses* of the component series whose results are evaluated by referring to the classical R^2 determination index.

We might refer to a Principal Components Regression (PCR, Chatterjee and Price, 1977). However, PCR usually shows the inconvenient of disregarding the peculiar aim of the analysis when building the principal components. In order to overcome this drawback, we propose a Regression based on Principal Components onto a Reference Subspace (PCAR, D'Ambra and Lauro, 1992).

This technique allows for explicitly considering the different role played by the component series and the aggregate one. Moreover, it is proved to fit closer than PCR and consequently provides the user with more stable results. From the geometrical point of view, PCAR is based on a PCA performed on the subspace spanned by the explanatory variables where the response variables have been previously projected. In our case, there is just one response variable so that this geometrical approach is not directly feasible.

Therefore, we finally take advantage from the Partial Least Squares technique (PLS1, Tenenhaus et al., 1995). The aim of PLS is exactly the same as that of PCAR but its approach is algorithmic rather than geometrical.

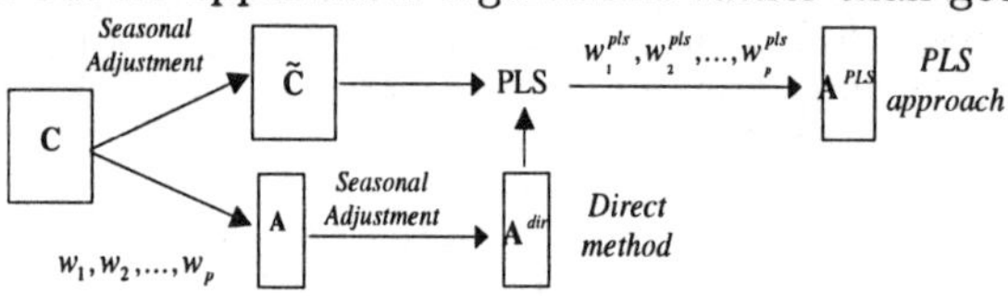

Figure 2: The supervised approach

In the first step, the PLS1 algorithm works on the cross-product matrix between $\tilde{\mathbf{C}}$ and A^{dir}, both column-centred and standardised. In the successive steps, the procedure is iterated by working on the residuals provided at the previous step. Therefore, the PLS1 regression coefficients refer to the residual matrices and need to be transformed by means of the weights and loading matrices (Martens and Næs, 1989) in order to be directly applicable to the seasonally adjusted component series and to be used for prediction purposes. Finally, the transformed regression coefficients represent the new weighing system (w_1^{pls}, w_2^{pls}, ..., w_p^{pls}) which allows to define $A^{pls}=w_1^{pls}\tilde{C}_1+w_2^{pls}\tilde{C}_2+ \ldots +w_p^{pls}\tilde{C}_p$.

4 The hybrid approach

This last approach is useful when the time series analyst has, for some reasons, to stick with the preassigned system of aggregation weights or when he/she is not able to choose between aggregating the original or the seasonally adjusted component series. In fact, instead of working on the definition of a different system of weights as the unsupervised and supervised approaches do, the hybrid approach aims at defining a *compromise* set of component series (C^{com}).

This task is accomplished (Fig. 3) by referring to a Procrustean rotation (Gower, 1975) of the seasonally adjusted components towards the originally observed ones. Procrustes Rotations are peculiar orthogonal transformations that allow to linearly constraint the configuration of $\tilde{\mathbf{C}}$ to that of $\mathbf{C}$ taken as the rotation reference. Therefore, they allow to model the seasonally adjusted components in order to match them, in terms of distances, as much as possible with the original ones.

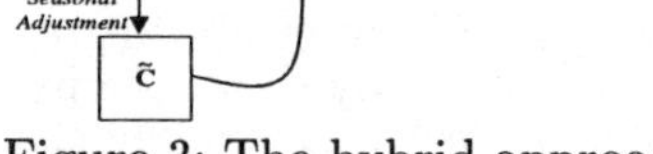

Figure 3: The hybrid approach

The goodness of fit of the Procrustean rotation is measured by means of a matching index whose minimum (0) indicates a perfect matching of the two sets of component series. From the compromise component series, the most coherent aggregate series may be found by any of the weighing systems previously defined.

5 Applications on real time series

The proposed approaches have been extensively tested, with satisfying results, as the off-line experimentation[2] of the ESPRIT project TESS by analysing several Spanish real socio-economic time series.

For the sake of brevity, hereinafter we limit to show the results obtained by working on the consumer prices index time series (training-set: from January 1980 to December 1996; test-set: year 1997). This series is yielded by combining 9 component series related to the sectors of manufactured food, non-manufactured food, other, culture, building, medicine, home tools, transportation, clothes. The aggregation weights are preassigned by the Bank of Spain and sum up to 1. In Table 1, we show this system of weights as well as the ones coming from our approaches.

	manu	*no-man*	*oth*	*cult*	*build*	*medic*	*home*	*trans*	*cloth*	α
Assig.	**0.1748**	0.1188	0.1526	0.0727	0.1028	0.0313	0.0668	**0.1654**	0.1148	0.973
CLU	0.1112	0.1103	0.1113	0.1112	0.1113	0.1113	0.1111	0.1113	0.1111	0.998
PCA	0.1115	0.1096	0.1114	0.1115	0.1112	0.1114	0.1114	0.1109	0.1111	0.998
PLS	**0.0888**	0.1077	0.1747	0.1228	0.1150	0.1113	0.1047	**0.0836**	0.0915	0.987

Table 1: Aggregation weights and their reliability

As it can be easily observed, the weights coming from PCA and CLU roughly assign the same importance to all component series. Differently, the preassigned and PLS weights strongly discriminate among the components. In particular, the most determining sectors for the Bank of Spain are manufactured food and transportation, while PLS (with R^2=0.9999) assigns them a weaker role. This is due to the fact that the models of these series do not seem to provide a meaningful statistical information with respect to the model of the aggregate. The last column in Table 1 shows that the alternative systems of weights increase the coherence of the aggregate with respect to its components in terms of reliability.

In Table 2, we evaluate the performance of the different systems of weights in terms of revisions and forecasts.

With respect to *revisions*, this is done by computing the average absolute percent revision (AAPR, Gomez 1999) index. Revisions are basically the differences between the earliest adjustment of a month's datum obtained when that month is the final month in the series and the later adjustment based on all future data available at the time of the diagnostic analysis. The AAPR index is a relative measure of this difference so that lower values indicate a better performance. Other measures, such as the root mean square revision (RMS), and the average absolute revision (AAR), may be used for the same purpose but are not expressed in percent terms.

It is worth noticing that, independently from the chosen system of weights, the indirect method is always preferable to the direct one as it constantly shows a lower AAPR. Beyond this, it is evident that all alternative weighing systems perform better than the preassigned one. In particular a time series analyst should choose the indirect method for seasonal adjustment and the system of weights yielded by the PLS method.

[2] The approaches proposed in this paper have been implemented in MATLAB 5.3 codes.

As it concerns the *forecasts*, we have studied the quality of the forecasts produced by the aggregate and the component series. We have firstly identified the ARIMA models for the training-set and then, keeping the models fixed, we have added one of the test-set observations at a time. Each time, we have calculated one-period-ahead forecasts and compared them with the actual values. The forecasting statistic we have used is the root mean square error of the out-of-sample 1-period-ahead forecasts (RMSE1, Gomez 1999).

		Assigned weights	CLU	PCA	Procrustes PCA	**Procrustes CLU**	**PLS**
	Revisions						
AAPR	Direct	0.0258	0.0228	0.0227	0.0227	0.0228	0.0178
	Indirect	0.0214	0.0186	0.0184	0.0184	0.0185	**0.0175**
	Forecasts						
RMS1	Direct	0.3407	0.2760	0.2747	**0.2745**	0.2758	0.2958
	Indirect	0.4192	0.3713	0.3698	0.3695	0.3711	0.3630

Table 2: Performances of the aggregation weights

The forecasting part of Table 2 shows how, in terms of forecasts, the time series analyst should prefer the direct method of seasonal adjustment performed on the compromise component series obtained from the hybrid approach (with a very good matching index equal to 0.00032) and with the weights coming from PCA (RMS1=0.2745).

These results seem to prove that the proposed approaches may represent a valid alternative to the classical ones as there exist cases where they show a better performance.

References

Chatterjee, S. and Price, B. (1977). *Regression Analysis by Example.* New York: Wiley, chapters 7 and 8.

Cronbach, L.J. (1984). *Essentials of Psychological Testing.* Harper & Row.

D'Ambra, L. and Lauro, N.C. (1992). Non Symmetrical Exploratory Data Analysis, *Statistica Applicata*, **4**, 4, 511-529.

Gomez, V. (1999). Minimisation of Revisions *Technical Report of TESS Project*, Department of Statistics and Operational Research, UPC, Barcelona.

Gower, J.C. (1975). Generalised Procrustes Analysis, *Psychometrika*, **40**, 33-51.

Hartigan, J.A. (1975). *Clustering Algorithms.* New York: Wiley.

Joliffe, I. (1986). *Principal Component Analysis.* New York: Springer-Verlag.

Martens, H. and Næs, T. (1989). *Multivariate Calibration.* Chichester: Wiley.

Tenenhaus, M. and Gauchi, J.P. and Menardo, C. (1995). Regréssion PLS et Applications, *Revue de Statistique Appliquée*, **53**, 7-63.

Some space-time models: an application to NO_2 pollution in an urban area

S. De Iaco [1] and D. Posa [2]

[1] Facoltà di Economia - Università di Chieti - Italy
[2] Facoltà di Economia - Università di Lecce - Italy

Abstract. In this paper a class of product-sum covariance models is introduced, in order to estimate and model realizations of space-time random fields, which are very common in environmental applications. Some constraints on the coefficients of this class of models are given in order to guarantee the positive definiteness condition. An overview of some classes of space-time covariance models and a short comparative study is presented; moreover, an application is considered.

Keywords.Space-time correlation, geostatistics

1 Introduction

In recent years a number of spatial temporal problems concerning air pollution modelling have been addressed in the atmospheric and statistical literature through the formalism of space-time random fields. In order to model spatial-temporal phenomena, time is simply considered as another "dimension"and instead of using an appropriate metric in space-time, the technique of separability is performed.

In this paper a comparative study among some classes of spatial temporal covariance models and the product-sum covariance model is made. Moreover, this last model has been used for modelling and prediction of hourly average concentrations of NO_2 in an urban area.

It is shown how the coefficients of the product-sum class of models are related to the "global"sill and "partial"spatial and temporal sills; moreover, some constraints on these sills are given in order to assure positive definiteness of the product-sum covariance model.

GSLIB routines (Deutsch and Journel, 1997) for kriging and cross-validation have been modified to use the product-sum spatial temporal model.

2 Some classes of spatio-temporal covariance models

In this section some classes of spatio-temporal covariance models, which have been used in the literature, are discussed, in order to compare them with the product-sum model. In order to distinguish between space and time, let $Z = \{Z(s,t), (s,t) \in D \times T\}$ be a second order stationary spatial-temporal random field, where $D \subset \Re^d$ and $T \subset \Re_+$, with expected value

$$E(Z(s,t)) = 0 \tag{1}$$

In this case, the covariance

$$C_{st}(h) = \mathrm{Cov}(Z(s+h_s, t+h_t), Z(s,t)) \tag{2}$$

where $h = (h_s, h_t)$, $(s, s + h_s) \in D^2$ and $(t, t + h_t) \in T^2$ and the variogram

$$2\gamma_{st}(h) = \mathrm{Var}(Z(s + h_s, t + h_t) - Z(s, t))$$

depends solely on the lag vector h, not on location or time. The function C_{st} in (2) must satisfy a positive-definiteness condition in order to be a valid covariance function.

1. *The metric model.* It is assumed (Dimitrakopoulos and Luo, 1993) that:
$$C_{st}(h_s, h_t) = C(a^2|h_s|^2 + b^2 h_t^2) \tag{3}$$
where the coefficients $a, b \in \Re$. Note that in (3) the same type of model is assumed for the spatial and temporal covariances, with possible changes in the range.
2. *The product model.* One of the simplest way to model a covariance in space-time is to keep the spatial and temporal domains separated (De Cesare et al., 1996). The product spatial-temporal covariance model is:
$$C_{st}(h_s, h_t) = C_s(h_s)C_t(h_t) \tag{4}$$
where the spatial dependence is separated by the temporal one. In (4) C_s is a positive-definite function in $\Re^d$ and C_t is a positive-definite function in $\Re$; admissible spatial covariance models and admissible temporal covariance models are readily available (Cressie, 1993), hence they can be combined in product form to give spatio-temporal covariance models. However, the class (4) is severely limited, since for any pair of spatial locations the cross-covariance function of the two time series always has the same shape. In fact, for any two fixed spatial lags h_1 and h_2
$$C_{st}(h_1, h_t) \propto C_{st}(h_2, h_t)$$
A similar result holds for any pair of time points and the cross-covariance function of the two spatial processes.
3. *The linear model.* Another type of separability involves adding spatial and temporal covariances (Rouhani and Hall, 1989), that is:
$$C_{st}(h_s, h_t) = C_s(h_s) + C_t(h_t) \tag{5}$$
For this model, covariance matrices of certain configurations of spatio-temporal data are singular (Myers and Journel, 1990): in this case, the covariance function is only positive semidefinite and it is unsatisfactory for optimal prediction.
4. *The nonseparable model.* A new approach that allows us to obtain classes of nonseparable, spatio-temporal stationary covariance functions has been derived by Cressie and Huang (1999). The authors assume that:
$$C_{st}(h_s, h_t) = \int e^{ih_s^T \omega} \rho(\omega, h_t) k(\omega) d\omega \tag{6}$$
where the following two conditions are satisfied:
   - for each $\omega \in \Re^d$, $\rho(\omega, .)$ is a continuous autocorrelation function and $k(\omega) > 0$;
   - the positive function $k(\omega)$ satisfies:
$$\int k(\omega) d\omega < \infty$$

3 The product-sum model

The product-sum covariance model can be obtained in the following way:

$$C_{st}(h_s,h_t) = k_1 C_s(h_s)C_t(h_t) + k_2 C_s(h_s) + k_3 C_t(h_t) \quad (7)$$

or equivalently, in terms of the semivariogram functions:

$$\gamma_{st}(h_s,h_t) = [k_2 + k_1 C_t(0)]\gamma_s(h_s) + [k_3 + k_1 C_s(0)]\gamma_t(h_t) - k_1\gamma_s(h_s)\gamma_t(h_t) \quad (8)$$

where C_s and C_t are covariance functions and γ_s and γ_t are the corresponding semivariogram functions. Note that $C_{st}(0,0)$ is the sill of γ_{st} ("global"sill), $C_s(0)$ is the sill of γ_s and $C_t(0)$ is the sill of γ_t ($C_s(0)$ and $C_t(0)$ are named "partial"sills); by definition: $\gamma_{st}(0,0) = \gamma_s(0) = \gamma_t(0) = 0$.

3.1 Estimation and modelling

From (7) it follows that:

$$k_1 C_s(0)C_t(0) + k_2 C_s(0) + k_3 C_t(0) = C_{st}(0,0) \quad (9)$$

Note also from (8) that:

$$\gamma_{st}(h_s,0) = [k_2 + k_1 C_t(0)]\gamma_s(h_s) \quad (10)$$
$$\gamma_{st}(0,h_t) = [k_3 + k_1 C_s(0)]\gamma_t(h_t) \quad (11)$$

In order to estimate and model $\gamma_s(h_s)$ and $\gamma_t(h_t)$ by $\gamma_{st}(h_s,0)$ and $\gamma_{st}(0,h_t)$, respectively, it is assumed that:

$$\begin{aligned} k_2 + k_1 C_t(0) &= 1 \\ k_3 + k_1 C_s(0) &= 1 \end{aligned} \quad (12)$$

Let H be the set of data locations, then the standard moment estimator for the spatial variogram at the vector lag r_s, with spatial tolerance δ_s, is:

$$\widehat{\gamma}_s(r_s) = \widehat{\gamma}_{st}(r_s,0) = \frac{1}{2|N(r_s)|}\sum[Z(s+h_s,t) - Z(s,t)]^2 \quad (13)$$

where the summation is over the set

$$N(r_s) = \{(s+h_s,t) \in H \text{ and } (s,t) \in H \text{ such that } ||r_s - h_s|| < \delta_s\},$$

and $|N(r_s)|$ is the cardinality of this set. Similarly

$$\widehat{\gamma}_t(r_t) = \widehat{\gamma}_{st}(0,r_t) = \frac{1}{2|M(r_t)|}\sum[Z(s,t+h_t) - Z(s,t)]^2 \quad (14)$$

where

$$M(r_t) = \{(s,t+h_t) \in H \text{ and } (s,t) \in H \text{ such that } ||r_t - h_t|| < \delta_t\},$$

and $|M(r_t)|$ is the cardinality of this set.
From (7) it is clear that $k_1 > 0$, $k_2 \geq 0$ and $k_3 \geq 0$ is a sufficient condition for positive definiteness. From (9), and (12) it follows that:

$$\begin{aligned} k_1 &= [C_s(0) + C_t(0) - C_{st}(0,0)]/C_s(0)C_t(0) \\ k_2 &= [C_{st}(0,0)) - C_t(0)]/C_s(0) \\ k_3 &= [C_{st}(0,0) - C_s(0)]/C_t(0) \end{aligned} \quad (15)$$

In order to model the spatial and temporal variograms it is necessary to choose the sills so that the numerators in (15) remain positive.

3.2 Some general comments

A brief comparative study among the classes of models previously described has been made in this section. Particularly:

- if the autocorrelation function ρ in (6) is purely a function of h_t, then the product covariance model is obtained;
- the product model and the linear model are easily obtained by the product-sum covariance model setting, respectively, $k_2 = k_3 = 0$ and $k_1 = 0$;
- the product-sum covariance model is more flexible than the nonseparable covariance model for estimating and modelling spatial-temporal correlation structures.

4 An application

The data collection network, planned to the standards and information provided by the Lombardy Region's Environmental Town Council, consists of 52 survey stations for hourly averages of NO_2 measured from the 1st to the 31st of October 1998.
In order to remove the seasonal effect, time series analysis has been carried out for each location by the standard technique of moving average estimation (Brockwell and Davis, 1987). Residuals have been generated for all stations after removing the seasonal component.
Figure 1(a) shows the sample temporal variograms of the original data and the residuals with the fitted model: note the periodic structure at 24 hours exhibited by the variogram of the original data. Figure 1(b) shows the sample spatial variogram of the residuals with the fitted model.

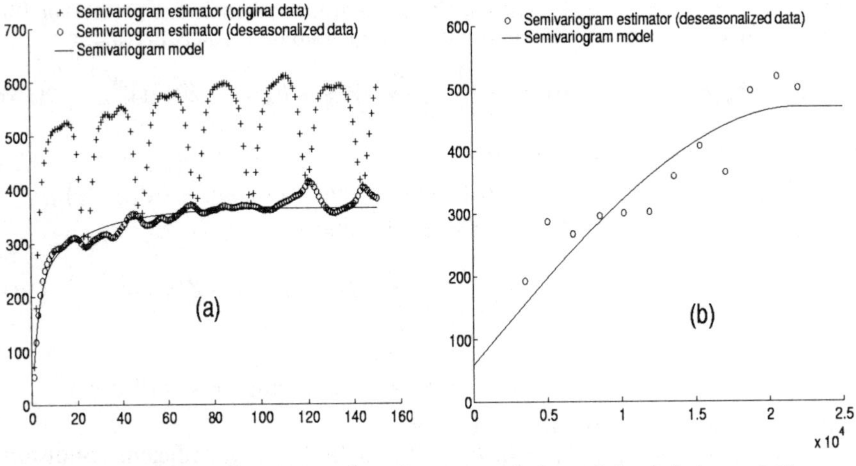

Figure 1. Sample temporal variograms of the original data and the residuals with the fitted model (a); sample spatial variogram of the residuals with the fitted model (b).

The "global" sill value $C_{st}(0,0)$ of γ_{st} has been estimated through the spatial-temporal variogram surface of the residuals (see Figure 2). Note that $C_{st}(0,0)$ is needed to solve (15).

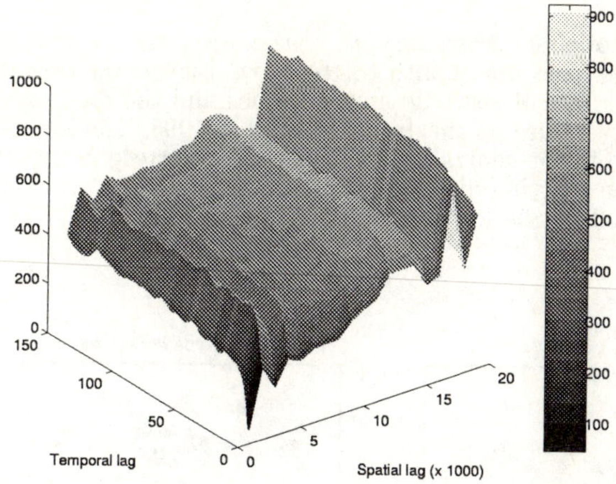

Figure 2. Spatial-temporal variogram surface of the residuals.

4.1 Spatial-temporal prediction

The hourly averages of NO_2 measured during October 1998 and the product-sum variogram model given in (8) have been used to predict the hourly averages of NO_2 at the monitoring stations available on the 1st of November 1998, by ordinary kriging. The daily averages of the predicted hourly values have been compared to the daily averages of the true hourly values (Figure 3); these last values have not been used throughout the above analysis. The correlation coefficient between true and predicted values is 0.9.

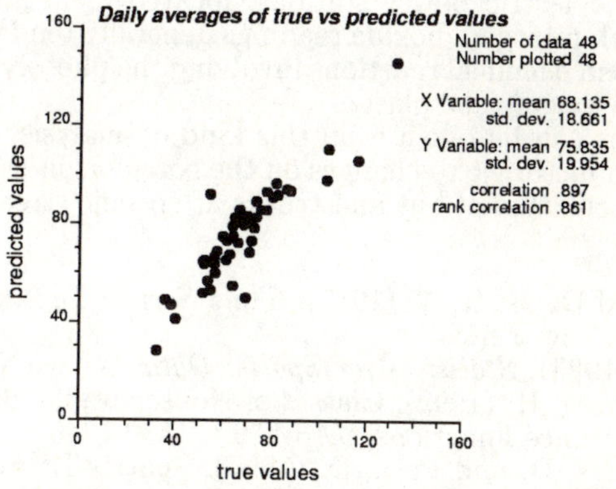

Figure 3. Daily averages of the true hourly values versus the daily averages of the predicted hourly values on the 1st of November 1998.

The predicted values were computed in the following way:

1. the residual product-sum variogram model was used to predict the residuals for the 1st of November 1998 at the same locations where the true values are available by ordinary kriging;
2. the seasonal component corresponding to October 1998, which was previously estimated, has been added to the predicted residuals in order to obtain the predicted values.

In the above modelling procedure, it is assumed that the trend component does not change from one month to the next. Figure 4 shows the postplots of the daily averages of the true hourly values and the daily averages of the predicted hourly values on the 1st of November 1998. The highest values are clearly localized in the central part of the graph, corresponding to the city of Milan, and some outlying districts, while the peripherical parts of the district are characterized by the lowest values.

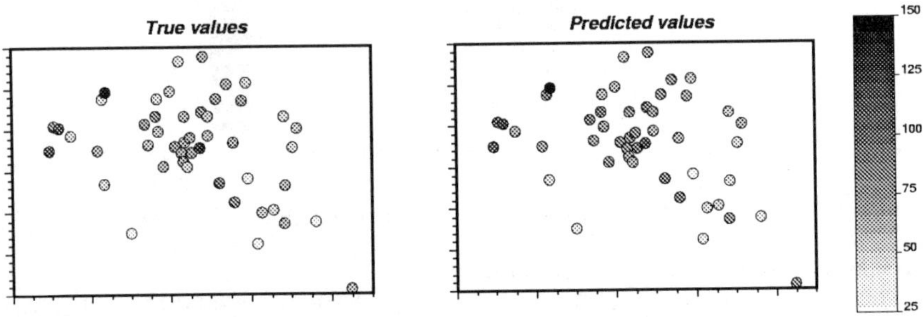

Figure 4. Postplots of the daily averages of the true hourly values and the daily averages of the predicted hourly values on the 1st of November 1998.

It is important to point out that NO_2 is caused by the oxidation of NO (nitric oxide) in the air: nitric oxide is generated by civil and industrial heating systems, thermoelectric power stations, located in Milan district, and motor vehicles. Peak nitrogen dioxide readings generally tend to occur in the evening, when certain chemical reactions involving the photolysis of NO_2 and formation of O_3 no longer take place.

Different decisions can be taken from this kind of analysis, from technological investments in factories, to changes on the corridors in the entrance city and to limit the circulation of bus and trucks at specific streets.

References

Brockwell, P. J. and Davis, R. A. (1987). *Time Series: Theory and Methods*, Springer-Verlag: New York.

Cressie, N. A. C. (1993). *Statistics for Spatial Data*, Wiley: New York.

Cressie, N. and Huang, H. (1999). Classes of Nonseparable, Spatio-temporal Stationary Covariance Functions, *JASA*, **94**, 1330-1340.

De Cesare, L., Myers, D. and Posa, D. (1996). Spatial-Temporal Modelling of SO_2 in Milan District, *Geostatistics Wollongong'96*, Kluwer Academic Publishers, **2**, 1031-1042.

Deutsch, C. V. and Journel, A. G. (1997). *GSLIB: Geostatistical Software Library and User's Guide*, Oxford Univ. Press: New York.

Dimitrakopoulos, R. and Luo, X. (1994). Spatiotemporal modelling: covariances and ordinary kriging systems, *Geostatistics for the next century*, Kluwer Academic Publishers, 88-93.

Myers, D. E. and Journel, A. G. (1990). Variograms with Zonal Anisotropies and Non-Invertible Kriging Systems, *Math. Geology*, **22**, 779-785.

Rouhani, S. and Hall, T. J. (1989). Space-time Kriging of Groudwater Data. *Geostatistics*, Kluwer Academic Publishers, 2, 639-651.

SLICE: generalised software for statistical data editing and imputation

Ton de Waal

Department of Statistical Methods, Statistics Netherlands, PO Box 4000, 2270 JM Voorburg, Netherlands

Abstract. Statistical offices have to face the problem that data collected by surveys or obtained from administrative registers generally contain errors. Another problem they have to face is that values in data sets obtained from these sources may be missing. To handle such errors and missing data efficiently, Statistics Netherlands is currently developing a software package, called SLICE (Statistical Localisation, Imputation and Correction of Errors). SLICE will contain several edit and imputation modules. Examples are a module for automatic editing and a module for imputation based on tree-models. In this paper I describe SLICE, hereby focussing on the above-mentioned modules.

Keywords. automatic data editing, computer-assisted editing, Fellegi-Holt paradigm, imputation, statistical data editing

1 Introduction

Statistical offices have to face the problem that data collected by surveys or obtained from administrative registers generally contain errors. Errors in statistical data sets may arise for a variety of reasons. For instance, the respondent may have misunderstood a question, the respondent may have made a mistake when giving the answer, or the respondent's database may contain errors. Alternatively, errors may have been introduced at the statistical office, e.g., while transferring data from a paper questionnaire to the computer system. Another problem that has to be faced by statistical offices is that values in data sets obtained from surveys or administrative registers may be missing. A value may be missing in a data set, e.g., because the respondent did not understand the corresponding question, the respondent forgot to answer the question, or the respondent did not know the answer. A value may also be missing because at the statistical office this value was not entered when transferring the data from a paper questionnaire to the computer system.

Incorrect data and missing data in a data set have to be treated in such a way that statistical analyses based on this data set give reliable results. Incorrect data can be treated by first identifying the errors and subsequently correcting these errors. This process is referred to as data editing. Missing values can be treated by estimating these values and filling in these estimates in the data set. This is referred to as imputation.

To handle errors and missing values in data sets efficiently, Statistics Netherlands is currently developing a software package called SLICE (Statistical

Localisation, Imputation and Correction of Errors). SLICE will contain several edit and imputation modules. Examples of such modules are a module for automatic data editing and a module for imputation based on tree-models. SLICE itself is planned to become a module of Blaise, the integrated survey processing system developed by Statistics Netherlands (see *Blaise Reference Manual,* 1998, and the *Blaise Developer's Guide,* 1998).

The aim of the present paper is to give a description of the automatic editing module and the imputation module of SLICE. For an overview of other editing software I refer to Pierzchala (1995), and for an overview of other imputation software to Hox (1999).

The remainder of the paper is organised as follows. Section 2 discusses automatic data editing and SLICE. Section 3 describes the present imputation module in SLICE, and the future imputation module based on so-called WAID tree-models. Section 4 concludes the paper with a short discussion.

2 Automatic data editing and SLICE

Traditionally, incorrect data in a data set are edited 'manually'. That is, each faulty record is modified manually, whereas the record has been checked either manually or by means of a computer program. In the latter case where records are checked by means of a computer program one also says that the data are edited in a computer-assisted manner. Data are modified, e.g., on the basis of subject-matter knowledge, by contacting the respondent again, by comparing the respondent's data to his data from previous years, or by comparing the respondent's data to data of similar respondents. Correcting data 'manually' is a time and money consuming process.

It has long been recognised, however, that it is not necessary to correct all data in every detail (see e.g. Granquist, 1995, and Granquist & Kovar, 1997). A reason why it is not necessary that all errors are removed from the data set is that published results are usually aggregated data, such as totals or means. Small errors will often cancel out when they are aggregated. Moreover, when the data have been obtained by means of a sample of the population there will always be a small error in the results, even when all collected data are completely correct. This implies that for data obtained by means of a sample an error in the results due to 'noise' in the data is acceptable as long as this error is negligible in comparison to the sampling error. These observations allow one to automate the editing process to some extent.

Three steps can be distinguished when editing data automatically. The first step is to identify the erroneous records. Once a record has been identified as incorrect, its erroneous, or implausible to be more precise, values have to be identified. These erroneous values are set to missing. The final step is to impute plausible values for the missing values, both for the values that were actually missing in the original data set as well as for the fields that were set to missing because their original values were considered incorrect.

To establish whether a record is erroneous edit checks have to be specified. A record is considered erroneous if one or more edit checks specified by subject-matter specialists are violated. If all specified edit checks are satisfied for a particular record, this record is considered correct. If records are edited automatically a record that is considered correct is not modified, whereas a record that is considered erroneous is always modified in such a way that all edit checks are satisfied.

A well-known paradigm for identifying erroneous fields in an erroneous record has been proposed in an influential paper by Fellegi and Holt (1976). According to this Fellegi-Holt paradigm the data of a record should be made to satisfy all edit checks by changing the values of the fewest possible number of variables. Over the years this paradigm has been generalised slightly, because it was realised that the values of some variables are more trustworthy than the values of other variables. Therefore, to each variable a weight, the so-called confidence weight, can be assigned. Such a confidence weight is a positive number that expresses the confidence one has in the correctness of the values of this variables. A high confidence corresponds to a variable of which the values are considered trustworthy, a low confidence weight to a variable of which the values are considered not so trustworthy. According to the generalised Fellegi-Holt paradigm the data of a record should be made to satisfy all edit checks by changing the values of the variables with the smallest possible sum of confidence weights.

To illustrate the Fellegi-Holt paradigm we assume that the following two edit checks have been specified by subject-matter specialists:

$$P + C = T \tag{1}$$

and

$$C/T \geq 0.6, \tag{2}$$

where P is the profit of an enterprise, C its costs and T its turnover. Suppose furthermore that a record contains the following data: P=755, C=125 and T=200. This record violates edit check (1). Hence, the record is considered erroneous.

Now, suppose that the confidence weights of P, C and T are all equal to 1. According to the Fellegi-Holt paradigm the data of the record should be made to satisfy all edit checks by changing the values of the fewest possible number of variables. Preferably, only one value should be changed. By trial and error we can find in our example that there is only one way to modify a single value such that all edit checks are satisfied. Namely, the value of P should be modified to 75.

In this example the Fellegi-Holt paradigm yields exactly one optimal solution, i.e. a set of variables with a minimum sum of confidence weights, to modify the data In other cases, however, application of the (generalised) Fellegi-Holt paradigm may yield several optimal solutions to modify the data. In such cases additional criteria may be used to select the best option.

The (generalised) Fellegi-Holt paradigm is a very natural principle to determine implausible fields. However, the resulting problem of finding those fields with the smallest possible sum of confidence weights that can be modified in such a way that all edit checks are satisfied is quite complicated. This so-called error localisation problem can be formulated as a mathematical programming problem. In fact, the error localisation problem is so complicated that most algorithms have either been designed to solve this problem for categorical data only or for numerical data only.

In the current prototype version of SLICE a module for automatic editing of numerical data only is implemented. This module, called CherryPi, was developed in the spirit of GEIS, the Generalized Edit and Imputation System of Statistics Canada (see Kovar and Whitridge, 1990). The basic principle to identify the implausible fields in an erroneous record is the generalised Fellegi-Holt paradigm that has been described above.

In the (near) future the implemented algorithm is likely to be replaced by an alternative algorithm that is currently being developed at Statistics Netherlands. This new algorithm is based on searching a large binary tree for optimal solutions to the error localisation problem. The new algorithm has several advantages in comparison to the old one. First, the new algorithm can be extended to solve the error localisation for a mix of numerical data and categorical data. At the moment of writing this paper two similar algorithms exist: one for numerical data (cf. Quere, 2000) and one for categorical data (cf. Daalmans, 2000). We hope to combine these two algorithms into a single algorithm that can handle both types of data simultaneously within the next few months. Second, the new algorithm is easier to understand and implement than the old one. Third, for numerical data the new algorithm is clearly faster than the old algorithm (cf. Quere, 2000).

3 Imputation and SLICE

Considering the limited size of the present paper and the vast literature on imputation, it is impossible to give even a brief overview of imputation methods that have been developed. Instead, I only discuss the present imputation module of SLICE and the imputation module that is planned to be implemented in SLICE in the near future. For two overview papers on imputation techniques I refer to Kalton & Kasprzyk (1986) and Kovar & Whitridge (1995).

The present imputation module of SLICE is an improved version of the imputation module of CherryPi. In CherryPi regression imputation is used to impute for missing values. For each variable a subject-matter specialist has to select a predictor variable. The subject-matter specialist also has to determine the constant term and the regression coefficient, e.g. by means of statistical analysis. If after regression imputation a resultant record still fails one or more edit checks, the imputed values are modified slightly in order to satisfy all edit checks. See De Waal and Wings (1999) for more information on this regression imputation module. In SLICE a similar regression imputation module has been implemented. The main difference is that the number of predictor variables is not limited to only one like in CherryPi. Instead, SLICE allows the use of an arbitrary number of predictor variables.

At Statistics Netherlands we are currently developing a new, more advanced, imputation module together with University of Southampton, Office for National Statistics in the UK, Statistics Finland and Instituto Nacional de Estatística de Portugal. These institutes all participate in an European project called AUTIMP. This project is partly financed by the 4th Framework Programme of the European Commission. One of the aims of AUTIMP is the development of imputation software based on automatic interaction detection (AID) trees, cf. Sonquist, Baker and Morgan (1971). The developed imputation software is planned to become a module in SLICE. The algorithm to construct AID trees has been developed by University of Southampton. Because the algorithm gives lower weights to outliers while constructing the tree, the technique is referred to as weighted automatic interaction detection (WAID).

In general, a tree–based model is a set of classification rules defined in terms of the values of a set of (typically categorical) predictor variables. It is typically constructed by successively splitting a "training" data set into subsets that are increasingly more homogeneous with respect to a response variable of interest until a stopping criterion is met.

The WAID technique that is developed is suitable for both categorical and numerical data. To use the WAID methodology for imputing missing values in a data set, first the missing data patterns in this data set have to be determined. For certain missing data patterns, or parts of missing data patterns, predictor variables are selected, and WAID trees are grown using a complete data set. The 'leaves' of these trees form homogeneous subsets of records. After generation of the trees, and hence the homogeneous clusters of records, the data set with missing values is supplied to the computer program.

To impute a record with missing values, we first determine which WAID tree corresponds to the missing data pattern of this record. Subsequently, we determine the homogeneous leaf corresponding to this record by examining the values of the predictor variables. The records from that leaf are used to impute the missing data in the record under consideration. The imputation software will support several imputation methods, e.g., a random donor record can be selected from the homogeneous leaf, or (in case of numerical data) a mean of the records in the homogeneous leaf may be imputed. For the case where only numerical predictor variables are used we will also support nearest neighbour hot deck imputation.

4 Discussion

In this paper I have described the present and planned automatic editing and imputation modules of SLICE. The fact that I have only described automatic editing and imputation, may give the false impression that Statistics Netherlands plans to rely solely on these techniques for checking and correcting its statistical data. I want to emphasise here that this definitely is not the case. In particular, at Statistics Netherlands we consider automatic editing to be just one step in the editing process, not as the entire editing process. In our view the ideal edit strategy consists of a combination of selective editing (where only the most suspicious and influential records are selected for computer-assisted editing), automatic editing and macro-editing (where records are only selected for computer-assisted editing when aggregated totals based on these records are suspicious, or when they are outliers in the distribution of all records). In this setup automatic editing is only applied to the records in the so-called non-critical stream, i.e. the records with only non-influential errors. Influential errors always have to be corrected by subject-matter specialists.

In comparison with the traditional computer-assisted approach the new combined approach requires substantially less resources for cleaning the data while the quality of the data can be maintained. Moreover, the timeliness of publication of the statistical data clearly is improved. For more information on the view of Statistics Netherlands on the ideal editing strategy I refer to De Waal, Renssen & Van de Pol (2000).

References

Blaise Reference Manual (1998). Department of Statistical Informatics, Statistics Netherlands, Heerlen.

Blaise Developer's Guide (1998). Department of Statistical Informatics, Statistics Netherlands, Heerlen.

Daalmans, J. (2000). Automatic editing of categorical data (working title). Statistics Netherlands, Voorburg.

De Waal, T. and H. Wings (1999). From CherryPi to SLICE. Report, Statistics Netherlands, Voorburg.

De Waal, T., R. Renssen and F. Van de Pol (2000). Graphical macro-editing: possibilities and pitfalls. Report, Statistics Netherlands, Voorburg.

Fellegi, I.P. and D. Holt (1976). A systematic approach to automatic edit and imputation. *Journal of the American Statistical Association*, **71**, pp. 17-35.

Granquist, L. (1995). Improving the traditional editing process. In: *Business Survey Methods* (ed. Cox, Binder, Chinnappa, Christianson & Kott), John Wiley & Sons, Inc., pp. 385-401.

Granquist, L. and J. Kovar (1997). Editing of survey data: how much is enough?. In: *Survey Measurement and Process Quality* (ed. Lyberg, Biemer, Collins, De Leeuw, Dippo, Schwartz & Trewin), John Wiley & Sons, Inc., pp. 415-435.

Hox, J.J. (1999). A review of current software for handling missing data. *Kwantitatieve Methoden*, **62**, pp. 123-138.

Kalton, G. and D. Kasprzyk (1986). The treatment of missing survey data. *Survey Methodology*, **12**, pp. 1-16.

Kovar, J. and P. Whitridge (1990). Generalized edit and imputation system; overview and applications. *Revista Brasiliera de Estadistica*, **51**, pp. 85-100.

Kovar, J. and P.J. Whitridge (1995). Imputation of business survey data. In: *Business Survey Methods* (ed. Cox, Binder, Chinnappa, Christianson & Kott), John Wiley & Sons, Inc., pp. 403-423.

Pierzchala, M. (1995). Editing systems and software. In: *Business Survey Methods* (ed. Cox, Binder, Chinnappa, Christianson & Kott), John Wiley & Sons, Inc., pp. 425-441.

Quere, R. (2000). Automatic editing of numerical data. Report, Statistics Netherlands, Voorburg.

Sonquist, J.N., E.L. Baker and J.A. Morgan (1971). Searching for structure. Institute for Social Research, University of Michigan.

Improved PCB inspection: computational issues

Dee Denteneer

Philips Research Laboratories Eindhoven, Prof. Holstlaan 4, 5656 AA Eindhoven, the Netherlands

Abstract. Process control on Printed Circuit Boards (PCB's) involves measurement of the volumes of a large number of small heaps of solder paste deposit, that are used to fix components, such as IC's, on the board. Surface estimation is crucial to their measurement but a traditional approach based on local polynomials was a major contributor to measurement variation. In this paper we show that the principles pioneered in Cleveland and Grosse can be successfully applied to reduce measurement variation. In addition, we discuss some new computational issues that arise in such an industrial application of local fitting: speeding up computations by reordering the data, fast adaptation, and some spatial design issues.

Keywords. Printed Circuit Board, local fitting, spatial design, kd-tree

1 Introduction

Manufacture of printed circuit boards (PCB's) involves the deposition of a large number of small heaps of solder paste that are used to fix the components on the PCB. Roughly, the process is as follows. Firstly, heaps of moist solder paste are deposited on the board. Secondly, the components (e.g. IC's) are placed on the board and each pin of each component must be positioned in such a heap of solder paste. Finally, the paste is dried to fix the components.

Typically, the number of heaps of solder paste deposit is in the order 1000-5000, the areas of the heaps range from $1mm^2$ to $.1mm^2$, and their heights range from $.1mm$ to $.2mm$.

The volume of solder paste in each heap is crucial to board quality: too much paste may cause components to short-circuit and too little paste may result in loose components. Hence, accurate measurement of these volumes is essential for process control in PCB manufacture. Typically, the measurement is performed after the deposition of the paste and before the placement of the components, so that the results can also be employed in a reject (and re-wash) policy that prevents expensive components from being placed on defective boards.

One way to carry out the measurements is optically. The spot on the PCB where the solder paste is located is illuminated with a laser beam and the absolute height of the top of the heap is computed from the amount of reflected light. However, to estimate the volume in each heap of solder paste deposit requires measurement of the height relative to the PCB surface. Consequently, the surface height must be estimated at a number of *target* locations: the locations of the paste dots, and surface estimation based on additional *reference* measurements on the PCB surface is necessary.

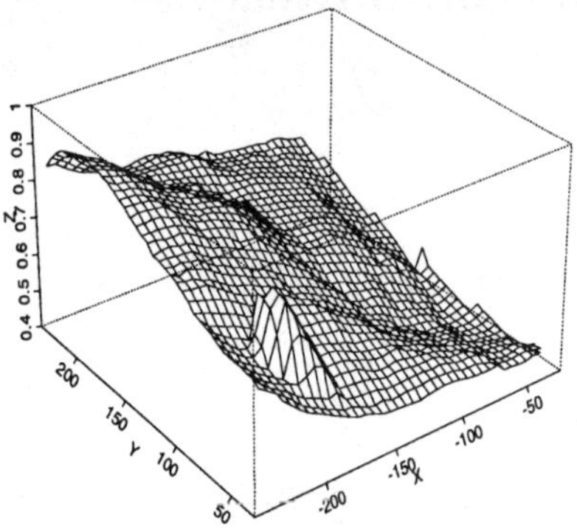

Fig. 1. Example warpage of PCB in inspection machine: interpolation of reference measurements

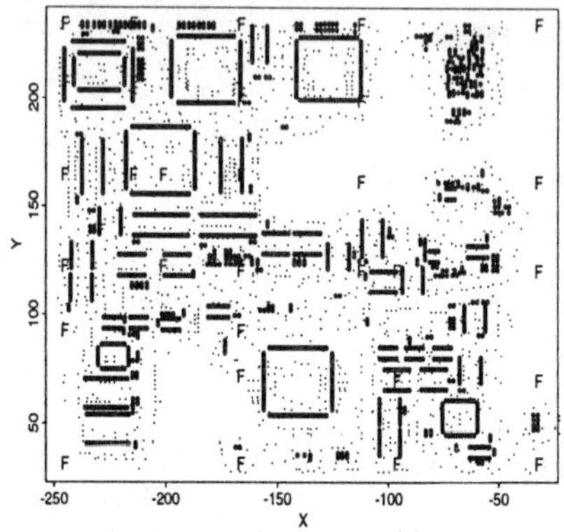

Fig. 2. Example layout of PCB: shown are 1998 target locations (big dots), 1697 reference locations (small dots), and 34 fit locations (denoted by F)

To compute this surface height, the following procedure was traditionally used at each target location. The height of the surface of the PCB was measured at a number of reference locations, near the target location. Then a local, quadratic polynomial was fitted to these reference measurements and the height of the surface was estimated by evaluating this polynomial at the target location. This procedure, however, was subject to occasional problems. Both noise peaks in the measurement of the height at the reference locations and unfortunate spatial arrangement of these reference locations can lead to substantial errors. To remedy these problems, this original procedure was extended with the sophisticated principles pioneered in Cleveland and Grosse (1991):

- For each target location, employ a large number, rather than a small number, of reference measurements. This can be implemented without taking extra reference measurements by borrowing reference locations from neighbouring target locations.
- Reduce bias by employing distance weights in the fitting procedure.
- Increase robustness, by using an iterative fit which down-weights reference measurements with large residuals.
- Fit at a limited number of *fit* locations only, as fitting at each target location would be computationally prohibitive.
- Compute the heights at the target locations by 'blending' these fits. Blending is the spatial generalisation of interpolation by means of a cubic polynomial, see Cleveland and Grosse for details.

Figures 1 and 2 illustrate the set-up. Thus Figure 1 shows an example surface of a PCB in the inspection machine. It was made by interpolating reference measurements. It shows the magnitude of the board warpage, here approximately .6mm, which is sizable as compared to the heights of the dots of solder paste deposit. The latter are not shown here, but are in the order of .1 mm. Also, visible is noise in the centre area of the PCB, which makes it clear that smoothing is needed. Figure 2 displays the target locations (big dots), the reference locations (small dots), and the fit-locations (denoted by F) on the surface of the PCB. As usual, there are a large number of target locations and reference locations: 1998 and 1697, respectively. The number of fit-locations is small by computational necessity and equals 34 in Figure 2.

The proposed procedure was extensively tested against the original procedure on the quantity of primary interest: estimated volumes of solder paste. A convenient testbed was provided by empty PCB's, i.e. boards with no solder paste on them. Here, all estimated volumes must be zero, so that there is an objective ground truth for comparison. There were two general findings. Firstly, there is a statistical improvement in that the variability of the volume estimates is reduced: the interquartile range of the error distribution is typically reduced by 1/3. Secondly, the occurrence of outliers is strikingly reduced. Clearly, the occurrence of outliers and their size is highly dependent on the type of PCB tested. However, a reduction of the maximum error by a factor 6 was not uncommon.

2 Computational issues

The principles of local fitting can be successfully applied in the current situation. Yet there are also some differences with the situation in which local fitting is traditionally employed, and we now turn to these. These differences are mainly computational and derive from the fact that, usually, surface estimation is invoked as an exploratory tool to investigate a single data set. In the current application, however, surface estimation is used as an automatic tool to be applied to a sequence of PCB's under the time constraints inherent to a production process. Apart from a greater focus on computational speed, the current application differs in that some preprocessing can be performed, which would not be useful in the standard context.

The following subsections survey some of the differences and our stance on them.

2.1 Selection of reference locations

Aim of the selection of the reference locations is to have accurate estimates at the target locations with the lowest number of reference measurements: less measurements results in a faster machine. Note that there are two benefits to a reduction of the number of reference measurements. Firstly, it reduces measurement time for the machine. Secondly, it reduces the execution time of the algorithms.

Strictly speaking, the design issue involves the selection of areas on the PCB, rather than locations: any area on the bare copper on the surface area, that is larger than a given minimal area, suits as a reference location. The traditional selection algorithm works roughly as follows. For each target location divide up the space around this location into four quadrants. Then these quadrants are searched in turn for the closest area of bare copper that exceeds the minimum size. These areas are then used as reference locations. Usually, 8 to 10 reference locations are associated with each target location. The set of reference locations then is the union of the reference locations found for all target locations.

This yields a large set of reference locations, and reduction can be necessary for reasons of speed. Moreover, as fitting is done only at a modest number of fit-locations, it seems more appropriate to derive reference locations from fit-locations rather than from target locations. Our implementation then reduces the given set of reference measurements by using this intuition.

To this end, we use the Delaunay triangulation of the reference locations; see e.g. Ripley (1981) for an excellent introduction to Delaunay triangulation and Voronoi (or Dirichlet) tessellation. Very briefly, the Voronoi tessellation associates a tile with each point in a collection of spatial points. The tile is the area closer to this point than to any other point in the collection. The Delaunay triangulation then connects the points that are neighbours in the Voronoi tessellation. This defines a neighbouring relation on a collection of planar points and we can speak of first neighbours: points that are directly connected in the triangulation, second neighbours: points that are indirectly connected in the triangulation via one intermediary point, etc.

Our algorithm first computes the Delaunay triangulation of the set of reference locations. Then for each fit-location, it selects the reference location closest to it, as well as the reference locations that are upto k-th neighbours of this initial point. All points so selected are retained in the collection of reference locations, all other points are deleted. Thus it is possible to reduce the collection of reference locations by an appropriate choice of k

In a representative board (see Figure 2) taking the reference locations that are upto third and fourth neighbours substantially reduces the number of reference locations to approximately 40% and 65% respectively of the original size. Experiments with simulated data and Gaussian noise indicate this reduction can be carried out without sacrificing much accuracy: i.e. with only 40% of the original reference points the squared error increased with approximately 50%.

This approach suffers from a lack of granularity: taking third neighbours reduces the set of reference locations to 40% and taking fourth neighbours reduces to 65%. However, there is nothing in between so that it is difficult to achieve a prescribed number of reference locations. We have currently resolved this issue using randomisation: use all up to k-th nearest neighbours, and select randomly from the $k+1$-th nearest neighbours until the target number of reference locations is attained. Clearly, alternative approaches that select from the set of $k+1$-th neighbours based on variance measures will be more appropriate and are currently being investigated.

2.2 Selection of fit locations

The fit locations are not optimally placed with respect to the reference locations: a lot of the vertices in Figure 2 are located at the edges of the estimation area. This is a general feature of kd-trees and derives from the fact that they are built by recursively dividing the smallest rectangle that encloses all data points. However, this feature is a disadvantage for local fitting: to fit at these edge-vertices we must extrapolate out of the data. This extrapolation will adversely affect the precision of the fit, whence of the surface that is estimated by interpolation of the fit. It is intuitively plausible that it is better to avoid this extrapolation and to locate the vertices closer to the observations.

It is possible to improve the accuracy of the surface estimate by relocating the fit locations on the edge of the estimation area toward the centre. We have devised an algorithm that does this, yet retains the kd-structure of the fit locations, which is convenient for interpolation. The algorithm is based on the following, one dimensional observation. Let $\hat{c}(x)$ be a function on $[0,1]$, say, obtained by interpolating local linear fits at V_1 and V_2 with a cubic spline, where $0 \leq V_1 < V_2 = 1$. The the average variance of $\hat{c}(x)$,

$$\int_0^1 \operatorname{var}(\hat{c}(x))dx \tag{1}$$

# Meas	# Target	# Vert	α	Efficiency	Speed-up
1697	1998	34	.05	1.10	3.5
1127	1247	18	.08	1.14	3.2
4499	2545	34	.02	1.09	4.0

Table 1. Efficiency: squared error of surface estimation with original kd-tree divided by squared error of surface estimation with modified kd-tree. Speed-up: execution time for surface estimation with reordered data divided by execution time for surface estimation with data in original order.

can be analytically computed and is a polynomial in V_1. Consequently, the average variance can be minimised with respect to V_1, e.g. by a grid search.

This example shows that it is possible to optimally locate the edge vertex, V_1, which would be located at 0 with the usual kd-tree. Our algorithm then generalises this observation to higher dimensions, roughly in the following way. It visits both edges in all dimensions of the estimation area. For each vertex on the current edge, the average efficiency is computed for a series of shifts that moves this vertex into the estimation area. Here, the efficiency is computed for the projected data, i.e. the coordinate values that correspond to the current dimension. Finally, that shift is retained that minimises the total, i.e. over all vertices on this edge, average efficiency.

Table 1 gives an overview of the gains in efficiency that were achieved in a number of simulations with Gaussian noise, for a variety of PCB's. Here efficiency is defined as the average squared error of the surface estimate with the original kd-tree divided by the average squared error of the surface estimate using the relocated vertices.

2.3 Reordering the data

The association of reference measurements to fit locations is computationally the most demanding operation in local fitting. This comes about because, traditionally, a large array is linearly searched, and the time required for this operation dominates the total time needed for surface fitting. However, huge speed-ups can be achieved by the following observations:

- For each fit location, store pointers so that only a subinterval of the large array need be searched.
- Reorder the reference measurements such that the sum of the lengths of these subintervals is small.

Intuitively, the task is to order spatial data so that points that are close in Euclidean sense are also close in this linear order. Standard techniques that come to mind are e.g. Euclidean distance to an origin, order induced by a minimum spanning tree (see Friedman and Rafsky, 1981) or by a Peano type plane filling curve (e.g. see Mandelbrot, 1977).

Table 1 stresses the relevance of such a reordering strategy for the current application. The last column gives an overview of the speed-ups that were obtained for a series of representative boards, where speed-up is defined as the computational time for surface estimation for the restricted search procedure with pre ordering strategy divided by the computational time for this operation without a pre ordering strategy. The speed-ups that were achieved are considerable. To this it can be added that for very large boards, with 25000 reference points, speed-ups with a factor 25 have been measured.

2.4 Fast adaptation

Often, reference measurements in some areas of the PCB have relatively low noise as compared to the measurements taken in other areas. Hence, some form of adaptation of the bandwidth of the surface estimator is required. Again, the quest for speed precludes the use of approaches as in Loader (1999), and novel, computationally inexpensive, approaches are called for.

Some adaptivity, without extra computational cost, can be achieved by exploiting the iterated approach to the least squares problem. The approach decides, after the first iteration, to one of three possibilities: iterated reweighted least squares, ordinary least squares, or ordinary least squares with reduced bandwidth.

3 Conclusion

In this paper we have shown that surface estimation is important for process control on PCB's and that the sophisticated principles for local fitting as pioneered in Cleveland and Grosse (1991) constitute a natural and successful extension of the engineering approach. Although this approach was found to be very successful, questions concerning the relative performance of this approach as compared to other techniques for surface estimation, such as kriging, have been left unanswered. This is a relevant subject for further research.

Part of this paper is devoted to spatial design issues for surface estimation, i.e. accurate estimation of the surface with the least number of measurements. This involves both the selection of fit locations and the selection of reference locations. These operation are mutually dependent: the best fit locations depend on the measurement locations, and, conversely, the best measurement locations depend on the fit locations. In our approach, we have only marginally exploited this mutual dependency by shifting the fit locations after selecting the final set of reference locations. However, there is no compelling reason to stop here. It is also possible to iterate the algorithm in order to further improve upon the selected fit and reference locations. Again, this could be a fruitful subject for further research.

From the applied point of view, however, the thing to go after is undoubtedly the measurement noise. A better understanding of the noise associated with the reference measurements, and its dependence on size and material, pays off immediately in both selection of reference measurements and fitting procedures: just avoid the places that you know will cause trouble. Moreover, a better understanding of the noise will give further structure and direction to the research questions to be investigated.

References

Cleveland, W.S. and Grosse, E. (1991). Computational Methods for Local Regression, *Statistics and Computing* **1**, pp. 47-62.

Friedman, J. H. and Rafsky, L. C. (1981). Graphics for the Multivariate Two-Sample Problem. *Journal of the American Statistical Association*, **76**, pp. 277-287.

Loader, C. (1999). *Local Regression and Likelihood.* New York: Springer.

Mandelbrot, B. (1977). *Fractals, form, chance and dimension*, W.H. Freeman and Co.: San Fransisco.

Ripley, B.D. (1981). *Spatial Statistics.* New York: Wiley.

Optimization of the antithetic Gibbs sampler for Gaussian Markov random fields

Johannes M. Dreesman

Governmental Institute of Public Health Service Lower Saxony,
Roesebeckstrasse 4-6, D-30449 Hannover, Germany

Abstract. The efficiency of Markov chain Monte Carlo estimation suffers from the autocorrelation of successive iterations, which is typical for this sampling method. In order to improve the efficiency, antithetic methods attempt to reduce this autocorrelation or even introduce negative autocorrelation. In this paper the antithetic method is adopted to Gibbs sampling of the spatial correlation structure of Gaussian Markov random fields and a rule for the optimal choice of the antithetic parameter is developed. The antithetic Gibbs sampler turns out to perform much better than the classical Gibbs sampler and could compete with i.i.d. sampling, which indeed is usually intractable for this kind of application.

Keywords. MCMC, Antithetic parameter, Variance reduction, Output diagnostics, MCMCML estimation

1 Simulating Gaussian Markov random fields

Consider the Gaussian Markov random field (GMRF) which is specified on the regular lattice $L := \{(i,j) : i = 1,\dots,I, j = 1,\dots,J\}$ by the set of local conditional distributions

$$X_{i,j}|\{x_{r,s} : (r,s) \neq (i,j)\} \sim N\left(\eta_{i,j}, \tau^2\right) \text{ for all } (i,j) \in L, \quad (1)$$

$$\text{where } \eta_{i,j} := \mu + (x_{i-1,j} + x_{i+1,j} + x_{i,j-1} + x_{i,j+1} - 4\mu) \cdot \beta. \quad (2)$$

Furthermore the restriction $|\beta| < 0.25$ is imposed on the model in order to obtain a proper joint distribution for $\mathbf{x} := (X_{1,1},\dots,X_{I,J})$, which is multivariat Gaussian with nondiagonal covariance structure.

Direct simulation of $\mathbf{x}$ can be performed by means of Cholesky factorization of the covariance matrix, which requires about $(I \cdot J)^3/6$ floating point operations. On the other hand, Markov chain Monte Carlo (MCMC) is easy to apply. The Gibbs sampler is the natural algorithm for this purpose. It resamples $\mathbf{x}$ componentwise from the local conditional distributions given by (1) and (2).

MCMC enables us to estimate the expectation $E(f(\mathbf{x}))$ by the sample mean $\bar{f}_M(\mathbf{x}) := \frac{1}{M}\sum_{m=1}^{M} f(\mathbf{x}^{(m)})$, where m is indexing the iterated MCMC-realizations of the entire field. Since MCMC takes the current realization of the field as starting point for generating a new one, in general the successive values of $f(\mathbf{x})$ are not independent. Usually they show positive autocorrelation. In order to reduce the autocorrelation of successive realizations and thus reduce the variance of the estimator $\bar{f}_M(\mathbf{x})$, Barone & Frigessi (1990) suggested the introduction of an antithetic parameter, which counteracts the autocorrelation.

2 The antithetic Gibbs sampler for GMRFs

Compared with the ordinary Gibbs sampler for GMRFs, the innovation of the antithetic Gibbs sampler is that in iteration m the field's component $X_{i,j}^{(m)}$ is resampled from

$$N\left((1+\psi)\cdot\eta_{i,j}^{(m)} - \psi\cdot x_{i,j}^{(m-1)}, (1-\psi^2)\cdot\tau^2\right) \tag{3}$$

instead of $N\left(\eta_{i,j}^{(m)}, \tau^2\right)$, where the constant $\psi \in (-1,1)$ is denoted as antithetic parameter and $\eta_{i,j}^{(m)}$ is obtained from the current realization of $\mathbf{x}$ as described in (2). The subtraction of $\psi \cdot x_{i,j}^{(m-1)}$ in (3) induces a negative autocorrelation component.

The antithetic Gibbs sampler includes the ordinary Gibbs sampler as a special case for $\psi = 0$, but is no Gibbs sampler anymore for $\psi \neq 0$. Therefore, the more general concept of Metropolis Hastings (MH) algorithms is helpful for analyzing the samplers equilibrium properties.

Let $\pi(\mathbf{x})$ be the distribution of interest and $q(\mathbf{x}'|\mathbf{x})$ the proposal density from which the MH step generates a new candidate state $\mathbf{x}'$, given the current state $\mathbf{x}$. Then

$$a(\mathbf{x}'|\mathbf{x}) = \min\left(\frac{\pi(\mathbf{x}')q(\mathbf{x}|\mathbf{x}')}{\pi(\mathbf{x})q(\mathbf{x}'|\mathbf{x})}, 1\right) \tag{4}$$

is the probability of accepting the candidate $\mathbf{x}'$ as the new state and $1-a(\mathbf{x}'|\mathbf{x})$ is the probability that the old state $\mathbf{x}$ is taken as new state again. The ordinary Gibbs update is known to be a single component MH update with acceptance probability identically 1 since $\frac{q(\mathbf{x}'|\mathbf{x})}{q(\mathbf{x}|\mathbf{x}')} = \frac{\pi(\mathbf{x}')}{\pi(\mathbf{x})}$ (see, e.g., Gilks et al., 1995). By constructing $q(\mathbf{x}'|\mathbf{x})$ and $q(\mathbf{x}|\mathbf{x}')$ from (3) it can be shown that the ratio $\frac{q(\mathbf{x}'|\mathbf{x})}{q(\mathbf{x}|\mathbf{x}')}$ for the antithetic Gibbs step is independent from ψ, which is due to the correction factor $(1-\psi^2)$ of the conditional variance. Thus, like the classical Gibbs sampler, the antithetic Gibbs sampler is also a MH sampler with stationary distribution $\pi(\mathbf{x})$ if proposals are accepted with probability 1 (Green & Han, 1992).

Besides these equilibrium properties, Green & Han (1992) could give strong results by means of theory of vector autoregressive processes how the antithetic method reduces the asymptotic variance of the sample mean of any linear function $f(\mathbf{x})$.

3 Markov chain Monte Carlo maximum likelihood estimation

Maximum likelihood (ML) estimation of the spatial dependence parameter β of GMRFs usually is intractable. This is caused by the normalizing term in the joint density. It contains the determinant of the high dimensional spatial covariance matrix and cannot be omited since it depends on

β. Therefore, several approximations to the ML estimator have been proposed, which are easy to calculate but considerably less efficient than ML estimation (see, e.g., Besag, 1977). More recently Geyer (1991) proposed the Markov chain Monte Carlo maximum likelihood (MCMCML) method, which enables ML estimation of β even if the likelihood contains an unknown normalizing term which depends on β. Consider a family of densities $\{f_\beta\}$ of the form $f_\beta(\mathbf{x}) = \frac{1}{z(\beta)} h_\beta(\mathbf{x})$, where h_β is known and the normalizing term $z(\beta) = \int h_\beta(\mathbf{x}) d\mathbf{x}$ is intractable. Geyer (1991) suggested to maximize the log-likelihood-ratio $lr(\beta, \beta^*|\mathbf{x}) := \log\left(\frac{L(\beta|\mathbf{x})}{L(\beta^*|\mathbf{x})}\right) = \log\left(\frac{h_\beta(\mathbf{x})}{h_{\beta^*}(\mathbf{x})}\right) - \log\left(\frac{z(\beta)}{z(\beta^*)}\right)$ for observed $\mathbf{x}$ and arbitrary β^*, $f_{\beta^*} \in \{f_\beta\}$, instead of the likelihood. He showed that $\frac{z(\beta)}{z(\beta^*)}$ could be expressed as $E_{\beta^*}\left(\frac{h_\beta(\mathbf{x})}{h_{\beta^*}(\mathbf{x})}\right)$ and estimated by $\frac{1}{M}\sum_{m=1}^{M} \frac{h_\beta(\mathbf{x}^{(m)})}{h_{\beta^*}(\mathbf{x}^{(m)})}$, $\mathbf{x}^{(m)} \sim f_{\beta^*}$. The sample $\left(\mathbf{x}^{(m)}\right)_{m=1,\ldots,M}$, $\mathbf{x}^{(m)} \sim f_{\beta^*}$, is generated by MCMC, which is possible even if the normalizing term of f_{β^*} is unknown. Thus, the MCMCML estimator is obtained by maximizing

$$\widehat{lr}_M(\beta, \beta^*|\mathbf{x}) := \log\left(\frac{h_\beta(\mathbf{x})}{h_{\beta^*}(\mathbf{x})}\right) - \log\left(\frac{1}{M}\sum_{m=1}^{M} \frac{h_\beta(\mathbf{x}^{(m)})}{h_{\beta^*}(\mathbf{x}^{(m)})}\right), \ \mathbf{x}^{(m)} \sim f_{\beta^*}, \quad (5)$$

with respect to β. For the GMRF (1) and (2), $\widehat{lr}_M(\beta, \beta^*|\mathbf{x})$ reduces to

$$(\beta - \beta^*) \cdot g(\mathbf{x}) \ - \log\left(\frac{1}{M}\sum_{m=1}^{M} \exp\left((\beta - \beta^*) \cdot g(\mathbf{x}^{(m)})\right)\right), \ \mathbf{x}^{(m)} \sim f_{\beta^*}, \quad (6)$$

where

$$g(\mathbf{x}) = \sum_{(i,j)\in L} (X_{i,j} - \mu) \cdot (X_{i+1,j} - \mu) + \sum_{(i,j)\in L} (X_{i,j} - \mu) \cdot (X_{i,j+1} - \mu). \quad (7)$$

Therefore, for MCMCML estimation of β we have to rely on a sampler for GMRFs which is easy to implement and shows good convergence properties for the statistic $g(\mathbf{x})$. The goodness properties of the antithetic Gibbs sampler are determined by the choice of the antithetic parameter ψ, which interacts with the field's spatial dependence parameter β. Since $g(\mathbf{x})$ is no linear function of $\mathbf{x}$, the theoretical results of Green & Han (1992) do not apply and we conducted a simulation study in order to find the optimal value of ψ.

4 Simulation study

For each combination of $\beta \in \{0.01, 0.1, 0.2, 0.24\}$ and $\psi \in \{-0.95, -0.9, -0.8, \ldots, 0.8, 0.9, 0.95\}$ two simulations of a GMRF on a 16×16 torus lattice were carried out. The run length of each simulation was 10000 sweeps over the whole field after a burn-in period of length 100.

The implementation of the antithetic Gibbs sampler is determined by two issues: the single variable update, which follows straightforwardly from equations (1), (2) and (3), and the visiting schedule of the variables. We employed a visiting schedule that makes use of the fields spatial dependence structure. Like a chessboard, the field was partitioned into two so called coding sets so

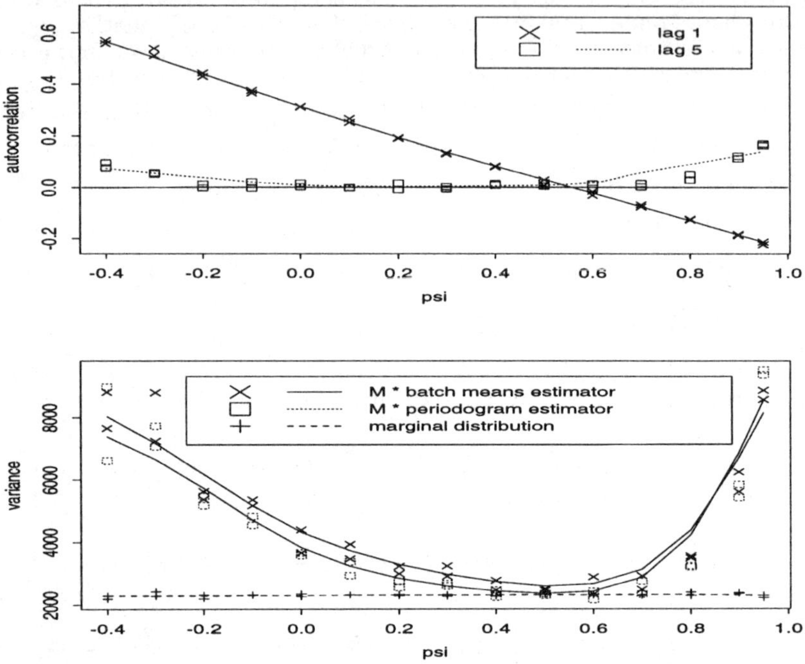

Fig. 1. Output diagnostics for antithetic Gibbs sampling of GMRF with $\beta = 0.2$. Lag-1 and lag-5 autocorrelation structure of $g(\mathbf{x})$ (top graph). Comparison of $M \cdot \widehat{var}(\bar{g}_M(\mathbf{x}))$ and $\widehat{var}(g(\mathbf{x}))$ (bottom graph).

that within each set, conditionally on the other set, the variables are mutually independent (Besag, 1974). Thus, updating both coding sets is identical with one sweep over the whole field. The coding sets were updated alternately and within each coding set the variables were updated quasi simultaneously. This allows efficient computation with array programming languages like S-Plus and has a high potential for parallel processing. Besag et al. (1995) could show that this updating schedule leads to the desired equilibrium properties of the Gibbs sampler.

In order to measure the goodness properties of the antithetic Gibbs sampler, the simulation output $g(\mathbf{x}^{(m)})$, $m = 1, \ldots, 10000$, was analyzed by means of several diagnostic statistics calculated with CODA (Best et al., 1995). Figure 1 demonstrates this analysis for $\beta = 0.2$, which is a typical value in applications. The top graph shows the empirical lag-1 and lag-5 autocorrelation of the two simulation outputs obtained for each ψ. The lag-1 autocorrelation declines almost linearly with increasing ψ. For $\psi = 0$ the autocorrelation is about 0.35. It becomes 0 for $\psi = 0.5$. The lag-5 autocorrelation first decreases with increasing ψ, reaches its minimum near 0 for $\psi = 0.2$ and increases again with ψ even though the lag-1 autocorrelation decreases further.

The bottom graph shows how the variance of $\bar{g}_M(\mathbf{x}) := \frac{1}{M}\sum_{m=1}^{M} g(\mathbf{x}^{(m)})$ depends on ψ. For each simulation, the graph presents two estimates of $var(\bar{g}_M(\mathbf{x}))$, multiplied by M, with smooth curves fitted to them. One set of estimates is obtained by the batch means method, where the simulation output is divided into subsequences (batches) of equal lengths and the sample means of these subsequences are calculated. From the observed variance of the batch means, which are asymptotically independent, an estimator for $var(\bar{g}_M(\mathbf{x}))$ is derived (cf. Geyer, 1992).

Let $\varrho\left(g(\mathbf{x}^{(m)}), g(\mathbf{x}^{(m+l)})\right)$ denote the lag-l autocorrelation of $\left(g(\mathbf{x}^{(m)})\right)$. Then the second estimator in the bottom graph uses the property

$$M \cdot var\left(\frac{1}{M}\sum_{m=1}^{M} g(\mathbf{x}^{(m)})\right) \approx var(g(\mathbf{x})) \cdot \sum_{l=-\infty}^{\infty} \varrho\left(g(\mathbf{x}^{(m)}), g(\mathbf{x}^{(m+l)})\right) \quad (8)$$

and estimates the components on the right hand separately, using the periodogram to estimate the autocorrelation structure (Geweke, 1992). The batch means estimator gives systematically higher values than the periodogram estimator but they both lead to the choice of $\psi_{opt} = 0.5$ being the most efficient by eye sight.

In addition, the bottom graph gives estimates for $var(g(\mathbf{x}))$, the variance of the marginal distributions. For $\psi_{opt} = 0.5$ it becomes obvious that $M \cdot var\left(\frac{1}{M}\sum_{m=1}^{M} g(\mathbf{x}^{(m)})\right) \approx var(g(\mathbf{x}))$. This corresponds to (8) together with the fact that the autocorrelations for lags $l \neq 0$ are about zero as can be observed for the lag-1 and the lag-5 autocorrelation from the top graph. Consequently, for ψ_{opt} our statistic is as efficient as if i.i.d. sampling of the field were possible. On the other hand, the relative efficiency of the classical Gibbs sampler ($\psi = 0$), which is obtained by calculating the ratio of the respective variances, is about 60% (cf. Roberts, 1996).

In the same way, ψ_{opt} was determined for the other values of β and we found a strong relationship between β and ψ_{opt} which could be expressed by the linear function

$$\widehat{\psi_{opt}} = 2.767 \cdot \beta \quad (9)$$

and is at least valid over the range $0.01 \leq \beta \leq 0.24$. Additional convergence diagnostics, not presented here, also revealed good convergence properties of the antithetic Gibbs sampler if $\widehat{\psi_{opt}}$ was taken for ψ.

5 Conclusions

We conclude that for simulating Gaussian Markov random fields the antithetic Gibbs sampler provides a substantial efficiency gain if the antithetic parameter is chosen according to (9). For a parametrization of the field which is typical for applications, the antithetic Gibbs sampler competes with i.i.d. sampling. Without considerably more complicated implementation, it performs much better than the pure Gibbs sampler, which has only about 60% efficiency.

In some recent papers the antithetic Gibbs sampler was applied to related problems and also turned out to produce a significant variance reduction; see e.g. Frigessi et al. (1999) and Barone & Sebastiani (1999).

References

Barone, P. and Frigessi, A. (1990). Improving stochastic relaxation for Gaussian random fields. *Probability in the Engineering and Informational Sciences* **4**, 369-389.

Barone, P. and Sebastiani, G. (1999). Over-relaxation methods and Metropolis-Hastings coupled Markov chains for Monte Carlo simulation. Available from `http://www.iac.rm.it/~barone/mcmcnode.ps`.

Besag, J.E. (1974). Spatial interaction and the satistical analysis of lattice systems. *Journal of the Royal Statistical Society, B* **36**, 192-225.

Besag, J.E. (1977). Efficiency of pseudolikelihood estimation for simple Gaussian fields. *Biometrika* **64**, 616-618.

Besag, J.E., Green, P.J., Higdon, D. and Mengersen, K. (1995). Bayesian computation and stochastic systems. *Statistical Science* **10**, 3-66.

Best, N.G., Cowles, M.K. and Vines, S.K. (1995). CODA Manual version 0.30. Cambridge: MRC Biostatistics Unit.

Frigessi, A., Gasemyr, J. and Rue, H. (1999). Antithetic coupling of two Gibbs sampler chains. Preprint statistics No. 13/1999, Norwegian university of science and technology, Trondheim.

Geweke, J. (1992). Evaluating the accuracy of sampling based approaches to the calculation of posterior moments. In *Bayesian Statistics* **4** (eds. J.M. Bernardo, J. Berger, A.P. Dawid & A.F.M. Smith), 169-194. Oxford University Press.

Geyer, C.J. (1991). Markov chain Monte Carlo maximum likelihood. In *Computing Science and Statistics: Proc. 23rd Symp. Interface* (ed. E.M. Keramidas), 156-163.

Geyer, C.J. (1992). Practical Markov chain Monte Carlo. *Statistical Science* **7**, 473-511.

Gilks, W.R., Richardson, S. and Spiegelhalter, D. (1996). Introducing Markov chain Monte Carlo. In *Markov Chain Monte Carlo in Practice* (eds. W.R. Gilks, S. Richardson & D.J. Spiegelhalter), 1-16. London: Chapman & Hall.

Green, P.J. and Han, X.L. (1992). Metropolis methods, Gaussian proposals, and antithetic variables. In *Stochastic Models, Statistical Methods and Algorithms in Image Analysis* (eds. P. Barone, A. Frigessi & M. Piccioni), 142-164. Berlin: Springer.

Roberts, G.O. (1996). Markov chain concepts related to sampling algorithms. In *Markov Chain Monte Carlo in Practice* (eds. W.R. Gilks, S. Richardson & D.J. Spiegelhalter), 45-57. London: Chapman & Hall.

Computing zonoid trimmed regions of bivariate data sets

Rainer Dyckerhoff

Seminar für Wirtschafts- und Sozialstatistik, Universität zu Köln, 50923 Köln, Germany

Keywords. Multivariate trimming, algorithm, zonoid depth, convex sets

1 Introduction

In data analyis an important task is to identify sets of points that are central in a data set $\{x_1, \ldots, x_n\} \subset \mathbb{R}^d$. A set of points that is central in some sense is called a *trimmed region*. In the univariate case a trimmed region is, e.g., the closed interval between two proper quantiles. Nolan (1992) and Massé and Theodorescu (1994) introduced concepts of trimmed regions, based on a multivariate analogue of the quantile function, suggested by Tukey (1975) and Eddy (1983). These concepts can be seen as generalizations of the univariate interquantile intervals.

Another approach is to define trimmed regions via a properly chosen data depth. Data depth is a concept that measures the "depth" of a point in a given data set. Many such notions of data depth have been proposed by, e.g., Mahalanobis (1936), Tukey (1975), and others. A trimmed region can be defined as the set of points that have a depth of at least α.

In this paper we focus on the so-called zonoid trimming, introduced by Koshevoy and Mosler (1997). The α-zonoid trimmed region is defined as the set of all convex combinations of the data points, where an additional restriction is imposed on the convex coefficients. Zonoid trimming is related to zonoid data depth: The α-trimmed region equals the set of all points that have zonoid depth at least α. Zonoid trimming has been applied in cluster analysis, see Hoberg (2000).

To use a notion of trimmed regions in practical applications it has to be computationally tractable. For the halfspace trimming nice algorithms have been developed by Ruts and Rousseeuw (1996) and Johnson, Kwok and Ng (1998).

Although an efficient algorithm for computing the zonoid data depth had been developed by Dyckerhoff, Koshevoy and Mosler (1996), there was no exact algorithm to compute zonoid trimmed regions. Here, we present an algorithm that allows to compute arbitrary zonoid trimmed regions of a bivariate data set with complexity $O(n^2 \log n)$. Thus, this algorithm has the same complexity as the algorithm proposed by Ruts and Rousseeuw (1996), which solved the same problem for the halfspace trimming. Unlike the algorithm of Ruts and Rousseeuw (1996) our algorithm can deal with three or more collinear points. A generalization of the algorithm can even cope with the problem of two or more points being equal.

In this paper we write S^{d-1} for the unit sphere in $\mathbb{R}^d$, the transpose of a vector $p \in \mathbb{R}^d$ will be denoted by p'.

2 Zonoid trimming

Zonoid trimming was first introduced by Koshevoy and Mosler (1997). The zonoid trimmed regions are defined as follows.

Definition 1. Let $x_1, x_2, \ldots, x_n \in \mathbb{R}^d$. For $\alpha \in [0,1]$ the α-zonoid trimmed region is given by

$$D_\alpha(x_1, \ldots, x_n) = \left\{ \sum_{i=1}^n \lambda_i x_i \, : \, \sum_{i=1}^n \lambda_i = 1, \, 0 \le \lambda_i, \, \alpha\lambda_i \le \frac{1}{n} \text{ for all } i \right\}.$$

Thus, $D_\alpha(x_1, \ldots, x_n)$ is the set of all convex combinations of the x_i, where the coefficients λ_i, are subject to the additional restriction $\alpha\lambda_i \le \frac{1}{n}$.

For each α the zonoid trimmed region is a convex polytope. As can be seen from the definition, the zonoid trimmed regions are decreasing in α. For $\alpha = 1/n$ the zonoid trimmed region equals the convex hull of the data points, for $\alpha = 1$ the trimmed region is a singleton, namely the set containing only the mean of the data points. Furthermore the zonoid trimmed regions are affine equivariant. If the data are transformed by an affine transformation, the zonoid trimmed regions are transformed in the same way.

For $\alpha > 0$ the zonoid trimmed regions can also be written in the alternative form

$$D_\alpha(x_1, \ldots, x_n) = \left\{ \frac{1}{n\alpha} \sum_{i=1}^n \lambda_i x_i \, : \, \sum_{i=1}^n \lambda_i = n\alpha, \, 0 \le \lambda_i \le 1 \text{ for all } i \right\}.$$

From now on we will write shortly D_α instead of $D_\alpha(x_1, \ldots, x_n)$.

3 Extreme points of zonoid trimmed regions

Recall that the *support function* $h_C : S^{d-1} \to \mathbb{R}$ of a compact convex set $C \subset \mathbb{R}^d$ is defined by $h_C(p) := \max\{p'x : x \in C\}$. The algorithm for computing zonoid trimmed regions relies on the following proposition.

Proposition 1. *Let C be a convex polytope in $\mathbb{R}^d$ and $x_0 \in C$. Then, x_0 is an extreme point of C if and only if there exists a direction $p \in S^{d-1}$ such that the equation*

$$p'x = h_C(p), \quad x \in C \tag{1}$$

has the unique solution x_0.

The points which are unique maximizers of the support function can be generated in an easy way. Let H be the set of all directions p for which there exist $i \ne j$ such that $p'x_i = p'x_j$. For a direction $p \in S^{d-1} \setminus H$ we denote by $r_p(x_i)$ the position of x_i in the *increasing* ordered sequence of the $p'x_j$, $j = 1, \ldots, n$. $r_p(x_i)$ is well-defined since the $p'x_j$, $j = 1, \ldots, n$, are pairwise distinct. Thus, $p'x_i > p'x_j$ if and only if $r_p(x_i) > r_p(x_j)$.

Now denote by $x_{p,\alpha}$ the point

$$x_{p,\alpha} = \frac{1}{n\alpha} \sum_{i=1}^n \lambda_i^p x_i \,,$$

where

$$\lambda_i^p = \begin{cases} 1, & \text{if } r_p(x_i) > n - [n\alpha], \\ n\alpha - [n\alpha], & \text{if } r_p(x_i) = n - [n\alpha], \\ 0, & \text{if } r_p(x_i) < n - [n\alpha]. \end{cases}$$

Here $[n\alpha]$ denotes the integer part of $n\alpha$. Obviously, $x_{p,\alpha} \in D_\alpha$. The next theorem tells that the points $x_{p,\alpha}$ are exactly the extreme points of D_α.

Theorem 1. *Let $x_1, \ldots, x_n \in \mathbb{R}^d$ be pairwise distinct. Then the set of extreme points of D_α is given by*

$$\{\, x_{p,\alpha} \,:\, p \in S^{d-1} \setminus H \,\}\,. \tag{2}$$

4 The algorithm

In the bivariate case the enumeration of the points $x_{p,\alpha}$ can efficiently be achieved by means of a so-called *circular sequence*, see, e.g., Goodman and Pollack (1980) or Edelsbrunner (1987).

Consider n points in the plane $\mathbb{R}^2$. Each direction $p \in S^1 \setminus H$ defines a permutation of the x_i, $i = 1, \ldots, n$, namely the permutation π_p for which

$$p' x_{\pi_p(1)} < p' x_{\pi_p(2)} < \ldots < p' x_{\pi_p(n)}\,.$$

The permutation is well-defined, since $p \notin H$. If we let p rotate through the unit circle we get a sequence of permutations. The permutation changes when p passes over a value in H. The change that takes place can be deduced from the data: For a direction p_0 in H there may be several sets of collinear points which lie on parallel straight lines that are othogonal to p. When p passes over p_0 the order of each of these sets is reversed. In Figure 1 the points x_1, x_2 as well as the points x_3, x_4, x_5 lie on two parallel lines. When p changes from p_1 to p_2 the order of the points x_1, x_2 as well as the order of the points x_3, x_4, x_5 is reversed.

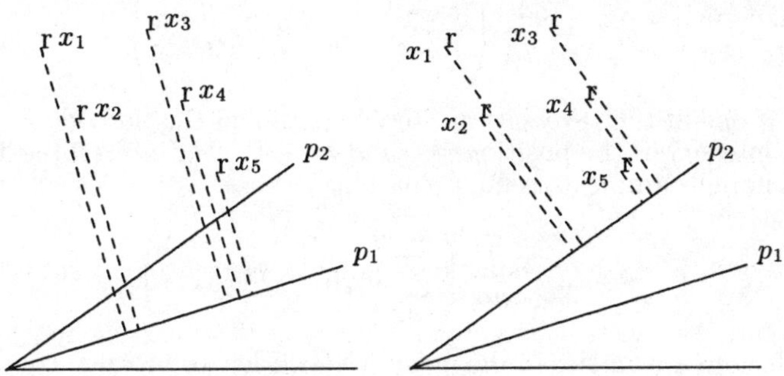

Fig. 1. Illustration of a step in the circular sequence

The permutation π_{-p}, corresponding to the direction $-p$, is exactly the reverse permutation of π_p, i.e., $\pi_{-p}(i) = \pi_p(n+1-i)$. Thus, to construct all permutations of the circular sequence it is not necessary to rotate p trough the full unit circle, but only through the upper half of the unit circle.

The circular sequence enables us to compute the points $x_{p,\alpha}$ in an efficient way, as described in the following algorithm.

Step 1: We store all data points in an array. This array is sorted according to the order defined by

$$(x_1, x_2) < (y_1, y_2) \iff (x_1 < y_1) \text{ or } (x_1 = y_1 \text{ and } x_2 > y_2)\,.$$

We further set up two arrays ORDER and RANKS which store the current values $\pi_p(i)$ and $r_p(i)$, respectively. This arrays are initialized with the values 1 to n.

Step 2: For each pair of points x_i, x_j we compute the angle between the first coordinate axis and the normal vector of the line passing through x_i and x_j. We always choose the angle in the interval $[0, \pi)$. This angle is stored in a record together with the indices i and j. We store these records in an array ANGLE. This array is sorted according to the value of the angle.

Step 3: We collect all succesive entries of the array ANGLE which correspond to the same angle. From these entries we construct the sets of collinear points, described above. This can easily be done, since x_i and x_j are in the same set if and only if one of the collected entries corresponds to the indices i and j. In the situation depicted in Figure 1 we would have four entries. The corresponding pairs would be $(1,2)$, $(3,4)$, $(3,5)$ and $(4,5)$ which would result in the two sets $\{1,2\}$ and $\{3,4,5\}$.

Step 4: For each set of collinear points reverse their order in the arrays ORDER and RANKS. This is best illustrated by an example. Suppose the situation in Figure 1 where x_1, x_2 and x_3, x_4, x_5 lie on parallel lines. Then the array ORDER is updated in the following way

		1	2			3	4	5	...	$\longrightarrow$			2	1			5	4	3	...

and the array RANKS analogously.

Step 5: Let $k_- = [n\alpha]+1$, $k_+ = n - [n\alpha]$ and $\delta = (n\alpha - [n\alpha])$. Check if one of the reversals in Step 4 contained the position k_+ (if $n\alpha$ is not an integer) or the positions k_+ and $k_+ + 1$ (if $n\alpha$ is an integer). In this case, compute a new extreme point $x_{p,\alpha}$:

$$x_{p,\alpha} = \frac{1}{n\alpha}\left(\sum_{i=k_+ +1}^{n} x_{\pi_p(i)} + \delta x_{\pi_p(k_+)}\right).$$

Check if one of the reversals in Step 4 contained the position k_- (if $n\alpha$ is not an integer) or the positions k_- and $k_- - 1$ (if $n\alpha$ is an integer). In this case, compute a new extreme point $x_{-p,\alpha}$:

$$x_{-p,\alpha} = \frac{1}{n\alpha}\left(\sum_{i=1}^{k_-} x_{\pi_p(i)} + \delta x_{\pi_p(k_-)}\right).$$

Step 6: If there are entries in the array ANGLE left to process, continue with Step 3, otherwise stop.

The algorithm generates two sequences of extreme points, namely the $x_{p,\alpha}$ and the $x_{-p,\alpha}$, which are both generated in counterclockwise order. Both sequences can be joined to give the sequence of all extreme points in counterclockwise order.

The complexity of the algorithm is determined by the sorting in Step 2. Since there are $\binom{n}{2}$ pairs of points, this results in a complexity of $O(n^2 \log n)$.

5 Applications and a real data example

Zonoid trimming can be applied in cluster analysis. A common approach to cluster analysis is to define a function which measures the inhomogeneity of the clusters. A cluster criterion can then be defined as the sum of the inhomogeneities of the clusters. Under all possible partitions of the data set in k clusters one has to find the partition which yields the minimum value of the cluster criterion. Hardy and Rasson (1982) use the volume of the convex hull of a cluster as a measure of inhomogeneity. However, this approach has some drawbacks. Adding points which lie inside a cluster, does not change the value of the cluster criterion. To overcome this drawback Hoberg (2000) has generalized this approach. As a measure of inhomogeneity he uses a weighted average of the volumes of zonoid trimmed regions of a cluster. Since dispersed data leads to larger zonoid trimmed regions this concept is a sensible inhomogeneity measure. The evaluation of the cluster criterion for a given partition requires the computation of a large number of zonoid trimmed regions. Thus, an efficient algorithm, like the one proposed above, is needed to make this approach computationally tractable.

To give an application of the algorithm to real data consider the heptathlon data set of Hand et al. (1994). This data set consists of the results of 25 heptathletes in the 1988 Olympic summer games. Here we consider their 100 metres time (in seconds) and their short put distances (in metres). Figure 2 shows the data as well as the zonoid trimmed regions for $\alpha = \frac{1}{25}, \frac{2}{25}, \ldots, \frac{24}{25}$.

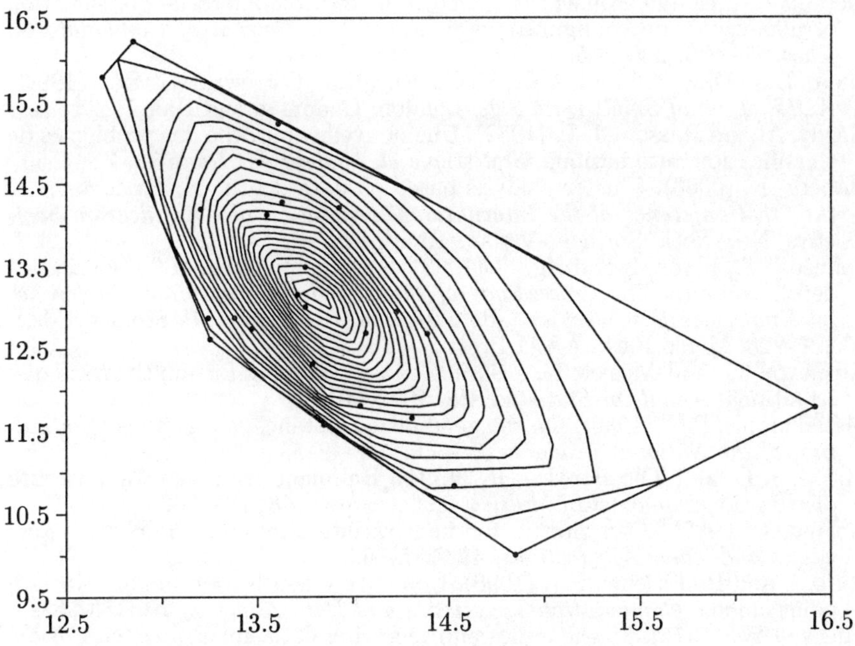

Fig. 2. Zonoid trimmed regions for the heptathlon data set

6 Concluding remarks

We have generalized the above algorithm to compute zonoid trimmed regions of bivariate discrete distributions with finite support. This generalized algorithm can also cope with the problem of two or more points being equal.

The algorithm as presented here is confined to bivariate data sets. Since the above algorithm uses the concept of a circular sequence there is no obvious generalization to the case of dimension greater than two. Nevertheless, an approximate algorithm can be constructed. Theorem 1 says that the set of extreme points consists of all points $x_{p,\alpha}$ where p runs through all directions. In the approximate algorithm, instead of checking all relevant directions, only a random subset of all directions is considered. For each chosen direction an extreme point is constructed. Thus, this approximate algorithm computes a subset of the extreme points of D_α, which yields an inner approximation of the α-zonoid trimmed region.

References

Dyckerhoff, R., Koshevoy, G. and Mosler, K. (1996). Zonoid data depth: Theory and computation. In: *COMPSTAT 96 – Proceedings in Computational Statistics*, 235–240. Heidelberg: Physica-Verlag.

Edelsbrunner, H. (1987). *Algorithms in Combinatorial Geometry.* Berlin: Springer-Verlag.

Eddy, W.F. (1983). Set-valued orderings for bivariate data. In: *Stochastic Geometry, Geometric Statistics, and Stereology* (R.V. Ambartzumian, W. Weil, Eds.), 79–90. Leipzig: Teubner-Verlag.

Goodman, J.E. and Pollack, R. (1980). On the combinatorial classification of nondegenerate configurations in the plane. *Journal of Combinatorial Theory A*, **29**, 220–235.

Hand, D.J., Daly, F. Lunn, A.D., McConway, K.J., Ostrowski, E., Eds. (1994). *A Handbook of Small Data Sets.* London: Chapman and Hall.

Hardy, A. and Rasson, J.-P. (1982). Une nouvelle approche des problèmes de classification automatique. *Statistique et Analyse des Données*, **7**, 41–56.

Hoberg, R. (2000). Cluster analysis based on data depth. In: *Proceedings of the 7th Conference of the International Federation of Classification Societies*, New York: Springer-Verlag. To appear.

Johnson, T., Kwok, I. and Ng, R. (1998). Fast computation of 2-dimensional depth contours. In: *Proceedings of the Fourth International Conference on Knowledge Discovery and Data Mining* (R. Agrawal, P. Stolorz, Eds.), 224–228. Menlo Park: AAAI Press.

Koshevoy, G. and Mosler, K. (1997). Zonoid trimming for multivariate distributions. *Annals of Statistics*, **25**, 1998–2017.

Mahalanobis, P.C. (1936). On the generalized distance in statistics. *Proceedings of the National Academy of India*, **12**, 49–55.

Massé, J.-C. and Theodorescu, R. (1994). Halfplane trimming for bivariate distributions. *Journal of Multivariate Analysis*, **48**, 188–202.

Nolan, D. (1992). Asymptotics for multivariate trimming. *Stochastic Processes and Their Applications*, **42**, 157–169.

Ruts, I. and Rousseeuw, P.J. (1996). Computing depth contours of bivariate point clouds. *Computational Statistics and Data Analysis*, **23**, 153-168.

Tukey, J.W. (1975). Mathematics and picturing data. In: *Proceedings of the 1974 International Congress of Mathematicians, Vancouver*, **2**, 523–531.

Outlier resistant estimators for canonical correlation analysis

P. Filzmoser[1], C. Dehon[2] and C. Croux[2]

[1] Department of Statistics, Probability Theory, and Actuarial Mathematics, Vienna University of Technology, Wiedner Hauptstr. 8-10, A-1040 Vienna, Austria

[2] ECARES, Université Libre de Bruxelles, CP139, Av. F.D. Roosevelt 50, B-1050 Bruxelles, Belgium

Abstract. Canonical correlation analysis studies associations between two sets of random variables. Its standard computation is based on sample covariance matrices, which are however very sensitive to outlying observations. In this note we introduce, discuss and compare different ways for performing a robust canonical correlation analysis. Two methods are based on robust estimators of covariance matrices, the others on projection-pursuit techniques.

Keywords. Canonical correlation, minimum covariance determinant estimator, projection-pursuit, robustness.

1 Introduction

The aim of Canonical Correlation Analysis (CCA) is to identify and quantify the relations between a p-dimensional random variable $\boldsymbol{X}$ and a q-dimensional random variable $\boldsymbol{Y}$. Herefore we look for linear combinations $a^t\boldsymbol{X}$ and $b^t\boldsymbol{Y}$ of the original variables having maximal correlation. Expressed in mathematical terms, CCA seeks for vectors $\alpha \in I\!R^p$ and $\beta \in I\!R^q$ such that

$$(\alpha, \beta) = \underset{a,b}{\operatorname{argmax}} |\operatorname{Corr}(a^t\boldsymbol{X}, b^t\boldsymbol{Y})|. \tag{1}$$

The resulting univariate variables $U = \alpha^t\boldsymbol{X}$ and $V = \alpha^t\boldsymbol{X}$ are then called the *canonical variates* and can be used for dimension reduction and graphical display. Note that the vectors α and β are only determined up to a constant factor by definition (1). The first canonical correlation ρ is defined as the absolute value of the correlation between the two canonical variates, which equals the maximum attained in (1).

Higher order canonical variates and correlations are defined as in (1), but now under the additional restriction that a canonical variate of order k, with $1 < k \leq \min(p, q)$, should be uncorrelated with all canonical variates of lower order. Due to space limitations, we restrict attention to a first order canonical analysis.

The above CCA problem (1) has a fairly simple solution (see e.g. Johnson and Wichern, 1998, Chapter 10). Denote by Σ the population covariance matrix of the random variable $\boldsymbol{Z} = (\boldsymbol{X}^t, \boldsymbol{Y}^t)^t$. We decompose Σ as

$$\Sigma = \begin{pmatrix} \Sigma_{xx} & \Sigma_{xy} \\ \Sigma_{yx} & \Sigma_{yy} \end{pmatrix}.$$

The vectors α and β are now the eigenvectors corresponding to the largest eigenvalues of the matrices

$$\Sigma_{xx}^{-1}\Sigma_{xy}\Sigma_{yy}^{-1}\Sigma_{yx} \quad \text{and} \quad \Sigma_{yy}^{-1}\Sigma_{yx}\Sigma_{xx}^{-1}\Sigma_{xy}. \tag{2}$$

Both of the above matrices have the same positive eigenvalues and the largest one equals the squared first canonical correlation.

For estimating the unknowns α, β, and ρ one typically computes the sample covariance matrix $\hat{\Sigma}$ from a sample $z_1, \ldots, z_n$, with $z_i = (x_i^t, y_i^t)^t \in I\!R^p \times I\!R^q$. Computing eigenvectors/values of the empirical counterparts of the matrices in (2) results then immediately in estimates of the canonical variates and correlations. Since the classical estimator of a covariance matrix is very vulnerable with respect to outlying observations, also the eigenvalues and -vectors based on $\hat{\Sigma}$ will be very sensitive, as was shown in the context of CCA by Romanazzi (1992). In Section 2 of this paper an approach based on robust estimators of the covariance matrix is outlined and illustrated with a real data example. Two other approaches to robust CCA, which are in the spirit of projection-pursuit, are proposed and discussed in Section 3. They are compared by means of a modest stability study in Section 4. A more comprehensive study of the different approaches is part of current research of the authors.

2 Robust CCA based on robust covariance matrices

2.1 Using the minimum covariance determinant estimator

The obvious way for robustifying CCA is to estimate Σ robustly and to compute eigenvectors/values from the estimated version of (2) in the usual way. Theoretical results for this approach have been obtained by Croux and Dehon (2000). As robust covariance estimators one could use M-estimators as in Karnel (1991), but it is known that these estimators have poor robustness properties in higher dimensions. A more appropriate choice is the *Minimum Covariance Determinant* (MCD) estimator of Rousseeuw (1985). The MCD estimator is obtained by looking for that subset of size h of the data which has the smallest value of the determinant of the empirical covariance matrix computed from it. Maximal robustness is obtained for $h \approx n/2$. The resulting estimator is then nothing else but the covariance matrix computed over that optimal subset. An efficient algorithm for computing the MCD estimator has been proposed by Rousseeuw and Van Driessen (1999).

Robust covariance matrix estimators can routinely be used in multivariate statistics. Filzmoser (1999) applied them for robust factor analysis of geostatistical data.

To illustrate the usefulness of a robust CCA, we applied the MCD-based approach to the "Diabetes data" (Andrews and Herzberg 1985, data set 36, page 215). For a group of $n = 76$ normal persons, the variables Glucose intolerance (X_1), Insulin Response to Oral Glucose (X_2), Insulin Resistance (X_3), Relative Weight (Y_1) and Fasting Plasma Glucose (Y_2) were measured. It is of medical interest to establish a relation between the X and the Y variables. The classical estimators for the eigenvectors are $\hat{a}^{cl} = (-0.32, 0.47, -1.04)^t$ and $\hat{b}^{cl} = (0.98, 0.06)^t$, compared to $\hat{a}^{rob} = (-0.11, 0.46, -1.06)^t$ and $\hat{b}^{rob} = (0.83, 0.36)^t$ for the robust estimates. From these estimates, a scatter plot of the scores of x_i and y_i $(1 \leq i \leq n)$ on the first canonical variates for X

and Y is constructed in Figure 1. Corresponding regression fits are indicated and the estimates for ρ are given by $\hat{\rho}^{cl} = 0.50$ and $\hat{\rho}^{rob} = 0.71$. From this, we conclude that the robust estimate found a relation between the X and Y variables which is well followed by a huge majority of the data. The classical approach tries to find an association between X and Y which is valid for all data points, outliers included, and thereby leading to a much weaker first canonical correlation. It was also verified that the robust first order canonical variates takes 97% of the total correlation between X and Y into account (defined as the first eigenvalue divided by the sum of all eigenvalues of the estimated matrices in (2)), while this was only 80% for the classical approach.

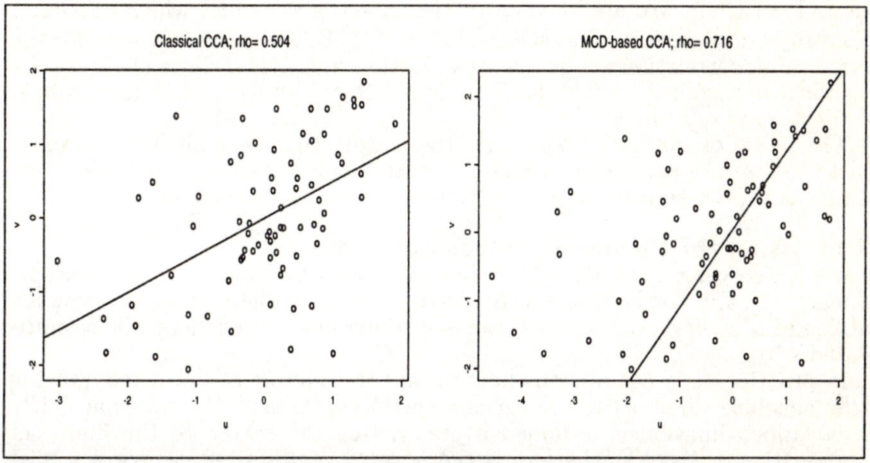

Fig. 1. Scatter plot of the first canonical variate b^tY versus a^tX for the classical estimator (left) and the MCD-based estimator (right).

2.2 Using the signs of the observations

The MCD-estimator requires that the number of observations is twice as much as the number of variables present in the data set. But in practical applications (e.g. in psychology, signal processing, etc.) it occurs often that one has a lot of variables but only a limited number of observations.

As an alternative, sign-based covariance matrices (Visuri, Koivunen, and Oja, 2000) can be used. For computing the sign of an observation x_i (and similarly for y_i), we first need to compute the spatial median μ_X of the data cloud $x_1, \ldots, x_n$. The spatial median is defined as

$$\mu_X = \underset{\mu}{\text{argmin}} \sum_{i=1}^{n} ||x_i - \mu||,$$

with $||\cdot||$ the Euclidean norm. The sign of x_i is then the projection of x_i on a unit sphere centered at μ_X:

$$S(x_i) = \frac{x_i - \mu_X}{||x_i - \mu_X||}.$$

Note that the signs are vector valued. By projecting observations on this unit sphere the influence of outliers is heavily reduced, leading to an outlier resistant procedure. Afterwards, ordinary covariance matrices are computed from these signs. Since fast iterative algorithms to compute μ_X and μ_Y exist and the signs can be computed in $O(n)$ time, the resulting procedure will be extremely fast. However, the statistical efficiency of the method can be quite low, since a lot of information is lost by only taking the direction of the data points into account.

3 Robust CCA based on projection-pursuit

Projection-Pursuit (PP) techniques for CCA start from the initial definition (1) of CCA. We are looking for two directions a and b which maximize the projection-pursuit index $|\mathrm{Corr}(a^t\boldsymbol{X}, b^t\boldsymbol{Y})|$. Taking as an estimate for the population correlation an ordinary correlation coefficient, yields of course the classical non robust approach. The idea is therefore to work with a robust projection-pursuit index.

The methods outlined below have the feature that they allow to compute just the first few canonical variates, without using an estimate of Σ. For high dimensions, where $p+q$ is huge, this is an important advantage.

3.1 Using the Spearman correlation index

Since the correlation index in (1) is between two univariate variables, a simple Spearman rank correlation can be used to measure the correlation between a^tX and a^tY. By working with ranks, the influence of outliers will be mitigated.

In practice, it is not obvious how to find the vectors α and β maximizing (the absolute value of) the Spearman correlation index. A simple and fairly good approximation is obtained by restricting the search to the finite set $\{(a_i, b_j)|1 \leq i, j \leq n\}$, where a_i is the normed vector $x_i - \mu_X$, μ_X a robust location measure of the X-population (e.g. the spatial median defined in Section 2.2 or the coordinate-wise median) and b_j the normed $y_j - \mu_Y$.

Although the non-parametric nature of the Spearman correlation is very appealing, the computational complexity of $O(n^3 \log n)$ may become prohibitive for bigger sample sizes.

3.2 Using robust alternating regressions

Application of the alternating regressing technique to CCA was already proposed by Wold (1966). Its use is motivated by the observation that, for a given α,

$$\beta = \underset{b}{\mathrm{argmax}}\, |\mathrm{Corr}(\alpha^t\boldsymbol{X}, b^t\boldsymbol{Y})|.$$

But then it follows from standard results on multiple regression, that β is proportional to the regression coefficient b in the model

$$\alpha^tX = b^tY + \gamma_1 + \varepsilon_1. \tag{3}$$

In the same way, for a given β, the optimal α equals (up to a scalar term) the parameter a in the regression model

$$b^tY = a^tX + \gamma_2 + \varepsilon_2. \tag{4}$$

Start now with an initial value α^0. (This can for example be obtained by performing a robust principal components analysis on the data matrix formed by $x_1, \ldots, x_n$.) According to (3), we get a first β^1 by regressing the univariate $X^t\alpha^0$ on the Y variables. Afterwards, using (4), an updated α^1 is obtained by regressing $Y^t\beta^1$ on X. This procedure is then iterated until convergence. The estimated regression coefficients are normalized in each step, and computing a bivariate (robust) correlation coefficient between the estimated canonical variates yields an estimator of ρ.

To be outlier resistant, the regression estimators in the above alternating regression scheme need to be robust. Since they are computed several times, a fast, but reliable estimator should be chosen. We propose to use a weighted L_1-estimator, as was motivated by Croux and Filzmoser (1998) in an application of alternating regressions to two-way tables.

4 Stability experiment

In this section all proposed methods of Section 2 are compared by a small statistical experiment. We generated a data set $Z = \{z_1, \ldots, z_n\}$ with $z_i = (x_i, y_i) \in I\!R^p \times I\!R^q$ from a specified $N(0, \Sigma)$ distribution, and with $n = 30$, $p = 2$, and $q = 3$. Since we are interested in the behavior of the estimators when outliers are present, we added randomly generated noise from $N(0, 50I_p)$ to the first 5 observations. This means that 16% of all observations are contaminated.

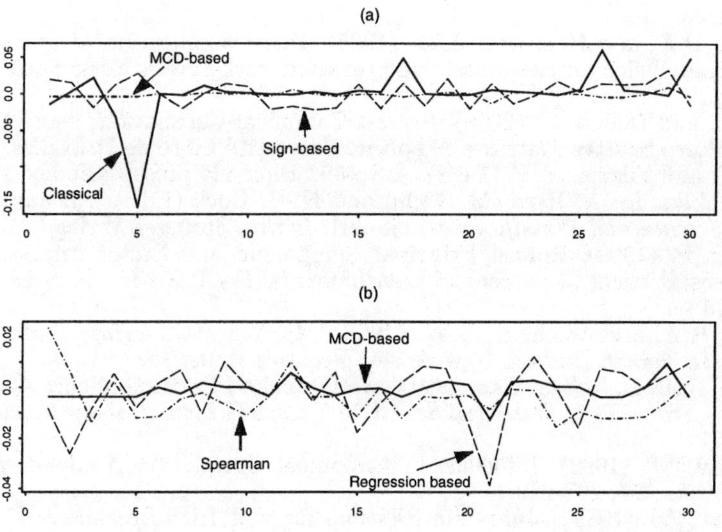

Fig. 2. Empirical influence functions for (a) the classical method, the MCD-based method and the Sign-based method; (b) the Spearman correlation based method, the robust alternating regression estimator and the MCD-based method.

A tool for assessing the robustness of an estimator is the *empirical influence function* (EIF). Here we compute the EIF of every observation z_i on the estimator $\hat{\rho}$ of the the first canonical correlation coefficient. By definition

$$\text{EIF}(z_i, \hat{\rho}) = \hat{\rho}(Z \setminus \{z_i\}) - \hat{\rho}(Z),$$

so it measures the effect of deleting the observations z_i on the estimator. In the spirit of robustness, we do not want single points to have a too high influence on the estimator. Figure 2a plots $\text{EIF}(z_i, \hat{\rho})$ versus the index of each observation for the classical estimator, the MCD-based estimator and the Sign-based estimator. The non robustness of the classical estimator appears by the extremely high value for the fifth point. The MCD-based estimator seems to be more robust for this example than the Sign-based method: the empirical influence function is indeed flatter. Figure 2b compares the EIF of the MCD with the two projection pursuit procedures. The MCD-based method remains the most robust, closely followed by the Spearman and the robust alternating regression method. Note the different scale of the vertical axis in Figure 2a and 2b.

As a final conclusion, making a choice between the available robust procedures is quite difficult at the present state. More theory needs to be developed and more practical experience is necessary. On the basis of several experiments we performed, it looks as if the MCD-based procedure is performing quite well in cases were the number of observations is high enough. Otherwise, the robust alternating regression method is an alternative.

References

Andrews, D.F. and Herzberg, A.M. (1985). Data: a collection of problems from many fields for the student and research worker. New York: Springer-Verlag.

Croux, C. and Dehon, C. (2000). Robust Canonical Correlations using High Breakdown Scatter Matrices. Preprint, Université Libre de Bruxelles.

Croux, C. and Filzmoser, P. (1998). A Robust Biplot Representation of Two-way Tables. In: A. Rizzi, M. Vichi, and H.-H. Bock (Eds.): *Advances in Data Science and Classification*, 355-361. Berlin: Springer-Verlag.

Filzmoser, P. (1999). Robust Principal Component and Factor Analysis in the Geostatistical Treatment of Environmental Data. *Environmetrics*, **10**, 363-375.

Johnson, R.A. and Wichern, D.W. (1998). *Applied Multivariate Statistical Analysis: Fourth Edition.* New Jersey: Prentice Hall.

Karnel, G. (1991). Robust Canonical Correlation and Correspondence Analysis. *The Frontiers of Statistical Scientific Theory & Industrial Applications*, 335-354.

Romanazzi, M. (1992). Influence in Canonical Correlation Analysis. *Psychometrika*, **57**, 237-259.

Rousseeuw, P.J. (1985). Multivariate Estimation with High Breakdown Point. In: W. Grossmann et al. (Eds.): *Mathematical Statistics and Applications, Vol. B*, 283-297. Dordrecht: Reidel.

Rousseeuw, P.J. and Van Driessen, K. (1999). A Fast Algorithm for the Minimum Covariance Determinant Estimator. *Technometrics*, **41**, 212-223.

Visuri, S., Koivunen, V., and Oja, H. (2000). Sign and Rank Correlation Matrices. *Journal of Statistical Planning and Inference.* To appear.

Wold, H. (1966). Nonlinear Estimation by Iterative Least Squares Procedures. In: F.N. David (Ed.): *A Festschrift for J. Neyman*, 411-444. New York: Wiley and Sons.

Graphical and phase space models for univariate time series

Roland Fried

Department of Statistics, University of Dortmund, 44221 Dortmund, Germany

Abstract. There are various approaches to model time series data. In the time domain ARMA-models and state space models are frequently used, while phase space models have been applied recently, too. Each approach has got its own strengths and weaknesses w.r.t. parameter estimation, prediction and coping with missing data. We use graphical models to explore and compare the structure of time series models, and focus on interpolation in e.g. seasonal models.

Keywords. Time series analysis, ARMA models, phase space embedding, conditional independence, missing data

1 Introduction

New technical possibilities for data collection and data storage allow to collect huge amounts of data subsequently in time. In ecology, e.g., the resulting time series often contain a lot of missing data and outliers. Hence, iterative and approximative procedures have to be applied for model specification (Delicado and Justel, 1999, Lewellen and Vessey, 1999, Fried, 2000).

In time series analysis ARMA and state space models are widely used. Furthermore, there are specialized methods such as structural time series models. It is hard to decide without further knowledge which model class should be applied. The problems mentioned above may result in high computational costs and an increase of the risk of model misspecification.

Especially for multivariate time series, the use of the concept of conditional independence and graphical models has been suggested repeatedly (e.g. Smith, 1992, Kjaerulff, 1995). The graphical visualization enhances the comprehension of the model structure and facilitates the communication between researchers and statisticians. Structural equivalences and differences of models can be recognized in this way (Wermuth and Lauritzen, 1990).

Here we use graphical models to compare univariate ARMA models and phase space models. An application of the latter class to time series in intensive care was given in Bauer et al. (1998).

2 Graphical models

Graphical interaction models have become an important tool for analyzing multivariate data. They are based on the concept of conditional independence (Dawid, 1979). Hypothetical relationships between the variables are illustrated by graphical means: The vertices of the graph represent the variables, whereas the edges indicate relationships between them. Directed edges (*arrows*) express directed influences, while undirected edges (*lines*) stand for symmetrical associations. Introductions to graphical models and proofs of the following statements can be found in Cox and Wermuth (1996) and in Lauritzen (1996). In the following let $\{Y_t : t \in V\}$ be a family of random variables with joint probability distribution P and joint observation space $\mathcal{X}$. The variables are often denoted briefly with their indices, i.e., small letters.

2.1 Graphical models on undirected graphs

An undirected graphical model can be specified through a system of subsets ("cliques") $\{C \subset V\}$ and factor potentials ψ^C for each of these subsets . These are non-negative functions on $\mathcal{X}$ which depend on $y \in \mathcal{X}$ only through the components y_C corresponding to C. The system $\{C \subset V\}$ can be visualized in form of a graph G with a vertex for each random variable $Y_t, t \in V$, and a line connecting all pairs (Y_s, Y_t) belonging to a clique.

Under the simplifying basic assumption that P has a positive continuous density w.r.t. a product measure μ, P factorizes according to G, i.e.,

$$P = f \cdot \mu \qquad \text{where} \qquad f(y) = \prod_C \psi^C(y_C),$$

iff P has the global Markov property w.r.t. the conditional independence graph G. This means, that $A, B \subset V$ are conditionally independent given $S \subset V$, whenever S separates A from B in G, i.e., any path (finite sequence of connected variables) between variables $a \in A$ and $b \in B$ necessarily contains at least one $s \in S$. This permits to relate the concept of conditional independence to the pleasant concept of graph separation.

2.2 Graphical models on directed graphs

Since time is considered to have a direction, directed graphical models appear more natural for time series analysis. We assume that P has a strictly positive density w.r.t. a product measure μ. A directed graphical model can be specified via a recursive factorization of P based on a partial ordering of the variables. We have to specify kernels $k^v : \mathcal{X}_v \times \mathcal{X}_{pa(v)} \mapsto \mathrm{R}^+$, $v \in V$, $\int_{\mathcal{X}_v} k^v(y_v, y_{pa(v)})\mu_v(dy_v) = 1$, such that

$$P = f \cdot \mu, \qquad \text{where} \qquad f(y) = \prod_{v \in V} k^v(y_v, y_{pa(v)})$$

and $pa(v)$ is the set of parents of v, i.e., the set of larger elements w.r.t. the applied ordering. A visual representation of this model in form of a directed acyclic graph ("dag") G is achieved by drawing a vertex for each random variable $Y_t, t \in V$, and a directed edge (arrow) pointing towards t for each pair (Y_s, Y_t) with $s \in pa(t)$. The *moral graph* G^m of a dag is the undirected graph which is derived from G by marrying parents, i.e., connecting all pairs of variables with a common child, and neglecting directions.

Under the stated assumptions P admits a recursive factorization iff it has the directed global Markov property with respect to G. This means, that $A, B \subset V$ are conditionally independent given $S \subset V$ whenever S separates A and B in $(G_{An(A \cup B \cup S)})^m$, the moral graph attained from G after eliminating all variables which are not parent of any of the variables in $A \cup B \cup S$ or elements of this union themselves. If P has these equivalent properties, it factorizes w.r.t. the moral Graph G^m (see section 2.1).

3 Time series models

Graphical models can be used to inspect the structures and properties of time series models. We consider especially ARMA-models and phase space models. Let $y_1, \ldots, y_N$ be a time series arising from a stochastic process $\{Y_t : t \in Z\}$.

3.1 ARMA-models

An ARMA model formally resembles a multiple regression. A stationary stochastic process $\{Y_t : t \in Z\}$ is called an ARMA(p,q) process with non negative integers p and q, if we have for every $t \in Z$

$$Y_t = \phi_1 Y_{t-1} + \ldots + \phi_p Y_{t-p} + \theta_1 W_{t-1} + \ldots + \theta_q W_{t-q} + W_t, \qquad (1)$$

where $\phi_1, \ldots, \phi_p, \theta_1, \ldots, \theta_q$ are unknown coefficients measuring the influence of the past variables on Y_t. The variables $W_t, t \in Z$, are assumed to be i.i.d. $N(0, \sigma^2)$-distributed random variables ("shocks"). The orders p and q of the process can be identified via the sample autocorrelations and partial autocorrelations. For $q = 0$ the process is called AR(p)-process and for $p = 0$ it is called MA(q)-process (for details see Box et al., 1994).

A graphical representation of an ARMA(p,q)-process can be found by drawing a (dashed) vertex for any variable Y_t (W_t) belonging to the period of interest. An arrow headed towards Y_t has to be drawn from all variables Y_i and W_j which appear in the equation for Y_t. This is due to the recursive factorization of the joint probability distribution implied by equation (1). Figure 3.1 shows graphs corresponding to AR(1)-, AR(2)- and MA(1)-processes. For AR(p) processes we can neglect the unobservables W_t from the graph, as W_t only has influence on Y_t.

Some basic properties of these models can be recognized easily from the corresponding moralized graphs also shown in Figure 3.1. In the AR(1)-model each value is conditionally independent of the past (future) given its direct precessor (successor). For the AR(2)-model we need to know the two previous (subsequent) observations. For a ARMA(p,q) model with $q > 0$ no such conditional independence exists as there always is a path between any two variables via unobservables W_t only. In consequence, we immediately see that optimal predictions are functions of the p recent observations in the AR(p)-model, whereas all past observations have to be included when $q > 0$.

Fig.1. Graphical representations of ARMA models

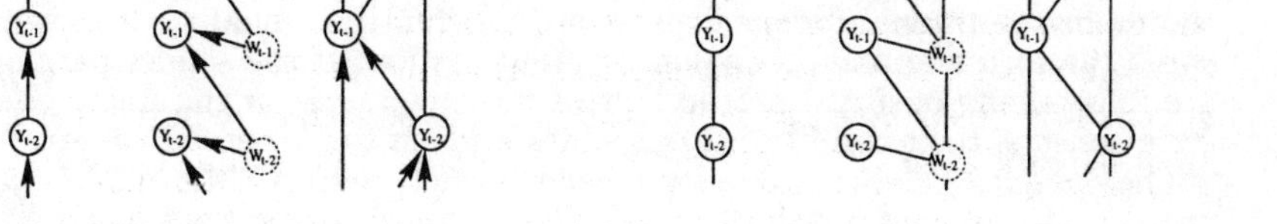

3.2 Phase space models

Phase space models are based on the phase-space embedding of the time series in a Euclidean space (Packard et al., 1980, Takens, 1980):

$$\mathbf{y}_t := (y_t, y_{t+T}, y_{t+2T}, \dots, y_{t+(m-1)T}) \in \mathrm{R}^m, t = 1, \dots, n-(m-1)T.$$

Here, m and T are suitably chosen non negative integers, namely the dimension of the embedding and the time delay. By this embedding, statistical techniques originally developed for multivariate data such as multivariate outlier identifiers (Becker and Gather, 1999) can be transferred into the time series context. Bauer et al. (1999) recommended to choose $T = 1$, and m corresponding to the partial autocorrelation structure of the process.

We put assumptions for linear stationary phase space models with embedding dimension m (PS(m)) in concrete terms as follows. We define the cliques of an undirected graphical model as the subsets of variables corresponding to each of the phase space vectors. We get a graphical representation of $\mathbf{Y}_t \sim N(\boldsymbol{\mu}, \Sigma)$ by connecting all variables via undirected edges. The graphs of the resulting time series models for the embedding dimensions $m = 2$ and $m = 3$ are the same as the moral graphs for the AR(1) and the AR(2) model respectively in Figure 3.1. In consequence, the conditional independence structure implied by a phase space model with embedding dimension m as defined above is the same as for an AR(m-1)-process.

4 Parameter restrictions

In ARMA models with high orders, e.g. in seasonal models, often some of the coefficients are set to zero. The aim is to eliminate missing influences from the model to avoid overparameterization by parameter restrictions. We could e.g. be interested in the AR(4) model

$$Y_t = \phi_1 Y_{t-1} + \phi_4 Y_{t-4} + W_t,$$

i.e., $\phi_2 = \phi_3 = 0$ for describing quarterly measured data.

For defining missing interactions in the PS formulation, however, we would use restrictions such as $\sigma^{i,j} = 0$ for $|i-j| = h \leq m-1$, where the $\sigma^{i,j}$ denote the concentrations, i.e., the elements of the inverse of Σ. In the example mentioned above we would specify $\sigma^{i,j} = 0$ for $|i-j| = 2, 3$. At first sight this assumption seems more complex as we are more familiar with coefficients in linear models than with concentrations.

For comparing both types of restrictions we look at the corresponding graphical representations (Figure 2). We recognize several facts. First, the conditional independence structures are not the same, the restrictions do not lead to the same subclasses of AR(4)-models. Second, the structure is much more complicated for the restrictions on the coefficients in the AR model than for the restrictions on the concentrations in the PS approach. We get much more conditional independences for the latter possibility. In general, the more conditional independences are inherent in a model, the easier become the formulae for prediction or interpolation: If the model parameters are known, the optimal interpolator of missing values in the mean squared error sense is their conditional expectation given the observed values.

The complex conditional independence structure in ARMA-models causes problems when a lot of values are missing. A couple of methods has been suggested to overcome this problem. Methods such as the factorization method,

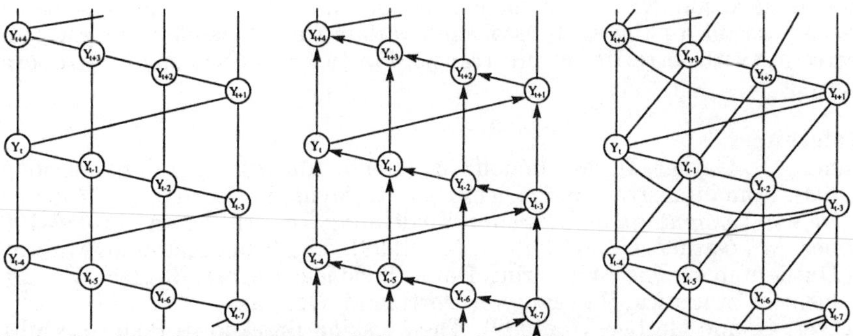

Fig. 2. Graphical interaction models for a restricted PS model (left) and a restricted AR model (middle) as well as the corresponding moralized graph.

the EM algorithm, Kalman filtering using a state space formulation, and filling in the holes with arbitrary values and treating them as additive outliers can be found in the literature (see, e.g., Ferreiro, 1987, and Gomez et al., 1999). In case of many missings an iterative procedure starting with simple estimators for filling the gaps and using "good" estimators thereafter may enhance the results (Delicado and Justel, 1999).

In comparison, for a PS model the assumed conditional independences result in more weights equal to zero in formulae for interpolation. However, maximum likelihood estimation of the model parameters is difficult when the graph is not triangulated. For such "non-decomposable" models explicit formulae for the maximum likelihood estimators usually do not exist. Dempster (1972) showed that the MLE of the covariance matrix of a multivariate normal distribution with zero restrictions for some of the concentrations has the same elements as the sample covariance matrix **S** at the non restricted positions. We can adapt this result to the sample of phase space vectors using $S_{ij} = \hat{\gamma}(i-j)$, the matrix of the sample autocovariances. Of course this is not equivalent to maximum likelihood estimation since our sample consists of dependent vectors. Nevertheless, some experiments with seasonal data show that this simple approach may sometimes outperform advanced statistical procedures based on the ARMA approach (Chen and Liu, 1993) w.r.t. the interpolation of a large amount of missing data. We used a FORTRAN subroutine written by Wermuth and Scheidt (1977) for our calculations.

5 Discussion

Graphical models can be used to illustrate critical model assumptions and influential variables for prediction and interpolation. They may be also useful for the identification and validation of a model in an iterative, computer-intensive data-based model search strategy.

Unrestricted autoregressive models of fixed order have the same conditional independence structure as appropriately defined stationary phase space models. However, as autoregressive models are naturally represented by directed graphs and phase space models by undirected graphs, parameter restrictions in form of missing interactions lead to distinct subclasses of both model types. Missing interactions in ARMA models correspond to zero weights of past variables, while missing interactions in PS models mean partial uncorrelatedness. Because of the underlying symmetry the subclasses of phase space models have simpler conditional independence structures. This may be especially useful for the analysis of multivariate time dependent data, e.g. in regression with correlated errors, transfer function models or spatial-temporal analysis.

References

Bauer, M., Gather, U. and Imhoff, M. (1998). Analysis of high dimensional data from intensive care medicine. In: R. Payne, P. Green (eds.): *Proceedings in Computational Statistics.* Heidelberg: Physica Verlag, pp. 185-190.

Bauer, M., Gather, U. and Imhoff, M. (1999). The Identification of Multiple Outliers in Online Monitoring Data. Technical Report 29/1999, Department of Statistics, University of Dortmund, Germany.

Becker, C. and Gather, U. (1999). The Masking Breakdown Point of Multivariate Outlier Identification Rules. *J. Amer. Statist. Assoc.*, **94**, 947-955.

Box, G.E.P., Jenkins, G.M. and Reinsel, G.C. (1994). *Time Series Analysis. Forecasting and Control. Third edition.* Englewood Cliffs: Prentice Hall.

Chen, C. and Liu, L.-M. (1993). Joint Estimation of Model Parameters and Outlier Effects in Time Series. *J. Americ. Statist. Assoc.*, **88**, 284-297.

Cox, D.R. and Wermuth, N. (1996). *Multivariate Dependencies.* London: Chapman and Hall.

Dawid, A.P. (1979). Conditional Independence in Statistical Theory. *J. Roy. Statist. Soc.* B, **41**, 1-31.

Delicado, P. and Justel, A. (1999). Forecasting with Missing Data: Application to Coastel Wave Heights. *J. Forecasting*, **18**, 285-298.

Dempster, A. (1972). Covariance Selection. *Biometrics*, **28**, 157-175.

Ferreiro, O. (1987). Methodologies for the Estimation of Missing Observations in Time Series. *Statistics & Probability Letters*, **5**, 65-69.

Fried, R. (2000). Exploratory Analysis and a Stochastic Model for Humusdisintegration. *Environmental Monitoring and Assessment*, (to appear).

Gomez, V., Maravall, A. and Peña, D. (1999). Missing Observations in ARIMA models: Skipping approach versus additive outlier approach. *J. Econometrics*, **88**, 341-363.

Harvey, A.C. and Pierse, R.G. (1984). Estimating Missing Observations in Economic Time Series. *J. Amer. Statist. Assoc.*, **79**, 125-131.

Kjaerulff, U. (1995). dHugin: a computational system for dynamic time-sliced Bayesian networks. *Int. J. Forecasting*, **11**, 89-111.

Lauritzen, S.L. (1996). *Graphical Models.* Oxford: Clarendon Press.

Lewellen, R.H. and Vessey, S.H. (1999). Analysis of Fragmented Time Series Data Using Box-Jenkins Models. *Commun. Statist. - Sim.*, **28**, 667-685.

Packard, N.H., Crutchfield, J.P., Farmer, J.D. and Shaw, R.S. (1980). Geometry from a time series. *Physical Review Letters*, **45**, 712-716.

Smith, J.Q. (1992). A Comparison of the Characteristics of Some Bayesian Forecasting Models. *Int. Statist. Rev.*, **60**, 75-87.

Takens, F. (1980). Dynamical Systems and Turbulence. In: *Vol. 898 of Lecture Notes in Mathematics.* Berlin: Springer-Verlag.

Wermuth, N. and Lauritzen, S.L. (1990). On Substantive Research Hypotheses, Conditional Independence Graphs and Graphical Chain Models. *J. Roy. Statist. Soc.* B, **52**, 21-72 (with discussion).

Wermuth, N. and Scheidt, E. (1977). Fitting a Covariance Selection Model to a Matrix. *Applied Statistics*, **26**, 88-92.

The use of the Tweedie distribution in statistical modelling

Robert Gilchrist and Denise Drinkwater

STORM, University of North London, London, N7 8DB. U.K.

Abstract. This paper discusses the estimation of the parameters of the so-called Tweedie distribution, $T_p(\mu, \sigma^2)$. Two special cases are considered, namely the Compound Poisson ($1 < p < 2$) and the Stable form ($p > 2$). The former is appropriate for data with a non-zero probability of zero observations and the latter is appropriate for data with a large dispersion. Our models will assume that we have data Y_i, $i = 1, ..., N$, with differing means μ_i, with common p and σ^2. The $T_p(\mu_i, \sigma^2)$ distribution can be characterised by $Var(Y_i) = \sigma^2 \mu_i^p$, $i = 1, ..., N$. In general, we shall model the μ_i in terms of explanatory variates x_{ij}, $i = 1, ..., N, j = 1, ..., m$. We discuss how it is straightforward to construct the maximum likelihood estimates of p, μ_i, and σ^2 in a GLM oriented computer package. The Tweedie distribution is used to model the alcohol consumption of British 16 and 17 year olds and randomised quantile residuals are used to validate the modelling.

Keywords. Tweedie, Compound Poisson, Consumption Data, Alcohol, GLM, GLIM

1 Introduction

In a recent paper (Gilchrist and Drinkwater, 1999), we discussed the use of the so-called Tweedie Compound Poisson distribution in the modelling of 'consumption' data. Such data are characterised by having a non-zero density for positive values, but with a point mass at the origin. Typical data which we have considered includes the consumption of alcohol by teenagers and the use of packaging materials by UK industrial companies. It might be appropriate to use our approach to model data on, e.g. rainfall, although some caution is needed for such data, as our assumption will be that all observations are independent.

The Tweedie distribution is a special case of the exponential dispersion family, as discussed by Jørgensen (1997). As such, the distribution fits nicely into the class of distributions which can be modelled in a generalised linear modelling package. We used GLIM4 (Francis *et al.*,1993).

For estimation purposes, the Tweedie distribution can be most easily characterised as the member of the exponential family for which the variance $Var(Y) = \sigma^2 \mu^p$, where $\mu = E(Y)$, and σ^2 and p are (unknown) parameters.

This paper initially discusses computational issues related to the estimation of these parameters for the Compound Poisson case ($1 < p < 2$). We show that it is straightforward to find their maximum likelihood estimates. We also discuss the residual analysis for such models. Raw residuals from a Tweedie model are highly skewed, so other approaches are needed.

This paper also discusses the extension of the estimation technique to the case where $p > 2$. The density of this extended form of the distribution is effectively of the same form as for the Tweedie Compound Poisson case , but new computational problems arise with the evaluation of the infinite sums. This distribution (a so-called Stable distribution) can be considered useful to provide a family to model the dispersion in a so-called Mean and Dispersion Additive Model (Rigby and Stasinopoulos, 1996).

2 Modelling with the Tweedie distribution

The so-called Tweedie distribution perhaps dates back to Tweedie (1947). There is a full discussion in Jørgensen (1997). We here distinguish between the cases where $1 < p < 2$ and $p > 2$ (note that $p = 2$ corresponds to the Gamma distribution).

The density of the so-called reproductive form of the Tweedie distribution, denoted by $\mathrm{T}_p(\mu, \sigma^2)$, can be written in the following notations (Jørgensen, 1997):

For the Compound Poisson Tweedie, where $y \geq 0$, $1 < p < 2$,

$$P_Y(y, \theta, \lambda, k) = C_p(y, \lambda) \exp[\lambda(\theta y - \kappa_p(\theta)] \tag{1}$$

$$\text{where } C_p(y,\lambda) = \tfrac{1}{y}\textstyle\sum_{r=1}^{\infty} \dfrac{\lambda^r \kappa_p^r(\frac{-1}{\lambda y})}{\Gamma(kr) r!}, \qquad y > 0,$$

$$= 1 \qquad y = 0.$$

For the Stable Tweedie, where $y > 0$, $p > 2$,

$$P_Y(y, \theta, \lambda, k) = \lambda C_p^*(y, \lambda) \exp[\lambda(\theta y - \kappa_p(\theta)] \tag{2}$$

$$\text{where } C_p^*(y,\lambda) = \tfrac{1}{\pi y}\textstyle\sum_{r=1}^{\infty} \dfrac{\Gamma(1-kr)}{r!} \lambda^r \kappa_p^r(\tfrac{-1}{\lambda y}) \sin(r\pi k)$$

In this notation, θ and k are used where $\mu = \kappa_p'(\theta)$, $p = (k+2)/(k+1)$, with $\kappa_p(\theta) = \frac{1}{2-p}[\theta(1-p)]^{(p-2)/(p-1)}$. The 'scale' σ^2 is given by $\sigma^2 = 1/\lambda$.

In the regression context here considered, the θ (and associated μ) are different for each y (i.e. we have θ_i and μ_i where $\mu_i = \kappa_p'(\theta_i), i = 1, ..., N$).

In general, we shall have N independent observed observations y_i, $i = 1, ..., N$, so there are potentially $N + 2$ parameters to estimate. However, we shall usually be in a linear modelling environment where we wish to model the unknown $\mu_i = E(Y_i)$ in terms of m parameters β_j, $j = 1, ..., m$. More specifically, as usual, we assume $g(\mu_i) = \sum_j x_{ij}\beta_j$, for known explanatory variates x_{ij}, $i = 1, ..., N$, $j = 1, ..., m$, where $g(.)$ is a known function. The function g is referred to as the 'link' function and might typically be the identity or logarithmic function. Thus, in modelling with a Tweedie distribution we shall need to estimate the $m + 2$ parameters σ^2, p and β_j, $j = 1, ..., m$.

3 Estimation of the parameters of the Tweedie compound Poisson distribution

The Tweedie Compound Poisson distribution can be thought of as being generated from a (Poisson) random number of events (uses), with each use following a Gamma distribution. Moreover, the mean of independent Tweedie variables also follows the Tweedie distribution (the *convolution* theorem). The physical interpretation and convolution theorem of the Tweedie distribution would seem to indicate that it might provide a reasonable approximation to reality for mean consumption (e.g. mean daily consumption of alcohol). Our estimation approach is direct and straightforward to implement in GLIM. The method appears not to have been used by others and is as follows:

The maximum likelihood estimate for λ (it is convenient for us to estimate $\lambda = 1/\sigma^2$) is given by

$$\lambda^{-1} = \frac{\frac{1}{k}\sum_{i=1}^{N} \mu_{y_i}^{\frac{k}{k+1}} + \sum_{i=1}^{N-n} \frac{y_i}{\mu_{y_i}} \mu_{y_i}^{\frac{k}{k+1}}}{\sum_{i=1}^{N-n} \frac{A(y_i)}{B(y_i)}}, \tag{3}$$

where

$$A(y_i) = \sum_{1}^{\infty} \frac{\{[\lambda(k+1)]^{k+1} y_i^k\}^r}{k^r \Gamma(rk)(r-1)!} \tag{4}$$

and

$$B(y_i) = \sum_{1}^{\infty} \frac{\{[\lambda(k+1)]^{k+1} y_i^k\}^r}{k^r \Gamma(rk) r!}. \tag{5}$$

Thus our maximum likelihood procedure is to first fix p and then find the maximum likelihood estimates of the μ_i. (A major bonus of the GLM formulation is that these estimates of the μ_i can be found without knowledge of σ^2). Then with some initial starting values for λ (e.g. the mean residual deviance), equation 3 can be used iteratively to find the m.l.e. of λ. The size of the (profile) likelihood can then be calculated for this value of p. This is then repeated in suitably sized steps over the possible range of p (i.e. $1 < p < 2$) and the overall maximum of the (profile) likelihood found, finally giving us the maximum likelihood estimates of p, σ^2 and the μ_i, $i = 1, 2, ..., N$. The algorithm is fast and seems to work well.

4 Example: Teenage alcohol consumption data

As an example of the use of the Compound Poisson Tweedie, we now discuss the modelling of teenage alcohol consumption data. A rationale for the use of this distribution for 'consumption' data comes from one of its characterisations. Using alcohol consumption as an example, consider a person who has a (Poisson) number of events on which s/he consumes alcohol, with a Gamma distribution for the amount drunk on each occasion. Then the resulting overall consumption by the individual follows the Tweedie distribution.

The alcohol consumption data were collected in 1986/7 from the 1970 longitudinal birth cohort study 16-17 year 'Youthscan' survey (see Crawley and While 1996) . Data were collected on diet, body size, activity and a number of cross-sectional demographic, personal and lifestyle variables of both the individual and the parent of the individual. There were 1545 individuals (Males=658, females=887).

The data in which we have a special interest are the reported alcohol intakes (in grams per day) collected from the dietary diary. 26 explanatory variables (outlined above) were included in the analysis.

An important characteristic of the data Y_i is that there is a non-zero probability that Y_i is zero, with a continuous density for positive Y_i.

The data were analysed for all respondents together with gender as an explanatory variable. The data contained $n = 882$ non-drinkers (57%).

4.1 Compound Poisson Tweedie model results

For the Tweedie model the parameter estimates for the best linear predictor (with log link) are shown in Table 1.

Table 1: Parameter Estimates for the Tweedie Model

parameter	estimate	s.e.
constant	0.128	0.185
eats out	0.516	0.129
smoker	0.287	0.135
has 3 or more take-aways per week	0.298	0.105
female	-0.637	0.151
watches TV 1-2hr	-0.350	0.120
watches TV >2hr	-0.279	0.097
Not dieting	-0.501	0.095
modern family/affluent housing	0.235	0.086
Living in Scotland	-0.352	0.133
Female smoker	0.520	0.181
Female 17 year old	0.522	0.142

In the above table, the parameters in the linear predictor are factors indicating that an individual has the trait in question. As an illustration of the use of these parameters, consider 'eats out'. A teenager who eats out has a contribution of 0.516 to his/her linear predictor (where consumption is exp(linear predictor)). A teenager who does not eat out has no such contribution to his/her linear predictor. Thus it may be observed that the consumption of alcohol increases if the individual eats out. If we consider throughout that the other variables are kept fixed, someone who eats out drinks 67% (exp(0.516) * 100) more.

Similar discussion applies to other variables; positive parameters indicating increased alcohol consumption and negative parameters indicating lower alcohol consumption. We also calculated the following 95% confidence intervals for p and σ^2 directly from the profile likelihood, using a χ^2 approximation (see, e.g. Francis *et al.*; 1993, page 289).

Table 2: Parameter Estimates and Confidence Intervals for p and σ^2

Parameter	Estimate	Confidence Interval.
p	1.41	1.37, 1.46
σ^2	3.47	3.40, 3.54

4.2 Diagnostics

Model diagnostics were considered by way of a residual analysis and by calculation of Cook's Distances (Cook and Weisberg, 1982.). Raw residuals from a Tweedie distribution are highly skewed and the modified deviance residuals, r^* suggested by Jørgensen (1997) proved unsatisfactory. However, the characterisation of the $T_p(\mu, \sigma^2)$ distribution for $1 < p < 2$ as a Poisson sum of Gamma densities allows us easily to construct the cumulative distribution function (c.d.f) of $T_p(\mu, \sigma^2)$, $1 < p < 2$. Hence we were able to calculate the so-called random quantile residuals (see e.g. Dunn and Smyth, 1996). The normalised QQ plot (Fig 1) appeared quite linear, indicating an acceptable fitting model.

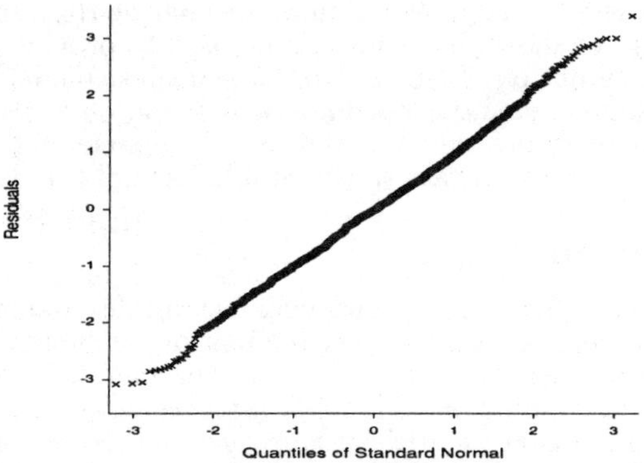

Fig. 1. Normal probability plot of residuals

We also successively removed observations with the largest observed Cook's distances and refitted the model. We checked the significance of these observations using an approximate χ^2_1 distribution and found that no observations had any significant influence on the model.

5 Estimation of the parameters of the Tweedie stable distribution

For $p > 2$ the Tweedie distribution can be considered useful to provide a family to model the dispersion in a so-called Mean and Dispersion Additive Model (Rigby and Stasinopoulos, 1996). Thus, when jointly modelling the mean and dispersion in a GLM context, an appropriate density from the GLM family is required for the dispersion modelling. For $p > 2$, the Tweedie distribution is effectively of the same form as for the Compound Poisson case ($1 < p < 2$) and hence essentially the same GLM algorithm may be used. Thus we again fix p (but $p > 2$) and estimate the μ_i within the GLM package. The only difference from the Compound Poisson case is that we need to replace $\Gamma(kr)$ by $\Gamma(1 - kr)\sin(r\pi k)$ in the infinite sum required to estimate σ^2. This makes convergence less rapid. We note that, when $p = 3$, the Tweedie distribution corresponds to the Inverse Gaussian distribution; in this case, the m.l.e estimate of σ^2 is given by the Deviance/N, which enables us to confirm the validity of our approach. Further numerical results will be given in the full oral presentation.

6 Conclusion

We have shown that the Tweedie $T_p(\mu, \sigma^2)$ distribution for $p > 1$ offers a flexible approach for two types of data, namely data with a non-zero probability of zero observations and data with large dispersion. Indeed, the combination of the Poisson, Gamma and Tweedie (Compound Poisson and Stable) provides a family with $Var(Y) = \sigma^2\mu^p$, for $p \geq 1$. Moreover it is well known (see, e.g. Jørgensen 1997) that there is no well-defined density of this family for $0 < p < 1$, so by combination with the Gaussian case ($p = 0$), our approach indicates how to model all well-defined densities within a GLM, for all $p \geq 0$. For the case with a non-zero probability of zero observations we have developed a user-friendly algorithm for estimating the parameters of the distribution and have shown how this can be used to provide understanding of the alcohol consumption of British 16/17 year olds. We have investigated the use of randomised quantile residuals and found that they indicate an acceptable fit. Furthermore we have indicated how to extend our technique from the Compound Poisson case to the Stable form of the density.

Acknowledgement

Our thanks are due to Dr Helen Crawley for helpful discussions on the data. Denise Drinkwater was supported by an ESRC Research studentship during the period of this research.

References

Cook, R.D. and Weisberg, S.(1982). *Residuals and Influence in Regression.* London: Chapman & Hall.

Crawley, H.F. and While, D. (1996). Parental smoking and the nutrient intake and food choice of British teenagers aged 16-17 years. *J of Epidemiology and Community Health.* **50:** 306-312.

Dunn, P.K. and Smyth, G.K. (1996). Randomized Quantile Residuals. *J.Comput. Graph. Statist.*, **5**, 236-244.

Francis, B., Green, M., Payne, C.(eds)(1993). *GLIM4 Manual*, Oxford: Oxford University Press.

Gilchrist, R. (2000). Regression models for data with a non-zero probability of a zero response. *Communications in Statistics Theory and Methods* 29(9). (to appear).

Gilchrist, R. and Drinkwater, D. (1999). Fitting Tweedie models to data with probability of zero responses. In *Statistical Modelling, Proceedings of the 14th International Workshop on Statistical Modelling*, eds. Friedl, H, Berghold, A, Kauerman, G., p207-214.

Jørgensen, B. (1997). *The Theory of Dispersion Models.* London: Chapman & Hall.

Rigby, R. and Stasinopoulos, D.(1996). Mean and Dispersion Additive Models. In: *Statistical Theory and Computational Aspects of Smoothing.* eds. Hardle, W. and Schimek, M.G., pp215-230. Heidelberg:Physica Verlag.

Tweedie, M.C.K. (1947). Functions of a statistical variate with given means, with special reference to Laplacian distributions. *Proc. Cambridge Phil. Soc*, **49:** 41-49.

Predictive dimension: an alternative definition to embedding dimension

Dominique Guègan[1] and Francesco Lisi[2]

[1] University of Reims UPRESA 6056 - CREST-ENSAE, Paris - France
[2] Department of Statistics - University of Padova - Italy

Abstract. In this paper we propose an alternative definition to the embedding dimension that we call predictive dimension. This dimension does not refer to the number of delayed variables needed to characterize the system but to the best predictions that can be obtained for the system. This kind of definition is particularly useful in a forecasting context because it leads to the same value of the traditional embedding dimension for chaotic time series and it is always finite for stochastic ones.

Keywords. Chaos, embedding dimension, prediction

1 Introduction

The ability of forecasting chaotic time series is strictly connected to a state space reconstruction problem. In this context, according to Takens theorem, the basic idea consists in creating a multidimensional state space from a univariate time series. When this operation leads to a diffeomorphism, the main features of the system are preserved and we are able to well predict its future evolution. To this purpose, the reference procedure is the so-called delay coordinates method, that uses delayed variables for the space reconstruction (Sauer *et al.*, 1991). The dimension of the reconstructed state space is called embedding dimension.

Several methods for embedding dimension estimation have been proposed in literature (Cutler, 1993; Takens, 1996). A lot of them identify the embedding dimension as the value beyond which some invariant feature of the system does not change when the embedding dimension increases. Some examples of these invariant features are the correlation dimension (Grassberger and Procaccia, 1983) and the ergodic density in the zero-one explosive method (Bosq *et al.*, 1999). Most of these methods, however, have a number of problems (Smith, 1992; Ramsey and Yuan, 1989). For example, they work unsatisfactory when data are contaminated by noise. In fact, the classical definition of embedding dimension does not have any meaning for stochastic processes, for example a white noise, because in this case it is not finite. The consequence is that for the computation of the embedding dimension we have to know that the system is deterministic. Furthermore, most of these methods work only in low-dimensional state spaces, because when the true dimension of the space increases linearly, the number of needed observations increases exponentially.

The aim of this work is looking for a statistic that allows us to estimate some kind of "embedding dimension" both in a fully deterministic context and in a stochastic one, which avoids the mentioned problems. To this purpose, we propose the concept of predictive dimension which is based on prediction and leads to the same results of the embedding dimension for deterministic systems but it is univocally defined also for stochastic processes.

2 The predictive dimension

From the state space reconstruction perspective the embedding dimension can be thought of as the number of delayed variables that we need to obtain a diffeomorphism of the original system.

Predictive dimension, instead, represents the number of lagged variables needed to obtain the best prediction for the system evolution. The underlying idea is that the best prediction is achieved when the "correct" dimension is selected; in this case a further increase of the space dimension will not lead to a significant improvement. This means also that, when the dimension increases, the mean absolute forecasting error (MAE) will decrease up to reaching some stable value. This approach is similar to that of Cao (1997), who simply looks for the number of delays that minimizes MAE. In general, however, this approach leads to overestimation of the correct dimension, with a risk of overfitting in the forecasting step. Thus, to avoid overestimation a stopping rule, based on the size of the reduction error, has to be introduced.

Let us now specify in more detail the definition of the predictive dimension. Suppose to have a time series $\{Y_t\}_{t=1}^{T}$, $Y_t \in \Re$; cut it in two parts: the first nf observations, and the last $T - nf = nt$. Then, based on the first part, make a number of one-step-ahead predictions on the second part with a "good" method. In this paper we use the nearest neighbours method (Farmer and Sidorowich, 1987; Casdagli, 1992; Guègan and Mercier, 1997), which is based on two parameters: the number of neighbours k, and the embedding dimension d. This method is quite general because it approximates a nonlinear dynamics by local linear models. The choice of nf is not very important provided that nf is not too small.

For each couple of parameters (d, k) belonging to a set of possible values we can obtain the one-step-ahead prediction errors $e_t(d,k) = \left|\hat{Y}_t(d,k) - Y_t\right|$, $(t = nf+1, ..., nf+nt)$, where $\hat{Y}_t(d,k)$ is the nearest neighbours prediction of Y_t made in $t-1$

For a set $\mathbf{S}$ of possible values of k and for a fixed d consider the prediction errors $e_t(d,k)$.

Using the nearest neighbours predictor, the predictive dimension is given by the smallest d for which

$$R(d) = |e(d) - e_{\min}| \leq \varepsilon * \sigma \qquad d \in \{0, 1, 2, ..., D\}$$

where σ is the standard deviation of the data, $\varepsilon > 0$ is a constant and

$$e(d) = \min_{k \in \{\mathbf{S}\}} \left\{ \frac{1}{nt} \sum_{t=1}^{nt} |e_t(d,k)| \right\},$$

nt being the total number of one-step-ahead predictions. Finally

$$e_{\min} = \min_{d \in \{0,1,...,D\}} e(d)$$

For $d = 0$ the prediction is simply the mean of the data; in this case we have not a true nearest neighbours predictor.

This criterion chooses the value of d which leads to the best prediction neglecting improvements less than $\varepsilon * \sigma$. The parameter ε permits to follow the parsimonious principle, avoiding in this way overfitting problems. The choice of ε is quite subjective. Throughout the paper we have always chosen $\varepsilon = 0.05$, because this value resulted to give good performances.

In this framework we expect that the predictive dimension coincides with the classical embedding dimension for deterministic (chaotic) dynamical systems. For stochastic systems, instead, we expect that it coincides with the order of an autoregressive representation of the model. In general, it differs from an identification problem because only autoregressive structures are considered.

Furthermore, with this formulation all processes should have finite predictive dimension. In fact, in practice, there will always be an order d such that a local linear model of order greater than d produces only marginal and not significant predictive improvements. In particular, for a white noise we expect the estimated predictive dimension to be equal to zero because no prediction, based on a finite structure, is better than its average and thus no delayed variables are present in the estimated model.

Before starting the simulations we want to stress that, while predictive dimension can be considered as a different way for embedding dimension estimation, it does not allow to discriminate between stochastic and deterministic systems. Thus predictive dimension has to be seen as a complementary and not completely substitutive tool with respect to the classic approach based, for example, on the correlation dimension.

3 Simulation results

The method has been validated by simulation. To this purpose chaotic systems, noiseless and noisy, as well as linear and nonlinear stochastic processes have been considered.

For deterministic systems we chose the usual chaotic parameters, while for the stochastic ones we tried several values of parameters. In more detail, the predictive dimension has been calculated for 200 series of length $T = 1100$ generated by each process.

For the estimation $nt = 100$ one-step-ahead predictions, based on the first $nf = 1000$ observations, have been performed. These predictions allow us to estimate the predictive dimension by minimizing $R(d)$ with the parameters $\varepsilon = 0.05$ and $D = 10$. For each series the value of the estimated predictive dimension recorded and the final results are then expressed as percentage of the estimated predictive dimension which assumed a fixed value in the set $\{0, 1, ..., D\}$.

For the class of the chaotic systems two classical maps have been selected: the logistic map, $X_{t+1} = 4X_t(1 - X_t)$, and the Hénon system, $X_{t+1} = 1 - 1.4X_t^2 + Y_t$; $Y_{t+1} = 0.3X_t$. For this latter $X-$component has been considered.

Both clean and noisy data were taken into account. For noisy data an additive gaussian noise, with variance proportional to the variance of the data, was added.

For clean data the method performed excellently, selecting always the correct dimension. For noisy data it continues to work quite well up to a noise-to-signal ratio of 5% that seems to be a bound for a correct estimation. For example, with this level of noise, for the logistic map the predictive dimension has been estimated to be equal to one 93.5% of times. The complete results for the Hénon map are given in table 1. However, it is worth noting that in several papers the percentage of noise is given by the ratio between the standard deviations. If standard deviations are considered the percentage of noise for which the prediction dimension gives good results is, obviously, much higher than 5%. Among stochastic processes, we considered first a

Table 1. Results, expressed in percentage, for the Hénon map

Noise	d=0	d=1	d=2	d=3	d≥4
0%	0	0	100	0	0
1%	0	0	100	0	0
5%	0	0	94.5	5.5	0
10%	0	5	68	27	0

white noise. As expected, the predictive dimension is finite and equal to zero; for example, for a uniform white noise it resulted to be zero with a rate of 97% and for a gaussian white noise with a rate of 94.5%.

When dealing with autoregressive (AR) stochastic systems the predictive dimension resulted to be, essentially, the order of the autoregressive model. For example, for an AR(2) with $\phi_1 = 0.2$ and $\phi_2 = 0.7$ the predictive dimension was equal to two 99% of times . Obviously, the performances depend also on the intensity of the relation, that is on the value of the parameters. This point is illustrated in table 2, that shows that when the parameter of an AR(1) process decreases in absolute value, also the rate of correct selected orders does the same. Nonlinear autoregressive models, such as Threshold

Table 2. Results, expressed in percentage, for AR(1) models: $X_t = \phi X_{t-1} + \xi_t$

Parameters	d=0	d=1	d≥2
$\phi = 0.9$	0	100	0
$\phi = 0.7$	0.5	99.5	0
$\phi = 0.5$	9.5	90.5	0
$\phi = 0.3$	53.5	46.5	0

Autoregressive, behave similarly to linear AR models. In our simulations, for example, the series generated by the SETAR(1;1,1)

$$\begin{array}{lll} X_t = 0.9X_{t-1} + \xi_t, & \text{if} & X_{t-1} \leq 0.5 \\ X_t = 0.2X_{t-1} + \xi_t, & \text{if} & X_{t-1} > 0.5 \end{array}$$

led to a prediction dimension equal to one 100% of times.

For not purely autoregressive stochastic processes the situation is not always completely clear. In the case of moving average (MA) processes predictive dimension systematically overestimates the order of the model. This is not strange because we are using an autoregressive structure. For this reason, at first glance it might seem that, because an autoregressive structure is used and MA processes do not possess a finite autoregressive representation, predictive dimension might be infinite as well. However this is not the case and, in fact, predictive dimension never reaches the maximum value D. This is due to the fact that, from a predictive point of view, a finite truncation of an infinite sum is a sufficiently accurate approximation. Furthermore in general, we observe a PD that is close to the value at which the autocorrelation function becomes not significant.

A similar situation arises for nonlinear processes such as the bilinear ones. Also in this case there is not direct relation with the order of the process and we simply interpret PD as the number of lagged variables we need to obtain the best prediction in terms of local linear autoregression. In any case, however, in all considered cases, the predictive dimension resulted to be strictly less than D.

With respect to ARCH models, such as, for example, the ARCH(1)

$$X_t = h_t^{1/2}\xi_t, \qquad h_t = a_0 + a_1 X_{t-1}^2,$$

we obtained a behaviour similar to that of a white noise for X_t, and similar to an AR for X_t^2. This is not surprising because ARCH models generate uncorrelated signals and can be thought of as AR processes with respect to the squares of the observations.

Although, throughout the paper we used $T = 1100$, in several situations some hundreds of observations are enough. As an example, we considered the logistic map and the AR(1) process and we estimated the predictive dimension, again for 200 series, using the same parameters of the previous cases, but with a sample size $T = 250$ ($nf = 200$ and $nt = 50$). The results are reported in table 3 where we can see that predictive dimension still performs satisfactory.

Table 3. "Small" sample results (nt=200) for the logistic map and the AR(1) model.

Logistic	d=0	d=1	d=2	d≥3
0%	0	100	0	0
5%	0	96.4	4	0
AR(1)				
$\phi = 0.9$	0	100	0	0
$\phi = 0.7$	2	97	1	0
$\phi = 0.5$	5.9	93.1	1	0
$\phi = 0.3$	38	59.1	2.9	0

4 Conclusion

In this paper an alternative definition to embedding dimension has been proposed and its validity has been shown by simulations.

From a practical and computational point of view this method is interesting because:

i) it can be applied both in a deterministic and in a stochastic context, giving always a finite predictive dimension. While this feature is not useful to test for chaos, it is very pleasant in the forecasting step;

ii) in a fully chaotic context it essentially estimates the embedding dimension, giving good results also in the noisy case, at least when the noise is not too large. In this sense it can be also thought of as a competitive way to estimate embedding dimension, with the further appeal that it works quite well with relatively short time series avoiding the problems highlighted by Ramsey *et al.* (1990) concerning the number of points needed to correctly estimate the correlation dimension;

iii) being a prediction-based procedure, it is particularly useful when the final goal is forecasting. Our approach is similar to that proposed by Cao (1997). However, since Cao did not use any stopping rule, his method leads very often to too much large embedding dimensions, as is the case of Soofi and Cao (1999), who applied this method to financial time series, obtaining embedding dimension estimates up to 56. Dimensions which are so high would require so many data to be completely unuseful, and in fact they were not (and probably nobody will not be) able to obtain any interesting results.

Finally, an example of a successful application of this method can be found in Bordignon and Lisi (2000). To predict the daily mean discharge of an important italian river in a possibly chaotic context, Bordignon and Lisi used PD to select the number of delayed variables. This allowed the authors to significantly improve the prediction with respect to linear models.

References

Bordignon, S. and Lisi, F. (2000). Nonlinear analysis and prediction of river flow time series. To appear in *Environmetrics.*

Bosq, D., Guégan, D. and Leorat G. (1999). Statistical estimation of the embedding dimension of a dynamical system. *International Journal of Bifurcations and Chaos*, **9**, pp. 645-656.

Casdagli, M. (1992). Chaos and deterministic versus non-linear modelling. *Journal of Royal Statistical Society B*, **54**, pp. 303-328.

Cao, L. (1997). Practical method for determining the minimum embedding dimension of a scalar time series. *Physica D*, **110**, pp. 43-50.

Cutler, C. (1993). A review of the theory and estimation of fractal dimension. In: H. Tong (Ed.) *Dimension, Estimation and Models.* Singapore: World Scientific.

Farmer, J.D. and Sidorowich, J.J. (1987). Predicting chaotic time series. *Physical Review Letters*, **59**, pp. 845-848.

Grassberger, P. and Procaccia, I. (1983). Measuring the strangeness of strange attractors. *Physica D*, **9**, pp. 189-208.

Guégan, D. and L. Mercier (1997). Prediction in chaotic time series: method and comparisons using simulations. In: (eds.) Prochazka A., Hlir J.U. and Sovka, P. *Signal analysis in prediction I, ECSAP-97*, pp. 215-218.

Ramsey, J.B. and Yuan, H. (1989). Bias and error bias in dimension calculation and their evaluation in some simple models. *Physics Letters A*, **134**, pp. 187-197.

Ramsey, J.B., Sayer, C.L. and Rothman, P. (1990). The statistical properties of dimension calculation using small data sets: some economics applications. *International Economic Review*, **31**, pp. 991-1020.

Sauer, T., Yorke, J.A. and Casdagli, M. (1991). Embedology. *Journal of Statistical Physics*, **65**, pp. 570-616.

Smith, L.A. (1992). Estimating dimension in noisy chaotic time series. *Journal of Royal Statistical Society B*, **52**, pp. 329-351.

Soofi, A.S. and Cao, L. (1999). Nonlinear deterministic forecasting of daily Peseta-Dollar exchange rate. *Economics Letters*, **62**, pp. 175-180.

Takens, F. (1996). Estimation of dimension and order of time series. In: *Progress in Nonlinear Differential Equations and their Applications*, **19**, 405-422. Birkhäuser: Verlag Basel.

Post-stratification to correct for nonresponse: classification of ZIP code areas

Mark Huisman

Department of Statistical Methods, Statistics Netherlands, P.O. Box 4000, 2270 JM Voorburg, The Netherlands.

Abstract. Presently used weighting procedures for the Dutch Labor Force Survey are suspected not to correct sufficiently for bias due to nonresponse. In this paper, a new correction procedure is proposed that is based on post-stratification. Starting from small geographical units, i.e., ZIP code areas, homogeneous clusters of individuals are sought. For this purpose a two-stage strategy for cluster-analyzing the large number of ZIP code areas is presented. Because hierarchical clustering procedures cannot be applied to large numbers of objects, initial clusters of objects are sought by categorizing the cluster variables, in the first stage of the procedure. In the second stage, these initial clusters are further combined with an hierarchical procedure, resulting in a final classification. The two-stage procedure is applied to classify the ZIP code areas in the Netherlands with respect to socio-economic variables.

Keywords. Two-stage cluster analysis, large data sets, post-stratification, labor market.

1 Introduction

The Dutch Labor Force Survey (DLFS) is one of the major surveys of Statistics Netherlands. In the past years, however, the response rates were only 40 to 50%. The large amount of nonresponse results in less precise estimates of the main population characteristics which the DLFS is supposed to measure (e.g., the unemployment rate). Moreover, the nonresponse is suspected to be selective with respect to the position of the respondents on the labor market. This means that the results of the DLFS may be biased due to systematic differences between respondents and nonrespondents, and procedures are needed to correct for the selectiveness of the nonresponse.

A well-known and frequently used correction method is *post-stratification* (Bethlehem, 1988). With this technique, the population is divided into strata on the basis of auxiliary information that is related to the nonresponse. Identical weights are assigned to all elements in each stratum, and the weighted cases are analyzed. To reduce the nonreponse bias as much as possible, homogeneous strata have to be constructed. The presently used weighting procedures for the DLFS are suspected not to be completely succesful in reducing the bias sufficiently, because the strata are not strongly related to variables that influence the response. Therefore, a new stratification is sought. In this paper, we present a procedure consisting of three steps, to construct a new stratification for the DLFS (cf. Huisman, 1999). These three steps are selection of *auxiliary information*, construction of *strata*, and computation of *adjustment weights*.

The new stratification for the DLFS is based on small geographical areas, i.e., ZIP code areas, which are assumed to be homogeneous with respect to the position of the inhabitants on the labor market. We seek homogeneous clusters of ZIP code areas, which will serve as post-strata to correct for the nonresponse in the DLFS. In the first step of the procedure, a set of cluster variables has to be collected, on which the clustering of the ZIP code areas should be based. The selection of relevant variables to be used in the geographical post-stratification procedure is described in Section 2.

In the second step, strata are constructed with an hierarchical cluster analysis based on the variables selected in the first step. Problems arise, however, with an hierarchical cluster analysis, because existing techniques and software cannot handle the large data set that contains the auxiliary information selected for every ZIP code area. To overcome this problem, a two-stage strategy to cluster-analyze large data sets was proposed by Huisman & Bethlehem (1999). We use this strategy to cluster the ZIP code areas, and both the process and the resulting stratification are presented in Section 3.

The third and final step in the construction of a new stratification to correct for nonresponse in the DLFS, consists of computing the adjustment weights in each new stratum. The new weighting scheme leads to new estimates of population characteristics, which should be compared with unweighted results and the results of the presently used weighting scheme. In this paper, however, we only present the description of the procedure to create a new stratification. The merits of the new weighting scheme for the DLFS is topic of future research. In Section 4 some general remarks are made about the use of the clusters of ZIP code areas for nonresponse correction in the DLFS.

In this paper, the focus of attention is on the clustering process described in Section 3. Existing computer software are not capable of cluster-analyzing large data sets. Therefore we present the two-stage procedure, together with an algorithm to cluster-analyze large data sets, described by Huisman & Bethlehem (1999), and use this procedure to classify the ZIP code areas. The paper ends with a discussion of the results of the classification.

2 Selecting cluster variables

One of the most important correction techniques for nonresponse is *adjustment weighting* (cf. Bethlehem, 1999). Every respondent in the survey is assigned a weight, and estimates of population characteristics are obtained by processing the weighted observations. Post-stratification is a frequently used weighting technique in which the population is divided into strata based on auxiliary information, i.e., variables that have been measured in the survey. Population and sample distributions of these variables are compared to determine whether nonresponse resulted in a selective sample.

Correction for nonresponse with post-stratification is most effective when the strata are homogeneous with respect to the variables of interest in the survey and/or to the probability of response (Bethlehem, 1988). The reduction of the bias will, however, only be complete when the nonresponse is *ignorable*, i.e., independent of the main variables given the auxiliary (cluster) variables. In this case the data are *Missing At Random* (MAR, cf. Little & Rubin, 1987). The nonresponse in the DLFS, is suspected to be nonignorable. This means that the data are not missing at random and missingness is related to the main variables in the survey. Therefore, we build a stratification based on auxiliary variables that are strongly related to the topic of interest, i.e., a respondent's position on the labor market. In this way the strata will be homogeneous with respect to both variables of interest and probability to respond, and nonresponse bias will be reduced substantially, although the bias cannot be completely removed.

The actual number of auxiliary variables plays an important role in the correction process. When it is unknown to what extent and in which way the nonrespons resulted in a selective sample, a large set of variables is most effective in reducing the nonresponse bias. A major disadvantage of a large set of variables, however, is the existence of (almost) empty strata when all variables and (a large part of) their interactions are included in the weighting scheme. These empty cells are caused by crossing the auxiliary variables in traditional schemes to construct strata. When there are no observations in a stratum, adjustment weights cannot be computed. This means that not all interactions between the variables can be used, or that even some variables cannot be used in the correction procedure, and information is thrown away.

Moreover, in traditional weighting schemes, auxiliary information is only useful when it has been (completely) measured in the survey and the population distribution is known (Bethlehem, 1999). This leaves out correction for unobserved (unknown) causes of nonresponse. By constructing strata in another way both problems may be solved. When strata are constructed by clustering the population, register information that is not available in the sample may be used, and all information in the cluster variables is taken into account. Still, we are bound to the restriction that clusters should be large enough to contain both (enough) population and survey members.

Data from registrations
Huisman (1999) described the 22 selected variables and the composition of the data set on which the cluster analysis is based. It contains information on the 397,692 ZIP code areas in the Netherlands related to labor market, and come from combined registers at Statistics Netherlands. Some of the variables are highly correlated. This can lead to unwanted results, because the common factor of the highly correlated variables would get a higher weight than the individual variables, and some (latent factors of) variables would have a larger influence on the clustering than others. Therefore, the variables are combined in five factors by means of a principal components analysis.

The factors are labeled the *family factor* (number of inhabitants, women, children, married persons, working persons, families with children, and average family size), the *unemployment factor* (number of persons not born in the Netherlands, divorced persons, persons receiving an unemployment benefit, persons registered at an unemployment office, the degree of urbanization of the ZIP code area), the *elderly factor* (number of elder persons, widows and widowers, persons not living in a family), the *income factor* (average income of inhabitants, average earnings of the working persons), and the *social security factor* (number of persons with a disability benefit, average value of the benefits). By combining the original variables some information is lost, but still 75% of the variance in the register data is explained by these factors.

3 Clustering ZIP code areas

The ZIP code areas are combined into homogenous clusters with an *hierarchical cluster analysis* based on the scores of the five factors. In an hierarchical process clusters are formed by grouping objects into bigger and bigger clusters, starting from the situation in which each cluster contains a single object, until all objects are grouped into a single cluster. In each step of the process, distances $d(C_g, C_h)$ between clusters C_g and C_h are computed, and two clusters for which the distance is smallest are combined. Because homogeneous clusters are sought, the method of Ward was used. This method combines clusters that result in the smallest increase in the within-cluster sum of squares (cf. Romesburg, 1984):

$$d(C_g, C_h) = \frac{n_g n_h}{n_g + n_h} \sum_{j=1}^{p} (\bar{F}_j^{(g)} - \bar{F}_j^{(h)})^2,$$

where $\bar{F}_j^{(g)}$ is the mean of factor j in cluster C_g and n_g the number of objects in cluster C_g. The method favors the grouping of homogeneous clusters with respect to variation in the factors. The agglomerative process is stopped when new clusters become too heterogeneous. This is determined by inspecting the distances for which clusters are combined in each step (i.e., the fusion distances), which reflect the homogeneity of the clusters.

3.1 An algorithm for hierarchical clustering

Hierarchical cluster methods depend on the computation of a matrix of fusion distances $d(C_g, C_h)$ in every step of the procedure. When the number of objects to be clustered is large, computations and storage of the distance matrix in a computer poses difficulties in time required to perform the computations and/or available memory. Keeping the distance matrix in memory is out of the question for a data set of several thousand records or larger. Keeping the data set in memory and recomputing distances, however, takes too much time even with modern-day computers. Therefore Huisman & Bethlehem (1999) presented a hybrid algorithm that partly stores distance information, and partly recomputes distances.

The basic idea of the algorithm is that for each cluster C_g only two values are stored, i.e., the index of the cluster for which the index to C_g is minimal, and the minimum fusion distance between the two clusters. To obtain candidate clusters for merging, only the minimum value of the minimum fusion distances has to be determined, and for the new cluster all distances to the other clusters have to be computed. Still, the data set with register information on the 397,692 ZIP code areas is much too large to be cluster-analyzed within a reasonable time span, and other strategies must be used (cf. Romesburg, 1984).

3.2 Two-stage cluster analysis

Hierarchical procedures have the advantage that different solutions are embedded in each other and therefore comparable, and that the number of clusters to extract does not have to be specified beforehand (Romesburg, 1984). To profit from these advantages in large data sets, Huisman & Bethlehem (1999) proposed an hierarchical cluster analysis in two stages. In the first stage, initial clusters are formed by categorizing the standardized cluster variables. Objects with the same categorized scores are combined in initial clusters, and the mean factorscores are used as input in the second stage.

In the second stage, an hierarchical cluster analysis is performed on the reduced data set containing the means of the initial clusters. The method of Ward is used to construct homogenous clusters with small within-cluster variances. With the hybrid algorithm described above, the reduction of the data set in the first stage does not have to be too large. A small number of initial clusters is undesirable, because the marginal distributions of the cluster variables may be too much distorted which results in a bad classification.

Huisman & Bethlehem (1999, pp. 13–15) derived rules to determine the categorization in the first stage. These are:

1. Define categories such that the number of categories per variable results in the largest reduction of the data set, given a maximum number of initial clusters.
2. Define categories such that the marginal distributions of the categorized variables best resemble the original marginal distributions.
3. Define categories with equal widths, except for tail categories.

In the first stage, it is important that the reduction is as large as possible with minimal loss of information. To determine the final, optimal clustering of ZIP code areas, measures of homogeneity based on the within-cluster sum of squares are used (Jansen & Bethlehem, 1979).

3.3 Results of the two-stage analysis

Huisman (1999) and Huisman & Bethlehem (1999) described in detail the analysis of the ZIP code data. In this subsection, only the main results are presented. In the first stage of the analysis, bounds are determined to categorize the five factors (e.g., the categorized value of factor j for object i equals 3 if $-1 < F_{ij} \leq -0.5$). All five factors have nine categories (bounds -1.5, -1, -0.5, 0, 0.5, 1, 1.5, and 3), and the first four get and extra tenth outlier-category (bound 5). This categorization reduces the data set to 15,506 initial clusters, which can be clustered within a reasonable time span, and results in the smallest loss of information.

To determine the optimal number of clusters in the second stage of the analysis, the fusion distances, (i.e., the distances for which clusters are combined), are inspected. Sudden changes in these distances reflect an optimal number of clusters. Also homogeneity measures (depending on the amount of variance accounted for by the clustering) can be used to determine the number of clusters (cf. Jansen & Bethlehem, 1979). Huisman (1999) presented clusterings with 15 to 20 final clusters, in which the clusters account for 50% to 45% of the variance in the data. However, one optimal clustering does not exist, because there are several substantial jumps in the fusion distances and the homogeneity measures, for different numbers of clusters. To determine the best clustering to be used in the post-stratification, more research is needed.

4 Determining adjustment weights

When an suitable number of strata (clusters) is found, we have to compute the adjustment weights in each stratum. In post-stratification, the correction weight c_k for element k in stratum U_g is defined by (simple random sampling)

$$c_k = \frac{N_g}{\sum_{i \in u_g} 1/\pi_i} = \frac{N_g}{N}\frac{n}{n_g},$$

where the sum is over all sample elements i in stratum U_g with inclusion probability π_i, which equals n/N in case of simple random sampling. This means that for each stratum both the number of inhabitants of the ZIP code areas and the number of respondents of the DLFS are needed. In this stage of the process of constructing the stratification, the problem of small strata emerges. When a stratum is too small, the number of respondents may be (almost) zero and the weighting procedure cannot be used.

On the other hand, the existence of some very large strata also is undesirable. These strata consists of too many persons and correction will not be optimal because the strata are too heterogeneous. Splitting the large strata into smaller ones is preferred. Huisman (1999), however, showed that extracting more clusters also results in extracting more small ones.

The results of the two-stage cluster analysis showed some large clusters with average scores on the five factors and some (very) small strata, which have more extreme (high or low) scores. The small strata are very homogeneous and probably very useful for the nonresponse correction. A solution for the problem of small strata, is *collapsing* (some of) them. This is not part of the cluster analysis and is based on ad hoc decisions. However, the influence of such, possibly hetrogeneous rest-strata may be large, and needs to be studied further.

5 Discussion and future research

Post-stratification to correct for nonresponse in the DLFS will lead to less biased results, especially when strata are related to the target variables and the cause of nonresponse. We presented a procedure to construct strata, based on a two-stage cluster analysis of ZIP code areas. A large set of auxiliary variables were collected, and the information available in the data was used optimally to construct homogeneous strata.

Determining the optimal number of clusters to extract in the ZIP code data is a diffcult task. The two-stage cluster procedure does not provide strict rules and criteria outside the procedure may determine the (minimal) number of strata. Moreover, with a large number of clusters, the problem of small strata will emerge. Collapsing these strata is necessary for the correction procedure to succeed, but needs more study.

The two-stage cluster analysis provides a clustering of ZIP code areas which seem useful in a post-stratification procedure. The actual computation of the adjustment weights, however, was not performed yet. The merits of the stratification have to be tested by comparing the results of the new weighting scheme with the scheme presently used for the DLFS and the unweighted results. This is topic of further study.

References

Bethlehem, J.G. (1988). Reduction of nonresponse bias through regression estimation. *Journal of Official Statistics*, **4**, 251-260.

Bethlehem, J.G. (1999). *Weighting adjustment for ignorable nonresponse.* Paper presented at the International Conference on Survey Nonresponse, October 28-31 1999, Portland, Oregon.

Huisman, M. (1999). *Poststratificatie: Het clusteren van postcodegebieden in Nederland* [in Dutch: Post-stratification: Clustering ZIP code areas in the Netherlands]. (Internal report 5543-99-RSM.) Voorburg: Statistics Netherlands.

Huisman, M. & Bethlehem, J.G. (1999). *Hierarchical clustering of large data sets: Classification of ZIP code areas.* (Research paper no. 9937.) Voorburg: Statistics Netherlands.

Jansen, M.E. & Bethlehem, J.G. (1979). *An application of cluster analysis to geographical classification.* (Research report 9613-79-M1.) Voorburg: Statistics Netherlands.

Little, R.J.A. & Rubin, D.B. (1987). *Statistical Analysis with Missing Data.* New York: Wiley.

Romesburg, H.C. (1984). *Cluster Analyses for Researchers.* London: Wadsworth, Inc.

Approximate Bayesian inference for simple mixtures

K. Humphreys and D.M. Titterington

Department of Statistics, University of Glasgow, Glasgow, G12 8QQ, Scotland.

Keywords. Mixtures, Bayesian, Recursive, Variational approximations

1 Introduction

Exact likelihoods and posterior densities associated with mixture data are computationally complex because of the large number of terms involved, corresponding to the large number of possible ways in which the observations might have evolved from the different components of the mixture. This feature is partially responsible for the need to use an algorithm such as the EM algorithm for calculating maximum likelihood estimates and, in Bayesian analysis, to represent posterior densities by a set of simulated samples generated by Markov chain Monte Carlo; see for instance Diebolt and Robert (1994).

However, Markov chain Monte Carlo methods are clearly time-consuming, samples of parameter values are required to be stored and there are risks to be run as to whether or not convergence has occurred. It may therefore be of value to consider the use of non-iterative approximate methods that are easy to use and that may be quite adequate for practical purposes. One such approach, referred as a *variational approach*, has recently come to prominence in the neural-computing literature, for instance in MacKay (1997) and Ghahramani and Beal (2000). The purposes of this paper are to present the variational approach, to illustrate the approach on a very simple mixture problem and compare its performance with the ideal Bayesian approach and alternative approximate methods from the statistical literature, to suggest modifications of the variational approach, and to indicate the scope for the use of the approximate methods in the context of more complicated problems.

2 Methods for approximate Bayesian inference

2.1 The variational approach

Suppose that θ represents the set of parameters in the model and that z represents the set of values missing from the data, which in the case of mixture problems corresponds to the mixture-component indicators of the observations. Suppose also that x represents the data that are observed. Then Bayesian inference centres on the posterior distribution of θ, given x, which may in theory be obtained as the marginal distribution from the joint distribution of θ and z, given x; this is the key to the MCMC approach of Diebolt and Robert (1994). In the variational approach, this joint posterior density is approximated by a distribution $q(z,\theta)$, chosen to maximise

$$\int \sum_{\{z\}} q(z,\theta) \ln \frac{p(x,z,\theta)}{q(z,\theta)} \, d\theta \,, \tag{1}$$

equivalent to the negative of the Kullback-Leibler divergence between the exact and approximate joint distributions of θ and z, given x. In the variational approach $q(z,\theta)$ is assumed to take a factorised form, and typically the factors corresponding to the parameters turn out to have the same distributional form as the conjugate family that would obtain were there no missing data; the appropriate values of the (hyper-)parameters are obtained by optimisation. We illustrate this below.

2.2 The variational approach for a simple mixture problem

The approach has been applied to several complex models where computational difficulties arise for the ideal Bayesian approach. Examples are hidden Markov models (Mackay, 1997), mixtures of factor analysers (Ghahramani and Beal, 2000) and multi-layer networks (Barber and Bishop, 1998). Attias (1999) and Ghahramani and Beal (2000) consider mixture models with unknown numbers of components. It is enlightening to illustrate the approach on simpler problems and to compare it to alternative approximate methods. We consider the simplest mixture problem, in which we have a mixture of two known distributions. Suppose that x contains a series of observations $x_1,..,x_n$, generated from a series of underlying independent random variables $z_1,..,z_n$, contained in z, which represent the two mixture components and take possible values 0 and 1. We let f_{0i}, f_{1i} denote the known conditional densities evaluated at the observation x_i given the (unknown) components 0 and 1, respectively, and we parameterise the model by θ, which represents the probability of an observation originating from component 1. We assign a Beta$(a^{(0)}, b^{(0)})$ prior distribution to θ, so that

$$p(x,z,\theta) \propto \theta^{(a^{(0)}-1)}(1-\theta)^{(b^{(0)}-1)} \prod_{i=1}^{n} \theta^{z_i}(1-\theta)^{(1-z_i)} f_{1i}^{z_i} f_{0i}^{(1-z_i)} . \tag{2}$$

We use an approximating conditional distribution $q(z,\theta)$, which factorises as $q(z,\theta) = \{\prod_{i=1}^{n} q_i^{(z)}(z_i)\}q^{(\theta)}(\theta)$. To maximise (1) we can use the following iterative procedure. In turn, we (i) optimise $q^{(\theta)}(\theta)$ for fixed $\{q_i^{(z)}(z_i); i = 1,..,n\}$ and (ii) optimise $\{q_i^{(z)}(z_i); i = 1,..,n\}$ for fixed $q^{(\theta)}(\theta)$. The terms in (1) which involve θ and $q^{(\theta)}(\theta)$ are

$$\int q^{(\theta)}(\theta) \log \left\{ \frac{\theta^{(a^{(0)}-1)}(1-\theta)^{(b^{(0)}-1)} \left[\prod_{i=1}^{n} \theta^{r_{1i}}(1-\theta)^{r_{0i}}\right]}{q^{(\theta)}(\theta)} \right\} d\theta , \tag{3}$$

where $r_{ji} = q_i^{(z)}(z_i = j), j = 0,1$. The above expression is equivalent to the negative of a Kullback-Leibler divergence, which is non-positive, hence the optimal choice for $q^{(\theta)}(\theta)$, when $\{q_i^{(z)}(z_i); i = 1,..,n\}$ is fixed is

$$\theta \sim \text{Beta} \left[a^{(0)} + \sum_{i=1}^{n} r_{1i}, b^{(0)} + \sum_{i=1}^{n} r_{0i} \right] . \tag{4}$$

The terms in (1) which involve a specific z_i and $q_i^{(z)}(z_i)$ are

$$\sum_{j=0}^{1} r_{ji} \log \left(\frac{f_{ji}\phi_j}{r_{ji}} \right) , \tag{5}$$

where $\phi_1 = \exp\{\int q^{(\theta)}(\theta)\log\theta\, d\theta\}$ and $\phi_0 = \exp\{\int q^{(\theta)}(\theta)\log(1-\theta)\, d\theta\}$. Consequently the optimal $q_i^{(z)}(z_i)$ $(i = 1,..,n)$ is defined by $r_{ji} \propto f_{ji}\phi_j$. it is natural to initialise the iterative procedure by taking $r_{1i} = f_{1i}a^{(0)}/\{f_{1i}a^{(0)} + f_{0i}b^{(0)}\}$ and $r_{0i} = f_{0i}b^{(0)}/\{f_{1i}a^{(0)} + f_{0i}b^{(0)}\}$. The variational approximation for the posterior distribution of θ thus turns out to be a Beta distribution, which is the conjugate distribution for this prior, and is obtained by solving a bank of nonlinear equations that involve digamma functions. There is no explicit formula for the optimal values of the hyper-parameters in the variational approximation. This is true for other families of models to which the approximation has been applied. The relationship between the variational Bayes algorithm and the EM algorithm is discussed by Attias (1999). Both algorithms are guaranteed to converge.

2.3 Recursive methods

There are in fact alternative ways of deriving Beta approximations for our simple mixture problem, in the form of recursive approximations. We consider two methods, quasi-Bayes and probabilistic-editor. For a more complete discussion of these methods the reader is referred to Chapter 6 of Titterington, Smith and Makov (1985). We let $a^{(i)}$, $b^{(i)}$ be the values of the hyper-parameters of the (approximate) Beta distribution after processing observation i. If we let $w_{1i} = f_{1i}a^{(i-1)}/\{f_{1i}a^{(i-1)} + f_{0i}b^{(i-1)}\}$ and $w_{0i} = f_{0i}b^{(i-1)}/\{f_{1i}a^{(i-1)} + f_{0i}b^{(i-1)}\}$ $(i = 1,..,n)$, then the exact marginal distribution of θ given only observation x_1 is

$$p(\theta|x_1) = w_{11}\text{Beta}\,[a^{(0)} + 1, b^{(0)}] + w_{01}\text{Beta}\,[a^{(0)}, b^{(0)} + 1]\,. \qquad (6)$$

The exact marginal posterior distribution builds up as a weighted average of Beta distributions. As each observation i is processed, the recursive methods replace the posterior distribution of θ by an approximation based on a single Beta distribution with hyper-parameters $a^{(i)}$, $b^{(i)}$. Formula (6) is used repeatedly with weights w_{11} and w_{01} replaced by w_{1i} and w_{0i} and hyper-parameters $a^{(0)}$, $b^{(0)}$ replaced by $a^{(i-1)}$, $b^{(i-1)}$. Computations are thus made manageable; the updating of the hyper-parameters is trivial. The drawback of these methods is that they are *recursive*, in that the final result depends on the order in which the data are processed.

2.3.1 Quasi-Bayes

Suppose the z were known. As the first observation was processed, the exact marginal posterior distribution of θ would be $\text{Beta}\,[a^{(0)} + \Delta_1, b^{(0)} + (1-\Delta_1)]$, where $\Delta_i = z_i$. Similarly, as subsequent observations (i=2,..,n) were processed Δ_i and $1-\Delta_i$ would be added to the values of the hyper-parameters. The z_i are of course not known. The idea of the quasi-Bayes method is to replace the values of Δ_i by their expectations, based on the values of the hyper-parameters of the (approximate) Beta distribution before i is processed. For the mixture of two known distributions $a^{(i)} = a^{(i-1)} + f_{1i}a^{(i-1)}/\{f_{1i}a^{(i-1)} + f_{0i}b^{(i-1)}\}$. The quasi-Bayes method is known to provide consistent estimators for simple mixture models.

2.3.2 The probabilistic editor

In this approach the Δ_i are chosen so that the first two moments of the (single) Beta distribution approximation are identical to those of the mixture distribution (based on the distribution of θ after processing the previous observation, which is itself an approximation of the true posterior distribution at the processing of the second and subsequent observations). For the mixtures of two known distributions problem, after observation i is processed the hyper-parameters take the values

$$a^{(i)} = \frac{\mathbf{E}(\theta)\{\mathbf{E}(\theta) - \mathbf{E}(\theta^2)\}}{\mathbf{E}(\theta^2) - \{\mathbf{E}(\theta)\}^2}\ , \quad b^{(i)} = a^{(i)}\{\frac{1}{\mathbf{E}(\theta)} - 1\}\ , \tag{7}$$

where $\mathbf{E}(.)$ denotes expectation based on the posterior distribution of θ, when it has prior distribution Beta $\left[a^{(i-1)}, b^{(i-1)}\right]$.

2.3.3 A recursive variational approach

It is easy to construct a recursive method based on the variational approach by optimising (3) and (5) (with n=1) for each observation as it is processed. The advantage of this method is that it requires the solution of n single equations as opposed to the solution of a (single) system of n nonlinear equations, which is necessary in the non-recursive variational approach.

3 Numerical results

We applied the various methods to the simple mixture of two known Normal densities with means of 3 and 5; both have unit variance. We generated a sample of size n=50 using $\theta = 0.65$. In Figure 1 we have plotted density estimates of the posterior distributions, based on a Beta prior distribution for θ with hyper-parameter values $a^{(0)}$=$b^{(0)}$=1. It is time consuming to calculate the posterior density of the mixing weight exactly, so Gibbs sampling was used. For the recursive methods we have plotted the posterior density for just one particular ordering of the data. For this example, and for other examples of mixture models which we have considered, the probabilistic-editor approach turned out to be superior to all the other approximations, including the variational approximation, even taking into account the influence of the recursive nature of the method. It is the only method which does not significantly under-estimate the spread of the posterior distribution. The probabilistic editor is also less sensitive to the order in which the data are processed than are the other recursive methods. This is evident from the results presented in Table 1; the standard deviations of the first two moments of the approximations taken over 1000 different orderings of the data are relatively small for this method.

We can calculate the probability that observation i belongs to a particular mixture component using

$$\mathrm{pr}(z_i = 1|x_1, .., x_n) = \int \mathrm{pr}(z_i = 1|x_i, \theta)\mathrm{pr}(\theta|x_1, .., x_n)d\theta\ . \tag{8}$$

The estimator (8) based on the probabilistic editor estimator of the posterior distribution (of θ given x_1,..,x_n) will clearly be closer to the exact estimator than those based on the other posterior distribution approximations. It is not usually computationally feasible to calculate (8). For classification based on the means of the posterior distribution estimates, the variational method performs as well as the probabilistic editor. The performances of the quasi-Bayes and recursive variational methods are sensitive to the ordering of the

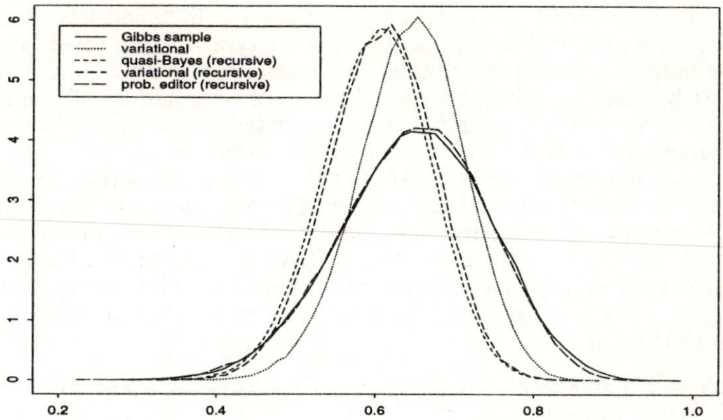

Fig. 1. Density estimates for the one parameter mixture model

	mean	variance
Gibbs sampling	0.6500	0.0088
variational	0.6502	0.0043
quasi-Bayes	0.6278	0.0044
	(0.0381)	(0.0002)
probabilistic editor	0.6527	0.0091
	(0.0034)	(0.0004)
recursive variational	0.6338	0.0043
	(0.0454)	(0.0002)

Table 1. First two moments of the Beta distribution approximations (means and, in parentheses, standard deviations of the first two moments based on 1000 different data orderings are given for the recursive methods)

data. For recursive computation of probabilities for classifying observations to mixture components we can use

$$\mathrm{pr}(z_i = 1|x_1, .., x_n) \propto f(x_i = 1|x_1, .., x_{n-1})\mathrm{pr}(z_i = 1|x_1, .., x_{n-1}) \,. \qquad (9)$$

For our simple mixture model, using (9), the recursive variational method is not more accurate than the other recursive methods. In the next section we draw attention to a case where the recursive variational method is superior to other methods of approximation.

4 Further developments and summary

Our simple example consisted of an underlying independence process $\{z_i; i{=}1,..,n\}$ determining the (independence) distribution of the variables $\{x_i; i{=}1,..,n\}$, where each z_i consists of a single binary (Bernoulli distributed) latent variable parameterised by θ, which is univariate and has a Beta prior. The approximation methods described in Section 2 extend easily to the case where each z_i consists of several binary latent variables; such models are popular in the neural computing literature. The probabilistic editor is the only method which does not easily accommodate the case where z_i consists of latent variables with more than two categories. If z_i consists of a single categorical latent variable with k states it is typically assumed that z_i is multinomially distributed and that θ, which has dimension k-1, has a Dirichlet prior distribution. The probabilistic editor method would ideally match the "exact" and approximate marginal posterior distributions on 2(k-1) moments; however there are only k (unknown) hyper-parameters, so for $k > 2$, except in special circumstances, this is not possible. For a few examples we have matched all first moments and one second moment, and have found that, for small k, this approach can offer high quality approximations for the marginal posterior distributions. This procedure is however somewhat ad-hoc.

We have considered each method for a hidden (first-order) stationary Markov chain for $\{z_i;\ i=1,..,n\}$. In this case recursive estimation is based on the relationship

$$\text{pr}(z_{i-1} = j, z_i = k | x_1, .., x_i) \propto \text{pr}(z_{i-1} = j | x_1, .., x_{i-1}) \phi_{jk} \text{pr}(x_i | z_i = k) , \tag{10}$$

where ϕ_{jk} denotes the (stationary) conditional probability of a transition out of state j into state k. We use this result in the same spirit as do Holst and Lindgren (1991). Even for binary z_i, if there are several (independent) realisations of this process, the probabilistic editor is unwieldy and we have found that our recursive adaptation of the variational Bayes approach provides approximations of the marginal distributions which are superior to those provided by other methods. We will report this result in detail in Humphreys and Titterington (2000). There is scope, in complicated mixture problems, for the use of the variational approximation and its recursive adaptation.

Acknowledgement
This research was supported by a grant from the UK Engineering and Physical Sciences Research Council.

References

Attias, H. (1999) Inferring parameters and structure of latent variable models by variational Bayes. In *Proc. 15th Conference on Uncertainty in Artificial Intelligence.*

Barber, D. and Bishop, C.M. (1998) Ensemble learning for multi-layer networks. In M.I. Jordan, M.J. Kearns, S.A. Solla (eds.) *Advances in Neural Information Processing Systems*, 10. MIT Press, 1998.

Diebolt, J. and Robert, C.P. (1994) Estimation of finite mixture distributions through Bayesian sampling. *J.R. Statist. Soc. B.* 56, 363-375.

Ghahramani, Z. and Beal, M.J. (2000) Variational inference for Bayesian mixture of factor analysers. In S.A. Solla, T.K. Leen and K.-R. Müller (eds.) *Advances in Neural Information Processing Systems*, 12. MIT Press (to appear).

Holst, U. and Lindgren, G. (1991) Recursive estimation in mixture models with Markov regime. *IEEE Transactions on Information Theory*, 37, 6, 1683-1690.

Humphreys, K. and Titterington, D.M. (2000) Some examples of recursive variational approximations. In preparation.

MacKay, D.J.C. (1997) Ensemble learning for hidden Markov models. Technical report, Cavendish Laboratory, University of Cambridge.

Titterington, D.M., Smith, A.F.M. and Makov, U.E. (1985) *Statistical Analysis of Finite Mixture Distributions.* New York: Wiley.

Correlated INAR(1) process

Nikolai Kolev and Delhi Paiva

Department of Statistics, IME, University of São Paulo, C.P. 66281, 05315-970 São Paulo, SP, Brazil

Abstract. We introduce the concept of correlated integer valued autoregressive process of order 1. It is based on equi-correlated binary responses, instead of independent responses involved with the usual INAR(1) process. Results related to the extended Steutel and van Harn operator are presented, their correlation structure, and additional properties are shown. A procedure for conditional likelihood estimation of the parameters of the model is proposed. The case of Poisson innovations illustrates the considered processes.

Keywords. Autoregressive model, conditional maximum likelihood, correlated binomial thinning, random summation, simulation.

1 Introduction

Al-Osh and Alzaid (1987) define the following INAR(1) model

$$X_t = \alpha * X_{t-1} + \epsilon_t, \quad t = 0, 1, 2, \ldots \tag{1}$$

where X_{t-1} and ϵ_t are independent sequences of non-negative integer valued random variables (r.v.). X_{t-1}'s have Poisson distribution of parameter λ, and ϵ_t's are Poisson distributed with parameter $(1-\alpha)\lambda$, with $\alpha \in (0,1)$. The operator $\alpha * X_{t-1}$ is defined as the random sum $\sum_{i=1}^{X_{t-1}} Y_i$, where $\{Y_i\}$ is a sequence of independent identically distributed (i.i.d.) Bernoulli r.v.'s with parameter α, i.e. $P(Y=1) = 1 - P(Y=0) = \alpha$. The process (1) is stationary. This type of processes have been studied by McKenzie (1986). He derived various properties for specific marginal distributions (geometric and negative binomial), and extended the concept to ARMA-like models. The mixture operator $\alpha*$ in (1) is known as *binomial thinning*, and was introduced by Steutel and van Harn (1979). It replaces the scalar multiplication in the continuous AR(1) model.

An immediate extension of INAR(1) process can be obtained by substituting the Bernoulli variables involved in the operator $\alpha*$ by other non-negative integer-valued r.v.'s (with finite variance and mean $\alpha < 1$). To emphasize the difference, we denote these operators by $\alpha\bullet$ instead of $\alpha*$. Latour (1998) studied a corresponding p-order generalized process and named it GINAR(p).

The proposed here models suit for describing statistical data which may be expressed in terms of counts, taken sequentially in time, and which are correlated. In modelling counting processes with dependence, it is natural to consider the process X_t as composed by survivors in the previous periods, X_{t-1}, with an addition represented by some new arrivals (innovation), which form the process ϵ_t, of certain discrete distribution.

The paper is organized as follows. In Section 2 we briefly discuss distributional questions, related to our generalization of the Steutel and van Harn operator. In Section 3 we introduce correlated INAR(1) processes and outline its basic properties. Estimation of parameters and specific results when the innovation sequence ϵ_t is Poisson distributed illustrate our considerations. At the end we apply this methodology to study real mortality data observed during a two years period in the state of São Paulo, Brazil, and show some simulation results.

2 Correlated binomial thinning

We generalize the operator $\alpha *$ in (1) by the extended operator

$$\alpha \bullet_\rho X = \sum_{i=1}^{X} Y_i, \tag{2}$$

The r.v.'s $Y_i, i = 1, 2, \ldots$ are identically distributed, $P(Y = l) = \pi_l, \sum_{l=0}^{\infty} \pi_l = 1$, with $E(Y) = \alpha < 1$. They are also equi-correlated, i.e. $corr(Y_i, Y_j) = \rho > 0$ for $i \neq j$. The probability generating function (PGF), $P_Z(s)$, of the random sum $Z = \sum_{i=1}^{X} Y_i$ is given by

$$P_Z(s) = \rho \sum_{l=0}^{\infty} \pi_l P_X(s^l) + (1-\rho) P_X(P_Y(s)), \quad |s| < 1,$$

where $P_X(.)$ and $P_Y(.)$ denote the PGF's of X, and Y_i's respectively. The mean and the variance are given by $E(Z) = E(X)E(Y)$ and

$$Var(Z) = E(X)Var(Y) + Var(X)[E(Y)]^2 + \rho Var(Y)E[X(X-1)].$$

If Y_i's are independent, then $\rho = 0$, and we have $P_Z(s) = P_X(P_Y(s))$ and $Var(Z) = E(X)Var(Y) + Var(X)[E(Y)]^2$, i.e. the classical results.

We assume further that Y_i's are equi-correlated identically Bernoulli distributed with mean α. To obtain explicit formula for the k-th order operation

$$\underbrace{\alpha \bullet_\rho \cdots \alpha \bullet_\rho}_{k-\text{times}} X =^d \alpha^k \bullet_\rho X, \tag{3}$$

we consider $P_k(s)$, the PGF of the random sum (3). It is given by

$$P_k(s) = 1 - (1-\rho+\rho\alpha)^k + \sum_{i=0}^{k} \binom{k}{i} (\alpha\rho)^{k-i} (1-\rho)^i P_X(1-\alpha^i+\alpha^i s), \quad |s| < 1.$$

This expression is a mixture of the PGF's $P_X(1-\alpha^i+\alpha^i s)$ for $i = 0, 1, \ldots, k$, since the sum of the coefficients in front of the PGF's equals to one. Each PGF $P_X(1-\alpha^i+\alpha^i s)$ represent corresponding random summation as in the independent case, when applying a binomial thinning operator of parameter α^i, $i = 0, 1, 2, \ldots, k$.

3 CINAR(1) process

Definition. An integer-valued time series X_t is called an CINAR(1) process (it means *correlated integer-valued autoregressive process of order 1*), if

$$X_t = \alpha \bullet_\rho X_{t-1} + \epsilon_t, \quad t = 0, 1, 2 \ldots, \tag{4}$$

where the operator $\alpha\bullet_\rho$ is defined by (2), and the Y_i's are equi-correlated identically Bernoulli distributed r.v.'s with $corr(Y_i, Y_j) = \rho > 0$ for $i \neq j$. The ϵ_t are i.i.d. r.v.'s with finite mean μ and variance σ^2, independent of X_{t-1}.

For example, CINAR(1) processes can be observed as the number of patients at specific point of time, or as the number of calls waiting in a queue for service at certain moment. In many cases the assumption of independence (as in INAR(1) process) for counting time series seems unreasonable.

3.1 Properties

CINAR(1) and INAR(1) processes have similar properties. In particular,

$$X_t =^d \alpha^k \bullet_\rho X_{t-k} + \sum_{j=0}^{k-1} \alpha^j * \epsilon_{t-j}, \quad k = 1, \ldots, t.$$

In this summation $\alpha^j \bullet_\rho \epsilon_t$ transformation is the same as $\alpha^j * \epsilon_t$, since the sequence ϵ_t is i.i.d. In particular, the following equality in distribution for the considered CINAR(1) is valid:

$$X_t =^d \alpha^t \bullet_\rho X_0 + \sum_{j=0}^{t-1} \alpha^j * \epsilon_{t-j}.$$

The auto-covariance structure of (4) is represented by the function

$$\gamma(k) = Cov(X_{t-k}, X_t) = \alpha^k Var(X_{t-k}) + \sum_{j=0}^{k-1} \alpha^j Cov(X_{t-k}, \epsilon_{t-j}) = \alpha^k \gamma(0).$$

The mean of the process X_t is given by

$$E(X_t) = \alpha^t E(X_0) + \mu \sum_{j=0}^{t-1} \alpha^j.$$

Moreover, if $E(X_0) = \frac{\mu}{1-\alpha}$, then CINAR(1) has a constant mean. The variance of the process, $Var(X_t)$, has the form

$$\alpha^t \left[B^t Var(X_0) + (1 - B^t)E(X_0) + (B^t - \alpha^t)(E(X_0))^2\right] + \sigma^2 \sum_{j=1}^{t} \alpha^{2(j-1)},$$

where $B = \rho + (1 - \rho)\alpha$.

The second-order stationarity in CINAR(1) is not valid, as it holds for the INAR(1) process. Therefore, a CINAR(1) process is non-stationary.

3.1 Parameter estimation

The CINAR(1) process (4) has the Markovian property $P(X_t|x_{t-1}, x_{t-2}, \ldots) = P(X_t|x_{t-1})$. The likelihood function for the sample $\mathbf{x} = (x_0, x_1, \ldots, x_N)$ is $L_N(\mathbf{x}, \theta) = P_\theta(X_0 = x_0) \prod_{t=1}^{N} P_\theta(x_{t-1}, x_t)$, where $\theta \in \Theta$ is the vector of unknown parameters and $P_\theta(x_{t-1}, x_t)$ are the transition probabilities of the corresponding Markov chain.

We neglect the dependency on the initial value x_0, and consider the conditional maximum likelihood (ML) estimate for θ, given $X_0 = x_0$. We get it by maximizing the corresponding conditional log-likelihood function over the parameter space Θ. We apply the ML method for the CINAR(1) process with Poisson distributed innovations ϵ_t of parameter λ. Then the unknown vector parameter is $\theta = (\alpha, \rho, \lambda)$. The conditional distribution of X_t given x_{t-1} is *correlated binomial-Poisson mixture*, (see Luceño, (1995) for the correlated binomial distribution), represented by

$$P(X_t = n|x_{t-1} = m) = (1-\rho)e^{-\lambda} \sum_{j=0}^{\min\{m,n\}} \frac{\lambda^{n-j}}{(n-j)!} \binom{m}{j} \alpha^j (1-\alpha)^{m-j}$$
$$+\rho e^{-\lambda} \frac{\lambda^n}{n!}(1-\alpha) + \rho e^{-\lambda} \frac{\lambda^{n-m}}{(n-m)!} \alpha \mathbf{1}_{\{n \geq m\}},$$

where $\mathbf{1}_{\{.\}}$ is the indicator function of the event in the subscript.

We denote the conditional ML estimator by $\hat{\theta} = (\hat{\alpha}, \hat{\rho}, \hat{\lambda})$, and obtain it as solution of the system of equations $\frac{\partial l_N}{\partial \alpha} = 0$, $\frac{\partial l_N}{\partial \rho} = 0$ and $\frac{\partial l_N}{\partial \lambda} = 0$, where $l_N = log\ L_N$. For example, $\frac{\partial l_N}{\partial \alpha} = 0$ gives the equation

$$\sum_{t=1}^{N} \sum_{j=0}^{x_{t-1}} (1-\rho) \frac{\lambda^{x_t - j}}{(x_t - j)!} \binom{x_{t-1}}{j} [\alpha^{j-1} (1-\alpha)^{x_{t-1}-j-1} (j - \alpha x_{t-1})]$$
$$= \rho \sum_{t=1}^{N} \left[\frac{\lambda^{x_t}}{x_t!} + \frac{\lambda^{x_t - x_{t-1}}}{(x_t - x_{t-1})!} \mathbf{1}_{\{x_t \geq x_{t-1}\}} \right]. \qquad (5)$$

The ML estimator $\hat{\theta}$ can be obtained by using Newton-Raphson method. The initial values of the parameters α and λ can be obtained by an estimation procedure, based on minimization of the sum of squared deviations about the conditional expectation, developed by Klimko and Nelson (1978).

The conditional expected value of a CINAR(1) process with Poisson innovations is $E(X_t|x_{t-1}) = \alpha x_{t-1} + \lambda$.

The conditional least square estimators for α and λ are values that minimize the function $F(\alpha, \lambda, x_{t-1}) = \sum_{t=1}^{N} [X_t - E(X_t|x_{t-1})]^2$. The standard procedure gives

$$\hat{\alpha} = \frac{\sum_{t=1}^{N} x_t x_{t-1} - \frac{1}{n} \sum_{t=1}^{N} x_t \sum_{t=1}^{N} x_{t-1}}{\sum_{t=1}^{N} x_{t-1}^2 - \frac{1}{n} (\sum_{t=1}^{N} x_{t-1})^2} \text{ and } \hat{\lambda} = \frac{1}{n} \sum_{t=1}^{N} x_t - \frac{1}{n} \hat{\alpha} \sum_{t=1}^{N} x_t. \qquad (6)$$

4 An application

The number of the daily dying people registered with respiratory disease in the PROAIM (Programa de Aprimoramento das Informações de Mortalidade no Municipio de São Paulo) during the years 1996 and 1997 is studied. For the observed data we apply an INAR(1) and a CINAR(1) processes with Poisson innovations, and estimate their respective parameters.

The initial estimates of the parameters α and λ are computed in accordance with formulas (6) and are given in Table 1. The initial estimate of the parameter ρ of the CINAR(1) model is obtained as solution of equation (5), for the already found initial estimates of α_0 and λ_0. These values of α_0, ρ_0 and λ_0 are used as starting values for the Newton-Raphson algorithm. The ultimate estimates are given in Table 1, correspondingly. We were able to get these estimates for about 1000 iterations of the procedure proposed by the OX program (see Doornik, 1996).

Let us discuss the estimated values of the parameters for the CINAR(1) process. According to our model, α is the probability of death of an individual. Within a year it increases by 0.0905. This indicates some significant increase in the average air-pollution in São Paulo over the same period of time, confirmed by other observations and studies too. The mean number of patients entered daily in the PROAIM (represented by the parameter λ) during the years 1996 and 1997 is 6.9809 and 5.3356, correspondingly. This is a little bit surprising result, nevertheless it is confirmed by the data. One would expect just the opposite, since a higher air-pollution could cause a higher number of patients.

Similar conclusions can be made from detailed interpretation of the estimation of the parameters α and λ of the INAR(1) process.

Table 1: *CINAR(1) and INAR(1) estimated parameters for the respiratory disease data in the years 1996 and 1997.*

Initial values				CINAR(1)			INAR(1)	
year	α_0	ρ_0	λ_0	$\hat{\alpha}$	$\hat{\rho}$	$\hat{\lambda}$	$\hat{\alpha}$	$\hat{\lambda}$
1996	0.5041	0.9987	9.4927	0.4581	0.2677	6.9808	0.4839	7.9427
1997	0.5775	0.9999	7.6177	0.5486	0.1536	5.3356	0.5357	6.7167

The significant positive estimate of ρ indicates that there exists dependence between the binary variables Y_i involved in the CINAR(1) construction. The estimated value of ρ for the 1997 year is less than that for 1996. This shows that the correlation between deaths decreases when the pollution increases.

On Fig.1a, and Fig.1b we show simulated CINAR(1) and INAR(1) processes for the year of 1996 with parameter's values as given in Table 1. Fig.1c, and Fig.1d illustrate the corresponding simulation results for the year of 1997.

One can observe that the use of CINAR(1) model is preferable, because it represents higher concentration of the simulated observations than the concentration represented by the INAR(1) model. And this behavior actually corresponds to the observed data. The difference comes from the stationary properties of the INAR(1), and the non-stationarity of CINAR(1) process.

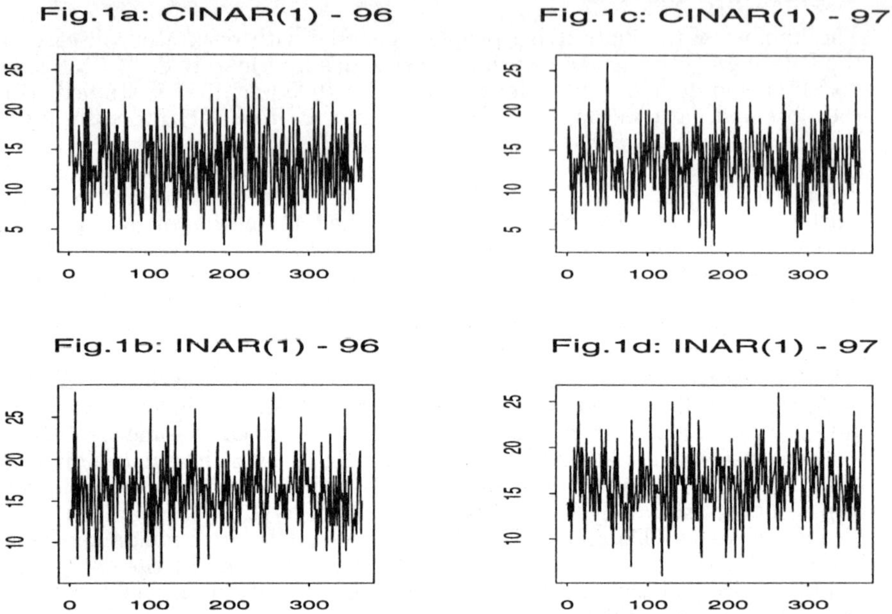

Figure 1: *Simulating series of CINAR(1) and INAR(1) processes for the data of respiratory disease during the years 1996 and 1997.*

References

Al-Osh, M.A. and Alzaid, A.A. (1987). First-order integer-valued autoregressive (INAR(1)) process. *Journal of Time Series Analysis*, **8**, pp. 261-275.

Doornik, J.A. (1996). *OX. An Object - Oriented Matrix Programming Language.* London: International Thompson Business Press.

Latour, A. (1998). Existence and stochastic structure of a non-negative integer-valued autoregressive process. *Journal of the Time Series Analysis*, **19**, pp. 439-455.

Luceño, A. (1995). A family of partially correlated Poisson models for overdispersion. *Computational Statistics & Data Analysis*, **20**, pp. 511-520.

Klimko, L. and Nelson, P. (1978). On conditional least squares estimation for stochastic processes. *The Annals of Statistics*, **6**, pp. 629-642.

McKenzie, E. (1986). Autoregressive moving-average process with negative-binomial and geometric marginal distributions. *Advances of Applied Probability*, **18**, pp. 679-705.

Steutel, F.B. and van Harn, K. (1979). Discrete analogues of self-decomposability and stability. *The Annals of Probability* **7**, pp. 893-899.

Confidence regions for stabilized multivariate tests

Siegfried Kropf

Coordination Centre for Clinical Trials, University of Leipzig, Prager Str. 34, D-04317 Leipzig, Germany

Abstract. Stabilized multivariate tests as proposed by Läuter (1996) and Läuter, Glimm and Kropf (1996, 1998) are becoming more and more interesting for clinical research. In this paper we consider tests for the comparison of two independent groups and investigate the corresponding confidence regions by numerical methods. The confidence regions reflect the properties of the tests from another perspective. The construction method is, however, applicable only for small dimensions. It turns out that – due to their involved shape – these confidence regions are difficult to handle.

Keywords. Confidence region, stabilized multivariat tests, treatment effects

1 Background

The difficulties of traditional parametric or nonparametric multivariate tests to cope with high-dimensional data, when the samples are small or moderate, have been the reason for the development of alternate tests (O'Brien, 1984, Wei & Lachin, 1984, Tang, Geller & Pocock, 1993), often summarized as multiple endpoint tests. The main idea was to restrict the search for differences among groups to special directions in the space of the p-dimensional observations assuming a 'similar behaviour' of the p variables. Whereas these tests are distribution-free or parametric with only asymptotically known distribution, Läuter and co-workers (Läuter, 1996; Läuter, Glimm & Kropf, 1996, 1998; Läuter, Kropf & Glimm, 1999) have proposed a class of exact parametric tests based on the theory of spherical distributions (Fang & Zhang, 1990). Many of these new proposals have the same basic principle as the earlier proposals for multiple endpoint tests. The p-dimensional data vectors $\boldsymbol{x}$ are transformed into linear scores z with a data-dependent weight vector $\boldsymbol{d}$, $z = \boldsymbol{d}'\boldsymbol{x}$, and then the scores are analyzed in standard tests for univariate data. The special feature of the new proposals are restrictions for the derivation of the weight vector which ensure that the vector of the score values from all sample elements has a spherical distribution and, hence, the usual parametric tests with the scores are exact level α tests, though the scores are no longer normal nor independent from each other. These restrictions are mild enough to adapt the tests to different real data situations. That makes these tests quite promising for the data analysis in the field of medicine and in many other fields.

Here we consider tests for the comparison of means from two independent samples and we want to investigate the corresponding confidence regions, which has not been investigated before, though it is common practice in the analysis of clinical trials to complete test results with estimators of the effect size.

2 Statistical model and stabilized tests

We consider the standard two-sample problem with independent p-dimensional normal sample vectors

$$\boldsymbol{x}_j^{(k)} \sim N_p\left(\boldsymbol{\mu}^{(k)}, \Sigma\right) \quad , \quad k=1,2;\, j=1,\ldots,n^{(k)} . \tag{1}$$

If $\boldsymbol{\mu}$ denotes the difference of both means (e.g., the treatment effect in a clinical trial), $\boldsymbol{\mu} := \boldsymbol{\mu}^{(1)} - \boldsymbol{\mu}^{(2)}$, then we want to test the null hypothesis $\mathrm{H}_0 : \boldsymbol{\mu} = \boldsymbol{0}$ against $\mathrm{H}_1 : \boldsymbol{\mu} \neq \boldsymbol{0}$ or a more specialized alternative hypothesis, where $\boldsymbol{0}$ denotes the null vector. Theorem 3 in Läuter, Glimm and Kropf (1996) shows that this can be done by calculating the score values $z_j^{(k)} = \boldsymbol{x}_j^{(k)\prime}\boldsymbol{d}$, $k=1,2;\, j=1,\ldots,n^{(k)}$, for each sample element and comparing them in the usual two-sample t test

$$t = \frac{\bar{z}^{(1)} - \bar{z}^{(2)}}{s_z}\sqrt{\frac{n^{(1)} \cdot n^{(2)}}{n^{(1)} + n^{(2)}}}$$

with $\bar{z}^{(k)} = \frac{1}{n^{(k)}}\sum_{j=1}^{n^{(k)}} z_j^{(k)}$ $(k=1,2)$ and $s_z^{\,2} = \frac{1}{n^{(1)} + n^{(2)} - 2}\sum_{k=1}^{2}\sum_{j=1}^{n^{(k)}}\left(z_j^{(k)} - \bar{z}^{(k)}\right)^2$,

provided that the weight vector $\boldsymbol{d}$ depends on the data only through the total sums of products matrix $\boldsymbol{W} = \sum_{k=1}^{2}\sum_{j=1}^{n^{(k)}}(\boldsymbol{x}_j^{(k)} - \bar{\boldsymbol{x}})(\boldsymbol{x}_j^{(k)} - \bar{\boldsymbol{x}})'$ with $\bar{\boldsymbol{x}} = \frac{1}{n^{(1)} + n^{(2)}}\sum_{k=1}^{2}\sum_{j=1}^{n^{(k)}}\boldsymbol{x}_{(j)}^{(k)}$

and that $\boldsymbol{d}'\boldsymbol{W}\boldsymbol{d} \neq 0$ with probability 1. The null distribution of the score-based t test is the usual t distribution with $n^{(1)} + n^{(2)} - 2$ degrees of freedom despite of the pre-processing with a data-dependent weight vector $\boldsymbol{d}$. The power of the resulting tests depends on a suitable choice of an algorithm for deriving $\boldsymbol{d}$ in accordance with the expected data structure.

Proposals for such tests are the standardized sum test (SS test) with the weight vector $\boldsymbol{d} = (1/\sqrt{w_{ii}})_{i=1,\ldots,p}$ (w_{ii} are the diagonal elements of $\boldsymbol{W}$) and the principal component test (PC test), where the weight vector $\boldsymbol{d}$ is derived from the eigenvalue problem $\boldsymbol{W}\boldsymbol{d} = \lambda\,\mathrm{Diag}(\boldsymbol{W})\boldsymbol{d}$. Both tests are scale invariant (which is a desirable feature of the test in most situations). The SS test is appropriate if we expect equally directed effects of approximately equal relative magnitude in all variables. The PC test should have a good power if the p variables are approximately determined by the same (unobservable) latent variable, however, each with a specific scale factor and each with a specific degree of disturbance by independent noice variables. The type I error is maintained exactly, even if these additional assumptions are not true.

The final t test can be *one-sided or two-sided*. In the latter case, however, it should be ensured that all weight coefficients are non-negative to make the test result interpretable. If this is not originally given (as for the SS test), it can be enforced by taking the absolute values from all vector components.

Another useful modification consists in a procedure for the *selection of variables*. After having calculated raw scores in the usual way (e.g., SS scores), the correlation coefficients between the score and the p original variables are determined. Then the weight coefficients of variables that have a non-significant correlation to the score are set to zero. In that way, ineffective variables are excluded from the analysis.

These test versions are available as SPSS and SAS macros (using the matrix language in both systems and a graphical surface for SPSS) and can be obtained from the author by request per email (Siegfried.Kropf@KKSL.Uni-Leipzig.de, cf. also Kropf & Läuter, 1998).

3 Construction of the confidence region

The basic model (1) for the stabilized multivariate tests is the same as in conventional multivariate parametric statistic. Hence, we could apply the usual confidence ellipsoid closely related to Hotelling's T^2 test, or we could construct a rectangular region by combining the Bonferroni-corrected univariate confidence regions, i.e., each at a confidence level $1-\alpha/p$. In the latter case, we would have a well interpretable confidence region, but we would totally ignore the correlation structure of the variables and loose power that way. With the confidence ellipsoids, we would utilize the correlations, however, we would disregard the special philosophy of stabilized multivariate statistics. Particularly, we would not utilize our prior assumptions about the special data structure.

There are two possibilities to keep the link to the stabilized tests. The first one is simply to invert the test and to determine a confidence region C for $\boldsymbol{\mu}$ in the p-dimensional space. A vector $\boldsymbol{\mu}^* \in \Re^p$ is included in C if the shifted null hypothesis $\mathrm{H}_{\boldsymbol{\mu}^*}: \boldsymbol{\mu} = \boldsymbol{\mu}^*$ is not rejected. Let us denote the test statistic of the stabilized test (including the derivation of the weight vector with all possible modifications, the calculation of the scores and the final t test) by $T = T(\boldsymbol{x}_1^{(1)},\ldots,\boldsymbol{x}_{n^{(1)}}^{(1)},\boldsymbol{x}_1^{(2)},\ldots,\boldsymbol{x}_{n^{(2)}}^{(2)})$ and the corresponding t quantile by t_0. Then the confidence region C is defined by

$$\boldsymbol{\mu}^* \in C \quad \Leftrightarrow \quad T(\boldsymbol{x}_1^{(1)}-\boldsymbol{\mu}^*,\ldots,\boldsymbol{x}_{n^{(1)}}^{(1)}-\boldsymbol{\mu}^*,\boldsymbol{x}_1^{(2)},\ldots,\boldsymbol{x}_{n^{(2)}}^{(2)}) \le t_0 . \qquad (2)$$

Unfortunately, the function T is complicated with respect to the argument $\boldsymbol{\mu}^*$, even for such a simple test version as the SS test without further modifications. Therefore, the subsequent results have been obtain in a purely numerical way by a systematical scanning of the $\Re^p$ in the neighbourhood of the vector $\overline{\boldsymbol{x}}^{(1)} - \overline{\boldsymbol{x}}^{(2)}$ using a SAS macro. Because of the necessity to present the result in a graphical way, this version is obviously restricted to small dimensions p.

Another useful objective could be to find a confidence interval for the score used in the test. This would eliminate the restriction to low dimensions of the sample vectors. It is not necessary for the test to interpret the score by means of its weights, but it makes it easier to understand the results. Of course, this is a score with random coefficients. Nevertheless, it can be used to characterize the elements of the population. A problem is, however, that in the above test version the weight vector changes when the shift vector $\boldsymbol{\mu}^*$ in (2) is altered, because the total sums of products matrix $\boldsymbol{W}$ is then also dependent from $\boldsymbol{\mu}^*$. To circumvent this problem, Läuter (yet unpublished) proposed to derive the scores from the usual sums of products matrix which is based on the sample means (as in the proposal by O'Brien) instead of $\boldsymbol{W}$. Then the resulting tests are no longer exact, but one can find test versions that are at least conservative. However, the enormous variety of test modifications as indicated in Section 2 is lost.

4 Examples of numerically derived confidence regions

In this section we use data from Hothorn (cf. Kropf, Hothorn & Läuter, 1997) to demonstrate typical features of the confidence regions corresponding to the stabilized multivariate tests, thus also describing the tests through another perspective. A potentially toxical substance has been administered to Wistar rats and then some parameters of liver function have been observed. We only use the control group and the low dose group with each six animals, and we consider only three paramaters (Table 1).

Table 1. Example data from a toxicological experiment with Wistar rats (data from Hothorn)

Control group			Low dose group		
AP	ASAT	ALAT	AP	ASAT	ALAT
0.29	1.82	0.71	0.28	2.27	0.47
0.24	2.30	0.46	0.27	2.73	0.89
0.23	1.99	0.45	0.25	1.79	0.55
0.22	1.90	0.31	0.34	2.31	0.77
0.29	1.94	0.18	0.36	2.32	0.99
0.30	1.81	0.20	0.33	2.19	0.87

The mean differences of the three variables AP, ASAT and ALAT are -0.043, -0.308 and -0.372, respectively. Figure 1 shows the two-dimensional confidence

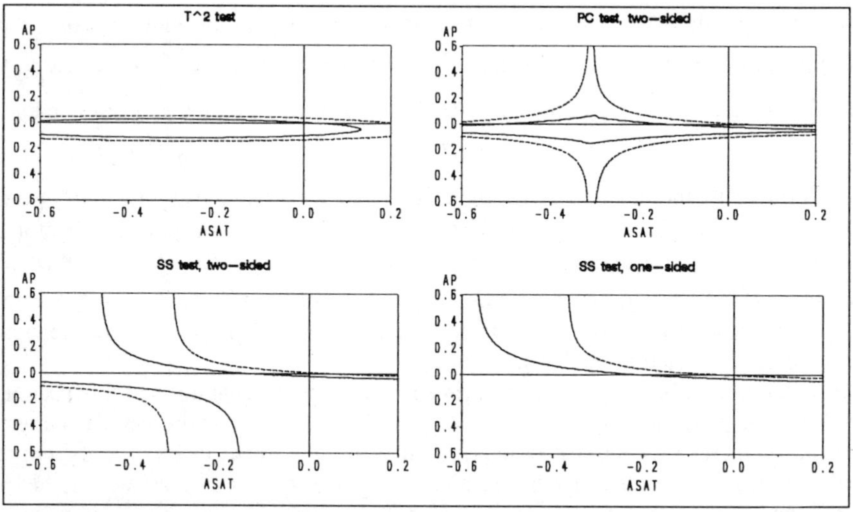

Fig. 1. Two-dimensional confidence regions (AP, ASAT) for the example with different tests (solid line – confidence level 0.95, dashed line – confidence level 0.99)

regions for AP and ASAT. The classical T^2 test has the well-known ellipsoid as corresponding confidence region. The PC test and the SS test have large or even infinite confidence regions. However, they better utilize the structure of the data for delimiting from the null hypothesis. Therefore, the origin is contained in the 95% region for the T^2 test, not in those for the two-sided PC test and SS test. The fourth part of the figure illustrates that when considering one-sided tests, the origin is even outside the 99% region.

Figure 2 shows the confidence region for the variables AP, ASAT and ALAT.

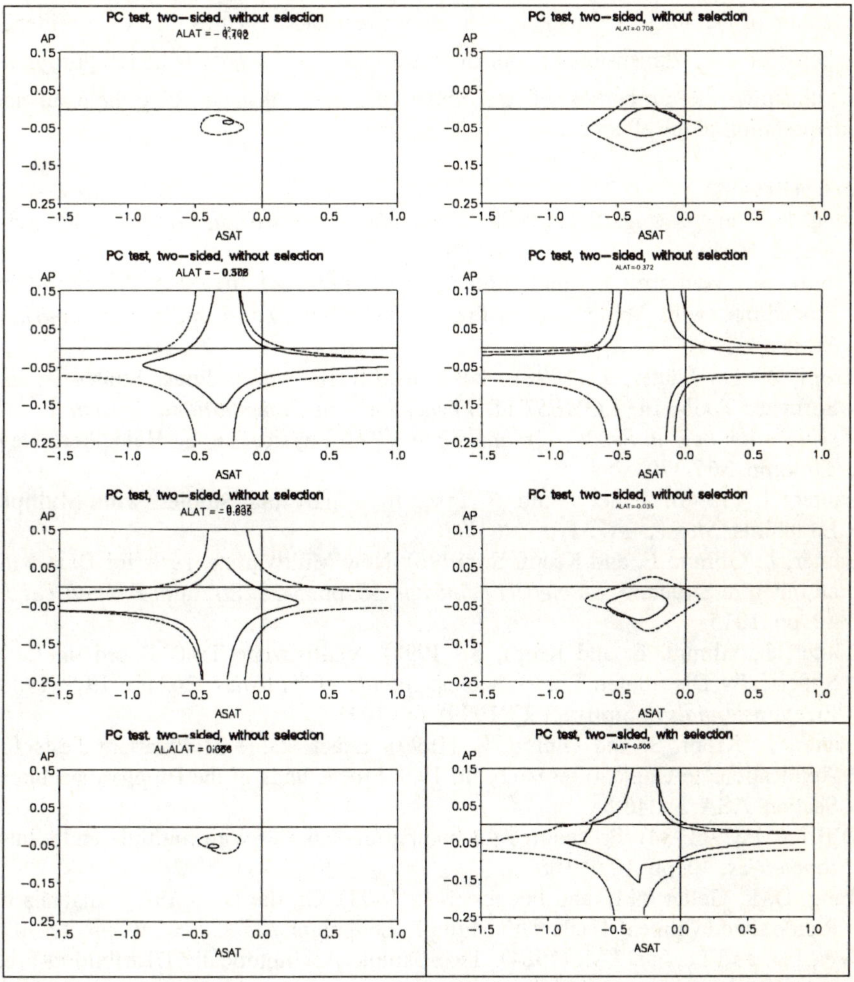

Fig. 2. Three-dimensional confidence region for the example with the two-sided PC test, presented as a series of two-dimensional plots with the variables AP and ASAT for different values of ALAT. Note that the ALAT values are not equally spaced and that the last plot shows a version with selection of variables (solid line – confidence level 0.95, dashed line – confidence level 0.99)

The three-dimensional region is presented as a series of two-dimensional AP – ASAT plots (slices) for varying fixed values of ALAT. The first seven plots of the figure are one series without selection of variables, the last plot shows the counterpart of the third plot in a test with selection.

As can be seen from the figure, the confidence region has a center around the empirical mean difference vector and has infinite 'tentacles', parallel to the three co-ordinate axes. With these three variables, the PC test gives a highly significant P-value of 0.006, the T^2 test only a non-significant P-value of 0.077. Nevertheless, the envolved shape of the confidence region prevents its easy interpretation. With selection of variables, the shape is even more irregular. Thus, this kind of confidence regions for the p-dimensional parameter vector $\mu := \mu^{(1)} - \mu^{(2)}$ is more appropriate to illustrate the properties of the underlying tests than to give help in the interpretation of the effect size.

References

Fang, K.T. and Zhang, Y.T. (1990). *Generalized Multivariate Analysis.* Springer-Verlag, Berlin.

Kropf, S., Hothorn, L. and Läuter, J. (1997). Multivariate Many-to-One Procedures, with Application in Pre-Clinical Trials. *Drug Information Journal*, **31**, pp. 433-447.

Kropf, S. and Läuter, J. (1998). Stable Multivariate Procedures. Strategies and Software Tools. In: *COMPSTAT. Proceedings in Computational Statistics. 13th symposium held in Bristol, Great Britain, 1998*. Physica-Verlag Heidelberg New York, pp. 365-370.

Läuter, J. (1996). Exact *t* and *F* Tests for Analyzing Studies with Multiple Endpoints. *Biometrics*, **52**, pp. 964-970.

Läuter, J., Glimm, E. and Kropf, S. (1996). New Multivariate Tests for Data with an Inherent Structure. *Biometrical Journal*, **38**, pp. 5-22. Erratum: *Biometrical J.*, **40**, pp. 1015.

Läuter, J., Glimm, E. and Kropf, S. (1998). Multivariate Tests Based on Left-Spherically Distributed Linear Scores. *Annals of Statistics*, **26**, pp. 1972-1988. Erratum: *Annals of Statistics* **27**(1999), pp. 1441.

Läuter, J., Kropf, S. and Glimm, E. (1999). Exact Stable Multivariate Tests for Applications in Clinical Research. In: 1998 Proceedings of the Biopharmaceutical Section, ASA, pp. 46-55.

O'Brien, P.C. (1984). Procedures for comparing samples with multiple endpoints. *Biometrics*, **40**, pp. 1079-1087.

Tang, D.-I., Geller, N.L. and Pocock, S.J. (1993). On the Design and Analysis of Randomized Clinical Trials with Multiple Endpoints. *Biometrics*, **49**, pp. 23-30.

Wei, L.J. and Lachin, J.M. (1984). Two-Sample Asymptotically Distribution-Free Tests for Incomplete Multivariate Observations. *JASA*, **79**,Theory and Methods Section, pp. 653-661.

Comparison of stationary time series using distribution-free methods

Elizabeth Ann Maharaj

Department of Econometrics and Business Statistics, Monash University, Caulfield Campus, P.O. Box 197, Caulfield East, Vic. 3145, Australia

Abstract. In this paper we propose distribution-free procedures based on the moving blocks bootstrap for differentiating between two stationary time series that are not necessarily independent. A chi-square type statistic and a Kolmogorov-Smirnov type statistic, each of which are based on the differences between the autocorrelations and the differences between the partial autocorrelations of the two series, are constructed. Monte Carlo studies carried out to assess the tests, show that they perform reasonably well. The tests are applied to real financial time series

Keywords. Moving blocks bootstrap, Stationary time series, Autocorrelations, Partial autocorrelations

1 Introduction

Discriminating between two or more time series has useful applications in various fields. For example, in medicine, discriminating between different sectors of brain wave recordings; in geology, discriminating between the waveforms of earthquakes and nuclear explosions; in finance, differentiating between performance of certain stocks and shares. A further motivation for discriminating between time series is efficient forecasting. It is well known that better estimates are obtained by pooling similar data sets. These similar data sets, which in this case would be stationary time series, can be identified on the basis of the techniques for differentiating between them. Forecasts can then be obtained.

In this paper we propose distribution-free procedures based on the moving blocks bootstrap for testing for differences between the generating processes of two stationary time series that are not necessarily independent. This contrasts with existing tests, such as those by De Sousa and Thomson (1982), Basawa *et al.* (1984), Maharaj (1996), Maharaj (1998) and Guo (1999), all of which first require fitting models, or those by Jenkins (1961), Coates and Diggle (1986), Swanepoel and Van Wyk (1986), Diggle and Fisher (1991), and Timmer *et al.* (1999), that first require estimating spectra. All these tests, except the one in Maharaj (1998), are applicable only to independent time series. The test statistics considered, namely, a chi-square type statistic and a Kolmogorov-Smirnov type statistic, are based on the differences between the autocorrelations, as well as the differences between the partial autocorrelations of the two stationary time series at each lag. None of these test statistics requires the fitting of models to the series, or the estimating of spectra from the series or from the fitted models, as is the case with the existing tests. Furthermore these tests are not restricted by the condition that the two series must be independent, as is the case with most of the existing tests.

2 Moving blocks bootstrap

The bootstrap procedure of Efron (1979) for estimating unknown sampling distributions of statistics is based on independent and identically distributed (*iid*) observations. However if the observations are not independent, as is the case with time series observations, the situation may be totally different. Singh (1981) observed that the *iid* resampling scheme of the usual bootstrap method failed to capture the dependence structure even for the sample mean of dependent data. Since then, there have been several attempts in the literature to modify and extend Efron's idea to dependent data. In most cases, the modifications take into account certain special properties of the specific models under consideration.

A general bootstrap procedure for weakly stationary data which is free of specific modelling, has been formulated by Kunsch (1989). A similar procedure has been suggested independently by Liu and Singh (1992). This procedure which is referred to as the moving blocks bootstrap method, does not require one to first fit a parametric or semi-parametric model to the dependent data. The procedure works for arbitrary stationary processes with short range dependence. Moving blocks bootstrap samples are drawn from a stationary dependent data set as follow: Let $X_1, X_2, \ldots, X_T$ be a sequence of stationary dependent random variables with common distribution function F for each X_i. Let θ_F be the parameter of interest and θ_T be its estimator based on $X_1, X_2, \ldots, X_T$. Let $B_1, B_2, \ldots, B_{T-b+1}$ be the moving blocks, where b is the length of each block. B_j stands for the jth block consisting of b consecutive observations starting from x_j, that is, $B_j = \{x_j, x_{j+1}, \ldots, x_{j+b-1}\}$. k independent block samples are drawn with replacement from $B_1, B_2, \ldots, B_{n-b+1}$. All observations of the k-sampled blocks are then pasted together in succession to form a bootstrap sample. The number of blocks k is chosen so that $T \cong kb$. The bootstrap estimator θ_T^* is then obtained from this sample. With the moving blocks bootstrap, the idea is to choose a block length b, large enough so that the observations more than b units apart are nearly independent. We simply cannot resample from the individual observations because this would destroy the correlation that we are trying to capture. However by sampling the blocks of length b, the correlation present in observations less than b units apart, is retained.

It has been shown that the moving blocks bootstrap is asymptotically valid for a wide range of statistics that belong to the class of smooth functionals (Kunsch (1989), Liu and Singh (1992)). In all these cases, it is assumed that the block length $b \rightarrow \infty$ at a rate such that $b/T \rightarrow 0$ as $T \rightarrow \infty$. Furthermore it has been shown asymptotically that the optimal block length for estimating standard errors is of order $O\left(T^{1/3}\right)$. However in practice, one has to decide on the constant associated with $T^{1/3}$. This issue was addressed by Hall *et al.* (1995). They suggested some data driven procedures for estimating the standard errors of specific statistics. Buhlmann and Kunsch (1999) suggested a fully data driven procedure to determine the optimal block length for estimating the standard errors of general statistics. While we will not attempt to determine the optimal block lengths for the statistics used in our proposed tests, we will select a range of appropriate block lengths based on the performance of the tests.

Variations of the moving blocks bootstrap, in particular subsampling of blocks have been discussed by Politis *et al.* (1997). However this procedure is far too computationally intensive and will not been considered here.

3 Hypothesis testing procedure

It is known that the complete probabilistic structure of a stochastic process $\{Z_t, t = 1, 2, \ldots, T\}$ is determined by the set of distributions of all finite collections of Z's. Much of the information in these joint distributions can be described in terms of the means, variances and covariances. If $\{Z_t, t = 1, 2, \ldots, T\}$ is a stationary[1] process, then the mean is constant, and the covariance $Cov(Z_t, Z_s) = \gamma_{|t-s|}$ depends only on the time difference $|t-s|$ and not on the actual times. The set $\{\gamma_k, k = 0, 1, 2, \ldots\}$, where $k = |t-s|$, is called the autocovariance function. A corresponding function is the autocorrelation function $\{\rho_k, k = 0, 1, 2, \ldots\}$. Since the autocorrelation function is a useful device for describing time series processes, we use it in the construction of two of the test statistics. Like the autocorrelation function, its companion function, the partial autocorrelation function $\{\phi_{kk}, k = 1, 2, \ldots\}$ which measures the dependence of Z_{i+k} on Z_i with the dependence $Z_{i+k-1}, Z_{i+k-2}, \ldots, Z_{i+1}$ removed, is also a useful device for describing time series processes. We use it in the construction of the other two test statistics.

Let $\{x_t\}$ and $\{y_t\}$, be two stationary time series that have some logical connection and let $\{X_t\}$ and $\{Y_t\}$ be their respective generating stochastic processes. Consider the following hypotheses: H_0: These is no difference between $\{X_t\}$ and $\{Y_t\}$; H_1: These is a difference between $\{X_t\}$ and $\{Y_t\}$. The proposed test statistics, each of which is of the form of a metric are $W_1 = \sum_{k=1}^{m} (r_{xk} - r_{yk})^2$, $W_2 = \max_{1 \le k \le m} |r_{xk} - r_{yk}|$, $W_3 = \sum_{k=1}^{m} (q_{xk} - q_{yk})^2$ and $W_4 = \max_{1 \le k \le m} |q_{xk} - q_{yk}|$, where r_{xk} and r_{yk} are the estimated autocorrelations at lag k of x_t and y_t respectively, q_{xk} and q_{yk} are the estimated partial autocorrelations at lag k of $\{x_t\}$ and $\{y_t\}$ respectively, and m is the number of lags under consideration.

The bootstrap provides a first-order asymptotic approximation to the distribution of the test statistic under the null hypothesis. Hence, the null hypothesis can be tested by comparing the test statistic to a bootstrap-based critical value. This is equivalent to comparing a bootstrap-based *p-value* to a nominal significance level α.. The bootstrap-based *p-value* is obtained by the following steps: (1) Consider the series $x_1, x_2, \ldots, x_T$. Sample with replacement k times from the set $\{B_1, B_2, \ldots, B_{T-b+1}\}$. This produces a set of blocks $\{B_1^*, B_2^* \ldots, B_k^*\}$, which when laid end-to-end forms a new time series, that is the bootstrap sample $x_1^*, x_2^*, \ldots, x_T^*$, of length T. Do the same for the series $y_1, y_2, \ldots, y_T$. (2) The autocorrelations r_{xk}^*, r_{yk}^* and the

[1] Stationarity here shall implies weak stationarity

partial autocorrelations q^*_{xk}, q^*_{yk}, $k = 1, 2, \ldots, m$ are obtained from the bootstrap samples, and the four test statistics W_1^*, W_2^*, W_3^* and W_4^* are then determined. The steps (1) and (2) are then repeated J times. The empirical distribution of the J values of each statistic W_i^* is the bootstrap estimate of the distribution of the corresponding statistic W_i, $i = 1, 2, 3, 4$. The bootstrap-based *p-value* p_i^* is an estimate of the *p-value* that would be associated with each test statistic W_i, and it is obtained as follows: $p_i^* = \#(W_i^* > W_i)/J$, $i = 1, 2, 3, 4$. For a nominal significance level α, H_0 is rejected if $p_i^* < \alpha$.

4 Simulation study

4.1 Outline

To estimate the size of each test for both $T = 50$ and $T = 200$, two series were simulated from each of the following processes: AR(1), $\phi = 0, 0.1, 0.5, 0.9$; MA(1), $\theta = 0.1, 0.5, 0.9$; ARMA(1,1), $\phi = -0.6$, $\theta = 0.3$. These models were chosen, so that both first and second order processes, as well as a range of parameter values would be considered. Estimates of the size were obtained at the 10%, 5% and 1% levels of significance. Estimates of power for both $T = 50$ and $T = 200$, at the 10%, 5% and 1% levels of significance were obtained by applying the tests to series generated from the following pairs of processes: (AR(1), $\phi = 0$; AR(1) $\phi > 0$) and (AR(1) $\phi = 0$; AR(2) $\phi_1 = 0$, $\phi_2 > 0$). In order to observe the effect of different block lengths on the performance of each test, block lengths $b = T^k$, $0.4 < k < 0.8$ in increments of 0.1 were considered. This range of k values was selected to ensure that the blocks did not contain too few observations or too many observations. To assess the sensitivity of the tests to different lag lengths, the test statistics were calculated for lags $\ell = T^k$, $0.4 < k < 0.8$ in increments of 0.1. Each Monte Carlo trial used 100 simulations, while a total of 500 bootstrap replications were generated for each Monte Carlo trial. All programming was done in *Gauss*.

4.2 Discussion

For both $T = 50$ and $T = 200$, each of the tests W_1 and W_2 performed fairly similarly at different lag and block lengths, in terms of size, for most of the case considered. As the process from which the series were generated, tended towards non-stationarity, namely, AR(1), $\phi = 0.9$, the size of each of these tests , especially for the smaller block lengths, was overestimated. For most of the other processes considered, these tests were generally valid. Each of the tests W_3 and W_4 also performed fairly similarly at different lag and block lengths, in terms of size, and they were generally valid in most cases. However there were relatively fewer instances when the sizes of W_3 and W_4 were overestimated as compared to that of the tests W_1 and W_2. For both $T = 50$ and $T = 200$, for all four tests, power calculations for both scenarios showed that power decreased as the number of lags that were included, increased. In all cases, for each block length considered, the highest power was achieved for the smallest number of lags ($T^{0.4}$) considered. Also the tests generally tended to have better power for the block lengths $T^{0.5}$, $T^{0.6}$ and $T^{0.7}$ than for the very small block length $T^{0.4}$ or the very large block length $T^{0.8}$.

Hence it is seems to be quite clear that varying the block and the lag lengths does have an influence on the power performance of the test. Power performance was generally better for the Kolomogrov-Smirnov type tests (W_2 and W_4) than the chi square type tests (W_1 and W_3). As expected, the power of the tests increased with increasing series length. Some simulation results for $T = 200$ are given in Table 4.1. All other results can be obtained from the author on request.

Table 4.1 Power of the tests $b = T^{0.4}$; $\ell = T^{0.4}$ and $T^{0.8}$ at the 5% level of significance

Process	$T = 200$ $b = 24$							
	W_1		W_2		W_3		W_4	
AR(1) ϕ	ℓ=8	ℓ=69	ℓ=8	ℓ=69	ℓ=8	ℓ=69	ℓ=8	ℓ=69
0.0	0.02	0.03	0.06	0.01	0.06	0.05	0.05	0.03
0.2	0.15	0.08	0.17	0.10	0.14	0.07	0.17	0.14
0.4	0.61	0.23	0.80	0.64	0.66	0.36	0.82	0.69
0.6	0.99	0.63	1.00	0.99	0.99	0.86	1.00	1.00
0.8	1.00	0.81	1.00	1.00	1.00	0.99	1.00	1.00

5 Application

To gauge the performance of the tests on observed time series, we considered the number of dwelling units financed by all lenders (banks and other institutions) in three states of Australia (Victoria (VIC), Queensland (QLD), and South Australia (SA)) from January 1978 to March 1998. The natural logarithmic transformation was taken for each series[2], and a graph revealed that the series for SA and QLD appeared to move in tandem while the pattern of the series for VIC appeared to differ to some extent from the other two. First differencing of each of these series appeared to render them stationary. The tests for block lengths $b = T^{0.5}$, $T^{0.6}$ and $T^{0.7}$ and lag $\ell = T^{0.4}$ were applied to these differenced series. The results which are non-significant for the pair (SA; QLD), for all four test, significant for the pairs (SA ;VIC), and (QLD;VIC) mainly for the tests W_3 and W_4 , would appear to be consistent with the observations made from the graph of the series. Furthermore when the parametric test from Maharaj (1998) was applied to these series, similar results were obtained. The p-values of the tests W_3 and W_4 are shown in Table 5.1.

Table 5.1 P-values of tests for $T = 242$; $\ell = T^{0.4}$; $b = T^{0.5}$, $T^{0.6}$ and $T^{0.7}$

	SA/VIC			SA/QLD			VIC/QLD		
	b=16	b=27	b=46	b=16	b=27	b=46	b=16	b=27	b=46
W_3	0.00	0.00	0.00	0.81	0.75	0.54	0.02	0.01	0.00
W_4	0.00	0.00	0.00	0.95	0.94	0.82	0.01	0.00	0.00

6 Concluding remarks

The tests appear to perform reasonably well for both short and long series for certain block and lag lengths. The advantage that these tests have over the existing

[2] Source: Australian Bureau of Statistics (web address: gopher://gopher.abs.gov.au:70/11/PUBS/finan/56090)

tests is that they do not require fitting models to the series or estimating spectra from the series. Furthermore they are not restricted by the condition that the two series under consideration must be independent, as is the case with most other tests.

References

Basawa, I.V., Billard, L. and Srinivasan, R. (1984). Large-sample tests of homogeneity for time series models. *Biometrika,* **71**, p. 203-206.

Buhlmann, P. and Kunch, H.R. (1999). Block length selection in the bootstrap for time series. *Computational Statistics and Data Analysis*, **31**(3), p. 295-310.

De Souza, P. and Thomson, P.J. (1982). LPC distance measures and statistical tests with particular reference to likelihood ratio. *Institute of Electrical and Electronics Engineers, Transmissions on Acoustics, Speech and Signal Processing,* **30**, p. 304-315.

Diggle, P.J. and Fisher, N.I. (1991). Nonparametric comparison of cumulative periodograms. *Applied Statistics*, **40**, p. 423 -434.

Efron, B. (1979). Bootstrap methods: Another look at the jackknife. *Annals of Statistics*, **7**, p. 1-26.

Guo, J.H. (1999). A non-parametric test for the parallelism of two first-order autoregressiive processes. *Australian and New Zealand Journal of Statistics*, **41**, p. 59-65.

Hall, P., Horowitz, J.L. and Jing, B.Y. (1995). On blocking rule for the bootstrap with dependent data. *Biometrika*, **82** (3), p. 561-574.

Jenkins, G.M. (1961). General considerations in the analysis of spectra. *Technometrics*, **3**, p. 133-166.

Kunsch, H.R. (1989). The jacknife and bootstrap for general stationary observations. *Annals of Statistics*, **17** (3), p. 1217-1241.

Liu, R.Y. and Singh, K (1988). *Exploring the Limits of the Bootstrap*. (ed. R. Lepage and L. Billard), John Wiley and Sons.

Maharaj, E.A. (1996). A significance test for classifying ARMA models. *Journal of Statistical Computation and Simulation*, **54**, p. 305-331.

Maharaj, E.A. (1998). Classification of stationary time series. In: *Proceedings in Computational Statistics 1998 - Short Comminications and Posters*. (ed. R. Payne and P. Lane), Bristol, August 1998.

Politis, D.N., Romano, J.P.and Wolf, M. (1997). Subsampling for hetroskedastic time series. *Journal of Econometrics*, **81**, p. 281-318.

Swanepoel, J.W.H. and Van Wyk, J.W.J. (1986). The Comparison of two spectral density functions using the bootstrap. *Journal of Statistical Computation and Simulation*, **24**, p. 271-282.

Singh, K. (1981). On the asymptotic accuracy of Efron's bootstrap. *Annals of Statistics*, **9**, p. 1187-1195.

Timmer, J., Lauk, M., Vach, W. and Lucking, C.H. (1999). A test for a difference between spectral peak frequencies. *Computational Statistics and Data Analysis*, **30** (1), p. 45-55.

Generation of Boolean classification rules

Roger J. Marshall

Department of Community Health, University of Auckland, New Zealand

Abstract. An algorithm to generate a class of Boolean classification rules is described. The algorithm is implemented in search partition analysis software (SPAN), a program designed to find an optimal binary data partition. Some comments on the relationship of the procedure with tree-based search procedures are discussed.

Keywords. Boolean classifiers, trees, decision rules

1 Introduction

Classification and regression trees are now well known statistical tools for data exploration and to develop classification rules. Other non-hierarchical search partitioning methods are less well known. One, search partition analysis (SPAN), is a procedure that I have proposed (Marshall, 1986,1995,1999). It has been applied primarily in medical problems. It is based on an algorithm to generate a binary partition from a class of Boolean classifiers in disjunctive normal form:

$$A = I_1 \cup I_2 \cup \ldots \cup I_q \tag{1}$$

where I_i is the conjunction of p_i attributes from a pool of m potential predictive attributes. I will refer to q as the *order* of the expression (1).

For example, an optimal classification rule with $q = 3$ and $p_1 = p_2 = p_3 = 2$ for predicting death within 28 days for a person experiencing a stroke is (Broad et al, 2000)

$$A = (C \cap I) \cup (C \cap S) \cup (D \cap I) \tag{2}$$

where C, S, I and D are respectively the *attributes*: loss of consciousness, institutionalised pre-stroke, severe motor deficit and requires help dressing. This rule was found by generating the possible Boolean expressions of the form (1) that can be generated from a pool of, in this case, $m = 12$ of the best predictors of death.

Although the methodology has been published (Marshall, 1986, 1995, 1999), the algorithm to enumerate the class of possible expressions of the form (1) has not. It can be presented as a combinatoric problem: the enumeration of the ways of selecting q groups of objects, with replacement, from a pool of m objects such that there are p_i objects in group i. The pool of m objects is the set of attributes, say T_m, that are to be combined in the rule. In order that the expression does include redundant I_i's, it is necessary to ensure that, for $p_j \leq p_i$, all the objects of I_j are not the same objects as those in I_i, for if they are $I_i \cup I_j = I_i$ and the order of the expression is no longer q.

2 The lock and key enumeration algorithm

Suppose the m objects of T_m are labelled $1, 2, \ldots, m$. Let I_i denote the set of objects in group i. If, for the moment, the requirement that *all* objects in one group are not contained in another is ignored, that is, "embedding" $I_j \subseteq I_i$ *is* allowed, there are $\binom{m}{p_i}$ possibilities for each I_i. We identify each possibility by a key K_i, where $K_i = 1, \ldots, \binom{m}{p_i}$ and the objects of the q groups are determined by a set keys $K_1, \ldots, K_q$. By generating a set of keys, the actual objects in each group can be obtained by "unlocking" the combination of each key to obtain I_i.

2.1 Unlocking a key

We discuss generating the keys in a moment. First, consider how to unlock a key K to obtain the combination $I = (i_1, \ldots, i_p)$ of a group of p objects from m. There are $\binom{m}{p}$ combinations and suppose the keys correspond to an ordering in which i_p is incremented the fastest and i_1 the slowest. Thus $I = (1, 2, \ldots, p)$ corresponds to $K = 1$, and $I = (m-p+1, m-p+2, \ldots, m)$ corresponds to $K = \binom{m}{p}$ and $i_1 < i_2 < \cdots < i_p$. Given i_1 there are $L_{i_1} = \binom{m-i_1}{p-1}$ possibilities for the remaining $p-1$ objects. Thus, if we define

$$S_i = \sum_{j=1}^{i} L_j$$

with $S_0 = 0$ and find an i such that $S_{i-1} < K \leq S_i$ then the first digit is $i_1 = i$. The remaining combination, $i_2, \ldots, i_p$ is now a permutation of the digits $i+1, \ldots, m$ and $K - S_{i-1}$ is the key to unlock the remaining combination. Thus with K , m and p reset to $K - S_{i-1}$, $m - i$, and $p - 1$ respectively we proceed in the same way to find the second digit and so on.

2.2 Generating keys

To generate the key combinations $K_1, \ldots, K_q$ consider first the case when $p_1 = \ldots = p_q = p$. There are $c = \binom{m}{p}$ possible values for each K_i and, to avoid embedding, these have to be distinct. Therefore the problem is that of enumerating the $\binom{c}{q}$ ways to select q objects from c, each enumeration satisfying $K_1 < K_2 < \cdots < K_q$. When the p_i's are not all equal, the keys do not necessarily have to satisfy the full inequality $K_1 < K_2 < \cdots < K_q$ unless $p_{i-1} = p_i$ for some i and it is then necessary to ensure that $K_{i-1} < K_i$. Subject to these constraints the sequence of keys can be generated by the simulated nested loop device of Gentleman (1975). Except in the special case $p_1 =, \ldots, = p_q = p$ the algorithm will generate some I_j that are embedded, and in this case it is necessary to check for embedding.

3 Size of the search

Let $N_{q,m}$ denote the number of expressions of the form of (1), that is, $N_{q,m}$ is the number of ways of selecting q groups of objects from a pool of m objects, with no embedding.

Consider first, the special case $p_1 = p_2 = \ldots = p_q = p$. Then, there are $c = \binom{m}{p}$ possible groups and $N_{q,m} = \binom{c}{q}$ ways of picking q of them.

When the p_i's are not all equal then there is no simple general formula for $N_{q,m}$. Consider first $q = 2$ and suppose $p_1 \geq p_2$. Let $c_i = \binom{m}{p_i}$ for $i = 1, 2$ and $c_{12} = \binom{p_1}{p_2}$. There are c_1 to select elements of I_1 and c_2 ways for I_2. However the latter must exclude the c_{12} ways in which I_2 may be embedded in I_1. Hence

$$N_{2,q} = c_1(c_2 - c_{12})/e!$$

where e is the number of times a p_i is repeated, that is, either 1 or 2. Note that $e = 2$ when $p_1 = p_2$ and the formula coincides with that given by $N_{q,m}$ above.

For the case $q = 3$ define, as above, $c_{jk} = \binom{p_j}{p_k}$ and, additionally, $c'_{jk} = \binom{m - p_k}{p_j - p_k}$, then it can be shown by extending the inclusion/exclusion argument above that

$$N_{q,3} = (c_1c_2c_3 - c_1c_3c_{12} - c_1c_2c_{13} - c_1c_2c_{23} + c_1c_{12}c_{13} + c_3c'_{13}c'_{23})/e!$$

where e is the number of times a p_i is repeated, either 1,2 or 3. Table 1 gives $N_{q,m}$ for some values of m and $p_1, \ldots, p_q$.

I have been unable to derive a general formula for $N_{q,m}$ when $q \geq 4$ and when the p_i's are unequal.

Table 1. $N_{q,m}$, the number of ways of selecting q groups of objects from m.

$p_1, \ldots, p_q$	8	12	16	20
1	8	12	16	20
2	28	66	120	190
3	56	220	560	1140
1,1	28	66	120	190
2,1	168	660	1680	3420
3,1	280	1980	7280	19380
2,2	378	2145	7140	17955
3,2	1400	13860	65520	213180
3,3	1540	24090	156520	649230
1,1,1	56	220	560	1140
2,1,1	440	2970	10840	29070
2,2,1	1680	17820	87360	290700
2,2,2	3276	45760	280840	1125180

4 Practical aspects

In practice, to generate decision rules of the form (1), one would try different values of $p_1, \ldots, p_q$. Specifically, combinations in which $q \leq Q$ and $p_i \leq P_i$. Obviously from Table 1, to make the search manageable $P_1, \ldots, P_Q$ need to be small, say 2, 2.

Also, one can consider interchanging $\cup$ and $\cap$ in (1), so admitting combinations in which the the complement of the decision rule is in normal disjunctive form and written in terms of the complements of the elements of T_m. Once the lock and key combination is generated, two partitions of the data ensue.

4.1 Expanding a search recursively

A recursive procedure to allow more complex rules to be generated, but without doing an exhaustive search, was suggested in Marshall(1995). In this procedure, an initial search is done with fixed $P_1, \ldots, P_Q$ and the best rule determined. The rule is then used to create an attribute which is added to the attribute set for a subsequent search.

In terms of the analogy used in the lock and key algorithm, this procedure is tantamount to adding an extra object, the attribute generated on the first search, to the pool, T_m, of m objects and again generating q groups of objects. The number of ways of doing this, now $N_{q,m+1}$, will include the $N_{q,m}$ that are already generated plus new combinations that have the new object in at least one of the groups. Only the latter are of interest and there are $N_{q,m+1} - N_{q,m}$ of them. I have been unable to devise an algorithm which generates just these, however, it is easy and quick to generate the $N_{q,m+1}$ combinations and just immediately discard those which do not contain the new attribute.

4.2 Complexity penalising

In the recursive approach, the expression that are generated may become quite complex, in the sense that q and p_i may become large. Suppose a criterion, G, is used to choose the "best" classification rule. For instance reduction in mean square error. In practice G may reach a plateau as expression complexity increases. A mechanism for offsetting G against increasing complexity is desirable, to choose simple and effective decision rules.

There are different ways to assess the complexity of a Boolean expression. One is to choose the number of distinct elements that comprise it, for example, four in (2). Another, which takes account of the complexity of the complement of A in (1) is $c = q + q' - 1$ where q' is the order of $\bar{A}$ when it is expressed in disjunctive normal form. For example, the complement of (2) is

$$\bar{A} = (\bar{C} \cap \bar{D}) \cup (\bar{C} \cap \bar{I}) \cup (\bar{S} \cap \bar{I})$$

so that the complexity is $c = 5$.

Since A and $\bar{A}$ are two sides of the same coin, this definition of complexity seems most reasonable.

Complexity penalising can be achieved formally by using $G' = G - \beta c$ to select a decision rule, where β is a penalty parameter which can be arbitrarily imposed. In practice, however, the upper envelope of a plot of G versus c provides an effective means to select an optimal rule. Typically there are initial rapid gains in G with increasing c, but these fall away at some point that may provide an optimal rule.

5 Contents of the attribute set: regular rules

In Marshall(1986,1999) is was argued that, to generate rational classifiers, rules of the form of (1) should be in terms of attributes that are "positive" with respect to the outcome variable. In effect this requires that T_m consists of a pool of positive attributes. For example, to establish the decision rule (2), T_m was a set of 12 attributes which included the four adverse features U, C, S, I that are in rule (2) plus eight other adverse pre-stroke attributes: unable to dress, unable to walk, aphasia, incontinence and so on. I have termed this kind of rule *regular*.

Where it is anticipated that interactions in which an adverse factor may become beneficial in combination with other factor, it is possible to include *both* an attribute and its complement in T_m. For example, if in the stroke example, it was felt that *not* being institutionalised might also, in some circumstances, present a risk on stroke outcome, both I and $\bar{I}$ could be included in T_m. The benefits of this, in the author's experience are dubious. Typically one ends up with a regular rule; the inclusion of complements only serve to make the search unnecessarily long.

6 Comparison to trees

One of the features of tree-based classification rules is that, typically, the rules that are generated are not regular. That is, when splits are made in terms of the presence or absence of an attribute, the routes to terminal nodes are generally combinations of some attributes that are present and some that are absent. Interpretation of the meaning of these combinations can often be difficult (Marshall, 2000).

However, in the process of amalgamating terminal nodes to form a decision rule, it is possible that the rule so formed, after Boolean simplification, *may* be regular. For example, Herman et al(1995) created a tree for classifying patients at high diabetes risk. There are four high risk terminal nodes which, when unioned, give the rule

$$(EO)(\bar{E}OS)(E\bar{O}H)(E\bar{O}\bar{H}G) \tag{3}$$

where O, E, H, S and G are respectively obesity, elderly (over 65), hypertension, sedentary and impaired glucose tolerance. Expression (3) reduces to simply

$$(EO)(OS)(EH)(EG)$$

which is regular. Also the reduced rule emphasises that, for example, being obese and sedentary is sufficient to satisfy (3), it is not necessary to be age under 65 (as the element $\bar{E}OS$ seems to imply). That is, some of the combinations are misleading and redundant.

On the other hand, because of the so-called replication problem (Pagallo and Haussler, 1990) it generally requires a complex tree to represent a relatively simple regular rule. For example, to represent (2) by a tree requires a tree with 11 terminal nodes.

These consideration, and others (Marshall, 2000) suggest that the rules derived by SPAN are often more interpretable, at least in the medical domain, than tree based rules. However, one could consider a hybrid of the two approaches by allowing splits of a tree based on combinations of the form (1). The SPAN software (mentioned below) allows this to be done. The benefits of this kind of approach, if any, over the direct non-hierarchical approach remain to be demonstrated.

7 Software

The algorithm describe is implemented in SPAN software, which can be obtained from URL http://www.auckland.ac.nz/mch/span.htm. SPAN is a Windows 95, 98 or NT program, which, as well as implementing the methods described above, also allows tree based analysis, with node splits that can be of the canonical form (1).

Generally, the time required to generate Boolean expressions is negligible compared to the time spent applying the expression to the data. SPAN has an built in algorithm to collapse data into grouped form which gives substantial gains in performance. For example, in a data set of 50,000 observations and with $m = 12$ and a binary outcome variable, the data collapse to only about 500 distinct data combinations, making the search very fast.

References

Broad, J., Marshall, R.J. and Bonita, R. (2000). The prediction of death soon after a stroke: A case study comparing four statistically derived prediction rules. *Journal of Clinical Epidemiology* (submitted).

Gentleman, J.F. (1975). Algorithm AS 88. Generation of all ${}^nC_r$ combinations by simulating nested Fortran DO loops. *Applied Statistics*, **24**, 374-376.

Herman, W.H., Engelgau, M.M., Smith, P.J., Aubert, R.E. and Thompson, T.J. (1995). A new and simple questionnaire to identify people at increased risk for undiagnosed diabetes. *Diabetes Care*, **18**,382-387.

Marshall, R.J. (1986). Partitioning methods for classification and decision making in medicine. *Statistics in Medicine*, **5**, 517- 526.

Marshall, R.J. (1995). A program to implement a search method for identification of clinical subgroups. *Statistics in Medicine*, **14**, 2645-2660.

Marshall, R.J. (1999). Classification to ordinal categories using a search partition methodology with an application in diabetes screening. *Statistics in Medicine*, **20**, 2723-2736.

Marshall, R.J. (2000). The use of classification and regression trees in clinical epidemiology. *Journal of Clinical Epidemiology* (in press)

Pagallo, G. and Haussler, D. (1990). Boolean feature discovery and empirical learning. *Machine Learning*, **5**, 71-99.

A statistical package based on Pnuts

Junji Nakano[1], Takeshi Fujiwara[2], Yoshikazu Yamamoto[3] and Ikunori Kobayashi[3]

[1] Center for Development of Statistical Computing, The Institute of Statistical Mathematics, 4-6-7 Minami-azabu, Minato-ku, Tokyo 106-8569, Japan
[2] The Graduate University for Advanced Studies, 4-6-7 Minami-azabu, Minato-ku, Tokyo 106-8569, Japan
[3] Faculty of Engineering, Tokushima Bunri University, Shido 1314-1, Okawa, Kagawa 769-2193, Japan

Abstract. We are developing a statistical system named Jasp in order to utilize recent advanced computational environments. We design Jasp language based on Pnuts, a script language written in and for Java language. Pnuts is a functional language without type declaration, and is easy to use for tentative and experimental work. We add tools for statistical analysis, and object oriented syntax mainly to bundle related functions. Besides a Jasp language window, Jasp user interface also has a graphical user interface window to show the history of analysis and to operate the system by pop-up menus. These two windows are tightly connected and can be used alternatively. Jasp is realized by client/server approach, and one client can execute calculations on more than one servers and can perform distributed computing. Jasp is able to use programs written in foreign languages such as C, C++ and Fortran.

Keywords. Distributed computing, Graphical user interface, Java, Object oriented approach

1 Introduction

We are building a new statistical system based on a script language Pnuts, which is implemented in Java language and enables full interaction with Java environments. Our system aims to adopt new technologies developed recently, such as object oriented approach, GUI (Graphical User Interface), Internet and distributed computing .

We have already many good statistical systems, for example, S (Chambers and Hastie, 1992) and XploRe (Härdle *et al.*, 1999), to name a few. They have been used widely in many fields for a long time, are sufficiently reliable, and are also sophisticated by adopting new technologies such as GUI and Internet usage. Many of them, however, acquired such modern facilities by adding new components on their original basic designs, which strongly reflect the technologies of the time when they were first designed. In other words, their new facilities are limited by their original designs.

Computer technologies have been developing explosively in these years both in hardware and software. Our aim is to make a statistical system which can use full abilities of these new technologies without limitation by making a completely new design. Such an attempt was difficult for a statistician before, because of the tremendous complexity of programming work. As a lot of good tools have become available recently for GUI and network programming at cheap cost, statisticians can do advanced programming work themselves relatively easily by using them. Among such tools, Java is though to be one of the most promising system programming languages.

Java has many new characteristics such as object oriented programming, platform independency, good libraries for network abilities(TCP/IP), and GUI and graphics support. It has also advanced Internet functions (RMI, Remote Method Invocation) and an interface to other languages (JNI, Java Native Interface). Although the calculation speed is still slow compared with traditional compiler languages, its advanced features are very attractive. This is the reason why many recent statistical system projects, for example, Omegahat (Chambers *et al.*, 2000), use it.

We also hope to make a new statistical system in Java language, named Jasp (JAva based Statistical Processor). To realize functional abilities of Jasp language, we use a script language Pnuts (Tomatsu, 2000) as the base language. We add statistical tools and object oriented mechanism to Pnuts for Jasp language, which are suitable for complex statistical work. We describe Jasp language more precisely in section 2.

As GUI is so popular in recent computational environments, GUI which is easy to use both for naive and professional users is strongly required for any statistical system. In section 3, we propose a way to connect GUI and CUI tightly, to record the history of analysis on the GUI and to operate Jasp by pop-up menus.

In section 4, we explain the client/server approach in Jasp. This enables the quick operation from Web browsers and distributed computing on several servers. We note that a similar approach was adopted in quantlet client/server technique of XploRe (Härdle *et al.*, 2000).

In section 5, we show that JNI is used in Jasp to import existing programs written in C, C++ and Fortran languages by slightly modifying them.

2 Jasp language

In statistical packages, language or macro for describing statistical procedures is important especially for advanced or frequent users. Statistical language should be simple and intuitive for tentative programming work, appropriate to write large and complex programs and able to reuse existing codes repeatedly and smoothly.

It is known that simplicity and flexibility of programming for statistical analysis can be achieved by functional language interpreters like S or XploRe. It is also admitted that object oriented languages are useful to write reusable and rigid programs for large systems. For example, C++ is mainly used for building complex systems. Considering these facts, we decided to design new functional and object oriented language for statistics, which is as simple as possible but suitable enough to modern computational environments.

We use Java language for building our statistical system, because it adopts many modern advanced features of computer science in it. Java language is well designed and easier to use than C++ language, and a pure object oriented compiler language. We think, however, it is still difficult for statisticians, who are not professional programmers, to use it as a daily tool for statistical analysis.

We decide to use Pnuts as the basis of Jasp language, because Pnuts is much simpler than Java and still can enjoy the merits of Java. Pnuts is a script language interpreter written in Java and has simple functional syntax and built-in language extension facility. Pnuts can use Java classes directly and seamlessly.

As Pnuts is a general purpose language and lacks desirable functions for statistical work, we add such facilities to Pnuts, and name it Jasp language. As Pnuts can use Java classes easily, we use Java classes freely available for basic statistical work: Jampack (Stewart, 2000) for calculating matrices, Ptplot (Lee and Hylands, 2000) for drawing graphics, and Colt (Hoschek, 2000) for computing probability distributions and generating random numbers. We make Jasp interpreter by adding some programs on Pnuts to use these classes effectively. Jasp language is basically a functional language as Pnuts is. Jasp function has no type declaration and is easy to use without any care for variable types. A simple example of Jasp function is:

```
function ols(y,x){
  coeff = (x.trans * x).inv * x.trans * y
  return coeff
}
```

We can write all statistical procedures by defining Jasp functions. When we write a lot of functions for our particular analysis, we often notice that they should be bundled in some way for reusing them in future work. We also know that some functions are related to a particular kind of data set. These characteristics are expressed by object oriented approach so well that we add this structure to Jasp language as one of the basic properties. We recommend to use this mechanism to define classes for assembling closely related functions and variables. In Jasp language, Java classes and Jasp classes are used almost similarly. This means that we can extend Jasp by both Jasp and Java languages. Basic system extensions should be realized by Java classes, and advanced statistical classes, for example, linear regression analysis or principle component analysis, should be written in Jasp language. An example of Jasp class definition is:

```
jaspclass LinearRegression {
  method LinearRegression(y,x){
    this.beta = ols(y,x)
    this.forecast = x * beta
  }

  function ols(y,x){
    coeff = (x.trans * x).inv * x.trans * y
    return coeff
  }
}
```

Note that above defined function is used without any modification in this class definition, and can not be called from outside the object. Variables prefixed by `this.` and all methods are accessible anywhere. The method whose name is the same as the class name is a constructor of the class. All Jasp functions also can be thought as constructors. In a class definition, a class can inherit another class like other object oriented languages.

3 Data based GUI

Terminal windows on which users give their commands by the language, or CUI (Character User Interface) were popular before. Recently, however, GUIs are preferred, especially by beginners or non-frequent users, because GUI is easier to operate than CUI. CUI requires heavy study of the language and memorizing commands. GUI uses pop-up menus, which show available procedures on the screen and work as a kind of help system. However, when users become familiar to a statistical system, learn the language and want to do complex analysis by it, they feel frustration due to much mouse clicking, and need to use the language on CUI or to write programs. For these reasons, it is desirable that GUI and CUI (or language) can be used alternatively without difficulty.

We design Jasp UI (User Interface) in such a way that CUI and GUI correspond as much as possible. In Jasp session, we usually have both a CUI terminal window and a GUI window. A CUI window is divided into upper and lower sub windows. We can operate Jasp fully by evaluating Jasp programs written in the upper sub window, and the result is printed as a series of characters in the lower sub window. When they are evaluated, top-level assignment sentences produce icons of assigned variables on a GUI window. A GUI window is divided into three parts. The upper left window is to record icons of variables assigned at the top-level in evaluating commands. For example, when we read data from a file by the command

```
data = read("japan.dat")
```

in the CUI window, an icon which expresses variable `data` appears on the upper left GUI window. The lower left window shows available functions and Jasp classes. The right window shows tables and graphs of results by command evaluation.

The GUI window is designed to use Jasp functions as much as possible by mouse operations. In the upper left window, icons of variables which express objects are arranged according to the generated orders. It records and shows the history of analysis clearly and intuitively. When we specify an icon by a mouse pointer, a pop-up menu appears, which shows a list of available methods for the object. By choosing one of menu items, we can send the message to the object, and have results. This GUI operation can generate corresponding Jasp command in a CUI window. Functions in the lower left windows also can be executed by mouse operation.

An example of Jasp UI is shown in Fig. 1.

4 Client/server approach and distributed computing

Jasp UI is designed for Internet usage. Jasp consists of a client UI program and a server interpreter program. To start Jasp really means to start Jasp UI, which starts Jasp server usually on the same machine. However, Jasp server can be on a remote computer, and the user does not need to know where the Jasp server is located. Usual data files are thought to be on the computer where UI is executed. It is reasonable that Jasp server works on remote powerful computers and Jasp UI runs on local small personal computers. In Jasp, even when network is not available and client and server are on the same computer, they still communicate by the internal TCP/IP connection.

Jasp UI can be used on WWW browsers which support Java applets. As Jasp applet on a less powerful local computer is just a client program and real calculations are performed on the powerful remote server, we can expect quick responses even when we need much calculation. In this situation, some limitations are set for security. For example, files on the local computer can not be read or written directory. In Jasp commands, however, we can use Internet URL (Uniform Resource Locators) for specifying the file we want to read.

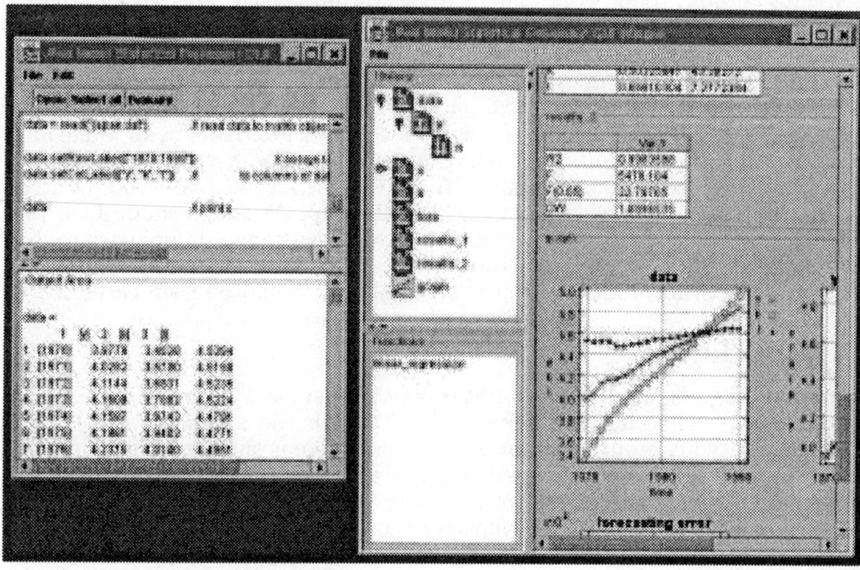

Fig. 1. An example of Jasp UI

Today, as we can use many powerful computers connected by network, it is natural to perform calculations simultaneously on them (Nakano, 1998). Although some distributed computing technologies for this purpose are available, for example, MPI (Message Passing Interface) and PVM (Parallel Virtual Machine), they are too general and difficult to use for statisticians. We realize simple commands for distributed computing mainly for executing calculations independently on remote computers. We have to note that Java applet can communicate only with programs on the computer from which Java applet was transferred. This means that client UI applet can use only the server on the Web server computer. We call the server to which the client connects firstly as a main server, and let a main server communicate with other remote servers. Thus communications are performed among servers and between a client and a main server. Functions we realize for this purpose are simple ones: a function to copy data between servers and a function to execute functions on a specified remote server. They are easy to use and sufficient for statistical work such as grid search for maximizing complicated functions or large simulation experiments. We use Java RMI to implement these functions.

5 Foreign language interface

Although many good statistical systems are available, not a few statisticians still write their programs for statistical research in traditional languages, because they have used such languages for a long time and/or need huge amount of calculation by high performance computers on which modern statistical systems do not work. Their programs implement advanced statistical techniques developed by themselves but are usually difficult to use because of the lack of good UI. To merge their programs into general purpose statistical systems is an important task. Therefore, foreign language interface is also indispensable for a statistical package. We use JNI mechanism to import procedures written in Fortran, C and C++ languages. It helps to make Java classes which execute foreign language programs without much modification. We have imported some programs in TIMSAC (TIMe Series And Control) package (Akaike and Nakagawa, 1988) into Jasp. Although main Java programs can be used commonly on all operating systems, different shared libraries of foreign language programs are required for each operating systems. It damages platform independency of Jasp, and should be used carefully.

6 Conclusion

Java language is used for building many statistical systems including our system Jasp. The characteristics of Jasp are on the language, the UI and the distributed computing abilities. Jasp language is simple, easy to use and enough powerful. We add statistical parts and object oriented programming syntax on Pnuts, which is a script language and can use Java classes smoothly. Our GUI is designed to have tight connections with CUI, and they can be used simultaneously. This UI is achieved by a client/server approach, and is available from Web browsers as a Java applet. Client/server approach is also useful for distributed computing on several servers. Jasp language has simple functions for realizing it.

The current version of Jasp runs on Linux, Sun Solaris and Microsoft Windows. Our system is still on the way of developing. Although the basic design principles are decided, details may change in future versions.

Jasp will be available freely from our Web site (`http://jasp.ism.ac.jp/`).

References

Akaike, H. and Nakagawa, T. (1988). *Statistical Analysis and Control of Dynamic Systems.* Dordrecht: Kluwer Academic Publishers.

Chambers, J.M. and Hastie, T.J. (ed.) (1992). *Statistical Models in S.* Pacific Grove: Wadsworth.

Chambers, J.M., Lang, D.T. *et al.* (2000). Omegahat. (`http://www.omegahat.org/`)

Härdle, W., Klinke, S. and Müller, M. (1999). *XploRe – Learning Guide.* Berlin: Springer. (`http://www.xplore-stat.de/`)

Härdle, W., Kleinow, T. and Tschernig, R. (2000). Web quantlets for time series analysis. SFB 373 Discussion Paper No. 1, 2000. (`http://sfb.wiwi.hu-berlin.de/`)

Hoschek, W. (2000) Colt. (`http://nicewww.cern.ch/ hoschek/colt/`)

Lee, E.A. and Hylands, C. (2000). Ptplot. (`http://ptolemy.eecs.berkeley.edu/java/ptplot/`)

Nakano, J. (1998). Graphical user interface for statistical software using Internet. In: *COMPSTAT98 Proceedings in Computational Statistics*, 407-412. Heidelberg: Physica-Verlag.

Stewart, P. (2000). Jampack. (`ftp://math.nist.gov/pub/Jampack/Jampack/AboutJampack.html`)

Tomatsu, T. (2000) Pnuts. (`http://javacenter.sun.co.jp/pnuts/`)

Generalized regression trees [1]

Marlon Núñez

Departmento de Lenguajes y Ciencias de la Computación, Universidad de Málaga, Campus de Teatinos s/n, Málaga 29071, Spain

Abstract. At present regression trees tend to be accurate, however they can be incomprehensible to experts. The proposed algorithm Economic Generalized Regression (EGR) induces regression trees that are more logical and convenient. EGR uses domain knowledge. The domain knowledge contains "*is-a*" hierarchies and *cost* associated to each variable. After generating several subtrees from training examples, EGR selects the best one according to a user-defined balance between accuracy and average classification cost. The user can define the degree of economy and generalization. This information will influence directly on the quality of search that the algorithm must undertake.

Keywords. Regression trees, domain knowledge, cost-sensitive learning.

1 Introduction

Regression trees, similar to classification trees are used when learning a relation between variables and a continuous class. Their inner nodes are labeled with a test on the value of a variable, and their leaves are labeled with a function prescribing a value to the class.

A regression tree actually implements a function $y(x_1, x_2, ..., x_n)$ of continuous or discrete variables. In the basic CART (Breiman, et al, 1984) algorithm and other CART-based algorithms (Le Blanc & Tibshirani, 1998), (Morimoto, et al, 1997) for construction of regression trees, the class value in the leaves is estimated as a constant function. In other CART-based algorithms (Karalic, 1992), the class is a linear function of continuous variables.

This paper shows that a regression tree can be both accurate and comprehensible at the same time. This goal can be reached by taking into account not only the data but also the domain knowledge. EGR uses domain knowledge. The domain knowledge contains "*is-a*" hierarchies and *cost* associated to each variable. In many domains these two aspects are important.

The term "cost" should be interpreted in its most abstract sense. Cost may be measured in many different units, such as monetary units (dollars), temporal units (seconds), or abstract units of utility. In medical diagnosis, cost may include such things as the quality of life of the patient, in so far as such things can be (approximately) measured. There are a number of references in which cost is

---

[1] This work has been partially supported by project FACA number PB98-0937-C04-01 of the CICYT, Spain. FACA is a part of the FRESCO project.

associated to problems by induction of decision trees. In Núñez (1988, 1991), and Tan (1991), each test (attribute, measurement, feature) may have an associated constant cost. For example, in medical diagnosis, a blood test has a cost. In Turney (1995) the cost of performing a certain test may be conditioned by the circumstances surrounding the test. The cost of performing a certain test on a given patient may be conditioned by the previous tests that have been chosen for the patient. For example, a group of blood tests ordered together may be cheaper than the sum of the costs of each test considered by itself, since the tests share common costs, such as the cost of collecting blood from the patient. In Breiman et al. (1984) there is a cost of misclassification, where there is a cost of assigning a prediction to a class i, when it actually belongs to class *j*. EGR deals with a not analyzed kind of cost analysis in regression trees learning. EGR selects variables by analyzing a constant cost associated to attributes and, minimizes the average classification cost of the regression tree on testing examples. This technic is an efficient way to learn knowledge in a cost-sensitive regression problem.

Domain knowledge in a regression tree reduces the number of leaves. There are problems in which there is a hidden model behind the data. In those cases, a single domain-driven "rule" may be as descriptive as several CART "rules". In Karalic (1997), an Inductive Logic Programming approach is proposed for using linear regression to deal with numerical data and domain knowledge.

2 EGR algorithm

The proposed EGR algorithm grows a maximal regression trees using domain knowledge on training examples, by recursively making splits based on the selection of the best cost/benefit ratio in every split. Then it constructs several subtrees based on the maximal tree. The best subtree is the one with the best accuracy and average classification cost measured with testing examples. EGR algorithm, presented in Table 1, can generate a tree with "rules" like:

IF (Age = 25-33) AND (Degree-of-Employee = *engineer*) AND (IQ-Test = 120-140)
THEN Salary ➔ (μ = 50000, σ = 3000, π=85%)

For the example above, suppose that the variable IQ-TEST has an associated cost, because knowing the results of the test implies a spending of time and money.

2.1 Growing the maximal tree

The selection of the best variable is based on a cost/benefit analysis of variables. The benefit is calculated analyzing the mixture of the data in every possible split (see figure 1). EGR transforms a continuous class into a discrete class by finding its components, and then, calculates the gain of information of the variable. The algorithm calculates the mean μ, standard deviation σ, and weight π of the components of the mixture, $[(\mu_1,\sigma_1,\pi_1), (\mu_2,\sigma_2,\pi_2)... (\mu_n,\sigma_n,\pi_n)]$, using Tanner method (Tanner, 1962) and some variations (Titterington, et al 1987).

Table 1. EGR algorithm

```
EGR (Examples, DomainKnowledge, MajorityThresholdInterval)
  Distribute Examples in TrainingExamples (66.66%) and TestingExamples (33.33%)
  MaximalTree = GrowMaximalTree (TrainingExamples, DomainKnowledge)
  SetOfSubtrees = GenerateSubtrees (TestingExamples, MaximalTree)
  SelectBestSubtree(SetOfSubtrees)
```

The procedures and functions used in EGR are:

```
GrowMaximalTree (TrainingExamples, DomainKnowledge)
  Find μ,σ,π of every component of mixture of the continuous class of the examples
  If there is only a component in the mixture
    Label the leaf with (μ,σ,π) of component
    Return the leaf
  Else
    Select variable in accordance with a cost/benefit criterion
    Initialize AbstractValuesList with abstract values associated to the selected variable
    Let LastValidTree = nil
    Let Subtree =GrowMaximalTree (examples re-writing the first abstract values)
    If the Subtree below that branch is consistent and complete
      Let LastValidTree = Subtree. Then try recursively a better generalization than
      the one stored, performing a union with another element in AbstractValuesList
    Else
      Return LastValidTree

GenerateSubtrees (TrainingExamples, MaximalTree, MajorityThresholdInterval)
  For Threshold = Min to Max of MajorityThresholdInterval step ϕ%
    For every node of MaximalTree
      Find μ,σ,π of every component of mixture in class of the TrainingExamples
      If the majority component has a π (weight) greater than Threshold
        Label the leaf with (μ,σ,π) of majority component and add it to Subtree.
    Insert Subtree in ListOfSubtrees
  Return ListOfSubtrees

SelectBestSubtree(TestingExamples, SetOfSubtrees)
  For every Subtree in SetOfSubtrees
    Measure relative mean square error of Subtree based on TestingExamples
    Measure average Classification Cost of Subtree based on TestingExamples
  Select the subtree with the best balance (user-calibrated) of error and classification cost.
```

The benefit of the variable is its gain of information (ΔI) as a function of the weights π of components found before and after splitting. The gain of information is the difference between the initial mixture entropy in the class (before using the variable) and the pondered mixture entropy in the class after splitting with that variable. The cost of a variable is the quantification of the degree of difficulty that implies knowing the value of that variable (i.e. a clinical test costs money and time).

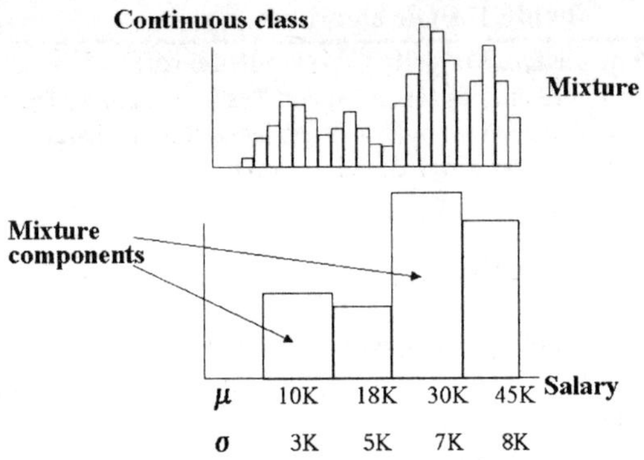

Figure 1: Entropy of a mixture

The selected variable is the one with the best cost/benefit relation. This cost/benefit function is defined as the ratio between the cost and the information efficiency of the variable: *cost* / ($2^{\Delta I}$-1). Figure 2 illustrates how the information efficiency $2^{\Delta I}$-1 is calculated from entropy of mixtures.

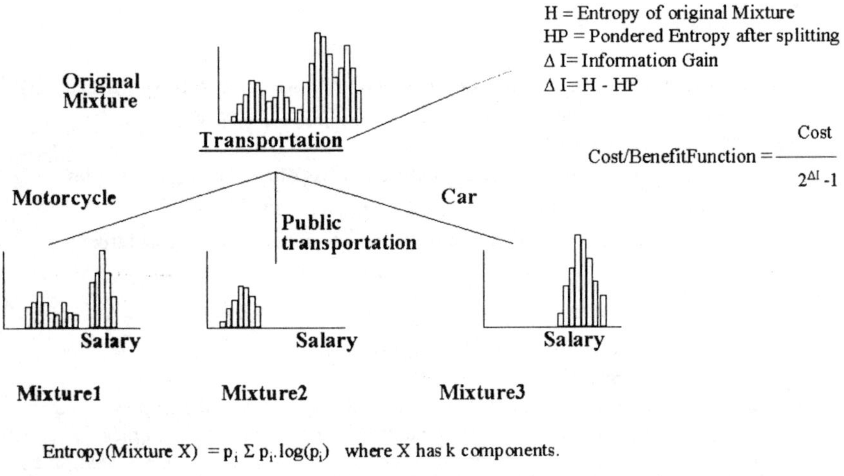

Figure 2: Illustration of cost/benefit analysis of an attribute (i.e. transportation)

Analyzing taxonomies in domain knowledge

While constructing the maximal tree, EGR analyses the possible branches of the selected variable (i.e. Degree-of-Employee). The purpose is to generalize its original values (i.e. civil-engineer, computer-engineer, electrician, and plumber), climbing the taxonomy of abstract values *engineer* and *technician* (i.e. civil-engineer "is-an" *engineer*; electrician "is-a" *technician*, etc.).

It initializes an internal list (AbstractValuesList) with the more general abstract values of is-a hierarchy. It then constructs a subtree according to that abstract value (i.e. Degree-of-Employee = *engineer*). If the subtree below that branch is consistent and complete the algorithm stores it in a temporary place (LastValidTree), and tries a better generalization than the one stored, performing a logic union (OR) with another value (i.e. Degree-of-employee = *engineer* OR *electrician*). If this attempt does not work, it will at least have the stored subtree (LastValidTree).

Continuous variables
Continuous variables are processed as taxonomic values. Firstly, the algorithm tries an interval (i.e. age: 25-28). If the subtree bellow this value is consistent and complete the algorithm tries a union with another interval (Age = 25-28 or 28-33, namely Age = 25-33). As it was explained before, if this attempt does not work, it will at least have the stored subtree (Age = 25-28).

Algorithm EGR proceeds by growing the maximal tree until it is not possible to grow it any further. The leaves of the maximal tree are labeled with the parameters of the majority component (μ_i, σ_i ,π_i).

2.2 Pruning and testing

The maximal tree is then used to build several subtrees by applying a majority threshold in every node. Namely, EGR generates a number of subtrees changing this majority threshold in a user-defined interval. At every node of MaximalTree, EGR finds μ, σ, π of every component of mixture in class of the TrainingExamples. If the majority component has a π (weight) greater than Threshold, then it labels the leaf with (μ,σ,π) of majority component. Once the subtrees are generated, EGR measures the relative mean square error and average classification cost in every subtree using testing examples. The best subtree is the one with the best user-defined balance of error and classification cost.

3 Experimentation

EGR was compared to CART in terms of *relative mean squared error* (RE^{ts}) and average classification cost. Generally speaking, CART is slightly better in terms of accuracy, but EGR generates tree with better (lower) average classification cost.

In Boston classical data, for instance, CART obtained 0.28 of RE^{ts}, while EGR got 0.34. Regarding classification cost analysis, EGR trees achieved much lower average classification cost than CART trees (up to 82% of saving). For this later analysis we assigned costs for knowing the values of variables. This result is not surprising because EGR has been specially designed to cost-sensitive problems.

CART selected the variable RM in the first split; however, in the CART overall analysis on *importance of variables,* the best ranked is LSTAT. It is interesting to mention that EGR selected LSTAT as the first variable for splitting.

EGR has been used for learning regression trees that predicts time of occurrence of discrete alarms in a computer network (Núñez, 2000). The Relative Mean Squared Error of EGR was slightly greater compared to CART (0.09 and 0.08). Experts in network management analyzed the generated rules. Regression trees were easier to understand since EGR took into account taxonomies and associated cost. Unfortunately, understandability measures also involve subjective factors difficult to be contrasted among methods.

4 Conclusions

Generally speaking, the main characteristics of EGR are the following:

- It does not require variables to be selected in advance, because it uses a method for determining the best variables for splitting.
- It generates regression trees with nodes of multiple branches.
- It uses mixture analyses of the continuous class in the learning phase, as well as, in the pruning phase.
- EGR regression tress are more understandable and convenient without sacrificing accuracy, because it uses domain knowledge.

Acknowledgement. Thanks to Raúl Fidalgo for developing some of the modules, and testing the final application extensively.

References

Breiman, L., Friedman, J.H., Olsen, R.A., and Stone, C.J., (1984). *Classification and Regression Trees,* Wadsworth Int. Group, Belmont, California.

Karalic, A. (1992). Employing linear regression in regression tree leaves. *Proceedings of European Congress on Artificial Intelligence*, Vienna, Austria.

Karalic A, (1997). First Order Regression. *Machine Learning*, Vol. 6, 26 (2/3):147-176, Kluwer Academic Publishers.

Le Blanc, M. and Tirshirani, R. (1998). Monotone Shrinkage of Trees. *Journal of Computational and Graphics Statistics,* Volume 7, Number 4.

Morimoto, Y., Ishii, H. and Morishita, S. (1997). *Efficient Construction of Regression Trees with Range and Region Splitting.* VLDB 1997.

Núñez, M. (2000). Learing Patterns of Behavior by Observing System Events. To appear in *Proceedings of European Conference of Machine Learning.*

Núñez, M. (1991). The Use of Background Knowledge in Decision Tree Induction. *Machine Learning*, Vol. 6, Number 3, Kluwer Academic Publishers.

Núñez, M. (1988). Economic Induction: A Case Study, *Proceeding of Third European Working Session on Learning,* Pitman Publishing, London.

Tan, M. (1991). Learning a cost-sensitive internal representation for reinforcement learning. In *Proceedings of IWML-1991.* Morgan Kaufmann.

Tanner, W.F. (1962). Component of the Hypsometric Curve of the Earth. J. Geophis. Res.

Titterington, D.M., Smith, A.F.M. and Makov, U.E. (1987). *Statistical Analysis of Finite Mixture Distributions*, John Wiley & Sons.

Turney, P.D. (1995). Cost-sensitive classification: Empirical evaluation of a hybrid genetic decision tree induction algorithm, *Journal AI Research*, **2**, 369.

Generalized linear mixed models: An improved estimating procedure

Jian-Xin Pan[1] and Robin Thompson[2]

[1] Centre for Medical Statistics, Department of Mathematics, Keele University, Staffordshire ST5 5BG, UK
[2] Statistics Department, IACR-Rothamsted, Harpenden, Hertfordshire, AL5 2JQ, UK

Abstract. The analytically intractable integrated-likelihood in the generalized linear mixed models (GLMM) is approximated in terms of Gauss-Hermite (GH) quadrature. Maximizing the approximated likelihood leads to an improvement over the existing penalized quasi-likelihood (PQL) estimation. The bias caused by the PQL estimation can be eliminated by adding GH quadrature nodes.

Keywords. Gauss-Hermite quadrature; Generalized linear mixed models; Penalized quasi-likelihood estimates

1 Introduction

Generalized linear mixed models (GLMM) are an extension of the generalized linear models (GLM) in which random effects are incorporated into the linear predictor. Such extensions of the GLM can be used to accommodate correlation and overdispersion often observed among discrete outcomes. The GLMM hence broadens extensively the application ranges of the GLM to a variety of disciplines where either binary or binomial data with within-subject correlation are commonly encountered. Incorporating random effects into the linear predictor of the GLM, however, complicates the estimation problems considerably. To obtain the maximum likelihood estimates (MLE) of the parameters, for example, we have to integrate out random effects from the models, which except for a few special cases is analytically intractable. In the literature most of the technical efforts are mainly based on the Laplace approximation. The 1st-order Laplace approximation leads to the so-called *penalized quasi-likelihood* (PQL) estimates (Breslow and Clayton, 1993). Several analogous procedures were proposed, e.g., by Lee and Nelder (1996).

A key feature of the PQL estimation is that it is easily implemented by iteratively fitting a linear mixed model to a modified dependent variable. A recently developed GENSTAT procedure GLMM (Payne *et al*, 1995), implementing the restricted maximum likelihood (REML) techniques (see, e.g., Patterson and Thompson, 1974) in each iteration, provides an easy access to this approach, which makes this approximation increasingly popular in the use of the GLMM. Although the PQL procedure has been demonstrated to work reasonably well for certain data, it could produce seriously biased estimates in particular if the underlying random effects parameter values are large, see, e.g., Goldstein and Rasbash (1996). To reduce the serious biases caused by the PQL estimates, Goldstein and Rasbash (1996) suggested applying an adjusted 2nd-order Laplace approximation to the integrand in the integrated likelihood, and the resulting estimates, namely the 2nd-order PQL estimates, are more accurate than the ordinary PQL estimates in certain circumstances. However, the bias-corrected estimate yields less satisfactory inference in the GLMM in particular when the number of observations in each cluster is small or the variance components of the random effects are large. Recently, some numerical approximation methods such as Markov Chain Monte Carlo (MCMC) techniques and Monte Carlo EM (MCEM) algorithm were proposed to calculate the estimates in the models. These numerical approximation procedures, however, are computationally intensive.

In this paper an improved estimation approach with less computational effort is proposed to reduce the bias caused by the PQL estimates. The key idea is to make an appropriate transformation for random effects and then apply Gauss-Hermite (GH) quadrature to approximate the transformed likelihood. The parameters in the GLMM are then estimated by iteratively fitting a linear mixed model to a modified dependent variable. More specifically, the fixed effects in the models are estimated using iteratively weighted least squares (IWLS) and the variance components of random effects are estimated using REML techniques. Thus the proposed approach takes advantage of the computational flexibility of the PQL approach and the optimal property of the GH approximation. In contrast, the PQL approach is a special case of our proposed method if only one quadrature node is utilized. In fact, the improved PQL approach can be thought of as a higher-order Laplace approximation and the number of quadrature nodes corresponds to the order of Laplace approximation. Numerical results show that this method can satisfactorily eliminate the biases of the PQL estimates by successively adding GH quadrature nodes, in particular when the variance component of the random effects is large.

2 Improved Penalized Quasi-Likelihood Estimation

Let y_{ij} be the response for the jth observation in subject i, where $j = 1, 2, \cdots, n_i$ and $i = 1, 2, \cdots, N$. Suppose $\boldsymbol{x}_{ij}$ is a $p \times 1$ vector of covariates and $\boldsymbol{\beta}$ is the regression coefficient. Given an unobserved random effect b_i, the observations y_{ij} in the ith subject are independent and follow *a regular exponential distribution family* $f(y_{ij}|b_i) = \exp\{(a_{ij}/\phi)(y_{ij}\xi_{ij} - d(\xi_{ij}) + c(y_{ij}))\}$, where a_{ij} is a given constant, ϕ is a known overdispersion scalar, and ξ_{ij} is called a *natural parameter*. Obviously, the conditional expectation and variance of y_{ij} given b_i are of the forms $u_{ij} = E(y_{ij}|b_i) = d'(\xi_{ij})$ and $v_{ij} = \text{var}(y_{ij}|b_i) = \phi a_{ij}^{-1} d''(\xi_{ij})$, respectively, where $d'(\cdot)$ and $d''(\cdot)$ denote the 1st- and 2nd-order derivatives of $d(\cdot)$. The random effects b_i are usually assumed to follow a normal distribution with mean 0 and variance θ, i.e., $b_i \sim N(0, \theta)$. The GLMM introduces a *link function* $g(\cdot)$, which links the conditional expectation u_{ij} to the *linear predictor* $\eta_{ij} = \boldsymbol{x}_{ij}^{\tau}\boldsymbol{\beta} + b_i$, i.e., $g(u_{ij}) = \eta_{ij}$ or $u_{ij} = h(\eta_{ij})$, where $\boldsymbol{x}_{ij}^{\tau}$ stands for the transpose of $\boldsymbol{x}_{ij}$ and $h(\cdot)$ is the inverse function of $g(\cdot)$. Furthermore, the conditional variance v_{ij} can be written as $v_{ij} = \phi a_{ij}^{-1} v(u_{ij})$ where $v(\cdot)$ is known as a *variance function*.

Suppose $\boldsymbol{y}_i = (y_{i1}, y_{i2}, \cdots, y_{in_i})^{\tau}$ is the observation vector of the ith subject. Since the subject observations $\boldsymbol{y}_1, \boldsymbol{y}_2, \cdots, \boldsymbol{y}_N$ are independent, the log-likelihood function of the GLMM must be of the integrated form

$$l(\boldsymbol{\beta}, \theta) = \sum_{i=1}^{N} \log \int f(\boldsymbol{y}_i|b_i) f(b_i) db_i \tag{1}$$

where $f(\boldsymbol{y}_i|b_i) = \prod_{j=1}^{n_i} f(y_{ij}|b_i)$ and $f(b_i)$ is the density function of b_i, i.e., $N(0,\theta)$. Except for a few special cases, the integral involved in (1) is analytically intractable and thus good approximations to the integrated likelihood should be sought. Rewrite the integrand in (1) as $\exp\{-\kappa(b_i)\}$, where $\kappa(b_i) = -l_i + b_i^2/2\theta + (1/2)\log(2\pi\theta)$ and $l_i = \sum_{j=1}^{n_i} \log f(y_{ij}|b_i)$. Let $\kappa'(\cdot)$ and $\kappa''(\cdot)$ be the 1st- and 2nd-order derivatives of $\kappa(\cdot)$. Denote $\tilde{b}_i$ as the minimizer of $\kappa(b_i)$, i.e., $\tilde{b}_i = \arg\min_{b_i}\{\kappa(b_i)\}$, then $\tilde{b}_i$ must be the solution to the equation $\kappa'(b_i) = 0$. The PQL approach (Breslow and Clayton, 1993) is to expand the integrand in the 1st-order Taylor series about $\tilde{b}_i$, which leads to inaccurate estimates in particular when the variance component θ is large. Instead of using Taylor expansion, we apply GH quadrature approximation to the integral in (1). The GH quadrature is commonly used for numerical integration in statistics because of its relation to the Gaussian density. However, it samples integration nodes from a Gaussian distribution irrespective of the range of integrand. To allow the integration nodes to be sampled in a reasonable range, certain transformation for the integration variable should be applied (see, e.g., Liu and Pierce, 1994). Let $a_i = \tilde{\tau}_i^{-1/2}(b_i - \tilde{b}_i)$, where $\tilde{\tau}_i = [\kappa''(\tilde{b}_i)]^{-1}$. Dividing and multiplying the integrand by the density of the normal distribution $N(0,1)$ allows the log-likelihood to be approximated by

$$l(\boldsymbol{\beta},\theta) \approx \sum_{i=1}^{N} \log\Big[\sum_{k=1}^{m}\Big(\frac{\omega_k}{\pi^{1/2}}\Big)\exp\Big\{\tilde{l}_i^{(k)} - \frac{1}{2\theta}\tilde{b}_i^{(k)2} + c_k^2 + \frac{1}{2}\log\frac{\tilde{\tau}_i}{\theta}\Big\}\Big] \tag{2}$$

where $\tilde{l}_i^{(k)} = \sum_{j=1}^{n_i} \log f(y_{ij}|\tilde{b}_i^{(k)})$ and $\tilde{b}_i^{(k)} = \tilde{b}_i + (2\tilde{\tau}_i)^{1/2}c_k$. The GH quadrature node c_k ($k = 1, 2, \cdots, m$) is the kth zero of Hermite polynomial with degree m, and the weight ω_k depends on the number m and Hermite polynomial with degree $m-1$ evaluated at c_k, see, e.g., Davis and Rabinowitz (1975, Chapter 2) for more details. The Equation (2) is known as GH *log penalized quasi-likelihood* and the estimates maximizing Equation (2) are called GH *penalized quasi-likelihood* (GH-PQL) estimates.

In the spirit of the PQL estimates, the variance component θ is assumed to be fixed first. Let $\widehat{\boldsymbol{\beta}}$ be the estimate of the fixed effects $\boldsymbol{\beta}$ maximizing $l(\boldsymbol{\beta},\theta)$ for fixed θ, i.e., $\widehat{\boldsymbol{\beta}} = \max_{\boldsymbol{\beta}}\{l(\boldsymbol{\beta},\theta)\}$. According to the definition of $\tilde{b}_i$, if $\hat{b}_i$ is the value of $\tilde{b}_i = \tilde{b}_i(\boldsymbol{\beta},\theta)$ with $\boldsymbol{\beta}$ replaced by $\widehat{\boldsymbol{\beta}}$, i.e., $\hat{b}_i = \tilde{b}_i(\widehat{\boldsymbol{\beta}},\theta)$, then the estimates $\widehat{\boldsymbol{\beta}}$ and $\hat{b}_i$ jointly maximize $l_P^{gh}(\boldsymbol{\beta}, b_i)$, the right-hand side of the Equation (2) but with $\tilde{l}_i^{(k)}$ and $\tilde{b}_i^{(k)}$ replaced by $l_i^{(k)}$ and $b_i^{(k)}$, respectively, where $l_i^{(k)} = \sum_{j=1}^{n_i} \log f(y_{ij}|b_i^{(k)})$, $b_i^{(k)} = b_i + (2\tau_i)^{1/2}c_k$, and τ_i is of the same form but with $\tilde{b}_i$ replaced by b_i. Differentiating $l_P^{gh}(\boldsymbol{\beta}, b_i)$ with respect to $\boldsymbol{\beta}$ and b_i leads to the score equations for the mean parameters

$$\frac{\partial l_P^{gh}}{\partial \boldsymbol{\beta}} = \sum_{i=1}^{N}\sum_{k=1}^{m} \omega_{i,k}^{*} \sum_{j=1}^{n_i} \frac{(y_{ij} - u_{ij}^{(k)})\boldsymbol{x}_{ij}}{\phi a_{ij}^{-1} v(u_{ij}^{(k)}) g'(u_{ij}^{(k)})} = \mathbf{0} \tag{3}$$

and

$$\frac{\partial l_P^{gh}}{\partial b_i} = \sum_{k=1}^{m} \omega_{i,k}^{*} \Big[\sum_{j=1}^{n_i} \frac{(y_{ij} - u_{ij}^{(k)})}{\phi a_{ij}^{-1} v(u_{ij}^{(k)}) g'(u_{ij}^{(k)})} - \frac{1}{\theta} b_i^{(k)}\Big] = 0, \tag{4}$$

where $u_{ij}^{(k)} = h(\eta_{ij}^{(k)})$, $\eta_{ij}^{(k)} = \boldsymbol{x}_{ij}^{\tau}\boldsymbol{\beta} + b_i^{(k)}$, and the weight $\omega_{i,k}^*$ is of the form

$$\omega_{i,k}^* = \frac{\omega_k \exp\{l_i^{(k)} - b_i^{(k)2}/(2\theta) + c_k^2\}}{\sum_{k=1}^m \omega_k \exp\{l_i^{(k)} - b_i^{(k)2}/(2\theta) + c_k^2\}}.$$

We now employ the Fisher scoring algorithm to iteratively calculate the solutions to Equations (3) and (4). Let $\delta_{ij}^{(k)} = \{\phi a_{ij}^{-1} v(u_{ij}^{(k)})[g'(u_{ij}^{(k)})]^2\}^{-1}$ and $\gamma_{ij}^{(k)} = \{\phi a_{ij}^{-1} v(u_{ij}^{(k)}) g'(u_{ij}^{(k)})\}^{-1}$, and $\boldsymbol{W}_i^{(k)} = \mathrm{diag}(\delta_{i1}^{(k)}, \cdots, \delta_{in_i}^{(k)})$ and $\boldsymbol{R}_i^{(k)} = \mathrm{diag}(\gamma_{i1}^{(k)}, \cdots, \gamma_{in_i}^{(k)})$ be the diagonal matrices with size $n_i \times n_i$. Also, we denote $\boldsymbol{u}_i^{(k)} = (u_{i1}^{(k)}, \cdots, u_{in_i}^{(k)})^{\tau}$, $\boldsymbol{X}_i = (\boldsymbol{x}_{i1}, \boldsymbol{x}_{i2}, \cdots, \boldsymbol{x}_{in_i})^{\tau}$, and $\boldsymbol{1}_{n_i}$ stands for the vector of 1's with length n_i. Furthermore, we form large vectors and matrices in lexicographic order as $\boldsymbol{y} = (\boldsymbol{y}_1^{\tau}, \cdots, \boldsymbol{y}_N^{\tau})^{\tau}$, $\boldsymbol{b} = (b_1, \cdots, b_N)^{\tau}$, $\boldsymbol{u}^{(k)} = (\boldsymbol{u}_1^{(k)\tau}, \cdots, \boldsymbol{u}_N^{(k)\tau})^{\tau}$, $\boldsymbol{X} = (\boldsymbol{X}_1^{\tau}, \cdots, \boldsymbol{X}_N^{\tau})^{\tau}$, $\boldsymbol{J} = \mathrm{diag}(\boldsymbol{1}_{n_1}, \cdots, \boldsymbol{1}_{n_N})$, $\boldsymbol{W}^{(k)} = \mathrm{diag}(\omega_{1,k}^* \boldsymbol{W}_1^{(k)}, \cdots, \omega_{N,k}^* \boldsymbol{W}_N^{(k)})$, and $\boldsymbol{R}^{(k)} = \mathrm{diag}(\omega_{1,k}^* \boldsymbol{R}_1^{(k)}, \cdots, \omega_{N,k}^* \boldsymbol{R}_N^{(k)})$. For starting values of $\boldsymbol{\beta}$ and $\boldsymbol{b}$, we define a working vector $\boldsymbol{Z} = \boldsymbol{X\beta} + \boldsymbol{Jb} + \boldsymbol{W}^{-1}(\boldsymbol{y}^* - \boldsymbol{u}^*)$ where $\boldsymbol{y}^* = \sum_{k=1}^m \boldsymbol{R}^{(k)}\boldsymbol{y}$, $\boldsymbol{u}^* = \sum_{k=1}^m \boldsymbol{R}^{(k)}\boldsymbol{u}^{(k)}$ and $\boldsymbol{W} = \sum_{k=1}^m \boldsymbol{W}^{(k)}$. Then the solutions to Equations (3) and (4) can be approximately written as the iterative solutions to the system

$$\begin{pmatrix} \boldsymbol{X}^{\tau}\boldsymbol{WX} & \boldsymbol{X}^{\tau}\boldsymbol{WJ} \\ \boldsymbol{J}^{\tau}\boldsymbol{WX} & \boldsymbol{J}^{\tau}\boldsymbol{WJ} + \frac{1}{\theta}\boldsymbol{I}_N \end{pmatrix} \begin{pmatrix} \widehat{\boldsymbol{\beta}} \\ \hat{\boldsymbol{b}} \end{pmatrix} = \begin{pmatrix} \boldsymbol{X}^{\tau}\boldsymbol{WZ} \\ \boldsymbol{J}^{\tau}\boldsymbol{WZ} + \frac{1}{\theta}(\boldsymbol{b} - \boldsymbol{b}^*) \end{pmatrix}, \tag{5}$$

where $\boldsymbol{b}^* = \sum_{k=1}^m \boldsymbol{b}^{(k)}$ with $\boldsymbol{b}^{(k)} = (\omega_{1,k}^* b_1^{(k)}, \cdots, \omega_{N,k}^* b_N^{(k)})^{\tau}$, and the working dependent variable $\boldsymbol{Z}$ and the weighted matrix $\boldsymbol{W}$ are updated at each iteration. Obtaining the solution $\widehat{\boldsymbol{\beta}}$ to Equation (5) is equivalent to solving the system $(\boldsymbol{X}^{\tau}\boldsymbol{V}^{-1}\boldsymbol{X})\widehat{\boldsymbol{\beta}} = \boldsymbol{X}^{\tau}\boldsymbol{V}^{-1}\boldsymbol{Y}$, where $\boldsymbol{V} = \boldsymbol{W}^{-1} + \theta\boldsymbol{JJ}^{\tau}$ is a modified weighted matrix and $\boldsymbol{Y} = \boldsymbol{X\beta} + \boldsymbol{Jb}^* + \boldsymbol{W}^{-1}(\boldsymbol{y}^* - \boldsymbol{\mu}^*)$ is a modified working dependent variable to $\boldsymbol{Z}$ with $\boldsymbol{b}$ replaced by $\boldsymbol{b}^*$. The solution $\hat{\boldsymbol{b}}$ to Equation (5) must be of the form $\hat{\boldsymbol{b}} = \theta\boldsymbol{J}^{\tau}\boldsymbol{V}^{-1}(\boldsymbol{Y} - \boldsymbol{X}\widehat{\boldsymbol{\beta}}) + (\boldsymbol{b} - \boldsymbol{b}^*)$. In other words, the fixed effects $\boldsymbol{\beta}$ can be estimated by using IWLS technique. This also suggests that an approximate covariance matrix for the estimate $\widehat{\boldsymbol{\beta}}$ is of the form $(\boldsymbol{X}^{\tau}\boldsymbol{V}^{-1}\boldsymbol{X})^{-1}$ if $\boldsymbol{X}$ is of full rank. Compared with Breslow and Clayton's (1993) PQL equations, the estimates have similar forms but the working dependent variable $\boldsymbol{Y}$ and the weighted matrices $\boldsymbol{W}$ and $\boldsymbol{V}$ are updated at each iteration in terms of GH quadrature nodes and weights. In particular, if only one integration node is utilized, the estimates are identical to the PQL estimates.

In the discussion above we assumed that the variance component θ of the random effects is known. In practice, however, it is unknown and needs to be estimated simultaneously with the fixed effects $\boldsymbol{\beta}$. In the spirit of the PQL estimate, we treat the working dependent variable $\boldsymbol{Y}$ as normally distributed with covariance matrix $\boldsymbol{V}$, except that the dependence of the weighted matrix $\boldsymbol{W}$ on θ is ignored when calculating derivatives. Then at each iteration of (5), the REML estimate of θ must be the solution to the estimating equation

$$\frac{\partial l_R^{gh}}{\partial \theta} \approx \frac{1}{2}\Big\{(\boldsymbol{Y} - \boldsymbol{X}\widehat{\boldsymbol{\beta}})^{\tau}\boldsymbol{V}^{-1}\frac{\partial \boldsymbol{V}}{\partial \theta}\boldsymbol{V}^{-1}(\boldsymbol{Y} - \boldsymbol{X}\widehat{\boldsymbol{\beta}}) - \mathrm{tr}\Big(\boldsymbol{P}\frac{\partial \boldsymbol{V}}{\partial \theta}\Big)\Big\} = 0, \tag{6}$$

where $\boldsymbol{P} = \boldsymbol{V}^{-1} - \boldsymbol{V}^{-1}\boldsymbol{X}(\boldsymbol{X}^\tau\boldsymbol{V}^{-1}\boldsymbol{X})^{-1}\boldsymbol{X}^\tau\boldsymbol{V}^{-1}$. Thus, at each iteration we simultaneously solve Equations (5) and (6) for obtaining updated estimates of $\boldsymbol{\beta}$, $\boldsymbol{b}$ and θ. These procedures are repeated until convergence is obtained.

3 A simulation

To compare our methods with the PQL estimates, we conducted simulation studies and analyzed several well-known practical examples. To save space, we only present our simulation studies here. We follow McGilchrist's (1994) simulation model, a logistic-normal mixed model, where the random effect b_1, $b_2, \cdots, b_{15}$ are i.i.d samples from the normal distribution $N(0, \theta)$. Given the random effects $b_1, b_2, \cdots, b_{15}$, we generate independent binomial observations $y_{ij} \sim \text{binomial}(6, \pi_{ij})$ with $\text{logit}(\pi_{ij}) = \beta_0 + \beta_1 x_{ij} + b_i$ $(i = 1, 2, \cdots, 15; j = 1, 2)$, where $\pi_{ij} = \Pr\{y_{ij} = 1|b_i\}$, $x_{i1} = 2i - 16$ and $x_{i2} = 2i - 15$. The true values of parameters set in the simulation are (a) $\beta_0 = 0.2$, $\beta_1 = 0.1$ and $\theta = 1$ (b) $\beta_0 = 0.2$, $\beta_1 = 0.1$ and $\theta = 2$, and the simulation is conducted 100 runs for each set of the true values. Table 1 presents the averaged values of the GH-PQL estimates over 100 runs for the fixed effects β_0 and β_1 and the variance component $\theta^{1/2}$. The associated estimated standard errors of the estimates are also given in parentheses. In Table 1, the estimates PQLm represent the GH-PQL estimates with m integration nodes. The GLM estimates are the classical estimates without random effects and the MLE estimates are calculated directly using numerical integration with 20 GH quadrature nodes (Pan and Thompson, 1998).

Table 1. *Averaged values of the estimates over 100 runs of simulation*

	1st set of parameters			2nd set of parameters		
Method	β_0	β_1	$\theta^{1/2}$	β_0	β_1	$\theta^{1/2}$
True	.200	.100	1.000	.200	.100	1.414
GLM	.148(.162)	.086(.020)	–	.185(.162)	.076(.020)	–
MLE	.176(.176)	.100(.022)	.844(.200)	.234(.189)	.096(.023)	1.295(.243)
PQL1	.165(.291)	.094(.034)	.881(.287)	.217(.378)	.089(.044)	1.248(.335)
PQL2	.179(.308)	.102(.036)	.963(.306)	.224(.389)	.092(.045)	1.291(.342)
PQL5	.171(.305)	.098(.036)	.928(.289)	.216(.397)	.093(.046)	1.332(.346)
PQL10	.175(.304)	.098(.036)	.928(.289)	.230(.398)	.094(.046)	1.336(.346)
PQL20	.175(.304)	.098(.036)	.928(.289)	.217(.399)	.093(.046)	1.338(.347)
PQL40	.175(.304)	.098(.036)	.928(.289)	.217(.399)	.093(.046)	1.338(.347)

Table 1 shows that the classical logistic regression without random effects, produces under-estimates of the regression coefficients β_0 and β_1 because the within-subject correlation of the data is ignored. The ordinary PQL approach, i.e., PQL1, takes the correlation of the data into account and yields more accurate estimates than the GLM. However, the PQL estimates are biased in particular as the variance component of the random effects is large. The biases caused by the PQL estimates can be reduced by adding Gauss-Hermite quadrature nodes, and the improvement over PQL1 is significant in particular as the variance component of the random effects is large, see, e.g., the PQL2, PQL5, PQL10 and PQL20 estimates in Table 1. Also, the estimated standard errors become more stable as we add integration nodes.

4 Discussion

Regarding the choice of the number of GH quadrature nodes, we suggest that the integration nodes should be added successively until relatively stable estimates are obtained. A single choice for the number of integration nodes could lead to inaccurate estimates. When the underlying variance components of the random effects are small, the ordinary PQL technique could perform satisfactorily and the GH-PQL approach with more integration nodes may not be necessary. An interesting fact is that, when the data are Gaussian and the model has the identity link, our approach completely reduces to the REML method for normal linear mixed models. Hence the proposed approach may be viewed as a natural extension of REML estimation. In principle, the procedure proposed in this paper can also be extended to multivariate random effects. However, the GH quadrature approximation is only appropriate for lower-dimensional integrals as the computational effort in the approximation increases exponentially with the dimension of the integral. For multivariate random effects, it is expected that the Quasi-Monte Carlo (QMC) integration approximation which involves less computational effort could be used to replace GH quadrature. The key idea behind QMC method is to choose specific integration nodes scattered uniformly on the integration domain, see Pan and Thompson (1998) for more details.

Acknowledgments

This research was supported by a grant from Agriculture and Fisheries Department, Scottish Office. A substantial part of J. Pan's work was conducted when he was in Rothamsted. We thank referees for their comments which led to worthwhile improvements.

References

Breslow, N. E. and Clayton, D. G. (1993). Approximate inference in generalized linear mixed models. *Journal of the American Statistical Association*, **88**, 9-25.

Davis, P. J. and Rabinowitz, P. (1975). *Methods of Numerical Integration.* New York: Dover.

Goldstein, H. and Rasbash, J. (1996). Improved approximation for multilevel models with binary responses. *Journal of the Royal Statistical Society, Series A* **159**, 505-513.

Lee, Y. and Nelder, J. A. (1996). Hierarchical generalized linear models (with discussions). *Journal of the Royal Statistical Society, Series B* **58**, 619-678.

Liu, Q. and Pierce, D. A. (1994). A note on Gauss-Hermite quadrature. *Biometrika*, **81**, 624-629.

McGilchrist, C. A. (1994). Estimation in generalized mixed models. *Journal of the Royal Statistical Society, Series B* **56**, 61-69.

Pan, J. X. and Thompson, R. (1998). Quasi-Monte Carlo EM algorithm for MLEs in generalized linear mixed models. *COMPSTAT: Proceedings in Computational Statistics, 1998* (edited by Payne R. and Green P.); Physica-Verlag, 419-424.

Patterson, H. D. and Thompson, R. (1974). Recover of interblock information when block sizes are unequal. *Biometrika*, **58**, 545-554.

Payne, R. W., Lane, P. W., Digby, P. G. N., Harding, S., A., Leech, P. K., Morgan, G. W., Todd, A. D., Thompson, R., Tunnicliffe Wilson, G., Welham, S. J. and White, R. P. (1995). *Genstat 5 Release 4 Reference Manual.* Oxford: Oxford University Press.

The stochastic dimension in a dynamic GIS

Edzer J. Pebesma, Derek Karssenberg and Kor de Jong

Utrecht Centre for Environment and Landscape Dynamics, Faculty of Geographical Sciences, Universiteit Utrecht, P.O. Box 80.115, 3508 TC Utrecht. e-mail: e.pebesma@geog.uu.nl

Abstract. Coping with random fields in a time-dynamic geographic information system (GIS) increases the computational burden and storage requirements with a large amount, and calls for a number of custom functions to enable easy analysis of the resulting random components, as well as specialised output reporting functions. This paper addresses the computational and implementation issues when a Monte Carlo approach is taken, and shows some results from a rainfall-runoff model running within a GIS.

Keywords. Geographical information systems, Monte Carlo, temporal GIS, stochastic modelling, geostatistics

1 Introduction

Geographical information systems (GIS, Burrough and McDonnel, 1998) liberate the end-user from worrying about looping over all spatial entities by providing a set of functions that operate on maps as a whole. Such functions operate either point-wise, meaning that a result value exclusively depends on the values in a set of maps at the same location, or they operate in a spatial neighbourhood, meaning that values at other spatial locations contribute as well. A point-wise operation can be the addition of two maps,

```
sum.map = a.map + b.map
```

a neighbourhood operation the calculation of distances to a point location:

```
distances.map = distanceto(point.map)
```

Writing GIS operations on maps as algebraic expressions was first proposed by Tomlin and Berry (1979), and they named it 'map algebra'.

Beyond the standard map algebra, PCRaster (Wesseling et al., 1996) extends the set of spatial-only, static functions with a set of functions that have a temporal dimension, by adding a loop over a (discretised) time axis: spatial time series (stacks of maps) as well as aggregates over the iteration period can now be processed and analysed. The thus obtained dynamic GIS (implemented as a concise modelling language) has shown to be an easy tool for fast implementation of a wide range of spatio-temporal processes, ranging from distributed hydrological models to ecological dispersion or urban growth models (cellular automata). Easy control over the modelling process enables users with little expertise in GIS or computer programming to evaluate many similar but alternative models, by simply trying them out.

Although some of the functions provided by PCRaster may contain some form of randomness (e.g. a function returning a map with standard normal *i.i.d.* variates), all functionality regards deterministic computation, meaning that uncertainty with respect to model input variables cannot be handled. In the practice of environmental modelling however, users are often confronted with highly uncertain model inputs, and want to be able to assess how input errors propagate to the output of their GIS models.

Today, freely available geostatistical software tools (e.g. Deutsch and Journel, 1998; Pebesma and Wesseling, 1998) provide the means for a fairly straightforward modelling and simulation of spatial fields for both discrete (nominal) and continuous variables, using the indicator simulation formalism for the former and Gaussian random fields for the latter. Using these maps as input to GIS models is far from easy though, and it may prevent large groups from doing so for that reason.

This paper discusses the concept of a stochastic (dynamic) GIS, and proposes a set of functions that should be added to the GIS to provide error estimates of model output, coming from model input error. In a case study regarding the analysis of stream flow and surface runoff during a rain storm, we analyse how uncertainty on infiltration capacity influences the model output, being spatial distribution of runoff and temporal variation of river flow at the catchment outlet.

2 Stochastic GIS

2.1 A Monte Carlo approach

Spatial operations of a map algebra GIS can be generalised as follows:

$$\{r\} = f(A, B, C, ...) \tag{1}$$

with $\{r\}$ a set of one or more maps where the output is written to, with $f(\cdot)$ a simple or compound operation, and with input maps A, B, C,... (possibly including some constant maps). In a dynamic GIS (1) is evaluated every time step, and the operation may involve an update of some maps for each time step. The challenge in a stochastic GIS is to characterise the joint (that is, multivariable) distribution function of $\{r\}$ over space and time, given the joint distributions of A, B, C,... . Analytic approaches (Heuvelink, 1998) have addressed analytical solutions to (1) for the case where only non-spatials were stochastic. The only feasible approach to cases that include random spatial or spatio-temporal fields appears to be a full spatio-temporal Monte Carlo (MC) simulation (Heuvelink, 1998).

The most general implementation for an MC approach considers evaluation over the stochastic, temporal and spatial domain in the order:

```
for m in MC-Sample
 for t in TimeSteps
   < evaluate r >
```

(evaluation of r trivially involves the looping over all spatial locations). For a full analysis of the results, all output of each MC run has to be stored. This can be a full spatial time series for each MC run, or a single map (e.g. a state variable at the last time step, or a map aggregate over time) or a non-spatial time series (read from a location in a map, or aggregated over space), or an aggregated (scalar) value over either of the latter two.

Moving the MC loop inside the time or space loop may be more efficient in some respects, but only works in highly simplified cases, e.g. in absence of temporal and spatial correlation of the model input.

2.2 The 'stochastic' dimension

Taking the 'randomness' as a new dimension, in the sense of spatial or temporal dimensions, can be justified from both a conceptual and an operational point of view. From a conceptual point of view, geographical space, or the three-dimensional space where we live in, is captured by traditional GIS (although most emphasis is traditionally put to the two 'horizontal' dimensions). A temporal GIS adds to this the temporal dimension, to capture changes of the spatial settings over time. These four dimensions can thus be used to represent everything we know. To add to this a representation of things we do not know, for instance by probability density functions (PDFs), we need at least one more dimension. It may be argued that *every* point in space/time that needs a PDF adds one dimension, but usually easier representations are sought (for instance by assuming stationarity).

From an operational point of view, the obvious regular discretisation of a 3D-space+time 'block' is a four-dimensional array. Using MC analysis, this array is replicated a number of times equal to the MC sample size. The obvious discretisation of such a stochastic 3D-space/time block is a five-dimensional array: The added index identifies the MC sample element.

The consequence of adding a stochastic dimension is that the modeller has to take one more aspect into account when choosing the spatial discretisation (raster map cell size) and temporal discretisation (time step): the MC sample size. The total computational burden (and, as we will see in the worst case, storage requirements) is of order:

<*map cells*> × <*time steps*> × <MC *sample size*>

A real danger will be that naive users choose a small sample MC size because they prefer high-resolution maps, and end up with highly inaccurate *estimates* of model output distributions.

3 Analysis of the Monte Carlo output

To obtain the complete results from the MC analysis, the full output has to be retained: for instance the cumulative probability density function of r at a given time and location is estimated from the ranked MC sample values at this time and location. Also, this output may serve as input to another function, say $g(\cdot)$:

$$p = g(r, ...)$$

An example of $g(\cdot)$ is spatial aggregation of output of the function $f(\cdot)$, which should be applied to every MC sample element (Heuvelink and Pebesma, 1998; Pebesma and Heuvelink 1999).

To summarise all marginal cumulative probability density functions, at the cost of loosing information of the joint density of r, one can use two approaches that differ strongly with respect to storage requirements:

1. collect a set of percentiles, e.g. the 5-,10-,25-,50-,75-,90-,and 95-percentile,
2. collect the sum, sum of squares, sum of cube squares, ... and the frequency of sample elements being above a set of pre-defined thresholds.

The second option gives the possibility of estimating for each location and time the mean, variance, skewness and kurtosis, and probabilities of exceeding thresholds.

Estimation of the percentiles of a distribution requires that the full set of MC sample elements is stored, because they need to be ranked before percentiles can be estimated. It should be noted here that for very high or low percentiles some gain is obtained by only storing the n tail values, with $n \approx qN$, with q the quantile estimated and with N the MC sample size. In general however, the full output distribution is of interest, eliminating this advantage.

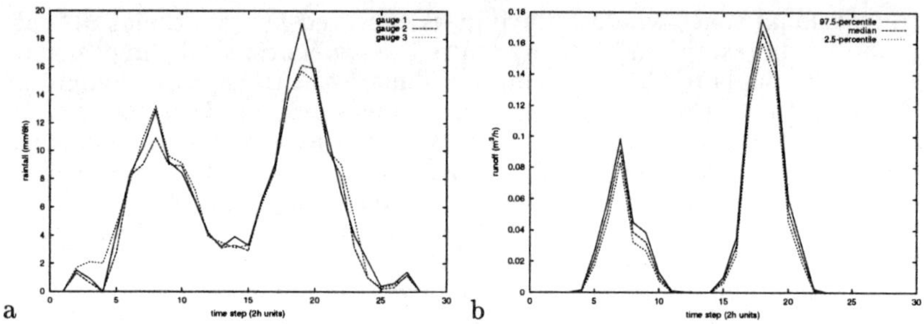

Fig. 1. (a) Rainfall, measured at the three gauges (see Fig. 2b) in mm/6h; (b) runoff at the catchment outlet: 2.5- and 97.5-percentiles and median value in m^3/h

The second group of summary variables has the attractive property that they can be collected *without* storing the complete set of MC sample elements: they can be obtained by tracking all the necessary sums during the MC simulation, and after each MC run the previous results can be discarded.

4 Case study: rainfall-runoff simulation

Rainfall-runoff modelling can be useful for predicting river floods, soil moisture contents, soil erosion, and eventually landscape evolution, in undulating or hilly catchments. A simple rainfall-runoff simulation model for the 0.42 km^2 Catsop catchment (Limburg, The Netherlands) was implemented in PCRaster. It evaluates for each time step: (1) when rainfall intensity exceeds the soil infiltration capacity, the soil becomes saturated and excess rainfall will run off downhill; (2) surface runoff is routed through a drainage network and will either re-infiltrate in subsequent non-saturated cells, or leave the catchment at its outlet.

Rainfall at each location is obtained by reading the rainfall time series data (Fig. 1a) from the nearest rainfall gauge (Fig. 2b) at that moment. The local drain direction map defining the runoff network is derived from the elevation map (Fig. 2a); and infiltration is modelled as a function of soil texture (Fig. 2b). Mean values for infiltration were 2.8 (clay), 8.3 (loam) and 19.0 (sand) [mm/6h]. Standard deviations were taken as one third of the mean value, and the semivariance between infiltration at two sites a distance h ($h > 0$) apart within the same soil texture class was modelled as an exponential variogram

$$\gamma(h) = \sigma^2(s)(1 - 0.9\exp(-h/50)),$$

with $\sigma^2(s)$ the soil-dependent variance (variance and spatial correlation information were obtained from Loague and Kyriakidis, 1997).

Simulation of infiltration capacity was done by sequential Gaussian simulation (Deutsch and Journel, 1998). Negative simulated values were reset to a zero value. To increase the MC sampling efficiency, Latin hypercube sampling (Pebesma and Heuvelink, 1999) was applied. A sample of 1000 infiltration capacity maps was obtained to study the effect of unknown infiltration capacity on spatial patterns of runoff (Fig. 3) and time series of runoff at the catchment outlet (Fig. 1b).

The results show that the uncertainty increases with runoff levels, but that the errors are small compared to the modelled runoff levels. Actual runoff

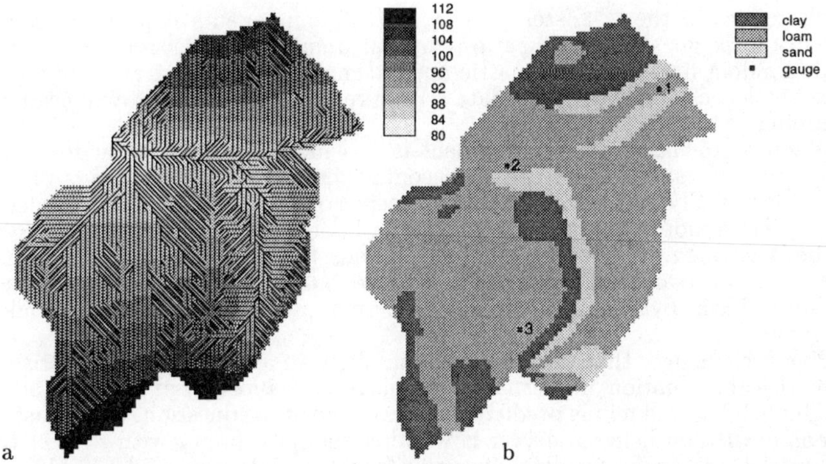

Fig. 2. (a) map of altitude (grey shades; units m above sea level) and local drain directions (drawn lines); (b) map of soil texture and location of rain gauges; mapped area is 1 km × 0.8 km

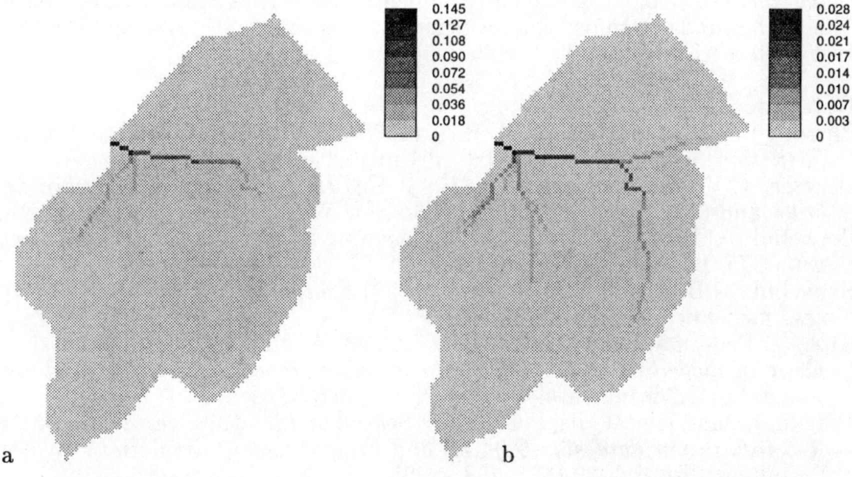

Fig. 3. (a) Median runoff at time step 20; (b) width of the empirical 95% confidence interval at time step 20

measurements should be used to assess whether the errors in infiltration as they are modelled here can account for the model prediction errors. In addition, infiltration measurements should have been used for a more realistic assessment of spatial errors in infiltration.

5 Discussion

A stochastic dynamic GIS can be obtained when one dimension is added to a GIS that provides a temporal dimension. We do not yet provide such an environment in the PCRaster project, but all components required (a map algebra package; stochastic spatio-temporal simulation of discrete or continuous random fields, and stochastic output analysis functions) are available. The bottleneck is a convenient data structure that allows higher dimensional mapping.

When a stochastic GIS environment is provided, the question arises how a suitable MC sample size can be recommended to non-expert users. This depends trivially on which results (e.g., means or tail percentiles) are needed, but general guidelines such as they exist for bootstrapping methods still have to be developed. In addition, simple methods to experimentally determine the accuracy (sampling error) of estimated MC sample statistics, such as repeated Latin hypercube sampling (Pebesma and Heuvelink, 1999) should be provided.

Bigger challenges than handling random fields in a GIS are the comparison and critical evaluation of alternative GIS *model structures*, by analysing model residuals (observed minus predicted values). One of the questions a stochastic dynamic GIS can help to answer is whether the uncertainty with respect to GIS model input can completely account for the variation in model residuals.

Acknowledgments
This paper was written while the first author was a visiting scholar at the Department of Geological and Environmental Sciences at Stanford University. The Netherlands Organisation for Scientific Research (NWO) supported this visit with a travel stipend.

References

Burrough, P.A., and McDonnell, R.A. (1998). *Principles of Geographical Information Systems.* Oxford: Oxford University Press.

Deutsch, C.V. and Journel, A.G. (1998) *GSLIB Geostatistical Software Library and User's guide, second edition.* New York: Oxford University Press.

Heuvelink, G.B.M. (1998), *Error Propagation in Environmental Modelling with GIS*, London: Taylor & Francis.

Heuvelink, G.B.M. and E.J. Pebesma (1999) *Spatial aggregation and soil process modelling.* Geoderma 89, 47-65.

Kros, J., Pebesma, E.J., Reinds, G.J., Finke, P.A. (1999) *Uncertainty assessment in modelling soil acidification at the European scale: a case study.* Journal of Environmental Quality 28 (2), 366-377.

Loague, K., and Kyriakidis, P.C. (1997) *Spatial and temporal variability in the R-5 infiltration data set: Déjà vu and rainfall-runoff simulations.* Water Resources Research 33 (12), 2883-2895.

Pebesma, E.J., and Wesseling, C.G. (1998) *Gstat, a program for geostatistical modelling, prediction and simulation.* Computers & Geosciences, 24(1), 17-31. Software at `http://www.geog.uu.nl/gstat/`

Pebesma, E.J., Heuvelink, G.B.M. (1999) *Latin hypercube sampling of Gaussian random fields*, Technometrics 41(4), 303-312.

Tomlin, C.D., and Berry, J. (1979) *A Mathematical Structure for Cartographic Modeling and Environmental Analysis.* In: Proceedings of the ACSM, 269-283.

Wesseling, C.G., Karssenberg, D., Van Deursen, W.P.A. and Burrough, P.A. (1996) *Integrating dynamic environmental models in GIS: the development of a Dynamic Modelling language.* Transactions in GIS 1, pp 40-48. Software at `http://www.geog.uu.nl/pcraster/`

A robust version of principal factor analysis

G. Pison[1], P. J. Rousseeuw[1], P. Filzmoser[2] & C. Croux[3]

[1] Department of Mathematics and Computer Science, Universitaire Instelling Antwerpen, Universiteitsplein 1, B-2610 Antwerpen, Belgium.
[2] Department of Statistics and Probability Theory, Vienna University of Technology, Wiedner Hauptstrasse 8-10, A-1040 Vienna, Austria.
[3] ECARE and Institut de Statistique, Université Libre de Bruxelles, CP-139, Avenue Roosevelt 50, B-1050 Bruxelles, Belgium.

Abstract. Our aim is to construct a factor analysis method that can resist the effect of outliers. We start with a highly robust initial covariance estimator, after which the factors can be obtained from maximum likelihood or from principal factor analysis (PFA). We find that PFA based on the minimum covariance determinant scatter matrix works well. We also derive the influence function of the PFA method. A new type of empirical influence function (EIF) which is very effective for detecting influential data is constructed. If the data set contains fewer cases than variables, we estimate the factor loadings and scores by a robust interlocking regression algorithm.

Keywords. Factor Analysis, Influence Function, Outliers, Robustness

1 The factor analysis model

One aim of a factor analysis (FA) is to summarize the correlation structure of some observed variables $X_1, X_2, \ldots, X_p$. For this purpose it introduces $k < p$ unobservable or latent variables $\Phi_1, \ldots, \Phi_k$ which are called *factors*, and which are linked with the original variables through the equation

$$X_j = \mu_j + \Lambda_{j1}\Phi_1 + \Lambda_{j2}\Phi_2 + \ldots + \Lambda_{jk}\Phi_k + \varepsilon_j \tag{1}$$

for each $1 \leq j \leq p$. The error variables $\varepsilon_1, \ldots, \varepsilon_p$ are assumed to be independent, with $\varepsilon_j \sim N(0, \psi_j)$ where $\psi_1, \ldots, \psi_p$ are called the *specific variances.* The coefficients Λ_{jl} are called the factor *loadings*, and they are collected into the matrix of loadings $\boldsymbol{\Lambda}$.

Using the vector notations $\boldsymbol{X} = (X_1, \ldots, X_p)^t$, $\boldsymbol{\Phi} = (\Phi_1, \ldots, \Phi_k)^t$, and $\boldsymbol{\varepsilon} = (\varepsilon_1, \ldots, \varepsilon_p)^t$, the usual conditions on factors and error terms can be written as $E(\boldsymbol{\Phi}) = E(\boldsymbol{\varepsilon}) = 0$, $\mathrm{Cov}(\boldsymbol{\Phi}) = I_p$, and $\mathrm{Cov}(\boldsymbol{\varepsilon}) = \boldsymbol{\Psi}$, with $\boldsymbol{\Psi}$ a diagonal matrix containing on its diagonal the specific variances. Furthermore, $\boldsymbol{\varepsilon}$ and $\boldsymbol{\Phi}$ are assumed to be independent.

In factor analysis, one needs to estimate the matrices $\boldsymbol{\Psi}$ and $\boldsymbol{\Lambda}$ (the latter is only specified up to an orthogonal transformation). Let us denote the covariance matrix of X as $\boldsymbol{\Sigma}$. (In case that $X_1, \ldots, X_p$ are standardized versions of the originally measured variables, $\boldsymbol{\Sigma}$ represents a correlation matrix.) It follows from (1) that

$$\boldsymbol{\Sigma} = \boldsymbol{\Lambda}\boldsymbol{\Lambda}^t + \boldsymbol{\Psi} \tag{2}$$

under the model assumptions. In classical factor analysis the matrix $\boldsymbol{\Sigma}$ is estimated by the sample covariance matrix $\boldsymbol{S}$. Afterwards one tries to decompose $\boldsymbol{S}$ as in (2) to obtain the estimators for $\boldsymbol{\Lambda}$ and $\boldsymbol{\Psi}$. Typically $\boldsymbol{S}$ cannot be decomposed exactly as in (2), so we must resort to an approximate decomposition. Many methods have been proposed for this, of which maximum likelihood estimation (MLE) and the principal factor analysis (PFA) method are the most frequently used.

It is, however, well known that outliers can heavily influence the estimation of $\boldsymbol{\Sigma}$ and hence also the parameter estimates. Our aim is to construct a factor analysis method that can resist the effect of outliers. Therefore a robust covariance matrix estimator needs to be used. Kosfeld (1996) used an M-estimator for this purpose. Here we will use the Minimum Covariance Determinant (MCD) estimator (Rousseeuw, 1985) because it has a higher breakdown value. The MCD looks for the subset of h observations having the smallest determinant of its covariance matrix. Typically, $h \approx 3n/4$. The MCD estimator of $\boldsymbol{\Sigma}$ is then the covariance matrix computed from that subset, and the MCD estimator of $\boldsymbol{\mu}$ is its mean. The MCD estimator has good efficiency properties (Croux and Haesbroeck, 1999). Moreover, a fast algorithm for the MCD has recently been developed (Rousseeuw and Van Driessen, 1999).

The resulting robust loadings $\boldsymbol{L}_n^r$ and specific variances $\boldsymbol{P}_n^r$ allow us to obtain robust factors $\boldsymbol{F}_n^r$. They describe the correlation or covariance between the uncontaminated data.

2 Empirical study

Some empirical studies with outliers are carried out to investigate their effect on classical and robust FA.

2.1 Sensitivity analysis

We investigate the sensitivity of factor analysis to outliers and small errors. Therefore we compare the sensitivity of classical maximum likelihood estimation (CLAS.MLE), principal factor analysis (CLAS.PFA), and their MCD-based versions on the stock price data set of Johnson and Wichern (1998), with $n = 100$ observations and $p = 5$ variables. The variables are standardized by their average and standard deviation.

We first estimate $k = 2$ factors based on the classical and robust correlation matrices, yielding the loadings $\boldsymbol{L}_n^{(0)} \in \mathbb{R}^{5\times 2}$ and unique variances $\boldsymbol{P}_n^{(0)} = (P_1^{(0)}, \ldots, P_5^{(0)})$. For the sensitivity analysis we add a noise matrix $(err^{(s)})$ and a matrix $(xout^{(s)})$ which causes n_{out} data points to become outliers. The elements of the noise matrix are distributed according to $N(0, (0.05)^2)$. The outlier matrix $xout^{(s)}$ is mainly zero, except for n_{out} elements. We generate only one outlying entry per outlying object, which is generated from the normal distribution $N(10, (0.05)^2)$. The disturbed data sets $\boldsymbol{X}^{(s)}$ are thus generated as $\boldsymbol{X}^{(s)} = \boldsymbol{X}^{(0)} + err^{(s)} + xout^{(s)}$ for $s = 1, \ldots, m$. Fitting this model yields estimates $\boldsymbol{L}_n^{(s)}$ and $\boldsymbol{P}_n^{(s)}$ for $m = 1000$ simulated samples. Since the loadings matrix is only determined up to an orthogonal matrix, we consider the $p \times p$ matrix $\boldsymbol{A}_n^{(s)} = \boldsymbol{L}_n^{(s)}(\boldsymbol{L}_n^{(s)})^t$ instead. We will compare the elements $a_{ij}^{(s)}$ of $\boldsymbol{A}_n^{(s)}$ with the undisturbed entries $a_{ij}^{(0)}$ of the matrix $\boldsymbol{A}_n^{(0)} = \boldsymbol{L}_n^{(0)}(\boldsymbol{L}_n^{(0)})^t$ by computing the mean squared error (MSE) of the estimates:

$$MSE(a_{ij}) = \frac{1}{m}\sum_{s=1}^{m}\left(a_{ij}^{(s)} - a_{ij}^{(0)}\right)^2$$

for $i, j = 1, \ldots, p$. The average MSE is $MSE(\boldsymbol{A}) = \frac{1}{p^2}\sum_{i=1}^{p}\sum_{j=1}^{p} MSE(a_{ij})$. Similarly, for the square root of the unique variances P_j we compute

$$MSE(P_j) := \frac{1}{m}\sum_{s=1}^{m}\left(\sqrt{P_j^{(s)}} - \sqrt{P_j^{(0)}}\right)^2$$

where $j = 1, \ldots, p$ and the average MSE is $MSE(\boldsymbol{P}) = \frac{1}{p}\sum_{j=1}^{p} MSE(P_j)$.

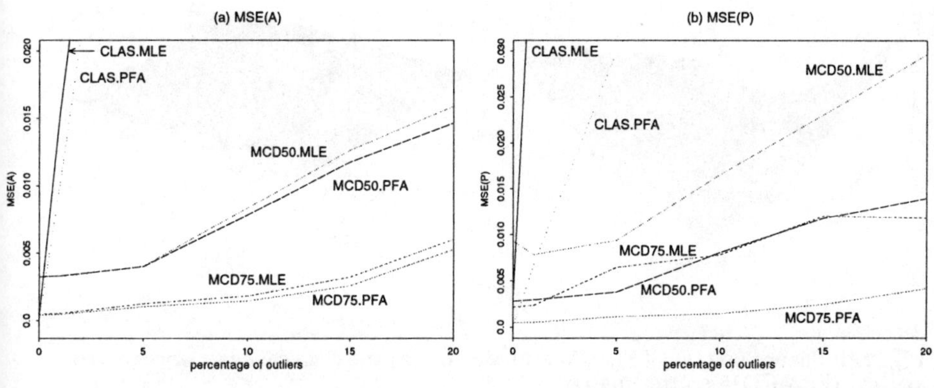

Fig. 1. Sensitivity of factor analysis on the stock price data: (a) $MSE(\mathrm{A})$ versus the fraction of outliers; (b) $MSE(\mathrm{P})$ versus the fraction of outliers.

Figure 1 shows the average MSE versus the fraction of outliers (here, 0% to 20%). We see that the MSE's of CLAS.PFA and CLAS.MLE are much higher than those based on the robust correlation matrix. Comparing MCD50 ($h \approx 0.5n$) and MCD75 ($h \approx 0.75n$), we find that FA using MCD75 systematically yielded a lower MSE than the corresponding method based on MCD50. Because MCD75 also has a higher efficiency than MCD50, we will work with MCD75 from now on.

2.2 Monte Carlo study

We start from fixed parameter values, i.e. $\boldsymbol{\Lambda}$ and a diagonal matrix $\boldsymbol{\Psi}$. Then we construct data sets $\boldsymbol{X}^{(s)}$ as follows: $\boldsymbol{X}^{(s)} = \boldsymbol{\Lambda}\boldsymbol{\Phi}^{(s)} + \boldsymbol{\varepsilon}^{(s)} + \mathrm{Out}^{(s)}$. For each s we generated the matrix of factor scores $\boldsymbol{\Phi}^{(s)}$ from $N(0,1)$, the entries $\varepsilon_{ij}^{(s)}$ of the noise term $\boldsymbol{\varepsilon}^{(s)}$ from $N(0, \psi_j)$, and the outlying term $\mathrm{Out}^{(s)}$ from $N(10, (0.05)^2)$.

Fitting the factor analysis model to $\boldsymbol{X}^{(s)}$ gives the estimates $\boldsymbol{L}_n^{(s)}$ and $\boldsymbol{P}_n^{(s)}$ for $s = 1, \ldots, 1000$. These estimates are compared to the true $\boldsymbol{\Lambda}$ and $\boldsymbol{\Psi}$ by computing the MSE.

From the simulations (with $n = 100$, $p = 5$, and $k = 2$) we conclude that PFA based on MCD is more robust than MLE based on MCD. Therefore, we will focus on the MCD75.PFA technique.

3 The influence function of PFA

3.1 The theoretical influence function

We have derived the influence functions $IF(x, \boldsymbol{P}, G)$ and $IF(x, \boldsymbol{LL}^t, G)$ of the PFA method. In Figure 2 we plot the influence function of the first specific variance.

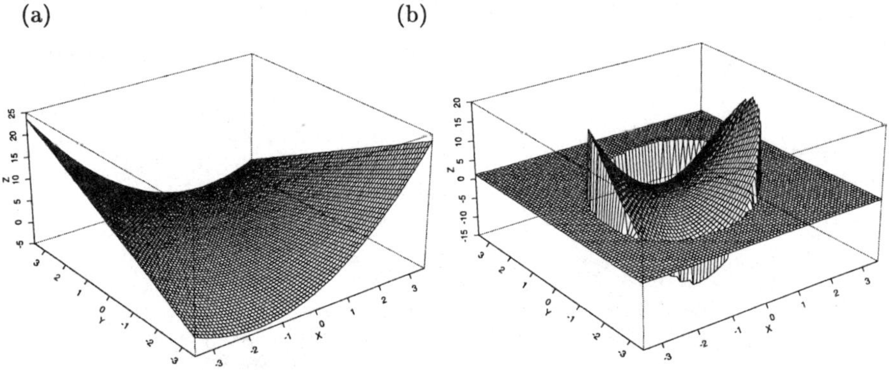

Fig. 2. Influence function $IF(x, P_1, G)$ based on (a) the classical covariance matrix and (b) the MCD75 scatter matrix.

We see that for the PFA based on the classical covariance matrix, the influence functions of ψ_j and $(\boldsymbol{\Lambda}\boldsymbol{\Lambda}^t)_{ij}$ are unbounded. This shows that an outlying x can have an arbitrarily large effect on the PFA method. On the other hand the influence function of MCD-based PFA is bounded, so an outlier x has a bounded effect on the robust PFA. Inside the elliptical central region of the x-distribution the IF looks like that of the classical PFA. Outside, the influence function is a constant for the MCD covariance matrix, and it becomes zero when working with the MCD correlation matrix.

3.2 The empirical influence function

In the previous subsection, we computed the influence functions in the population case, where we know the true underlying distribution G. In the empirical setting we only have a sample $\boldsymbol{X}_n \in \mathbb{R}^{n \times p}$ without knowing G. However, the unknown G depends only on the parameters $\boldsymbol{\mu}$ and $\boldsymbol{\Sigma}$, which we will replace by estimates $\boldsymbol{T}(\boldsymbol{X}_n)$ and $\boldsymbol{S}(\boldsymbol{X}_n)$ in the formula of the influence function. The resulting *empirical influence function* (EIF) is then evaluated in each data point x_i $i = 1, \ldots, n$ to measure their effect on the principal factor analysis. Our aim is to detect the most influential observations x_i by comparing the EIF(x_i).

We can construct the EIF of the classical PFA (e.g. of $\boldsymbol{P}_n^c$) and of the robust PFA (e.g. of $\boldsymbol{P}_n^r$). For $\boldsymbol{T}(\boldsymbol{X}_n)$ and $\boldsymbol{S}(\boldsymbol{X}_n)$ we take the classical estimates $(\boldsymbol{T}_n^c, \boldsymbol{S}_n^c)$ or the robust estimates $(\boldsymbol{T}_n^r, \boldsymbol{S}_n^r)$. This yields three ways to define the EIF:

- Tanaka and Odaka (1989) computed $EIF(x_i; \boldsymbol{P}_n^c; \boldsymbol{T}_n^c, \boldsymbol{S}_n^c)$. This approach often masks outliers because $\boldsymbol{T}_n^c$ and $\boldsymbol{S}_n^c$ break down.
- Substituting the robust $\boldsymbol{T}_n^r$ and $\boldsymbol{S}_n^r$ in the robust IF yields $EIF(x_i; \boldsymbol{P}_n^r; \boldsymbol{T}_n^r, \boldsymbol{S}_n^r)$. This function illustrates the fact that an outlying x_i has only a small effect on $\boldsymbol{P}_n^r$, which is natural because we constructed $\boldsymbol{P}_n^r$ for this purpose.
- Substituting the robust $\boldsymbol{T}_n^r$ and $\boldsymbol{S}_n^r$ in the classical IF yields $EIF(x_i; \boldsymbol{P}_n^c; \boldsymbol{T}_n^r, \boldsymbol{S}_n^r)$. This is the most useful, because $\boldsymbol{T}_n^r$ and $\boldsymbol{S}_n^r$ are not affected by outliers. Ideally, we would like to have $EIF(x_i; \boldsymbol{P}_n^c; \boldsymbol{\mu}, \boldsymbol{\Sigma})$ for the true $\boldsymbol{\mu}$ and $\boldsymbol{\Sigma}$ of the parent distribution, but in the presence of outliers the $\boldsymbol{T}_n^r$ and $\boldsymbol{S}_n^r$ are good approximations to these parameters.

In practice, to detect the most influential data points x_i (i.e. points that would strongly affect the classical PFA) we therefore recommend to compute the $EIF(x_i; \boldsymbol{P}_n^c; \boldsymbol{T}_n^r, \boldsymbol{S}_n^r)$, which avoids the masking effect.

4 Example

The aircraft data set (Gray 1985) consists of $n = 23$ single-engine aircraft built in the years 1947-1979 with $p = 5$ variables. Applying the MCD to these data indicates that cases 14 and 22 are outliers. Let us now compute the empirical influence functions $EIF(x_i; P_j)$ and an overall value $\|EIF(x_i; \boldsymbol{P})\| = \sqrt{|EIF(x_i; P_1)|^2 + \ldots + |EIF(x_i; P_5)|^2}$ in the 23 observations x_i for the different versions of the EIF considered above. Figure 3 plots $\|EIF(x_i; \boldsymbol{P})\|$ versus the case number i. We see that the outlying cases 14 and 22 have a

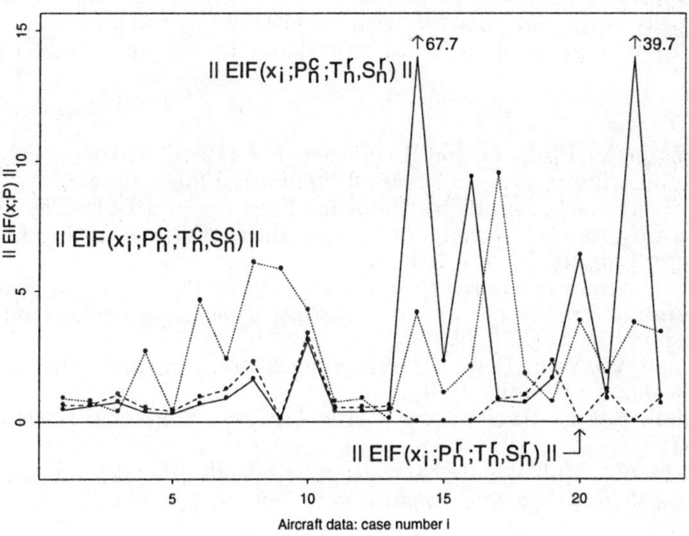

Fig. 3. The empirical influence functions $\|EIF(x_i; \mathrm{P})\|$ evaluated in 23 aircraft.

relatively small $\|EIF(x_i; \boldsymbol{P}_n^c; \boldsymbol{T}_n^c, \boldsymbol{S}_n^c)\|$. This is because $\boldsymbol{T}_n^c$ and $\boldsymbol{S}_n^c$ try to fit all the data points, so $\boldsymbol{S}_n^c$ becomes too large. Secondly, using the robust estimates $\boldsymbol{P}_n^r$, $\boldsymbol{T}_n^r$ and $\boldsymbol{S}_n^r$ leads to $\|EIF(x_i; \boldsymbol{P}_n^r; \boldsymbol{T}_n^r, \boldsymbol{S}_n^r)\| = 0$ for cases 14 and 22. This illustrates the robustness of $\boldsymbol{P}_n^r$ but does not help to detect the influential points. The only function that clearly shows the influential points is $\|EIF(x_i; \boldsymbol{P}_n^c; \boldsymbol{T}_n^r, \boldsymbol{S}_n^r)\|$, which takes on huge values for cases 14 and 22. In this example the outliers for $\boldsymbol{S}_n^r$ are also the most influential points for PFA. This does not always have to be the case.

5 Robust interlocking regression

When the data set contains fewer cases than variables, we can not compute the MCD covariance matrix. In this case an approach called "Factor Analysis using Interlocking Regressions" (FAIR) can be used. This method estimates the unknown parameters (factor scores and loadings) directly, without passing via an estimate of the covariance matrix.

Let the variables already be standardized to have zero location and unit spread. We denote the vector of unknown parameters as $\boldsymbol{\theta} = (\boldsymbol{\Phi}_1, \ldots, \boldsymbol{\Phi}_n, \boldsymbol{\Lambda}_1, \ldots, \boldsymbol{\Lambda}_p)$. The FAIR estimator is defined by:

$$\hat{\boldsymbol{\theta}}_{FAIR} = \operatorname*{argmin}_{\boldsymbol{\theta}} \sum_{i=1}^{n} \sum_{j=1}^{p} w_i(\boldsymbol{\theta}) v_j(\boldsymbol{\theta}) |x_{ij} - \hat{x}_{ij}(\boldsymbol{\theta})|$$

with $w_i(\boldsymbol{\theta}) = \min(1, \chi^2_{k,0.95}/\mathrm{RD}_i^2)$ for $i = 1, \ldots, n$ where the RD_i are robust distances computed from the collection of score vectors. Analogously, the set of column weights v_j is defined using the loading vectors. The FAIR estimator is extensively discussed in Croux, et al (1999). It can cope with many outlying cells in the data matrix, without loosing too much efficiency. A robust interlocking regression algorithm is available to compute the FAIR estimator.

References

Croux, C., Filzmoser, P., Pison, G. and Rousseeuw, P.J. (1999). Fitting Factor Models by Robust Interlocking Regression. Submitted for publication.

Croux, C. and Haesbroeck, G. (1999). Influence Function and Efficiency of the Minimum Covariance Determinant Scatter Matrix Estimator. *Journal of Multivariate Analysis*, **71**, pp. 161–190.

Gray, J.B. (1985). Graphics for Regression Diagnostics. In: *American Statistical Association Proceedings of the Statistical Computing Section*, pp. 102–107. Washington, D.C.: ASA.

Johnson, R.A. and Wichern, D.W. (1998). *Applied Multivariate Statistical Analysis.* New Jersey: Prentice Hall.

Kosfeld, R. (1996). Robust Exploratory Factor Analysis. *Statistical Papers*, **37**, pp. 105–122.

Rousseeuw, P.J. (1985). Multivariate Estimation with High Breakdown Point. In: *Mathematical Statistics and Applications, Vol. B*, pp. 283-297. Dordrecht: Reidel.

Rousseeuw, P.J. and Van Driessen, K. (1999). A Fast Algorithm for the Minimum Covariance Determinant Estimator. *Technometrics*, **41**, pp. 212-223.

Tanaka, Y. and Odaka, Y. (1989). Influential Observations in Principal Factor Analysis. *Psychometrika*, **54**, pp. 475–485.

TESS: system for automatic seasonal adjustment and forecasting of time series

Albert Prat[1], Victor Gomez[2], Ignasi Solé[1] and Josep M. Catot[1]

[1] Department of Statistics, UPC, Av. Diagonal 647, 08028 Barcelona, Spain
[2] Ministerio de Economia y Hacienda, Madrid, Spain

Abstract. TESS, SysTEm for Automatic Seasonal Adjustment and Forecasting of Time Series, is an ESPRIT IV project (Number 29.741) that started on January 1999 and finished on June 2000. More information can be found in http://esl.jrc.it/tess

The principal objective of the project was the production of a system for automatic seasonal adjustment and forecasting of time series, with special attention to the problem of aggregation. In order to achieve this goal, we enhanced the pre-existing software FORCE4/R (1997), based on SEATS and TRAMO (Maravall and Gomez (1992), Gomez and Maravall (1992)), with the methods required to perform the above stated functions in an optimal way. Furthermore, the system offers a homogenous interactive environment employing visualisation techniques.

Keywords: Time Series, Knowledge Based Systems, ARIMA models, TRACE, TRAMO, SEATS, Seasonal Adjustment.

1 Introduction

TESS, SysTEm for Automatic Seasonal Adjustment and Forecasting of Time Series, is an ESPRIT IV project (Number 29.741) that started on January 1999 and finished on June 2000. More information can be found in http://esl.jrc.it/tess

In this paper after a general introduction we will concentrate on the description of the functionalities, give an example of application and present the conclusions.

TESS is a software package designed to work under WINDOWS95/98 operating system for univariate time series analysis. There are 4 main programs that can be invoked for that purpose within TESS. These are: TRAMO (Time series Regression with Arima noise, Missing observations and Outliers), SEATS (Signal Extraction in Arima Time Series), TRACE (1999) (TRend And Cycle Estimation) and SSAG (1999) (State Space form for AGgregation) (1999).

It performs a fairly complete statistical treatment of time series, be that in routine large-scale applications or, for example, in careful analysis for short-term economic policy, control and monitoring. Although other frequencies are possible, our attention centres mainly on monthly and quarterly series. The statistical treatment we have in mind includes short-term forecasting, interpolation of missing values, seasonal adjustment, deciding whether to forecast or seasonally adjust the aggregate or its component series, estimation of short-term and long-term trends, estimation of the business cycle, estimation of special effects and removal of outliers, perhaps for a large number of series.

TESS uses a methodology that solves all the previous problems with an internally consistent model-based approach. The TRAMO, SEATS and TRACE programs available in TESS can be applied in an entirely automatic way. The use of signal extraction techniques in ARIMA models (even regression variables, missing observations, and outliers) facilitates statistical inference. In this way it is possible to provide precise answers to many questions of applied interest in short-term data analysis.

Many of the series generated by TRAMO, SEATS and TRACE together with different models and diagnostics obtained can be stored in the system for further analysis.

2 Functionalities

To perform the above complete statistical treatment of time series, TESS provides the following type of functionalities:

2.1 Visualisations aspects

The graphic interface and the graphic system that TESS incorporates have been specially designed to make the use of TESS very user-friendly. The user interface allows the user to interact with the system to perform all the functionalities that TESS offers, in a homogenous way. TESS provides plots of many of the series generated by TRAMO, SEATS and TRACE. The graphical possibilities allow the user to visually understand the main characteristics of the results of the different functions of the system.

2.2 Data base system

The data are organised in domains. A domain is composed of time series with the same seasonality or some other common characteristics. It is possible to create several levels within each domain (industry, sector, etc.), in a very flexible way.

TESS allows to aggregate series, assign weights to the components, duplicate series, edit series, import/export data, including multiple time series, automatic update of the database when new data become available, and also store and retrieve models, parameters, trends, and cycles, etc.

The time series manager window works in a way very similar to the explorer window in Microsoft Windows©. The window is splitted into two parts: the left side allows to open the tree structure and go deeper inside a time series to show properties of the model, forecasts, etc. The right side shows the time series when a domain is selected, its beginning year and the seasonality. All these operations can be performed with individual or aggregated time series.

2.3 Configuration of the pre-existing software

TESS allows a flexible configuration of the main parameters of TRAMO, SEATS and TRACE for modelling and forecasting, seasonal adjustment and trend and cycle estimation.

The systems include several default configurations created by the authors. Furthermore, a user can create his/her personal configurations for each time series and can save these configurations for future use.

2.4 Time series modelling and forecasting

For modelling and forecasting, TESS uses a DLL to access the program TRAMO (Time Series Regression with ARIMA Noise, Missing Observations and Outliers).

TRAMO is a program for estimation and forecasting of regression models with possibly nonstationary (ARIMA) errors and any sequence of missing values. The program interpolates these values, identifies and corrects for several types of outliers, and estimates special effects such as Trading Day and Easter and, in general, intervention variable type of effects. Fully automatic model identification and outlier correction procedures are available.

Intervention analysis is performed in a very user-friendly way.

2.4.1 Forecasting a large number of series and monitoring the forecast error

TESS allows for batch forecasting of a group of series. The system has been prepared to run automatically. The system can import automatically many time series, each with its previously selected configuration or with the default configuration, obtain forecasts and export the results to a file in only one step. In this case, TESS works like a black box, where we introduce time series and obtain forecasts.

Another feature related with forecasting is the automatic control of the forecast errors. Two control charts are used in order to monitor the forecast errors given by a model, and provide alarm signals that point out possible model's changes. These charts are the EWMA (see Page (1954 and 1961) and Barnard (1959)), and CUSUM (see Roberts (1959) and Hunter (1986)).

2.5 Treatment of special time series in the automatic modelling

When automatically forecasting a great number of time series, a test is performed in order to identify time series with intermittent demand (i.e. many zeroes followed by a period of demand). An improved version of the method developed by Croston (1972) is used to forecast this type of series.

TESS also offers the possibility to model very short time series (more than 6 values), this is made possible, in automatic way, by an extension of TRAMO. The method used to forecast short series is a sort of exponential smoothing, in which airline models with different seasonalities are specified and estimated.

2.6 Seasonal adjustment

SEATS is a program for estimation of unobserved components in time series following the so--called ARIMA--model--based method. Trend, seasonal, irregular, and transitory components are estimated and forecasted with signal extraction techniques applied to ARIMA models. The standard errors of the estimates and forecasts are obtained and the model--based structure is exploited to answer questions of interest in short--term analysis of the data. When used for seasonal adjustment, TRAMO preadjusts the series to be adjusted by SEATS.

TESS can integrated the use of TRAMO/SEATS for one or more time series using different configurations of SEATS. It can also create and use different configuration for each time series and save the results of seasonal adjustment together with a variety of plots.

2.7 Trend and cycle estimation

TESS integrates the new software TRACE. This is a program for the estimation of trends and cycles. The filters used by the program are fixed ARMA filters of three types. First, low--pass filters to estimate trends which are the two--sided forms of the well known Butterworth filters of electrical engineering, like the Hodrick--Prescott filter. Second, band—pass filters based on Butterworth filters to estimate cycles. Third, trend-cycle filters which are the finite versions of the Wiener--Kolmogorov filters corresponding to the trend in a canonical decomposition of an airline model. These last filters can also be used for seasonal adjustment assigning appropriate values to the parameters,.

The main purpose of TRACE is the estimation of trends, with various degrees of smoothness, and business cycles. To this end, it is strongly recommended that the program be always used as a second step in a two--step procedure. The first step in this procedure should consist of the estimation of the trend--cycle component by means of the application of a model--based method, like the one implemented in the TRAMO and SEATS programs. In this way, undesirable effects, like the generation of spurious cycles, are avoided.

The TRACE program is structured so as to be used both for in--depth analysis of a few series or for automatic routine applications to a large number of series. TESS interfaces TRACE in a user-friendly way allowing for:

- Trend a cycle estimation using differents configuration of TRACE.
- Possibility of creating and using different configuration types for each time series.
- Possibility of multiple time series selection and trend and cycle estimation.
- Save low pass (trend) and band pass (cycle).

2.8 Seasonal adjustment and forecasting of aggregated time series

Suppose that the series to be seasonally adjusted, X_t, is the sum of component series

$$X_t = X_{1t} + X_{2t} + \ldots + X_{kt}$$

Then, in addition to the series obtained by direct adjustment of the X_t, a second seasonally adjusted version of the series can be obtained by summing the adjusted component series. This second approach is called indirect adjustment.

A solution to the problem of aggregation of time series for seasonal adjustment is to choose between direct and indirect adjustment using a criterion based on the minimization of data revisions.

The revision statistics used in TESS are the average absolute percent revision (AAPR) and the absolute average revision (AAR). These statistics are calculated for both the aggregate series and its components. The criterion chooses the minimum AAPR or AAR value. The formulas are:

$$AAPR = 100\frac{1}{M}\sum_{t=T+1}^{M}\frac{\left|A_{t|N} - A_{t|t}\right|}{\left|A_{t|t}\right|} \qquad AAR = \frac{1}{M}\sum_{t=T+1}^{M}\left|A_{t|N} - A_{t|t}\right|$$

where $A_{t|N}$ is the final adjustment and $A_{t|t}$ the concurrent adjustment.

A new routine has been added to the TESS system. This routine (SSAG) computes both the average absolute percent revision (AAPR) statistic and the root mean square error of the one-step-ahead forecasts (RMSE1) . The two tests are based on the minimisation of revisions (AAPR) or (AAR) and the agreement of forecast (RMSE1). The last test is used in TESS to select whether it is better to use the aggregate (direct) or its components (indirect) to forecast. The formula is:

$$RMSE1 = \sqrt{\frac{1}{N-T}\sum_{t=T+1}^{N}\left(Y_t - Y_{t/t-1}\right)^2}$$

where Y_t is the real data and $Y_{t/t-1}$ denote the forecast of Y_t based on the data upto t-1.

3 TESS applied to an example

In the oral presentation of this paper, we will give details of the utilisation of the TESS system in two test cases: the AIRLINE model and Spanish Consumer Price Index

The first one is a typical example of univariate time series modelling. It is a time series used by Box-Jenkins (Box et al (1994)) to explain seasonal models. The test case will show how you can perform a seasonal adjustment and estimate a trend and cycle.

The second one is an example of the aggregation problem. The Spanish Consumer Price Index (IPC) is an aggregate of several economic variables. The test case will show how does TESS make a seasonal adjustment, obtain forecasts and discriminates between working on the aggregate time series or on its single components (direct and indirect methods).

This example of application can be also find in Prat et al. (1999).

4 Conclusion

There is a clear need for forecasting and seasonal adjustment methods for a broad spectrum of European organisations. From industrial and commercial companies to national statistical institutes, all have the real need of using these methods. It is also clear that the usage of these methods is a potential improving competiveness for industrial and commercial companies.

TESS was designed to perform modelling, forecasting, seasonal adjustment and trend and cycle estimation from time series. It is endowed with the most recent available methodologies and deals with the problem in a user-friendly way.

It suits both advanced and non-expert users. The non-expert will find software that produces forecasts of time series automatically taking into account all the possible errors that might affect the series. The expert will find an adaptable system that allows modifying its configuration in the most suitable way.

Professional statisticians will find a system that allows them to work with constantly updating series in an automatic way. Last, but not least, users of TRAMO, SEATS and TRACE will find that all the procedures are available within the software while the Graphical User Interface is much easier to handle.

Acknowledgements

TESS is partially funded by ESPRIT IV and the CICYT, grant TIC99-1280-CE.

References

Barnard, G.A. (1959). Control charts and stochastic processes. Journal of the Royal Statistical Society, Series B, **21**, pp. 239-271.

Box, G.E., Jenkins, G.M., Reinsel, G.C. (1994). Time series analysis, forecasting and control. Third Edition. Holden-Day, San Francisco.

Croston, J.D. (1972). Forecasting and stocks control for intermittent demands. Operational Research Quarterly, Vol. **23** N. 3, pp. 289-303.

Gomez, V. and Maravall, A. (1992). Time Series Regression with ARIMA Noise and Missing Observations – Program TRAM. Eui Working Paper Eco N. 92/81, Department of Economics, European University Institute.

Hunter, J.S. (1986). The exponentially weighted moving average. Journal of Quality Technology, **18**, pp. 203-210.

Maravall, A. and Gomez, V. (1992). Signal Extraction in ARIMA Time Series – Program SEATS. Eui Working Paper Eco N. 92/65, Department of Economics, European University Institute.

Page, E.S. (1954). Continuous inspection schemes. Biometrika, **41**, pp. 100-114.

Page, E.S. (1961). Cumulative sum charts. Technometrics, **3**, pp. 1-9.

Prat, A., Sole, I., Catot, J. and Lores, J. (1998). FORCE4/R, A new software product for forecasting and seasonal adjustment. NTTS'98, New Techniques and Technologies for Statistics, Sorrento, 1998. Contrib. paper, pp. 429-434.

Prat, A. et al. (1999). TESS/JRC-UPC/5/1.1 P deliverable. It's a document describing two relevant test cases. Available from the authors

Roberts, S.W. (1959). Control chart test based on geometric moving averages. Technometrics, **1**, pp. 239-250.

References to programmes

FORCE4/R (1997). Software. It is a forecasting, modelling and signal extraction software that was developed by UPC as part of FORCE4, an ESPRIT IV project number 20.704.

TRACE (1999). Software (DLL). Trend and Cycle Estimation, done by Victor Gomez and integrated into TESS by UPC.

SSAG (1999). Software (DLL). State Space form for Aggregation, done by Victor Gomez and integrated into TESS by UPC.

The BADAME project

Hana Rezanková and Luboš Marek

Department of Statistics and Probability, University of Economics, Prague, 13067, Czech Republic

Abstract. This paper describes a project, which aims at the creation of a database of indicators and models of the Czech economy. Datasets with time series in a form convenient for analysis are prepared and analysis results are published on a website. Information about the statistical computational environment is also provided.

Keywords. BADAME project, Czech economy, time series analysis, data sources, statistical computational environment, SAS system

1 Introduction

BADAME is an abbreviation of the Base of Data and Models of the Economy of the Czech Republic. This word means "we research" in the Czech language. Project BADAME is financed by the Ministry of Education, Youth and Sports of the Czech Republic. The project team consists of researchers and teachers of the University of Economics, Prague (UEP) and is coordinated by Jirí Ivánek, vice-rector for development and information systems of UEP.

The aim of this project is the creation of a user-friendly environment for economic research. It concentrates on microdata, macrodata and metadata concerning the Czech economy, methods and software for their analysis. These data can be used by researchers, teachers and students of the University of Economics. The project can serve as a model for other universities and research institutes.

The whole project is composed of the following six specific projects:

- implementation of macroeconomic models for the Czech Republic;
- evaluation of financial market models and corresponding software applications;
- databases of enterprises and models for evaluation of enterprises and branches;
- databases of statistical indicators and statistical models;
- evaluation of methods and software applications for knowledge discovery in databases;
- technical support of the project.

Results of activities of project participants are published on the project website (http://badame.vse.cz), where some parts of the environment are available to all Internet users.

One of the main results is the installation of databases with indicators of individual enterprises, which are only locally available at university. In our paper we mainly concentrate on the specific project concerning the creation of a database of statistical indicators and statistical models.

2 The BADAME project and statistics

The range of the fourth project is greater than its name suggests. We would like to provide both data and the statistical computational environment to users. We consider the following parts and their components.

Data and models: a "mirror" of the Czech Statistical Office Website, a page with links to web pages containing datasets concerned with the Czech Republic, and a database of economic time series and results of their analyses.

Teaching statistics: "mirrors" of some English textbooks on statistics, a page with links to web pages containing textbooks on statistics and lists of them, and our own interactive textbook on statistics in Czech and partially in English.

Statistical software: updating statistical packages (SPSS, STATGRAPHICS), installation of the SAS System, description of these software products, and a page with links to websites of software firms.

The choice of statistical packages is influenced by the tradition in the Czech Republic (which software was used on the mainframes and which one on the first type of personal computers) and by the fact of which firms have their representations in our country. The price of the site license is also very important for our university of more than 13,000 students.

Each package usually includes some method, algorithm or graphical presentation which is not included in others packages. Therefore it is useful both to have several different statistical packages for data analysis and provide information about their possibilities.

Cooperation in other projects: presentation of web page SYSEL (Czech name for suslik) about the statistical computational environment containing information about statistical websites, software, literature, data and teaching on the Internet.

3 Data sources on the Internet

We have been interested in data about the Czech Republic which can be found on the Internet and how datasets are presented on web pages. There are many datasets with indicators about the Czech economy on the Internet. *The Czech Statistical Office* (www.czso.cz) provides some tables. However most of them cover only a short period. Tables can be displayed mostly in a spreadsheet form.

Websites of ministries and other institutions of the Czech Republic also provide some data, e.g. *the Ministry of Finance of the Czech Republic* (www.mfcr.cz) shows macroeconomic forecast on its website and *the Czech National Bank* (www.cnb.cz) publishes monetary indicators on the Internet. There are some other interesting sources, e.g. the commercial website *PATRIA* (www.patria-finance.com/default.asp).

Further, we can find indicators on foreign websites, for example:

- *Business Central Europe* (www.bcemag.com/_bcedb/stat_main.htm) – a database, which includes basic macroeconomic indicators of Central European countries (the chosen indicator is displayed for each of the countries or all indicators are shown for the chosen country – for several time periods),

- *InfoNation* (www.un.org/Pubs/CyberSchoolBus/infonation/e_infonation.htm) – up to four indicators for chosen countries (members of the United Nations) can be displayed,
- *EIA* (www.eia.doe.gov/emeu/iea/contents.html) – *Energy Information Administration*, includes tables for individual indicators (number of habitants, GDP, energy consumption), rows represent individual countries and columns represent years,
- *United Nations – Global Statistics* (www.un.org/Depts/unsd/global.htm) with the database FAOSTAT (Food and Agriculture Organizations of the United Nations), chosen tables are displayed and can be downloaded,
- *The World Bank Group* (www.worldbank.org/data) – makes it possible to study a publication with data and charts about individual countries.

4 Time series analysis

The main purpose of our specific project is time series analysis with respect to the Czech economy and its presentation on the website. We met some serious problems during data acquisition and time series analysis. The problems are the following:

Most Czech economic time series are very short. The small number of observations is due to the fact that the Czech economy is a very young one. Price liberalization was started in January 1992 and many of the economic series are recorded from this date. For this reason, we cannot use some of the usual methods (e.g. Box-Jenkins approach), because ARIMA model construction requires at least 60 observations.

Some observations in longer time series are not comparable with later observations. The series before the year 1990 were realized in a socialist economy. Methods for investigation of indicators were changed several times. The long economic time series are often "contaminated" by many events such as the separation of our state into the Czech and Slovak Republics, price liberalization, political decisions and other interventions. Therefore it is very difficult to recalculate them into a usable form.

Most indicators are not available in electronic form, or they are available in this form for a very short period. If some tables exist, they contain data just for presentation and they are not convenient for computer analysis. Therefore we have studied only some time series, which we have prepared in the electronic form from Statistical Yearbooks and other publications.

The fact that comparability is a serious problem is clear from Fig 4.1. This is a quarterly series of GDP (Gross Domestic Product). Methodology changes in the first quarter 1994 have caused the step in this time series.

We have analyzed Czech economic time series of different categories - macroeconomy, external trade, environment and agriculture, prices, labor and social statistics, industry and construction. The main focus is on basic series such as GDP, unemployment, inflation rate, producer price indices, consumer price indices, external trade, population and so on. Most series are recorded as annual or quarterly ones and a few only as monthly series.

Web users can see the source data, time series plot, tentative series model and small horizon forecasting. Datasets are in Excel format and every visitor can download them. We update the series continuously and then recalculate the indicators, model and forecasting, of course.

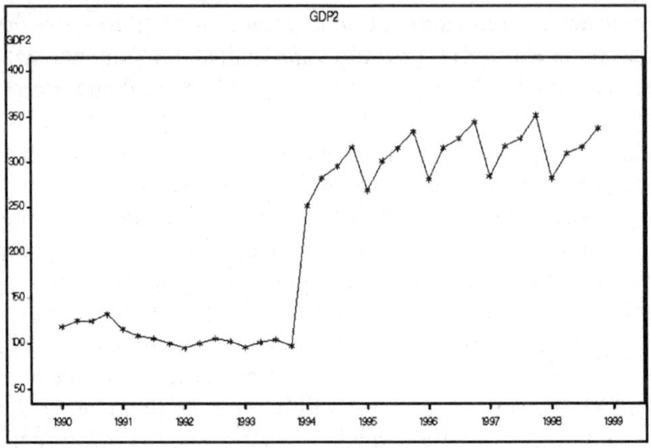

Fig 4.1. GDP plot

Different methods such as decomposition, weighted averages methods, exponential smoothing methods, Box-Jenkins methodology and intervention analysis have been used for time series analysis. The method applied depends on the actual series observations.

We have performed the analysis with the SAS System. The forecasting procedure in the SAS System is very suitable for these analyses, because it includes the most advanced time series methods. However, some other statistical package (SPSS) could be also used.

An example of what is offered with respect to time series analyzed is shown in Fig 4.2.

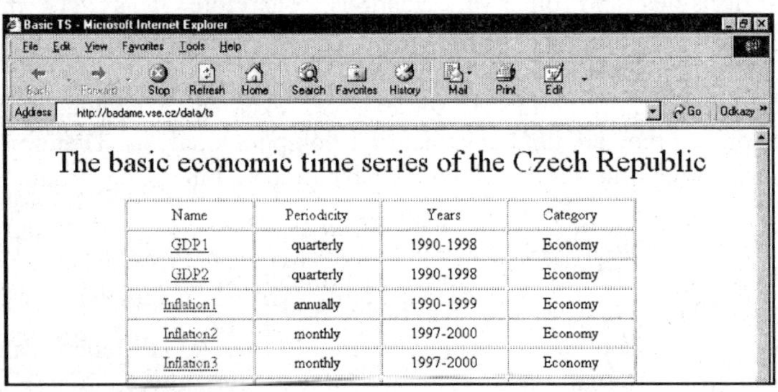

The basic economic time series of the Czech Republic

Name	Periodicity	Years	Category
GDP1	quarterly	1990-1998	Economy
GDP2	quarterly	1990-1998	Economy
Inflation1	annually	1990-1999	Economy
Inflation2	monthly	1997-2000	Economy
Inflation3	monthly	1997-2000	Economy

Fig 4.2. Offer of time series

All columns of the table are generated from an Access database dynamically. The database contains such information as name, comments (both in English and Czech) and the names of files with data, analysis results and graphs for each time series (graphs are in JPEG format). Therefore both English and Czech web page scan be generated. For each time series, the output is generated in the form which is displayed in Fig 4.3.

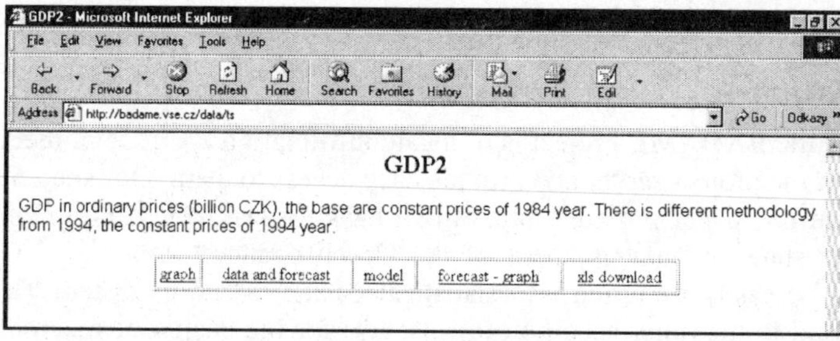

Fig 4.3. Links to data and results presentation

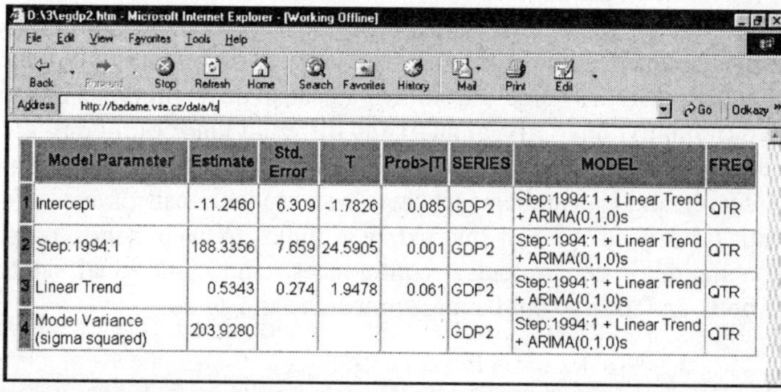

	Model Parameter	Estimate	Std. Error	T	Prob>|T|	SERIES	MODEL	FREQ
1	Intercept	-11.2460	6.309	-1.7826	0.085	GDP2	Step:1994:1 + Linear Trend + ARIMA(0,1,0)s	QTR
2	Step:1994:1	188.3356	7.659	24.5905	0.001	GDP2	Step:1994:1 + Linear Trend + ARIMA(0,1,0)s	QTR
3	Linear Trend	0.5343	0.274	1.9478	0.061	GDP2	Step:1994:1 + Linear Trend + ARIMA(0,1,0)s	QTR
4	Model Variance (sigma squared)	203.9280	.	.	.	GDP2	Step:1994:1 + Linear Trend + ARIMA(0,1,0)s	QTR

Fig 4.4. Model estimated

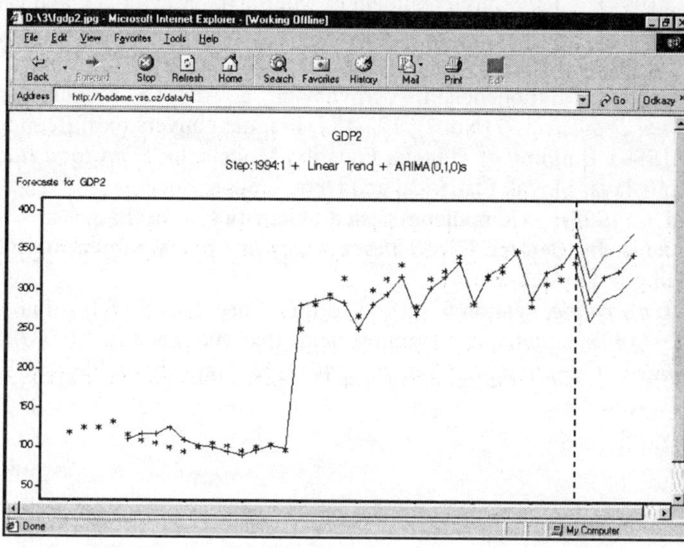

Fig 4.5. Final model and forecasts

The appropriate model has been chosen with regard to R-square and other statistics which exist in the SAS System. The parameters estimated, standard errors, t-statistics and other results are obvious from Fig 4.4. The graph in Fig 4.5 shows the final model and forecasts.

5 Conclusion

The aim of the BADAME project is to locate information about data, methods and software on the project server and provide easy access to them. Our specific project described in this paper provides data in two basic forms: tables and graphs which present the state and development of the Czech economy and Excel data files which can be easily imported into statistical packages and analyzed. These data files are used in the other specific projects, e.g. for the design of macroeconomic models. They can be also used for the evaluation and comparison of time series procedures in statistical packages.

We consider adding more time series to our database and updating them. We would like to perform some parts of time series representation dynamically. However this approach has some difficulties, because different types of time series are presented (monthly, quarterly, annual) and different kinds of models are used.

Cooperation in other projects is also useful. The web page SYSEL is located on the project server and researchers, teachers and students can obtain information about statistical sources on the Internet, e.g. links to web pages of statistical offices, associations, departments, conferences and individual statisticians, websites of book and journal publishing companies and so on.

References

Box, G.E.P. and Jenkins, G.M. (1970). *Time series analysis, forecasting and control.* San Francisco: Holden Day.

Lenz, H.-J. (1994). A Rigorous Treatment of Microdata, Macrodata, and Metadata. In: *COMPSTAT94 Proceedings in Computational Statistics*, 357-362. Heidelberg: Physica-Verlag.

Marek, L. (1998). Exponenciální vyrovnávání a ARIMA modely. In: *Acta Oeconomica Pragensia*, **6** (No. 2), 173-177. Prague: University of Economics.

Marek, L. (1999). Building of Transfer Function Models. In: *Výpo tová štatistika*, 64-68. Bratislava: Slovak Statistical and Demographic Society.

Müller, M. (1998). Computer-Assisted Statistics Teaching in Network Environments. In: *COMPSTAT98 Proceedings in Computational Statistics*, 77-88. Heidelberg: Physica-Verlag.

SAS/ETS User's Guide, version 6, second edition. Cary, NC: SAS Institute Inc.

Talbot, M. (1998). Statistics Training and the Internet. In: *COMPSTAT98 Proceedings in Computational Statistics*, 461-466. Heidelberg: Physica-Verlag.

Improving Statistics Canada's cell suppression software (CONFID)

Dale A. Robertson

Statistics Canada, Tunney's Pasture, Ottawa ON, K1A 0T6 Canada

Abstract. The need for cell suppression software is discussed. A brief review of the theoretical framework is given. A recent modification to improve the treatment of the common respondent problem in CONFID (software in use at Statistics Canada) will be described.

Keywords. Disclosure control, cell suppression, mathematical programming

1 Introduction

Statistical agencies have to grapple with a trade-off between the amount of information they release, and the privacy of the individuals who provide this information. In the past various bureaus have devised their own techniques. It is only recently that general use software has appeared. ACS, Sande (1995), is a complete suite of cell suppression software produced by the original creator of CONFID. The multi-national ARGUS project, Waal and Willenborg (1999), aims to produce disclosure avoidance software for both tabular data and microdata files. The problems that arise in the area are often those of operations research, which tends to be outside the realm of the expertise of the typical statistician.

In cell suppression a particular problem arises because the same respondent contributes to different tabulation cells. An improvement in the treatment of this problem will be described.

1.1 Cell suppression

Cell suppression is a disclosure avoidance technique used in magnitude tables. These are typically produced in the economic statistics area. Business parameters such as production volume, sales, ... are tabulated. Detailed multi-level standard classification systems (industrial coding schemes, geography...) are used, leading to complex collections of simple tables, related by common cells. Responses vary widely in size. A single entity can provide most of the cell total, even with many contributors. The identities of the respondents in a cell are generally known while the absence of a certain activity (a zero cell) is also known. All variables should be considered bounded. Often by their nature, the tabulated quantities are non-negative and some estimate of response can be made by a knowledgeable user (say within 50%).

The resulting problem is of significant scale. Hundreds of thousands of observations may be aggregated into perhaps tens of thousands of cells. These cells must be inspected for disclosure and protective measures taken.

In the tabulations, some cells are intrinsically sensitive: release of their value is a disclosure. These cell values must be concealed or rendered uncertain. Simply hiding them (X in place of the value) won't do. Cell values are related by linear equations that permit estimation of the suppressed cells (residual disclosure). Study of bounded variables subject to linear equations falls into the general domain of linear programming (LP) and this well-established methodology will be useful.

1.2 Prevention of residual disclosure

Preventing residual disclosure requires the suppression of the values of other cells beside the intrinsically sensitive ones, the "complementary suppressions." One can see three questions then arising.

ξ What are suitable criteria for deciding upon the sensitivity of cells?
ξ How can one choose complementary suppressions?
ξ How can one be sure that a pattern is in fact adequate?

It is convenient to begin with the last question. Consider the following table.

X	X	204
X	X	9
204	9	213

If that is all we are given, not much can be done. Add the condition that the tabulated quantity cannot by its nature be negative, and the above is equivalent to a table of ranges.

195-204	0-9	204
0-9	0-9	9
204	9	213

Adding a plausible estimate for the hidden cells to about 50% of their value will further modify this to

196-201	3-6	204
3-6	3-6	9
204	9	213

This principal is clearly general. Any table system with hidden cells, where the cells are subject to certain bounds, is equivalent (via a LP calculation) to a table of correlated ranges for the hidden cells. We would not want these ranges to be overly precise. One must decide upon suitable protection ranges for sensitive cells. An adequate pattern will supply sufficient ambiguity for the sensitive cells, and ensure that implicitly published aggregations are non-sensitive. Note that because of the correlated ranges, naive measures of pattern quality can be quite misleading

2 Theory

A theoretical framework was set up for this problem by Sande (1977), with important contributions by Cox (1980).

The theory gives a way to measure sensitivity in a quantitative manner. This sensitivity defines the ambiguity a sensitive cell needs, and allows one to set up effective procedures for evaluating or generating suppression patterns. The sensitivity is found by evaluating a weighted sum of the responses in the cell. Sande (1977) observes that the rule should be subadditive. This is the inequality

$$S(C_1+C_2) \leq S(C_1) + S(C_2).$$

Here C_1 and C_2 are two arbitrary tabulation cells of sensitivities $S(C_1)$, $S(C_2)$ and $S(C_1+C_2)$ is that of their union. The direction of the inequality captures the notion that aggregation provides protection in general. An immediate consequence is that non-sensitive cells ($S <= 0$) can never combine to make a sensitive cell ($S > 0$). Cox (1980) gives conditions on the weights that are necessary and sufficient for sub-additivity to hold.

Another important relationship that holds under cell aggregation is

$$S(C_1) - T(C_2) \leq S(C_1+C_2)$$

Here C_1 is a sensitive cell, united with another cell C_2, in an attempt to produce a non-sensitive union. The above limits the effect of C_2 and captures the intuitive notion that a more sensitive cell will require a larger complement for its protection. (The above implies a natural normalization for S, as yet undetermined up to a multiplicative factor.)

The importance of the behaviour under cell union arises because the union of all the suppressed components in a published total is an effectively published cell. If this union is sensitive then one has not provided proper protection. An effective complementary suppression renders the union non-sensitive.

The common rules in use obey the above natural inequalities. It is recommended that any proposed rule follow the two inequalities.

From sub-additivity, given a sensitive cell C_1, it is clear that a complement C_2 whose sensitivity is sufficiently negative,

$$|S(C_2)| = -S(C_2) \geq S(C_1)$$

will ensure that the resulting union is non-sensitive.

However such a cell if it exists (and it may not) would generally be large, of the same order as C_1 or greater. This would be much more suppression than is needed, leading to unacceptable patterns.

Rather one must focus on the other inequality, and see if can be made precise. If so then one can use a (generally much smaller) cell with $T(C_2)$ comparable in size to $S(C_1)$. ($S(C_1)$ is in general a few percent of $T(C_1.)$)

One cannot determine the exact sensitivity of a union without obtaining the data for the cells, constructing the new response sequence, and evaluating the formula for S. However the continuity inequality becomes sharp if the proposed complement has no common respondents with the sensitive cell, and its responses are small compared to the large responses of the sensitive cell. If one can solve the problem of common respondents, then one can use the size of the cell as a rule for choosing complements. This permits the separation of the process of table construction from that of pattern generation, a very desirable goal for a practical implementation.

Various approaches have been proposed to solve the common respondent problem. One technique was that of "protection capacity", Jewett (1993). Cells with respondents in common are less effective complements. A protection capacity is assigned which reflects the power of the cell to protect the sensitive one. It is our understanding that ACS, Sande (1995), also treats the common respondent problem, but we are unaware of the details of the method.

The approach discussed here involves detecting "all" sensitive unions of cells at table construction time. Sensitive unions are added as objects to protect. Thus ultimately will ensure that all effectively published aggregations are safe.

For this to be feasible, the formulation must allow the representation of combinations of cells as objects to protect. This prevents the use of the network flow models, which cannot handle unions. A more general code, capable of handling full LPs, is required.

3 The practical treatment of unions

Detecting "all" sensitive unions suggests combinatorial explosion, but things are not nearly as bad as one might first think. It turns out the number of unions to test and to retain is not nearly as large as first suspected.

Although non-obvious combinations of cells can sometimes be isolated, we are not going to solve the general problem here. We will only worry about unions in which all the component cells lie along a single axis and are components of a non-sensitive total. (If the total is suppressed, then one cannot trivially obtain the total of the hidden cells.) So the task is to check relevant subsets of a non-sensitive total for sensitivity.

Fortunately one doesn't need to examine all subsets. A little thought, using the fact that two non-sensitive cells can never combine to produce a sensitive cell, drastically reduces the number of cases to consider.

All the sensitive cells in the published total will always be suppressed. Thus the first union to check is that of all the sensitive cells. If this is non-sensitive, then no sensitive combinations can be formed by adding other cells. If it is sensitive, then add it to the list of objects to be protected, and more suppression will be forced.

This was the historical level of treatment of common respondents in CONFID. The idea can clearly be extended. When the set of sensitive cells is itself sensitive, proceed to check all combinations of it with single non-sensitive cells. Again if all of these are non-sensitive there is nothing further to do. If one is sensitive put it into the protection list and check all combinations of this union with a second member of the set of non-sensitive cells etc.

One can imagine all the possibly interesting combinations to be arranged as a tree of non-uniform depth and 2^n nodes (n is the number of non-sensitive cells in the total). The root of this tree corresponds to the set of all sensitive cells in the total. The first level down has the combinations of the root set with a single non-sensitive cell, the next any with two non-sensitive cells, etc. A depth-first search can explore this tree. The search tree can be drastically pruned, as descent stops as soon as a non-sensitive combination is met. In addition the sub-additivity bound can often be used to deduce non-sensitivity of a combination without calculation.

3.1 The algorithm

The algorithm to examine the subsets can be informally described as follows. One needs to maintain two sets of cells A, and B. Sens(T) is a function which, given a set T, will evaluate the sensitivity of the union of the cells contained in T. Then the algorithm is

Initialize: Set A to the set of all sensitive cells contained in the total.
Set B to the set of all non-sensitive cells in the total.

Then invoke the procedure traverse (A,B) where traverse is the following

```
procedure traverse(SS, NN)
     real sSen
  sSen = Sens(SS)
  if (sSen <= 0.0 )   ! non-sensitive stop this probe
     return
     else  ! record and descend
     add SS to the list of objects to protect
     while NN is non-empty
           remove an arbitrary cell C from NN
           if (sSen + Sens (C )> 0 ! otherwise must be non-sensitive
                         add C to SS
                traverse(SS,NN)
                remove C from SS
           end if
     end while
  end procedure traverse
```

Note that it is not necessary to make local copies of the sets if one uses a stack to store the set elements. One only needs local stack pointers which correspond to the current instances of A and B

4 Results

Initial results were encouraging. The number of evaluations and combinations was not excessive, and the overall effect was modest.

In a case using agricultural data, the number of sensitive unions of sensitive cells was 73. Turning on the full combination analysis resulted in a total number of sensitive combinations of 122. This means that most tree searches quit right at the root, and few go deeper than the first level. On the other hand, at least one case of significant expansion was observed. The tree traversal added 119 more calls of the sensitivity evaluation function. The whole table building step is not expensive and the resources consumed are negligible. The extra amount of work is insignificant.

Of much more interest is the consequences for the suppression phase. Again results were very encouraging. The chief worry was the production of many more complements. This would be difficult to justify to users, who have difficulty understanding the concepts. In fact one ended up with the same number of complementary suppressions in either run (555 complementary suppressions to protect 2000 sensitive cells.) The patterns were not identical as the suppressed total rose slightly more than 1%, from 457000 to 462000. This change should be acceptable given the greater degree of protection, which results. Again the extra work was not significant.

References

Cox, L. (1980). *Suppression Methodology and Statistical Disclosure Control.* Jour. Amer. Stat. Ass. **75**, no.370 (Theory and Methods Section) 377-385.

Cox, L. (1979). *Linear Sensitivity Measures in Statistical Disclosure Control.* Proc. Amer. Stat. Ass. Social Statistics Section. 634-639

Cox, L. & Sande, G. (1979). *Techniques for Preserving Statistical Confidentiality.* Proc. 42nd session Int.Stat.inst.

Jewett, R. (1993). *Disclosure Analysis for the 1992 Economic Census.* Working paper, U.S.B.C.

Sande, G. (1977). *Towards Automated Disclosure Analysis for Establishment Based Statistics.* Working paper, Statistics Canada

Sande, G. (1984). *Statistical Journal of the United Nations. ECE 2* 34-41

Sande, G. (1995). *ACS documentation.* Sande & Associates, 600 Sanderling Ct. Secaucus NJ, 07094 U.S.A.

Sande, G. (1999). *Blunders Official Statistical Agencies make While Protecting the Confidentiality of Business Statistics.* Technical Note

The multimedia project MM*STAT for teaching statistics

Bernd Rönz, Marlene Müller and Uwe Ziegenhagen

Humboldt University Berlin, Faculty of Economics and Business Administration, Institute for Statistics and Econometrics, Spandauer Straße 1, 10178 Berlin, Germany

Keywords. Interactive computational statistics, multimedia teaching, internet bases teaching

1 Introduction

The multimedia project MM*STAT was developed to have an additional tool for teaching statistics. There are some important facts which influenced this development. First, teaching statistics for students in socio-economic sciences must include a broad spectrum of applications of statistical methods in these fields. A pure theoretical presentation is generally considered by the students to be tedious. Second, in practice no statistical analysis is carried out without a computer. Thus, teaching statistics must include the acquisition of computational capabilities. Third, statistics has become more and more complicated over time, because of increasingly complex data structures, statistical methods and models. Thus, an ever-increasing special knowledge of statistics is required and has to be taught. Fourth, notwithstanding these high demands on teaching statistics, the available lecture time, especially for the introductory courses, has remained constant over the years or has even been cut down.

To ensure an effective education in statistics, there is clearly a need to update and to revise the current teaching methodology. Blackboard work, overhead transparencies and textbooks indisputably remain essential components. *Additional* teaching and learning tools can help to integrate the aforementioned requirements. New multimedia technologies are a challenge to extend the traditional methods of teaching statistics by new techniques, especially by computer-aided components. In this regard, the internet will have a major influence on teaching statistics.

The natural succession of statistics lectures incorporates three important dimensions (Derby, N., Härdle, W., Rönz, B., 1999): presentation and explanation of important statistical methods and models with their assumptions (first dimension); deepening the subjects by means of more background information and demonstrating the application of statistical methods by elementary classroom examples as well more elaborate statistical analysis (second dimension); reinforcing previous ideas within the process of presenting new ones (third dimension). The multimedia project MM*STAT implements these three dimensions of teaching statistics in a HTML based computer teaching and learning tool. It is a flexible tool to support teaching and learning statistics in introductory courses via the internet or from a CD-ROM. It can be used in the class room, but is mainly used as a tool for reviewing the lecture subjects by the students at their own place. MM*STAT is intended as an add-on to the traditional classroom lecture. It allows the students – in an individual way – to get more detailed information about statistical terms and methods and to practice statistics with real data sets. In particular, the interactive examples which are integrated into MM*STAT provide a variety of different data sets.

At present MM*STAT is only available in German, being written for and tested by students at the Faculty of Economics and Business Administration, Humboldt University Berlin, Germany. If all components are well implemented it will be translated into English. In the following some of the main characteristics of MM*STAT will be pointed out.

2 Details of MM*STAT

The style of MM*STAT can be considered as a composition of filing cards, allowing fast access to each card and opening up to 10 cards at the same time. Main components of MM*STAT are lecture units, additional information and examples.

2.1 The lecture units

Upon entering the electronic statistics course from the logo screen of MM*STAT the first card shows the table of lecture contents.

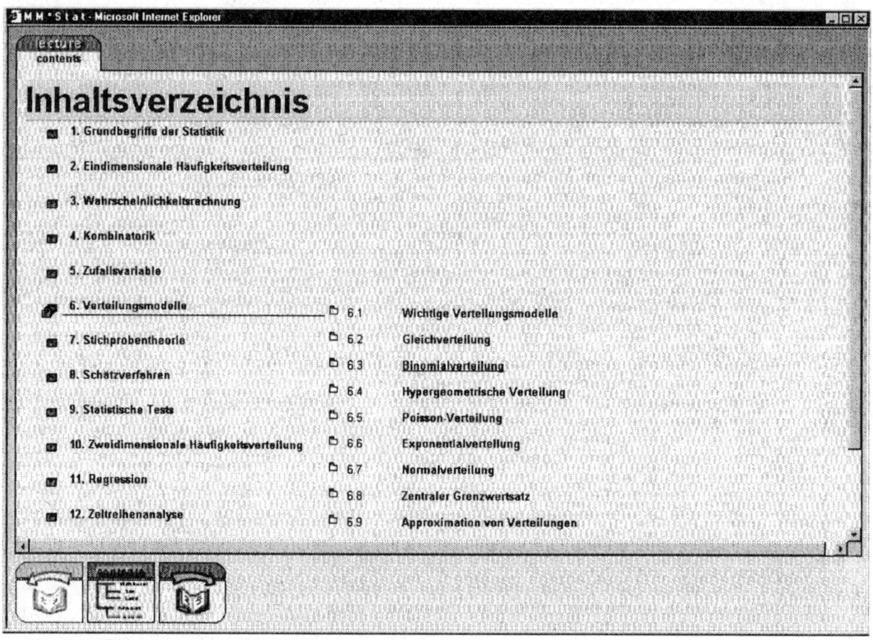

Fig 1. Lecture contents of MM*STAT

From here the user can choose the lecture unit to be reviewed. Because MM*STAT is designed as a teaching and learning tool for an introductory statistics course, it contains the common and well-known topics: basic concepts of statistics, one-dimensional frequency distributions, basics of probability calculus, combinatorics, random variables, probability distributions, sampling theory, estimation methods, hypothesis testing, two-dimensional frequency distributions, regression analysis, time series analysis.

Each "lecture" filing card consists of an outline of the lecture, providing the basic concept of the method considered, definitions, important formulas, graphics, fundamental requirements of the application of the statistical method or model considered. "Forward" and "Backward" icons allow the user to move along the lessons' sequence which represents the first dimension of teaching.

2.2 More information

If required, more information on the statistical method explained in the lecture unit is provided. In this case an "information" button appears on the lower right of the "lecture unit" screen. More information should deepen the understanding of statistical methods and can be for example the derivation of the method or of a special formula, additional textual explanations, comments on the main applications. In this sense, more information belongs to the second dimension of teaching statistics (cf. Section 1). The filing tabs "lecture" and "information" enable the user to switch at any time between the lecture unit card and the information card, while leaving the other card to be previewed again.

2.3 Examples

Examples are essential for understanding the statistical methods given in the lecture unit and are considered as part of the second dimension of teaching statistics. MM*STAT provides three types of examples: fully explained examples, enhanced examples and interactive examples. Of course, not every lecture unit is connected with examples or with each type of example. What is available for the selected lecture unit can be seen by the buttons in the lower right corner of the lecture unit screen. Each type of example has its own background colour which is again distinct from the background colour of the lecture unit. The filing tabs enable the user to switch at any time between the lecture unit card, the information card and the example card if the mentioned cards have previously been opened.

Fully explained examples: Fully explained examples are directly related to the content of the lecture unit. They start out with the statement of the problem to be solved and the illustration of the data, continue with the exact argumentation for the statistical method to be applied and the choice of the relevant formulas given in the lecture unit, show the calculations and finally give an interpretation of the results.

Enhanced examples: All that was said about the fully explained examples holds for the enhanced examples, too. In addition, enhanced examples are characterised by at least one of the following criteria: (1) They refer not only to the content of the lecture unit just considered, but also to the previous ones. (2) They compare different versions of a statistical method (e.g. confidence interval estimates for the population mean with known and unknown population variance respectively; hypothesis testing based on a small sample and on a large sample). (3) They comprise various statistical methods (e.g. range, quartiles, mean, standard deviation, histogram and boxplot for an income distribution). (4) They contain the application of the same statistical method on different variables or on the breakdown of a variable by a factor (e.g. on household expenditure for food and for clothes; on the overall sampling distribution of age and the age distribution broken down according to sex).

Interactive examples: The fact, that learning by doing is most effective in the learning process, is not new and naturally applies to statistics, too. However, its translation into action is often difficult, especially if the number of students in an introductory course is large. Therefore, an important feature of MM*STAT is its interactive capability embodied in the interactive examples. This kind of example ensures that the student can actively participate in the learning process. Interactive examples enable the students to study the effects of changes of various aspects of the problem at hand and to get an intuitive understanding how the statistical method works.

Let us mention some of the implemented interactive examples here: Study the same data set as in the fully explained or enhanced example, but now with the opportunity to select other variables; choose among different data sets; carry out the same statistical procedure, but with a different sample size; change the number of classes or the interval width in the case of grouped data and show the effect on the parameters and on the resulting histogram; change the parameters of distributions; transform a variable and compare the resulting distribution to that of the original variable; repeated confidence interval estimations with the same or changing confidence level and/or sample size; change the significance level and/or the sample size for hypothesis tests; repeated regression and time series estimations.

2.4 Reinforcing previously learned statistical concepts

In the progress of a statistics course it is very often helpful to reinforce statistical concepts and methods already introduced earlier in the course. That is the third dimension of teaching statistics. In an electronic statistics course the user must be enabled to resort to a previous statistical idea without starting from the very beginning.

In MM*STAT this can be practised in different ways: To recall a special term or definition a glossary is available. Within a lecture unit, important terms are hyperlinked to the glossary. A small symbol at the end of each definition in the glossary is a direct link to the lecture unit in which the term is explained in detail.

If the user wants to completely reread a lecture then there are two opportunities: If the desired lecture unit is the previous one than the "backward" icon can be used. If the desired lecture unit is not the previous one then the selection must be made via the “contents” card.

Sometimes a hyperlink to statistical concepts and methods already introduced earlier in the course can be found in the lecture unit. This hyperlink enables the user to open that lecture unit in addition and to switch to and from by clicking on the tabs.

2.5 Some other features of MM*STAT

Some other features of MM*STAT worth mentioning are:

Multiple choice questions are included at the end of each section so that the student has the opportunity to check his/her knowledge. The correctness of the response to the questions is automatically checked and displayed. Furthermore, the correct answers can be displayed so that the student has a feedback to his/her decisions.

A special filing card called *bookmark* can be opened from the last entry on the table of contents. This card contains a list of the filing cards last viewed. These filing cards can be reopened either separately or all together. This feature enables continuous work with MM*STAT.

In a few lecture units *sound sequences* are available for audio explanations of statistical facts.

Help pages, concerning the handling of MM*STAT and the required software configuration to run MM*STAT, are accessible from the start page of MM*STAT.

3 Technical requirements of MM*STAT

MM*STAT is based on HTML 4.0 combined with JavaScript and Cascading Style Sheets (CSS). HTML is used for the text of the filing cards and to realise hyperlinks. Formulas and graphical illustrations are included as GIF or JPEG bitmaps. CSS documents record the internal structure of the MM*STAT documents and comprise for instance background designing, typographical attributes of the filing cards and the embodiment of links. JavaScript programs (ECMA script dialect), inserted into the HTML documents, enable a comfortable and innovative navigation through all sections of MM*STAT, dynamically fit in multimedia objects (e.g. sound sequences) and improve the overall visual impression (e.g. by MouseOver effects). The JavaScript program in the basic document of MM*STAT manages the whole filing card system and integrates the filing cards requested by the user.

The technical minimum requirements are: Pentium 90 or comparable processor, CD-ROM drive (for the CD version of MM*STAT), internet connection (for the MM*STAT online version), Microsoft Internet Explorer 4.0 (or higher) or Netscape Navigator 4.5 (or higher), Java plugin 1.1 and swing class library, sound card and audio/video plugin for the used webbrowser.

A special feature of MM*STAT is its interaction capability. This allows the user to study statistical methods under varying application conditions or by using different data sets without any additional software overhead. To include the interactive examples, the XploRe Quantlet technology was used. This is a client/server concept which was especially developed for statistical computing over the internet. In this architecture the graphical user interface of the program, the XploRe Quantlet Client (XQC), is separated from the computational part, the XploRe Quantlet Server (XQS).

The XQC is a Java Applet and can therefore be integrated into HTML documents. On one hand its flexible configuration makes it possible to provide the inexperienced user with a menu driven software by allowing only certain changes of parameters, the selection of variables or data sets and executing the statistical method. On the other hand it permits the experienced user to edit the program and to execute the modification.

The XQS is a statistical software environment which provides a variety of statistical methods as well a matrix oriented programming language. The communication between client and server is realised by a special protocol (MD*Crypt), using TCP/IP which is available on each internet capable computer.

For a detailed explanation of the XQC/XQS technology see Härdle, Kleinow, Tschernig (2000) and Kleinow, Thomas (1999).

4 Comments of the students

MM *STAT has been used via the internet in the class for three terms (since the summer term 1999) and was sold as CD-ROM to the students. MM*STAT met with a very good response from the students. An inquiry of the students attending the class in the winter term 1999 revealed that 72 percent already work with the tool. Further results for those who used the tool were:

Medium		Frequency of usage	
CD-ROM version	48%	more than once per week	4 %
via internet	40 %	once per week	20 %
both versions	12 %	more infrequent	76 %

75 % of the users said that MM*STAT was helping them to better understand the introductory statistics course. This positive reaction is an encouragement for us to further improve the teaching and learning tool MM*STAT.

References

Derby, N., Härdle, W., Rönz, B. (1999). The three dimensions of multimedia teaching of statistics. In: *Discussion Paper No. 76*, Sonderforschungsbereich 373, Humboldt University Berlin

Härdle, W., Kleinow, T. & Tschernig, R. (2000). Web quantlets for time series analysis. In: *Discussion Paper No. 1*, Sonderforschungsbereich 373, Humboldt University Berlin

Kleinow, T. & Thomas, M. (1999). Computational Resources for Extremes. *Discussion Paper*, Sonderforschungsbereich 373, Humboldt University Berlin.

Project-URL: http://www.mm-stat.de

Projection pursuit approach to robust canonical correlation analysis

M. Rosário de Oliveira and João A. Branco

Department of Mathematics and Applied Mathematics Centre. Instituto Superior Técnico, Technical University of Lisbon, Av. Rovisco Pais, 1049-001 Lisboa, Portugal

Abstract. Projection pursuit techniques are used to build new robust estimators for the parameters of the canonical correlation model. A simulation study shows that for non-ideal data these estimators can perform as well as other robust estimators. However, they can have much higher breakdown points. This advantage makes these estimators the right choice for use with real data, where potential outlying observations are very frequent.

Keywords. Canonical correlation analysis, projection pursuit, robust estimation

1 Introduction

Very frequently two groups of variables are measured on the same sample units, and it is of great practical interest to explore any association that might exist between the two sets of variables. Canonical correlation analysis (CCA), developed by Hotelling (1936), is the usual method for studying the interrelations between two sets of variables. The most direct approach is to consider two sets of variables: $\mathbf{x} = (x_1, ..., x_p)^t$ and $\mathbf{y} = (y_1, ..., y_q)^t$ $(p \leq q)$ and find linear combinations of the variables in each set $(u, v) = (\boldsymbol{\alpha}^t\mathbf{x}, \boldsymbol{\beta}^t\mathbf{y})$ that have maximum correlation. The new variables (u, v) are called canonical variates. Considering that Σ (the overall covariance matrix of $\mathbf{z}^t = (\mathbf{x}^t, \mathbf{y}^t)$) can be partitioned, in the usual way, it can be shown that the canonical variates and correlations can be determined by the eigenvectors and eigenvalues of the matrix $\Sigma_{11}^{-1}\Sigma_{12}\Sigma_{22}^{-1}\Sigma_{21}$. In fact, the eigenvectors are the $\boldsymbol{\alpha}$'s, the eigenvalues (λ) are the square of the canonical correlations (ρ) and the $\boldsymbol{\beta}$'s are known functions of the $\boldsymbol{\alpha}$'s.

As happens with other multivariate methods, estimation in CCA can be heavily affected by the presence of atypical or outlying observations. Non-ideal data can lead to instability of the coefficients of the canonical variates, complicating and confusing the interpretation of these variates. Hence, it is convenient to use robust methods since these are capable of accommodating discrepant observations in the estimation process. Robust estimates of canonical coefficients and canonical correlations have been suggested by Karnel (1991).

The object of this paper is to propose alternative robust methods of estimation in CCA using projection pursuit (PP). In section 2 the new estimators are introduced, as well as the transformations needed for the implementation of the algorithm described in section 3. In section 4 artificial data from a simulation study are used to compare various methods of estimation. A global performance measure is proposed and used to evaluate the various estimation options. Section 5 is devoted to the study of the empirical breakdown point and the main conclusions are presented in section 6.

2 Robust projection pursuit estimators in CCA

PP is a general strategy of data projection that looks for projections which reveal interesting aspects of the data (Huber, 1985). Using these projections we can transform a difficult problem, defined in a high dimension space, into a more manageable problem in a lower dimension space (usually one or two dimensions). PP is based on the optimization of a function called projection index (PI).

With this idea in mind CCA can be seen as a special case of PP. In fact, finding linear combinations of $(\mathbf{x}, \mathbf{y})$ that maximize the correlation between them is equivalent to looking for projections of the data in directions $(\boldsymbol{\alpha}, \boldsymbol{\beta})$ that maximize a certain objective function PI (correlation). In order to obtain robust estimators we have to choose a robust estimator of the PI. Li and Chen (1985) used a similar approach for the robustification of principal components taking the PI as a robust estimator of the variance. In our case, we need joint estimators of the mean and the variance of samples in a bidimensional space to build the PI. Two different robust estimators are considered: Huber's proposal-2[1] (Huber, 1981) and the minimum covariance determinant estimator (MCD) (Rousseeuw and Van Driessen, 1999).

In order to accommodate the restrictions associated with the CCA it is convenient to make three kinds of transformations of the data: (i) transform the data in order to obtain restrictions of the form $\boldsymbol{\alpha}^{*t}\boldsymbol{\alpha}^* = 1$ and $\boldsymbol{\beta}^{*t}\boldsymbol{\beta}^* = 1$ instead of $\boldsymbol{\alpha}^t\Sigma_{11}\boldsymbol{\alpha} = 1$ and $\boldsymbol{\beta}^t\Sigma_{22}\boldsymbol{\beta} = 1$, (ii) project the observations in a space orthogonal to the directions previously estimated in order to satisfy restrictions of the form: $\boldsymbol{\alpha}_i^{*t}\boldsymbol{\alpha}_j^* = 0$ and $\boldsymbol{\beta}_i^{*t}\boldsymbol{\beta}_j^* = 0$ for $i \neq j$ instead of the form: $\boldsymbol{\alpha}_i^t\Sigma_{11}\boldsymbol{\alpha}_j = 0$ and $\boldsymbol{\beta}_i^t\Sigma_{22}\boldsymbol{\beta}_j = 0$ for $i \neq j$ (iii) use polar coordinates to include the restrictions $\boldsymbol{\alpha}^{*t}\boldsymbol{\alpha}^* = 1$ and $\boldsymbol{\beta}^{*t}\boldsymbol{\beta}^* = 1$ in the function to be optimized.

2.1 Transformations

The first transformation is of the form: $(\mathbf{x}^*, \mathbf{y}^*) = \left(K^{-\frac{1}{2}}U^t\mathbf{x}, L^{-\frac{1}{2}}V^t\mathbf{y}\right)$ where U, K, V and L are given by the spectral decompositions $\Sigma_{11} = UKU^t$ and $\Sigma_{22} = VLV^t$. This same transformation is used in the classical CCA to carry out the eigen analysis of a non-symmetric matrix by changing the problem to an eigen analysis of a symmetric matrix (Cooley and Lohnes, 1971). The process can be viewed as a transformation of the original spaces in Euclidean spaces taking advantage of the nice properties of orthogonality in Euclidean spaces.

The second type of transformations projects the observations into spaces orthogonal to the canonical coefficients vectors already estimated. This will ensure that all estimated canonical coefficients, except the vectors that define each pair of canonical variates, are orthogonal with each other.

Changing to polar coordinates, the final transformation, ensures that there is no need to consider additional restrictions in the optimization routine to obtain vectors with norm equal to one (*i.e.* $\boldsymbol{\alpha}^{*t}\boldsymbol{\alpha}^* = 1$ and $\boldsymbol{\beta}^{*t}\boldsymbol{\beta}^* = 1$).

[1] A special case of an M-estimator.

3 Algorithm

The main steps of the algorithm are as follows:

1. Estimate Σ in some robust manner, *e.g.* using an M-estimator or any of the estimators listed in Table ??. Perform the spectral decomposition of Σ_{11} and Σ_{22}.
2. Make the transformation: $(\mathbf{x}^*, \mathbf{y}^*) = \left(K^{-\frac{1}{2}}U^t\mathbf{x}, L^{-\frac{1}{2}}V^t\mathbf{y}\right)$ where $\Sigma_{11} = UKU^t$ and $\Sigma_{22} = VLV^t$.
3. Maximize the PI, that is, obtain $\max PI\left(\boldsymbol{\alpha}^*, \boldsymbol{\beta}^*\right)$ where $\boldsymbol{\alpha}^*$ and $\boldsymbol{\beta}^*$ are written in polar coordinates. As a result of the maximization process $\hat{\boldsymbol{\alpha}}_1^*$ and $\hat{\boldsymbol{\beta}}_1^*$ are obtained.
4. Build two orthonormal bases, A and B where $A = [\hat{\boldsymbol{\alpha}}_1^*|A^r]$, $B = [\hat{\boldsymbol{\beta}}_1^*|B^r]$ and A^r, B^r are bases of the spaces orthogonal to $\hat{\boldsymbol{\alpha}}_1^*$ and $\hat{\boldsymbol{\beta}}_1^*$.
5. Project the observations $\mathbf{x}^*$ into the space orthogonal to $\hat{\boldsymbol{\alpha}}_1^*$ and $\mathbf{y}^*$ into the space orthogonal to $\hat{\boldsymbol{\beta}}_1^*$. This is equivalent to the transformation: $(\mathbf{x}^{**}, \mathbf{y}^{**}) = \left((A^r)^t\,\mathbf{x}^*, (B^r)^t\,\mathbf{y}^*\right)$.
6. Compute $\max PI\left(\boldsymbol{\alpha}^{**}, \boldsymbol{\beta}^{**}\right)$. As a result of the maximization process $\hat{\boldsymbol{\alpha}}_2^* = A^r\boldsymbol{\alpha}_1^{**}$ and $\hat{\boldsymbol{\beta}}_2^* = B^r\boldsymbol{\beta}_1^{**}$ are obtained.
7. Update the orthonormal bases in the following way: $A = [\hat{\boldsymbol{\alpha}}_1^*\ \hat{\boldsymbol{\alpha}}_2^*|A^r]$ and $B = [\hat{\boldsymbol{\beta}}_1^*\ \hat{\boldsymbol{\beta}}_2^*|B^r]$. Update the transformations in (5) and maximize the PI, as in (6), to obtain the next pair of robust estimates of the canonical coefficients. Repeat this procedure until all pairs of canonical coefficients are estimated.
8. Rewrite the robust estimates in the original data space:
$$\left(\hat{\boldsymbol{\alpha}}_i, \hat{\boldsymbol{\beta}}_i\right) = \left(UK^{-\frac{1}{2}}\hat{\boldsymbol{\alpha}}_i^*, VL^{-\frac{1}{2}}\hat{\boldsymbol{\beta}}_i^*\right), \quad i = 1, \cdots, p. \tag{1}$$
9. Obtain the scores: $(\hat{u}_i, \hat{v}_i) = \left(\hat{\boldsymbol{\alpha}}_i^t\mathbf{x}, \hat{\boldsymbol{\beta}}_i^t\mathbf{y}\right)$. Choose a robust estimator of the correlation to estimate $\rho_i = Corr\left(u_i, v_i\right)$ for $i = 1, \cdots, p$.

To complete the algorithm we have to choose robust estimators for various parameters: (i) Σ (step 1), (ii) correlation that defines the PI (step 3), (iii) canonical correlations (step 9). In the estimation of Σ and PI two different estimators are considered: an M-estimator (Huber's proposal-2) and the MCD estimator. For the estimation of the canonical correlations only an M-estimator is considered.

4 Simulation study

4.1 Simulation conditions

For testing the performance of the proposed PP estimators, data was generated according to three types of distribution: (i) Normal($\mathbf{0}$,Σ) (NOR), (ii) symmetric contaminated normal: SCN=$0.9\mathcal{N}(\mathbf{0}, \Sigma)$+$0.1\mathcal{N}(\mathbf{0}, 9\Sigma)$ and (iii) a T-distribution with 3 degrees of freedom (T_3).

We have used two groups of three variables ($p = q = 3$). Samples of size 500 were used and for each type of distribution and each method of estimation 500 replications were generated.

The various methods of estimation used are indicated in table 1.

Table 1. Methods of estimation.

	Class	Mar(0.05)	Mar(0.10)	Mar/Mar	MCD/Mar	Mar/MCD	MCD/MCD	MMar
Σ	Classical	Maronna	Maronna	Maronna	MCD	Maronna	MCD	Maronna
PI	–	–	–	Maronna	Maronna	MCD	MCD	Maronna

4.2 Measuring the performance of each method of estimation

Each method of estimation produces $2p$ vectors of canonical coefficients and p canonical correlations. It would be desirable to have a global measure that could give some indication of how good these estimates are when compared with the true values of the corresponding parameters. This measure should not depend on the generated samples, except through the estimates in the study.

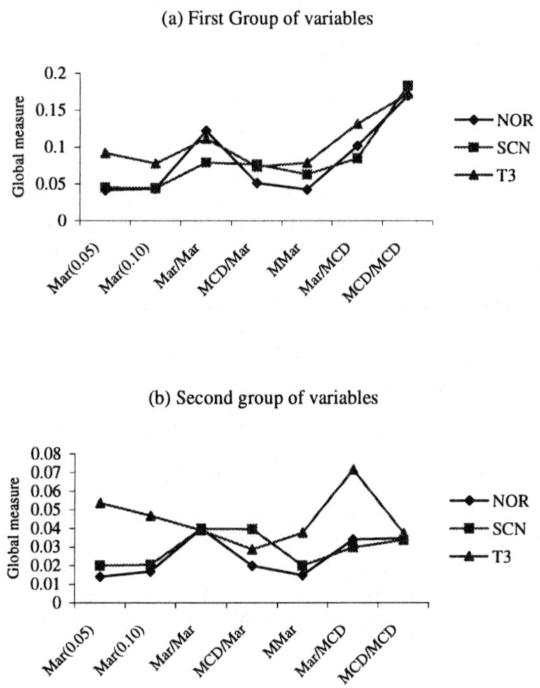

Fig.1. Performance of various estimation methods.

A possible measure associated with the estimates in the first group is given by:

$$\sum_{i=1}^{p} \frac{\hat{\lambda}_i \int_{\Re^p} (\hat{\boldsymbol{\alpha}}_i^t \mathbf{x} - \boldsymbol{\alpha}_i^t \mathbf{x})^2 \phi(\mathbf{x}) d\mathbf{x}}{\sum_{i=1}^{p} \hat{\lambda}_i},$$

where

$$\int_{\Re^p} (\hat{\alpha}_i^t \mathbf{x} - \alpha_i^t \mathbf{x})^2 \phi(\mathbf{x}) d\mathbf{x} = E\left\{ \left(\hat{\alpha}_i^t \mathbf{X} - \alpha_i^t \mathbf{X}\right)^2 \mid \mathbf{X} \sim \mathcal{N}(\mathbf{0}, \Sigma_{11}) \right\},$$

and $\hat{\lambda}_i / \sum_{i=1}^{p} \hat{\lambda}_i$ is a weight attached to the ith canonical variate.

4.3 Results

The simulation results show that the robust estimators are better than the classical estimators, under various contamination conditions.

The method MMar is a refinement of Mar/Mar, but it will not be explained here. It gives results of the same level as MCD/Mar and is better than Mar/Mar, Mar/MCD and MCD/MCD (Figures 1a and 1b).

MCD/MCD is generally inferior to the other methods, mainly in the first group variables, but it has high empirical breakdown point as can be seen in the next section. The less satisfactory results are associated with the T_3 distribution.

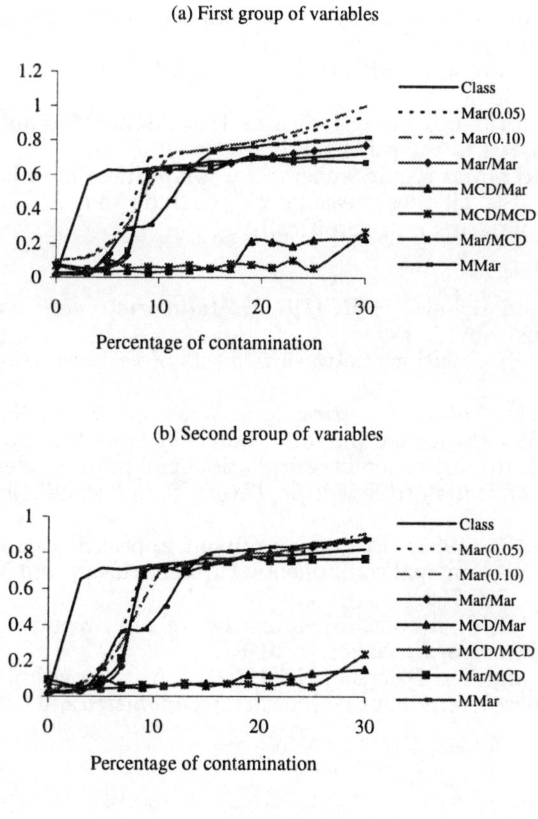

Fig.2. Empirical breakdown point.

5 Empirical breakdown point

In order to obtain the empirical breakdown point samples of size 200 were generated from a normal distribution ($p = q = 3$) with zero mean and a simple covariance matrix such that: $\boldsymbol{\alpha}_1 = (3,0,0)^t$, $\boldsymbol{\alpha}_2 = (0,\sqrt{6},0)^t$, $\boldsymbol{\alpha}_3 = (0,0,\sqrt{3})^t$, $\boldsymbol{\beta}_1 = (\sqrt{10},0,0)^t$, $\boldsymbol{\beta}_2 = (0,\sqrt{10},0)^t$, $\boldsymbol{\beta}_3 = (0,0,\sqrt{10})^t$, $\lambda_1 = 0.9$, $\lambda_2 = 0.6$ and $\lambda_3 = 0.3$. For each sample a fixed percentage of the observations were replaced by the point $(10,10,10,10,10,10)^t$.

Figures 2a and 2b show that when the initial estimate of Σ is of Maronna type all the associated breakdown points are low, even for the PP estimators. This is not surprising since the breakdown point of the Maronna estimators is less then or equal to $1/(p+q)$ (Maronna, 1976). But if Σ is estimated by MCD (a high breakdown point estimator) the breakdown point for the PP estimators shows a noticeable increase even when PI is a Maronna estimator. Other interesting property of the procedure is the equivariance of the estimators.

6 Conclusions

The simulation study shows that the PP methods (with MCD as initial estimate of Σ) have higher breakdown point than the usual procedure based on a robust covariance matrix (Mar(0.05) and Mar(0.10)).

Among all PP estimators MCD/Mar and MCD/MCD have the highest breakdown point.

The global performance index indicates that MCD/Mar and MMar are superior as compared to the other methods.

MCD/MCD and MMar require substancial computing time when compared with MCD/Mar. For all these reasons we conclude that MCD/Mar is the estimator which performs best empirically.

References

Cooley, W. W. and Lohnes, P. R. (1971). *Multivariate data analysis.* John Wiley and Sons, New York.

Hotelling, H. (1936). Relations between two sets of variates. *Biometrika*, **28**, 321-377.

Huber, P. J. (1981). *Robust Statistics.* John Wiley and Sons, New York.

Huber, P. J. (1985). Projection pursuit. *Annals of Statistics*, **13**, 435-475.

Karnel, G. (1991). Robust canonical correlation and correspondence analysis. *The Frontiers of Statistical Scientific Theory & Industrial Applications*, **2**, 335-354.

Li, G. and Chen, Z. (1985). Projection-pursuit approach to robust dispersion matrices and principal components: Primary theory and Monte Carlo. *JASA*, **80**, 391, 759-766.

Maronna, R. A. (1976). Robust M-estimators of multivariate location and scatter. *The Annals of Statistics*, **4**, 51-67.

Rousseeuw, P. J. and Van Driessen, K. (1999). A fast algorithm for the minimum covariance determinant estimator. *Technometrics*, **41**, 212-223.

A fast algorithm for highly robust regression in data mining

Peter J. Rousseeuw[1] and Katrien Van Driessen[2]

[1] Department of Mathematics and Computer Science, University of Antwerp (UIA), Universiteitsplein 1, B-2610 Wilrijk, Belgium
[2] Faculty of Applied Economics, University of Antwerp (UFSIA), Prinsstraat 13, B-2000 Antwerp, Belgium

1 Introduction

Data mining aims to extract previously unknown patterns or substructures from large databases. In statistics, this is what robust estimation and outlier detection were constructed for, see e.g. Rousseeuw and Leroy (1987). Our goal is to construct algorithms which allow us to compute robust results in a data mining context. Such algorithms thus need to be fast, and able to deal with large data sets.

One often needs to fit a linear model $y_i = x_{i1}\theta_1 + \cdots + x_{ip}\theta_p + e_i$ ($i = 1, \ldots, n$) to n data points of the form $(\boldsymbol{x}_i, y_i) = (x_{i1}, \ldots, x_{ip}, y_i)$, with $x_{ip} = 1$ for regression with an intercept term. Many estimators of $\boldsymbol{\theta} = (\theta_1, \ldots, \theta_p)$ break down in the presence of outliers. For data with several explanatory variables it is no longer possible to detect outlying observations by visual inspection of a scatterplot. In higher dimensions, it takes a robust method to *detect* the outliers by means of residuals based on a robust regression or robust distances of $\boldsymbol{x}_1, \ldots, \boldsymbol{x}_n$.

Several methods for robust regression have been proposed, such as the least median of squares (LMS) method of Rousseeuw (1984). The LMS is defined by minimizing $(r^2)_{h:n}$ where $(r^2)_{1:n} \leq (r^2)_{2:n} \leq \cdots \leq (r^2)_{n:n}$ are the ordered squared residuals, and where $h = [(n + p + 1)/2]$. Here p is again the number of coefficients, including the intercept term. The LMS attains the highest possible breakdown value, namely $([(n - p)/2] + 1)/n$. For a survey of positive-breakdown methods and some substantive applications, see (Rousseeuw 1997).

An alternative to the LMS is the least trimmed squares (LTS) estimator, which was also proposed in (Rousseeuw 1984, page 876). Its objective is to minimize

$$\Sigma_{i=1}^{h} (r^2)_{i:n} \tag{1}$$

where $(r^2)_{1:n} \leq \cdots \leq (r^2)_{n:n}$ are the ordered squared residuals. This is equivalent to finding the h-subset with smallest least squares objective function. The LTS regression estimate is then the least squares fit to these h points. The breakdown value of LTS with $h = [(n + p + 1)/2]$ equals that of the LMS. Moreover, LTS regression has several advantages over LMS. Its objective function is more smooth and its statistical efficiency is better (Rousseeuw 1984). For a discussion of algorithms for LMS and LTS see (Visek 1996) and the references cited therein.

Until now the LTS estimator has been applied less often than the LMS because it was perceived as being harder to compute. However, in this paper we will construct a new LTS algorithm which is substantially *faster* than all existing LMS algorithms. The new LTS algorithm can deal with a sample size n in the tens of thousands or more. It makes positive-breakdown robust regression and the corresponding outlier detection method available as tools for data mining. This should make the LTS the method of choice.

2 Basic idea and the C-step

One of the keys of the new algorithm is the fact that starting from any approximation to the LTS regression coefficients, it is possible to compute another approximation yielding an even lower objective function.

Property 1. Consider a data set $(\boldsymbol{x}_1, y_1), \ldots, (\boldsymbol{x}_n, y_n)$ consisting of p-variate $\boldsymbol{x}_i$ (typically with $x_{ip} = 1$) and a response variable y_i. Let $H_1 \subset \{1, \ldots, n\}$ with $|H_1| = h$, and put $Q_1 := \sum_{i \in H_1} (r_1(i))^2$ where $r_1(i) = y_i - (\hat{\theta}_1^1 x_{i1} + \hat{\theta}_2^1 x_{i2} + \cdots + \hat{\theta}_p^1 x_{ip})$ for all $i = 1, \ldots, n$ where $\hat{\boldsymbol{\theta}}_1 = (\hat{\theta}_{11}, \ldots, \hat{\theta}_{p1})$ is any p-dimensional vector. Now take H_2 such that $\{|r_1(i)|; i \in H_2\} := \{|r_1|_{1:n}, \ldots, |r_1|_{h:n}\}$ where $|r_1|_{1:n} \leq |r_1|_{2:n} \leq \cdots \leq |r_1|_{n:n}$ are the ordered absolute values of the residuals, and compute the least squares (LS) fit $\hat{\boldsymbol{\theta}}_2$ of the h observations in H_2. This yields $r_2(i)$ for all $i = 1, \ldots, n$ and $Q_2 := \Sigma_{i \in H_2}((r_2(i))^2$. Then

$$Q_2 \leq Q_1.$$

Applying the above property to H_1 yields H_2 with $Q_2 \leq Q_1$. In our algorithm we will call this a *C-step*, where C stands for 'concentration' since H_2 is more concentrated (has a lower sum of squared residuals) than H_1.

Repeating C-steps yields an iteration process. If $Q_2 = Q_1$ we stop; otherwise we apply another C-step yielding Q_3, and so on. The sequence $Q_1 \geq Q_2 \geq Q_3 \geq \ldots$ is nonnegative and hence must converge. In fact, since there are only finitely many h-subsets there must be an index m such that $Q_m = Q_{m-1}$, hence convergence is always reached after a finite number of steps. (In practice, m is often below 10.) This is not sufficient for Q_m to be the global minimum of the LTS objective function, but it is a necessary condition.

This provides a partial idea for an algorithm:

> *Take many initial choices of H_1 and apply C-steps to each until convergence, and keep the solution with lowest value of (1).* (2)

3 Construction of the FAST-LTS algorithm

3.1 Creating initial subsets $\boldsymbol{H_1}$

In order to apply the algorithmic idea (2) we first have to decide how to construct the initial subsets H_1. Let us consider the following two possibilities:

(a) Draw a random h-subset H_1.

(b) Draw a random p-subset J, and compute $\hat{\boldsymbol{\theta}}_0 :=$ the coefficients of the hyperplane through J. If J does not define a unique hyperplane (i.e., when the rank of X_J is less than p), extend J by adding random observations until it does. Then compute the residuals $r_0(i) := y_i - \hat{\boldsymbol{\theta}}_0 \boldsymbol{x}_i^t$ for $i = 1, \ldots, n$. Sort them into $|r_0(\pi(1))| \leq \cdots \leq |r_0(\pi(n))|$ and put $H_1 := \{\pi(1), \ldots, \pi(h)\}$.

Option (a) is the simplest, whereas (b) starts like the PROGRESS algorithm (Rousseeuw and Leroy 1987).

When the data set does not contain outliers or deviating groups of points, it makes little difference whether (2) is applied with (a) or (b). But to be realistic, we have to consider contaminated data sets, for example data with one big group of normally distributed data and a smaller, outlying group of normally distributed observations. Empirical trials showed that option (b) performs well, while approach (a) fails because each random subset H_1 contains a sizeable number of points from the majority group as well as from the minority group, which follows from the law of large numbers. When starting from a bad subset H_1 the iterations will not converge to the good solution. On the other hand, the probability of a p-subset without outliers is much higher, which explains why (b) yields many good initial subsets H_1 and hence a robust result. Therefore, from now on we will always use (b).

3.2 Selective iteration

Each C-step calculates LS regression coefficients based on h observations, and the corresponding residuals for all n observations. Therefore, reducing the number of C-steps would improve the speed. The fact that this can be done without losing the effectiveness of the algorithm is based on empirical trials which show that often the distinction between good (robust) solutions and bad solutions already becomes visible after two or three C-steps, see also (Rousseeuw and Van Driessen 1999). Therefore, from now on we will take only two C-steps from each initial subsample H_1, select the 10 different subsets H_3 with the lowest sums of squared residuals, and only for these 10 we continue taking C-steps until convergence.

3.3 Nested extensions

For a small to moderate sample size n the above algorithm does not take much time. But when n grows the computation time increases, due to the LS regressions on h-subsets and the n residuals that need to be calculated each time. To avoid doing all the computations in the entire data set, we will consider a special structure. When $n > 1500$, the algorithm generates a nested system of subsets which was introduced in (Rousseeuw and Van Driessen 1999). Iterations are carried out within these "nested extensions".

3.4 Intercept adjustment

One other feature of the new algorithm that has not yet been mentioned is the intercept adjustment. This is a technique which decreases the LTS objective value of a fit. Each initial p-subset and each C-step yields an estimate $\tilde{\boldsymbol{\theta}} = (\tilde{\theta}_1, \ldots, \tilde{\theta}_{p-1}, \tilde{\theta}_p)$ where $\tilde{\theta}_1, \ldots, \tilde{\theta}_{p-1}$ are the slopes and $\tilde{\theta}_p$ is the intercept. The corresponding LTS objective value is given by

$$\Sigma_{i=1}^{h}(r^2(\tilde{\boldsymbol{\theta}}))_{i:n} = \Sigma_{i=1}^{h}((y_i - x_{i1}\tilde{\theta}_1 - \cdots - x_{i,p-1}\tilde{\theta}_{p-1} - \tilde{\theta}_p)^2)_{i:n} \tag{3}$$

The adjusted intercept $\tilde{\theta}'_p$ is then calculated as the exact, univariate LTS location estimate of $\{t_i = y_i - x_{i1}\tilde{\theta}_1 - \cdots - x_{i,p-1}\tilde{\theta}_{p-1}; i = 1, \ldots, n\}$, i.e. $\tilde{\theta}'_p = \underset{\mu}{\operatorname{argmin}} \Sigma_{i=1}^{h}((t_i - \mu)^2)_{i:n}$ which, by construction, yields the same or a lower objective value (3).

For a given univariate data set, the LTS estimator is the mean of the h-subset with the smallest variance. In other words, any contiguous h-subset

$t_{j:n}, \ldots, t_{j+h-1:n}$ has a mean $\bar{t}^{(j)}$ and a sum of squares $SQ^{(j)} = \Sigma_{i=j}^{j+h-1}(t_{i:n} - \bar{t}^{(j)})^2$ and we want to minimize this $SQ^{(j)}$. The complete univariate algorithm can be found in (Rousseeuw and Leroy 1987, pages 171-172).

4 Performance of FAST-LTS

To get an idea of the performance of the overall algorithm, we start by applying FAST-LTS to some small regression data sets taken from (Rousseeuw and Leroy 1987). The first column of Table 1 lists the name of each data set, followed by n and p, where n is the number of observations and p stands for the number of coefficients including the intercept term. We stayed with the default value of $h = [(n+p+1)/2]$. The next column shows the number of starting p-subsets used in FAST-LTS, which is usually 500 except for two data sets where the number of possible p-subsets out of n was fairly small, namely $\binom{12}{3} = 220$ and $\binom{18}{3} = 816$, so we used all of them.

Table 1. Performance of the FAST-LTS algorithm on some small data sets.

data set	n	p	starts	best h-subset found	time (seconds)
Heart	12	3	220	1 2 4 5 6 7 11 12	1.3
Phosphor	18	3	816	1 2 3 4 6 7 11 12 14 15 18	3.5
Coleman	20	6	500	2 5 6 7 8 9 11 13 14 15 16 19 20	4.5
Wood	20	6	500	2 3 9 10 11 12 13 14 15 16 17 18 20	4.3
Salinity	28	4	500	2 3 4 6 7 12 14 15 17 18 19 20 21 22 26 27	4.2
Aircraft	23	5	500	1 5 6 7 8 9 10 11 13 14 15 17 20 23	4.1
Delivery	25	3	500	2 5 6 7 8 10 12 13 14 15 17 21 22 25	3.1

The next entry in Table 1 is the result of FAST-LTS, given here as the final h-subset. By comparing these with the slower but exact LTS algorithm of Agulló (1997) it turns out that these h-subsets do yield the exact global minimum of the objective function. The next column shows the running time of FAST-LTS, in seconds on a Sun SparcStation 20/514. We may conclude that for these small data sets FAST-LTS gives very accurate results in little time.

Let us now try the algorithm on larger data sets. Because outliers in x-space are very often the cause for a regression fit to break down, we will generate data with many bad leverage points to illustrate the performance of the FAST-LTS algorithm. Each data set is generated according to

$$y_i = x_{i,1} + x_{i,2} + \cdots + x_{i,p-1} + x_{i,p} + e_i$$

in which $e_i \sim \mathrm{N}(0,1)$ is the error term, $x_{i,j} \sim \mathrm{N}(0,100)$ for $j = 1,\ldots,p-1$ are the nontrivial explanatory variables, and $x_{i,p} = 1$ is the intercept term. We have introduced outliers in the x direction by replacing a percentage of the $x_{i,1}$ by values that are normally distributed with mean 100 and variance 100. For each data set, Table 2 lists n and p followed by the percentage of outliers. The algorithm always used 500 starts.

For all data sets FAST-LTS yielded the robust result, i.e. corresponding to the first distribution. The computation times were quite low for the given values of n and p. Even for a sample size as high as 50,000 a few minutes

Table 2. Performance of the FAST-LTS algorithm on larger data sets, with time in seconds.

n	p	outliers	starts	time
100	2	40%	500	10
	3	40%	500	12
	5	40%	500	35
500	2	40%	500	115
	3	40%	500	125
	5	40%	500	140
1,000	2	35%	500	70
	5	35%	500	95
	10	35%	500	160
10,000	2	40%	500	95
	5	40%	500	125
	10	40%	500	200
50,000	2	40%	500	185
	5	40%	500	250

suffice, whereas no previous algorithm we know of could handle such large data sets.

Note that the FAST-LTS algorithm is affine equivariant: when the data are translated or subjected to a linear transformation, the resulting $\hat{\boldsymbol{\theta}}_{\text{full}}$ will transform accordingly. The program FAST-LTS has been thoroughly tested and can be obtained free of charge from the authors.

5 Example

In this section we will illustrate the diagnostic plot (see, e.g., Rousseeuw 1997) of robust residuals $r_i/\hat{\sigma}$ versus robust distances $\text{RD}(\boldsymbol{x}_i)$, given by

$$\text{RD}(\boldsymbol{x}_i) = \sqrt{(\boldsymbol{x}_i - T)C^{-1}(\boldsymbol{x}_i - T)}.$$

Here T is the robust estimate of multivariate location and C is the robust scatter matrix obtained from the minimum covariance determinant (MCD) method of (Rousseeuw 1984). We can compute the MCD with the algorithm of (Rousseeuw and Van Driessen 1999), which we apply to the set of $\boldsymbol{x}_i$ (without using the y_i).

Example. In their report, Odewahn et al. (1998) describe the Digitized Palomar Sky Survey (DPOSS) data they are working on at the California Institute of Technology. The data consist of many celestial objects or light sources (mainly stars and galaxies) for which nine characteristics in the blue, red, and near-infrared bands are measured. S.C. Odewahn gave us access to a part of this database, from which we extracted the 56,744 stars for which all the characteristics in the blue colour (the F band) were available. We considered MAperF, the aperture magnitude (which is a kind of brightness measure) as response variable and regressed it against the other 8 characteristics of the colour band F. These characteristics describe the size of a light source and the shape of the spatial brightness distribution in a source. Figure 1a plots the standardized LS residuals versus the classical Mahalanobis distances. Some isolated outliers in the y-direction as well as in x-space fall outside this plot.

(We zoomed in to have a better view of the large majority of the data.) According to Figure 1a, most data follow the same linear trend.

(a) (b)

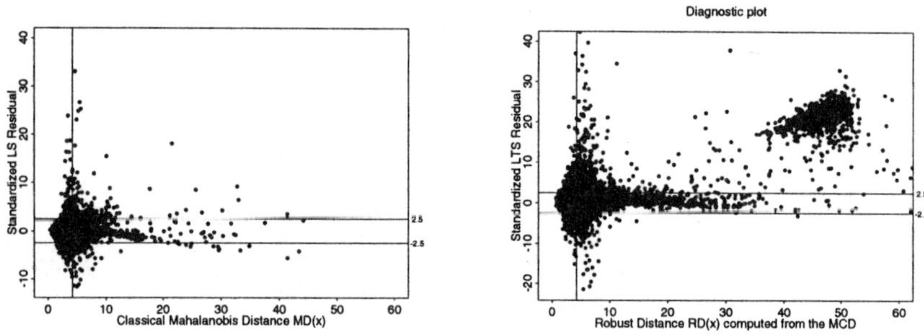

Fig. 1. DPOSS data: regression of MAperF on 8-dimensional x_i. (a) Plot of LS residual versus Mahalanobis distance $MD(x_i)$; (b) diagnostic plot of LTS residual versus robust distance $RD(x_i)$.

However, Figure 1b tells a very different story. It plots the LTS residuals obtained by FAST-LTS versus the robust distances $RD(\boldsymbol{x}_i)$ based on the MCD. This diagnostic plot shows us that there is a rather large group of bad leverage points, i.e. observations with both large robust residuals and large robust distances. These were invisible in Figure 1a because they attract the LS fit as well as the classical Mahalanobis distances. It turns out that this deviating substructure consists of giant stars.

References

Agullo, J. (1997). *Computación de estimadores con alto punto de ruptura.* Ph.D. Thesis. University of Alicante, Spain.

Odewahn, S.C., Djorgovski, S.G., Brunner, R.J. and Gal, R. (1998). Data From the Digitized Palomar Sky Survey. Technical Report. California Institute of Technology.

Rousseeuw, P.J. (1984). Least Median of Squares Regression. *Journal of the American Statistical Association*, **79**, pp. 871–880.

Rousseeuw, P.J. (1997). Introduction to Positive-Breakdown Methods. In: G.S. Maddala and C.R. Rao (Eds.): *Handbook of Statistics*, Vol. **15**: Robust Inference. Amsterdam: Elsevier, pp. 101–121.

Rousseeuw, P.J. and Leroy, A.M. (1987). *Robust Regression and Outlier Detection.* New York: John Wiley.

Rousseeuw, P.J. and Van Driessen, K. (1999). A Fast Algorithm for the Minimum Covariance Determinant Estimator. *Technometrics*, **41**, pp. 212–223.

Visek, J.A. (1996). On High Breakdown Point Estimation. *Computational Statistics*, **11**, pp. 137–146.

Optimal classification trees

Petr Savický[1], Jan Klaschka[1] and Jaromír Antoch[2]

[1] Institute of Computer Science, Academy of Sciences, Pod Vodárenskou věží 2, 182 07, Prague, Czech Republic
[2] Department of Statistics, Charles University, Sokolovská 83, 186 75 Prague, Czech Republic

Abstract. Classification and regression trees have been traditionally grown by recursive partitioning, i.e. by a top-down search for "locally optimal" splits. The "local", or "one-step", optimization of splits can to some extent, using the present power of computer hardware, be substituted by the full optimization of whole trees. In this paper, two bottom-up optimization algorithms are outlined and first experimental experience is presented.

Keywords. Classification trees, cost-complexity measure, optimal trees, algorithms, bottom-up procedures.

1 Introduction

Since the sixties, when the first tree-based methods appeared, recursive partitioning (binary or k-ary) has kept the position as a state-of-art method to search for tree-structured classification models, though it has been known, that "local" optimization of individual splits does not yield, in general, "globally" optimal trees. In their classical monograph on classification and regression trees (section 2.5.8), Breiman *et al.* (1984) advocate recursive partitioning (compare also with scepticism of Gelfand *et al.*, 1991): *"At this stage of computer technology, an overall optimal tree growing procedure does not appear feasible for any reasonably sized data set."*

The aim of this paper is to show that with the *current* power of computer technology, optimal tree growing is possible *to some extent.* We present two novel algorithms that allow us, when the number of predictors is not too large, to construct trees with the smallest classification error in training data within the class of all possible binary classification trees of a specified size. The size is specified in different ways in the two algorithms. Our limited experimental experience suggests that such trees may have good generalization properties for problems with a complex relationship between the dependent variable and the predictors.

Our work was inspired by both classical (namely CART by Breiman *et al.*, 1984) and recent developments (e.g., Siciliano, 1998) in the tree-based methodologies, as well as by recent works on RAM-resident solutions to complex data-analytical problems (e.g., Pijls & Bioch, 1999).

2 Basics and notation

We consider tree-structured classification models with $P \geq 1$ binary $\{0,1\}$-valued predictors (independent variables) $X_1, \ldots, X_P$ and $K \geq 2$ classes $C_1, \ldots, C_K$ (an extension to nonbinary predictors via, e.g., dummy variables is also possible).

Data of an individual case consist of a *predictor vector* $\mathbf{x} = (x_1, \ldots, x_P) \in \{0,1\}^P$ and a *class label* $C \in \{C_1, \ldots, C_K\}$. The set $\{0,1\}^P$ is the *predictor space*. Let *cubes* be such subsets $B = A_1 \times \cdots \times A_P$ of the predictor space which, for each index i, have either $A_i = \{0\}$, or $A_i = \{1\}$, or $A_i = \{0,1\}$. Further, let $\mathcal{B}_P$ denote the set of all 3^P cubes $B \subseteq \{0,1\}^P$. The *dimension* of the cube $B = A_1 \times \ldots \times A_P$ is the number of those indices i for which $A_i = \{0,1\}$. The cube B is a *subcube* of the cube B' if $B \subseteq B'$. Cubes of dimension 0 are singletons. For data set $\mathcal{L}$ and cube B, let $\mathcal{L}_B$ be the set of those cases from $\mathcal{L}$ whose predictor vectors belong to B.

The optimization algorithms will be described in terms of assigning trees to cubes. A binary classification tree for a cube B is a description of a partition of B into its subcubes, together with a function that is a constant from $\{C_1, \ldots, C_K\}$ on each of the used subcubes. For a tree T, the function is denoted by f_T and its domain by D_T. More specifically, a binary classification tree T is either

(i) a *trivial* tree, when D_T is a cube and $f_T \in \{C_1, \ldots, C_K\}$ is a constant function defined on D_T;
(ii) a *compound* tree, i.e. a pair of trees $T = (T_L, T_R)$, such that D_{T_L} and D_{T_R} are disjoint and $D_{T_L} \cup D_{T_R}$ is a cube. Then, $D_T = D_{T_L} \cup D_{T_R}$ and f_T equals f_{T_L} on D_{T_L} and f_{T_R} on D_{T_R}.

The tree diagram corresponding to a trivial tree consists of only one node, which is a leaf and also the root. The root of a tree diagram corresponding to a compound tree is a split on the predictor that is constant on the two subcubes and nonconstant on the whole cube.

We say that a predictor vector $\mathbf{x} = (x_1, \ldots, x_P) \in B$ is classified by T to a class C, if $f_T(\mathbf{x}) = C$. The size of a trivial tree T is $|T| = 1$. For a compound tree $T = (T_L, T_R)$, let $|T| = |T_L| + |T_R|$. Note that $|T|$ is equal to the number of leaves of the tree diagram corresponding to T.

The set of all trees T for a cube B, i.e. trees satisfying $D_T = B$, will be denoted $\mathcal{T}(B)$. Let $\mathcal{T}$ be a shorthand for $\mathcal{T}(\{0,1\}^P)$. By $\mathcal{T}_m(B)$ (leaving argument B, when B is the whole predictor space) we denote the set of all trees for cube B whose size is equal to m.

Let a data set $\mathcal{L}$, cube $B \in \mathcal{B}_P$ and a classifier $T \in \mathcal{T}(B)$ be given. Let $N_i > 0$ for $i = 1, \ldots, K$ be the number of those cases from $\mathcal{L}$ that belong to class C_i. Let $N_{ij}(B)$ for $i, j = 1, \ldots, K$ be the number of those cases from $\mathcal{L}_B$ that belong to class C_i, and are classified into C_j by T. For $i, j = 1, \ldots, K$ we denote by Z_{ij} the cost of classifying (i.e., misclassifying for $i \neq j$) one case from class C_i to C_j. Let $\pi_1, \ldots, \pi_K$ be the prior probabilities of classes. (See, e.g., Breiman *et al.*,1984, for various modes of prior probabilities usage.) The misclassification cost $R(T|\mathcal{L}_B)$ of the classifier T on $\mathcal{L}_B$ is defined as

$$R(T|\mathcal{L}_B) = \sum_{i=1}^{K} (\pi_i / N_i) \sum_{j=1}^{K} Z_{ij} N_{ij}(B).$$

Note, that for $T = (T_L, T_R) \in \mathcal{T}(B)$, $T_L \in \mathcal{T}(B_L)$ and $T_R \in \mathcal{T}(B_R)$, the equality $R(T|\mathcal{L}_B) = R(T_L|\mathcal{L}_{B_L}) + R(T_R|\mathcal{L}_{B_R})$ holds.

3 Tree optimization algorithms

Traditional tree-growing by recursive partitioning consists of a search for the best split of the predictor space in the root, followed by a search for the best split of the obtained subcubes, etc. At the moment a split for a cube is being chosen, the properties of further splits of the considered subcubes are typically unknown. (For a rare example of an attempt to "look more steps ahead", see Friedman, 1979.)

The optimization approach proposed in this paper is characterized, instead, by searching for the best tree among *all possible trees* of some size. The algorithms build the optimal trees in a bottom-up manner. In fact, first to complete an optimal tree, we assign (as an intermediate step) an optimal tree to each cube in $\mathcal{B}_P$. At the moment a split of a cube is being chosen, the optimal trees for all subcubes of the cube are already known.

While optimizing a split, the algorithms can optimize the two candidate subtrees independently due to the following simple fact.

Proposition 1. Let B be a cube, $\mathcal{L}$ a data set and m an integer. Let the misclassification cost $R(T|\mathcal{L}_B)$ be minimized in $\mathcal{T}_m(B)$ by $T^* = (T_L, T_R)$ where $|T_L| = m_L$ and $|T_R| = m_R$. Let $B_L = D_{T_L}$ and $B_R = D_{T_R}$. Then T_L and T_R minimize $R(T|\mathcal{L}_{B_L})$ in $\mathcal{T}_{m_L}(B_L)$ and $R(T|\mathcal{L}_{B_R})$ in $\mathcal{T}_{m_R}(B_R)$, resp.

3.1 Cost-complexity minimization (Algorithm I)

We have designed and implemented a novel algorithm that, based on a learning data set $\mathcal{L}$, can find, for a given *complexity parameter* $\alpha \geq 0$, a classification tree minimizing the *cost-complexity measure*

$$R_\alpha(T|\mathcal{L}) = R(T|\mathcal{L}) + \alpha|T|.$$

Note that the cost-complexity-based pruning in CART – see Breiman *et al.* (1984) – minimizes the same quantity. However, the search for the best tree in CART is confined to the set of *trees obtained by pruning from one specific tree*, while our algorithm looks for the tree minimizing the cost-complexity measure in the set of *all possible trees.*

Assume, we have $R_\alpha(T^*|\mathcal{L}) = \min_{T\in\mathcal{T}} R_\alpha(T|\mathcal{L})$, where $\alpha > 0$ and $|T^*| = m$. Then, at the same time, we have $R(T^*|\mathcal{L}) = \min_{T\in\mathcal{T}_m} R(T|\mathcal{L})$. Thus, by varying the α-values, we can calculate $\min_{T\in\mathcal{T}_m} R(T|\mathcal{L})$ for some values of m, along with finding the trees that realize the minima.

The trees minimizing $R_\alpha(T|\mathcal{L})$ for some α have to satisfy the following geometrical condition.

Proposition 2. For all m, if $\mathcal{T}_m$ is nonempty, let $\varphi(m) = \min_{T\in\mathcal{T}_m} R(T|\mathcal{L})$. Let H be the convex hull of the set $\{(m, y); \varphi(m) \text{ well-defined}, y \geq \varphi(m)\}$. Let T^* minimize $R(T|\mathcal{L})$ within $\mathcal{T}_m$. Then, there is an $\alpha \geq 0$ such that T^* minimizes $R_\alpha(T|\mathcal{L})$ within $\mathcal{T}$, if and only if the point $(m, \varphi(m))$ lies on the boundary of H.

Let us call an ordering of $\mathcal{B}_P$ a *subcube-consistent* order, if every cube is preceded by all its proper subcubes.

Algorithm I

Input: Description of the variables, data set $\mathcal{L}$, α.
Output: A tree T^* minimizing $R_\alpha(T|\mathcal{L})$.
Description:

- For each cube B in $\mathcal{B}_P$ containing at least one training case, where the cubes are taken in any subcubc-consistent order, do:
  1. Choose a class $C \in \{C_1, \ldots, C_K\}$ such that $R_\alpha(T|\mathcal{L}_B)$ of the trivial tree T with $D_T = B$ and $f_T = C$, is minimized.

2. For every pair B_L,B_R of disjoint subcubes of B, such that both contain at least one case from $\mathcal{L}$ and $B = B_L \cup B_R$ (if any exists), consider the tree $T = (T_L, T_R)$, where T_L and T_R are the trees assigned to B_L and B_R.
3. Among the trees considered in the two preceding steps, choose a tree T which minimizes $R_\alpha(T|\mathcal{L})$ (breaking possible ties deliberately). Store the information about which tree has been selected. The tree is guaranteed to minimize R_α in $\mathcal{T}(B)$.

- After finishing the above loop, trace the tree T^* assigned to $B = \{0,1\}^P$ using the stored information. This is the output.

Note that by setting the parameter α close to zero, the tree minimizing the size among trees with the lowest possible misclassification cost is found.

3.2 The cost minimization for a given size (Algorithm II)

The second algorithm can find for any given $M \geq 1$ a sequence of trees minimizing the misclassification cost $R(T|\mathcal{L})$ on the learning data set $\mathcal{L}$ within the sets $\mathcal{T}_m$ for $m = 1, 2, \ldots, M$. Unlike the former algorithm, the new algorithm can yield trees that do not fulfill the condition from Proposition 2. We pay for this advantage with increased space and time complexity.

The algorithm is similar to Algorithm I. The difference is that Algorithm II selects and stores a cost-minimal tree for each of the sizes $1, 2, \ldots, M$, if they exist, for each cube with at least one training case.

The cubes are processed in the same order as by Algorithm I. The description differs only in parts 2 and 3, which should be replaced by:

2′. For every pair B_L,B_R of disjoint subcubes of B such that both contain at least one case from $\mathcal{L}$ and $B = B_L \cup B_R$ (if any exists) and every pair m_L, m_R such that $m_L + m_R \leq M$, consider tree $T = (T_L, T_R)$, where T_L and T_R are the trees assigned to B_L for size m_L and to B_R for size m_R, resp. The resulting tree is a candidate for size $m_L + m_R$.

3′. Use the tree from step 1 for size 1. For each $m = 2, 3, \ldots, M$, choose among the candidates for size m from step 2′ the tree minimizing $R(T|\mathcal{L})$ (breaking possible ties deliberately). Store the information about which tree has been selected for each size.

The M trees assigned to $\{0,1\}^P$ are the output of the algorithm.

3.3 Software and algorithmic complexity

Algorithms I and II have been implemented in C++. The present versions are simple command line applications designed for research purposes only, see `http://www.cs.cas.cz/~savicky/trees`. The program for generating the experimental data may be found at the same site.

The space required by Algorithm I is $O(3^P K)$ with a moderate constant depending on the implementation. For Algorithm II, the space requirement is $O(3^P(K + 5M))$ with a similar constant. The time required by both algorithms is linear in the corresponding space requirement.

For example, for the PWN dataset described below, we have $P = 10$, $K = 2$. On a Pentium III under Linux, Algorithm I with 50 different values of α, 1000 training and 10000 testing cases required 9 seconds and 1.6MB of RAM. For the same data, Algorithm II with $M = 50$ different sizes required 12 seconds and 58.2MB of RAM.

4 Generalization issues and numerical experience

Experiments with three types of simulated data, aimed at assessing the generalization properties of trees constructed by our algorithms, are reported.

4.1 Distributions for generating the experimental data

1. **Decimal digits (DD).** A benchmark task introduced by Breiman *et al.* (1984). Each of the digits $y \in \{0, 1, \ldots, 9\}$ is coded by a fixed vector $(\xi_1(y), \ldots, \xi_7(y)) \in \{0, 1\}^7$, representing a combination of segments set ON and OFF on the LED display. Each case has a class label $y \in \{0, 1, \ldots, 9\}$ and predictor vector $(x_1, \ldots, x_7) \in \{0, 1\}^7$, where $\text{Prob}(x_i = \xi_i(y)) = 0.9$ for $i = 1, \ldots, 7$.
2. **High and low digits (HLD).** A variation of DD. Data are generated in the same way as above, but digits $0, \ldots, 9$ are transformed into class label 0 for "low" digits $0, \ldots, 4$, and 1 for "high" digits $5, \ldots, 9$.
3. **Parity with noise (PWN).** The dependent variable Y is defined as parity of $\{0, 1\}$-valued "hidden" variables $\xi_1, \ldots, \xi_5$, i.e. $Y = 1$ if $\sum_{i=1}^{5} \xi_i$ is odd, $Y = 0$ otherwise. Each case has a value y of Y (class label), and vector $(x_1, \ldots, x_{10})$, a realization of manifest binary variables $(X_1, \ldots, X_{10})$, where $\text{Prob}(X_i = \xi_i) = 0.978$ for $i = 1, \ldots, 5$ and $\text{Prob}(X_i = Y) = 0.6$ for $i = 5, \ldots, 10$. Note that Y is equal to the parity of $x_1, \ldots, x_5$ with probability approx. 0.9. Each of the five "misleading" predictors $X_6, \ldots, X_{10}$ alone gives a 60% prediction of Y, but even by combining them, one cannot achieve prediction comparable with the parity of $x_1, \ldots, x_5$.

4.2 Experimental setting and results

We applied Algorithms I and II to 100 data sets for DD and HLD and 10 data sets for PWN. Each set of training data contained 1000 cases. We used prior probabilities estimated from data, and penalized equally each misclassified case, so that R was simply equal to the misclassification rate.

Each learning data set $\mathcal{L}$ was split into subsets $\mathcal{L}_1$ and $\mathcal{L}_2$ of 500 cases each. Using $\mathcal{L}_1$, we constructed (i) by Algorithm I, trees $T_1, \ldots, T_J$ for J values of α in a geometric sequence ($J = 500$ for DD and HLD, $J = 50$ for PWN), (ii) by Algorithm II, trees $T'_1, \ldots, T'_{50}$ of sizes from 1 to 50. Then, we chose trees T^* and $T^{*\prime}$ from among $T_1, \ldots, T_J$ and $T'_1, \ldots, T'_{50}$, resp., so as to minimize misclassification cost on $\mathcal{L}_2$. Finally, we assessed the misclassification cost of trees T^* and $T^{*\prime}$, applying them to a test data set of size 10000 (one test set for all training sets of the same kind).

In parallel, we analyzed the same data by CART, using $\mathcal{L}_1$ for tree-growing and $\mathcal{L}_2$ for pruning, and the test set of 10000 cases for the assessment of misclassification cost of the resulting trees. As a splitting criterion, twoing was applied in the DD experiment, and Gini measure in the other experiments. We tried out both 0 SE and 1 SE rules for pruning.

The results of the experiments are summarized in Table 1. In the DD experiment, there was no or unconvincing improvement of the CART results (twoing, 1 SE rule) by our algorithms, while in the HLD experiment the superiority of our results was clear, but not too impressive. The reason probably is in that the problem DD was not hard and the problem HLD was not *too* hard for CART. In the PWN experiment, there was no doubt that our algorithms were able do disclose the hidden pattern, while CART "got lost".

Tab. 1. Results of experiments. (Q1, Q3 are the 1st and 3rd quartiles.)

Experiment		Optimization				CART			
		Alg. I		Alg. II		0 SE		1 SE	
		R %	size	R %	size	R %	size	R %	size
Decimal	min	25.31	10	25.30	14	24.85	10	25.03	10
digits	Q1	26.34	22	26.29	22	26.49	22.25	27.17	12.25
	median	26.93	28	26.78	27	26.88	30	27.98	17
(100 data	Q3	27.64	33.75	27.38	32	27.48	36	29.00	20.75
sets)	max	29.90	39	28.82	49	29.88	50	30.18	42
High/low	min	13.22	5	13.22	5	13.50	7	13.73	6
digits	Q1	14.16	9.25	14.01	10	14.42	13	14.95	8
	median	14.76	13	14.66	13	14.90	18	15.64	10
(100 data	Q3	15.24	18	15.17	18	15.39	27	16.80	15
sets)	max	17.16	27	17.16	48	17.73	40	18.33	32
Parity with	min	10.06	30	10.06	32	32.67	6	33.76	4
noise	Q1	10.06	32	10.06	32	33.72	8.50	34.47	6
	median	10.60	32	10.56	32	34.89	41.50	35.00	7.50
(10 data	Q3	12.05	34.50	11.63	34.75	35.44	180.75	36.08	17.25
sets)	max	13.89	36	12.17	38	36.19	204	37.69	90

5 Conclusions

In some cases, bottom-up strategies can be more successful than the traditional top-down "one-step" search for the best split. In order to establish a method of more general applicability based on our algorithms, one has to look for various ways to find a heuristic compromise between the full tree optimization and the traditional top-down methods. Moreover, the simple construction/selection strategy using two halves of the training data should be supplemented by other well-known approaches, e.g. cross-validation.

Acknowledgements
This research was supported by grants 201/00/1482 (for the first and second authors) and 201/00/0769 (for the third author) from the Grant Agency of the Czech Republic. We are grateful to Prof. N. C. Lauro, who kindly enabled us to perform part of calculations, using the program CART v.1.1, at the Dept. of Statistics at the Univ. *Federico II* in Naples.

References

Breiman L., Friedman J.H., Olshen R.A. and Stone C.J. (1984). *Classification and Regression Trees*. Belmont CA: Wadsworth.

Friedman J.H. (1979). A tree-structured approach to nonparametric multiple regression. In: *Smoothing techniques for curve estimation. Lecture notes in math.* **757**, (eds. Gasser, T. & Rosenblatt, M.), pp. 5-22. Berlin: Springer-Verlag.

Siciliano, R. (1998). Exploratory versus decision trees. In: *COMPSTAT 98, Proceedings in Computational Statistics* (eds. R. Payne & P. Green), pp. 113-124. Heidelberg: Physica-Verlag.

Pijls, W. and Bioch, J.C. (1999). Mining frequent itemsets in memory-resident databases. Manuscript, `http://www.few.eur.nl/few/people/pijls/`

Gelfand, S.B., Ravishankar, C.S. and Delp, E.J. (1991). An iterative growing and pruning algorithm for classification tree design. *IEEE Trans. on Pattern Analysis and Machine Intelligence* **13**, *2*, pp. 163-174.

GAM spline algorithms: a direct comparison

Michael G. Schimek

Karl-Franzens-University of Graz, IMI, A-8010 Graz, Austria

Abstract. For some while backfitting has been the numerical technique associated with generalized additive models. More recently alternatives such as relaxed iterative projection or penalized likelihood fitting have emerged. What they have in common is the use of spline methodology and S code to execute them. Here we check these algorithms for performance differences in the S-Plus environment. The main results: They do not differ much in standard situations, however, vary much under concurvity (near-singularity).

Keywords. Backfitting, generalized additive models, penalized likelihood, relaxed iterative projections, singularity, splines, S-Plus

1 Introduction

Since the introduction of generalized additive models (abb. GAM, Hastie & Tibshirani, 1986) until recently the one and only estimation technique has been backfitting. Now we have a number of alternative algorithms at hand. Linton & Nielsen (1995) proposed a marginal integration method, Schimek (1996) a relaxed iterative projection technique, and Marx & Eilers (1998) a penalized likelihood procedure. In the nineties GAMs have become popular among data analysts. Algorithms proposed so far can differ in asymptotic terms, even more with respect to numerical stability, (finite) sample size and smoothness properties of the estimated regression curves. Practitioner's choice seems to be determined rather by software availability than statistical theory. But this is no surprise as the literature does not really help to decide for a specific numerical procedure. For our goal of giving at least some guidelines to practitioners which GAM algorithms to use, we restrict our attention to a single but popular smoothing concept and such algorithms which converge (in asymptotic sense) to the same result. Otherwise any comparative result would be prune to misinterpretation.

Our study is confined to splines. Theoretical properties such as linearity and symmetry have made these shrinking smoothers widely used. In addition, a variety of GAM algorithms is available for them. As a result we cannot include the marginal integration method which is based on a kernel smoother (implemented in the XploRe package). Here we consider standard backfitting, relaxed iterative projection (abb. RIP) and penalized likelihood fitting. For non-normal data the above mentioned procedures become part of the so-called local scoring algorithm, similar to Fisher scoring in generalized linear models (abb. GLM). The comparison is based on S-Plus code provided by the authors of the algorithms and adapted to work with the known S-Plus routines for GAMs (such code is not available for the marginal integration method apart from its different nature and asymptotics).

2 Algorithms for GAMs

The evaluation of a GAM requires the solution of a linear system of normal equations $A\mathbf{g} = \mathbf{b}$, where $\mathbf{g}$ is unknown (i.e. the regression curves). Let the elements of the matrix A be a_{ij} and of the vector $\mathbf{b}$ be b_i. Because of the large size of the equation system it is usually solved by iterative numerical techniques. First we discuss two algorithms which solve such systems iteratively.

For non-normal data (within the GLM framework) we have an additional complication: There the local scoring algorithm (a scoring weight matrix and an adjusted dependent variable are evaluated in each iteration step) is required, built around backfitting or the RIP (i.e. the latter form the inner loop of this two-loop algorithm).

Finally a more direct penalized likelihood procedure is considered which avoids iterations over two loops. It is less flexible with respect to the choice of scatterplot smoothers in GAMs, because restricted to so-called P-splines.

2.1 Backfitting

By standard backfitting we understand Gauss-Seidel iterations (not to confuse with Jacobi iterations - see Schimek & Turlach, 2000), developed for solving non-singular linear systems. Gauss-Seidel backfitting is defined in the kth step by

$$v_i^{(k)} = [b_i - \sum_{j=1}^{i-1} a_{ij} v_j^{(k)} - \sum_{j=i+1}^{N} a_{ij} v_j^{(k-1)}]/a_{ii}$$

with iterative solutions $v_i^{(k)}$, $i = 1, 2, \ldots, N$ and starting values $v_i^{(0)}$ (usually $v_i^{(0)} = 0$). The idea is to determine estimates for the regression covariates in a successive manner, taking advantage of the specific structural features of the estimation problem. This algorithm is most effective in combination with certain smoothing splines, because we can avoid the explicit calculation of the smoother matrices of the p covariates. Convergence results can be obtained under the assumption of full rank of A. Convergence depends on the eigenvalues of the iteration matrix.

2.2 Relaxed iterative projection fitting

Different from backfitting which is equation-oriented, the RIP method is column-oriented. Thus we assume the matrix $A = (\mathbf{a}_1, \mathbf{a}_2, \ldots, \mathbf{a}_n)$ to consist of column vectors $\mathbf{a}_i$ for $i = 1, 2, \ldots, n$. Let us have a linear space $sp(A)$ generated by the columns of A and $\mathbf{b} \in sp(A)$. We define two real sequences, one is (μ_j) with

$$\left(\mathbf{b} - \sum_{i=1}^{j} \mu_i \mathbf{a}_i, \mathbf{a}_j \right) = 0, \quad j = 1, 2 \ldots,$$

where $\mathbf{a}_i$ are the column vectors of A as defined above. The other sequence is (s_{ik}) defined by $s_{ik} = \sum_j \mu_j, \quad j = i + nk, \quad k = 1, 2, \ldots$. In the j-th iteration step μ_j is determined by the orthogonal projection of the previous residual component $\mathbf{u}_{j-1}$ onto the dimension $\mathbf{a}_j$ and the coefficients μ_j can be calculated by dot (inner) products. Hence $\mu_j = (\omega \mathbf{u}_{j-1}, \mathbf{a}_j)/(\mathbf{a}_j, \mathbf{a}_j)$, where ω is a relaxation parameter $(1 \leq \omega \leq 1.8)$ and $\mathbf{u}_{j-1} = \mathbf{b} - \sum_{i=1}^{j-1} \mu_i \mathbf{a}_i$. Because the above s_{ik} tend to the x_i for $k \to \infty$, each element x_i of the solution vector $\mathbf{x} = (x_1, x_2, \ldots, x_n)$ and k sufficiently large can be obtained by $x_i = \sum_j \mu_j, \quad j = i + nk, \quad k = 1, 2, \ldots$ The necessary condition is that the residual components $\mathbf{u}_j$ tend to zero, which is always true. Indeed convergence does not depend on special features of the system matrix A, such as positive definiteness or diagonal dominance. Hence we can handle ill-conditioned systems. In standard applications they are rare (unless concurvity is present) but there is current research into GAMs for correlated or panel data where degeneracy is likely to occur. However, the RIP algorithm can always be applied instead of backfitting but convergence is slower. That is why a relaxation parameter concept has been introduced (Schimek, 1996). Increasing degeneracy requires larger parameter values. When applied to a non-singular system, the RIP algorithm converges to the same solution as backfitting.

2.3 Penalized likelihood fitting

As pointed out already for non-normal data, backfitting or RIP form the core of the local scoring algorithm. In Marx's & Eilers' (1998) P-GAM approach estimation is carried out simultaneously, for normal as well as non-normal data. For that purpose P-splines are adopted, which are B-splines with a difference penalty to avoid the messy problem of knot optimization. Comparison with the other algorithms still makes sense because we are not outside the scope of shrinking smoothers. Let B_j be known B-spline matrices (for covariates $j = 1, 2, \ldots, p$) and $\mathbf{B} = (I, \|B_1\|, \ldots, \|B_p\|)$. Further assume an unknown coefficient vector a_j associated with the B-spline bases and $\mathbf{a} = (a_0, a_1, \ldots, a_p)^T$. To solve $\mathbf{Ba} = \eta$ (η the predictor function of GLMs) iterations ($k = 1, 2, \ldots$)

$$\hat{\mathbf{a}}_{k+1} = (\mathbf{B}^T \hat{W}_k \mathbf{B} + P)^{-1} \mathbf{B}^T \hat{W}_k \hat{z}_k$$

are performed until convergence, where $\hat{W}_k$ is the scoring weight matrix, P a fixed blockdiagonal penalty matrix and $\hat{z}_k$ the adjusted dependent variable. This algorithm is derived from the penalized version of the log-likelihood of the regression model. According to Marx & Eilers (1998, p.208) iteration with P-splines is just Gauss-Seidel iteration. Hence in case of convergence and no degeneracy we can expect the same result as from the first two approaches in combination with local scoring.

3 The implementations in S-Plus

For a direct and fair comparison of all mentioned algorithms it is necessary having them all implemented within the function `gam()` of S-Plus (MathSoft, 1999). Hastie (1992) gives a few hints how this function works. The maybe one and only but time-consuming way to learn about the implementational details is to read the S code and help files of all concerned functions and to analyse how they interrelate. This is not trivial because of the object-orientation of the S language and the modular structure of the involved functions. The fact that a substancial part of S-Plus is coded in external Fortran and C routines (the source code cannot be accessed), makes it even harder to see what is going on. For instance, all the local scoring (backfitting) and smoothing is performed in a Fortran subroutine when each term in the predictor function is smooothed by a cubic spline (i.e. via function `s()`). A user-programmer cannot make any alteration in such a case. The primary motivation behind is computational speed.

When Marx & Eilers (1998) implemented their P-GAMs, for good reasons they restricted themselves to adapt only such S-Plus functions which are solely coded in S. All their own coding is also in S (no loading of external Fortran or C). A P-spline is activated via the function `ps()`.

The RIP algorithm described in Schimek (1996) was originally implemented in a self-contained C program. For this paper it was necessary to implement it in S (the core of the algorithm - large number of iterations - is still in C for performance reasons) with a new function `rip.gam()`.

Due to the page limit we cannot describe here the technical details of S-Plus for backfitting (without external source code), for penalized likelihood fitting, and for RIP fitting. For the comparisons in this paper all algorithms were embedded in the same S-Plus environment provided by MathSoft (version 2000). They were executed on a Pentium II platform under Windows 98.

4 Comparison of the algorithms

A study based on simulated data as well as on real data and the three algorithms in combination with cubic spline smoothers was carried out. The emphasis was on their comparative small sample behaviour, assuming non-singularity of the system matrix A (i.e. non-concurvity). What all three techniques have in common is the use of the local scoring concept. The "fitting"-related aspects can make the difference and are best studied on the simple additive regression model $Y = \sum_{j=1}^{p} g(x_j) + \epsilon$ with a response variable Y, covariates x_j and an error term ϵ under Gaussian errors. The sample size was always $n = 100$. All the samples for the variables x_j were drawn from $X \sim U(a, b)$, with $a = 0$ and $b = 1$. Sampling from a uniform distribution guaranteed varying sparseness of the design. The error samples were drawn from $\epsilon \sim N(\mu, {\sigma_\epsilon}^2)$, with $\mu = 0$ and $\sigma_\epsilon = \{0.01, 0.02, 0.03, 0.04, 0.05\}$, to have low to high noise conditions. The amount of smoothing was chosen via visual inspection aiming at a simultaneous optimal reconstruction of the known functions g. Criterion for a comparable amount of smoothing within the three techniques was the nonlinear (equivalent) degrees of freedom for each term (the standard approach in S-Plus for GAMs). For the problems with "automatic" smoothing parameter selection in GAMs see Schimek & Turlach (2000, p.321f).

First results for simulated GAMs (with up to three covariates) based on frequently applied functions g convinced us that all three techniques work equally well, even under large error variances. Hence we fixed $\sigma_\epsilon = 0.05$ and decided to take into account the specific characteristics of a cubic spline smoother, that is confront it with functions that can cause estimation problems in the boundaries or elsewhere due to a discontinuity. The selected functions are $g_1 = g(x_1) = x_1^3/2x_1^2$, $g_2 = g(x_2) = \sin(2\pi(1 - x_2)^2)$, and $g_3 = g(x_3) = 4 \mid x_3 - 0.5 \mid$. The top row of **Fig. 1** shows g_1 and g_2. g_3 (not displayed) is a V-shaped curve (i.e. with discontinuity).

Let us consider a model ($p = 2$) consisting of g_1 and g_2 in more detail. The obtained relative best results are virtually the same for the backfitting GAM and the RIP-GAM (thus only the RIP-GAM estimates are shown in the middle row of **Fig. 1**). There is a slight distortion of the asymptotic behaviour of g_1 (no surprise). The P-GAM results (bottom row of **Fig. 1**) differ with respect to g_1: There is more curvature to the left of the function (more smoothing results in singularity also seen in other examples studied). In such situations we are better off with the other two fitters. It is a general observation that certain combinations of equivalent degrees of freedom for

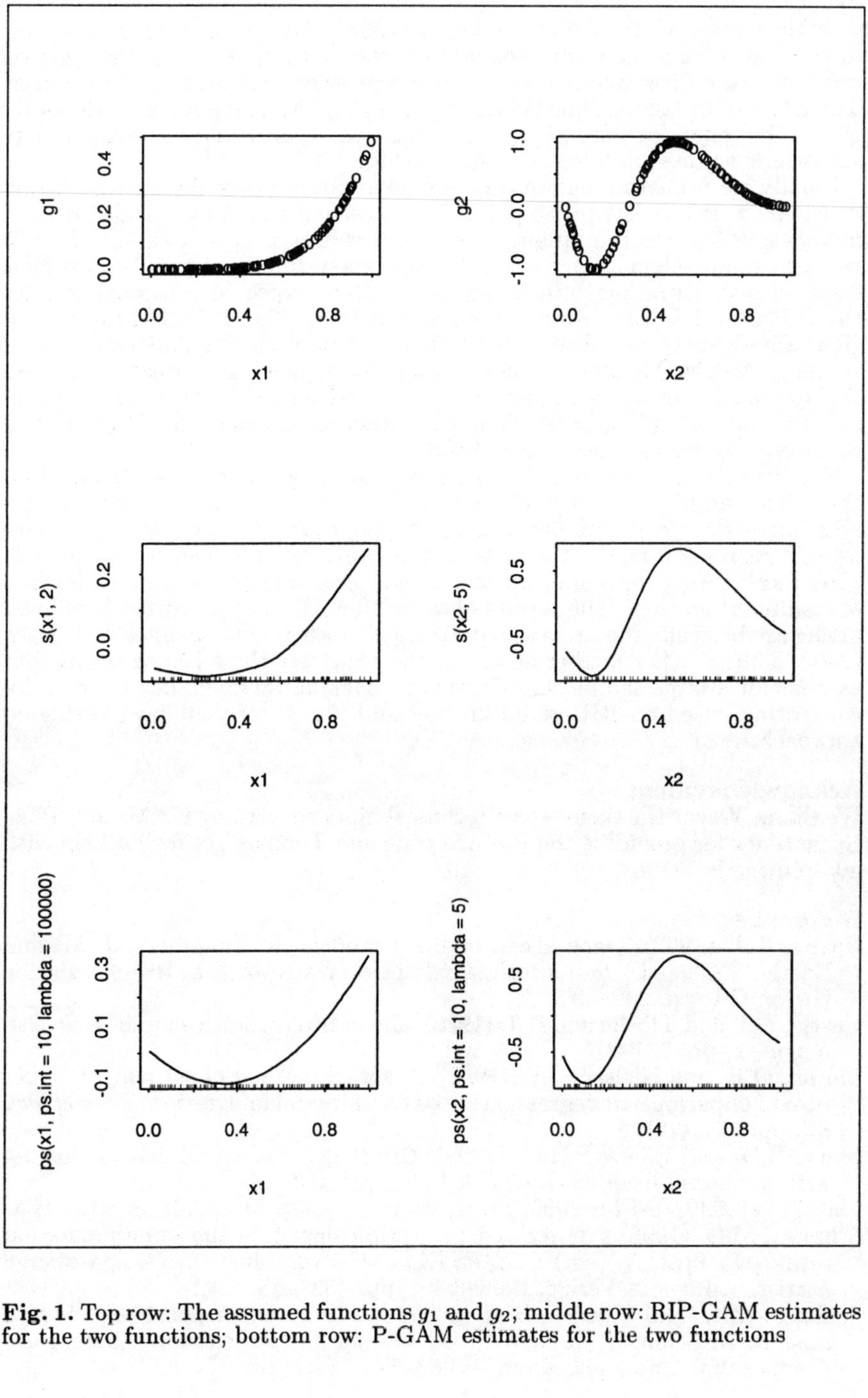

Fig. 1. Top row: The assumed functions g_1 and g_2; middle row: RIP-GAM estimates for the two functions; bottom row: P-GAM estimates for the two functions

the individual terms in backfitting or RIP-GAMs cannot be reproduced in P-GAMs.

In the model with g_3 and g_2 (i.e. another combination with $p = 2$) the discontinuity forces the cubic splines to oversmooth g_3, although the function estimates (not displayed here) are still reasonable in other areas of the curve. The effect of oversmoothing is larger in the P-GAM. Adaptation of the order of the difference penalty (P-splines offer additonal flexibility compared to smoothing splines) does not change anything.

Finally we want to mention a model having three covariates with a rank-deficient A. It consists of g_1, g_2, and g_1 a second time but calculated for a modified design (to the uniformly sampled values of x_1 a constant of 0.005 was added, introducing severe concurvity). The results in short: P-GAM fails, classical backfitting performs much better than expected (obviously due to the QR matrix decomposition adopted in S-Plus). The RIP-GAM does best (RIP algorithm is specifically contructed to cope with singularities).

From what we have experienced so far, we suggest to evaluate GAMs in the standard backfitting implementation of S-Plus unless there is reason to believe that the spline order should be other than cubic (use P-GAMs) or concurvity is present (use RIP-GAMs).

Finally we consider a practical example, the Irish Educational Transitions Data. The sample size is $n = 468$ (no missing values). We observe data sparseness but little correlation between the explanatory variables. In an additive logistic regression model the 0/1-outcome variable is *school leaving certificate*. Explanatory variables are *sex*, *verbal reasoning test score* and *father's occupational prestige*. The results obtained for RIP and penalized likelihood fitting are basically the same as reported for backfitting in Schimek & Turlach (2000, p.302ff). (For further details of the study see there.) So it seems that at least for a large sample size there is no substantial difference between local scoring based on RIP or backfitting and the direct penalized likelihood approach.

Acknowledgement

We thank Trevor Hastie for some technical hints concerning GAMs in S-Plus, Brian Marx for providing the P-GAM code and Thomas Yee for his help with a C-routine in S-Plus.

References

Hastie, T.J. (1992). Generalized additive models. In Chambers, J. M. and Hastie, T. J. (ed.) *Statistical models in S*, Wadsworth & Brooks, Pacific Grove, CA, pp. 249–307.

Hastie, T.J. and Tibshirani, R.J. (1986). Generalized additive models. *Statist. Scien.*, **1**, pp. 297–318.

Linton, O.B. and Nielsen, J.P. (1995). A kernel method of estimating structured nonparametric regression based on marginal integration. *Biometrika*, **82**, pp. 93–100.

Marx, B.D. and Eilers, P.H.C. (1998). Direct generalized additive modeling with penalized likelihood. *CSDA*, bf 28, pp. 193–209.

MathSoft (1999). S-Plus 2000. *Programmers's guide.* MathSoft, Seattle, WA.

Schimek, M.G. (1996). An iterative projection algorithm and some simulation results. In Prat, A. (ed.) *COMPSTAT '96. Proceedings in Computational Statistics.* Physica-Verlag, Heidelberg, pp. 453–458.

Schimek, M.G. and Turlach, B.A. (2000). Additive and generalized additive models. In Schimek, M.G. (ed.) *Smoothing and Regression. Approaches, Computation and Application.* Wiley, New York, pp. 277–327.

Markov Chain Monte Carlo methods for handling missing covariates in longitudinal mixed models

Ernst Schuster

Institute of Medical Informatics, Statistics and Epidemiology, University of Leipzig, Liebigstrasse 27, 04103 Leipzig, Germany.

Abstract. Handling missing covariates in longitudinal mixed effect models is demonstrated on a medical example.

Keywords. Markov Chain Monte Carlo, Missing Covariates, Longitudinal Mixed Models

1 Introduction

Regression models for longitudinal data with fixed and random effects for the covariates are investigated. One aim of statistical analysis is the determination of those covariates which have a significant influence on the response or which lead to a good model fitting.

Values of the response variable may be *missing at random* and we assume that the parameters of the data model and the parameters of the missingness mechanism are *distinct*, e.g. for the response variable the missing-data mechanism is *ignorable* (Schafer (1997)). If all covariates are complete, the observed-data likelihood can be utilized. But if covariates are missing it is intractable and multiple imputation is one possibility. Neither this method nor the EM algorithm, should be examined here. In my opinion the best way to handle missing covariates is the use of Markov Chain Monte Carlo (MCMC) methods to estimate the model parameters. As a result the missing covariates are estimated together with the whole model from a sample of the posterior distribution using non-informative priors for these. The analysis with complete covariates and the MCMC method with missing covariates is demonstrated on a medical example.

2 Statistical modelling and medical example

The examined model is a mixed effect regression model for repeated measurements with missing response values:

$$y_i = X_i\beta + u_i + \varepsilon_i$$

where y_i is an n_i column vector of the response variable for case i, X_i is an n_i x p design matrix with the values of the p covariates, β is a p column vector of regression coefficients assumed to be fixed, u_i are random intercepts, which are assumed to be independently distributed across subjects with $u_i \sim N(0, \sigma^2)$, ε_i is the within subject error with $\varepsilon_i \sim N(0, W)$, where W, for within, is a covariance matrix – n_i x n_i.

The aim of a prospective study was to evaluate the usefulness of clinical parameters, laboratory tests and immunogenetic markers as predictors of an erosive course of joint disease early in rheumatoid arthritis (RA). 96 patients with persistent oligio-or polyarthritis over a period of six weeks and a history of disease of less than 2 years were enrolled in the study. As one major outcome parameter of the destructive process, radiological evaluation of joint erosions was used. At study entry and at each scheduled visit, hand and feet radiographs were taken and scored using the Larsen score. As response for statistic modelling, **changes** in the Larsen score in the first (96), second (96), third (72) and fourth (54) year of observation were used. In parentheses the number of measured changes in the Larsen score is specified. Because the study is going on, the missing values of the response variable can be treated as ignorable.

Since the goal of the analysis was the identification of prognostic parameters, initial values of the covariates documented at study entry as well as the values obtained after 6 months of observation and therapy were used. In the preselected model (Table 1) only rheumatoid factor IgA (RfIgA) has two missing values. These missing values are also handled as ignorable.

3 Markov Chain Monte Carlo methods

MCMC methods are generic simulation methods. The unknown parameters and the missing values get prior distributions, mostly noninformative e.g. $N(0, 10^6)$ for location parameters and $Gamma(10^{-3}, 10^{-3})$ for each precision, which is the reciprocal of variance. Using Bayes' theorem the posterior density is proportional to prior times likelihood. In case of noninformative prior densities the posterior density results in similar estimations as likelihood function does. Therefore a sample from the posterior density is used, when the observed-data likelihood is intractable. The most important MCMC method is the algorithm of Metropolis and Hastings (Robert, C.P., Casella, G. (1999)) with the Gibbs sampler (Gilks, W.R., and Roberts, G.O. (1996)) as special case. The Gibbs sampler can be used if it is possible to sample from the conditional densities of each parameter given all other parameters and the data.

By using WinBUGS the joint posterior must be conjugate or log-concave. Otherwise, they are discretized. The Gibbs Sampler, after reaching stationary distribution, can be used for inference.

4 Modelling and results for the medical example

4.1 Complete covariates analysis and selection of the covariance structure

Mixed effect regression models were calculated with S-Plus 2000 for all cases with complete covariates. The stepwise term reduction began for a model with the initial values documented at study entry as well as the values obtained after 6 months of observation and therapy as covariates for fixed effects.

Table 1. Parameter-estimator (with p-value in parenthesis) for the pre-selected model, models with other covariance-structure and the BUGS-model

		Pre-selected model	Conditional independent + random slope	Conditional independent model	WinBUGS model
		estimator (p-value)	estimator (p-value)	estimator (p-value)	Estimator: median (95% credible interval)
Fixed effects	intercept	12.46 (<.0001)	13.34 (<.0001)	13.32 (<.0001)	13.31 (8.53, 18.27)
	DR4epi	2.87 (0.0087)	2.74 (0.0069)	2.77 (0.0059)	2.80 (0.84, 4.73)
	tint	-2.92 (0.0006)	-3.31 (0.0004)	-3.30 (0.0004)	-3.28 (-5.10, -1.56)
	Sex	-7.69 (0.0014)	-9.11 (0.0019)	-9.05 (0.0012)	-8.86 (-14.21,-3.72)
	RfIgA	0.0095 (0.0162)	0.0114 (0.0021)	0.0107 (0.0032)	0.0094 (0.0026,0.016)
	Sex:tint	2.11 (0.0245)	2.72 (0.0075)	2.70 (0.0080)	2.65 (0.75, 4.67)
Random effects: standard deviation	intercept	3.06	4.21	2.29	2.32 (0.35, 3.70)
	slope		0.74		
	Residual	7.08	6.97	7.10	7.18 (6.55, 7.90)
Corre-lations between time intervals	r_{12}	-0.237			
	r_{13}	0.127			
	r_{14}	0.158			
	r_{23}	0.007			
	r_{24}	0.102			
	r_{34}	0.556			
LR-Test to cond. ind. model: p-value		0.0406	0.3434	---	
AIC		2165	2168	2166	
BIC		2218	2206	2196	

Some interactions were also included. The stepwise deletion of parameters (biggest p-value first), that had no effect on the yearly increase of Larsen score leads to a preselected model. In this example the Akaike Information Criterion (AIC: -2 log-likelihood + p * 2) and the Bayesian Information Criterion (BIC: -2 log-likelihood + p * log(number of cases)) are minimized, too. Table 1 shows the preselected model in the first column. Conditional on the selected covariates and the random intercept the correlations between time intervals are nearby zero (exception r_{34}). Therefore it makes sense to use a conditional independent model (Table 1, third column). This is equivalent to the compound symmetry form for the total covariance matrix. The conditional independent model is worse for the log-ratio-test ($p=0.04$) but better in BIC. The fit of both models is equally good, therefore the more simple conditional independent model should be preferred, which is analysed with WinBUGS below. The second column shows that an additional random slope is not necessary, because AIC and BIC are worse.

4.2 Modelling with MCMC methods for missing covariates

Fig 1 shows a directed graph for the model created with WinBUGS. The graph is a special case of the general relation

$$\mathbf{y}_i \sim N(\mathbf{X}_i \beta + u_i , \sigma^2 I)$$

based on the following distributions and equations:

$$\mathbf{y}_i \equiv (larsd_{i1}, larsd_{i2}, larsd_{i3}, larsd_{i4})^T,$$

$$\mathbf{X}_i \beta \equiv \alpha_0 + b_1\, tint_k + b_2\, Sex_i + b_3\, Sex_i\, tint_k + b_4\, DR4epi_i + b_5\, RfIgA_i ,$$

$$\beta \equiv (\alpha_0 , b_1 , b_2 , b_3 , b_4 , b_5)^T, \qquad \mathbf{tint} = (1, 2, 3, 4)^T,$$

$$u_i \equiv \sigma_1 \alpha_i , \qquad \mu_{i,k} \equiv \mathbf{X}_i \beta + u_i ,$$

$$larsd_{ik} \sim Normal(\mu_{i,k}, \tau), \qquad larsd_{ik} \sim Normal(0, 10^{-6}) \text{ for missing } larsd_{ik}$$

$$\alpha_0 , b_1 , ..., b_5 \sim Normal(0, 10^{-6}), \tau \sim Gamma(10^{-3}, 10^{-3}),$$

$$\sigma_1 \sim Unif(0, 10), \qquad \alpha_i \sim Normal(0, 1),$$

$$RfIgA_{11} \sim Unif(0, 1000), \qquad RfIgA_{45} \sim Unif(0, 1000).$$

In contrast to the notation $N(\mu, \sigma^2)$, $Normal(\mu, \tau)$ means here and in BUGS normal distributed with expectation μ and precision τ, which is the reciprocal of the variance. The fixed effects are the constant α_0, the slope in time b_1, and the other factors b_2 to b_5 for the covariates. All these get as prior a noninformative normal distribution.

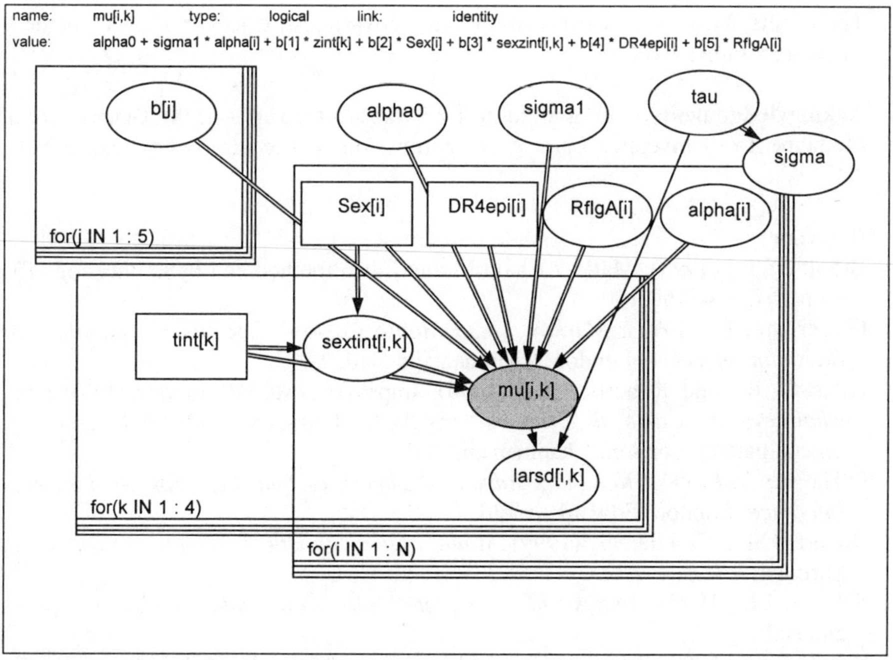

Fig 1. Graphical model of the rheumatoid arthritis data

The factor σ_1 is the standard deviation of the random intercept. Therefore the α_i are standard normal distributed. The factor σ_1 and both missing values of RfIgA get as prior a noninformative uniform distribution. The yearly change in Larsen score (larsd) is modelled as normal distributed. As prior for τ is chosen a noninformative conjugate distribution, namely a gamma distribution.

Table 1 shows in the last column the result of 50 000 iterations after a "burn in" of 1 million iterations. Because only two values of RfIgA were missing, the conditional independent model with complete covariates and the BUGS model are rather similar.

In the BUGS model, the progression of disease is indicated by a yearly baseline increase in the Larsen score of 13.31, that is described by the intercept. The influence of time indicates that this increase decreases by 3.28 yearly. The presence of an RA associated DRB1*04 allele had the strongest impact on progression resulting in an additional yearly increase in Larsen score of 2.8. In addition, the level of RFIgA measured at study entry had a significant influence on the yearly increase in Larsen score. There was a gender difference with men having a higher yearly increase of 8.86 compared to women. However the interaction with time interval shows that this increase decreases by 2.65 yearly.

A remarkable benefit of the MCMC methods is the ability to calculate credible intervals for random coefficients. A crucial advantage of MCMC methods is the possibility to calculate robust models by other distributions e.g. double exponential for the error term.

The results show that for data with missing covariates the use of MCMC methods can be recommended.

Acknowledgements. I am grateful to Dr Sylke Kaltenhäuser of the Department of Medicine IV, University Leipzig for permission to use the rheumatoid arthritis data.

References

Brooks, S.P., (1998). Markov Chain Monte Carlo method and its application. *The Statistician*, **47**, 69-100.

Gamerman, D. (1997). *Markov Chain Monte Carlo – Stochastic simulation for Bayesian inference*, London: Chapman and Hall.

Gilks, W.R., and Roberts, G.O. (1996). Improving MCMC mixing. In *Markov Chain Monte Carlo in Practice* (eds W.R. Gilks, S. Richardson and D.J. Spiegelhalter). London: Chapman and Hall.

O'Hagen, A. (1994). *Kendall's Advanced Theory of Statistics, Vol 2B, Bayesian Inference,* London: Edward Arnold.

Robert, C.P., Casella, G. (1999) *Monte Carlo Statistical Methods*, New York: Springer.

Schafer, J.L. (1997). *Analysis of Incomplete Multivariate Data*. London: Chapman and Hall.

Schuster, E. (1999). Markov Chain Monte Carlo Methods for Handling Missing Covariates in Multiple Regression Models. In: Friedl, H. Berghold, A. and Kauermann, G. (eds.) *Statistical Modelling – Proceedings of the 14th International Workshop on Statistical Modelling, Graz, Austria, July 19-23, 1999*, 651-655.

Spiegelhalter DJ, Thomas A, Best NG. (1999). *WinBUGS Version 1.2 UserManual. Cambridge, U.K.*

S-PLUS 2000 Guide to Statistics *(1999), MathSoft, Seattle.*

Tanner, M.A. (1996). *Tools for Statistical Inference: Methods for the Exploration of Posterior Distributions and Likelihood Functions.* Third Edition. New-York: Springer.

Robust Bayesian classification

Paola Sebastiani[1] and Marco Ramoni[2]

[1] Imperial College, London SW7 2B, UK
[2] The Open University, Milton Keynes, MK7 6AA UK

Abstract. This paper describes a method for learning robust Bayesian classification rules from incomplete data by producing interval-based classification. We provide two scoring methods to perform interval-based classification and a decision theoretic approach to choose the scoring method best suiting the problem at hand.

Keywords. Bayes classifier, missing data, interval probabilities

1 Introduction

Naive Bayes classifiers (NBCs) (Duda and Hart, 1973) are among the oldest supervised classification methods. The training step for a NBC consists of estimating the conditional probability distributions of each attribute given the class from a training set. Once trained, the NBC classifies a case, represented by a set of attribute values, by computing posterior probabilities of the classes via Bayes' Theorem and by assigning the case to the class with maximum posterior probability. When the training set has missing values, both the efficiency of the training process and the classification accuracy can be lost. Different solutions to handle this problem make restrictive assumptions on the missing data mechanism such as that data are *Missing at Random* (MAR) (Little and Rubin, 1987), so that the probability data are missing is independent of the unobserved data.

The need for a classification method robust with respect to the missing data mechanism is our motivation to introduce a robust version of the NBC that we call the *Robust Bayes Classifier* (RBC). The RBC produces interval-based classification rules by training the classifier with the *Robust Bayesian Estimator* (RBE) introduced in Ramoni and Sebastiani (2000) to learn conditional probabilities from incomplete data. The interval-based classification is based on a new propagation algorithm to compute posterior probability intervals containing all the class posterior probabilities that could be obtained from the exact computation of all possible completions of the training set. These intervals are then ranked by attaching a score to each of them and a new case is assigned to the class associated with the highest ranked interval. We provide two scoring methods for probability intervals and a general decision-theoretic criterion to select the best interval-scoring method by trading off accuracy and coverage. As a by-product, this decision theoretic criterion provides a principled way to asses the safety of adopting the MAR assumption for a training set.

2 The naive Bayes classifier

Suppose we have m attributes $A_1, \cdots, A_m$ and a set of q mutually exclusive and exhaustive classes c_j. We define by C the variable representing class membership as $C = c_j$. An NBC makes the assumption that attributes are independent, given the class and, therefore, is defined by the distribution of the variable C and by a set of conditional probability distributions of each attribute A_i, given class c_j. Each attribute A_i bears values $A_i = a_{ik}$. As shorthand, we will denote $C = c_j$ by c_j and $A_i = a_{ik}$ by a_{ik}. A consequence of the independence assumption underlying the classifier is that all these distributions can be estimated from a training set $\mathcal{D}$ locally, that is, independently from each other, thus making training very efficient.

Let $n(a_{ik}, c_j)$ be the frequency of cases in the training set $\mathcal{D}$ in which the attribute A_i appears with value a_{ik} and the class is c_j and let $n(c_j)$ be the frequency of cases in the training set with class c_j. When the training set $\mathcal{D}$ is complete, we have that $n(c_j) = \sum_k n(a_{ik}, c_j)$ and Bayesian estimates of $p(a_{ik}|c_j)$ and $p(c_j)$ using conjugate Dirichlet analysis are $p(a_{ik}|c_j) = (\alpha_{ijk} + n(a_{ik}, c_j))/\sum_k[\alpha_{ijk} + n(a_{ik}, c_j)]$ and $p(c_j) = (\alpha_j + n(c_j))/\sum_j[\alpha_j + n(c_j)]$ respectively, see Ramoni and Sebastiani (1999a). The quantity α_{ijk} can be regarded as frequency of the pair a_{ik}, c_j in an imaginary sample of size α. Similarly, the quantity α_j is the frequency of the classes c_j in the same imaginary sample. Once the classifier has been trained, we can use it for the classification of new cases. If we represent a case to be classified by a set of attribute values $e = \{A_1 = a_{1k}, \ldots, A_m = a_{mk}\}$, Bayes' theorem lets us calculate the posterior probability of each class c_j give e as

$$p(c_j|e) = p(c_j) \prod_{i=1}^{m} p(a_{ik}|c_j) / \sum_{h=1}^{q} p(c_h) \prod_{i=1}^{m} p(a_{ik}|c_h) \quad (1)$$

and the case is assigned to the class with maximum posterior probability. Next section describes a generalization of these results when the training set is incomplete.

3 The robust Bayes classifier

The robust Bayesian estimator introduced by Ramoni and Sebastiani (2000) is a novel approach that allows one to estimate the probabilities $\{p(c_j), p(a_{ik}|c_j)\}$ specifying the NBC without making any assumption about the missing data mechanism. This estimator is based on a new view of incomplete data: with no information on the pattern of missing data, an incomplete data set can only constrain the set of estimates that can be induced from the database. Hence, the robust Bayesian estimator returns estimates which are robust with respect to the missing data mechanism by providing interval-valued estimates containing all estimates that could be learned from all possible completions of the incomplete data. We denote by $[\underline{p}(a_{ik}|c_j)\ \overline{p}(a_{ik}|c_j)]$ the interval-valued estimate of $p(a_{ik}|c_j)$, and by $[\underline{p}(c_j)\ \overline{p}(c_j)]$ the interval-valued estimate of $p(c_j)$.

Once trained, the classifier must be used to label unclassifed cases by reasoning with probability intervals. Given a new case with attribute values $e = \{a_{1k}, ..., a_{mk}\}$, an NBC performs this task in two steps: first it computes the posterior probability of each class, given e, by Equation 1 and then it labels the case with the class having maximum posterior probability. Now, we need to perform these two steps by reasoning with probability intervals. Ramoni and Sebastiani (1999b) describe an exact algorithm for extending Equation 1 to probability intervals. The algorithm maintains the same computational complexity needed to evaluate Equation 1 and returns the probability interval $[\underline{p}(c_j|e); \overline{p}(c_j|e)]$ that contains all the values $p(c_j|e)$ we would obtain by the possible completions of the data. When the training set reports class membership of every case, it can be shown that $\overline{p}(c_j|e) = p(c_j)\prod_{i=1}^{m}\overline{p}(a_{ik}|c_j)/[p(c_j)\prod_{i=1}^{m}\overline{p}(a_{ik}|c_j)+\sum_{l\neq j}p(c_l)\prod_{i=1}^{m}\underline{p}(a_{ik}|c_l)]$, $\underline{p}(c_j|e) = p(c_j)\prod_{i=1}^{m}\underline{p}(a_{ik}|c_j)/[p(c_j)\prod_{i=1}^{m}\underline{p}(a_{ik}|c_j)+\sum_{h\neq j}p(c_h)\prod_{i=1}^{m}\overline{p}(a_{ik}|c_h)]$ and these boundary points become slightly more complex in the general case. Details are in Ramoni and Sebastiani (1999b).

The next step is to use these intervals to assign, unequivocally, a case with attribute values $e = \{a_{1k}, \ldots, a_{mk}\}$ to a class. We accomplish this task by associating each posterior probability interval $[\underline{p}(c_j|e)\ \overline{p}(c_j|e)]$ with a score $s(c_j|e)$ and by choosing the class with maximum score. This interval-based classification rule rests on the intuition that the score is a "meaningful" summary of the global information contained in the probability intervals. However, this is not the unique requirement. Since the standard NBC classifies cases on the basis of the posterior probabilities of the classes given the attribute values, we require that the set of scores associated with the probability intervals $[\underline{p}(c_j|e)\ \overline{p}(c_j|e)]$ defines a probability distribution. We propose two scoring systems having these properties: the *strong dominance score* and the *complete admissible score*, which are defined as follows.

Definition 1 Strong Dominance Score. Given a set of q consistent posterior probability intervals $[\underline{p}(c_j|e)\ \overline{p}(c_j|e)]$, we define the *strong dominance score* as: $s_d(c_j|e) = 1$ if $\underline{p}(c_j|e) > \overline{p}(c_h|e)$ for all $h \neq j$ and 0 otherwise.

The interval-based classification rule induced by the strong dominance score classifies a new case as c_j if and only if the probability $\underline{p}(c_j|e)$ is larger than the probability $\overline{p}(c_h|e)$, for any $h \neq j$. Strong dominance is a safe criterion since it returns the classification that we would obtain from all consistent completions of the training set $\mathcal{D}$. However, when the probability intervals are overlapping, the strong dominance score is not defined and we face a situation of undecidability. In order to increase the coverage of the classifier, we can weaken the criterion producing the strong dominance score by making the minimal assumption that all missing data mechanisms are equally possible, thus making all values within the intervals $[\underline{p}(c_j|e)\ \overline{p}(c_j|e)]$ equally likely. In this way, we summarize the interval into an average point by defining the score $s_u(c_j|e) = (1-k)\overline{p}(c_j|e) + k\underline{p}(c_j|e)$ where the quantity k is chosen so that the set of scores $\{s_u(c_j|e)\}$ is a probability distribution. As the scores $s_u(c_j|e)$ always lead to a decision, we define it a *complete-admissible score.*

Definition 2 Complete-Admissible Score. Given a set of q consistent posterior probability intervals $[\underline{p}(c_j|e)\ \overline{p}(c_j|e)]$, we define the quantity

$$s_u(c_j|e) = \overline{p}(c_j|e) - (\overline{p}(c_j|e) - \underline{p}(c_j|e))(1 - \sum_h \underline{p}(c_h|e))/\sum_h(\overline{p}(c_h|e) - \underline{p}(c_h|e))$$

a *complete-admissible score.*

It is worth noting that the classification based on the complete-admissible score subsumes that based on the strong dominance score because if $\underline{p}(c_j|e) > \overline{p}(c_h|e)$, for all $h \neq j$, then $s_u(c_j|e) > s_u(c_h|e)$ for all $h \neq j$. When the condition to apply the strong dominance score does not hold, the complete-admissible score lets us classify cases left unclassified by the strong dominance score. This strategy may result in an increased classification coverage at the price of a lower accuracy. The choice of the scoring system best suiting the problem at hand will then depend on the importance given to coverage and accuracy. This argument can be used formally by introducing missclassification costs and costs incurred for the inability to classify one case. We denote by RBC_d the classifier using strong dominance and by RBC_u the classifier using the complete-admissible score.

Let θ_d be the classification accuracy achieved by the RBC_d with coverage γ_d, and let θ_u be the classification accuracy achieved by the RBC_u, with 100% coverage. Suppose that the cost incurred for not being able to classify a case with attribute values e is a quantity C_i, while the cost for a wrong classification is C_w. Then, the expected cost incurred on using the RBC_d is $C(\text{RBC}_d) = C_w(1-\theta_d)\gamma_d + C_i(1-\gamma_d)$ if correct classification yields no cost. On the other hand, the expected cost incurred on using the RBC_u, is $C(\text{RBC}_u) = C_w(1-\theta_u)$. If we decide to use the system with minimum expected cost, the RBC_d is to be preferred to the RBC_u when $C(\text{RBC}_d) \leq C(\text{RBC}_u)$ which is true if and only if $\theta_u - \theta_d\gamma_d \leq (1-\gamma_d)(1-C_i/C_w)$. For example, if $C_i = C_w$, the best decision is to choose the RBC_d whenever $\theta_d\gamma_d \geq \theta_u$. In practical applications, the quantities θ_d, θ_u and γ_d can be estimated from the data available by running some cross validation experiment, as shown in the next section.

Suppose now the quantity θ_a is the accuracy of any other NBC_a trained on an incomplete data set by making the MAR assumption. We can use the same decision-theoretic approach to help one decide the best classifier and, as a by-product, the decision can be interpreted as an evaluation of the consequences of enforcing the MAR assumption.

4 Evaluation

The experimental evaluation was conducted on the ten databases reported in Table 1, available from the UCI Machine Learning repository (Blake, Keogh and Merz, 1998).

We compared the classification accuracy of the RBC and the standard NBC by running, on each data set D, a 5-fold cross validation experiment. For each database, we ran four tests: one based on using the NBC trained by removing the missing entries on each dependency A_i, C (NBC_m); one based on assigning the missing entries to a dummy state (NBC_*), one based on the strong dominance score (RBC_d) and one based on the complete-admissible score (RBC_u). The experiments were carried out using the implementation of the RBC_d and RBC_u in `RoC`, available at `http://kmi.open.ac.uk/projects/bkd`, and the implementation of the NBC_m and NBC_* in `Bayesware Discoverer`, available at `http://www.bayesware.com`.

We report two performance measures for each test: *accuracy* and *coverage*. The classification accuracy was estimated as average number of cases correctly classified in the test sets. The coverage was estimated as the frequency of cases classified. We also provide 95% confidence limits based on using a Normal approximation. Table 1 reports the results. The accuracy of RBC_d is overall better than all other methods and the accuracy gain ranges between 0.02% (D_2), in which there are only 6 missing entries in a data set of 699 cases, to 16.77% (D_6), in which data are heavily missing. This gain of accuracy is counter-balanced by a loss of coverage that can be as small as 6.51% in (D_6). The adoption of the complete-admissible score increases the

Database		NBC_m Accuracy	NBC_* Accuracy	RBC_d Accuracy	RBC_d Coverage	RBC_u Accuracy
D_1	Annealing	86.54±2.88	80.88±3.32	**97.53 ± 1.51**	49.12±4.87	96.73±1.24
D_2	B.Cancer	97.42±0.66	97.42±0.66	**97.49 ± 0.67**	99.65±5.23	97.23±0.67
D_3	Credit	84.88±1.30	84.88±1.30	**87.48 ± 1.72**	95.40±5.21	84.70±1.31
D_4	Cylinder	73.70±3.71	73.00±3.74	**91.71 ± 4.14**	31.30±6.97	74.26±0.67
D_5	Hepatitis	85.03±2.09	85.16±2.08	**90.50 ± 2.84**	76.45±9.73	85.55±2.08
D_6	Horse Colic	75.79±1.62	75.79±1.63	**92.56 ± 0.59**	6.51±2.05	77.73±1.61
D_7	L.Cancer	43.75±17.88	43.75±17.88	**46.67 ± 17.85**	93.75±8.66	43.75±17.88
D_8	Mushrooms	98.53±0.12	98.40±0.12	**99.04 ± 0.15**	98.88±1.53	98.70±0.11
D_9	Sick	91.60±0.51	90.87±0.53	**97.53 ± 0.34**	86.30±2.29	92.46±0.49
D_{10}	Vote	90.02±1.05	90.21±1.04	**92.05 ± 1.75**	94.94±6.47	90.21±1.04

Table 1. Classification accuracy and coverage. The figure in bold in each row is the maximum accuracy.

coverage to 100% at the price of reducing the accuracy, that in two cases (D_2 and D_3) is worse than the accuracy of the standard NBC. On the other hand, the accuracy gain of the RBC_u against NBC_m and NBC_* is significant in all the other data sets, and noticeably, it reaches 10.19% in the data set Annealing. This last case shows the danger of enforcing the wrong assumption on the missing data mechanism which produces a loss of 10% in the classification accuracy.

One way of looking at the results is to consider that strong dominance partitions the data into two parts. One part comprises the cases on which there is no classification ambiguity due to the missing data and the classification accuracy is only model-dependent. The remaining part comprises those cases which cannot be classified unless one is willing to make some assumption on the missing data mechanism. The classification accuracy on these cases of the other systems achieving 100% coverage is $\theta_{al} = (\hat{\theta}_a - \hat{\theta}_d\hat{\gamma}_d)/(1-\hat{\gamma}_d)$, where $\hat{\theta}_a$ is the (estimated) accuracy of the NBC_m, the NBC_* or RBC_u, while $\hat{\theta}_d$ and $\hat{\gamma}_d$ are the estimated accuracy and coverage of the RBC_d. Table 2 reports these accuracy values for the NBC_m, the NBC_*, and the RBC_u in the ten data sets used in this experiment. The sixth columns report the maximum cost ratio C_i/C_w to make the RBC_d the best classification system in terms of minimum expected costs and, for reference, the last two columns are the proportion of cases left unclassified by the RBC_d and size of each database. If the cost ratio is superior to the value reported, then the best classification system is that having the largest accuracy θ_{al} and it is highlighted in bold face in the table.

In two data sets (D_2 and D_3) the RBC_d is the best choice if $C_w \geq 4.44C_i$ and $C_w \geq 1.45C_i$, respectively. If these conditions on the costs C_i and C_w are not satisfied, then the NBC_m or, equivalently, the NBC_*, are the best classification systems. We note that in D_2, the complete-admissible score performs very poorly on the cases left unclassified by the strong dominance score, while the enforcement of the MAR assumption lets the standard NBC fully exploit the information provided by the available data and reaches an accuracy of 77.49%. This data set has, however, only 6 cases with missing entries. In D_3, none of the NBC_m, the NBC_* or RBC_u can achieve an accuracy larger than 50%, so that if the the cost incurred for a wrong classification is smaller than $1.45C_i$, a random assignment to classes of the cases left unclassified by the RBC_d is preferable. In the data set D_7, the accuracy θ_{al} is null for all systems, and hence the choice of the RBC_d is never under discussion. In the remaining

	Database	θ_{ml}	θ_{*l}	θ_{ul}	C_i/C_w	$(1-\gamma_d)100$	Size
D_1	Annealing	0.7593	0.6481	**0.9596**	0.0404	50.88	798
D_2	B.Cancer	**0.7749**	**0.7749**	0.2320	0.2251	0.35	699
D_3	Credit	**0.3096**	**0.3096**	0.2705	0.6904	4.60	598
D_4	Cylinder	0.6448	0.6549	**0.6631**	0.3369	68.70	512
D_5	Hepatitis	0.6782	0.6727	**0.6948**	0.3052	23.55	155
D_6	Horse Colic	0.7462	0.7462	**0.7670**	0.2330	93.49	368
D_7	L.Cancer	0	0	0	1.0005	6.25	32
D_8	Mushrooms	0.4190	0.5350	**0.6868**	0.3132	1.22	8124
D_9	Sick	0.4892	0.5424	**0.6052**	0.3948	15.70	2800
D_{10}	Vote	**0.5569**	0.5193	**0.5569**	0.4431	5.06	435

Table 2. Accuracy values for the NBC_m, the NBC_*, and the RBC_u

seven data sets, the RBC_u is always the second best choice, if the cost ratio C_i/C_w is superior to the value reported in the last column of the table. In D_1, for example, if the cost for not classifying a case is smaller than 25 times the cost for a wrong classification, the RBC_u is the best choice and achieves an accuracy 0.9596 on the cases left unclassified by the RBC_d. This is a gain of about 20% compared to the NBC_m that works under the MAR assumption. Again this result confirms that the MAR assumption on this data set has a negative effect of the classification accuracy and would not encourage one to enforce it.

The results show that classification based on the strong dominance criterion appears to be the most accurate at the price of, sometimes, low coverage. The use of the complete-admissible score improves coverage at the price of accuracy and yields better results than standard solutions except when the proportion of missing data is small. However, the results also show that there is no uniformly best classification strategy when some data are missing in the training set and the solution to adopt needs to take into account features of the data set and of the classification problem. The decision theoretic approach we provide is a principled way to help one choosing the best solution to his problem.

Acknowledgments This research is supported by the ESPRIT programme of the European Community, under contract EP29105.

References

Blake, C., Keogh, E. and Merz, C.J. (1998). *Repository of machine learning databases.* University of California, Irvine, Dept. of Information and Computer Sciences, http://www.ics.uci.edu/~mlearn/MLRepository.html.

Duda, R.O. and Hart, P.E. (1973). *Pattern Classification and Scene Analysis.* Wiley, New York, NY.

Little, R.J.A. and Rubin, D.B. (1987). *Statistical Analysis with Missing Data.* Wiley, New York, NY.

Ramoni, M. and Sebastiani, P. (1999a). Bayesian methods. In *Intelligent Data Analysis. An Introduction*, pages 129–166. Physica Verlag, Heidelberg.

Ramoni, M. and Sebastiani, P. (1999b). *Robust Bayes classifiers.* Technical Report KMi-TR-82, KMI, The Open University.
Available at http://kmi.open.ac.uk/techreports/KMi-TR-82.

Ramoni, M. and Sebastiani, P. (2000). Robust learning with missing data. *Machine Learning.* Available at http://kmi.open.ac.uk/techreports/KMi-TR-28.

Social science measurement by means of item response models

Klaas Sijtsma

Department of Methodology, FSW, Tilburg University, P.O. Box 90153, 5000 LE Tilburg, The Netherlands

Abstract. The basic ideas of measurement in the social and behavioral sciences is explained, followed by a discussion of item response theory, which supplies the family of modern statistical measurement methods. Four specialized topics in item response modeling are discussed, that are at the core of present-day research in item response theory.

Keywords. goodness-of-fit methods, item response theory, item selection, misfitting respondents, person ordering

1 The basic ideas of social science measurement

The measurement of hypothetical constructs, such as intelligence, introversion, and achievement in psychology, and attitudes and opinions in sociology and political science, has a long and impressive tradition. Well-known names in psychological measurement are Spearman (classical test theory, factor analysis), Thurstone (paired comparison, factor analysis), and Lord (classical test theory, item response theory). In sociology, Guttman (scalogram analysis, covariance structures) and Lazarsfeld (latent class analysis, latent trait models) immediately come to mind. Except for Spearman, each of these milestone researchers has contributed to what is nowadays known as item response theory (Van der Linden & Hambleton, 1997).

Item response models are used for the analysis of data from psychological and educational tests and from questionnaires which are used throughout the behavioral and social sciences. The results of such data analyses can be used for assessing the composition of the final test or questionnaire, estimating the reliability and the validity of the measurement values collected by a test or a questionnaire, and assigning measurement values to individuals who took a test or filled out a questionnaire. Once in their final form, tests are used for diagnosis aimed at the assignment of individuals to treatment or therapy (counseling, clinical setting) and remedial teaching (education) or at giving job advice. Also, the decision to admit pupils to certain schools and select people in jobs is often supported by test performance results. Questionnaires are often used in such diverse areas as opinion polls, demographic research, and marketing research where the measurement of preferences may be of primary interest.

Tests and questionnaires almost always consist of a number (J) of problems (e.g., mazes, building blocks, geometric figures which have to be rotated mentally), questions (essay questions or open-ended questions, multiple-choice questions), or statements (e.g., about ethical or political issues or about a respondent's behavior). In general, these building blocks are called "items". The items reflect the final operationalization of the hypothetical construct and are crucial to the success of measurement. The responses to items are assumed to contain information about a respondent's standing on the unobservable or latent trait measured by the complete set of items in the test or questionnaire.

Responses to items may be solutions to problems, choices, markings or written or oral reports. These responses are qualitative and as such they are not suited for statistical analysis. The necessary quantification is done by categorization of the responses, ordering of categories on the latent trait, and assigning ordered numbers to the categories which reflect the ordering on the latent trait. Thus, the quantification rests upon the idea: A higher item score reflects a higher standing on the latent trait. The statistical analysis by means of item response models of J item scores collected in a sample of N respondents has to point out whether the scoring rule made sense.

2 Item response models

Let X_j denote the random variable for the ordered score on item j ($j = 1, \ldots, J$), with realizations x_j, with $x_j = 0, 1, \ldots, m$. For $m \geq 2$, items are called polytomous items. Ordered polytomous scores may indicate the degree to which a respondent endorses an attitude statement such as "The present Dutch law on euthanasia is too liberal", but they may also indicate the rating of an essay on the exploitation of the tropical rain forests, written by high school students. Very common are dichotomous scores, $x_j = 0, 1$, where 0 indicates an incorrect answer and 1 a correct answer. Dichotomous scores are typical for multiple-choice items, but may also indicate agreement or disagreement with an attitude statement.

Furthermore, let $\mathbf{X}$ denote a vector with J item score variables and let $\mathbf{X} = \mathbf{x}$ denote the J realizations. Also, let θ denote the latent trait. We define $P(X_j = x|\theta)$ to be the conditional probability of obtaining a score of x on item j. This conditional probability describes the relationship of an item score to the latent trait, and is called the category response curve. For dichotomous items, this probability is defined as $P_j(\theta) \equiv P(X_j = 1|\theta)$, and is called the item response function. In general, we call category response curves and item response functions simply response functions.

In general, item response models rest on a number of assumptions. First, item response models can be multidimensional, which means that the items from a test simultaneously measure several traits, in which case $\boldsymbol{\theta}$ is a vector, or unidimensional, in which case θ is a single latent variable. Most item response models are unidimensional, which reflects the common practice that tests and questionnaires are required to measure one construct at a time. Second, it is mostly assumed that respondents do not improve or otherwise change their ability, attitude, an so forth, while taking the test or filling out the questionnaire. That is, the measurement procedure does not affect the traits it attempts to measure (cf. a cold thermometer that, when placed in warmer water, affects the water temperature read from its scale; the idea is that this should be avoided). Sometimes, item response models drop the assumption of local independence, but this is rather uncommon.

For item scores $x_j = 0, \ldots, m$ and $m \geq 1$, the assumptions of unidimensionality and local independence can be written as

$$P(\mathbf{X} = \mathbf{x}|\theta) = \prod_{j=1}^{J} P(X_j = x_j|\theta).$$

By integrating θ out, we obtain the J-variate distribution of the item scores,

$$P(\mathbf{X}=\mathbf{x}) = \int_{\theta} \prod_{j=1}^{J} P(X_j = x_j|\theta)dG(\theta), \tag{1}$$

where $G(\theta)$ is the cumulative distribution function of θ. Without further restrictions on the response functions, the cumulative distribution of θ, or both, the multivariate distribution $P(\mathbf{X} = \mathbf{x})$ is not restricted in any way. This is undesirable because in order to be able to test whether an item response model fits empirical data, observable consequences need to be derived from the model. In item response modeling, this is accomplished by placing restrictions on the response functions.

Two kinds of restrictions are possible. *Parametric* item response models assume a particular parametric function. For example, for dichotomous items the much used 3-parameter logistic model (Lord & Novick, 1968; part IV by A.L. Birnbaum; Lord, 1980) assumes that

$$P(X_j = 1|\theta) = \gamma_j + \frac{(1-\gamma_j)\exp[\alpha_j(\theta-\delta_j)]}{1+\exp[\alpha_j(\theta-\delta_j)]}, \tag{2}$$

where δ_j is a parameter which locates the item on the θ scale, α_j is a parameter which is monotonically related to the slope of the function at δ_j, and γ_j is the lower asymptote for $\theta \to -\infty$. Parameter δ_j gives us an impression about the difficulty of the item, α_j shows how well the item separates low θs from high θs, and γ_j gives the probability of a correct answer for low θ (e.g., when low-ability respondents guess for the correct answer on a multiple-choice item). *Nonparametric* models only place order restrictions on response functions. For example, for dichotomous items the model of monotone homogeneity (Mokken, 1971) assumes that $P_j(\theta_a) \leq P_j(\theta_b)$ whenever $\theta_a < \theta_b$ (for all pairs of unequal θs), and any item response function that has this property is accepted. Both the 3-parameter logistic model and the model of monotone homogeneity express the idea that a higher θ should lead to a higher probability of answering an item correctly. For item scores expressing preferences (score 1 means stimulus was chosen, and score 0 otherwise), item response models assume that response functions are bell-shaped (Post, 1992).

The estimation of parametric item response models mostly is done with maximum likelihood methods, but Bayesian methods are also used (Baker, 1992; Patz & Junker, 1999). Because of the variety of parametric item response models and estimation methods, we only mention conditional maximum likelihood estimation that is used, in particular, for estimating the 1-parameter logistic model (obtained from Equation 2 by setting γ_j to 0 and α_j to 1; see Fischer & Molenaar, 1995). Here, the likelihood of the data $\mathbf{X}_{N\times J}$ given the marginal distribution for the items (columns) is maximized using iterative algorithms, such as the Newton-Raphson. This yields estimates of the δs, which are assumed fixed when in the next step the θs are estimated. For estimating the 2- and 3-parameter logistic models (the 2-parameter logistic model is obtained from Equation 2 by setting γ_j to 0; see Lord, 1980), marginal maximum likelihood estimation is used. Here, a distribution for θ is assumed and the likelihood is integrated across this distribution. Next, the likelihood is solved for the item parameters, which in the final step are assumed fixed when the θs are estimated. For nonparametric models, that do not define parametric response functions with latent variables, estimation is relatively simple and can be done analytically, which usually does not pose any technical problems.

3 Some special topics

Due to space limitations, we will restrict ourselves to a brief discussion of four recent developments in item response theory, which together constitute the COMPSTAT 2000 symposium "Computer-intensive statistics in modern item response theory".

3.1 Goodness-of-fit methods

Usually, the univariate marginal distributions of the multivariate distribution of $\mathbf{X}$ (Equation 1) are used for investigating properties of response curves, for example, whether they comply with the shape described by Equation 2 (parametric) or whether they are nondecreasing functions of θ (nonparametric). The bivariate marginal distributions contain information about relationships between items and are used for evaluating the assumptions of unidimensionality and local independence. Other K-variate marginal distributions $(3 \leq K \leq J-1)$ contain information about simultaneous relationships between K items, but are almost never used for testing models due to the enormous complexity of the methods implied (Sijtsma & Junker, 1996).

There have been some attempts to reformulate item response models with continuous latent traits as latent class models with ordered latent classes; see, for example, Croon (1991). Here, the idea is that respondents do not have a score on a continuous θ, which indicates their level on that latent trait, but that someone with item score pattern $\mathbf{x}_i$ belongs to a discrete latent class q $(q = 1, \ldots, Q)$ with probability $P(q|\mathbf{x}_i)$. The number of ordered latent classes is limited and in this sense restricts the data structure. Van Onna (this symposium) formulated nonparametric item response models as ordered latent class models and constructed a Bayesian estimation algorithm for the model parameters, and also proposed goodness-of-fit methods.

3.2 Automated item selection from an item pool

In the context of nonparametric item response theory, three automated item selection procedures have been developed (similar procedures do not yet exist for parametric models). First, in the context of the model of monotone homogeneity a sequential clustering algorithm has been designed that starts with the subset of items (mostly, two items) from a pool of I items that best complies with the selection criterion. In each next step, an item is selected that together with the already selected items maximizes the selection criterion. This continues until all items are selected or no more items can be added that satisfy auxiliary selection criteria. The second procedure selects items stepwise according to a hierarchical clustering technique. In the first step, the best pair according to a particular formal criterion is selected; in the second step, either a third item is added to this pair or a second pair is selected; in the third step, either an item is added to the triplet already selected, to one of the two pairs already selected, or a third pair is selected; and this continues until finally all items are selected together in one set of J items. The researcher has to decide which intermediate clustering solution is his/her final solution. The third procedure searches for the subset of J items $(J = 2, 3, 4, \ldots, I)$ from the pool of I items that maximizes a particular formal criterion. Theoretically, this procedure tries all possibilities and may form several non-overlapping scales. In practice, algorithms have been designed to speed up the search from the enormous amount of possible scales. Van Abswoude (this symposium) recently has started a systematic comparison of the three item selection procedures. This comparison entails the effectiveness of each procedure for finding from simulated data with a known dimensionality structure the correct dimensionality and the correct assignment of items to subscales.

3.3 Using X_+ to order persons on θ

Psychologists and measurement practitioners may test children and adults, report the test results to these people, but perhaps also to school, companies and other clients. In all cases, professionals using tests and questionnaires often like to use and report simple scores, such as the number-correct or the number of points earned on a questionnaire, to be denoted $X_+ = \sum X_j$. When measurement instruments are used for ordering respondents, the ordering on X_+ should in a stochastic way reflect the ordering on θ. That is, the conditional cumulative distributions of θ should be higher for higher values of X_+: For integers $s_1 < s_2$, and for arbitrary real values t, we have that

$$P(\theta \le t|X_+ = s_1) \ge P(\theta \le t|X_+ = s_2). \tag{3}$$

For dichotomous items, Equation 3 is implied by unidimensional, locally independent item response models with monotonely nondecreasing item response functions. Most of the well known polytomous item response models, *theoretically* do not imply Equation 3 (Hemker, Sijtsma, Molenaar, & Junker, 1997). For three known classes of polytomous item response models, however, Van der Ark (this symposium) found in simulated *practically* relevant situations that for partial credit models and sequential ratio models Equation 3 was almost never violated, but that for graded response models a nonignorable number of violations occurred.

3.4 Adaptive test procedures and misfitting respondents

Item response theory has facilitated adaptive testing, a computerized testing procedure that adapts the difficulty level of the items stepwise to the estimated ability level of the examinee (Van der Linden & Glas, 1999). For example, after an examinee has received the first five items (presented on a computer screen; the examinee responds by pushing a button) the adaptive testing software selects the sixth item on the basis of the intermediate estimate of θ. The difficulty level of the sixth item corresponds to the estimated θ value. The result of this adaptive testing procedure is a highly reliable estimated θ, measured with relatively few items. This testing procedure is important, in particular, in educational measurement, where huge numbers of items are needed for testing large numbers of pupils. The items are stored in a so-called item bank, from which the adaptive testing software can select large numbers of different or partly overlapping tests.

Possible problems during adaptive testing are, for example, that an individual examinee may have pre-knowledge of some of the more difficult items from the item bank and when presented with such items, obtains better results than when presented with other items of equal difficulty, but previously unknown to him/her. Also, due to extreme anxiety the examinee may fail on the first series of moderate to easy items, and only catch up on the real θ level later on. Meijer (this symposium) devised a method for detecting unusual item score patterns at the individual level, that accumulates consecutive deviation scores that are larger than a preset level, and that identifies an item score pattern as aberrant when the index is higher than a particular critical value. The idea comes from industrial quality control and has proved to be useful in computerized adaptive testing.

References

Baker, F.B. (1992). *Item Response Theory. Parameter Estimation Techniques.* New York: Marcel Dekker.

Croon, M.A. (1991). Investigating Mokken scalability of dichotomous items by means of ordinal latent class analysis. *British Journal of Mathematical and Statistical Psychology,* **44**, 315-331.

Fischer, G.H., and Molenaar, I.W. (1995). *Rasch Models. Foundations, Recent Developments, and Applications.* New York: Springer-Verlag.

Hemker, B.T., Sijtsma, K., Molenaar, I.W., and Junker, B.W. (1997). Stochastic ordering using the latent trait and the sum score in polytomous IRT models. *Psychometrika,* **62**, 331-347.

Lord, F.M. (1980). *Applications of Item Response Theory to Practical Testing Problems.* Hillsdale, NJ: Erlbaum.

Lord, F.M., and Novick, M.R. (1968). *Statistical Theories of Mental Test Scores.* Reading, MA: Addison-Wesley.

Mokken., R.J. (1971). *A Theory and Procedure of Scale Analysis.* Berlin: De Gruyter.

Patz, R.J., and Junker, B.W. (1999). A straightforward approach to Markov Chain Monte Carlo methods for item response models. *Journal of Educational and Behavioral Statistics,* **24**, 146-178.

Post, W.J. (1992). *Nonparametric Unfolding Models. A Latent Structure Approach.* Leiden: DSWO Press.

Sijtsma, K., and Junker, B. W. (1996). A survey of theory and methods of invariant item ordering. *British Journal of Mathematical and Statistical Psychology,* **49**, 79-105.

Van der Linden, W.L., and Glas, C.A.W. (1999). *Computer Adaptive Testing: Theory and Practice.* Dordrecht, the Netherlands: Kluwer.

Van der Linden, W.J., and Hambleton, R.K. (1997). *Handbook of Modern Item Response Theory.* New York: Springer-Verlag.

Multivariate DLMs for forecasting financial time series, with application to the management of portfolios

Andrew Simpson and Darren J. Wilkinson

Department of Statistics, University of Newcastle, Newcastle upon Tyne, NE1 7RU, England.

Abstract. This paper considers a Bayesian approach to the multivariate forecasting of financial time series based on dynamic linear models (DLMs). It is shown how a marginal posterior forecast distribution may be simulated, and how this may be used directly in order to implement a fully Bayesian decision-theoretic approach to the selection of optimal stock portfolios. This is briefly compared with more traditional approaches to portfolio selection.

Keywords: Bayes; DLMs; Finance; Forecasting; Multivariate.

1 Introduction

This paper considers the problem of Bayesian modelling and forecasting for multivariate financial time series. Multivariate models are necessary, as prices of related stocks exhibit dependencies between series, as well as the usual dependencies over time. The multivariate dynamic linear state space models of West & Harrison (1997) are often appropriate for explaining the evolution of log-prices. The problem of Bayesian inference for the underlying states and covariance matrices is discussed in the context of Bayes linear, EM and MCMC based algorithms. In order for these algorithms to work efficiently, a variety of Kalman filtering, smoothing and simulation-smoothing techniques are required, as naive approaches suffer from problems associated with slow mixing and convergence.

The use of the EM algorithm for state space models is illustrated in Koopman (1993). MCMC approaches to state space modelling are also becoming increasingly popular. The basic strategy is outlined in West and Harrison (1997), and applications to financial series are discussed in Pitt & Shephard (1999). The advantage of the MCMC approach is that it allows simulation from the full joint posterior distribution of the hidden states and future observables, thus allowing analyses of forecasting problems which integrate over all uncertainties in the model.

Traditional portfolio management selection and analysis developed in Markowitz (1959) requires point estimates of the mean and covariance matrix of the returns distribution. The underlying theory is based on a very simple model for price changes which fails to capture the dynamic nature of price evolution. Further, uncertainty about the first and second order moments can lead to a sub-optimal portfolio. Traditional approaches have relied on historic point estimates of the mean vector and covariance matrix, ignoring uncertainty. Recently the Bayes-Stein method has been shown to provide better point estimates than conventional sample estimators; Jorion (1986) and Board and Sutcliffe (1994), but estimation risk is still generally treated in a somewhat *ad hoc* manner. The MCMC approach outlined above allows more realistic modelling of price evolution, and simulation from the full joint posterior probability distribution for the future returns bypasses the need for calculating point estimates by directly integrating over all uncertainties in the model. Portfolio management strategies based on the full forecast distribution are contrasted with standard mean/variance approaches.

In section 2 the model for price evolution is discussed, together with basic computational strategies. In section 3, a brief overview of portfolio analysis is given, and adapted to the modelling framework of section 2. In section 4, application to an example portfolio management problem is discussed, and various strategies are compared. Brief conclusions are given in section 5.

2 Model for price evolution

Suppose that we are interested in forming a portfolio of p different stocks. Let the vector of prices at time t be given by the p-vector X_t. Our model will be developed for the log-price vector $Y_t = \log X_t$. It will be assumed that at each time t, Y_t is multivariate normal. Hence the components of X_t are log-normal, and our model is a discrete-time generalisation of the Black and Scholes model for stock price behaviour, allowing for noise and dynamic stochastic variation in interest rates. More precisely, we model Y_t according to the following multivariate dynamic linear model.

$$\begin{aligned} Y_t &= \mu_t + \nu_t, & \nu_t &\sim N(0,V) \\ \mu_t &= \mu_{t-1} + \tau_{t-1} + \omega_t, & \omega_t &\sim N(0,W) \\ \tau_t &= \tau_{t-1} + \zeta_t, & \zeta_t &\sim N(0,Z) \end{aligned}$$

Here μ_t is a vector representing the "true" underlying log-price at time t, τ_t is a vector representing the "trend" in the log-price at time t, and ν_t, ω_t and ζ_t are independent "noise" vectors with covariance matrices V, W and Z respectively, which will typically be unknown. The model is initialised with the prior specification $(\mu_0', \tau_0')' \sim N(m, C)$ for some vector m and matrix C, which is typically taken to be fairly diffuse. Kalman filtering and smoothing (retrospective estimation) techniques allow estimation of μ_t and τ_t for given V, W and Z, as discussed in West & Harrison (1997). These local computation techniques also have a Bayes linear interpretation, which is discussed in the context of sales forecasting in Goldstein & Wilkinson (2000). Implementation details are discussed in Wilkinson (1998).

In the case of unknown V, W and Z, the problem of fully Bayesian inference becomes less straightforward, but a range of techniques are possible. The simplest approach is to adopt independent (semi-conjugate) inverse Wishart priors for V, W and Z. One can then use the Kalman smoothing techniques indicated above as the E-step of an EM algorithm which locates the joint marginal posterior mode for V, W and Z. Such techniques are are straightforward extension of those described in de Jong (1989) and Koopman (1993). This algorithm also has an interpretation as an iterative Bayes linear procedure for variance matrix estimation, though this will be explored in detail elsewhere. The EM algorithm may be contrasted with the (more complex) non-iterative approach to Bayes linear covariance matrix inference introduced in Wilkinson and Goldstein (1996) and applied to the problem of sales forecasting in Wilkinson (1997). Such techniques lead to good point estimates for the covariance matrices underlying the DLM, but it is difficult to properly account for full joint posterior uncertainty without adopting a MCMC approach.

A naive Gibbs sampler for the above model performs very poorly due to exceptionally slow mixing and convergence. However, it is relatively straightforward to adapt the Kalman smoother techniques outlined above to produce a simulation-smoother which simulates from the joint posterior of all unknown states given the data together with V, W and Z. This may then be used as the major part of a block MCMC sampler for the DLM. This basic strategy is examined in West & Harrison (1997), and leads to a sampler with greatly improved mixing. If a satisfactory MCMC scheme can be developed, one can simulate directly from the full joint posterior forecast distribution for future (log) prices, automatically integrating over all posterior uncertainty for the states, together with V, W and Z. This distribution can be used directly for optimal portfolio selection.

3 Portfolio selection

3.1 Exploitation of the returns distribution

If an investor has exponential utility for money, $u(x) = 1 - e^{-\lambda x}$, then her utility for the uncertain monetary reward $X \sim N(a, b)$ is given by $u(X) = \mathrm{E}(u(X)) = \mathrm{E}(1 - e^{-\lambda X}) = 1 - e^{-\lambda(a - \frac{1}{2}\lambda b)}$, and so for fixed λ, the investor will maximise her utility by maximising $a - \frac{1}{2}\lambda b$. In traditional portfolio analysis (Markowitz, 1959), attention focuses on the distribution of *returns*, $R = (X_{T+1} - X_T)/X_T$ (denoting the current time by T). It is usually supposed that returns are jointly Gaussian; $R \sim N(\mu, \Sigma)$. Typically the vector μ and matrix Σ are estimated from the data, and then assumed known. If the proportion of wealth invested in each stock is given by the vector α, where $\alpha'\mathbf{1} = 1$, the return on an initial investment of M is $M\alpha' R \sim N(M\alpha'\mu, M^2\alpha'\Sigma\alpha)$, and hence utility is maximised by solving the quadratic program

$$\max \alpha'\mu - \frac{1}{2}\eta\alpha'\Sigma\alpha, \qquad \text{subject to} \qquad \alpha'\mathbf{1} = 1,$$

where $\eta = M\lambda$. Using a Lagrange multiplier, it is easy to show that the optimal α is given by

$$\alpha = \eta^{-1}\Sigma^{-1}\mu + \frac{1 - \eta^{-1}\mathbf{1}'\Sigma^{-1}\mu}{\mathbf{1}'\Sigma^{-1}\mathbf{1}}\Sigma^{-1}\mathbf{1}.$$

There are several problems with this approach. The most fundamental is the lack of dynamic model for price evolution. This problem has been recognised by Quintana & West (1987), who develop a Bayesian dynamic model for price evolution which is applied to the problem of portfolio selection in Quintana (1992). Other problems are related to the estimation risk associated with plug-in estimates (Jorion, 1986).

A radically different approach to portfolio selection is afforded by MCMC. Here, direct simulation from the full joint posterior returns distribution is possible, integrating over all uncertainties in the model. Using a collection of simulated return vectors $\{R^{(i)}\}$, $1 \le i \le N$ (such as successive simulated values for R from an MCMC scheme, after suitable burn-in period), and an investment strategy, α, one can approximate the utility associated with strategy α as follows. For each return vector $R^{(i)}$, the utility associated with the payoff is $1 - e^{-\eta\alpha' R^{(i)}}$, and the sample mean of the collection of such utilities $\{1 - e^{-\eta\alpha' R^{(i)}}\}$ converges to the utility associated with the investment strategy α (this follows from the ergodic theorem and the fact that the utility of a gamble is the expected utility of that gamble). The utility of a given strategy, α, can therefore be evaluated directly, and maximised with respect to α using standard numerical optimisation techniques. This is a computationally intensive procedure, but the costs for the optimisation stage are small compared to the cost of carrying out the MCMC run in the first place.

3.2 Obtaining the returns distribution from the DLM

Given the modelling framework elucidated in section 2, and data $x_1, x_2, \ldots, x_T$, a suitably implemented MCMC scheme will generate, *inter alia*, simulated values from the joint marginal posterior distribution for parameters V, W, μ_T, and τ_T. Denoting the collection of simulated values by $\{V^{(i)}, W^{(i)}, \mu_T^{(i)}, \tau_T^{(i)}\}$, $1 \leq i \leq N$, values can be simulated from the returns distribution using the fact that $Y_{T+1}|\cdot \sim N(\mu_T + \tau_T, V + W)$, and then that $R|\cdot = (\exp(Y_{T+1}) - x_T)/x_T$ (using the usual convention that $|\cdot$ denotes conditioning on all data and other parameters). This gives a collection of simulated return vectors $\{R^{(i)}\}$ which can be used as the basis of a portfolio selection strategy as described in section 3.1. Note that this returns distribution is non-Gaussian, invalidating the presented assumptions underlying the quadratic programming approach to portfolio selection, and hence a numerical optimisation procedure should be used. It should be noted however, that there are justifications for the Markowitz approach which are not dependent on a Gaussian returns distribution.

4 Example application

By way of illustration we shall consider a simple portfolio consisting of four U.S. retail banks: CountryWide Credit, Capital One Finance, Bank America and First Union. Their shares are traded on the NYSE and the daily closing prices of the 252 trading days in 1998 are used in our analysis. The Kalman filter/simulation-smoother described in section 2 is used to obtain simulated values from the joint posterior distribution of the state given the data, V, W and Z. These simulated values are then used as part of a MCMC scheme which also simulates new values for the parameter matrices V, W and Z. Simulated values from the joint posterior forecast distribution are produced at each iteration of the sampler, integrating over posterior uncertainty for V, W and Z. Running the sampler for 50,000 iterations and dropping the first 10,000 generally provides an approximate sample from the posterior distribution which satisfies standard MCMC convergence diagnostics. Semi-conjugate inverse Wishart priors were adopted for V, W and Z. The prior specification used for V and W was fairly weak, but that for Z was quite strong ($df = 30$), in order to keep the variability of the trend component of the model small. It is interesting to note that the variance of the state evolution (W) turns out to be greater than that for the observational noise (V), as is the case with many financial time series, in sharp contrast to the situation in many engineering applications of DLMs.

Figure 1 shows the time series plot of all four banks and the 5, 50 and 95 percentiles for the 50-step ahead marginals of the posterior forecast distribution. The vectors of simulated posterior future values were used to calculate vectors of returns for one- and fifty-step ahead forecasts, based on the last known data point. These returns were then used to calculate optimal portfolio weightings for varying levels of the risk aversion parameter, η. Figure 2 shows plots for four different scenarios. Note that negative values of α correspond to the adoption of short positions (selling shares you don't own). Plot A shows the results of simple optimisation using the Markowitz model, based

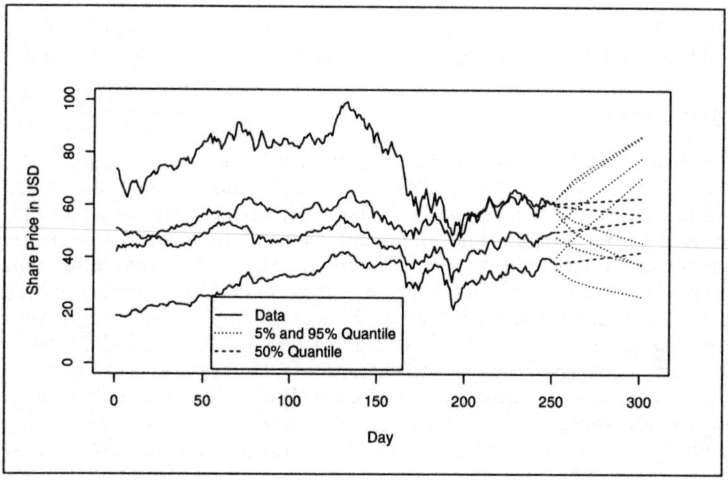

Fig. 1. Observed prices and pointwise predictive intervals for the four banks

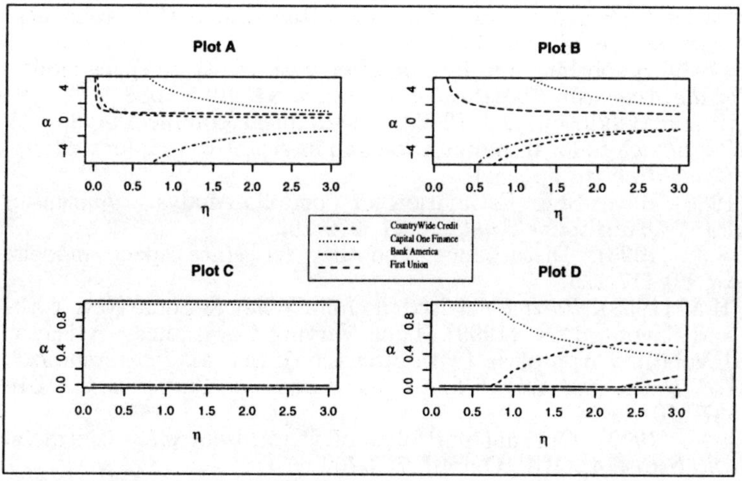

Fig. 2. Plot of portfolio weights for varying η

on observed returns. This gives extreme short positions for risk-neutral individuals. Plot B shows unrestricted optimisation of utility based on the joint one-step ahead forecast distribution. This again shows short positions being taken by risk-neutral individuals, though these are different to the Markovitz portfolios, due to more sophisticated modelling process. Plot C shows the effect of prohibiting short sales within the utility maximisation of the one-step ahead problem. Here no short positions are taken, and all individuals put all of their assets into Capital One Finance, which has the highest expected return. Plot D shows the optimal portfolios based on the fifty-step ahead forecast distribution. Here a more balanced portfolio is adopted by risk averse individuals, emphasising the fact that the optimal portfolio depends on the time frame adopted for re-balancing, in response to the differing volatilities.

5 Conclusions

This paper has illustrated a fully Bayesian approach to the multivariate forecasting of financial time series based on DLM theory. It has also been demonstrated that this can lead naturally to a decision-theoretic approach to portfolio selection and optimisation. This has a number of advantages over more simplistic approaches based directly on observed returns. In particular, this approach allows for dynamic changes in the price evolution process which is very important for accurate forecasting. The resulting framework is very flexible in the way it may be adapted and extended to accommodate additional structure as required. Obvious extensions of this work would allow for dynamic evolution of the variance matrices underlying the model. However, without careful prior specification and large amounts of data, problems due to the very weakly identifiable nature of the resulting model are likely to be considerable.

References

Board, J.L.G. and Sutcliffe, C.M.S. (1994). Estimation Methods in Portfolio Selection and effectiveness of Short Sales Restrictions:U.K. Evidence. *Management Science*, **40**, 516-534.

de Jong, P. (1989) Smoothing and interpolation with the state space model. *Journal of the American Statistical Association*, **84**, 1085-1088.

Goldstein, M. and Wilkinson, D.J. (2000). Bayes linear graphical models: A geometric approach to local computation and interpretive graphics. *Statistics and Computing* (to appear).

Jorion, P. (1986). Bayes-Stein Estimation for Portfolio Analysis. *Journal of Financial and Quantitative Analysis*, **21**, 279-292.

Koopman, S.J. (1993). Disturbance smoother for state space models. *Biometrika*, **80** 117-126.

Markowitz, H.M (1959). *Portfolio Selection*, John Wiley & Sons, New York.

Pitt, M.K. and Shephard, N. (1999). Time Varying Covariances: A Factor Stochastic Volatility Approach, (with discussion). *In J.M. Bernardo, J.O. Berger, A.P. Dawid and A.F.M. Smith (eds.) Bayesian Statistics*, **6**, OUP, Oxford, 547-570.

Quintana, J.-M. (1992). Optimal portfolios of forward currency contracts. *Bayesian Statistics*, **4**, OUP, Oxford, 753-762.

Quintana, J.-M. and West, M. (1987). Multivariate time series analysis: new techniques applied to international exchange rate data. *The Statistician*, **36**, 275-281.

West, M. and Harrison, J. (1997). *Bayesian Forecasting and Dynamic Models*, Second edition, Springer-Verlag, New York.

Wilkinson, D.J. and Goldstein, M. (1996). Bayes Linear Adjustment for Variance matrices. *Bayesian Statistics*, **5**, OUP, Oxford, 791-799.

Wilkinson, D.J. (1997). Bayes linear variance adjustment for locally linear DLMs. *Journal of Forecasting*, **16**, 329-342.

Wilkinson, D.J. (1998). An object-oriented approach to local computation in Bayes linear belief networks. In: *R. Payne and P. Green (editors) Proceedings in Computational Statistics 1998*, Physica Verlag, Heidelberg, 491-496.

An algorithm for the multivariate Tukey median

Anja Struyf[1,2] and Peter J. Rousseeuw[1]

[1] Department of Mathematics and Computer Science, University of Antwerp (U.I.A.), Universiteitsplein 1, B-2610 Wilrijk-Antwerpen, Belgium,

[2] Research Assistant with the FWO, Belgium.

Abstract. The halfspace location depth of a point $\boldsymbol{\theta}$ relative to a data set X_n is defined as the smallest number of observations in any closed halfspace with boundary through $\boldsymbol{\theta}$. As such, halfspace depth can be seen as a kind of multivariate ranking. The deepest location, i.e. the $\boldsymbol{\theta}$ with maximal halfspace depth, is a multivariate generalization of the median. Until now the deepest location could only be computed for bivariate data. In this paper, we construct an algorithm (called DEEPLOC) to approximate the deepest location in higher dimensions.

Keywords. Algorithm, Location depth, Multivariate median

1 Location depth

Tukey (1975) proposed the *halfspace depth* as a tool to visually describe bivariate data sets. In p dimensions, the halfspace location depth of a point $\boldsymbol{\theta}$ relative to a data set $X_n = (\mathbf{x}_1, \ldots, \mathbf{x}_n) \in \mathbb{R}^{p \times n}$ is denoted as $ldepth(\boldsymbol{\theta}; X_n)$. It is defined as the smallest number of observations in any closed halfspace with boundary through $\boldsymbol{\theta}$. In the univariate setting ($p = 1$), this definition becomes

$$ldepth_1(\theta; X_n) = \min\,(\#\{x_i \leq \theta\}, \#\{x_i \geq \theta\}).$$

The multivariate ldepth can then be seen as the smallest univariate ldepth of $\boldsymbol{\theta}$ relative to any projection of the data set onto a direction $\boldsymbol{u}$, since

$$ldepth(\boldsymbol{\theta}; X_n) = \min_{\|\boldsymbol{u}\|=1} \#\{i; \boldsymbol{u}'\mathbf{x}_i \leq \boldsymbol{u}'\boldsymbol{\theta}\}.$$

In words, $ldepth(\boldsymbol{\theta}; X_n)$ says how deep $\boldsymbol{\theta}$ lies in the data cloud. Several properties of the multivariate halfspace depth can be found in Donoho and Gasko (1992). Exact algorithms exist to compute the location depth of a point relative to a bivariate data set (Rousseeuw and Ruts 1996; Al-Furiah, Johnson and Ranka 1998). For higher dimensions, exact and approximate algorithms for $ldepth(\boldsymbol{\theta}; X_n)$ were proposed in (Rousseeuw and Struyf 1998).

Location depth is often described as multivariate ranking: points near the boundary of the data set have low ranks, and the rank increases as one gets deeper inside the data cloud. This can be visualized by means of the *ldepth regions* D_k given by

$$D_k = \{\boldsymbol{\theta} \in \mathbb{R}^p; ldepth(\boldsymbol{\theta}; X_n) \geq k\}.$$

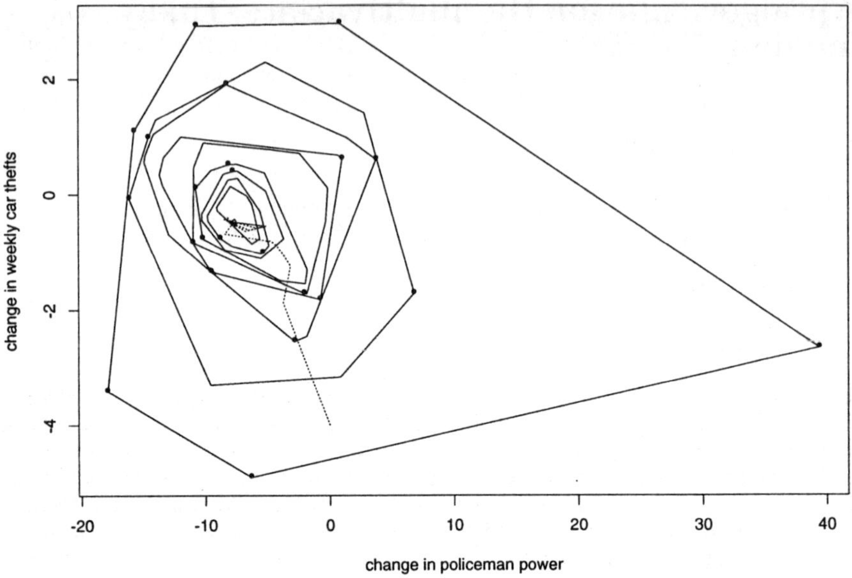

Fig. 1. Ldepth contours and algorithm steps (starting from a point with ldepth 1) for the New York crime data set.

All these ldepth regions are convex sets, and for each ldepth k it holds that $D_k \subseteq D_{k-1}$. The region D_1 equals the convex hull of X_n. The boundary of an ldepth region is called an *ldepth contour*. For bivariate data sets, ldepth contours can be computed with the algorithms of Ruts and Rousseeuw (1996) and Johnson, Kwok and Ng (1998). In Figure 1, the ldepth regions of the NY crime data set are drawn. This data set was taken from the DASL library (available at http://lib.stat.cmu.edu/DASL/) and gives the change (in percent) in policeman power and in weekly car thefts in New York City for 23 time periods in 1966-1967. Note that this data set contains one outlying observation, which only influences the outermost ldepth contour. The smallest ldepth region D_{k^*} (which is D_{10} in the example) contains all the points with maximal ldepth k^*. The center of gravity of this region is unique and has the maximal location depth, and can be seen as a generalization of the univariate median to higher dimensions. This location estimator is called the *Tukey median* or the *deepest location* and denoted as T^*. In two dimensions, the Tukey median can be computed in $O(n^2 \log^2 n)$ time with the algorithm of Rousseeuw and Ruts (1998). In this paper, we will construct an algorithm that computes an approximation for the Tukey median in every dimension. In Figure 1, the dashed line indicates the way through the data cloud which our algorithm follows to reach D_{10} starting from the point with coordinates $(0, -4)$. How the algorithm works is described in the next section.

A very useful property of halfspace depth is its *affine invariance*: for any affine transformation $g : \mathbb{R}^p \to \mathbb{R}^p : \mathbf{x} \to A\mathbf{x} + \boldsymbol{b}$, with $\boldsymbol{b} \in \mathbb{R}^p$ and $A \in \mathbb{R}^{p \times p}$ a nonsingular matrix, it holds that $ldepth(g(\boldsymbol{\theta}); g(X_n)) = ldepth(\boldsymbol{\theta}; X_n)$. Therefore the Tukey median is *affine equivariant*, i.e. $T^*(g(X_n)) = g(T^*(X_n))$. A major motivation for using the Tukey median as a multivariate location estimator is its robustness, which can be measured by means of the *breakdown value* ε^*. Donoho and Gasko (1992) have proved that $\varepsilon^*(T^*; X_n) \geq \frac{n}{p+1}$ for any sample in general position. This means that we can replace $\frac{n}{p+1}$ observations from X_n without moving the resulting Tukey median T^* outside of a bounded region. When the original data set X_n was drawn from any angularly symmetric distribution, and in particular from an elliptically symmetric distribution, the breakdown value $\varepsilon^*(T^*; X_n)$ tends to 1/3 in any dimension. In other words, when at least 67% of the data points come from such a distribution then the deepest location remains reasonable, no matter what the other data points are like. Very recently, Bai and He (1999) obtained the asymptotics of the Tukey median in arbitrary dimension.

Other notions of location depth and multivariate medians have been proposed. For a survey with applications see Liu, Parelius and Singh (1999).

2 Computing the deepest location in p dimensions

2.1 Algorithm

A naive algorithm for the p-variate Tukey median can easily be formulated, but it would be very time consuming. It can be proved that every ldepth region D_k is the intersection of hyperspaces whose boundary hyperplanes are connecting p observations. Hence there are $C(p, C(p, n)) = O((n^p)^p) = O(n^{2p})$ possible vertices of ldepth regions. Computing the exact ldepth of each candidate vertex requires $O(n^{p-1} \log n)$ time with the algorithm of Rousseeuw and Struyf (1998). Constructing the deepest contour D_{k^*} in this way would therefore take $O(n^{3p-1} \log n)$ time. For bivariate data sets ($p = 2$) the naive approach would take $O(n^5 \log n)$ time, but Rousseeuw and Ruts (1998) have constructed the program HALFMED which computes the exact bivariate deepest location in $O(n^2 \log^2 n)$ time.

Table 1 lists execution times of the algorithm HALFMED and our new approximate algorithm DEEPLOC, both on a Sun SparcStation 20/514. The data sets were generated from the bivariate standard normal distribution. The table also lists the exact halfspace depth of the location estimate found by each of the programs. Since DEEPLOC only works with a subset of directions $\boldsymbol{u}$ to approximate the depth, also this approximate depth is included. We see that the approximate depth is very close to the exact depth. For real data sets, DEEPLOC performs equally well.

In this section we will explain the basic ideas of the algorithm DEEPLOC for higher dimensions. More details can be found in Struyf and Rousseeuw (2000). DEEPLOC starts from an initial point, and then takes steps in carefully selected directions to increase the ldepth. After several steps, it approaches the deepest location.

To reduce the number of steps that must be taken, it is best to take an initial point which is already centrally located. For this purpose we use the coordinatewise median $M_1 = (\operatorname*{med}_{i=1}^{n} x_{i1}, \ldots, \operatorname*{med}_{i=1}^{n} x_{ip})$ since this location estimator is easy to compute, and usually has a fairly good halfspace depth relative to the data set. If one insists that the starting point for an affine equivariant estimator should itself be affine equivariant (which M_1 is not), we may as well start from the sample mean (which is). Since the ldepth has convex contours, there is no risk of ending in a suboptimal local maximum.

After computing a starting point, we construct m directions $\boldsymbol{u} \in \mathbb{R}^p$ with $||\boldsymbol{u}|| = 1$ which are randomly drawn from 4 predetermined classes of direc-

n	HALFMED time	HALFMED exact depth	DEEPLOC time	DEEPLOC approx. depth	DEEPLOC exact depth
20	0.13	8	0.4	8	8
40	1.3	17	2.1	17	17
60	3.9	27	3.3	27	27
80	10	36	4.5	36	35
100	20	45	5.7	45	44
150	65	70	8.8	70	70
200	150	90	12	90	89
300	580	140	19	141	140
400	1558	189	27	189	189
500	3057	240	32	240	240

Table 1. Comparison of execution times (in seconds) and the depth of the location estimate found by HALFMED and DEEPLOC, for different sample sizes n.

tions. By default $m = 1000$ directions are computed, but this number may be changed by the user, for example to let m increase with p. The univariate ldepth of M_1 relative to the projection of X_n on each of these m directions is then computed, and the set U_{move} of directions $\boldsymbol{u}$ which yield the same lowest $\#\{i; \boldsymbol{u}'\mathbf{x}_i \leq \boldsymbol{u}'M_1\}$ is stored. We consider these as directions in which the ldepth might still improve. We will try to increase the univariate ldepth relative to these directions, in order to improve the ldepth relative to the set of all directions. For this we compute the average $\boldsymbol{u}_{move}$ of the directions that yielded the smallest ldepth, and make sure that $\boldsymbol{u}_{move} \neq \mathbf{0}$. This moving direction is computed as

$$\boldsymbol{u}_{move} = \frac{1}{|U_{move}|} \sum_{\boldsymbol{u} \in U_{move}} \boldsymbol{u}.$$

Then we take a step away from M_1 in the direction $\boldsymbol{u}_{move}$. Intuitively, this technique is similar to the idea of the steepest ascent. The ldepth that is attained by the deepest location relative to X_n must be at least $\lceil \frac{n}{p+1} \rceil$ (see Donoho and Gasko 1992). If $ldepth_1(M_1; \boldsymbol{u}'_{move}X_n) < \lceil \frac{n}{p+1} \rceil$ then we take a step exactly large enough to reach a point M_2 which has univariate ldepth $\lceil \frac{n}{p+1} \rceil$ in direction $\boldsymbol{u}_{move}$. Otherwise, we take a step such that the univariate ldepth of the resulting point M_2 in direction $\boldsymbol{u}_{move}$ is 1 unit larger than the ldepth of M_1 in the same direction. Step 2 repeats this procedure, starting from M_2. We iterate until a point M_l has been found with maximal ldepth $n/2$, or until the algorithm has not been able to improve the ldepth during the previous *ntry* iteration steps.

Sometimes the algorithm starts wavering instead of moving towards the deepest location along a straight path. To avoid spending too much time by moving back and forth, we have included some extra features. If no ldepth improvement has been made during *nalt* consecutive steps then we consider some other directions, namely the directions connecting some previous M_j to the current M_l.

The default values for the constants *ntry* and *nalt* used in this algorithm are $10(p+1)$ and $4(p+1)$ respectively. These values were chosen empirically, based on their good performance on our data sets (we used real data sets as well as generated samples).

Remark. We have chosen to keep the same set of directions $\boldsymbol{u}$ throughout the algorithm, instead of generating other directions after each step. One reason for this is to save computation time, and another is to avoid fluctuations of the ldepth estimate between two steps due to evaluating other directions.

2.2 Time complexity and number of iteration steps

The algorithm described in the previous section has a time complexity of

$$O(kmn \log n + kpn + mp^3 + mpn)$$

where k is the number of steps taken by the program. This is very good, because it only goes up like $n \log n$ for fixed k and m. Further on, we will see that it is reasonable to impose an upper bound on k, which has little effect on the accuracy.

Table 2 gives the actual times for several data sets, on our Sun SparcStation. Each time is the average over 3 different data sets for that combination of n and p. All data sets were generated from a standard normal distribution $N_p(\boldsymbol{0}, \boldsymbol{I})$, with outliers coming from the normal distribution with mean vector $\mu = (10, \ldots, 10)$ and covariance matrix $\Sigma = 3\boldsymbol{I}$. It suffices to generate the good data according to the unit covariance matrix because the results are the same if we would generate them from any other covariance matrix (i.e., correlations between the variables) due to the affine equivariance of the Tukey median. The table confirms that the computation time increases with n and

n	p	0% outliers	20% outliers	40% outliers
100	2	5.7	6.0	6.2
	3	8.2	8.2	12
	4	9.0	9.7	11
	5	13	12	17
1000	2	77	90	152
	3	116	141	93
	4	121	143	108
	5	136	177	142

Table 2. Computation time of DEEPLOC (in seconds) for randomly generated data sets of size n in dimension p, with and without outliers.

p, while the presence of outliers does not seem to influence the time very much. However, due to the factor k in the time complexity formula, which can vary substantially across data sets, it is difficult to see a pattern in the table. For example, the larger time for $n = 1000$, $p = 2$ and 40% outliers is due to one data set which required twice as many steps as the other data sets with the same characteristics. A lot of additional iteration steps usually lead to only minor increments of the ldepth. The largest ldepth improvements typically occur during the first few steps of the algorithm.

Therefore, it is a good idea to impose a limit *nstp* on the number of steps. We have tried to find an appropriate limit by looking at several generated data sets (gaussian as well as contaminated data), and the number of steps they required before the algorithm stopped. We found empirically that after

$nstp = 5n^{0.3}p$ steps most of the runs have reached the final ldepth. The actual worst-case time complexity of our algorithm then becomes

$$O(5n^{0.3}(mpn \log n + p^2 n) + mp^3 + mpn).$$

Very often the number of steps is limited by the other upper bound (*ntry* steps without ldepth improvement), and in those cases the computation time is less. Of course, these stopping rules are only suggestions, and may be changed by the user. The contaminated data sets seem to require more steps, whereas the main increments still take place during the first few steps. If the user wants to have more confidence about reaching the exact deepest location, we suggest to enlarge (for example, to double) the step limit, thereby increasing the computation time by at most that factor.

When we impose the step limit $nstp = 5n^{0.3}p$ on the DEEPLOC runs in Table 2, only for 3 data sets a (slightly) lower ldepth was obtained. For 20% of outliers, the ldepth was 1/1000 lower for a data set with $p = 3$. For 40% outliers, we lost 5/1000 and 6/1000 of ldepth for two bivariate data sets. All other ldepths remained the same, whereas the computation time was always lowered (on average it decreased by 10%).

The complete Fortran code of DEEPLOC can be downloaded from our website at `http://win-www.uia.ac.be/u/statis/index.html`, questions and remarks about the program can be directed to `astruyf@uia.ua.ac.be`.

References

Al-Furiah, I., Johnson, T. and Ranka, S. (1998). Calculating robust depth measures for large data sets. *DIMACS Workshop on External Memory Algorithms and/or Visualization.*

Bai, Z.-D. and He, X. (1999). Asymptotic distributions of the maximal depth estimators for regression and multivariate location. *Annals of Statistics,* **27**, to appear.

Donoho, D.L. and Gasko, M. (1992). Breakdown properties of location estimates based on halfspace depth and projected outlyingness. *The Annals of Statistics*, **20**, pp. 1803–1827.

Johnson, T., Kwok, I. and Ng, R. (1998). Fast computation of 2-dimensional depth contours. *Proc. 4th Int. Conf. on Knowledge Discovery and Data Mining*, pp. 224–228.

Liu, R.Y., Parelius, J. and Singh, K. (1999). Multivariate analysis by data depth: descriptive statistics, graphics and inference. *The Annals of Statistics*, **27**, pp. pp. 783–840.

Rousseeuw, P.J. and Ruts, I. (1996). Algorithm AS 307: Bivariate location depth. *Applied Statistics (JRSS-C)*, **45**, pp.516–526.

Rousseeuw, P.J. and Ruts, I. (1998). Constructing the bivariate Tukey median. *Statistica Sinica*, **8**, pp. 827–839.

Rousseeuw, P.J. and Struyf, A. (1998). Computing location depth and regression depth in higher dimensions. *Statistics and Computing*, **8**, pp. 193–203.

Ruts, I. and Rousseeuw, P.J. (1996). Computing depth contours of bivariate point clouds. *Computational Statistics and Data Analysis*, **23**, pp. 153–168.

Struyf, A. and Rousseeuw, P.J. (2000). High-dimensional computation of the deepest location. *Computational Statistics and Data Analysis*, to appear.

Tukey, J.W. (1975), Mathematics and the picturing of data. In: *Proceedings of the International Congress of Mathematicians*, **2**, pp. 523–531, Vancouver.

Analyzing and synthesizing information from a multiple-study database

The Diverse Populations Collaborative Group[1]

Keywords: Permutation Tests, Bootstrap Analysis, Data Mining, Computer-Intensive Methods

1 Introduction

An important public-health issue receiving much discussion in the literature today concerns whether there is an ideal relative weight associated with minimum mortality for all populations. Standards developed by WHO and U.S. organizations propose that ideal weights be determined by Body Mass Index, a single measure incorporating weight and height (BMI=$\frac{wt(kg)}{ht(m)^2}$). They recommend BMIs of ≤ 25 for all. To better understand the relationship between BMI and mortality in diverse populations and to discern whether a

	N	Deaths	CHD Deaths	Age Range	Baseline
NHIS	144,143	11,765	3,138	18-90	1988
NHANES I	14,235	4,566	1,416	24-77	1973
NHANES II	9,089	2,119	623	30-75	1978
Framingham	4,541	2,775	716	34-69	1958
Glostrup	10,198	1,101	275	29-80	1985
Honolulu	8,006	2,466	419	45-68	1965
Iceland	18,912	4,753	1,509	33-81	1968
Israel	10,059	3,473	1,097	39-74	1963
LRC	8,252	955	332	30-97	1972
Norway	48,646	3,576	1,009	35-49	1976
Puerto Rico	9,824	1,737	376	35-79	1965
Renfrew-Paisley	15,411	4,447	1,571	45-64	1974
Scottish Collaborative	7,011	2,080	758	21-75	1971
Tecumseh	4,864	1,001	362	18-92	1959
Yugoslavia	6,459	1,340	226	34-83	1964
HDFP	10,940	1,424	494	30-69	1973
LRC-CPPT	3,617	283	134	34-60	1972
MRFIT	11,559	933	397	35-58	1974
Total	345,766	50,794	14,852		

Table 1.: Studies comprising the diverse populations collaboration

single BMI of minimum mortality is appropriate, we examine person-level data from 18 studies.

Table 1 presents information on the 18 studies, including national samples, cohort studies with mortality followup, and clinical trials. Studies include both genders, African-American and Hispanic subgroups, and represent several different countries. Baselines begin in 1948 and continue through 1991, age groups vary, and participants were followed from 6 to 30 years. BMI information is available for all participants, but is based on self-reported weight

[1] A list of investigators is provided at the end of this paper. Correspondence: Daniel L. McGee, Ph.D., Department of Preventive Medicine and Epidemiology. Loyola University Medical Center, 2160 S. First Ave., Maywood, IL 60153, USA. E-mail: dmcgee@luc.edu

and height for one large dataset. Our final analytic sample contains almost 350,000 participants and over 50,000 deaths.

Most studies contain natural subgroups, including gender, race or ethnicity, area of residence, and treatment status for clinical trials. These subgroups are analyzed separately and referred to as "cohorts". The 18 studies provide 48 distinct cohorts.

A list of investigators and summary information on the studies (with references) are found at the collaboration website: *azteca.medctr.luc.edu/Collaborative.*

2 Assembling and analyzing data from the multiple studies

Our information derives from public access datasets and special datasets prepared by the collaborators. For the former, we used the formats provided by the sources, and for the latter, we tried to minimize the effort required to transmit information and requested only raw data involving necessary characteristics. Collaborators used whatever medium they found most convenient, providing floppy disks, tapes, CD roms, and information via FTP. Data formats varied also, from SAS and ASCII (flat) files to an Excel spreadsheet.

To develop a data warehouse for multiple studies, we needed to manage the documentation and translate variables to uniform formats. The large public datasets containing many variables require complex documentation, while the smaller datasets require a much simpler approach. Some public agencies offer documentation electronically, but many datasets still require the use of hardcopy instructions. To simplify access and reduce the need for multiple copies of documentation, we post as much information as possible on the World Wide Web.

We developed uniform storage methods using SAS to incorporate the datasets into the warehouse and establish files containing the information for our analyses. English measurements were converted to metric, to solve one of the simpler transformation problems. Smoking status transformations were difficult, in that the studies had developed their own coding systems for this variable, and many contained missing values. We have not been able to develop a mechanical procedure to overcome all of these problems, which makes the data transformation process labor intensive and time-consuming.

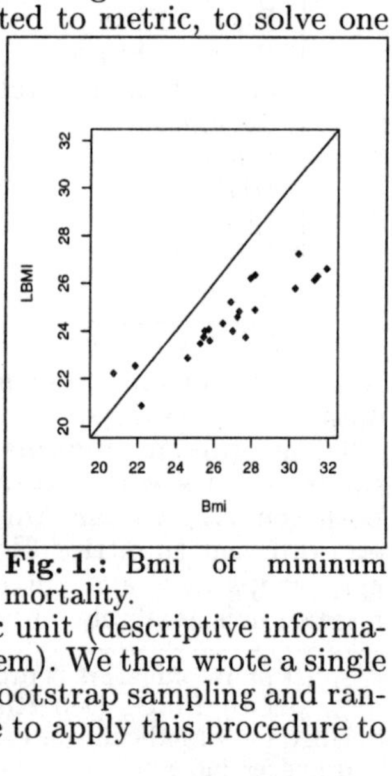

Fig. 1.: Bmi of mininum mortality.

To conduct our analyses, we created a single SAS file containing information from the 18 studies, which included variables to identify the studies and the specific cohorts from which the observations were derived. While using the SAS system to manage our data, we built an analytic capability using the statistical package STATA to conduct computer-intensive examinations of analytic methods. We first created a file of "meta-data," containing summary information on each analytic unit (descriptive information, including the location of data within the system). We then wrote a single program to perform various functions, including bootstrap sampling and randomization tests, and used the summary database to apply this procedure to all units.

In this report we present analyses to determine the shape of the BMI-mortality relationship for each cohort. For those cohorts that exhibit quadratic relationships, we also determine the BMI associated with minimum mortality.

3 Analytic algorithms

Analysis of the BMI-mortality relationship exhibits increases at both high and low levels of BMI. To model a non-monotonic relationship, some investigators have included a quadratic term in the model. This formulation, however, forces the relationship to be symmetric about a value, BMI_{min} (BMI of minimum mortality). But most studies that exhibit a non-monotonic relationship display asymmetry, and this can produce anomalous results when the model requires symmetry. When a symmetric model is used on an asymmetric relationship, the minimum may occur outside the range of available data or may be inconsistent with observed data (Goetghebeur 1995, Durazo–Arvizu 1997). We have demonstrated that data transformation provides one solution to the problem of asymmetry.

According to Cornfield (1961), a quadratic logistic model is needed when the independent variable is distributed normally with different means and variances for those who die and those who do not. When we required a transformation to meet these conditions, we often found it to be $LBMI = \frac{1}{BMI}$. (Some investigators suggest this to be a biologically meaningful measure of body fat (Nevill 1995)). We used this transformation in the following proportional-hazards model:

$$\begin{aligned}\lambda(t, LBMI, Age, Smoking) = \lambda_0(t) * \exp\{&\beta_1 Age + \beta_2 Smoking + \\ &\beta_3 LBMI + \beta_4 LBMI^2 + \\ &\beta_5(Smoking * LBMI) + \beta_6(Smoking * LBMI^2)\} \quad (1)\end{aligned}$$

We required an algorithm to select a best model to apply the *BMI*-mortality relationship uniformly to all our data. The choice of the model was a function of which parameters in Equation 1 were zero.

For our analyses, we assume that $\beta_1 \neq 0$ and $\beta_2 \neq 0$. Then, we fit four models to each cohort:

(i) $\beta_3 = \beta_4 = \beta_5 = \beta_6 = 0$ (iii) $\beta_3 \neq 0, \beta_4 \neq 0; \beta_5 = \beta_6 = 0$

(ii) $\beta_3 \neq 0; \beta_4 = \beta_5 = \beta_6 = 0$ (iv) $\beta_3 \neq 0, \beta_4 \neq 0, \beta_5 \neq 0, \beta_6 \neq 0$

Using generalized likelihood ratio statistics, we then placed each cohort into one of five mutually exclusive categories. (Model ii signifies a direct or an inverse relationship, depending on the sign of β_3).

Table 2 presents results from the application of our selection algorithm to the 48 cohorts. It includes analyses without transformations and shows only a modest qualitative difference between the use of raw and transformed data.

Results using LBMI	Results using BMI: None	Direct	Quadratic	Interaction	Total
None	8	0	2	0	10
Direct	0	1	0	0	1
Quadratic	2	2	27	3	34
Interaction	0	1	0	2	3
Total	10	4	29	5	48

Table 2.: Analytic results

For those cases in which the quadratic model is considered most appropriate, the estimated BMI of minimum mortality is as follows (provided that $\beta_4 > 0$):

$$BMI_{min} = -\frac{2\widehat{\beta}_4}{\widehat{\beta}_3} \tag{2}$$

When estimating the BMI_{min}, we find a large difference between the use of transformed and raw data. Figure 1 presents the estimated BMI_{min} for these two methods. With two exceptions, the BMI_{min} is much higher for BMI than for $LBMI$. Among studies selecting a quadratic model, the random effects average for the BMI_{min} is 27.8 for BMI and 24.3 for $LBMI$. This difference is large, and the value estimated using BMI is much higher than that suggested by published reports or an empirical examination of the data.

4 Two uses of computer intensive methods

If our algorithm is stable, we can expect similar results from new samples. To test for model stability, we selected 1,000 bootstrap samples (Efron 1982) from each of the 48 cohorts. We then applied our algorithm to determine the best model for each sample. The results of these 1,000 repetitions were tabulated and compared to those using original data. Figure 2 shows the percentage of time that agreement exists between bootstrap and original-data results. With a single exception, the model selected using the original data was the model most often selected using bootstrap samples. For the exception, the original model was quadratic; using bootstrap samples, an interaction model was selected 503 times, a quadratic model, 494 times.

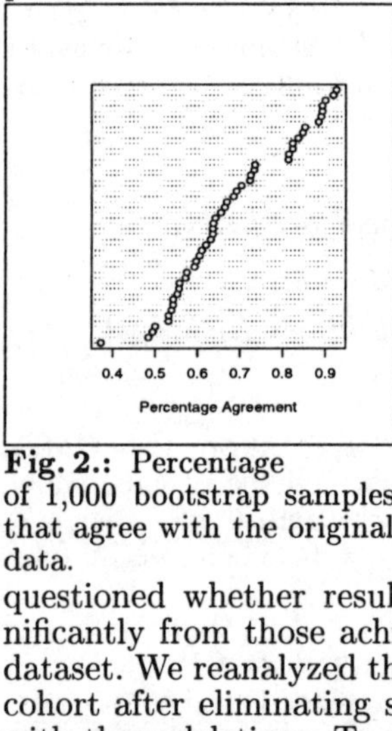

Fig. 2.: Percentage of 1,000 bootstrap samples that agree with the original data.

A common practice when analyzing the relationship between BMI and mortality is to eliminate smokers from the sample prior to analysis. We questioned whether results obtained after such deletions would differ significantly from those achieved after deleting random observations from the dataset. We reanalyzed the relationship between BMI and mortality for each cohort after eliminating smokers and did find that the best model changed with these deletions. To determine whether the exclusions produced significant changes in the relationship, we performed an approximate randomization test. This test randomly assigned smoking status to individuals from each cohort. The number of random participants designated smokers equaled the number of actual smokers within each cohort. These designated smokers were then excluded from the analysis. We conducted 1,000 repetitions of this simulation for all 48 cohorts. For each repetition, we eliminated a percentage of data chosen randomly prior to analysis and then selected the most appropriate model using our standard algorithm.

Table 3 compares outcomes involving the elimination of smokers and the randomization tests to those for the complete datasets. The top half compares results using the entire cohort with those excluding smokers from analysis. The bottom half presents results of the randomization test. For ease of comparison, the numbers are divided by 1,000. A total of 9,000 randomized simulations were conducted on the 9 cohorts that showed no relationship when all the data were included, and the following numbers were obtained: 8,041 agreed with the original finding, 1,071 showed inverse relationships, 63 showed direct relationships, and 825, quadratic relationships. In general, the results produced by random deletions were similar to those produced from the exclusion of smokers.

5 Discussion

We have discussed our data acquisition and management procedures, reported our algorithm for selecting the best fitting model to describe the BMI-mortality relationship, explained our use of bootstrapping to examine model sensitivity and our use of a randomization test to determine whether particular deletions lead to results different from random eliminations.

The results of our randomization test call into doubt the practice of eliminating smokers from analysis. The number of smokers within a cohort is often large, and deletions that may constitute more than half of the sample result in loss of power to discern true relationships. And with the sample having been diminished to such an extent, confusion exists concerning the proper population to which analytical results apply. Smokers themselves vary from study to study and do not share a set of common characteristics, which makes these eliminations from analysis problematic.

	None	Inverse	Direct	Quadratic	Total
	Results Using Original Data				
None	9	0	0	1	10
Direct	0	0	1	0	1
Quadratic	9	0	0	25	34
Interaction	0	0	0	3	3
Total	18	0	1	29	48
	Results After Random Deletions				
None	8.041	1.071	0.063	0.825	10
Direct	0.661	0	0.314	0.025	1
Quadratic	4.762	2.303	0.132	26.803	34
Interaction	0.004	0	0.002	2.994	3
Total	13.468	3.374	0.511	30.647	48

Table 3.: Results (number of cohorts obtained after deleting smokers (top) and from 1,000 simulations randomly deleting the same number of observations as there were smokers (bottom).

In our simulation study, we demonstrated that excluding smokers from analysis produces findings similar to those resulting from an equal number of random exclusions from the dataset. Based on these results, we caution against subsetting data without applying formal statistical procedures to test whether the omissions are biologically meaningful. If the type of random eliminations we performed above produce outcomes that differ from those resulting from the deletion of the specified data, there is evidence that such omissions are biologically meaningful and, hence, justified. Otherwise, regardless of widely accepted beliefs that certain subsets, such as smokers, will confound results, the omissions cannot be justified.

The concept of model uncertainty has gained some attention, particularly in the data-mining field (Glymour 1997). Several investigators have suggested that bootstrap sampling be used to examine the sensitivity of results from the application of various modeling algorithms (Gong 1982, Sauerbrei 1992). Most of the applications, however, involved much smaller numbers than does the present study. To our knowledge, our use of bootstrap sampling represents the largest medical study to examine model uncertainty using this methodology.

Acknowledgements This work was funded by a grant from the National Institutes of Health, DK52329.

References

Cornfield J, Gordon T, and Smith W. (1961). "Quantal respones curves for experimentally uncontrolled variables." *Bulletin of the International Statistical Institute*, **38**, 97–115.

Durazo-Arvizu R, McGee D, Li Z, and Cooper R. (1997). "Establishing the nadir of the body mass index-mortality relationship." *Journal Am Stat Assoc*, **92**, 1312–1319.

Efron B. (1982). The Jackknife, the Bootstrap and Other Resampling Plans. CBMS, **38**. SIAM, Philadelphia.

Glymour C, Madigan D, Pregibon D, Smyth P. (1997). "Statistical Themes and Lessons for Data Mining." *Data Mining and Knowledge Discovery*, **1**, 11–28.

Goetghebeur E and Pocock J. (1995). "Detection and estimation of a J-shaped risk-response relationships." *J.R. Statist. Soc. A*, **158, Part 1**, 107–121.

Gong G. (1982). "Some ideas on using the bootstrap in assessing model variability." In Heiner KW, Sacher RS, and Wilkinson JW, editors, *Computer Science and Statistics: Proceedings of the 14th Symposium on the Interface*, pages 169–173, Springer, New York.

Nevill AM and Holder RL. (1995). "Body mass index: a measure of fatness or leanness." *Bri J Nutr*, **73**, 507–516.

Sauerbrei W and Schumacher M. (1992). "A Bootstrap Resampling Procedure for Model Building: Application to the Cox Regression Model." *Statistics in Medicine*, **11**, 2093–2109.

Investigators, The Diverse Populations Collaboration:

Tecumseh Study: Dr. Victor Hawthorne

The Yugoslavia Cardiovascular Diseases Study: Dr. Djordje Kozarevic, Dr. Nikola Vojvodic

The Renfrew-Paisley Study and the Scottish Collaborative Study: Dr. Charles R. Gillis, Dr. David Hole, Ms. Carole Hart, Dr. Victor Hawthorne

The Glostrup Study: Dr. Torben Jorgensen, Dr. Troels Thomsen, Dr. Michael Davidson

The Israel Study: Dr. Uri Goldbourt, Ms. Shlomit Yaari

The Norwegian Counties Study: Dr. Yngve Haugstvedt, Dr. Aage Tverdal

The Iceland Study: Dr. Emil Sigurdsson

Loyola University Medical Center Ms. Guichan Cao, Dr. Richard Cooper, Dr. Ramon Durazo-Arvizu, Dr. Zhaohi Li, Dr. Youlian Liao, Dr. Daniel McGee, Dr. T. Elaine Prewitt, Dr. Christopher Sempos

Data from the following studies were obtained from the National Heart, Lung, and Blood Institute: The Framingham Heart Study, The Hypertension Detection and Followup Program, the Lipid Research Clinics Prevalence Study, The Multiple Risk Factor Intervention Trial, the Puerto Rico Heart Study, and the Honolulu Heart Study.

Data from the NHANES I, NHANES II, and NHIS studies were obtained from the National Center for Health Statistics.

For these studies the views expressed in this paper are those of the authors and do not necessarily reflect the views of these agencies.

Bootstrapping neural discriminant model

Masaaki Tsujitani[1] and Takashi Koshimizu[2]

[1] Faculty of Information Science & Arts, Osaka Electro-Communication University, Neyagawa, Osaka, 572-8530, Japan
[2] Product Development, Bayer Yakuhin, Ltd., Osaka, 532-8577, Japan

Abstract. In recent years, neural computing techniques have received increasing attention from statisticians. Statistical techniques are formulated in terms of the principle of the likelihood of neural network models, where the connection weights of the network are treated as unknown parameters in classification problems. Some comparisons with standard statistical techniques are included.

Key words: Bootstrap method, EIC, cross-entropy, bias correction, box plots

1 Introduction

In this article, the role of the bootstrap is highlighted for nonlinear discriminant analysis using a feed-forward neural network model when the data consist of a binary response and a set of explanatory variables. We illustrate that the information criterion based on the bootstrap method (Ishiguro et al.,1997) is shown to be favorable when selecting the optimum number of hidden units by thinking of a neural network as an approximation to an underlying model. It may provide a better bias estimate in certain cases, since analytic approximations are not used in the bootstrap method.

Introducing the maximum likelihood principle into the neural network model, the deviance allows us to test the goodness-of-fit of the neural network model. The deviance is, however, not even approximately distributed χ^2 in the case where ungrouped binary responses are available (Collett,1991). The number of degrees of freedom, which is required for the test for significance using the assumed χ^2 distribution for the deviance, is a contentious issue. No adequate distribution theory for the deviance exists. Consequently, the deviance on fitting a model to binary response data cannot be used as a summary measure of the goodness-of-fit test of a model. We can provide bootstrapping on the null distribution of the deviance with a specified significance level.

We additionally propose bootstrap methods for box plots of predicted values in order to identify outliers and examine distributional assumptions in neural network model fitting.

2 Neural discriminant model

A feed-forward neural network with a single hidden layer for two-class classification problems is shown in Fig 1. Each case has input variables $x_1,\ldots,x_I$ representing predictor variables, with the weight to hidden units being the α

terms, and the weights to the output unit being the β terms. The connection weight vector between the i-th unit in the input layer ($i = 1,\ldots I$) and the j-th unit in the hidden layer ($j = 1,\ldots,H$), is $\alpha_j=(\alpha_{0j}, \alpha_{1j},\ldots, \alpha_{Ij})^T$, where α_{0j} is a bias. Similarly, the weight vector connected to the output unit is $\beta=(\beta_0, \beta_1,\ldots, \beta_H)^T$, where β_0 is a bias.

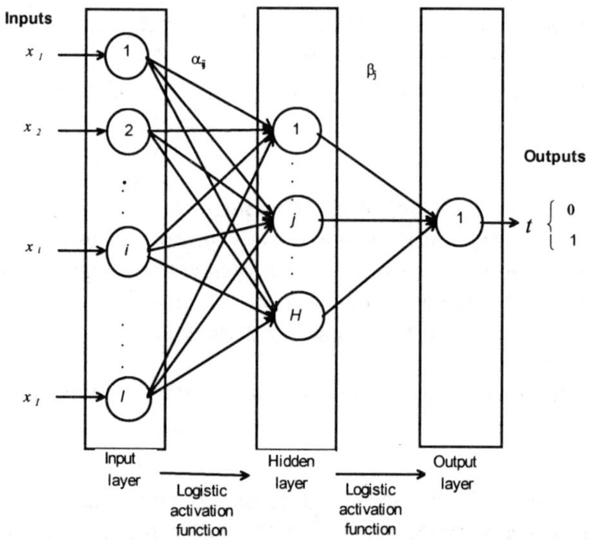

Fig 1 Feed-forward neural network with a single hidden layer

The input to the j-th hidden unit is a linear projection of the input vector $\boldsymbol{x}=(x_1,\ldots,x_I)$, i.e., $u_j = \sum_{i=0}^{I}\alpha_{ij}x_i$, $x_0 = 1$. When the logistic sigmoid function is used as $f(\cdot)$ and $g(\cdot)$, the functional representation of the output from the neural network for a given input vector $\boldsymbol{x}$ with $= (\alpha,\beta)(\{\alpha_{ij}\},\{\beta_j\}$ is

$$o(\boldsymbol{x};\theta) = g\left(\sum_{j=0}^{H}\beta_j\cdot f\left(\sum_{i=0}^{I}\alpha_{ij}x_i\right)\right) = \frac{1}{1+\exp\left(-\sum_{j=0}^{H}\frac{\beta_j}{1+\exp\left(-\sum_{i=0}^{I}\alpha_{ij}x_i\right)}\right)} \quad (1)$$

The number of parameters included in the function (1) is $p=H(I+1)+H+1$.

When an input vector, $\boldsymbol{x}^{<d>}=(x_1^{<d>},x_2^{<d>},\ldots,x_I^{<d>})$, $d=1,\cdots, D$, is given in two-class classification problems, an output of a classifier, $o^{<d>}=o^{<d>}(x;)$, is viewed as an estimator of the conditional probability that $t^{<d>}=1$. The log likelihood of the observed training sample of $\boldsymbol{t}=(t^{<1>},t^{<2>},\ldots,t^{<D>})^{\mathrm{T}}$ and $\boldsymbol{x}=(\boldsymbol{x}^{<1>},\boldsymbol{x}^{<2>},\ldots,\boldsymbol{x}^{<D>})^{\mathrm{T}}$ is written in the form,

$$\ln L(\boldsymbol{x},\boldsymbol{t};\theta) = \sum_{d=1}^{D}\left\{ t^{<d>}\ln o^{<d>} + \left(1-t^{<d>}\right)\ln\left(1-o^{<d>}\right)\right\}. \quad (2)$$

3 Bootstrapping

3.1 Determination of the optimum number of hidden units

In conventional statistics, various criteria have been developed for assessing the generalization performance. The *AIC* (*A*kaike *I*nformation *C*riterion) provides us with a decision as to which of several competing network architectures is best for a given problem. It is, however, pointed out that the use of the *AIC* may not be theoretically justified because the model is misspecified by thinking of a neural network as an approximation to an underlying model (Korn, 1999). We developed a generalization to the *AIC* for a misspecified model by bootstrapping.

Let the (original) training sample $X=\{X^{<1>},X^{<2>},\ldots,X^{<D>}\}$ with $X^{<d>}=(x^{<d>};t^{<d>})=(x_1^{<d>};x_2^{<d>},\ldots,x_I^{<d>};t^{<d>})$ be independently distributed in an unknown probability distribution F, where $(x_1^{<d>};x_2^{<d>},\ldots,x_I^{<d>})$ are I-dimensional explanatory variables and $t^{<d>}$ is a target value on the corresponding individual indicating the presence/absence. Let $\hat{F}$ be the empirical distribution, putting probability $1/D$ on each of the observed values $X^{<1>},X^{<2>},\ldots,X^{<D>}$. Generate B bootstrap samples $X^*=\{X^{<1>*},X^{<2>*},\ldots,X^{<D>*}\}$ from $\hat{F}$. We denote the b-th bootstrap sample by $X^*(b)$, b=1,…,B. For each bootstrap sample, fit a model and estimate parameters $\hat{\theta}\left(X^*(b)\right)$. They are sometimes called *nonparametric bootstrap* estimators because they are based on $\hat{F}$, the nonparametric estimator of the population F (Efron & Tibshirani,1993). The bootstrap pairs sampling algorithm is given as follows:

Step 1. The bootstrap estimator of bias C is given by

$$
\begin{aligned}
C^* &\equiv E_x\left[\ln L\left(X^*;\hat{\theta}\left(X^*(b)\right)\right)-\ln L\left(X;\hat{\theta}\left(X^*(b)\right)\right)\right] \\
&\cong \frac{1}{B}\sum_{b=1}^{B}\left[\ln L\left(X^*(b);\hat{\theta}\left(X^*(b)\right)\right)-\ln L\left(X;\hat{\theta}\left(X^*(b)\right)\right)\right]
\end{aligned}
\tag{3}
$$

where $\ln L(X^*(b);\hat{\theta}(X^*(b)))$ is the log likelihood on the bootstrap sample and $\ln L(X;\hat{\theta}(X^*(b)))$ is the log likelihood on the training sample given $\hat{\theta}(X^*(b))$.

Step2. The bootstrap bias correction of the log likelihood is then $\ln L(X;\hat{\theta}(X))-C^*$, and the *EIC* (*E*xtended *I*nformation *C*riterion) based information criterion is defined by

$$EIC=-2\ln L(X;\hat{\theta}(X))+2C^*. \tag{4}$$

3.2 Bootstrapping the log likelihood test statistic

Introducing the maximum likelihood principle into the neural network model, the deviance allows us to test the goodness-of-fit of the neural network model:

$$Dev=2\left[\ln L(\max)-\ln L(X;\hat{\theta}\right]=$$

$$= -2\sum_{d=1}^{D}\left\{t^{<d>}\ln\hat{o}^{<d>} + (1-t^{<d>})\ln\left(1-\hat{o}^{<d>}\right)\right\} \qquad (5)$$

where $\ln L(\boldsymbol{X};\hat{\theta})$ denotes the maximized log likelihood under some current neural network model, and the log likelihood for the saturated model $\ln L(\max)$ is zero. The degrees of freedom, D-p, equal the difference between the number of independent observations and the number of unknown parameters in the neural network model.

The deviance (5) is, however, not even approximately distributed χ^2 in the case where ungrouped binary responses are available. The number of degrees of freedom, which is required for the test for significance using the assumed χ^2 distribution for the deviance, is a contentious issue. No adequate distribution theory for the deviance exists. Consequently, the deviance on fitting a model to binary response data cannot be used as a summary measure of the goodness-of-fit test of a model.

From the above points of view, we employ bootstrapping the deviance (5) for the goodness-of -fit test:

Step 1. For the bootstrap sample $\boldsymbol{X}^*(b)$, the deviance (5) is computed by $Dev(b)=2[\ln L(\max)-\ln L(\boldsymbol{X}^*(b);\hat{\theta}^*(b))]$. This process is repeated independently B times.

Step 2. The value of the j-th order statistic Dev^* of the B replications can be taken as an estimate of the quantile of order $j/(B+1)$ for a specified significance level α.

Step 3. If the value of deviance (5) for the training sample is greater than the j-th smallest of its B bootstrap replications, the goodness-of-fit test has approximately a size of $\alpha=1-j/(B+1)$.

3.3 Excess error estimation

We can apply a prediction rule on the training sample. Let $e(F;\boldsymbol{X})$ be the error rate, the probability of incorrectly predicting the outcome of a new observation drawn from F, given a prediction rule on a training sample of $\boldsymbol{X}$. It is defined as the *actual error rate*, which is of interest for performance assessment of prediction rules. Let $e(\hat{F};\boldsymbol{X})$ be the error rate, the probability of incorrectly predicting the outcome on the sample drawn from the empirical distribution of the training sample, $\hat{F}$, which is called the *apparent error rate*. The usage of the training sample for both forming and assessing the prediction rule results in underestimating the error rate. The difference,

$$b(F) = E\left[\hat{e}(\hat{F};X) - e(F;X)\right] \qquad (6)$$

is bias, called the *excess error* of a given prediction rule. This excess error must be estimated when the training sample is small relative to the number of unknown parameters.

The bootstrap method can be used to estimate the excess error and actual error rate when fitting neural network models to data of classification problems. The bootstrap sampling algorithm is given as follows:

Step 1. The bootstrap estimate of the excess error in (6) is given by

$$\hat{b}(\hat{F}) = E\left[\hat{e}(\hat{F}^*; X^*) - \hat{e}(\hat{F}^*; X^*)\right] \cong$$
$$\cong \frac{1}{B}\sum_{b=1}^{B}\left\{\hat{e}(\hat{F}^*; X^*(b)) - \hat{e}(\hat{F}; X^*(b))\right\}, \tag{7}$$

where $\hat{e}(\hat{F}^*; \mathbf{X}^*)$ is the error rate on the bootstrap sample and $\hat{e}(\hat{F}; \mathbf{X}^*)$ is the error rate on the training sample given a prediction rule constructed with the bootstrap sample.

Step 2. Using bias correction (7), the bootstrap estimator of the actual error rate can be obtained as

$$\hat{e}(\hat{F}; \mathbf{X}) - \hat{b}(\hat{F}). \tag{8}$$

3.4 Box plots for predicted values

The box plots for predicted values estimated by using the bootstrap technique can be provided as follows:

Step 1. For each bootstrap sample, $b=1,\ldots,B$, maximize the log likelihood function (2) giving $\hat{\theta}_b^* = \left(\{\hat{\alpha}_{ij}^b\}, \{\hat{\beta}_j^b\}\right)$.

Step 2. Estimate the predicted value $\hat{o}^{<d>}(X^{<d>}; \hat{\theta}_b^*)$ for the d-th input vector $(x_1^{<d>}, x_2^{<d>}, \ldots, x_I^{<d>})$ of the training sample $X^{<d>}$ by used of $\hat{\theta}_b^*$. This process is repeated independently B times.

Step 3. Draw box plots that stretches from hinge to hinge, crossing it with a bar at the median. Then a "whisker" is drawn from each end of the box to display the variability of the median.

4 Application

We used the sample of developing kyphosis with three predictor variables, "Age of child", "The number of vertebrae" and "The beginning of the range" (Chambers et al., 1992). An examination of the value 19.59 of deviance (5) with two hidden units fits the data fairly well by comparison with the bootstrapped log likelihood test statistic 24.22 (α=0.05).

Table 1. Bootstrap estimates of the biases C^* and *EIC* values

Number of hidden units	*EIC* (C^*)	Log Likelihood	Number of apparent errors
1	93.88 (5.98)	-40.96	13
2	91.76 (9.76)	-36.12	9
3	95.36 (14.79)	-32.89	5
4	99.33 (20.30)	-29.37	2
5	99.90 (22.10)	-27.85	1

Table 1 summarizes the results after fitting a neural network model with one hidden unit to five hidden units. The optimum number of hidden units is determined as 2, where the number of apparent misclassification is 9. For the purpose of comparison, it is shown that the number of apparent misclassification errors for Fisher's linear discriminant analysis, logistic discriminant analysis and tree-based model (Chambers and Hastie, 1992) are 19,13, and 10, respectively. The bias estimates of the log likelihood, C^*, and the values of *EIC* are also given through refitting to each of B=400 bootstrap samples.

Fig 2 shows the box plots for the predicted values with two hidden units. Patients #16 and #43 have teach value t=0, and #77 has t=1. Patients #43 and #77 are misclassifications.

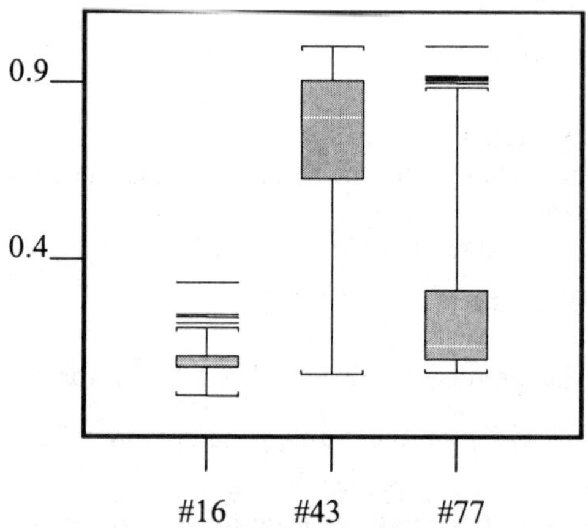

Fig 2 Box plots

References

Chambers, J.M. and Hastie, T.J. (1992). *Statistical Models in S*, Pacific Grove, CA:Wadsworth & Brooks/Cole Advanced Books & Software.

Collett, D. (1991). *Modeling Binary Data*. London:Chapman & Hall.

Efron, B. and Tibshirani, R.J. (1993). *An Introduction to the Bootstrap*. London:Chapman&Hall.

Ishiguro, M., Sakamoto, Y. and Kitagawa, G. (1997). Bootstrapping log likelihood and EIC, an extension of AIC," *Ann. Inst. Statist. Math.*, **49**, pp. 411-434.

Korn, U.A.O. (1999). Model selection in neural networks, *Neural Networks*, **12**, pp. 309-323.

An improved algorithm for robust PCA

Sabine Verboven, Peter J. Rousseeuw and Mia Hubert

Department of Mathematics and Computer Science, University of Antwerp (UIA),Universiteitsplein 1, 2610 Wilrijk, Belgium

Abstract. In Croux and Ruiz (1996) a robust principal component algorithm is presented. It is based on projection pursuit to ensure that it can be applied to high-dimensional data. We note that this algorithm has a problem of numerical stability and we develop an improved version. To reduce the computation time we then propose a two-step algorithm. The new algorithm is illustrated on a real data set from chemometrics.

Keywords. Orthogonal transformations, Principal component analysis (PCA), Projection pursuit (PP), Reflection algorithm

1 Introduction

There is a huge need for algorithms that perform well on high-dimensional data sets. In particular, principal component analysis (PCA) is often used to reduce the number of observed variables, making the data more manageable and interpretable. Classical PCA first rotates the data such that the new coordinate axes (i.e. the principal components) are the principal axes of the covariance matrix S, and then keeps a few components that together already explain a large proportion of the variance. To do this, one computes and ranks the eigenvalues (and the corresponding eigenvectors) of S.

However, it is well-known that the covariance matrix is highly attracted by anomalous observations. It is therefore necessary to apply robust methods that are resistant to outliers.

One approach is to robustify the covariance matrix (Devlin et al. 1975). The disadvantage of such methods is that they can only be applied if the number of observations n is larger than the number of variables p. To overcome this restriction, robust PCA based on projection pursuit (PP) has been proposed by Li and Chen (1985) and implemented by Croux and Ruiz (1996). The idea of PP is to search for the direction on which the projected observations have the largest robust scale. In subsequent steps, each new direction is constrained to be orthogonal to all previous directions. However, in high dimensions (large p) it appears that the latter algorithm has a numerical accuracy problem and consumes much computation time. This will be shown in Section 2. Therefore, in Section 3 we construct a modified version that is numerically much more stable. In Section 4 we introduce a fast two-step algorithm for the case $p > n$. Finally, Section 5 presents an example.

2 The algorithm of Croux and Ruiz

Let X be the $(n \times p)$ data matrix with rank $r \leq \min(n, p)$. The PP algorithm of Croux and Ruiz (1996), from now on called the C-R algorithm, constructs an r-dimensional subspace spanned by orthogonal unit vectors

$v^{(l)}$ $(l = 1, \dots, r)$ such that the robust scale $s^{(l)}$ of the projected observations on $v^{(l)}$ is maximal.

Starting with the original data one first looks for the direction $v^{(1)}$ such that the projection of the data on $v^{(1)}$ has maximal robust scale $s^{(1)}$. Once $v^{(1)}$ is found, the observations are projected onto the orthogonal complement of $v^{(1)}$. The next vector $v^{(2)}$ must then belong to this orthogonal complement and the robust scale $s^{(2)}$ of the projected observations on $v^{(2)}$ should be maximal. This loop is continued until r directions are found.

It has been proved by Li and Chen (1985) that the resulting method is highly robust, since it inherits the breakdown value of the robust scale estimator. Croux and Ruiz (1996) chose the median absolute deviation (MAD) and the biweight A-estimator (Hoaglin, Mosteller and Tukey 1983) in their actual implementation. Since A-estimators are more efficient than the MAD (Lax 1985), we will only work with A-estimators from now on.

To obtain a computationally feasible algorithm, Croux and Ruiz restricted the collection of directions that are investigated to all directions that pass through the origin and a datapoint. (We assume that the data have already been centered.)

Although this algorithm performs well in low dimensions, it lacks numerical stability in high dimensions, as will be shown in the following example.

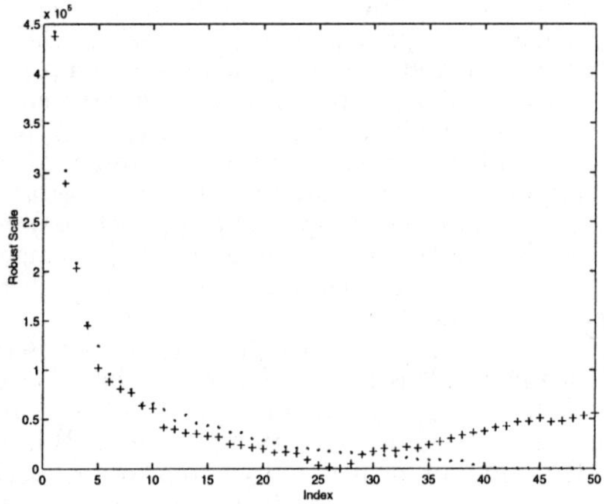

Fig. 1. The robust scales of the XRF data set computed with the C-R (+) and the RA (.) algorithm.

Energy Dispersive X-Ray Fluorescence (ED-XRF) is a fast non-destructive multi-element analysis technique. We will analyse a data set obtained from Pascal Lemberge containing 50 XRF spectra of aqueous solutions of metal salts at 1200 channels. The XRF data set is $\log_{10}$-preprocessed and mean-centered to lower the influence of the high background at the last channels (800 to 1200). In Figure 1 we see that the robust scales coming out of the C-R algorithm first decrease, as we expect from a PCA method, but from the 28th onwards they increase again.

3 An improved algorithm

The reason for this numerical instability is that all computations are performed in the high-dimensional space of dimension $p = 1200$. The succession of projections onto lower-dimensional spaces leads to an accumulation of round-off errors, and finally yields unreliable estimates. Therefore we propose an improvement of this algorithm through dimension reduction. Note that the basic principles of the C-R algorithm (the stepwise search for orthogonal directions $v^{(l)}$, the use of the robust A-estimator, the search in the direction of the data points) remain the same, and will therefore not be described again.

Suppose we have found the first direction $v^{(1)}$. We now apply an orthogonal data transformation $U^{(1)}$ such that $v^{(1)}$ is mapped onto the first basis vector $b_1^{(1)} = (1, 0, \ldots, 0)$. It turns out that the fastest way to do this is not by constructing a large orthogonal matrix but by means of a reflection. To construct this reflection $U^{(1)}$ we only need to compute the unit normal vector $n^{(1)} = (b_1^{(1)} - v^{(1)})/\|b_1^{(1)} - v^{(1)}\|$. Each data point $x_i^{(1)} := x_i$ is then transformed to

$$x_i^{(2)} = U^{(1)}(x_i^{(1)}) = x_i^{(1)} - 2\langle x_i^{(1)}, n^{(1)}\rangle n^{(1)}.$$

If we now want to project the data on the orthogonal complement of $U^{(1)}(v^{(1)})$ we simply have to remove the first coordinate of $x_i^{(2)}$ for every i. This reduces our space by one dimension. Moreover, all operations so far are numerically accurate. The reflection (i.e. the inner product) only requires elementary sums, and the projection itself needs no auxilary computations.

This procedure is repeated until $r = rank(X)$ directions have been found. At each step, we look for a new direction $v^{(l)}$ in the reduced $(p - l + 1)$-dimensional space. Then this vector $v^{(l)}$ is reflected onto the first basis vector $b_1^{(l)} \in I\!R^{p-l+1}$, and again the projections of the reflected data are obtained immediately by omitting their first coordinate.

In order to interpret the results, we have to retransform each direction $v^{(l)}$ to the original p-dimensional space. Note that this operation is again numerically stable since the inverse of a reflection is the reflection itself, so no matrix inversions are needed. We will refer to this approach as the *reflection algorithm* (RA).

In Figure 1 we see that the roots of the eigenvalues obtained by the reflection algorithm, i.e. the robust scales, are monotone decreasing.

If we now denote by P the $(p \times r)$-matrix that consists of the retransformed directions, we obtain a robust decomposition of X into a $(n \times r)$ score matrix and a matrix of loadings. Indeed, let $T = XP$, then $X = TP^t$. Finally, we can reduce the original number of variables p by only considering the first k columns of P. This type of algorithm benefits most from that situation, since it only needs to be run until $v^{(k)}$ is found instead of computing all r vectors $v^{(1)}, \ldots, v^{(r)}$. This saves a lot of computation time.

4 A two-step algorithm

Although the RA algorithm is more stable numerically than the C-R implementation it takes more computer time. Table 1 shows the computation time (in seconds on a 166Mhz PC running MATLAB) of the C-R algorithm and

RA on the XRF data set of Section 2, with $n = 50$ and $p = 1200$. The number k indicates the number of principal components that are actually computed. We see that RA is roughly 8 times slower than the C-R algorithm. Therefore, a faster algorithm is needed in such cases.

If $p > n$ we propose the following two-step algorithm. First, we reduce the data space to the affine subspace spanned by the n observations. This can be done fast and accurately by classical PCA, whereas we do not loose any information contained in the data. We thus decompose X as $X = T_1 P_1^t$. Secondly the RA algorithm is performed on the scores T_1, yielding $T_1 = T_2 P_2^t$. Finally we have $X = T_2 P_2^t P_1^t = T_2 P^t$ with $P = P_1 P_2$. This two-step algorithm will be denoted by RAPC, which stands for *reflection algorithm* on *principal components*. We see in Table 1 that the RAPC algorithm is much faster than RA and C-R.

Table 1. Computation times (in seconds) of the C-R algorithm, RA and RAPC on the high-dimensional XRF data set, for increasing numbers of principal components.

k	C-R	RA	RAPC
5	102.4	870.5	11.6
10	208.8	1977.5	14.8
15	349.8	2993.7	20.9
20	518.2	4052.8	26.7
25	671.4	5027.5	33.1
30	766.2	6021.1	39.1
40	994.9	8730.5	51.2
50	1388.6	9724.9	61.8

We have implemented the algorithms RA and RAPC in the MATLAB environment. The implementations can be downloaded from the website http://win-www.uia.ac.be/u/statis.

5 Example: Near-Infrared-Reflectance (NIR) Spectra of Biscuit Dough

We consider the well-known chemometrics data set of Osborne et al. (1984). It contains 40 NIR spectra of biscuit dough with measurements every 2 nanometers from 1200nm up to 2400nm. The data are first scaled using an inverse and a logarithmic transformation in order to eliminate drift and background scatter. Originally, the dataset consisted of 700 variables but the ends were discarded because of the lower instrumental reliability. As was done by Marx and Eilers (1999), we used the first differences to remove constants and sudden shifts. But contrary to previous analyses, we will leave the observation 23 (which is usually considered as an outlier on other grounds) in the data set. After this preprocessing we end up with a data set of 40 observations in 600 dimensions, to which we applied the RAPC algorithm.

Figure 2 shows the 'scree plot' of the robust eigenvalues (starting at the 7th eigenvalue) versus their index. We see that the robust eigenvalues decrease nicely. Often a scree plot helps us to decide how many principal components should be preserved. We should look at the 'elbow', the first place where there seems to be a kink in the line (Jackson 1991, p. 45). Here, the first twist is observed around 11, so we will retain $k = 12$ principal components.

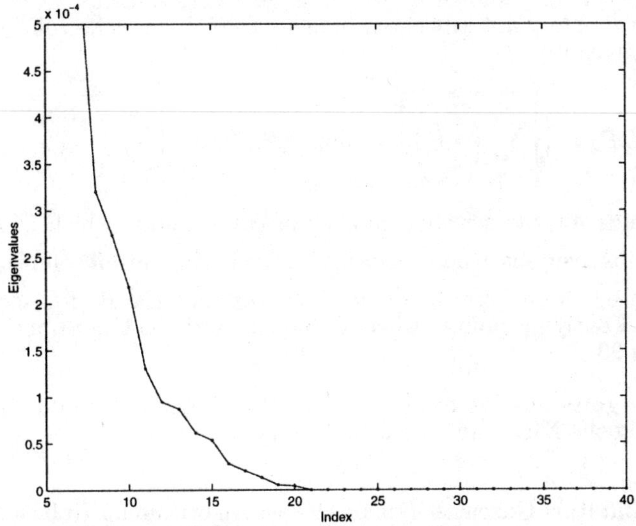

Fig. 2. Scree plot of the Biscuit NIR data set.

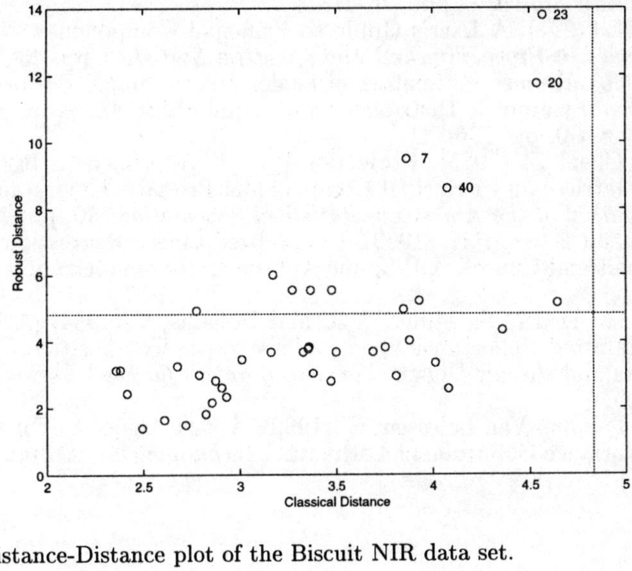

Fig. 3. Distance-Distance plot of the Biscuit NIR data set.

Next, we would like to identify the outliers in our data set. For this we will use the distance-distance plot of Rousseeuw and Van Driessen (1999) that shows for each data point its robust distance versus its classical distance. Denote by (T^C, s_j^C) and (T^R, s_j^R) the scores matrix and the square root of the jth eigenvalue, obtained by the classical PCA, and the RAPC method. Then for each data point i its classical distance (CD_i) and its robust distance (RD_i) are given by

$$CD_i = \sqrt{\sum_{j=1}^{k} \left(\frac{t_{ij}^C}{s_j^C} \right)^2} \quad \text{and} \quad RD_i = \sqrt{\sum_{j=1}^{k} \left(\frac{t_{ij}^R}{s_j^R} \right)^2}.$$

Here k stands for the selected number of components, which is 12 in our example. Observations whose distance exceeds the cut-off value $\sqrt{\chi^2_{k,0.975}}$ are pinpointed as outliers. In Figure 3 we see that the RAPC method detects several outlying points, whereas the classical PCA does not even find observation 23.

Acknowledgements We are grateful to Pascal Lemberge and Paul Eilers for providing the XRF and Biscuit data sets.

References

Croux, C. and Ruiz-Gazen, A. (1996). A Fast Algorithm for Robust Principal Components based on Projection Pursuit. *COMPSTAT: Proceedings in Computational Statistics 1996*, pp. 211-217, Heidelberg: Physica-Verlag.

Devlin, S., Gnanadesikan, R., and Kettenring, J. (1975). Robust Estimation and Outlier Detection with Correlation Coefficients. *Biometrika,* **62**, pp. 531-545.

Hoaglin, D.C., Mosteller, F., and Tukey, J.W. (1983). Understanding Robust and Exploratory Data Analysis. *New York: Wiley Series in Probability and Mathematical Statistics*, pp. 414-418.

Jackson, J.E. (1991). A User's Guide to Principal Components. *New York: Wiley Series in Probability and Mathematical Statistics*, pp. 45-46.

Lax, D. (1985). Robust Estimators of Scale: Finite-Sample Performance in Long-Tailed Symmetric Distributions. *Journal of the American Statistical Association*, **80**, pp. 736-741.

Li, G. and Chen, Z. (1985). Projection-Pursuit Approach to Robust Dispersion Matrices and Principal Components: Primary Theory and Monte Carlo. *Journal of the American Statistical Association*, **80**, pp. 759-766.

Marx, B.D. and Eilers, P.H. (1999). Generalized Linear Regression on Sampled Signals and Curves: A P-Spline Approach. *Technometrics*, **41(1)**, pp. 1-13.

Osborne, B.G., Fearn, T., Miller, A.R. and Douglas, S. (1984). Application of Near Infrared Reflectance Spectroscopy to the Compositional Analysis of Biscuits and Biscuit Dough. *Journal of Scientific Food Agriculture*, **35**, pp. 99-105.

Rousseeuw, P.J. and Van Driessen, K. (1999). A Fast Algorithm for the Minimum Covariance Determinant Estimator. *Technometrics,* **41**, pp. 212-223.

Applying techniques of dynamic programming to sequential mastery testing

Hans J. Vos

University of Twente, Faculty of Educational Science and Technology, P.O. Box 217, 7500 AE Enschede, The Netherlands

Abstract. The purpose of this paper is to derive optimal rules for sequential mastery testing, that is, deciding on mastery, nonmastery, or to continue testing and administering another random item. The framework of minimax sequential decision theory is used; that is, optimal rules are obtained by minimizing the maximum expected losses associated with all possible decision rules at each stage of testing.

Keywords. Sequential mastery testing, minimax sequential rules, backward induction

1 Introduction

In a sequential mastery test (SMT), the decision is to classify a student as a master, a nonmaster, or to continue testing and administering another random item. The main advantage of SMT (also denoted as tailored mastery testing) is to provide shorter tests for students who have clearly attained a certain level of mastery (or clearly nonmastery) and longer tests for those students for whom the mastery decision is not as clear-cut. One of the earliest approaches to SMT dates back to Ferguson (1969) using Wald's well-known sequential probability ratio test (SPRT).

As indicated by Ferguson (1969), three elements must be specified in advance in applying the SPRT-framework to SMT. First, two values p_0 and p_1 on the proportion-correct metric must be specified representing points that correspond to lower and upper limits of student's (unknown) true level of functioning at which a mastery and nonmastery decision will be made, respectively. Also, these two values mark the boundaries of the small region (i.e., indifference region) where we never can be sure to take the right classification decision, and, thus, in which testing will continue. Second, two levels of error acceptance α and β must be specified, reflecting the relative costs of the false positive (i.e., incorrect mastery decision) and false negative (i.e., incorrect nonmastery decision) error types. Third, Ferguson (1969) specified a maximum test length in order to classify within a reasonable period of time those students for whom the decision of declaring mastery or nonmastery is not as clear-cut (i.e., truncated procedure). It should be emphasized, however, that the SPRT-framework might also be applied to SMT if no maximum test length is specified.

A major drawback of the SPRT-framework is that it does not take costs of testing into account. The purpose of this paper, therefore, is to derive optimal rules for SMT that takes costs of testing explicitly into account. The framework of minimax sequential decision theory will be used (e.g., Lehmann, 1959).

2 Framework of minimax sequential decision theory

Three basic elements can be identified in minimax sequential decision theory. In addition to a psychometric model relating the probability of a correct response to student's true level of functioning and a loss function evaluating the total costs and benefits for each possible combination of classification outcome and true level of functioning, costs of administering one additional item must be explicitly specified in this approach. Doing so, maximum expected losses associated with the mastery and nonmastery decisions can now be calculated straightforward at each stage of testing. As far as the maximum expected loss associated with to continue testing concerns, this quantity is determined by averaging the maximum expected losses associated with each of the possible future decision outcomes relative to observing those outcomes (i.e., the posterior predictive distributions).

Optimal rules (i.e., minimax sequential rules) are now obtained by minimizing the maximum expected losses associated with all possible decision rules at each stage of testing using techniques of dynamic programming (i.e., backward induction). This technique starts by considering the final stage of testing (where the option to continue testing is not available) and then works backward to the first stage of testing.

3 Notation

In the sequel, following Ferguson (1969), a sequential mastery test is supposed to have a maximum test length of n ($n \geq 1$). Let the observed item response at each stage of testing k ($1 \leq k \leq n$) for a randomly sampled student be denoted by a discrete random variable X_k, with realization x_k. The observed response variables $X_1,...,X_k$ are assumed to be independent and identically distributed for each value of k, and take the values 0 and 1 for respectively incorrect and correct responses to the k-th item. Let the observed number-correct score be denoted by a discrete random variable $S_k = X_1 +...+ X_k$, with realization $s_k = x_1 +...+ x_k$ ($0 \leq s_k \leq k$).

Student's true level of functioning is unknown due to measurement and sampling error. Therefore, let student's true level of functioning be denoted by a continuous random variable T on the latent proportion-correct metric, with realization $t \in [0,1]$. Finally, a criterion level t_c ($0 \leq t_c \leq 1$) on T is required to be specified in advance by the decision-maker using methods of standard setting (e.g., Angoff, 1971). A student is considered a true nonmaster and true master if his/her true level of functioning t is smaller or larger than t_c, respectively.

It should be noted that the specification of a maximum test length in combination with the backward induction procedure guarantees that unusual response patterns like, for instance, 1111100000 in case of 10 items are properly taken into account.

4 Loss fuction, costs of testing, and psychometric model

The threshold loss function (e.g., Hambleton & Novick, 1973) is adopted as the loss structure involved. The choice of this loss function assumes that the total costs and benefits involved in the decision problem can be summarized by possibly different constants for each possible classification outcome. Let e represent the costs of administering one random item, then threshold loss can be formulated at each stage of testing k as follows:

Table 4.1. Table for threshold loss function at stage k of testing

True level of functioning / Decision	$T \leq t_c$	$T > t_c$
Declare nonmastery	ke	l_{01} + ke
Declare mastery	l_{10} + ke	ke

The loss parameters l_{ij} ($i = 0,1$; $i \neq j$) associated with the incorrect classification outcomes have to be empirically assessed, for which several methods, such as lottery methods, have been proposed in the literature (e.g., Luce & Raiffa, 1957).

Following Ferguson (1969), the binomial distribution will be adopted for the psychometric model. Its distribution is given at each stage of testing k by:

$$f(s_k \mid t) = \binom{k}{s_k} t^{s_k} (1-t)^{k-s_k} .$$

This modeling of response behavior corresponds to the assumption that all items have equal difficulty or are sampled at random from a large pool of items.

5 Optimal rules for the fixed-length mastery problem

Given response pattern $(x_1,...,x_k)$, first the minimax principle will be applied to the fixed-length mastery problem (i.e., mastery and nonmastery). Let $y = 0,1,...,k$ represent all possible values s_k can take, it then can easily be verified from Table 4.1 and the assumed binomial model that mastery is declared when s_k is such that:

$$\sup_{t \leq t_c} (l_{10} + ke) \sum_{y=s_k}^{k} \binom{k}{y} t^y (1-t)^{k-y} + \sup_{t > t_c} (ke) \sum_{y=0}^{s_k - 1} \binom{k}{y} t^y (1-t)^{k-y} <$$

$$\sup_{t \leq t_c} (ke) \sum_{y=s_k}^{k} \binom{k}{y} t^y (1-t)^{k-y} + \sup_{t > t_c} (l_{01} + ke) \sum_{y=0}^{s_k - 1} \binom{k}{y} t^y (1-t)^{k-y},$$

and that nonmastery is declared otherwise. Since the cumulative binomial distribution function is decreasing in t, and rearranging terms, it follows that mastery is declared when s_k is such that:

$$\sum_{y=s_k}^{k} \binom{k}{y} t_c^y (1-t_c)^{k-y} < 1/(1+R),$$

where $R = l_{10}/l_{01}$ denotes the loss ratio (i.e., false mastery relative to false nonmastery). If the above inequality is not satisfied, nonmastery is declared.

6 Derivation of minimax sequential rules

Let $d_k(x_1,...,x_k)$ denote the decision rule yielding the minimum of the maximum expected losses associated with the two classification decisions, and let the maximum expected loss associated with this minimum be denoted as $V_k(x_1,...,x_k)$. Minimax sequential rules can now be found by using the following backward induction computational scheme: First, the minimax sequential rule at the final stage of testing n is computed. Since the option to continue testing is not available at this stage of testing, it follows immediately that the minimax sequential rule is given by $d_n(x_1,...,x_n)$, and its associated maximum expected loss by $V_n(x_1,...,x_n)$.

To compute the maximum expected loss associated with the option to continue testing at stage (n-1) until stage 1, the risk $R_k(x_1,...,x_k)$ will be introduced at each stage of testing k ($1 \leq k \leq n$). Let the risk at stage n be defined as $V_n(x_1,...,x_n)$. Generally, given response pattern $(x_1,...,x_{k-1})$, the risk at stage (k-1) is then computed inductively as a function of the risk at stage k as:

$$R_{k-1}(x_1,...,x_{k-1}) = \min\{V_{k-1}(x_1,...,x_{k-1}), E[R_k(x_1,...,x_{k-1}, X_k) \mid x_1,...,x_{k-1}]\}.$$

The maximum expected loss associated with administering one more random item after (k-1) items have been administered, $E[R_k(x_1,...,x_{k-1}, X_k) \mid x_1,...,x_{k-1}]$, can then be computed as the expected risk at stage k of testing as:

$$E[R_k(x_1,...,x_{k-1}, X_k) \mid x_1,...,x_{k-1}] = \sum_{x_k=0}^{x_k=1} R_k(x_1,\ldots,x_k)*P(X_k = x_k \mid x_1,\ldots,x_{k-1}),$$

where $P(X_k = x_k \mid x_1,...,x_{k-1})$ denotes the posterior predictive distribution of X_k at stage (k-1) of testing. The minimax sequential rule at stage (k-1) is now given by: Administer one more random item if $E[R_k(x_1,...,x_{k-1}, X_k) \mid x_1,...,x_{k-1}]$ is smaller than $V_{k-1}(x_1,...,x_{k-1})$; otherwise, decision $d_{k-1}(x_1,...,x_{k-1})$ is taken.

7 Calculation of posterior predictive distributions

To compute $P(X_k = x_k \mid x_1,...,x_{k-1})$, a prior distribution must be specified representing our best prior beliefs concerning student's true level of functioning. The uniform prior on the standard interval [0,1] is taken as a prior in this paper. In combination with a binomial distribution for the psychometric model, it is then known (e.g., Lehmann, 1959) that $P(X_k = 1 \mid x_1,...,x_{k-1}) = (1+s_{k-1})/(k+1)$, and thus, that $P(X_k = 0 \mid x_1,...,x_{k-1}) = [1-(1+s_{k-1})/(k+1)] = (k-s_{k-1})/(k+1)$.

8 Simulation of different mastery testing strategies

In a Monte Carlo simulation the minimax sequential strategy will be compared with other existing approaches to mastery testing.

8.1 Description of the testing strategies used for comparison

The first comparison will be made with a conventional fixed-length mastery test (CT) in which student performance was recorded as proportion of correct answers.

The student was declared a master for answering 60% or more items correctly after completion of the test, whereas nonmastery was declared otherwise.

The second comparison will be made with Wald's SPRT procedure. The limits of the indifference region in which testing will continue were set at proportion-correct values p_0 and p_1 of 0.5 and 0.7, respectively, whereas the error rates α and β were each set equal to 0.1. For those students who could not be classified as either a master or nonmaster before the item pool was exhausted, a classification decision was made in the same way as in the CT procedure, using a mastery proportion-correct value of 0.6.

In order to make a fair comparison of the minimax sequential strategy with the two strategies described above, the criterion level t_c was set equal to 0.6. Furthermore, the losses l_{01} and l_{10} were assumed to be equal corresponding to the assumption of equal error rates in Wald's SPRT procedure. On a scale in which one unit corresponded to the costs of administering one random item (i.e., e = 1), l_{01} and l_{10} were each set equal to 150. Using the backward induction computational scheme discussed earlier, for given loss parameters and maximum test length n, a computer program called MINIMAX was developed to determine the appropriate decision at each stage of testing k for different values of s_k. A copy of the program MINIMAX is available from the author upon request.

8.2 Type of test and item response generation

The simulation study was conducted using a set of items that were perfect replications of each other. More specifically, it was assumed that each item could be described by a one-parameter logistic model (Rasch, 1960) with discrimination a of 1 and equal difficulty b of 0. This item pool reflected the choice of the binomial distribution for modeling response behavior in the minimax sequential procedure. Maximum test length varied (MTL) from 10 to 50 items.

Item responses for 1000 simulated students, drawn from a N(0,1) distribution, were generated for each item in the pool. For known ability of the simulated student and given item difficulty, first the probability of a correct answer was calculated using the one-parameter logistic model. Next, this probability was compared with a random number drawn from the uniform distribution in the range from 0 to 1. The item administered to the simulated student was scored correct and incorrect if this random number was less and greater than the probability of a correct answer, respectively.

Furthermore, a simulated student was supposed to be a "true" master if his/her ability used to generate the item responses was higher than a prespecified cut-off point on the N(0,1)-ability metric. Since a value of 0.6 on the proportion-correct metric corresponded after conversion with a value of 0 on the N(0,1)-ability metric, the cut-off point on the N(0,1)-ability metric was set equal to 0.

8.3 Results of the simulation study

Table 8.1 shows that the minimax sequential strategy yielded shorter average test lengths than the other two strategies at each MTL level, whereas the phi correlations between true and estimated mastery status (i.e., classification accuracy) were generally somewhat lower at each MTL level (see also Table 8.1). To examine which strategy is most efficient, therefore, we computed for the other two strategies the average test length at each MTL level for achieving the same phi correlation as under the minimax sequential strategy. Doing so, it turned out that the minimax sequential strategy was most efficient (i.e., combination of highest phi correlation and shortest average test length) at each MTL level.

Table 8.1. Average number of items to be administered and classification accuracy

	Maximum Test Length			Maximum Test Length		
Strategy	10	25	50	10	25	50
	Average Number of Items			Classification Accuracy		
CT	10	25	50	0.652	0.787	0.718
SPRT	8.64	14.47	17.42	0.584	0.723	0.718
MINIMAX	5.84	9.46	11.15	0.610	0.620	0.692

9 Discussion

It should be emphasized that the minimax sequential strategy is especially appropriate when costs of testing can be assumed to be quite large. For instance, when testlets (i.e., blocks of parallel items) rather than single items are considered. Also, the minimax sequential strategy might be appropriate in psychodiagnosis for making decisions concerning the effectiveness (e.g., effective or ineffective) of a new therapy (e.g., cognitive-analytic therapy) for patients suffering from some mental health problem (e.g., anorexia nervosa). In such clinical situations, costs of exposing a random patient to the new treatment are generally quite large.

References

Angoff, W.H. (1971). Scales, norms and equivalent scores. In: R.L. Thorndike (ed.), *Educational measurement*, 508-600. Washington: American Council on Education.

Ferguson, R.L. (1969). *The development, implementation, and evaluation of a computer-assisted branched test for a program of individually prescribed instruction*. Pittsburgh: University of Pittsburgh.

Hambleton, R.K. & Novick, M.R. (1973). Toward an integration of theory and method for criterion-referenced tests. *Journal of Educational Measurement*, **10**, 159-170.

Lehmann, E.L. (1959). *Testing statistical theory*. New York: Macmillan.

Luce, R.D. & Raiffa, H. (1957). *Games and decisions*. New York: Wiley.

Rasch, G. (1960). *Probabilistic models for some intelligence and attainment tests*. Copenhagen: Danish Institute for Educational Research.

The introduction of formal structure into the processing of statistical summary data

Andrew Westlake

Survey & Statistical Computing, 119 Florence Road, London N4 4DL, UK

Keywords. Multi-way table, Manipulation, Dissemination, Automation

1 Introduction

This paper reviews the technologies required for the production and publication of large volumes of statistical information, such as produced by National Statistical Offices (NSOs). We review both academic and commercial developments, and present these in the context of an analysis of the tools and structures needed.

Our ultimate objective is improvement in the accessibility and use of data from large or continuous surveys and inquiries. These are characterised as having a significant volume of data processing and output, often with complex processing to produce representative and comparable figures, but not great depth of analysis. Production will generally be in the form of large amounts of classified summary data, and analysis will be in the form of comment on trends perceived in headline summary figures. Some clients will extract small volumes of data of specific interest, while others will want to perform more detailed analysis.

We are looking for an approach to be applied after initial data entry and cleaning, which provides support for:

- Automation of recurrent processes (imputation, adjustments, standardisation, disclosure control, standard analyses)
- *Ad hoc* analyses by internal and external clients
- Exploration and discovery by potential users (is there an answer to my question?)
- Dissemination (on paper, by browsing, download, CD)

We conclude that useful progress has been made, but much remains to be done.

2 Structure

The data structure required for storing summary information is a **multi-way table**, a hypercube with multiple dimensions, and cells containing multiple values. Each **dimension** is a **classification structure**, with consolidation from detail to broad groups through various levels of aggregation in a hierarchical tree (more strictly, an acyclic graph). Each cell of the table contains multiple **measures**, the same in all cells, which can be counts, sums, means, or other expressions. Figure 1 shows these components in a hypothetical medical example, with classifications by time, location and disease. Note that the structure for time is not a simple hierarchy, because weeks and months do not map into each other.

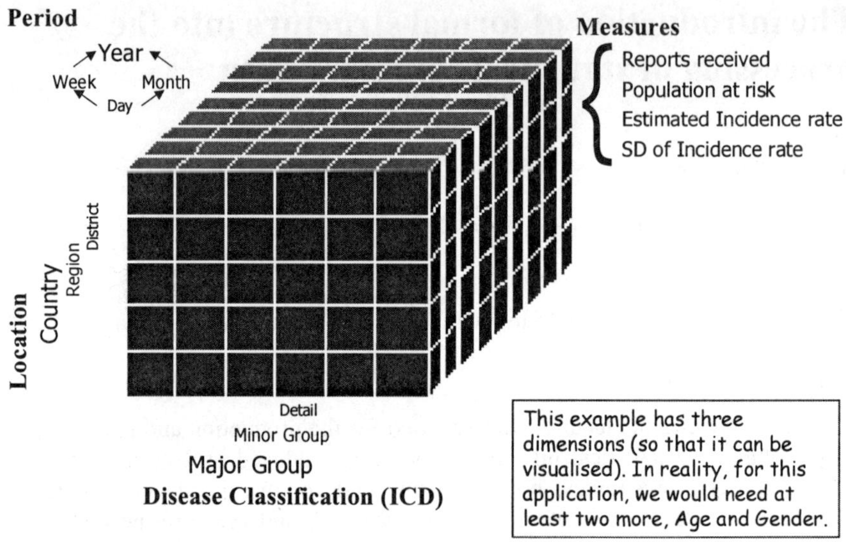

Fig. 1. A hypercube with dimensions and measures

Formal work on this structure has been carried out in the statistical database field from at least the early 1980's, and is found particularly in the SSDBM conference proceedings, with work by Rafanelli and Shoshani[i] on the Storm model being central. An alternative, more commercial, thread can be traced from work by Codd (who coined the term Olap – OnLine Analytical Programming) and has resulted in various tools, from database vendors (Oracle[ii], Ingress, MS SQL Server[iii]) and specialist firms, generally referring to Data Warehousing, or Data Cubes[iv]. This work is based on more traditional computer science views of databases and does not give much attention to the specific requirements of statistics, such as validity rules or contextual meta-data. Some special statistical products also exist, such as Beyond 20/20[v], SuperCross and PC-Axis.

From formal analysis of the nature and properties of the structure we can derive precise rules for specifying a table, plus consequential properties and behaviour. For example, clear rules are needed for identifying the underlying population elements that are described by a table (for ensuring that classification categories are exclusive and exhaustive) and for identifying different types of summary measure. This then leads to rules for computing aggregations of measures (for margins and consolidation) and for deriving additional measures (such as rates).

Automation is possible with a formal model of structure and functionality. An object model (consisting of structures, properties and methods) can be defined, and implemented as a set of programming objects with a suitable API (programming interface). This then provides the basis for application developers to implement specific tasks, or for the development of general-purpose packages. Using such tools it is possible to construct statistical production systems where the processes to be applied to statistical data are parameterised, using suitable meta-data.

Microsoft have proposed a standard API for use with OLAP products (an attempt at a broader standard through the OLAP Council did not succeed). The new Metaware project, funded by Eurostat[vi], is proposing to produce a production system linked to data warehouse products and building on the meta-data repository system Bridge, which was produced by the earlier IMIM project. A further project (MetaNet) has been proposed, to establish a network of organisations involved in statistical meta-data, with a view to establishing coherence between various standardisation efforts and disseminating information about these efforts.

3 Processing

Figure 2 shows the various components in a possible statistical processing system, plus the major flows of data. On the right is the database, holding all the data. This includes the source registers for sampling (the sampling frames) and the returned records from the surveys or inquiries. It also includes the various parameters that drive the processing of the data (such as the sampling rates within strata).

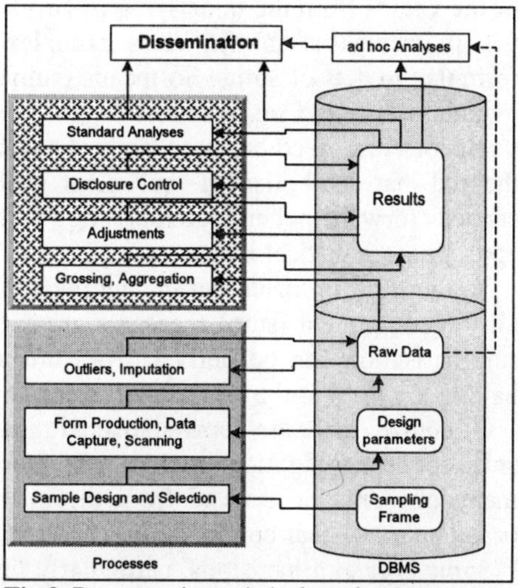

Fig 2. Processes in statistical production

There should also be explicit, structured storage of the various aggregated results of the inquiries, in the form of multi-way tables. This is logically part of the same database, but may be physically separate.

The boxes on the left of the diagram show processes applied to the data in the database. Those in the lower box are the ones currently implemented in most statistical processing systems. Some processes covered by the headings in the upper box are currently in place, but most of these are treated as tasks to be done manually outside the main statistical data processing system. We suggest that these should be implemented in a more formalised and standardised way, taking data from the database and returning derived values there.

In a traditional system, grossing and aggregation are tightly coupled to the specific reports needed for particular publications. The approach we are proposing here is to do as little aggregation as possible initially (i.e. to retain as much detail in classifications as is possible), thus retaining the option for much more flexible aggregation, whether for a number of different publications, or for analysis on demand. In the database literature there is considerable discussion about the physical aspects of storing aggregates, in particular about the efficiency trade-off between the space used to store more heavily aggregated data and the speed of retrieval from having the aggregations pre-calculated. However, logically, all aggregations derivable from the detailed classifications are available.

4 Manipulation

4.1 Aggregation

From a mathematical perspective, most of the processes applied to aggregate data are straightforward. The complex processes are the estimation of statistical precision (where complex sample designs are involved) and disclosure control (the avoidance of the disclosure of information about individual respondents).

Aggregation is simple (given knowledge of the types of measure involved), and most analysis in this context is little more than aggregation. Almost all aggregations are ultimately based on summation of more or less complex functions of the values from the underlying micro-data records (note that counts are sums of 0/1 indicator variables). More complex measures are obtained by applying formulae to sets of sums. So means (sum / count), percentages (restricted count / overall count) and ratios (sum1 / sum2) are all based on two sums, and standard deviations are based on three (sum of squares, sum and count). This refers back to the old statistical idea of 'sufficient statistics', the aggregates that need to be carried forward when summarising data, in order to be able to derive the appropriate statistical measures.

So the trick in managing the aggregation process for measures is to ensure that all the component (sum) measures are explicitly available. Then higher levels of summarisation can be derived from the more detailed aggregate tables, without having to return to the underlying micro-data records.

Of course, there are some statistical measures that are not based on sums. Some still support aggregation (such as minimum and maximum – and hence range), but many can only be derived from the micro-data – these are generally measures based on non-linear combinations of the micro-data values.

Some special dimensions, particularly time, also require care. For example, on a time dimension it is essential to differentiate between measures that represent aggregations over time (such as revenue) and those that represent cross-sectional observations (such as number of employees). Lenz and Shoshani[vii] have done important work on this topic.

4.2 Derivations

Grossing (the production of estimates for the whole population instead of just the sample) is done by applying factors representing the proportion of the population covered by the sample (either known or separately estimated factors, often ratio-based factors). More general adjustments are applied where the population sampled differs from the population for which results are to be reported. The most obvious example is in the National Accounts, where adjustment is made for known or estimated underreporting and concealment. Standardisation (such as adjustment to constant prices or the removal of seasonal effects) is similar, except that the data being adjusted may contribute to the factors to be applied. In general, all these operations involve the multiplication of measures in one multi-way table by factors from some other multi-way table, perhaps formed by aggregating the first.

5 Dynamic analysis

Traditional statistical publishing has concentrated on the production of a number of predefined tables, which are then committed to paper and fixed. However, developments in the Olap and data warehouse field have emphasised the dynamic nature of publication, with the aggregate data viewed as a resource to be explored by the user. The ubiquity of the Internet has reinforced this approach.

So we should view a multi-way table as the basis of a dynamic facility, through which a user can extract information of interest. This is similar to the idea of a *view* in a relational database. To support these dynamic views, various operations are required.

Selection, as in an RDBMS, is the operation to extract subsets of classifications and measures. There is also a need to reduce or increase detail in a classification, for example, moving from detailed to broad groups on a dimension, or showing the next level of detail for a particular group. These operations are called *roll-up* and *drill-down* in Olap terminology. Roll-up includes extraction of margins, the complete removal of some dimensions. A user may want to define new groupings within classifications, as combinations of existing groups.

The derivation of additional measures and classifications requires manipulation functionality, and also rules about the validity of the operation and the nature of any derived measure. With a production system we can assume that the derivation specifications are appropriate, but this is not reasonable for a system open to use by uncontrolled users. So we need to recognise the different types of measure in cells and the different types of operation between measures. It is then possible to warn a user can be warned if an operation could be invalid. Note that it is not in general possible to make absolute rules, since the user may have more information about the data than the system has.

As well as within cells, measure derivations can operate within and across classifications, and between tables. It seems likely that a general approach to this, based on operations between tables, can encompass all these cases. We need to define how the measures and dimensions of related but different tables can be combined, using repetition and distribution. Then the final step uses operations between measures within cells, as before. This will need rules to determine when tables are mutually conformant, which will draw heavily on explicit representation of the underlying population objects for the tables, and of the selection rules for the population and the measures.

6 Presentation and Dissemination

An important aspect of this approach to summary information is the separation of the internal, logical structure of the data from the external layout required for presentation. Additional functionality is required to support presentation, whether on screen, on paper, or in some other form. We need to map from logical structure to presentation layout, through the specification of rows, columns and pages, and to combine separate tables (in more ways than are needed for manipulation). When used in a browsing context the functionality should be dynamic, allowing the user to explore the data, as described previously. The 'Pivot Table' component of MS Office 2000 provides a good initial model for this, and Beyond 20/20 is more statistically oriented, though a little less flexible.

For dissemination it must be possible to generate different types of output, and with different packaging. While the full detail tables will be retained for internal use, exported tables will generally have less detail, so views will be turned into real tables. It is essential that disclosure control can be applied to exported tables, and to all other released material. Statistical reliability can be addressed by including appropriate measures in the tables.

For publication on paper, MS Office (or similar) products provide a good environment for generating reports and analyses (including automation options). Publication of static pages on the web is essentially the same problem but with different output formats and better navigation options, and, again, recent office products are addressing this issue. However, the web will also need dynamic versions, for exploration. Initial examples are provided by the ONS Statbase[viii] service, and the Beyond 20/20 web server. The Olap products generally address this as well, and a number of NSOs (such as The Netherlands[ix] and Hungary) are building systems using data warehouse products. Web access (or any other form of on-line access) has the important advantage of timeliness over any static dissemination method.

Users who need machine readable output should be able to export or download information. Spreadsheets provide a suitable form for simple cases, but full multi-way tables will be needed for in-depth analysis, which implies the availability of manipulation or browsing software for the user. Dissemination on CD, while less timely, addresses problems of scale and direct web access, but can simply reproduce the web content, providing static, dynamic and extract facilities through a browser.

References and Links

i Rafanelli M and Shoshani A (1990). STORM: A Statistical Object Representational Model. In Michalewicz Z, *Statistical and Scientific Database Management*, Springer-Verlag (1990), ISBN 3-540-52342-1.

ii Oracle Express OLAP: http://www.oracle.com/olap.

iii Microsoft SQL Server OLAP: http://www.microsoft.com/sql/productinfo/olapoverview.htm.

iv http://www.sybase.com/services/dwpractice/dimensional.html. This paper from Sybase gives a good flavour of the computer science approach to multidimensional data.

v Beyond 20/20 statistical dissemination server: http://www.ivation.com/index.htm.

vi Information about all the Eurostat research projects can be found at http://europa.eu.int/comm/eurostat/research

vii Lenz H-J and Shoshani A (1997), Summarizability in OLAP and Statistical Databases. In Hansen D and Ionnidis Y, *Proceedings of 9th SSDBM, IEEE Computer Society* (1997), ISBN 0-8186-7952-2.

viii ONS Statbase service: http://www.statistics.gov.uk/statbase/mainmenu.asp.

ix Statistics Netherlands Statline service http://statline.cbs.nl/

Validation of association rules by interactive mosaic plots

Adalbert Wilhelm and Heike Hofmann

Institut für Mathematik, Universität Augsburg, 86135 Augsburg, Germany

Keywords. Association rules, Data mining, Interactive graphics, MANET, Mosaicplots

1 Association rules

Association Rules have been proposed by Agrawal et al. (1993) in the context of market basket analysis. They were invented to provide an automated process, which could find connections among items, that were not known before, especially to answer questions like: "which items are likely to be bought together?". Typically, the data to be examined consists of customer purchases, i.e. a set of items bought by a customer over a period of time. The standard way of storing such data is the following: To be able to identify each customer the transactions are stored with unique numbers, the *transaction identification (TID)*. Beside that, we have a set of different *items*, the so-called *itemset* $\mathcal{I} = \{i_1, i_2, \ldots, i_m\}$. The data or *database* D is a set of purchases (*transactions*), where each transaction T includes a set of items, such that $T \subset \mathcal{I}$. A transaction T is said to *contain* a set of items X, if X is a subset of T.

An *association rule* is an implication of the form $X \to Y$, where X and Y are mutually exclusive itemsets (i.e. $X, Y \subset \mathcal{I}$ and $X \cap Y = \emptyset$).
An association rule $X \to Y$ holds with *confidence* $c = c(X \to Y)$, if $c\%$ of transactions in D that contain X also contain Y. $X \to Y$ has *support* s in the database D, if $s\%$ of transactions in D contain $X \cup Y$.

Most data mining software offer a procedure to generate all association rules with confidence and support that exceed some user-specified minimum thresholds *minsup* and *minconf*. The procedures are typically based on the a priori algorithm introduced by Agrawal & Srikant (1994). The key point of this algorithm lies in the fact that for any subset $X' \subseteq X$ of items its support is bounded from below by the support of the itemset X. In addition, the confidence for the rule $(X \setminus X') \to X'$ can be derived directly from the supports of the related itemsets using the equality

$$\text{confidence}\left((X \setminus X') \to X'\right) = \frac{\text{support}(X)}{\text{support}(X \setminus X')}.$$

Applications of association rules include market basket analysis, attached mailing in direct marketing, fraud detection and department store/shelf planning.

Let us consider an example originating from data that is available as one of the example data files for the SAS Enterprise Miner. For this data (in the following refered to as the *SAS Assocs Data*) the output for an association query with $minconf = 50$ and $minsup = 3$, limited by a maximum of 4 items

# items	conf	supp	count	rule
2	82.62	25.17	252	artichok → heineken
2	78.93	25.07	251	soda → cracker
2	78.09	22.08	221	turkey → olives
...				
3	95.16	5.89	59	soda & artichok → heineken
3	94.31	19.88	199	avocado & artichok → heineken
3	93.23	23.38	234	soda & cracker → heineken
...				
4	100.00	3.1	31	ham & corned beef & apples → olives
4	100.00	3.1	31	ham & corned beef & apples → hering
4	100.00	3.8	38	steak & soda & heineken → cracker
...				

Table 1. Examples of association rules as found in the SAS Assocs Data by the SAS Enterprise Miner Software. 1000 rules have been generated: 47 including 2 items, 394 with 3 items and 559 with 4 items.

per rule generated by the SAS Enterprise Miner consists of 1000 lines of the form shown in Table 1. Instead of concentrating on each transaction and the items bought in it, the statistical approach usually concentrates on the items themselves: Dummy variables are generated for each item, i.e. each item i_j $(1 \leq j \leq m)$ is coded in a binary variable X_j, taking ones and zeroes for "bought" and "not bought" respectively. The kth entry $X_j^{(k)}$ is 1, if item i_j is present in transaction t_k and 0 if not. Items bought together can be expressed as intersections of the corresponding item variables: $X_{j_1} \cap X_{j_2} \cap \ldots \cap X_{j_p}$, which we will call an *event*. The *frequency* with which two events X and Y occur together in the data corresponds to the support $s(X \cap Y)$, the *conditional frequency* of event Y given event X corresponds to the confidence $c(X \to Y)$.

In classical data analysis the frequency and the conditional frequency are maximum likelihood estimators of the probability of $X \cap Y$ and the conditional probability of Y given X, respectively. The conditions for an association rule given above can therefore be re-formulated for two disjoint events X and Y as $X \to Y \iff \begin{cases} P(X \cap Y) \geq s \\ P(Y|X) \geq c \end{cases}$ where s and c have the same meaning as before.

2 Visualizing association rules

Visualizing association rules aims at solving some major problems that come with association rules. First of all the rules found by automatic procedures must be filtered. Depending on what minimum confidence and what support is specified a vast amount of rules may be generated. Among those, however, not only "interesting and new" results - according to the principle of KDD - are found. In supermarket data for instance most purchases will also contain plastic or paper bags. The naive filtering approach that searches for the rules with highest confidence and/or highest support fails because it yields rules that are typically already known before and are thus not of interest. So, in many cases rules that just do not pass the thresholds can be economically better exploited and are therefore higher rated. By construction the a priori algorithm tends to prefer frequent events as right hand sidea of the rules. If $P(Y)$ is large, then it is very likely for any small event X, that $P(Y|X)$ is higher than the minimal confidence threshold. The meaning of rules found in this way, on the other hand, is more than questionable. Also, when dealing with different association rules which refer to the same response, it is of interest to examine, whether the populations described by the left hand side of a rule differ from each other. It well could be one and the same population, and different rules are just providing different descriptions for the same population. Therefore it is one goal of visualization to examine the impact of rules on one another and to find intersections amongst them. Quite often, slight variations of items are listed as different association rules: either the same items alternatively take right and left hand sides of rules or sequences occur.

Besides finding new and interesting results, datamining tools are to find *explicable* results. The interpretation of results therefore should be of great interest - if one can not explain a rule at first, we could use methods of cross-classification in the hope of finding clues within the data, which allow us to decide, whether a result has to be considered as a peculiarity of this particular data set or whether it can be generalized. Since support and confidence are equally important for the conclusions drawn from association rules any approach should visualize support and confidence of competing rules within one display. Figure 1 shows a standard visualization of association rules as a matrix of all left and right hand sides of rules. Left hand sides are the rows, right hand sides the columns of the matrix. Each rule, which fulfills $minsupp$ and $minconf$ is drawn as a square. The size of which is given by the actual support of the rule. This approach is rather unsatisfactory since

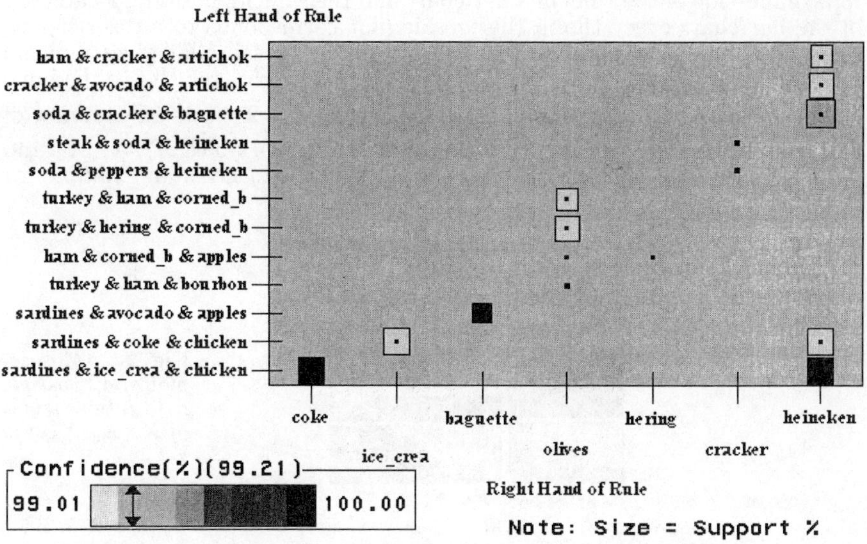

Fig. 1. *SAS Enterprise Miner: Visualisation of all 15 association rules with a minimum confidence of 99%.*

it reaches the space limits already for a small number of association rules. It also uses two different visual attributes to encode support and confidence of a rule: color for the confidence and size for the support. The ordering of colors is difficult and in no way unambiguous and depends heavily on the used scale of the color scheme. The encoding of the support by the size of the squares is visually problematic, since length of the square is used instead of area. For instance the rules *turkey&hering&cornedbeef* $\rightarrow$ *olives* and *ham&cornedbeef&apples* $\rightarrow$ *olives* have support 11.19% and 3.1%, respectively. The factor is approx. 4, whereas the areas differ with factor 16 which yields a lie factor as defined by Tufte (1983) of 400% .

Mosaic plots as introduced by Hartigan & Kleiner (1981) are a graphical analogue to multivariate contingency tables. They show the frequencies in a contingency table as a collection of rectangles whose areas represent the cell frequencies. Large areas therefore represent large cells. The shape of a tile is calculated during the (strictly hierarchical) construction. In classical mosaic plots alternately width and height of each bin is split according to the categories of the next variable included, in such a way, that the area of each bin is proportional to the number of cases falling into this specific intersection of variables. Thus, viewed statically the mosaic plot gives a graphical estimation of the joint distribution of the variables contained. Interactive features are necessary to turn the mosaic plot into a powerful exploration tool. In Hofmann (2000) the interactive features have been described for mosaic plots, the most essentials are *linked highlighting, rotating, navigating through dimensions, grouping categories,* and *variation of display.*

Linked highlighting provides the transition to conditional distributions and is essential for visualizing association rules. The basic approach is as follows: Combine all variables involved in the left-hand-side X of a rule as *explanatory* variables and draw them within one mosaicplot. Visualize the *response* Y, the right-hand-side of the rule, in a bar chart and then by highlighting a category in the bar chart every tile in the mosaic plot corresponds to an association rule. The amount of highlighting in a bin in the mosaic plot gives a visual measure for the frequency $S(X \cap Y)$, highlighting heights relative to the bin's height give the conditional frequency $\frac{S(X \cap Y)}{S(X)}$. Figure 2 shows an overview of all possible association rules involving four binary variables X_1, X_2, X_3 and Y. The bin in the bottom right corner ($x_1 \cap x_2 \cap x_3$) gives an example for

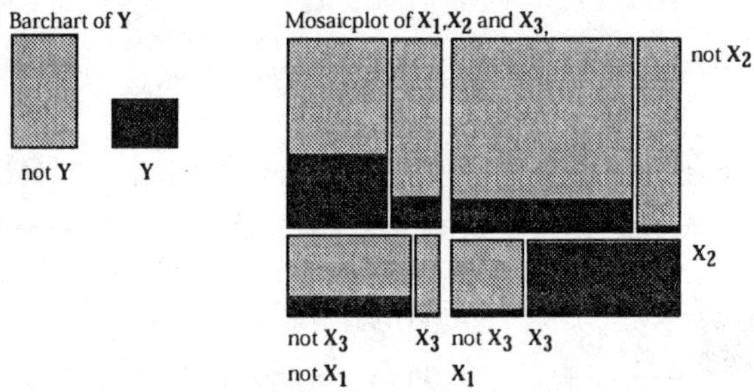

Fig. 2. *Mosaicplot of all possible association rules of X_1, X_2, X_3 and Y.*

a rule that passes most common thresholds. It has very high confidence (the highlighting almost fills this bin entirely), and also the support is relatively high (the bin itself, and therefore the amount of highlighting, is large). The leftmost bin in the upper row, (not $x_1 \cap x_2 \cap$ not x_3), represents a rule with a comparable support (this bin contains approximately the same amount of highlighting as the bin in the bottom right corner), yet the confidence of the corresponding rule is rather low (highlighting fills this bin to only one third, approximately). All the other possible rules have even lower confidence and their supports are also too small to be of interest. To make comparisons of

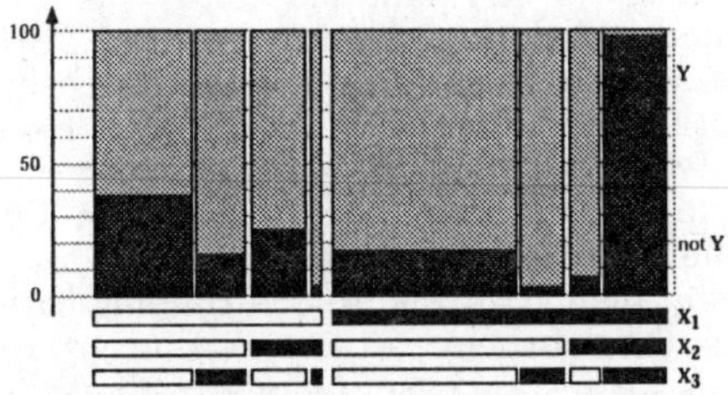

Fig. 3. *(Labelled) Doubledecker plot of variables X_1, X_2, X_3 and Y. Only one cell is almost filled entirely with highlighting - this hints to a "good" association rule.*

proportions of highlighting heights more easily, it is much better to align all bins. This yields mosaics of the following form: Starting from the basic rectangle, the width is divided according to the first variable's categories and their numbers of entries. Each of these bins is divided again and again horizontally in the same way according to the other variables. In MANET standard mosaic plots can be interactively rotated in this form by a simple mouse click and therefore we call these plots *rotated mosaic plots*. Highlighting splits the bins still vertically. Thus highlighting heights in a p dimensional mosaic plot of variables $X_1, \ldots, X_p$ show the conditional probabilities $P(Y|X_1, \ldots, X_p)$ when Y denotes the highlighting variable. In addition, to determine the exact combination that a bin represents more easily labels can be added underneath the mosaics (see figure 3). Each row of the labels corresponds to one variable, the white rectangles stand for "0"s, the blacks for "1"s. The first bin in fig. 3 therefore represents the combination of "no item bought" (all "0"s), the second bin contains all transactions, where only *corned beef* has been bought, and so on. This form of diagrams is also known as doubledecker plots introduced in Hofmann et al. (2000). This version of mosaic plots presents a good solution to the labelling problem of mosaics.

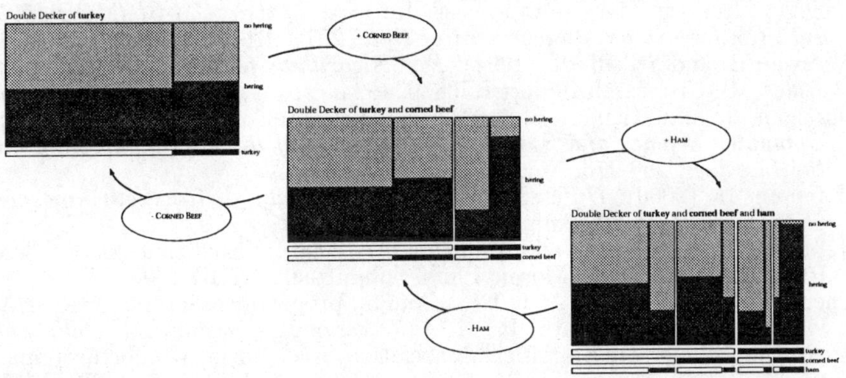

Fig. 4. *Sequence of Doubledecker plots. Highlighting shows* **hering** *buyers. From left to right variables are added to the display.*

Mosaic plots and Doubledecker Plots as well are constructed in a strictly hierarchical manner. One of their interactive features therefore deals with including and excluding variables. Statistically, the adding of variables corresponds to a further split of the underlying table. In terms of association rules it means, the addition of another item to the left hand side of a rule. The leftmost display in Figure 4 shows the doubledecker plot for the association rule **Turkey** → **Hering**. This rule is not significant since the highlighting proportions are almost equal. In the other plots, the variables **Corned Beef** and **Ham** have been added sequentially to the left hand side of the rule. The center plot suggests a strong rule **Turkey** & **Corned Beef** → **Hering**. The confidence of this rule is increased while the support almost remains the same when **Ham** is added to he rule, see right display in Figure 4. Grouping categories can be done for various reasons. One of them certainly is that groups belong logically together. Several items for instance can belong to the same class of products. This opens the way for including metainformation such as taxonomies to the data. Figure 5 shows an example of combining two items (**hering** and **sardines**) to a new class of products (**fish**).

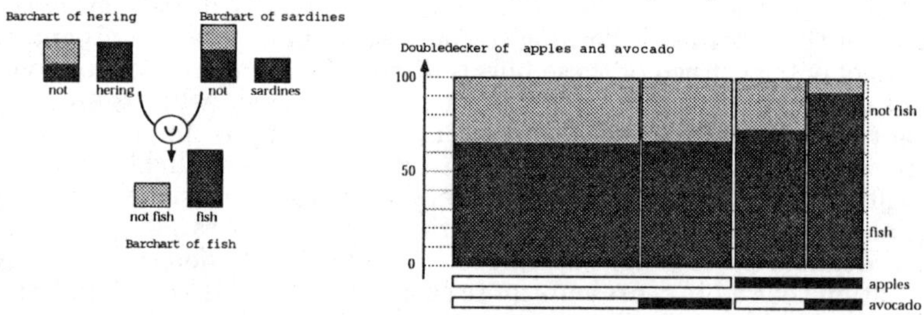

Fig. 5. *Barcharts for grouping* ***sardines*** *and* ***herings*** *to a new variable* ***fish****. The doubledecker on the right shows an association rule based on this new variable (apples&avocado → fish).*

References

Agrawal, R., Imielinski T. and Swami A. (1993). Mining Associations between Sets of Items in Massive Databases. In: *Proc. of the ACM-SIGMOD 1993 Int'l Conference on Management of Data*, 207 - 216, Washington D.C.

Agrawal, R. and Srikant R. (1994). *Fast Algorithms for Mining Association Rules.* IBM Research Report RJ9839, 1994.

Hartigan, J. and Kleiner, B. (1981). Mosaics for Contingency Tables. In: *Computer Science and Statistics: Proceedings of the 13th Symposium on the Interface*, 268-273, New York: Springer Verlag.

Hofmann, H. (2000). *Different levels of interactivity — Interactive mosaic plots.* Metrika, (forthcoming).

Hofmann, H., Siebes, A. and Wilhelm, A. (2000). Visualizing Association Rules with Interactive Mosaic Plots. Submitted to KDD 2000.

Theus, M. and Wilhelm, A. (1998). Counts, proportions, interactions — A view on categorical data. In: *ASA Proceedings - Section on Statistical Graphics*, American Statistical Association, Alexandria, VA, forthcoming.

Tufte, E. (1983). *The Visual Display of Quantitative Information.* Cheshire: Graphics Press.

Optimality models for PRAM

Leon Willenborg

Statistics Netherlands, Department of Statistical Methods, P.O. Box 4000, 2270 JM, Voorburg, The Netherlands

Abstract. The paper deals with the application of a technique to protect microdata against disclosure. This technique is called the Post Randomization Method (PRAM), and it works through a random perturbation of categorical variables, identifiers in fact, in a microdata set. In this paper two optimization models are formulated to arrive at nonsingular PRAM matrices, which are stochastic (Markov) matrices. The models presented are Nonlinear Programming (NP) models. The first model turns out to be equivalent to a Linear Programming (LP) problem. The second model, which is close to a geometric programming model, can be reformulated as a nonlinear optimization problem under linear constraints.

Keywords: Data perturbation, statistical confidentiality, microdata, nonlinear programming, geometric programming, linear programming

1 PRAM in a nutshell

The Post Randomization Method (PRAM) is described in Kooiman et al. (1997), Gouweleeuw et al. (1998), De Wolf et al. (1998) as an alternative technique for protecting microdata against disclosure. The technique can be seen as an analogue to noise addition. Whereas noise addition is applied to continuous variables, PRAM is applied to categorical variables. In microdata sets the idea is that PRAM should be particularly applied to identifiers in the microdata. In practice identifiers are usually categorical variables.

Given a microdata set M PRAM works as follows. Take an identifier v in M, which is a categorical variable which can take the values $a_1, \ldots, a_k$. We assume that a $k \times k$ Markov matrix $P = (p_{ij})$ is given. Now the records in M are taken one by one. If $v = i$ for a record r in M then with probability p_{ij} the value i for v is replaced by the value j. Note that the new value j—which may happen to be equal to the old value i—is generated independently of any other value in r or in another record in M.

So if the PRAM matrix P is given the execution of PRAM is easy. The problem to which this paper is devoted is the determination of a PRAM matrix, taking certain side conditions into account, such as the degree of safety obtained by the PRAM procedure and the nonsingularity of the PRAM matrix.

To make the discussion a bit more concrete consider for example the combination of values (in string notation)

$$u = u_1 u_2 u_3, \tag{1}$$

which is a combination of scores on the combination of variables $v_1 v_2 v_3$. Then the probability that the combination will remain unchanged after the application of PRAM is

$$p_{uu} = p_{u_1u_1}p_{u_2u_2}p_{u_3u_3}, \tag{2}$$

where $p_{u_ju_j}$ is the probability that category u_j of variable v_j remains the same after PRAM, i.e. an *invariance probability*. Expression (2) is an example of an *invariance probability for a combination*, in this case of u as defined in (1). Probability (2) is the product of three diagonal elements of three stochastic (sub)matrices, or PRAM (sub)matrices, P_{v_1}, P_{v_2} and P_{v_3} corresponding to the variables v_1, v_2 and v_3 respectively.

In general, for each identifier v we assume such a PRAM submatrix P_v to exist. The PRAM matrix is a stochastic matrix of block diagonal form, where the blocks on the diagonal are the PRAM submatrices P_v, for the identifiers v. In a PRAM matrix the diagonal entries are the invariance probabilities. So stated in our new terminology one aim of PRAM is to make the invariance probabilities p_{uu} of each unsafe combination (UC) u sufficiently small.

However, for statistical purposes (cf. Kooiman et al. (1997), De Wolf et al. (1998), Gouweleeuw et al. (1998) or Van den Hout (1999)) it is desirable that our PRAM matrix is also non-singular. This is ensured if we require that such a matrix is (row) dominant diagonal. This means that for each diagonal element p of a PRAM matrix P holds $p > 1 - p$ where $1 - p$ represents the sum of the off-diagonal elements in P in the same row as p. Of course, this condition is equivalent to the requirement that $p > 1/2$. Instead, we shall usually require for each invariance probability that $p \geq 1/2 + \epsilon_p$ for each diagonal element p, where $\epsilon_p > 0$ is a parameter to be specified by the data protector. From this it follows that for any combination c we have $p_{cc} > 1/2^{|c|}$, where $|c|$ denotes the *length* of c, i.e. is the number of different variables/values that define c. In particular this holds for an unsafe combination (UC) u, for which we have $1/2^{|u|} < p_{uu} \leq r_u$, where r_u is an upper bound for p_{uu} due to the safety requirements for u. This implies that the direct perturbative effect of PRAM applied to an UC u is limited due to the lower bound $1/2^{|u|}$.

Now, however, comes a secondary effect due to PRAM into play. This is the creation of "artificial" UCs. In other words PRAM makes sense not only because it changes UCs to other combinations but also because other combinations are changed to unsafe ones. So if an intruder sees an UC in a PRAM-ed file he cannot be sure that he is dealing with a true, unaltered UC or an artificial one, i.e. one that is the result of a perturbation.

Without this secondary effect PRAM would in fact be a useless method, since it would not protect UCs that have not changed after PRAM. Fortunately it does. So to phrase it a bit differently, the two pillars on which the effectiveness of PRAM rests are

1. **Outflow** The flow from each UC to other combinations should be large enough.
2. **Inflow** The flow from the other combinations to each UC should be sufficiently large.

In Model II (cf. Section 3) the second effect, the outflow, is taken into account explicitly and on an equal footing with the control of the outflow. Model I (cf. Section 2) only deals with the first effect, the inflow, explicitly; the second effect is either dealt with in a more casual way (in case of an unsophisticated approach), or explicitly (in a more sophisticated approach). In this paper we shall not deal extensively with the second step of Model I; see Willenborg (1999) for more details.

When we perturb a combination, a new combination will appear that can be classified as as an element in exactly one of the sets UC, SC or ZC. Here UC is the set of combinations that are unsafe in the original file M, SC the set of combinations that are safe in M and ZC the set of zero combinations, i.e. of combinations that do not occur in the original file M. Because the result of PRAM is that combinations (of values) are sent to other combinations of values, we can still partition the PRAM-ed combinations in terms of UC, SC or ZC. Of course, the meaning of this partitioning is different from that before the PRAM-ing. In other words, PRAM induces a flow among the sets UC, SC and ZC.

It should be stressed that we take ZC as an *approximation* to the set of structural zeros, i.e. the combinations that are impossible to occur or that happen not to occur in a population, and use it as a substitute for this set. If the sample size is not too small (compared to the number of cells in a table, corresponding to a combination of variables) this assumption is reasonable.

2 Model I: A two-step model

We assume a microdata set M to be given. The goal is to modify M by the application of PRAM, so as to arrive at a safe data set. We assume that, in principle, all identifiers in M will be PRAM-ed, that is, as far as they are involved in UCs. So there is no selection of identifiers to be PRAM-ed. We then talk about a PRAM matrix P for the super-variable v, constructed from all these identifiers. P is a block diagonal matrix, where each block is a stochastic matrix (a PRAM sub-matrix), which corresponds to one of the identifiers. If an identifier is not really PRAM-ed, the corresponding block matrix is the identity matrix.

2.1 Step 1: Diagonal entries

We start with a verbal description of the model. In formulating it we follow the general paradigm for disclosure protection: minimize information loss under the conditions of safety (and possibly other conditions). We translate this as follows into our setting.

First some words about the constraints. These are on the one hand inspired by safety requirements on the other by the non-singularity demand of the PRAM matrix to be constructed. For an UC u they require that

$$1/2^{|u|} < p_{uu} \leq r_u < 1, \tag{3}$$

where r_u is to be specified by the data protector, and which sets a level of protection for u. The first inequality in (3) is a result of the dominant diagonality constraint of the PRAM matrix which in the case of u_i requires that

$$p_{u_i u_i} > 1/2. \tag{4}$$

In order to complete our model we have to specify an object function, in this case measuring information loss. For this we take the product of the diagonal elements of the PRAM matrix, and we require this expression to be maximized, for the larger this expression is the closer the PRAM matrix is to the identity matrix. So the objective is to maximize the product of the entries on the main diagonal, i.e.

$$\max \prod_i p_{u_i u_i}, \tag{5}$$

where the product is taken over the set of all categories that appear in at least one UC (because the ones that do not occur in an UC can be taken equal to 1).

Now putting $x_i = -\log p_{u_i u_i}$ the optimization problem amounts to the following LP-problem

$$\min_x \iota' x \tag{6}$$

$$\text{s.t. } Jx \geq b$$
$$x < e, \tag{7}$$

where $\iota = (1, \ldots, 1)'$ the all-ones column vector of length equal to the number of variables x_i, J the incidence matrix of the UCs with $J_{jk} = 1$ if the k-th value is in the j-th UC and $J_{jk} = 0$ otherwise, b is a column vector of length equal to the number of UC's and with components $-\log r_u$ for each UC u, $e = \log 2\iota$. We can replace the final strict inequality in (7) by an inequality of the type $x \leq f$, for a vector f with $f < e$ and f close to e. So we arrive at our final formulation for the first step of Model I:

$$\min_x \iota' x \tag{8}$$

$$\text{s.t. } Jx \geq b$$
$$x \leq f.$$

It should be noted that we started with a Geometric Programming (GP) model, that turns out to be an LP-problem in disguise. No duality is needed to arrive at LP-model (8).

2.2 Step 2: Off-diagonal entries

The determination of the off-diagonal entries is necessary to control the inflow to the UCs. It can be done in two ways, one easy and the other somewhat more elaborate. The easy way is just to fill in the off-diagonal entries such that a stochastic matrix is obtained. In choosing the entries one can be guided by computational arguments, for instance. The second way is more complicated and is based on confidentiality arguments which are essentially the same as those used in Model II. For this reason we shall not elaborate them here. Refer to Willenborg (1999) for more details.

3 Model II: A one-step model

For the safety of the PRAM process it is important to control the inflow and outflow for the UCs (cf. Section 1). At the same time we should be wary that too many inconsistencies occur in the PRAM-ed microdata set, because these might help an intruder to undo a PRAM-ing. We translate this into the requirement that the inflow in ZC should be minimized. So information loss is measured here in terms of inflow to ZC.

We require, as in Model I, for the invariance probability p_{uu} for each u in UC that

$$1/2^{|u|} < p_{uu} \leq r_u < 1, \tag{9}$$

where, as in Model I, the threshold r_u should be chosen by the data protector. Condition (9) controls the degree of outflow for an UC u. In view of the nonsingularity requirement for the PRAM matrix we have the following general constraints:

$$p_{c_i c_i} > 1/2, \tag{10}$$

for each value c_i. Note that the requirement in (10) holds for all values c_i, i.e. all scores for the identifiers that appear in the combinations of variables that we have to consider, and not only for those appearing in a UC as in (4). Instead of (10) we will use for each c_i the following equivalent constraints:

$$\theta_{c_i} p_{c_i c_i}^{-1} \leq 1, \tag{11}$$

for a suitably chosen parameter $\theta_i > 1/2$.

To make sure that there is sufficient inflow for each UC u, we could require

$$\sum_{c \in UC \cup SC} v_c p_{cu} \geq \delta_u. \tag{12}$$

where v_c is the frequency of combination c in the original file M. However, in view of the format of GP-models the following constraints are considered for each u in UC instead:

$$\sum_{c \in UC \cup SC} (v_c p_{cu})^{-1} \leq \zeta_u. \tag{13}$$

Note that the replacement of (12) by (13) is in fact a replacement of a restriction on an arithmetic mean by a restriction on the corresponding harmonic mean. If we divide left-hand and right-hand side of (13) by ζ_u we get a restriction of the type as used in the formulation of the GP-model, if we substitute for each p_{cu} the product of the transition probabilities in the PRAM matrix.

As stated above, the goal is to apply PRAM in such a way that, under the constraints above, the flow to ZC is minimized. In other words, the aim is to

$$\min \sum_{c \in UC \cup SC} \sum_{d \in ZC} v_c p_{cd}, \tag{14}$$

under the constraints (9), (11) and (13). Summarizing the above, our Model II is the following GP-like problem:

$$\min \sum_{c \in UC \cup SC} \sum_{d \in ZC} v_c \prod_i p_{c_i d_j} \tag{15}$$

$$\text{s.t. } r_u^{-1} \prod_i p_{u_i u_i} \leq 1,$$

$$\theta_{c_i} p_{c_i c_i}^{-1} \leq 1,$$

$$\sum_{c \in UC \cup SC} (\zeta_u v_c)^{-1} \prod_i p_{c_i u_i}^{-1} \leq 1,$$

$$\sum_{c_j} p_{c_i c_j} = 1,$$

$$p_{c_i c_j} > 0.$$

where the inequalities are assumed to hold for all UC's (i.e. those involving u's) or all combinations (i.e. those involving c's), and the equalities hold for all c_i's. Due to these equalities the model 15 is not a pure GP-problem (hence the qualification GP-like model). In Avriel (1980) equality constraints in GP-models are briefly discussed, and it is clear that they are complicating factors in such models. No more than trial-and-error methods are suggested to cope with such models. It is not clear to the present author if more recent developments in the theory of GP-models have shed some light onto this issue.

In practice the number of terms in the object function can be very large, and there may be a need to reduce this number. One way to achieve this is to apply some global recoding first to the identifiers in the microdata set M in order to reduce the respective sizes of UC, SC and ZC. Another possibility is to only consider transitions from combinations c of $UC \cup SC$ to a d in ZC that differs in exactly one position from c. This is defendable because such transitions require only one value to change in a combination c. This is more likely to occur than two or more such changes in such a combination.

4 Concluding remarks

The present paper is essentially a summary of Willenborg (1999), to which the reader is referred (cf. also Willenborg and De Waal (2000)).

Model I has been implemented and some empirical experience has been gained with it but not enough to give a final verdict on its usefulness. Model II is more interesting than Model I since it controls both the inflow to and the outflow of the unsafe combinations. Research on Model II, however, is still going on at the time of writing.

Another way handle the approach underlying Model II is to first solve the flow problem at the level of the combinations, yielding transition probabilities for these combinations. In the second step one finds the transition probabilities at the level of the variables, using using linear regression for the logarithms of the transition probabilities.

References

Avriel. M. (ed.) (1981) *Advances in Geometric Programming*, New York: Plenum Press.

De Wolf, P.P., J.M. Gouweleeuw, P. Kooiman and L. Willenborg (1998) Reflections on PRAM, *Proceedings SDP98, Lisbon, 25–27 March.*

Gouweleeuw, J., P. Kooiman, L. Willenborg and P.-P. de Wolf (1998) The Post Randomization Method for Protecting Microdata, *Qüestiió*, **22**, 145–156.

Kooiman, P., L. Willenborg and J. Gouweleeuw (1997) PRAM: A Method for Disclosure Limitation of Microdata, Report, Department of Statistical Methods, Statistics Netherlands, Voorburg.

Willenborg, L. (1999) Optimization Models for PRAM Matrices, Report, Department of Statistical Methods, Statistics Netherlands, Voorburg.

Willenborg, L. and T. de Waal (1996) *Statistical Disclosure Control in Practice.* New York: Springer-Verlag.

Willenborg, L. and T. de Waal (2000) *Elements of Statistical Disclosure Control.* New York: Springer-Verlag.

Dealing with real ordinal data: recent advances in analyzing tied, censored, and multivariate observations

Knut M. Wittkowski

The Rockefeller University, General Clinical Research Center, New York, NY 10021, USA

Abstract. When ordinal data are inexact (tied, censored, multivariate), it is demonstrated that both asymptotic and "exact" rank tests (including the sign test) may yield liberal results. It is demonstrated that using the unconditional variance (without "correction for ties") for inexact data ensures that the probability of a "significant" result does not exceed the level. The proposed approach is based on the marginal likelihood principle. It is easily extended to multivariate ordinal data, allows for adjustment for confounding, is computationally feasible, and suggests criteria for the design of user interfaces.

Keywords. sign test, rank test, marginal likelihood, score, ties, exact, permutation, conditional, confounding, user interface

1 Introduction

Rank test (based on u-statistics) are widely accepted for the analysis of ordinal data in general, i.e. for data, where the functional relationship between the scale of the observations and their relevance in terms of the underlying concept is unknown. When such data are tied or censored, however, situations may arise, where ranks cannot be assigned in a unique fashion, and the discussion how to deal with such "inexact" data has recently again attracted attention.[Rayner et al., 1999]

For the sign test, first published in the early 18th century by John Arbuthnoth, this problem was originally identified in the 1940s. After some discussion,[Dixon et al., 1946] it was suggested that "tied" observations (no change) should always be discarded from the analysis.[Dixon et al., 1951] By the same argument, it was generally accepted in the 1950s that the variance of rank test should always be "corrected" (reduced) when observations are tied[Putter, 1955] or (interval) censored.[Kalbfleisch et al., 1973; Prentice, 1978; Schemper, 1983]

In the present paper, we will first present counterexamples showing that this treatment of ties may result in commonly used test statistics becoming extremely liberal. We will then introduce a distinction between "exact" and "inexact" ties (Section 3), show how to use this distinction when selecting test statistics (Section 4), and demonstrate that the proposed clarification resolves the anti-conservative behaviour of the currently suggested test statistic when dealing with ties. Finally, we will discuss statistical and computational relevance of the results.

2 Counterexamples

2.1 Sign test

The sign test (often named after McNemar's more recent publication [McNemar, 1947]) is often applied to paired data, when the distribution of differences cannot be assumed symmetric. (Otherwise, the paired t-test would be a robust alternative [Scheffé, 1959]). When the "correction" for ties is used, one may feel uncomfortable with a result being "significant", although only nine observations "improved" and one even "worsened", while 9,990 were (more or less) "unchanged". Assuming, for simplicity of the argument, that the differences follow a triangular distribution with a median of zero and a symmetric discretization interval around the origin (Fig. 1.), the "significance" of sign test results increases with the discretization bandwidth, i.e., with the inaccuracy of the measurements (Table 1.).

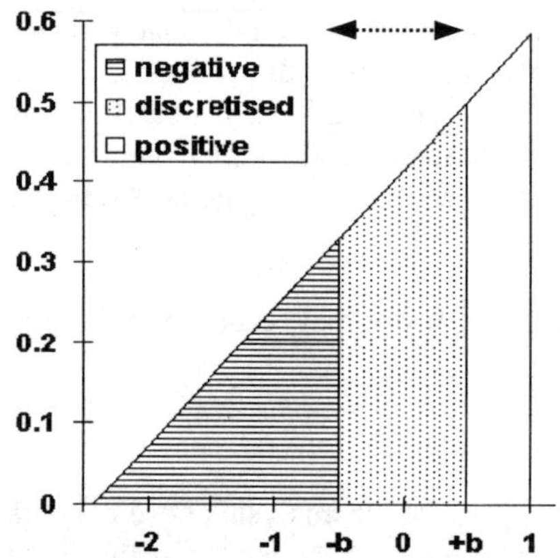

Fig. 1. Triangular distribution, discretized at zero with b=0.5

Table 1: Expected estimates and test statistics (n=25) by bandwidth b

b	p_0	p_+	$\frac{p_+}{1-p_0}$	T_{25}
0.00	0.00	0.50	0.50	-0.00
0.25	0.21	0.39	0.49	-0.06
0.50	0.41	0.27	0.46	-0.28
0.75	0.62	0.14	0.37	-0.78
0.90	0.75	0.06	0.23	-1.38
0.95	0.79	0.03	0.14	-1.68
0.99	0.82	0.01	0.03	-1.98
1.00	0.83	0	0	-2.07
1.50	0.93	0	0	-2.34
>2.41	1.00	0	-	-

2.2 Rank tests for tied/censored data

A similar situation arises when a linear rank test [Prentice, 1978] is used to compare, for instance, two samples (A:3,4; B:1,2,5,6,7; Table 2.). If either the highest observation (in group B) is censored after the highest group A observation, if the three highest observations in group B are tied (or, equivalently, all censored to "4^+"), or if non-informative observations are added, one would expect no effect on the between-group comparison. With both for the asymptotic and the "exact" version [Mehta et al., 1992] of the Prentice/Peto "log-rank" test, however, the test statistics increase (Table 2.; $T_{\text{"exact"}} = -\Phi^{-1}(p/2)$).

Table 2: Censorings, ties, and non-informative observations in linear rank tests

Grp A	Grp B	$T_{as.}$	$T_{\text{"exact"}}$
3,4	1,2,5,6	0.45	0.34
3,4	1,2,5,6,7	0.71	0.50
3,4	1,2,5,6,4^+	0.78	0.64
3,4	1,2,5,5,5	0.83	0.79

Grp A	Grp B	$T_{as.}$	$T_{\text{"exact"}}$
3,4	1,2,5,6	0.45	0.34
3,4	1,2,5,6, $1\times0^+$	0.47	0.37
3,4	1,2,5,6, $194\times0^+$	2.31	2.06
3,4,$18\times0^+$	1,2,5,6, $16\times0^+$	0.46	0.39

3 Two types of ties

To resolve the seeming discrepancy between the theoretical results seeming to indicate a specific treatment of ties as "optimal" and the counterintuitive consequences of using this "optimal" choice, one needs to consider the "nature" of ties. In genetics, for instance, "tied" observations can often be assumed to indicate identical phenomena (e.g., gene present or absent, [Rayner et al., 1999]) Often, however, a thorough formulation of the problem refers to an underlying continuous or unmeasurable factor, rather than the observed discretized variable. For instance, ties may be due to rounding of continuous variables (temperature, Fig. 1.) or to the use of discrete surrogate variables for continuous phenomena (parity for fertility).

4 Methods

4.1 Sign tests

The original sign test [Dixon et al., 1946] can be written as

$$T^* = \frac{(N_+ + p_{\mathrm{H}} N_0) - p_{\mathrm{H}} n}{\sqrt{p_{\mathrm{H}} q_{\mathrm{H}} n}} = \frac{(N_+ + p_{\mathrm{H}} N_0)/n - p_{\mathrm{H}}}{\sqrt{p_{\mathrm{H}} q_{\mathrm{H}}/n}} \sim_{\text{as.}} \mathrm{N}(0,1)$$

and the more recent versions [McNemar, 1947; Dixon et al., 1951] as

$$T = \frac{(N_+ + p_{\mathrm{H}} N_0) - p_{\mathrm{H}} n}{\sqrt{p_{\mathrm{H}} q_{\mathrm{H}} (n - N_0)}} = \frac{N_+/(n - N_0) - p_{\mathrm{H}}}{\sqrt{p_{\mathrm{H}} q_{\mathrm{H}}/(n - N_0)}} \underset{\{\text{if } p_{\mathrm{H}} = \frac{1}{2}\}}{=} \frac{N_+ - N_-}{\sqrt{N_+ + N_-}} \sim_{\text{as.}} \mathrm{N}(0,1)$$

The second form of each version is "normalized" in that the denominator is the (as.) standard deviation of the parameter estimated, $N_+ + p_{\mathrm{H}} N_0$ or $N_+ /(n - N_0)$, respectively. From the first form, it is more easily seen that T^*, based on the unconditional variance, is always smaller than T, based on the conditional variance. The difference in variance is the price to be paid for not having observed the positive and negative "signs" among the tied observations.

4.2 Rank tests

By means of the marginal likelihood (MrgL) principle, scores can be obtained even when not all observations can be ordered pairwise (Fig. 2,3) by averaging across all rankings compatible with the pairwise orderings (Fig. 3)

$$a_{jk} = \frac{1}{\operatorname{card} Q(\mathbf{x})} \sum_{\mathbf{r} \in Q(\mathbf{x})} a(r_{jk})$$

As in the case of the sign test above, the variance of the test statistic

$$T_{LR}^0 = \sum_{k=1}^{m_j} a_{1k} \Big/ \sqrt{\tfrac{m_1 m_2}{m_+} \operatorname{var}_{LR}^0}$$

is either based on all possible rankings

$$\operatorname{var}_{LR}^* = \frac{1}{m_+ - 1} \sum_{r=1}^{m_+} a(r)$$

or on the subset of compatible rankings:

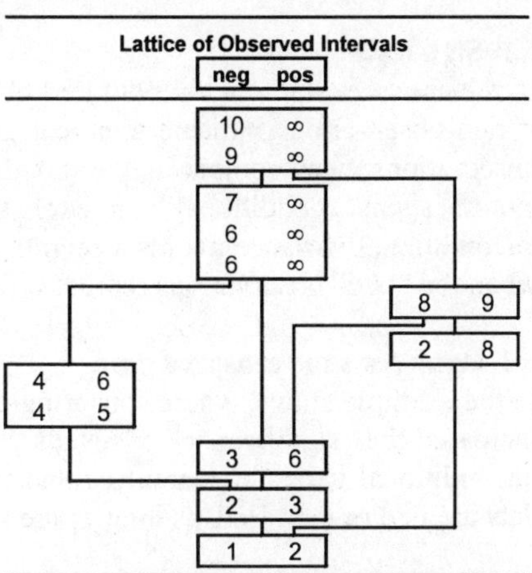

Fig. 2. Pairwise orderings for interval censored data (neg: time of last negative result; pos: time of first positive result; connecting lines indicate pairwise orderings).

$$\mathrm{var}_{LR} = \frac{1}{m_+ - 1} \sum_{j=1}^{2} \sum_{k=1}^{m_j} a_{jk}^2$$

Whether to base the test on the unconditional variance (T^*) or the conditional variance (T) depends on "nature" of the ties. If the ties are "exact", the conditional variance may be used. Otherwise, the unconditional variance should be used.

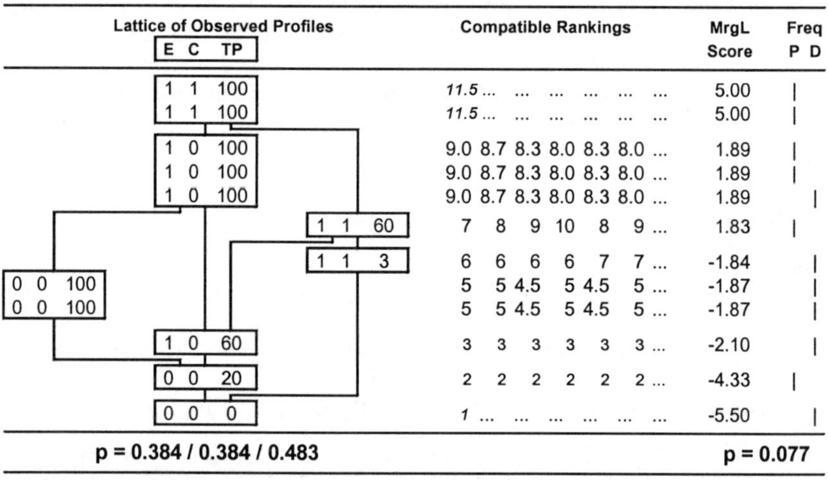

Fig. 3. Computation of MrgL scores for multivariate ordinal data (E,C: binary; TP: continuous). Note that the same scores are obtained for the data in Fig. 2.

5 Results

The distinction between "exact" and "inexact" ties resolves the above counter-examples within a framework of a unified generalization of rank tests to inexact (i.e., tied, censored, and multivariate) ordinal data.

5.1 Sign tests

In the above example with 9990 tied observation, the result remains "significant", if tied observations indicate identical genetic traits. In this case, eliminating tied observations when computing the (conditional) variance is reasonable, because the hypothesis is "conditional" on excluding ties. Otherwise, however, using the unconditional variance avoids a seeming "significant" result, as one would prefer, when the "tied" observations represent (unobservable) differences.

5.2 Rank tests for censored data

In the example above, where censoring or tying the largest observations in group B increased the "significance" of results "corrected" for ties, the proposed use of the unconditional variance guarantees that the test statistic remains invariant when the data are tied or censored, as long as the information content is not changed.

5.3 Rank tests for multivariate ordinal data

The use of the unconditional variance is also indicated for most applications where multivariate ordinal data are analysed in lieu of direct measurements of an underlying (latent) continuous factor.

For the sign test, the need for a distinction between "exact" and "inexact" ties has been demonstrated independently by Wittkowski [Wittkowski, 1989a; Wittkowski, 1998] and, more recently, Rayner and Best [Rayner et al., 1999]. While the different forms of the variance estimator [Wittkowski, 1988] have been linked to the corresponding types of ties earlier [Wittkowski, 1989b; Wittkowski, 1996], the concept is now being more widely accepted and we are now able to extend the results to multivariate data in general and to censored data in particular. We are also able to provide counterexamples explaining in detail why commonly used rank tests tend to be liberal when applied to "real" ordinal data.

The strikingly liberal results obtained when using "exact" (permutation) tests deserve special attention. It is commonly believed that a single permutation test can deal with ties "automatically" and is superior to both of the asymptotic tests. Such an assertion can easily seen to be mistaken by observing that (1) the asymptotical versions of rank tests are asymptotically "exact" and (2) the two versions of the sign test are not asymptotically equivalent for any $N_0/n \to p_0>0$ ($N\to\infty$). The above examples now demonstrate why "exact" tests may be less appropriate for inexact data. If the hypothesis is formulated with respect to the underlying continuous factor, it is not sufficient to consider the distribution of the test statistic conditional on the process, which lead to the discretized observations. The reason that the current permutation tests are not uniformly valid, is that the permutations do not preserve the structure of ties, censorings, or multivariate relationships. Thus, the currently used "exact" permutation tests are appropriate in general only for "exact" data. Luckily, this limitation has little practical relevance. Even for only moderate sample sizes, asymptotic tests (for both exact and inexact data) tend to be surprisingly similar to exact tests, unless designs are highly unbalanced. With small sample sizes, however, randomization is unlikely to eliminate confounding. Thus, "exact" tests excel primarily in the analysis of studies, where the effect of confounding variables is difficult to access.

Ever since the first spreadsheet-type knowledge-based user interface was introduced, attempts have been made to "simplify" the categorization of variables by scale level (nominal, ordinal, interval, and absolute) and scale type (discrete, discretized/continuous).[Wittkowski, 1985; Wittkowski, 1986] With the new examples, it now becomes clear that "to determine how to analyse data",[JMP 3.2, SAS Institute Inc.] it is not sufficient to classify variables simply by a simple "modeling type" as "Continuous, Ordinal, or Nominal". Rather, the distinction between type of ties, or, equivalently, discrete and discretized (continuous), is seen to apply to variables of all scale levels, proofing that more and better structured information is needed for user interfaces that allow for statistically adequate problem definition.[Wittkowski, 1991]

Finally, all these theoretical results would be of little relevance, if implementation of the procedure were not practical. Although computationally intensive in the case of multivariate data, however, recent advances in hardware and algorithms have made this approach feasible, even when adjustment for confounding or mediating variables is deemed necessary [Susser et al., 1998].

References

Dixon, W.J. and Massey, F.J.J. (1951). *An Introduction to Statistical Analysis*. New York: McGraw-Hill

Dixon, W.J. and Mood, A.M. (1946). The statistical sign test. *Journal of the American Statistical Association*, **41**, pp. 557–566

Kalbfleisch, J.D. and Prentice, R.L. (1973). Marginal likelihoods based on Cox's regression and life model. *Biometrika*, **60**, pp. 267–278

McNemar, Q. (1947). Note on the sampling error of the differences between correlated proportions or percentages. *Psychometrica*, **12**, pp. 153–157

Mehta, C. and Patel, N. (1992). *StatXact-Turbo User Manual*. Cambridge: CYTEL Software Corporation

Prentice, R.L. (1978). Linear rank tests with right censored data. *Biometrika*, **65**, pp. 167–179

Putter, J. (1955). The treatment of ties in some nonparametric tests. *Annals of Mathematical Statistics*, **26**, pp. 665–680

Rayner, J.C.W. and Best, D.J. (1999). Modelling Ties in the Sign Test. *Biometrics*, **55**, pp. 663–665

Scheffé, H. (1959). *The analysis of variance*. New York: Wiley

Schemper, M. (1983). A nonparametric k-sample test for data defined by intervals. *Statistica Neerlandica*, **37**, pp. 69–71

Susser, E., Desvarieux, M., Wittkowski, K.M. (1998). Reporting sexual risk behavior for HIV, a practical risk index and a method for improving risk indices. *American Journal of Public Health*, **88**, pp. 671-674

Wittkowski, K.M. (1985). *Ein Expertensystem zur Datenaltung und Methodenauswahl für statistische Anwendungen [An expert system for data management and method selection for statistical applications]*. Stuttgart: Technische Universität

Wittkowski, K.M. (1986). An expert system for testing statistical hypotheses. In: Boardman, T.J. (ed) *Computer Science and Statistics*, pp. 438–443. Washington: American Statistical Association

Wittkowski, K.M. (1988). Friedman-type statistics and consistent multiple comparisons for unbalanced designs. *Journal of the American Statistical Association*, **83**, pp. 1163–1170

Wittkowski, K.M. (1989a). An asymptotic UMP sign test for discretized data. *The Statistician*, **38**, pp. 93–96

Wittkowski, K.M. (1989b). A note on recommendations regarding the handling of ties in software for nonparametric statistical methods. *Statistical Software Newsletter*, **15**, pp. 107

Wittkowski, K.M. (1991). A structured visual language for a knowledge-based front-end to statistical analysis systems in biomedical research. *Computer Methods and Programs in Biomedicine*, **35**, pp. 59–67

Wittkowski, K.M. (1992). An extension to Wittkowski. *Journal of the American Statistical Association*, **87**, pp. 258

Wittkowski, K.M. (1996). A note on "Rahlfs, V.W. and Zimmermann, H., Nonparametric Methods in Standard Statistics Packages, The Case of the Friedman Analysis". *Computational Statistics and Data Analysis (Section IV)* **21**, pp. 116–117

Wittkowski, K.M. (1998). Versions of the sign test in the presence of ties. *Biometrics*, **54**, pp. 789–791

Testing for differences in location: a comparison of bootstrap methods in the small sample case

Karin Wolf-Ostermann

Institute of Health Economics and Clinical Epidemiology, University of Cologne, 50935 Cologne, Germany

Abstract. We compare several bootstrap methods for testing location differences in the two-sample case for small and moderate sample. We concentrate especially on the empirical shape of the underlying samples in order to allow a simple and obvious empirical identification of the properties (and limitations) of the used method. Our results show that there is an urgent need for more detailed empirical investigations of bootstrap properties in the finite sample case.

Keywords. adaptive test, bootstrap, kurtosis, location test, skewness, small sample

1 Introduction

A standard problem in statistical practice is the application of tests concerning differences in location. Since classical test procedures often require (strong) distributional assumptions like symmetry, equality of variances or special distributional models non parametric bootstrap methods for testing are often seen as a promising alternative. But although non parametric bootstrap methods are avoiding assumptions which limit other parametric or non parametric methods they are however not entirely assumption free. Using them we assume that the empirical distribution function is a good estimate of the unknown true distribution function. Therefore a given sample itself can influence the outcome of a bootstrap procedure seriously, regardless of probably favourable asymptotic properties of the bootstrap procedure.

There exist a multitude of different bootstrap procedures proposed for hypothesis testing. In many cases asymptotic properties of these bootstrap methods have been analysed and found comparable to classical test procedures. But there is often little knowledge about the performance of bootstrap tests in the small sample case since most of the theoretical results only hold if the sample size goes to infinity. We compare several bootstrap methods for testing location differences in the two-sample case to give some advice to the user of bootstrap methods to what extent one can rely on results. In Wolf-Ostermann (1996) we showed for small and moderate sample sizes that bootstrap location tests in the one-sample case are very sensitive to the empirical skewness of the underlying sample. We therefore developed adaptive bootstrap location tests which have good asymptotic and finite sample properties (Wolf-Ostermann 1998). We now adapt these modifications to the two-sample case and compare them to other approaches. We analyse location tests in the two-sample case because of its great practical relevance. In order to allow a simple and obvious empirical identification of the properties (and limitations) of an used method we concentrate especially on the empirical shape of the underlying samples.

2 Bootstrap procedures

Let $\mathbf{x}=(x_1,...,x_n)$ be a realisation of a random sample $\mathbf{X}=(X_1,...,X_n)$ from an unknown probability distribution F_X with mean μ_X and let $\mathbf{y}=(y_1,...,y_m)$ be a realisation of a random sample $\mathbf{Y}=(Y_1,...,Y_m)$ from an unknown probability distribution F_Y with mean μ_Y. We consider the two-sided test problem H_0: $\mu_X - \mu_Y = \mu_0$ (= 0) vs. H_1: $\mu_X - \mu_Y \neq \mu_0$. Beran (1986) proposed bootstrap techniques for hypothesis testing and proved the uniform consistency of simulated power functions. We use his approach in which the critical values of a test are approximated by the percentiles of the bootstrap test statistic distribution, and the unknown sample distribution is replaced by its empirical distribution. We compare the following non parametric bootstrap approaches:

- **standard bootstrap** (e.g. Efron & Tibshirani (1993)): This denotes the "original" non parametric bootstrap as it was introduced by Efron in 1979. It is the easiest and most straightforward way to implement a bootstrap test procedure.
- **adaptive trimmed bootstrap** (Wolf-Ostermann (1998)): This comprises two fine-adaptive bootstrap tests for a location parameter which depend on the empirical skewness and kurtosis of the underlying samples. We modify the given samples by asymmetrical trimming in order to get symmetrical empirical distributions with moderate tails. The trimming portion is determined by the empirical skewness and kurtosis of the samples. For the first procedure (*trim1*) we interpret the difference of trimmed means as an estimator of the true difference of means. For the second procedure (*trim2*) we take the trimmed samples as the given underlying samples.
- **adaptive bootstrap based on power functions** (Wolf-Ostermann (1998)): A fine-adaptive bootstrap test procedure in which the estimated power functions itself are used as a selection criterion to adapt the appropriate test. The test with the uniformly best estimated bootstrap power function is selected. Because of asymptotic considerations – the power functions of all included tests must be uniformly consistent estimates - is the approach based on the following tests: the standard bootstrap t-test and two adaptive trimmed tests as described before.
- **Efron's method II with identification of outlier bootstrap samples** (Jiménez-Gamero et al. (1998)): This bootstrap procedure disregards all bootstrap samples which are considered to be outlying bootstrap samples according to a discordancy criteria. This criterion is based on the number of distinct elements in the bootstrap sample. We adopt this method using algorithm 2 of Jiménez-Gamero et al. (1998) and leaving out all bootstrap samples which lie below the approximate 5%-quantile of the distribution of distinct elements contained in the bootstrap sample.

Results obtained by these bootstrap procedures will be characterised in the following way: standard bootstrap by *Tn-TSUR*, trimmed bootstrap by *T_trim1-TSUR* and *T_trim2-TSUR*, adaptive bootstrap based on power functions by

T_adap-TSUR and Efron's method II with identification of outlier bootstrap samples by *T_obs-TSUR*.

3 A Monte Carlo investigation

Following Young (1994) who demands empirical investigations of bootstrap properties in the finite sample case, the performance of the above mentioned bootstrap test procedures is analysed by a simulation study. All comparisons are carried out by considering significance levels and powers. The research concentrates on the case of small (and moderate) sample sizes which are of interest for the practical use of bootstrap tests. A bootstrap algorithm with equal selection probabilities – two-step uniform resampling (TSUR) – is used to estimate significance level and power of different bootstrap test procedures. The results are achieved using 1000 replicates for estimating the critical values and 1000 replicates for estimating the power functions.

We take the following different sample distributions with expectation μ denoting skewness (β_1[1]) and kurtosis (β_2) in a moment-based representation:

Normal *N(0,1)*:	$\mu = 0$	$\beta_1 = 0.0$	$\beta_2 = 3.0$
positive skewed Beta *+Beta(1,3)*:	$\mu = 0$	$\beta_1 = 1.4$	$\beta_2 = 3.1$
negative skewed Beta *-Beta(1,3)*:	$\mu = 0$	$\beta_1 = -1.4$	$\beta_2 = 3.1$
positive skewed Exponential *+Exp(-1,1)*:	$\mu = 0$	$\beta_1 = 2.0$	$\beta_2 = 9.0$
negative skewed Exponential *-Exp(-1,1)*:	$\mu = 0$	$\beta_1 = -2.0$	$\beta_2 = 9.0$

We analyse all possible combinations of these sample distributions, except those which are redundant because of symmetry. The combinations are representing typical situations of asymmetry and/or heavy tails. For each combination we draw 1000 independent samples of each distribution and perform the above mentioned bootstrap tests. The results presented below are "median results" of this 1000 samples although no serious differences could be seen in comparison to using means instead. The sample sizes of the two samples to be tested are equal and set to n = 20,50,99 while the nominal level of each test is $\alpha = 0.05$.

4 Results

The results of the simulations show to which empirical features of the underlying samples the particular bootstrap methods are sensible. Especially asymmetry of the empirical sample distributions is a crucial factor of influence. We found that the **estimated levels** of all bootstrap tests are very close to the nominal level of 5%, ranging from 4.7% to 5.2%. The **estimated power functions** are seriously influenced by asymmetrical distributed samples in the case of small samples. This is especially the case if we have the combination of positive and negative skewness. Roughly the analysed bootstrap test procedures can be divided in two groups. One group consists of the test procedures *Tn-TSUR*, *T_obs-TSUR* and *T_trim1-TSUR*, the other of *T_trim2-TSUR* and *T_adap-TSUR*. The tests of the first group are more seriously affected by the empirical shape of the underlying

[1] For simplicity we denote the moment-based skewness coefficient by β_1 instead of $\sqrt{\beta_1}$ as usual.

samples. This is especially true for the tests *Tn-TSUR* and *T_obs-TSUR*. Figure 4.1 shows the differences of estimated power functions in the case of two samples from exponential distributions. In table 4.1 a selection of estimated power function values is summarised.

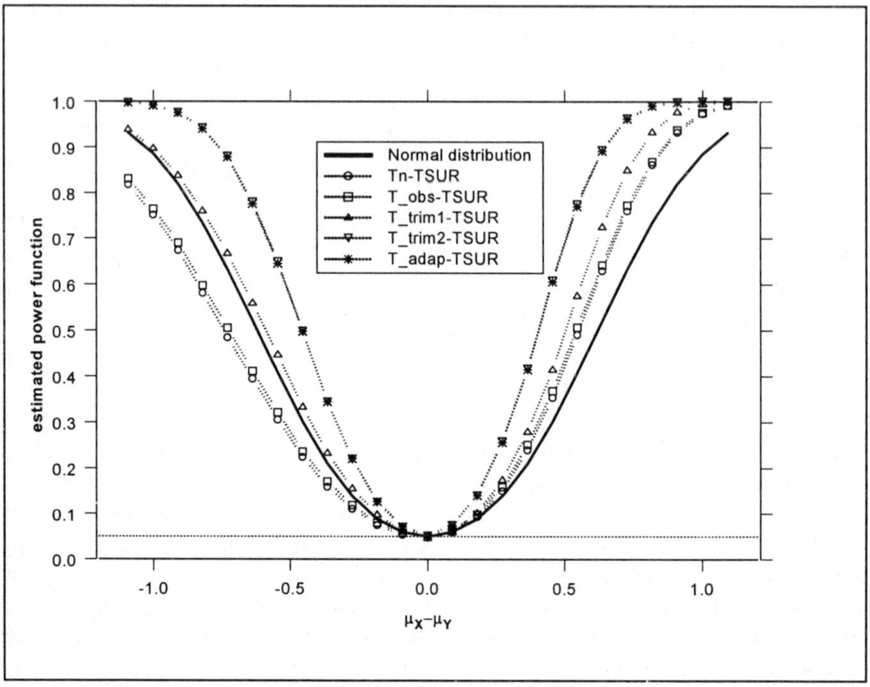

Fig. 4.1. Simulated power functions with $n = 20$, $\alpha = 0.05$ and F_X and F_Y being a positive and a negative skewed exponential distribution. ("Normal distribution" indicates the asymptotic power function of the classical t-test in the case of two normal distributed samples.)

In the case of two normal distributed samples all bootstrap test procedures perform well, the procedures *Tn-TSUR*, *T_obs-TSUR* and *T_trim1-TSUR* are nearly indistinguishable from the asymptotic power function of the classical t-test. This results remains nearly the same for two beta-distributed or two exponential-distributed samples as long as the skewness of both sampling distributions is (nearly) equal. The more the two samples differ in skewness the greater are the differences in the estimated power functions. Combining two samples of which only one has a skewed sampling distribution or two samples with both having positive (or negative) skewness but to a different extent leads to different results for the analysed bootstrap test procedures. For positive (or negative) skewness the bootstrap tests *Tn-TSUR*, *T_obs-TSUR* and – to a smaller extent – *T_trim1-TSUR* have no longer symmetric estimated power functions relating to μ_0 but tend to underestimate the true power function on one side of μ_0 and to overestimate it on the other side. Combining two samples with positive and negative skewed sampling distributions leads to similar results but the effects are more striking. In all model situations we simulated the bootstrap tests *T_trim2-TSUR* and *T_adap-TSUR* seem practically not affected by asymmetry of the underlying sample distributions. Their estimated power functions are symmetric to μ_0 and the estimated power values are remarkable greater than those of the other bootstrap tests.

sample distributions F_X / F_Y	*Tn-TSUR*		*T_obs-TSUR*		*T_trim1-TSUR*		*T_trim2-TSUR*		*T_adap-TSUR*	
N(0,1) /	73.8	8.6	74.9	9.2	74.5	9.7	83.8	10.0	83.1	10.1
N(0,1)	73.7	8.5	75.0	9.1	74.3	8.6	83.8	10.2	82.7	10.2
N(0,1) /	77.9	8.8	79.0	9.4	78.2	8.8	87.7	10.6	86.7	10.6
+Beta(1,3)	69.6	8.2	70.9	8.8	73.0	8.6	84.6	10.2	84.0	10.3
N(0,1) /	80.5	9.0	81.6	9.6	81.7	9.1	91.9	11.4	91.5	11.4
+Exp(-1,1)	67.2	8.2	68.5	8.8	72.5	8.8	88.6	10.9	88.5	11.9
+Beta(1,3) /	73.9	8.5	75.1	9.2	74.4	8.5	88.1	10.7	87.4	10.7
+Beta(1,3)	73.8	8.5	75.1	9.1	74.2	8.6	87.8	10.8	87.5	10.8
+Beta(1,3) /	64.8	7.8	66.4	8.4	74.0	9.0	84.5	10.3	83.7	10.3
- Beta(1,3)	81.4	8.9	82.4	9.6	83.8	9.0	90.8	11.0	89.9	11.0
+Exp(-1,1) /	74.3	8.9	75.4	9.6	74.8	9.0	95.9	12.7	95.7	12.6
+Exp(-1,1)	74.5	8.9	75.5	9.6	74.6	8.9	96.1	12.7	95.8	12.7
+Exp(-1,1) /	58.1	7.4	59.7	8.0	76.1	9.7	94.2	12.5	94.0	12.6
-Exp(-1,1)	86.2	9.2	86.9	9.8	93.3	10.1	99.1	14.0	99.0	13.9
+Beta(1,3) /	76.9	8.9	78.0	9.6	76.9	8.9	92.0	11.3	91.5	11.3
+Exp(-1,1)	71.4	8.6	72.7	9.3	72.6	8.7	92.8	11.6	92.3	11.5
+Beta(1,3) /	61.8	7.7	63.3	8.2	74.3	9.1	88.3	10.9	87.3	11.0
-Exp(-1,1)	84.1	9.1	85.0	9.8	88.5	9.5	95.5	12.0	95.0	11.8

Table 4.1. Estimated values (in %) of power functions at μ_0 = -0.8, -0.2 / 0.8, 0.2.

Comparing the ***(estimated) dispersion*** of the power function estimates we find that the adaptive bootstrap test procedures (*T_trim1-TSUR, T_trim2-TSUR, T_adap-TSUR*) tend to have a slightly greater spread in the middle part right and left of μ_0. The more the sample distributions are skewed and the more the two samples differ in skewness the greater is the (estimated) dispersion of the estimated power functions for all analysed bootstrap test procedures. Of course the estimated dispersion decreases with increasing sample sizes.

The described influence of sample skewness and/or sample kurtosis on the performance of the analysed bootstrap test procedures is depending on the sample sizes. The smaller the sample sizes are the more serious are the detected effects of the empirical shape of the sample distributions. With increasing sample sizes the effects tend to diminish. We have not yet investigated the influence of different sizes of the two samples in combination with the empirical shape of the sample distributions.

5 Conclusions

In this paper we briefly reviewed that favourable asymptotics of bootstrap test procedures is no guarantee of good behaviour for small or even moderate sample sizes. Standard non parametric bootstrap methods are often very sensitive to the empirical shape of the underlying samples. In general special regard must be paid to the problem of asymmetry which may lead to remarkable differences in power. The use of empirical skewness and kurtosis allows a simple identification of breakdown of the bootstrap location tests in the two-sample case. A suggestion to solve these difficulties may be the use of adaptive test procedures. The performance of the adaptive test procedures we developed and have analysed here is comparable to standard (bootstrap) test procedures in good-natured situations and much can be gained in disadvantageous situations. This is especially the case for practical applications where we are not interested in the "mean" performance of a special test procedure but in the performance for the samples being under consideration. Our results show that there is an urgent need for more detailed empirical investigations of bootstrap properties in the finite sample case also for many other fields of applications of bootstrap methods.

References

Beran, R. (1986). Simulated power functions. *Ann. Statist.*, **14**, pp. 151-173.

Efron, B. and Tibshirani, R.J. (1993). *An Introduction to the bootstrap*. New York: Chapman & Hall.

Jiménez-Gamero, M.D., Muñoz-García, J., Muñoz-Reyes, A. and Pino-Mejías, R. (1998). On Efron's Method II with Identification of Outlier Bootstrap Samples. *Computational Statistics*, **13**, pp. 301-318.

Wolf-Ostermann, K. (1996). The Influence of Skewness and Kurtosis on the Performance of Bootstrap Location Tests for Moderate Sample Sizes. In: *COMPSTAT 1996 Communications in Computational Statistics, Short Communications* (ed. A. Prat & E Ripoll), pp. 143-144.

Wolf-Ostermann, K. (1998). An Approach for Adaptive Bootstrap Testing Based on Power Functions. In: *COMPSTAT 1998 Communications in Computational Statistics, Short Communications* (ed. R. Payne & P. Lane), pp. 125-126.

Young, A. (1994). Bootstrap: More than a stab in the dark? *Statist. Science*, **9**, pp. 382-415.

Two principal points for location mixtures

Wataru Yamamoto[1] and Nobuo Shinozaki[2]

[1] Graduate School of Information Systems, University of Electro-Communications, Chofu 182-8585, Japan

[2] Faculty of Science and Technology, Keio University, Yokohama 223-8522, Japan

Abstract. Two principal points for location mixtures of symmetric or spherically symmetric distributions with equal proportions are investigated in this paper. In the univariate case, a sufficient condition on density functions is given for uniqueness. In the multivariate case, we give a lemma which enables us to compare candidates for two principal points geometrically and to restrict the region to search principal points. With this lemma, a subspace theorem is proved, which states that there exist two principal points in the linear subspace spanned by the component means. Further, a sufficient condition for uniqueness of two principal points is given for two component cases.

Keywords. principal points, k-means clustering, spherical distribution, location mixture, log-concave distribution

1 Principal points

Let $\boldsymbol{X}$ be a p-dimensional random variable. For a set of k points $\boldsymbol{\xi}_1, \ldots, \boldsymbol{\xi}_k$ in p-dimensional Euclidean space $\mathcal{R}^p$, the squared distance from $\boldsymbol{x} \in \mathcal{R}^p$ to this set is defined as

$$d^2\left(\boldsymbol{x}|\boldsymbol{\xi}_1, \ldots, \boldsymbol{\xi}_k\right) = \min_{1 \le j \le k} \left\{ \left\| \boldsymbol{x} - \boldsymbol{\xi}_j \right\|^2 \right\}, \tag{1}$$

where $||\cdot||$ stands for the Euclidean norm. We define the *mean squared distance* (MSD) as $\mathrm{E}\left[d^2\left(\boldsymbol{X}|\boldsymbol{\xi}_1, \ldots, \boldsymbol{\xi}_k\right)\right]$. Then the points $\boldsymbol{\xi}_j{}^*$, $j = 1, \cdots, k$ are called *k principal points of $\boldsymbol{X}$* (Flury, 1990), if they globally minimize MSD, i.e.

$$\mathrm{E}\left\{d^2\left(\boldsymbol{X}|\boldsymbol{\xi}_1{}^*, \ldots, \boldsymbol{\xi}_k{}^*\right)\right\} = \min_{\{\boldsymbol{\xi}_j\}} \mathrm{E}\left\{d^2\left(\boldsymbol{X}|\boldsymbol{\xi}_1, \ldots, \boldsymbol{\xi}_k\right)\right\}. \tag{2}$$

There is a closely related concept, *self-consistent points* (Tarpey and Flury, 1996). They are a set of points $\boldsymbol{y}_1, \ldots, \boldsymbol{y}_k$ which satisfy

$$\mathrm{E}\left[\boldsymbol{X}|\boldsymbol{X} \in C_i\right] = \boldsymbol{y}_i, \quad i = 1, \cdots, k,$$

where $C_i = \{x \in \mathcal{R}^p | \; \|x - y_i\| \leq \|x - y_j\|, \forall j \neq i\}, i = 1, \cdots, k$ are called *domains of attraction.* Flury (1993) shows that principal points are self-consistent and that the reverse is not always true. Thus self-consistent points are always candidates for principal points.

The properties of principal points for univariate distributions are investigated by several authors, including Flury (1990, 1993), Trushkin (1982), Tarpey (1994) and Zoppé (1995). Especially Trushkin (1982) shows that principal points are unique if the underlying distribution has log-concave density. However, it is not straightforward to extend these results to multivariate cases. A general result is obtained by Tarpey, Li and Flury (1995). They prove that k principal points for elliptical distributions are in the subspace spanned by the first principal components. We would like to call their theorem as the *principal subspace theorem.*

Principal points have much in common with k-means algorithm in cluster analysis. It would be interesting to study principal points of theoretical distributions that reflect group structures if developments in the theory of principal points are useful to understand cluster analysis better. However little is known about principal points when the distributions have structures such as location mixtures and the densities are not necessarily unimodal.

In this paper, we investigate two principal points for location mixtures of spherically symmetric distributions with equal weights

$$g(x) = \sum_{i=1}^{m} f(x - \mu_i) / m, \tag{3}$$

as a first step towards more general nonhomogeneous cases, and mainly treats the uniqueness and subspace properties of them.

2 Uniqueness for univariate case

First we state some results on uniqueness of two principal points for univariate distributions, which are given in Yamamoto and Shinozaki (2000a).

Lemma 1 . *Let a univariate probability density f be symmetric around the origin, and F be its cumulative distribution function. If the inequality*

$$\frac{c}{2F(c) - 1} > \frac{\int_c^\infty (x - c) f(x) \, dx}{1 - F(c)}, \quad \forall c \in \{c | 1/2 < F(c) < 1\} \tag{4}$$

holds, then f has two principal points uniquely and symmetrically around the origin,

$$\xi_1{}^* = \mathrm{E}[|X|], \quad \xi_2{}^* = -\xi_1{}^*. \tag{5}$$

Symmetric distributions with log-concave densities satisfy the inequality (4). Furthermore, t-distributions with degrees of freedom larger than two, whose densities are not log-concave, also satisfy it.

For distributions which do not have two principal points uniquely, see, for example, Flury (1990), Shimizu, Mizuta and Sato (1999), and Yamamoto and Shinozaki (2000a).

With Lemma 1, we have obtained the following lemma for two component location mixtures. (Yamamoto and Shinozaki, 2000a)

Lemma 2 . *Let F be a univariate probability distribution symmetric around the origin. If $c/\left(2F\left(c\right)-1\right)$ is nondecreasing for $c>0$ and F satisfies the condition (4), then the mixture distribution of the form*

$$g_m\left(x\right)=f\left(x+m\right)/2+f\left(x-m\right)/2 \tag{6}$$

also satisfies the condition (4) for arbitrary $m>0$.

We note that if $f\left(x\right)$ is unimodal, then $c/\left(2F\left(c\right)-1\right)$ is nondecreasing for $c>0$.

3 A subspace theorem

In this section, it is shown that there exist two principal points in the subspace $\mathcal{M}$ spanned by the component means $\boldsymbol{\mu}_1, \ldots, \boldsymbol{\mu}_m$ of the location mixture (3). For this purpose we introduce an operation on two points and give a lemma which enables us to compare MSD of two sets of self-consistent points geometrically. (Yamamoto and Shinozaki, 2000b)

Definition 1 *Let a and b be two real numbers with $a<b$. Contraction of a and b is the operation to replace a and b with $a+\delta$ and $b-\delta$ respectively, where δ is some number which satisfies $0<\delta<\left(b-a\right)/2$. (See Figure 1.)*

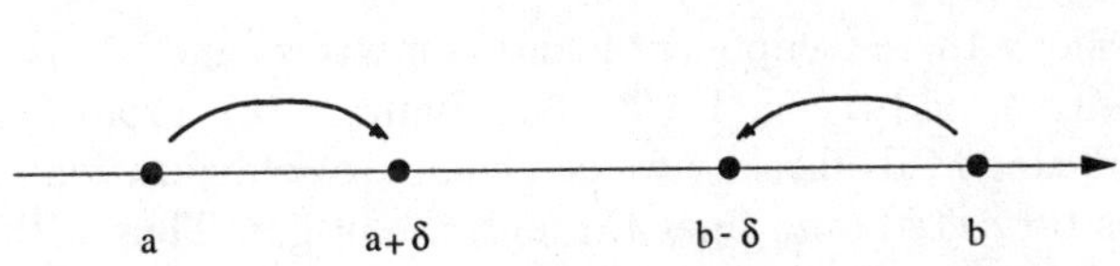

Fig.1. The operation of contraction of a and b

Lemma 3 *For a pair of points $\gamma_1, \gamma_2 \in \mathcal{R}^p$ on a line through the origin, we denote $\boldsymbol{a}=\left(\gamma_1-\gamma_2\right)/\left\|\gamma_1-\gamma_2\right\|$. Let $\boldsymbol{b}$ $(\neq \boldsymbol{a})$ be a vector with $\|\boldsymbol{b}\|=1$, and let P be an orthogonal matrix with $P\boldsymbol{a}=\boldsymbol{b}$. Suppose that the density is given by (3) and put $\xi_i=\boldsymbol{b}'\boldsymbol{\mu}_i$ and $\eta_i=\boldsymbol{a}'\boldsymbol{\mu}_i$, $i=1,\cdots,m$.*

If $\{\eta_i\}$ is obtained from $\{\xi_i\}$ by iterative applications of contraction of some two components, then

$$\mathrm{E}\left[d^2\left(\boldsymbol{X}|P\gamma_1,P\gamma_2\right)\right]\leq\mathrm{E}\left[d^2\left(\boldsymbol{X}|\gamma_1,\gamma_2\right)\right].$$

We note that two self-consistent points are always on a line through the origin (Tarpey, Li and Flury ,1995). Lemma 3 itself is useful to narrow the subset in which two principal points may lie as the following Example 1 shows.

With this lemma, we obtain the following theorem. (Yamamoto and Shinozaki, 2000b)

Theorem 1 *For m component location mixtures (3) of a spherically symmetric distribution, there exists a set of two principal points in the subspace spanned by component means $\boldsymbol{\mu}_1, \ldots, \boldsymbol{\mu}_m$.*

Although this theorem is given only for two principal points, it is similar to the principal subspace theorem of Tarpey, Li and Flury (1995). It allows us to search two principal points only in the linear subspace spanned by the mean vectors of component distributions of mixtures.

Remark 1 *Though Theorem 1 is stated for location mixtures with equal weights, it is also true for location mixtures with arbitrary weights.*

We can show that if $f_1(x) > 0$ for $^\forall x$ then two principal points are in $\mathcal{M}$. Further, for $m = 2$, we obtain the following Corollary, which is also true for arbitrary weight cases.

Corollary 1 *For the density (3) with $m = 2$, two principal points are always on the line through two component means $\boldsymbol{\mu}_1$ and $\boldsymbol{\mu}_2$.*

Example 1 . *Let us consider a three component mixture of spherically symmetric distributions*

Let us consider a three component location mixture, and put $\boldsymbol{\mu}_1 = (0, -1-\sqrt{3})$, $\boldsymbol{\mu}_2 = (-1, 1)$, and $\boldsymbol{\mu}_3 = (1, \sqrt{3})$. See Figure 2 (a). From Lemma 3, instead of calculating MSD directly we can compare configurations of the component means projected onto lines through the origin. Thus with a configuration on some line, we need to seek lines such that they have configurations which cannot be obtained from it by iterative applications of contraction.

We compare the configurations of component means projected onto each line with that onto the vertical axis. It can be easily seen that we need only to consider the region in which the configuration on a line has the second largest absolute value larger than $\sqrt{3}$. So the region to search for two principal points is restricted to the one which contains l_1, l_2, and l_3. l_4 cannot be a candidate. See Figure 2 (b).

4 Uniqueness for multivariate case

Combining two lemmas in Section 2 and Corollary 1, we have the following result.

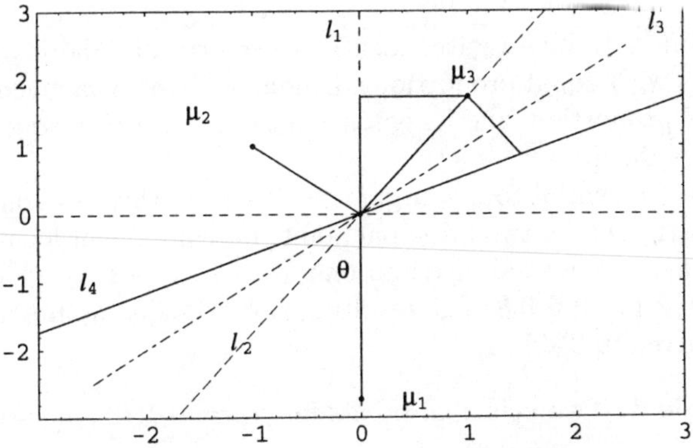

(a) Mean vectors and lines through the origin.

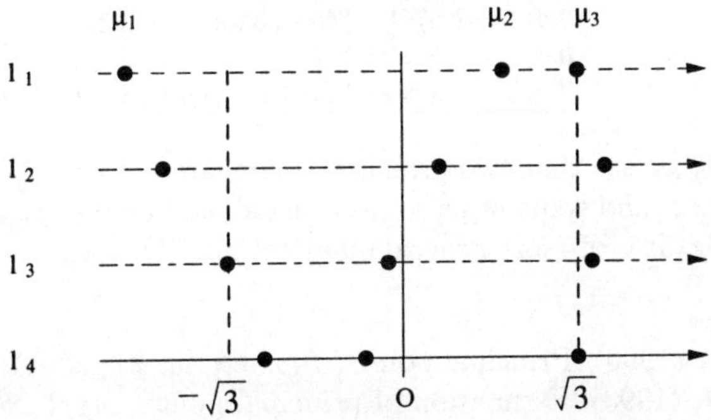

(b) Configurations on lines $l_1, \ldots, l_4$.

Fig.2. An example of three component mixture

Corollary 2 . *Let F be a spherically symmetric distribution on $\mathcal{R}^p$, and assume that the univariate marginal distribution of F has a density function. For one-dimensional marginal distribution F_1 if (a) $c/(2F_1(c) - 1)$ is nondecreasing for $c > 0$, and if (b) F satisfies the condition (4), then two component location mixtures of F have two principal points uniquely.*

We note that if a spherically symmetric distribution satisfies the condition of Corollary 2, then two principal points of its two component location mixtures can be obtained as

$$\xi_1^* = \mathrm{E}_{\mu_1,\mu_2}[\|X\|] \frac{\boldsymbol{\mu}_1}{\|\boldsymbol{\mu}_1\|}, \quad \xi_2^* = -\xi_1^*,$$

where $\mathrm{E}_{\mu_1,\mu_2}[\cdot]$ is the mean of the univariate marginal distribution on the line through $\boldsymbol{\mu}_1$ and $\boldsymbol{\mu}_2$.

5 Concluding remarks

This article mainly investigates location mixtures of spherically symmetric distributions with equal proportions. Uniqueness result is given only for the case of equal proportion, and its extension to the case of unequal proportions seems to be difficult.

Sphericity is essential to our argument. To show this, we conducted a numerical investigation of two principal points for component location mixtures of two dimensional normal distributions, with $\boldsymbol{\mu} = (-1, 0), (1, 0)$ and correlations $\rho = 0.2, 0.4, 0.6, 0.8$. The resulting two principal points and minimum MSD's are given in Table 1.

Table 1. Two principal points for $\boldsymbol{\mu}_1 = (-1, 0), \boldsymbol{\mu}_2 = (1, 0)$

ρ	min MSD	ξ_{11}^*	ξ_{12}^*	ξ_{21}^*	ξ_{22}^*
0.2	1.623	-1.162	-0.165	1.162	0.165
0.4	1.576	-1.148	-0.325	1.148	0.325
0.6	1.502	-1.134	-0.462	1.134	0.462
0.8	1.409	-1.126	-0.569	1.126	0.569

At a glance, we see that two principal points are not on a line through component means, and some other approaches should be necessary to investigate location mixtures of more general elliptical distributions.

References

Flury, B. D. (1990). Principal points, *Biometrika*, **77**, 33–41.

Flury, B. D. (1993). Estimation of principal points, *Appl. Statist.*, **42**, 139–151.

Shimizu, N., Mizuta, M., and Sato, Y.(1998). Some properties of Principal Points, *Japanese Jour. Appl. Statist.*, **27**, 1–16. (in Japanese)

Tarpey, T.(1994). Two principal points of symmetric, strongly unimodal distributions, *Statist. Probab. Letters*, **20**, 253–257.

Tarpey, T. and Flury, B.(1996). Self-consistency: a fundamental concept in statistics, *Statist. Sci.*, **11**, 229–243.

Tarpey, T., Li, L., and Flury, B. D.(1995). Principal points and self-consistent points of elliptical distributions, *Ann. Statist.*, **23**, pp. 103–112.

Trushkin, A. V.(1982). Sufficient conditions for uniqueness of a locally optimal quantizer for a class of convex error weighting functions, *IEEE Trans. Inform. Theory*, **IT-28**, 187–198.

Yamamoto, W. and Shinozaki, N.(2000a). On uniqueness of two principal points for univariate location mixtures, *Statist. Probab. Letters*, **46**, 33–42.

——(2000b). Two principal points for location mixtures of spherically symmetric distributions, *Jour. Japanese Statist. Soc.*, to appear.

Zoppè, A.(1995). Principal points of univariate continuous distributions, *Statistics and Computing*, **5**, 127–132.

Vector splines and other vector smoothers

Thomas W. Yee

Department of Statistics, University of Auckland, Private Bag 92019, Auckland, New Zealand

Abstract. Vector smoothers are nonparametric regression methods for smoothing a vector response $\mathbf{y}$ against a scalar x. Some theory and software details for two popular classes of vector smoothers are presented—one is based on splines and the other on local regression.

Keywords. Local regression, Nonparametric regression, Smoothing, Splines, S-PLUS, Vector generalized additive models, Vector spline

1 Introduction

Consider the *vector measurement model*

$$\mathbf{y}_i \;=\; \mathbf{f}(x_i) + \boldsymbol{\varepsilon}_i, \qquad E(\boldsymbol{\varepsilon}_i) = \mathbf{0}, \qquad \mathrm{Cov}(\boldsymbol{\varepsilon}_i, \boldsymbol{\varepsilon}_j) = \delta_{ij}\, \boldsymbol{\Sigma}_i \tag{1}$$

for data $(x_i, \mathbf{y}_i, \boldsymbol{\Sigma}_i)$, $i = 1, \ldots, n$, where $\mathbf{f}(x_i) = (f_1(x_i), \ldots, f_M(x_i))^T$ is a vector of M smooth functions to be estimated by the data. For example, we may measure the quantities diastolic (DBP) and systolic blood pressures (SBP) and X = age for a random sample of n people.

This smoothing problem is important because it forms the 'backbone' of the class of *vector generalized additive models* (VGAMs; Yee and Wild, 1996). The VGAM class is very large, for example, it includes univariate and multivariate distributions, categorical data analysis, time series, survival analysis, mixture models, generalized estimating equations, nonlinear least-squares problems, to name just a few. Software written by the author to fit VGAMs, called VGAM, is described in Section 4.

For the ordinary ($M = 1$) smoothing problem, there are several classes of smoothers common in the statistical community. Of these, the two most popular types are based on splines and local regression. We deal with these classes in the next two sections. Other classes include state-space models (Jones, 1993) and wavelets, but are not dealt with here.

Now we mention several notations and notes. Firstly, when smoothing for the VGAM class, the $\boldsymbol{\Sigma}_i$ in (1) are known, and we write $\mathbf{W}_i = \boldsymbol{\Sigma}_i^{-1}$. Secondly, tied observations $(x, \mathbf{y}_i, \boldsymbol{\Sigma}_i)$ $(i = 1, \ldots, T)$ may be replaced by a single observation $(x, \mathbf{y}_i^*, \boldsymbol{\Sigma}_i^*)$ where

$$\mathbf{y}_i^* = \left(\sum_{i=1}^{T} \boldsymbol{\Sigma}_i^{-1}\right)^{-1} \left(\sum_{i=1}^{T} \boldsymbol{\Sigma}_i^{-1}\mathbf{y}_i\right) \quad \text{and} \quad \boldsymbol{\Sigma}_i^* \;=\; \left(\sum_{i=1}^{T} \boldsymbol{\Sigma}_i^{-1}\right)^{-1}.$$

Consequently, we assume that $x_1 < x_2 < \cdots < x_n$. Thirdly, for fixed smoothing parameters, the two smoothers dealt with in the paper are linear smoothers. Fourthly, the vector $\mathbf{e}_j$ is a column vector of 0's but with a 1 in the jth position.

2 Vector smoothing splines

Fessler (1991) proposed the *vector (smoothing) spline* for fitting (1). It minimizes $S(f_1, \ldots, f_M) =$

$$\sum_{i=1}^{n} (\mathbf{y}_i - \mathbf{f}(x_i))^T \, \mathbf{W}_i \, (\mathbf{y}_i - \mathbf{f}(x_i)) \; + \; \sum_{m=1}^{M} \lambda_m \int_a^b \{f_m''(x)\}^2 \, dx \qquad (2)$$

over a Sobolev space of order 2. The quantity S fits naturally in the roughness penalty approach of Green and Silverman (1994). Fessler (1991) proposed an $O(n\,M^3)$ algorithm based on the Reinsch (1967) algorithm, and implemented it in his C program called VSPLINE. I originally modified Fessler's code, available at `http://www.eecs.umich.edu/~fessler/`, for statistical use and called it YEE-SPLINE.

Unfortunately, the Reinsch algorithm is known to be numerically unstable under certain conditions (Green and Silverman, 1994). To obtain a numerically more stable solution, and to integrate it with VGAM, I have developed a new $O(n\,M^3)$ algorithm based on the superior properties of B-spline basis functions (de Boor, 1978). I have called the new code YEE-SPLINE, and it is used in the author's S-PLUS functions `vsmooth.spline()` and `s()` for `vgam()`. The rest of this section is devoted to details of these.

2.1 Computation

The vector smoothing spline solution consists of component functions f_m that are spline functions. Hence we can write

$$f_m(x) \;=\; \sum_{j=1}^{n^*+2} \theta_{jm} \, B_j(x), \quad m = 1, \ldots, M,$$

where n^* is the 'effective' n (see later), θ_{jm} are coefficients and $B_j(x)$ are B-spline basis functions. Defining the $n \times (n^* + 2)$ matrix $\mathbf{B}$ and the $(n^* + 2) \times (n^* + 2)$ penalty matrix $\boldsymbol{\Omega}$ by

$$\left[(\mathbf{B})_{ij}\right] = B_j(x_i) \quad \text{and} \quad \left[(\boldsymbol{\Omega})_{ij}\right] = \int_a^b B_i''(x)\, B_j''(x)\, dx,$$

we can rewrite (2) as

$$(\mathbf{y} - \mathbf{B}_*\boldsymbol{\theta})^T \mathbf{W} (\mathbf{y} - \mathbf{B}_*\boldsymbol{\theta}) + \boldsymbol{\theta}^T \boldsymbol{\Omega}_* \, \boldsymbol{\theta} \qquad (3)$$

where $\mathbf{y} = (\mathbf{y}_1^T, \ldots, \mathbf{y}_n^T)^T$, $\mathbf{W} = \text{Diag}(\mathbf{W}_1, \ldots, \mathbf{W}_n)$, $\mathbf{B}_* = \mathbf{B} \otimes \mathbf{I}_M$, $\boldsymbol{\Omega}_* = \boldsymbol{\Omega} \otimes \text{Diag}(\boldsymbol{\lambda})$, and $\boldsymbol{\lambda} = (\lambda_1, \ldots, \lambda_M)^T$ is a vector of non-negative smoothing parameters. Note that $\boldsymbol{\Omega}$ is symmetric and has half-bandwidth 4 so that $\boldsymbol{\Omega}_*$ has half-bandwidth $3M + 1$.

Let $\boldsymbol{\theta} = (\theta_{11}, \ldots, \theta_{1M}, \ldots, \theta_{(n^*+2)1}, \ldots, \theta_{(n^*+2)M})^T$. Setting the derivative of (3) with respect to $\boldsymbol{\theta}$ to $\mathbf{0}$ gives the solution

$$\left(\mathbf{B}_*^T \mathbf{W} \, \mathbf{B}_* + \boldsymbol{\Omega}_*\right) \hat{\boldsymbol{\theta}} \;=\; \mathbf{B}_*^T \mathbf{W} \, \mathbf{y}.$$

The matrix $\mathbf{M} = \mathbf{B}_*^T \mathbf{W}\, \mathbf{B}_* + \mathbf{\Omega}_*$ is symmetric and has half-bandwidth $4M$, and its Cholesky decomposition $\mathbf{M} = \mathbf{U}^T\mathbf{U}$ can be easily computed. One then solves $\mathbf{U}^T \left(\mathbf{U}\hat{\boldsymbol{\theta}}\right) = \mathbf{B}_*^T \mathbf{W}\, \mathbf{y}$ by forward- and back-substitution to give $\hat{\boldsymbol{\theta}}$ and hence the solution $\hat{f}_1, \ldots, \hat{f}_M$ in $O(n^* \cdot M^3)$ operations. The influence matrix is

$$\mathbf{A}(\boldsymbol{\lambda}) \;=\; \mathbf{B}_* \left(\mathbf{B}_*^T \mathbf{W}\, \mathbf{B}_* + \mathbf{\Omega}_*\right)^{-1} \mathbf{B}_*^T \mathbf{W}.$$

The diagonal elements of this matrix (for *df* calculations) may be obtained by firstly applying the Hutchinson and de Hoog algorithm (1985) to $\mathbf{M}$ to get the central bands of its inverse. Then the central bands of $\mathbf{B}_* \mathbf{M}^{-1} \mathbf{B}_*^T$ are computed by performing block quadratic form type calculations. The diagonal of this matrix are the pointwise Bayesian variances. Finally, the diagonal elements of this matrix post-multiplied by $\mathbf{W}$ is computed; this is $\mathrm{Diag}(\mathbf{A})$.

2.2 vsmooth.spline()

The author's `vsmooth.spline()` is a stand-alone S-PLUS function for fitting vector smoothing splines. It has arguments:

```
> args(vsmooth.spline)
function (x, y, w, df = rep(5, M), spar = rep(0, M), all.knots = F,
          iconstraint = diag(M), xconstraint = diag(M),
          constraints = list("(Intercepts)" = diag(M), x = diag(M)),
          tol = 0.001, var.arg = T)
```

Written to operate as similar to the S-PLUS function `smooth.spline()` as possible, it calls the B-spline code of de Boor (1978) and segments of F. O'Sullivan's BART. By default, 5 degrees of freedom are assigned to each component function.

VGAM uses the routine `sknotl()` in BART to choose the 'effective' number of design points, n^*. For small data sets ($n < 50$), a knot is placed at every distinct data point x_i, so $n^* = n$. For larger data sets, the number of knots is chosen judiciously in order to keep the computation time manageable (if `all.knots=F`). In general, n^* is a piecewise exponential function of n, with $n^* = 200 + (n - 3200)^{\frac{1}{5}} \approx 200 \ll n$ when $n \geq 3200$.

Once n^* is chosen, the number of B-spline coefficients is $(n^* + 2)M$, and the number of knots is $n^* + 6$. Given $x_1 < x_2 < \cdots < x_n$, `sknotl()` chooses 3 boundary knots at both $\min(x_i)$ and $\max(x_i)$, and n^* knots at $x_{\mathrm{trunc}(1+(j-1)(n-1)/(n^*-1))}$ for $j = 1, \ldots, n^*$.

VGAM applies the Hutchinson and de Hoog algorithm (1985) using *rational* Cholesky. This is performed by a modification of the LINPACK subroutines `dpbfa` and `dpbsl`.

Another important feature of `vsmooth.spline()` is that it separates out the linear and nonlinear parts of each component function. The former is estimated first (based on LINPACK code), and the latter by smoothing on the residuals. This technique, called modified backfitting (Buja et al. 1989), is used in `gam()` (Hastie, 1992).

2.2.1 Matrix-band format

VGAM represents the $\mathbf{W}_i$ in *matrix-band* notation, which comprises a $n \times M^*$ matrix. The columns are enumerated by first transversing the diagonal, then the band above that, followed by the second band above the diagonal etc. For example, for $M = 3$, there are $M^* = 6$ columns enumerated as follows:

$$\mathbf{W}_i = \begin{pmatrix} 1 & 4 & 6 \\ & 2 & 5 \\ & & 3 \end{pmatrix}.$$

Why is such a format adopted? If $\mathbf{W}_i$ is diagonal then only the first M columns of `w` are needed. As well as reducing the size of `w` itself in most cases, the Cholesky decomposition of $\mathbf{W}_i$ can be stored in a similar representation, with M^* elements too. Another reason is that sometimes we want to input $\mathbf{W}_i$ into VGAM: if `w` is $M \times M \times n$ then `vsmooth.spline(..., w=w)` will result in an error, whereas if `w` being an $n \times M$ matrix will work. A final reason is that several VGAM models such as the bivariate logit model (see McCullagh & Nelder 1989, §6.5.6) can exploit this structure. Note when $M = 1$ the matrix-band format simplifies to a $n \times 1$ matrix. VGAM in this case allows `w` to be an ordinary n-vector.

2.2.2 Constraints-on-the-functions

It is sometimes very useful to constrain certain component functions to be equal or differ by a constant. For example, somebody might specify the difference between mean SBP and DBP to be constant over age.

`vsmooth.spline()` can do this (and other linear constraints) by allowing constraint matrices to be inputted. These are $\mathbf{B}_0$ and $\mathbf{B}_1$ in

$$\mathbf{f}_k(x_{ik}) = \mathbf{B}_0 \boldsymbol{\beta}_0 + \mathbf{B}_1 \mathbf{f}^*(x_i),$$

where the $\mathbf{B}_k$ are known $M \times r_k$ matrices, $r_k \leq M$. For example, if the component functions are parallel, then $\mathbf{B}_1 = \mathbf{1}_M$. The default, $\mathbf{B}_0 = \mathbf{B}_1 = \mathbf{I}_M$ corresponds to no constraints. $\mathbf{B}_0$ is inputted into `iconstraint=` and $\mathbf{B}_1$ is inputted into `xconstraint=`, or alternatively, both can be put into a list and assigned into `constraint=`.

Yee and Wild (1996) showed that the constrained vector spline problem is also a vector spline problem, but applied to adjusted responses and weight matrices.

3 Local regression

A second class of smoothers is based on the ideas of local regression (see, e.g., Fan and Gijbels, 1996). Let K be a kernel function, and $K_h(u) = K(u/h)/h$ where h is the bandwidth. Let

$$\mathbf{X}_x = \begin{pmatrix} 1 & (x_1 - x) & \cdots & (x_1 - x)^r \\ \vdots & \vdots & & \vdots \\ 1 & (x_n - x) & \cdots & (x_n - x)^r \end{pmatrix},$$

$\mathbf{X}_x^* = \mathbf{X}_x \otimes \mathbf{I}_M$, and $\mathbf{W}_x^* = \text{Diag}(K_h(x_1 - x)\mathbf{W}_1, \ldots, K_h(x_n - x)\mathbf{W}_n)$.

For the vector smoothing problem (1), an r-degree *vector polynomial kernel estimator* (VLPKE) is

$$\hat{\mathbf{f}}(x; r, h) = \hat{\boldsymbol{\beta}}_x^0, \tag{4}$$

where $\hat{\boldsymbol{\beta}}_x = (\hat{\boldsymbol{\beta}}_x^{0T}, \ldots, \hat{\boldsymbol{\beta}}_x^{rT})^T$ minimizes the quantity

$$\sum_{i=1}^{n} \left\{ \mathbf{y}_i - \sum_{j=0}^{r} \boldsymbol{\beta}_x^j (x_i - x)^j \right\}^T K_h(x_i - x)\, \mathbf{W}_i \left\{ \mathbf{y}_i - \sum_{j=0}^{r} \boldsymbol{\beta}_x^j (x_i - x)^j \right\}. \quad (5)$$

One can see that the weight matrices $\mathbf{W}_i$ are downweighted the further x_i is away from the target point x. This is the local regression idea, and the $M = 1$ case corresponds to the ordinary LPKE.

The solution to (5) is

$$\hat{\boldsymbol{\beta}}_x = \left(\mathbf{X}_x^{*T} \mathbf{W}_x^* \mathbf{X}_x^* \right)^{-1} \mathbf{X}_x^{*T} \mathbf{W}_x^* \mathbf{y}$$

so that (4) is equivalent to

$$\hat{\mathbf{f}}(x; r, h) = \left(\mathbf{e}_1^T \otimes \mathbf{1}_M^T \right) \hat{\boldsymbol{\beta}}_x.$$

Thus, to handle vector responses, the essential difference is that the usual weighted multiple linear regression is replaced by a weighted multivariate regression.

3.1 Derivative estimation

The extension of the local polynomial idea to estimate the νth derivative is straightforward. One can estimate $f_j^{(\nu)}(x)$ via the intercept coefficient of the νth derivative of the local polynomial being fitted at x, assuming $\nu \le r$. In general, for $\nu = 0, \ldots, r$,

$$\hat{\mathbf{f}}^{(\nu)}(x; r, h) = \nu! \left(\mathbf{e}_{\nu+1}^T \otimes \mathbf{1}_M^T \right) \left(\mathbf{X}_x^{*T} \mathbf{W}_x^* \mathbf{X}_x^* \right)^{-1} \mathbf{X}_x^{*T} \mathbf{W}_x^* \mathbf{y} \; = \; \nu! \, \hat{\boldsymbol{\beta}}_x^{\nu}.$$

Note that $\hat{f}_j^{(\nu)}(x; r, h)$ is not in general equal to the νth derivative of $\hat{f}_j(x; r, h)$.

3.2 Extensions

Welsh and Yee (in preparation) discuss several extensions of the basic formulation above. These include allowing each component function to have its own bandwidth and degree of polynomial, and the use of local (rather than global) bandwidths. Due to space restrictions, details are not presented here.

4 Discussion

The vector measurement model (1) presents the underlying smoothing problem for the VGAM class. This paper has outlined some of the essential ideas behind two classes of vector smoothers. For S-PLUS users, both are very relevant because they will form the basis of `s()` and `lo()` in the author's S function `vgam()`. For example, `vgam(y ~ lo(x1) + s(x2), multinomial)` could mean vector smoothing based on local regression on X_1 and vector spline smoothing on X_2, for a multinomial logit model. That is, for a categorical response Y taking values $1, 2, \ldots, M+1$, and $\mathbf{x} = (x_1, x_2)^T$,

$$\begin{pmatrix} \log\{p_1(\mathbf{x})/p_{M+1}(\mathbf{x})\} \\ \vdots \\ \log\{p_M(\mathbf{x})/p_{M+1}(\mathbf{x})\} \end{pmatrix} = \boldsymbol{\eta}_0 + \mathbf{f}_1(x_1) + \mathbf{f}_2(x_2),$$

where $\mathbf{f}_1$ is estimated by a VLPKE and $\mathbf{f}_2$ by a vector spline. Currently, only `s()` is implemented. The function `vgam()` can potentially fit hundreds of models/distributions; see
`http://www.stat.auckland.ac.nz/yee/vgam` for the current selection.

There are many details not mentioned here. For example, some of the most practically relevant properties include those of degrees of freedom, standard errors, and equivalent kernels. Some details of these can be found in Yee and Wild (1996).

There is much room for further work with the above. For example, the VLPKE can be extended to provide local likelihood estimates (Loader, 1999) of multivariate models. Furthermore, although only `s()` is currently available for `vgam()` one future project is to extend the LOESS code to handle the vector responses, and then incorporated into `vgam()` via `lo()`.

References

Buja, A., Hastie, T. and Tibshirani, R. (1989). Linear smoothers and additive models (with discussion) *Ann. Statist.*, **17**, 453–555.

de Boor, C. (1978). *A Practical Guide to Splines.* Springer: Berlin.

Fan, J. and Gijbels, I. (1996). *Local Polynomial Modelling and Its Applications.* Chapman & Hall, London.

Fessler, J.A. (1991). Nonparametric fixed-interval smoothing with vector splines. *IEEE Trans. Signal Process.*, **39**, 852–859.

Green, P.J. and Silverman, B.W. (1994). *Nonparametric Regression and Generalized Linear Models.* London: Chapman & Hall.

Hastie, T.J. (1992). *Generalized Additive Models.* In: Chambers, J.M. and Hastie, T.J. (eds.), *Statistical Models in S*, 249–307. Pacific Grove, CA: Wadsworth and Brooks/Cole.

Hutchinson, M.F. and de Hoog, F.R. (1985). Smoothing noisy data with spline functions. *Numerische Mathematik*, **47**, 99–106.

Jones, R.H. (1993). *Longitudinal Data with Serial Correlation: a State-space Approach.* New York: Chapman & Hall.

Loader, C. (1999). *Local Regression and Likelihood.* New York: Springer.

McCullagh, P. and Nelder, J.A. (1989). *Generalized Linear Models*, 2nd ed. London: Chapman & Hall.

Reinsch C. (1967). Smoothing by spline functions. *Numerische Mathematik*, **10**, 177–83.

Welsh, A.H. and Yee, T.W. (in preparation). Local regression for vector responses.

Yee, T.W. (in preparation). Vector splines.

Yee, T.W. and Wild, C.J. (1996). Vector generalized additive models. *J. Roy. Statist. Soc.* B **58**, 481–493.

Index of authors

Aguilera, A.M. 175
Alonso, A.M. 181
Ambroise, C. 187
Antoch, J. 427

Barcena, M.J. 193
Barnard, J. 199
Bartels, K. 205
Becker, C. 211
Beninel, F. 223
Bernaards, C.A. 217
Bertolino, F. 29
Biernacki, C. 223
Branco, J.A. 415
Bretagnolle, V. 223
Buzzigoli, L. 229

Cappelli, C. 235
Catot, J.M. 391
Conversano, C. 241
Corduas, M. 247
Cressie, N. 41
Croux, C. 301, 385

Darius, P. 253
Davies, P.L. 259
Davino, C. 265
De Falguerolles, A. 53
De Iaco, S. 271
De Jong, K. 379
De Waal, T. 277
Dehon, C. 301
Denteneer, D. 283
Dreesman, J.M. 289
Drinkwater, D. 313
Dyckerhoff, R. 295

Escabias, M. 175
Esposito, V. 265

Filzmoser, P. 301, 385
Fried, R. 307
Fujiwara, T. 361

Gilchrist, R. 313
Giusti, A. 229
Gomez, V. 391
Govaert, G. 187
Guègan, D. 319

Hand, D.J. 77
Hebrail, G. 87
Hofmann, H. 499
Hubert, M. 481
Huisman, M. 325

Karssenberg, D. 379
Klaschka, J. 427
Kobayashi, I. 361
Kolev, N. 337
Kooiman, P. 15
Koshimizu, T. 475
Kovac, A. 259
Kroese, A.H. 15
Kropf, S. 343

Lee, Y. 97
Lisi, F. 319
Lorimer, M. 161
Luetkepohl, H. 109

Maharaj, E.A. 349
Maravall, A. 121
Marek, L. 397
Marshall, R.J. 355
Michiels, S. 253
Mola, F. 241
Moreno, E. 29
Mugglin, A.S. 41
Müller, M. 409

Nakano, J. 361
Neuwirth, E. 131
Núñez, M. 367

Ottoy, J-P. 253

Paiva, D. 337
Pebesma, E.J. 379
Peña, D. 181
Pison, G. 385
Posa, D. 271
Prat, A. 391

Racugno, W. 29
Raeymaekers, B. 253
Ramoni, M. 445
Renssen, R.H. 15
Rezanková, H. 397
Robertson, D.A. 403
Romo, J. 181
Rönz, B. 409
Rosário de Oliveira, M. 415
Rousseeuw, P.J. 139, 385, 421, 463, 481
Rubin, D.B. 3

Sánchez, F.J. 121
Savicky, P. 427
Schimek, M.G. 433
Schuster, E. 439
Sebastiani, P. 445
Shannon, W.D. 235
Shinozaki, N. 523
Sijtsma, K. 451
Simpson, A. 457
Solé, I. 391
Solomin, A. 253
Stevens, R. 161
Struyf, A. 463

Thas, O. 253
Tsujitani, M. 475
Tusell, F. 193

Van Aelst, S. 139
Van Driessen, K. 421
Verboven, S. 481
Verbyla, A. 161
Vos, H.J. 487

Westlake, A. 493
Wilhelm, A. 499
Wilkinson, D.J. 457
Willenborg, L. 505
Wittkowski, K. M. 511
Wolf-Ostermann, K. 517

Yamamoto, W. 523
Yamamoto, Y. 361
Yee, T. 529

Ziegenhagen, U. 409

Index of keywords

Adaptive test 517
Aggregation weights 265
Alcohol 313
Algorithm 295, 463
Algorithms 427, 77
Antithetic parameter 289
Applets 253
Applications 87
ARFIMA models 247
ARIMA models 121, 391
ARMA models 307
Association rules 499
Autocorrelations 349
Automatic data editing 277
Automation 493
Autoregression 109
Autoregressive model 337

Backfitting 433
Backward induction 487
BADAME project 397
Bayes classifier 445
Bayes factor 29
Bayes 457
Bayesian methods 3
Best linear unbiased prediction 161
Bias correction 475
Binary trees 193
Biometrical variables 223
Biplot 53
Boolean classifiers 355
Bootstrap method 475
Bootstrap 109, 517
Bottom-up procedures 427
Box plots 475

Calibration estimator 15
Canonical correlation analysis 415
Canonical correlation 301
Cell suppression 403
Censored data 3
Chaos 319
Chemometrics 3
Classification trees 427
Clinical trials 3
Coherence 265
Compound Poisson 313
Compromise components 265
Computer-assisted editing 277
Conditional independence 307
Conditional maximum likelihood 337
Conditional 511
Confidence interval 109
Confidence region 343
Confidentiality 3
Confounding 511
Consumption data 313
Convex sets 295
Correlated binomial thinning 337
Cost-complexity measure 427
Cost-sensitive learning 367
Cross-entropy 475
Czech economy 397

Data mining 499, 87
Data perturbation 505
Data quality 77
Data sources 397
Data warehouses 87
Decision rules 355
Density estimation 259
Disclosure control 403
Dissemination 493
Distributed computing 361
DLMs 457
Domain knowledge 367

EIC 475
EM algorithm 187
Embedding dimension 319
Exact 511
Experiments 3

Factor analysis 217, 385
Fellegi-Holt paradigm 277
File matching 193
Finance 457
Forecasting 457

Fuzzy clustering 187

Generalized additive models 241, 433
Generalized bilinear models 53
Generalized linear model(s) 53, 97
Generalized non-linear models 53
Generalized regression estimator 15
Genetics 3
Geographical information systems 379
Geometric programming 505
Geostatistics 271, 379
GLIM 313
GLM 97, 313
Goodness-of-fit methods 451
Graphical user interface 361

HGLM 97
Hierarchical generalized linear model 97

Image reconstruction 3
Impulse response 109
Imputation 217, 277
Influence function 385
Interactive computational statistics 409
Interactive graphics 499
Interfacing software 131
Internet bases teaching 409
Interval probabilities 445
Intrinsic prior 29
Item response theory 451
Item selection 451

JAVA 253, 361
Joint modelling of mean and dispersion 97

Kd-tree 283
K-means clustering 523
Knowledge based systems 391
Kurtosis 517

Labor market 325
Large data sets 77, 325
Latent variables 3
Linear processes 181
Linear programming 229, 505
Linear regression 175
Linked tables 229
Load curve analysis 87
Local fitting 283
Local regression 529
Location depth 463
Location mixture 523
Location test 517
Log-concave distribution 523
Logistic regression 175
Long memory 247
Longitudinal mixed models 439

Manet 499
Manipulation 493
Marginal likelihood 511
Markov Chain Monte Carlo 41, 439
Markov random field 41
Matched tables 53
Mathematical programming 403
MCD 211
MCMC 289
MCMCML estimation 289
Microdata 505
Minimax sequential rules 487
Minimum covariance determinant estimator 301
Minimum mean distance 235
Misfitting respondents 451
Missing covariates 439
Missing data 193, 199, 217, 307, 445
Mixed models 161
Mixture model(s) 3, 187
Modality 259
Model selection 223
Model-based discriminant analysis 223
Models 77
Monte Carlo 379
Mosaicplots 499
Moving blocks bootstrap 349
Multidimensional item response theory 217
Multimedia teaching 409
Multiple imputation 193
Multiresolution analysis 259
Multivariate autoregression 41
Multivariate median 463
Multivariate trimming 295
Multivariate 457

Multi way table 493
MVE 211

Noncompliance 3
Nonlinear programming 505
Nonlinear regression 205
Nonnested models 29
Non-parametric regression 259, 529
Numerical consistency 15

Object oriented approach 361
Object Oriented 199
Official statistics 15
Optimal trees 427
Orthogonal transformations 481
Outlier detection 211
Outliers 121, 385
Output diagnostics 289

Partial autocorrelations 349
Penalized likelihood 433
Permutation 511
Person ordering 451
Phase space embedding 307
Post-stratification 325
Prediction intervals 181
Prediction 319
Predictive matching 193
Preliminary estimate 247
Principal component analysis (PCA) 481
Principal components 175
Principal points 523
Printed Circuit Board 283
Probability distribution 235
Projection pursuit (PP) 481
Projection- pursuit 301
Projection pursuit 415
Pruning 235
Psychometrics 217
Python 199

Random summation 337
Rank test 511
Recursive partitioning 241
Reflection algorithm 481
Regression depth 139
Regression trees 367
Relationship between populations 223
Relaxed iterative projections 133
Repeated weighting 15
Resampling methods 205
Residual maximum likelihood 161
Robust estimation 415
Robustness 139, 211, 301, 385

SAS system 397
Score 511
Seasonal adjustment 121, 265, 391
SEATS 391
Separable process 161
Sequential mastery testing 487
S-estimators 211
Sex determination 223
Shuttle algorithm 229
Sieve bootstrap 181
Sign test 511
Simplification 235
Simulation studies 205
Simulation 199, 337
Singular value decomposition 53
Singularity 433
Skewness 517
Small sample 517
Small-area counts 41
Smoothing spline 161
Smoothing 529
Space-time correlation 271
Spatial correlation 97
Spatial design 283
Spatio-temporal modelling 161
Specification tests 205
Spherical distribution 523
Splines 433, 529
S-Plus 433
SPLUS 529
Spreadsheets 131
Square tables 53
Stabilized multivariat tests 343
Stationary time series 349
Statistical computational environment 397
Statistical confidentiality 505
Statistical data editing 277
Statistical disclosure control 229
Statistical Education 131

Statistical methods 87
Statistics education 253
Statistics 77
Stochastic modelling 379
Strings 259
Supervised classification and prediction 241
Surveys 3

Teaching 253
Temporal correlation 97
Temporal GIS 379
Ties 511
Time series analysis 307, 397
Time series 121, 181, 391
TRACE 391
TRAMO 391
Treatment effects 343
Trees 355
Trees-based models 235
Trend estimation 121
Tweedie 313
Two-stage cluster analysis 325

User interface 511
Utility company 87

Variance reduction 289
Variogram 161
Vector generalized additive models 529
Vector spline 529
Vector 109

Zero restrictions 229
Zonoid depth 295